FORMULAS/EQUATIONS

Distance Formula The distance from (x_1, y_1) to (x_2, y_2) is $\sqrt{(x_2 - x_1)^2 + (y_2 - y_1)^2}$.

Midpoint Formula The midpoint of the line segment with endpoints (x_1, y_1) and (x_2, y_2) is $\left(\dfrac{x_1 + x_2}{2}, \dfrac{y_1 + y_2}{2}\right)$.

Standard Equation of a Circle The standard equation of a circle of radius r with center at (h, k) is
$$(x - h)^2 + (y - k)^2 = r^2$$

Slope Formula The slope m of the line containing the points (x_1, y_1) and (x_2, y_2) is
$$\text{slope } (m) = \frac{\text{change in } y}{\text{change in } x} = \frac{y_2 - y_1}{x_2 - x_1} \quad (x_1 \neq x_2)$$
m is undefined   if $x_1 = x_2$

Slope-Intercept Equation of a Line The equation of a line with slope m and y-intercept $(0, b)$ is $y = mx + b$

Point-Slope Equation of a Line The equation of a line with slope m containing the point (x_1, y_1) is $y - y_1 = m(x - x_1)$

Quadratic Formula The solutions of the equation $ax^2 + bx + c = 0$, $a \neq 0$, are $x = \dfrac{-b \pm \sqrt{b^2 - 4ac}}{2a}$

If $b^2 - 4ac > 0$, there are two unequal real solutions.

If $b^2 - 4ac = 0$, there is a repeated real solution.

If $b^2 - 4ac < 0$, there are two complex solutions that are not real.

GEOMETRY FORMULAS

Circle $r =$ Radius, $A =$ Area, $C =$ Circumference
$A = \pi r^2 \quad C = 2\pi r$

Triangle $b =$ Base, $h =$ Height (Altitude), $A =$ area
$A = \frac{1}{2}bh$

Rectangle $l =$ Length, $w =$ Width, $A =$ area, $P =$ perimeter
$A = lw \quad P = 2l + 2w$

Rectangular Box $l =$ Length, $w =$ Width, $h =$ Height, $V =$ Volume, $S =$ Surface area
$V = lwh \quad S = 2lw + 2lh + 2wh$

Sphere $r =$ Radius, $V =$ Volume, $S =$ Surface area
$V = \frac{4}{3}\pi r^3 \quad S = 4\pi r^2$

Right Circular Cylinder $r =$ Radius, $h =$ Height, $V =$ Volume, $S =$ Surface area
$V = \pi r^2 h \quad S = 2\pi r^2 + 2\pi rh$

CONVERSION TABLE

1 centimeter $\approx$ 0.394 inch	1 joule $\approx$ 0.738 foot-pound	1 mile $\approx$ 1.609 kilometers
1 meter $\approx$ 39.370 inches	1 gram $\approx$ 0.035 ounce	1 gallon $\approx$ 3.785 liters
$\approx$ 3.281 feet	1 kilogram $\approx$ 2.205 pounds	1 pound $\approx$ 4.448 newtons
1 kilometer $\approx$ 0.621 mile	1 inch $\approx$ 2.540 centimeters	1 foot-lb $\approx$ 1.356 Joules
1 liter $\approx$ 0.264 gallon	1 foot $\approx$ 30.480 centimeters	1 ounce $\approx$ 28.350 grams
1 newton $\approx$ 0.225 pound	$\approx$ 0.305 meter	1 pound $\approx$ 0.454 kilogram

The first person to invent a car that runs on water...

... may be sitting right in your classroom! Every one of your students has the potential to make a difference. And realizing that potential starts right here, in your course.

When students succeed in your course—when they stay on-task and make the breakthrough that turns confusion into confidence—they are empowered to realize the possibilities for greatness that lie within each of them. We know your goal is to create an environment where students reach their full potential and experience the exhilaration of academic success that will last them a lifetime. *WileyPLUS* can help you reach that goal.

Wiley**PLUS** is an online suite of resources—including the complete text—that will help your students:

- come to class better prepared for your lectures
- get immediate feedback and context-sensitive help on assignments and quizzes
- track their progress throughout the course

"I just wanted to say how much this program helped me in studying... I was able to actually see my mistakes and correct them. ... I really think that other students should have the chance to use *WileyPLUS*."

Ashlee Krisko, *Oakland University*

www.wiley.com/college/wileyplus

80% of students surveyed said it improved their understanding of the material. *

Prepare & Present

Create outstanding class presentations using a wealth of resources, such as PowerPoint™ slides, image galleries, interactive simulations, and more. Plus you can easily upload any materials you have created into your course, and combine them with the resources Wiley provides you with.

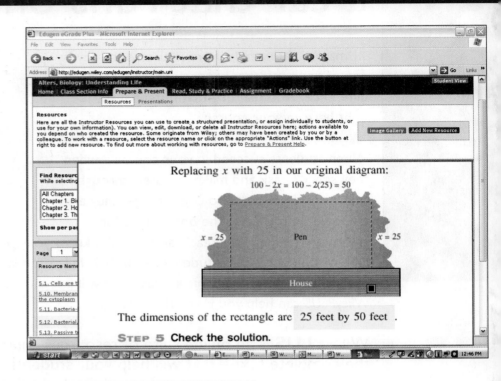

Create Assignments

Automate the assigning and grading of homework or quizzes by using the provided question banks, or by writing your own. Student results will be automatically graded and recorded in your gradebook. *WileyPLUS* also links homework problems to relevant sections of the online text, hints, or solutions— context-sensitive help where students need it most!

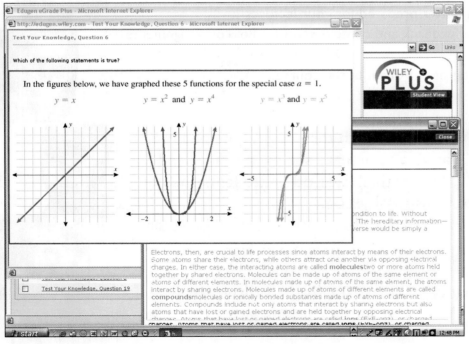

*Based on a spring 2005 survey of 972 student users of *WileyPLUS*

Track Student Progress

Keep track of your students' progress via an instructor's gradebook, which allows you to analyze individual and overall class results. This gives you an accurate and realistic assessment of your students' progress and level of understanding.

Now Available with WebCT and Blackboard!

Now you can seamlessly integrate all of the rich content and resources available with *WileyPLUS* with the power and convenience of your WebCT or BlackBoard course. You and your students get the best of both worlds with single sign-on, an integrated gradebook, list of assignments and roster, and more. If your campus is using another course management system, contact your local Wiley Representative.

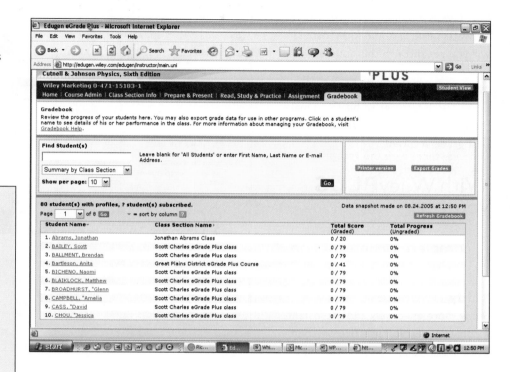

"I studied more for this class than I would have without *WileyPLUS*."

Melissa Lawler, *Western Washington Univ.*

For more information on what *WileyPLUS* can do to help your students reach their potential, please visit

www.wiley.com/college/wileyplus

76% of students surveyed said it made them better prepared for tests. *

You have the potential to make a difference!

Will you be the first person to land on Mars? Will you invent a car that runs on water? But, first and foremost, will you get through this course?

WileyPLUS is a powerful online system packed with features to help you make the most of your potential, and get the best grade you can!

With Wiley**PLUS** you get:

A complete online version of your text and other study resources

Study more effectively and get instant feedback when you practice on your own. Resources like self-assessment quizzes, tutorials, and animations bring the subject matter to life, and help you master the material.

Problem-solving help, instant grading, and feedback on your homework and quizzes

You can keep all of your assigned work in one location, making it easy for you to stay on task. Plus, many homework problems contain direct links to the relevant portion of your text to help you deal with problem-solving obstacles at the moment they come up.

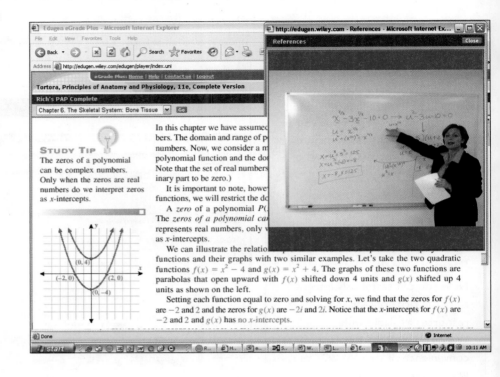

The ability to track your progress and grades throughout the term.

A personal gradebook allows you to monitor your results from past assignments at any time. You'll always know exactly where you stand.

If your instructor uses *WileyPLUS*, you will receive a URL for your class. If not, your instructor can get more information about *WileyPLUS* by visiting www.wiley.com/college/wileyplus

"It has been a great help, and I believe it has helped me to achieve a better grade."
Michael Morris, *Columbia Basin College*

69% of students surveyed said it helped them get a better grade. *

The Wiley Faculty Network

Where Faculty Connect

The Wiley Faculty Network is a faculty-to-faculty network promoting the effective use of technology to enrich the teaching experience. The Wiley Faculty Network facilitates the exchange of best practices, connects teachers with technology, and helps to enhance instructional efficiency and effectiveness. The network provides technology training and tutorials, including *WileyPLUS* training, online seminars, peer-to-peer exchanges of experiences and ideas, personalized consulting, and sharing of resources.

Connect with a Colleague

Wiley Faculty Network mentors are faculty like you, from educational institutions around the country, who are passionate about enhancing instructional efficiency and effectiveness through best practices. You can engage a faculty mentor in an online conversation at www.wherefacultyconnect.com

Participate in a Faculty-Led Online Seminar

The Wiley Faculty Network provides you with virtual seminars led by faculty using the latest teaching technologies. In these seminars, faculty share their knowledge and experiences on discipline-specific teaching and learning issues. All you need to participate in a virtual seminar is high-speed internet access and a phone line. To register for a seminar, go to www.wherefacultyconnect.com

Connect with the Wiley Faculty Network

Web: www.wherefacultyconnect.com

Phone: 1-866-4FACULTY

COLLEGE ALGEBRA

CYNTHIA Y. YOUNG
University of Central Florida

WILEY

John Wiley & Sons, Inc.

This book is dedicated to teachers and students

PUBLISHER	Laurie Rosatone
SENIOR ACQUISITIONS EDITOR	Angela Y. Battle
SENIOR DEVELOPMENT EDITOR	Ellen Ford
PROJECT EDITOR	Jennifer Battista
SENIOR PRODUCTION EDITOR	Sujin Hong
DIRECTOR OF MARKETING	Frank Lyman
MARKETING MANAGER	Amy Sell
COVER PHOTO	Photonica/Getty Images
SENIOR ILLUSTRATION EDITOR	Sigmund Malinowski
SENIOR PHOTO EDITOR	Tara Sanford
MEDIA EDITOR	Stefanie Liebman
DEVELOPMENT ASSISTANT	Justin Bow
DESIGN DIRECTOR	Harry Nolan
COVER DESIGNER	Harry Nolan
INTERIOR DESIGNER	Brian Salisbury

This book was set in 10/12 Times by Techbooks and printed and bound by Von Hoffmann. The cover was printed by Von Hoffmann.

This book is printed on acid free paper. ∞

To order books or for customer service, please call 1-800-CALL WILEY (225-5945).

ISBN-13 978-0-471-65958-7
ISBN-10 0-471-65958-4

Printed in the United States of America

10 9 8 7 6 5 4 3 2 1

About the Author

Cynthia Y. Young

EDUCATION

- Ph.D. **Applied Mathematics**   University of Washington
- M.S.   **Electrical Engineering**   University of Washington
- M.S.   **Mathematical Sciences**   University of Central Florida
- B.A.   **Mathematics Education**   University of North Carolina

RESEARCH

Associate Professor   Department of Mathematics, University of Central Florida
Sabbaticals/Fellowships   Naval Research Lab, Boeing, Kennedy Space Center
Principal Investigator   Turbulence Effects on LIDAR, Office of Naval Research

TEACHING

- University of Central Florida
- Shoreline Community College
- Courses taught: College Algebra, Trigonometry, Calculus, Differential Equations, Applied Boundary Value Problems, Advanced Engineering Mathematics, Special Functions, Optical Wave Propagation. Honors Courses: College Algebra, Calculus

AWARDS

2003–2004	Research Incentive Program	*University of Central Florida*
2002–2003	CAS Distinguished Researcher Award	*University of Central Florida*
2002–2003	University Professional Service Award	*University of Central Florida*
2001–2002	Teaching Incentive Program	*University of Central Florida*
2001	Young Investigator Award	*Office of Naval Research*
2001	Excellence Undergrad. Teaching Award	*University of Central Florida*
1993–1997	NPSC Four-Year Doctoral Fellowship	*Kennedy Space Center*

PERSONAL INTERESTS

- Her Labrador Retrievers' (Molly, Blue, and Ellie) Field Trials
- Tampa Bay Buccaneer Fan
- Golf
- Boating

In College Algebra it is difficult to get students to love mathematics, but you can get students to love *succeeding* at mathematics. Students come together in a College Algebra course with varied backgrounds and often resist change from their previous systems. As a result, a gap exists between students' high school and/or community college mathematics training and what is expected of them in a traditional college algebra course. I was compelled to write this text because I wanted to help students bridge that gap. My experiences as a high school, community college, and university teacher have taught me what students bring in the door, what causes them to struggle, and what methods will help them succeed.

We face many challenges in College Algebra classes: different learning styles, math anxiety, inappropriate placement, nontraditional students, and poor study habits. In spite of these challenges, I know students can succeed. In fact, I expect it. I *expect more* from my students because I know they can do it. I expect them to move beyond what they've learned in the past and to follow a more mature thought process. I expect them to learn to use a text to effectively support their classroom instruction and to ultimately gain confidence as they progress throughout the course. I sincerely hope this book meets your expectations, and I welcome your feedback.

Goals

How many times have we heard our students complain that they can't read the book? They might say, "I understand you in class, but when I get home I'm lost." I've tried several texts and even custom editions but could not find a text my students could read. So I decided to write my own book. I know there is no perfect text, but my overreaching goal was to *write a book that students could read without sacrificing rigor.* I believe I have accomplished this goal. The following principles represent how this book is designed to be an effective teaching and learning aid.

Encouraging Approach

Many people who take College Algebra (regardless of whether they are traditional or returning students) suffer from math anxiety. In order to conquer the math anxiety hurdle, the student has to experience success. Mathematics is a difficult subject to teach, even to students who like it. It requires an instructor and a textbook that present information in a clear, concise, inviting, and engaging format. Students typically do not read textbooks, so I have carefully chosen a tone that will encourage students and acknowledge their different learning styles.

Appropriate Rigor

Traditional College Algebra texts are rigorous. Although this text includes helpful student pedagogy and focuses on a balanced approach to skills and concepts, it also maintains the rigor appropriate for a true College Algebra course. This is because, for those students choosing to take additional mathematics courses, College Algebra is truly a foundation course. In fact, students will be asked to call upon the skills and concepts learned in College Algebra throughout their undergraduate mathematics curriculum. As such, mastery of skills cannot be at the expense of conceptual thought processes.

Overcoming Differences

Students have different learning styles. For example, some are visual learners whereas others respond better to audio methods. In addition, some students learn well with several examples; others learn more effectively with examples of what not to do.

Another challenge is that students come from different high school or community college backgrounds. Often, when I teach a topic a certain way, a student walks up after class and asks, "My high school teacher taught us this other way. Can I use that method?" This resistance to change makes it difficult for instructors to successfully lead students toward a more mature mathematical thought process. It is crucial for a College Algebra book to bridge the gap between what students know and what they need to know.

Topical Coverage

College Algebra topics are very standard (with the exception that some instructors opt not to cover sequences, series, and probability theory). The general topics in this text follow this convention. What differs are some of the approaches I use to solve them, such as word problems. In addition, some subtopics are presented in a slightly different order (e.g., moving complex numbers to the review section).

In this book a chapter on graphs precedes the chapter on functions and their graphs. Primarily this is to first introduce the Cartesian plane and graphing equations by point plotting with specific attention on lines and circles, and gaining familiarity and confidence with graphs before jumping into functions and their graphs. The second reason is that circles aren't functions and I wanted to cover circles and completing the square as a technique again before the following chapter covers quadratic functions where students will complete the square to find the vertex of the graph. Additionally, if the instructor does not choose to cover conics in a later chapter, the student will at least have seen a circle.

Innovative Pedagogy

I have developed the following features to address specific course issues. I hope that the benefits of these features help fill the gap that exists between class time and homework.

Skills and Conceptual Objectives

Every section opens with a list of both skills and conceptual objectives that students are expected to master in that section. This emphasizes and encourages not just the rote memorization of mathematical processes, but also the understanding of the concepts behind the content.

Parallel Words and Math

Most texts will present examples (equations on left) with brief running marginal annotations on the right. This text reverses that presentation so that an explanation in words is provided on the left and the mathematics is parallel to the right. This format reflects how students read naturally and makes it easier for them to *read* through examples.

Word Problems

Every College Algebra book I have ever taught from gives students a *procedure* for solving word problems. Such as, Step 1: read the problem, two or three times. I disagree with this approach. In my experience if a student reads the problem two or three times, he has absolutely no idea where to start and ends up staring at a blank piece of paper, intimidated and defeated. I believe that the more students get on their paper, the more confidence they gain. This book teaches students to read the problem once and write down what they are looking for. The second time through it tells them, *"Read until you get to a point where you can get something on your paper. Stop. Write or draw something, then continue."* Students gain confidence every time they draw or write something on their paper. Finally, students are instructed to run their final answers by the *Common Sense Department* to encourage critical thinking.

Common Mistakes

Some students learn better with lots of examples and repetition, and some learn more effectively with counterexamples. Traditional books will illustrate five different examples of how to do something and then put in a warning box with a totally different problem showing what not to do. It is difficult for students to bridge the two. I include what *NOT* to do on common mistakes with the *same* examples rather than waiting until later to post a Caution box. This helps students to avoid common mistakes.

Homework Exercises

There are *five* categories of homework exercises: *Skills, Applications, Catch the Mistake, Challenge,* and *Technology.* The Skills exercises strengthen students' ability to solve basic problems similar to worked examples. Once students have strengthened their skills and have gained confidence in the material, they can apply those skills in solving Application exercises. Then a Catch the Mistake set of exercises helps students who learn from counterexamples find mistakes and learn what not to do. For deeper conceptual understanding there are Challenge exercises that require a more mature analytical thought process. Finally, there are Technology exercises that make use of graphing calculators to complement analytic procedures.

Technology Using Graphing Calculators

College Algebra courses tend to fall into one of two camps: technology optional and technology required. Technology required is the heavy use of graphing calculators and/or computer software, and technology optional is the use of graphing calculators at the instructor's discretion. Both approaches have advantages and disadvantages. This book is well suited for the technology optional College Algebra course.

Features and Benefits at a Glance

All of the course issues discussed above reflect the gap that exists between course work and homework. The features in this book are designed to help instructors fill that gap.

Feature	Benefit to Student
Chapter opening vignette Chapter overview Organizational flow chart Supplement navigation chart	Preview what will be covered and how the topics are related. This manages students' expectations and improves their understanding. The supplement chart highlights available resources to develop students' study skills.
Skills and Conceptual Objectives	Emphasize the importance of conceptual understanding as well as skills.
Clear, concise, and inviting writing style, tone, and layout	Reduce math anxiety, encourage student success.
Parallel words and math	Increase students' ability to read the examples.
Common Mistake/ Correct Incorrect boxes	Demonstrate common mistakes so that the students strengthen their understanding (and avoid making the same mistakes).
Concept Checks and Your Turn exercises	Engage students during class time.
Catch the Mistake exercises	Encourage students to play the role of teacher and assess their understanding.
OOPS! Tying It All Together	Provide students with a fun and interesting way to apply what they have learned to real-life situations.
Chapter Review Review Exercises Practice Test	Improve student study skills, allow for student self-assessment and practice.

Supplements

Instructor Supplements

INSTRUCTOR'S SOLUTIONS MANUAL (ISBN: 0-471-77367-0)
- Contains worked out solutions to all exercises in the text.

POWERPOINT SLIDES
- For each section of the book a corresponding set of lecture notes and worked out examples are presented as PowerPoint slides. Available on the website www.wiley.com/college/young.

TEST BANK (ISBN: 0-471-77368-9)
- Contains a variety of questions and answers for every section of the text.

COMPUTERIZED TEST BANK (ISBN: 0-471-77775-7)
Electronically enhanced version of the Test Bank that
- provides varied question types as well as algorithmically-generated questions,
- allows instructors to freely edit, randomize, and create questions,
- allows instructors to create and print different versions of a quiz or exam,

- recognizes symbolic notation, and
- allows for partial credit if used within *WileyPLUS*.

BOOK COMPANION WEBSITE (WWW.WILEY.COM/COLLEGE/YOUNG)
- Contains all instructor supplements listed above along with a selection of personal response system questions.

WileyPLUS
- Provides additional resources for instructors, such as assignable homework exercises, tutorials, gradebook, integrated links between the online version of the text and supplements (see description above).

Student Supplements

DIGITAL VIDEO TUTOR (DVD) (ISBN: 0-471-70707-4)
- Streaming video of the author presenting chapter overviews, chapter summaries, and working selected examples step by step that are tied to specific textbook sections. DVD will be fully integrated with the textbook. An icon in the text indicates an example that has a video clip available. The Digital Video Tutor can be optionally packaged with the text.

STUDENT SOLUTIONS MANUAL (ISBN: 0-471-66280-1)
- Includes worked out solutions for all odd problems in the text.

BOOK COMPANION WEBSITE (WWW.WILEY.COM/COLLEGE/YOUNG)
- Provides additional resources for students, including web quizzes, video clips, and audio clips.

WileyPLUS
- Presents additional resources for students such as additional self-practice exercises, tutorials, integrated links between the online version of the text and supplements (see description above).

WileyPLUS

Expect More from Your Classroom Technology

Cynthia Young's *College Algebra* is supported by *WileyPLUS*—a powerful and highly integrated suite of teaching and learning resources designed to bridge the gap between what happens in the classroom and what happens at home. *WileyPLUS* includes a complete online version of the text, algorithmically generated exercises, all of the text supplements, plus course and homework management tools, in one easy-to-use website.

Organized around the everyday activities you perform in class, *WileyPLUS* helps you:

Prepare and Present: *WileyPLUS* lets you create class presentations quickly and easily using a wealth of Wiley-provided resources, including an online version of the textbook, PowerPoint slides, and more. You can adapt this content to meet the needs of your course.

Create Assignments: *WileyPLUS* enables you to automate the process of assigning and grading homework or quizzes. You can use algorithmically generated problems from the text's accompanying test bank, or write your own.

Track Student Progress: An instructor's gradebook allows you to analyze individual and overall class results to determine students' progress and level of understanding.

Promote Strong Problem-Solving Skills: *WileyPLUS* can link homework problems to the relevant section of the online text, providing students with context-sensitive help. *WileyPLUS* also features mastery problems that promote conceptual understanding of key topics and video walkthroughs of example problems.

Provide numerous practice opportunities: Algorithmically generated problems provide unlimited self-practice opportunities for students, as well as problems for homework and testing.

Support Varied Learning Styles: *WileyPLUS* includes the entire text in digital format, enhanced with varied problem types and video walkthroughs, to support the array of different student learning styles in today's classrooms.

Administer Your Course: You can easily integrate *WileyPLUS* with another course management system, gradebooks, or other resources you are using in your class, enabling you to build your course, your way.

WileyPLUS **includes a wealth of instructor and student resources:**

Digital Video Tutor: Presents chapter overviews, chapter summaries, and step-by-step walkthroughs of selected examples that are tied to specific textbook sections. Icons throughout the text direct the student to the video examples that the author works out "live." (Also available on DVD.)

Mastery Problems: Presented in a way that requires students to demonstrate complete understanding, or mastery, of the topics. They teach and assess at the same time, and provide students with hints and offer stepped out tutorials based on their input attempt at solving the problem.

Computerized Test Bank: Includes questions from the printed test bank with algorithmically generated problems.

Student Solutions Manual: Includes worked-out solutions for all odd-numbered problems and study tips.

Instructor's Solutions Manual: Presents worked out solutions to all problems.

PowerPoint Lecture Notes: In each section of the book a corresponding set of lecture notes and worked out examples are presented as PowerPoint slides that are tied to the examples in the text.

View an online demo at www.wiley.com/college/wileyplus or contact your local Wiley representative for more details.

The Wiley Faculty Network—Where Faculty Connect

The Wiley Faculty Network is a faculty-to-faculty network promoting the effective use of technology to enrich the teaching experience. The Wiley Faculty Network facilitates the exchange of best practices, connects teachers with technology, and helps to enhance instructional efficiency and effectiveness. The network provides technology training and tutorials, including *WileyPLUS* training, online seminars, peer-to-peer exchanges of experiences and ideas, personalized consulting, and sharing of resources.

Connect with a Colleague

Wiley Faculty Network mentors are faculty like you, from educational institutions around the country, who are passionate about enhancing instructional efficiency and effectiveness through best practices. You can engage a faculty mentor in an online conversation at www.wherefacultyconnect.com.

Participate in a Faculty-Led Online Seminar

The Wiley Faculty Network provides you with virtual seminars led by faculty using the latest teaching technologies. In these seminars, faculty share their knowledge and experiences on discipline-specific teaching and learning issues. All you need to participate in a virtual seminar is high-speed Internet access and a phone line. To register for a seminar, go to www.wherefacultyconnect.com.

Connect with the Wiley Faculty Network

Web: www.wherefacultyconnect.com
Phone: 1-866-4FACULTY

ACKNOWLEDGMENTS

I'd like to thank the entire Wiley team. To my editor and dear friend, Angela Battle, there are no words to express the gratitude in my heart. We share more than a middle name; this book is *ours*. To my developmental editor, Ellen Ford, thanks for helping me find my writing voice and for always finding the positive in every review. You kept me going. To Laurie Rosatone, publisher, thanks for all of the encouragement. To Jennifer Battista, project editor, thanks for all of your hard work coordinating supplements, recruiting reviewers, contributors, and class testers, and creating a community of colleagues. To Harry Nolan and Brian Salisbury for their creative and elegant cover and text design. To Sigmund Malinowski, Tara Sandford, and the rest of the photo and illustration team for developing exactly what I was thinking. To the media editor, Stefanie Liebman, videographer Antonio, and the advertising team for their support while filming the videos. To the production editor, Sujin Hong, who helped us all meet the deadlines. To Frank Lyman, Amy Sell, and the marketing team for your tireless efforts. And I'd especially like to thank Julia Flohr who inspired me to think about writing and all of the other Wiley sales reps who will help us share this book with instructors around the country.

To my husband, Dr. Christopher Laird Parkinson, thanks for making dinner so many nights while I typed. To our Labrador Retrievers (Molly, Blue, and Ellie) for snuggling on my feet while I worked on this book. To Bruce Spatz—our next lab will be named Wiley.

Dad, thanks for always reminding me that I was named after the only two teachers in the family. Mom, thanks for giving me confidence.

Writing this text has been the most humbling experience I've ever had. I have some very strong notions about what the book needs to do for students and instructors. The text development process has allowed me to hone my approach and test my convictions. Finding the balance between what instructors want and what students need has been challenging, but I think I've succeeded. The feedback I've received so far has been extremely gratifying.

I'd like to thank Lori Dunlop-Pyle for writing the Tying It All Together and OOPS! features at the end of every chapter and Pauline Chow for creating all of the Technology Tips throughout the text. I'd also like to thank all of the reviewers, accuracy checkers, class testers, contributors, and focus group participants. You have all contributed to this book. Your effort has made this text an ally for both students and instructors.

REVIEWERS

Ebrahim Ahmadizadeh, *Northampton Community College*
Margo Alexander, *Georgia State University*
David Arreazola, *Laredo Community College*
Kathy Autrey, *Northwestern State University*
Nicholas Belloit, *Florida Community College at Jacksonville*
Jean Bevis, *Georgia State University*
Ben Brink, *Wharton County Junior College*
Lee Ann Brown, *University of Central Oklahoma*
Connie Buller, *Metropolitan Community College*
Warren Burch, *Brevard Community College*
Elsie Campbell, *Angelo State University*
Sandra Campbell, *Armstrong Atlantic State University*
O. Pauline Chow, *Harrisburg Area Community College*
Suzanne Doviak, *Old Dominion University*
Gay Ellis, *Southwest Missouri State University*
Judith Fethe, *Pellissippi State Technical Community College*
Perry Gillespie, *Fayetteville State University*
Tammy Higson, *Hillsborough Community College*
Debra Hill, *University of North Carolina–Charlotte*
Laura Hillerbrand, *Broward Community College*
Joe Howe, *St. Charles Community College*
Nancy Johnson, *Manatee Community College*
David Keller, *Kirkwood Community College*
Susan Kellicut, *Seminole Community College*
Esmarie Kennedy, *San Antonio College*
Julie Killingbeck, *Ball State University*
Michael Lanstrum, *Cuyahoga Community College*
William Livingston, *Missouri Southern State University*
Stephanie Lochbaum, *Austin Community College*
Wanda Long, *St. Charles Community College*
Catherine Louchart, *Northern Arizona University*
Rudy Maglio, *Northeastern Illinois University*

James McGlothin, *Lower Columbia College*
M. Dee Medley, *Augusta State University*
Margaret Michener, *University of Nebraska–Kearney*
Debbie Millard, *Florida Community College at Jacksonville*
Linda Myers, *Harrisburg Area Community College*
Mihaela Poplicher, *University of Cincinnati*
Jonathan Prewett, *University of Wyoming*
William Radulovich, *Florida Community College at Jacksonville*
Stan Stascinsky, *Tarrant County College*
James Stein, *California State University–Long Beach*
Jacqueline Stone, *University of Maryland*
Matt Strom, *Minnesota State University–Mankato*
Todd Timmons, *University of Arkansas–Fort Smith*
Mark Turner, *Cuesta College*
Cari Van Tuinen, *Purdue University*
Kendall Walden, *Arkansas State University*
Natalie Weaver, *Daytona Beach Community College*
JanWehr, *University of Arizona*
Cheryl Whitelaw, *Southern Utah University*
Sandra Wray-McAfee, *University of Michigan— Dearborn*

CONTRIBUTORS

O. Pauline Chow, *Harrisburg Area Community College*
Lori Dunlop-Pyle, *University of Central Florida*

FOCUS GROUP PARTICIPANTS

Ray Abbasi, *Montgomery College*
Marwan Abu-Sawwa, *Florida Community College at Jacksonville*
Scott Adamson, *Chandler Gilbert Community College*
Margo Alexander, *Georgia State University*
Kathy Autrey, *Northwestern State University*

Ignacio Bello, *University of South Florida–Tampa*
Connie Buller, *Metropolitan Community College*
Nancy Carpenter, *Johnson County Community College*
James Chesla, *Grand Rapids Community College*
O. Pauline Chow, *Harrisburg Area Community College*
Scott Collins, *Pima Community College*
Judy Fethe, *Pellissippi State Technical Community College*
Laura Hillerbrand, *Broward Community College*
David Keller, *Kirkwood Community College*
Michael Montano, *Riverside Community College*
Shai Neumann, *Brevard Community College*
Ron Palcic, *Johnson County Community College*
William Radulovich, *Florida Community College at Jacksonville*
Mary Jane Sterling, *Bradley University*
Draga Vidakovic, *Georgia State University*

CLASS TESTERS

Lynn Cleaveland, *University of Arkansas*
Gay Ellis, *Southwest Missouri State University*
Joe Howe, *St. Charles Community College*
John Khoury, *Brevard Community College*
Julie Killingbeck, *Ball State University*
Michael Lanstrum, *Cuyahoga Community College*
Vic Perera, *Kent State University*
Mihaela Poplicher, *University of Cincinnati*
William Radulovich, *Florida Community College of Jacksonville*

SUPPLEMENT AUTHORS

O. Pauline Chow, *Harrisburg Area Community College*
Joe Howe, *St. Charles Community College*
Mark McKibben, *Goucher College*
Virginia Starkenburg, *San Diego City College*

Commitment to Accuracy

From the beginning, the editorial team and I have been committed to providing an accurate and error-free text. In this goal, we have benefited from the help of many people. In addition to the reviewer feedback on this issue and normal proofreading, we enlisted the help of several extra sets of eyes during the production stages and during video filming. I wish to thank them for their dedicated reading of various rounds of pages to help me insure the highest level of accuracy.

Elka Block, *Twin Prime Editorial*
Judith Fethe, *Pellissippi State Technical Community College*
Julie Killingbeck, *Ball State University*
Theodore Lai, *Hudson County Community College*
Michael W. Lanstrum, *Cuyahoga Community College*
Mark McKibben, *Goucher College*
Ann Ostberg, *Grace University*
Frank Purcell, *Twin Prime Editorial*
Marie Vanisko, *California State University—Stanislaus*
Donald Yee, *Essex County Community College*

TABLE OF CONTENTS

A NOTE FROM THE AUTHOR TO THE STUDENT

I wrote this text with careful attention to ways of making your learning experience more successful. If you take full advantage of the unique features and elements of the textbook, I believe your experience in College Algebra will be fulfilling and enjoyable. Let's walk through some of the special book features that will help you in your study of College Algebra.

Prerequisites and Review (Chapter 0)

A review of prerequisite knowledge (intermediate algebra topics) is included in Chapter 0 to provide a brush-up on knowledge and skills necessary for success in the course. In addition, reviews of prerequisite concepts are provided throughout chapters on an as-needed basis.

Clear, Concise, and Inviting Writing

Special attention has been given to presenting an engaging and clear narrative in a layout that is designed to reduce math anxiety in students.

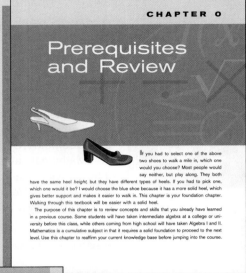

CHAPTER 0

Prerequisites and Review

If you had to select one of the above two shoes to walk a mile in, which one would you choose? Most people would say neither, but play along. They both have the same heel *height*, but they have different types of heels. If you had to pick one, which one would it be? I would choose the blue shoe because it has a more solid heel, which gives better support and makes it easier to walk in. This chapter is your foundation chapter. Walking through this textbook will be easier with a solid heel.

The purpose of this chapter is to review concepts and skills that you already have learned in a previous course. Some students will have taken intermediate algebra at a college or university before this class, while others coming from high school will have taken Algebra I and II. Mathematics is a cumulative subject in that it requires a solid foundation to proceed to the next level. Use this chapter to reaffirm your current knowledge base before jumping into the course.

CHAPTER 2

Graphs

In this chapter, we will review the Cartesian plane. We'll calculate the distance and midpoint between two points. We will then plot graphs of equations. We focus on graphing two types of equations: lines and circles.

Graphs

Cartesian Plane	Graphing Equations	Common Graphs
• Coordinates/Axes • Distance/Midpoint	• Point Plotting • Symmetry	• Lines • Circles

CHAPTER OBJECTIVES

- Plot points in the Cartesian plane and algebraically determine the distance and the midpoint between two points.
- Graph equations by point plotting.
- Determine if a graph has symmetry.
- Graph lines and circles.
- Determine equations of lines and circles.

NAVIGATION THROUGH SUPPLEMENTS

DIGITAL VIDEO SERIES #2

STUDENT SOLUTIONS MANUAL CHAPTER 2

BOOK COMPANION SITE
www.wiley.com/college/young

HIV infection rates

Circular orbits in the solar system

COKE 5-Minute 4:53 PM
Stock prices fluctuating throughout the day

The conversion between degrees Fahrenheit and degrees Celsius is a linear relationship.

Emperor penguins walking in a line, Weddell Sea, Antarctica

Graphs are used in many ways. There is only one temperature that yields the same number in degrees Celsius and degrees Fahrenheit. Do you know what it is? The penguins are a clue.

137

Chapter Introduction Flow Chart and Objectives

A flow chart and list of chapter objectives give you an overview of the chapter to help you see the big picture and the relationships between topics.

SECTION 2.3 Straight Lines

Skills Objectives	Conceptual Objectives

• Graph a line.
• Determine *x*- and *y*-intercepts.
• Find an equation of a line
  • given its slope and the *y*-intercept;
  • given its slope and a point;
  • given two points.
• Graph lines that are parallel or perpendicular to other lines.
• Solve application problems involving linear equations.
	• Classify lines as rising, falling, horizontal, and vertical.
	• Understand slope as a measure (rate) of change.
	• Associate two lines with the same slope with the graph of parallel lines.

Skills/Concepts Objectives

For every section, objectives are divided by skills *and* concepts so that you can learn the difference between solving problems and truly understanding concepts.

Examples

Examples pose a specific problem using concepts already presented and then work through the solution. These serve to enhance your understanding of the subject matter.

Concept Check

The Concept Check provides a periodic stopping point that asks you to stop and think about a related question to extend your conceptual understanding.

Your Turn

Immediately following an example, you are asked to solve a similar problem to reinforce and check your understanding. This feature helps build confidence as you progress in the chapter and is ideal for in-class activity and for preparing to do homework later.

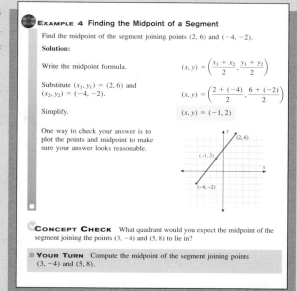

EXAMPLE 4 Finding the Midpoint of a Segment

Find the midpoint of the segment joining points $(2, 6)$ and $(-4, -2)$.

Solution:

Write the midpoint formula.
$$(x, y) = \left(\frac{x_1 + x_2}{2}, \frac{y_1 + y_2}{2}\right)$$

Substitute $(x_1, y_1) = (2, 6)$ and $(x_2, y_2) = (-4, -2)$.
$$(x, y) = \left(\frac{2 + (-4)}{2}, \frac{6 + (-2)}{2}\right)$$

Simplify.
$$(x, y) = (-1, 2)$$

One way to check your answer is to plot the points and midpoint to make sure your answer looks reasonable.

CONCEPT CHECK   What quadrant would you expect the midpoint of the segment joining the points $(3, -4)$ and $(5, 8)$ to lie in?

YOUR TURN   Compute the midpoint of the segment joining points $(3, -4)$ and $(5, 8)$.

COMMON MISTAKE

The most common mistake in calculating slope is writing the coordinates in the wrong order, which results in the slope being opposite in sign.
Find the slope of the line containing the two points $(1, 2)$ and $(3, 4)$.

 CORRECT

Label the points.
$(x_1, y_1) = (1, 2)$ $(x_2, y_2) = (3, 4)$

Write the slope formula.
$$m = \frac{y_2 - y_1}{x_2 - x_1}$$

Substitute the coordinates.
$$m = \frac{4 - 2}{3 - 1}$$

Simplify. $m = \frac{2}{2} = 1$

$$m = 1$$

INCORRECT

Label the points.
$(x_1, y_1) = (1, 2)$ $(x_2, y_2) = (3, 4)$

Apply the slope formula (**ERROR**).
$$m = \frac{4 - 2}{1 - 3}$$

The calculated slope is **INCORRECT** by a negative sign.
$$m = \frac{2}{-2} = -1$$

Common Mistake/ Correct vs. Incorrect

In addition to standard examples, some problems are worked both correctly and incorrectly to highlight common errors students make. Counterexamples like these are often an effective learning approach for many students.

Parallel Words and Math

This text reverses the common presentation of examples by placing the explanation in words *on the left* and the mathematics in parallel *on the right*. This makes it easier for students to read through examples as the material flows more naturally as is commonly presented in lecture.

In other words, if two lines in a plane are parallel, then their slopes are equal, and if the slopes of two lines in a plane are equal, then the lines are parallel.

WORDS	MATH
Lines L_1 and L_2 are parallel.	$L_1 \| L_2$
Two parallel lines have the same slope.	$m_1 = m_2$

STUDY TIP

Symmetry gives us information about the graph for "free."

Study Tips and Caution Notes

These marginal reminders call out important hints or warnings you should be aware of relating to the topic or problem.

CAUTION

Interchanging the coordinates can result in a sign error in the slope.

TECHNOLOGY TIP

To enter the graph of
$(x - 2)^2 + (y + 1)^2 = 4$, solve
for y first. The graphs of
$y_1 = \sqrt{4 - (x - 2)^2} - 1$ and
$y_2 = -\sqrt{4 - (x - 2)^2} - 1$
are shown.

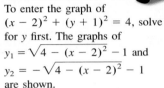

Plot1 Plot2 Plot3
\Y₁B√(4-(X-2)²)-

Technology Tips

These marginal notes provide instruction and visual examples using graphing and scientific calculators to solve problems.

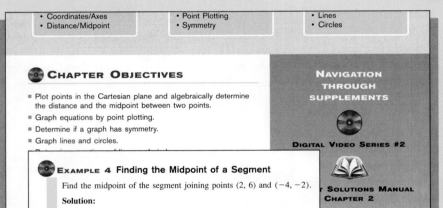

- Coordinates/Axes
- Distance/Midpoint
- Point Plotting
- Symmetry
- Lines
- Circles

CHAPTER OBJECTIVES

- Plot points in the Cartesian plane and algebraically determine the distance and the midpoint between two points.
- Graph equations by point plotting.
- Determine if a graph has symmetry.
- Graph lines and circles.

NAVIGATION THROUGH SUPPLEMENTS

DIGITAL VIDEO SERIES #2

SOLUTIONS MANUAL
CHAPTER 2

Video Icons

Video icons appear on all chapter overviews, chapter reviews, as well as selected examples and Your Turns throughout the chapter to indicate that a video segment is available for that element.

EXAMPLE 4 Finding the Midpoint of a Segment

Find the midpoint of the segment joining points $(2, 6)$ and $(-4, -2)$.

Solution:

Write the midpoint formula.

Substitute $(x_1, y_1) = (2, 6)$ and $(x_2, y_2) = (-4, -2)$.

Simplify.

One way to check your answer is to plot the points and midpoint to make sure your answer looks reasonable.

CONCEPT CHECK   Does a horizontal line have any x-intercepts or y-intercepts?

YOUR TURN   Determine the x- and y-intercepts (if they exist) for the lines given by the following equations:
a. $3x - y = 2$ **b.** $y = 5$

EXAMPLE 2 Graph, Classify the Line, and Determine the Slope

Sketch a line through each pair of points, classify the line as rising, falling, vertical, or horizontal, and determine its slope.

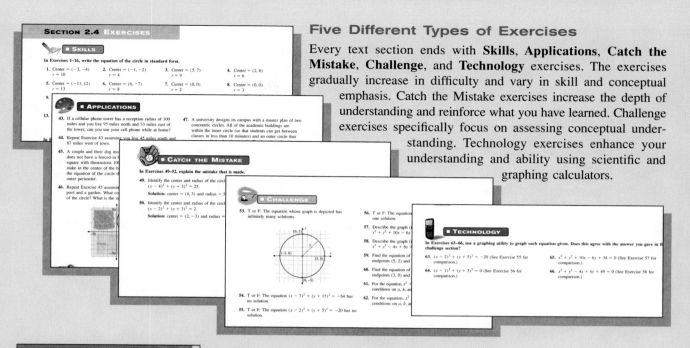

Five Different Types of Exercises

Every text section ends with **Skills**, **Applications**, **Catch the Mistake**, **Challenge**, and **Technology** exercises. The exercises gradually increase in difficulty and vary in skill and conceptual emphasis. Catch the Mistake exercises increase the depth of understanding and reinforce what you have learned. Challenge exercises specifically focus on assessing conceptual understanding. Technology exercises enhance your understanding and ability using scientific and graphing calculators.

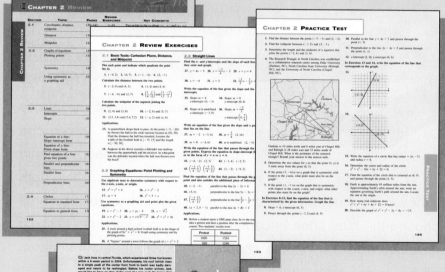

Chapter Review— Summary, Exercises, Practice Test

At the end of every chapter, a review chart organizes the topics in an easy-to-use one-page layout. This feature includes key concepts and formulas, as well as indicating relevant pages and review exercises so that you can quickly summarize a chapter and study smarter. Chapter Review Exercises are provided for extra study and practice. A Practice Test is also included to give you even more practice before moving on.

OOPS! and Tying It All Together

These unique end-of-chapter elements provide a fun and interesting way to apply what you have learned. OOPS! takes a situation in which doing math incorrectly results in a real-world mistake. Tying It All Together presents a question or problem that requires you to use more than one concept to solve the problem.

Prerequisites and Review

If you had to select one of the above two shoes to walk a mile in, which one would you choose? Most people would say neither, but play along. They both have the same heel *height,* but they have different types of heels. If you had to pick one, which one would it be? I would choose the blue shoe because it has a more solid heel, which gives better support and makes it easier to walk in. This chapter is your foundation chapter. Walking through this textbook will be easier with a solid heel.

The purpose of this chapter is to review concepts and skills that you already have learned in a previous course. Some students will have taken intermediate algebra at a college or university before this class, while others coming from high school will have taken Algebra I and II. Mathematics is a cumulative subject in that it requires a solid foundation to proceed to the next level. Use this chapter to reaffirm your current knowledge base before jumping into the course.

In this chapter you will review real and complex numbers, basic operations of polynomials, and factoring polynomials using the FOIL method. Simplifying rational expressions, exponents, and rationalization of denominators will be reviewed. To see how these topics relate to each other, look over the flowchart below. There will be a flowchart like this at the beginning of every chapter to help you understand how topics tie together and reinforce each other.

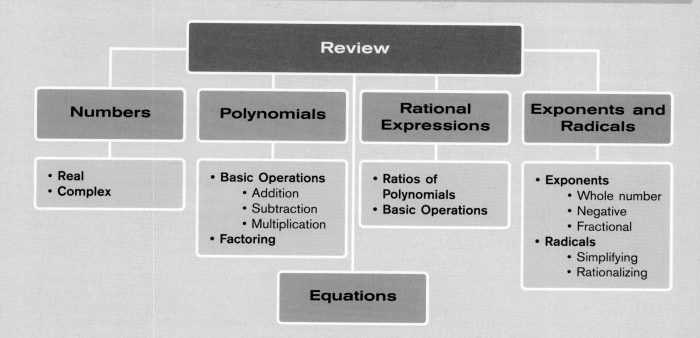

Review

Numbers
- Real
- Complex

Polynomials
- Basic Operations
  - Addition
  - Subtraction
  - Multiplication
- Factoring

Rational Expressions
- Ratios of Polynomials
- Basic Operations

Exponents and Radicals
- Exponents
  - Whole number
  - Negative
  - Fractional
- Radicals
  - Simplifying
  - Rationalizing

Equations

 ## CHAPTER OBJECTIVES

- Perform operations on polynomials.
- Multiply two binomials using the FOIL method.
- Factor polynomials.
- Perform operations on rational expressions.
- Simplify expressions with rational exponents.
- Solve simple equations.

NAVIGATION THROUGH SUPPLEMENTS

DIGITAL VIDEO SERIES #0

STUDENT SOLUTIONS MANUAL CHAPTER 0

BOOK COMPANION SITE
www.wiley.com/college/young

The Set of Real Numbers

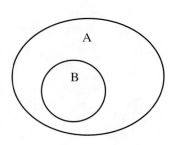

A **set** is a group or collection of objects called **members** or **elements** of the set. If *every* member of set B is also a member of set A, then we say B is a **subset** of A and denote it as $B \subset A$ or as shown on the left. For example, the starting lineup on a baseball team is a subset of the entire team. The set of **natural numbers**, $\{1, 2, 3, 4, \ldots\}$ is a subset of the set of **whole numbers**, $\{0, 1, 2, 3, 4, \ldots\}$. The set of whole numbers is a subset of the set of **integers**, $\{\ldots, -4, -3, -2, -1, 0, 1, 2, 3, \ldots\}$. The set of integers is a subset of the set of *rational numbers*, which is a subset of the set of *real numbers*.

The **set of real numbers** consists of subsets of *rational* and *irrational* numbers. **Rational numbers** are any integers or any fractions that are ratios of integers. In decimal form, numbers that repeat or terminate are rational, and those that do not are **irrational numbers**.

All of the following are examples of **real** numbers.

$$5, \ -17, \ \frac{1}{3}, \ \sqrt{2}, \ \pi, \ 1.37, \ 0, \ -\frac{19}{17}, \ 3.666, \ 3.2179\ldots$$

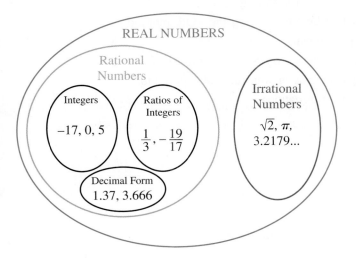

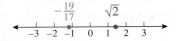

Typically real numbers are represented on a **real number line**. Two of the above examples of real numbers are written on the number line shown on the left.

EXAMPLE 1 Classifying Real Numbers

For the following set of real numbers:

$$-3, 0, \frac{1}{4}, \sqrt{3}, \pi, 7.51, \frac{1}{3}, -\frac{8}{5}, 6.66666$$

a. state the rational numbers.　　**b.** state the irrational numbers.

Solution:

a.　rational: $-3, 0, \frac{1}{4}, 7.51, \frac{1}{3}, -\frac{8}{5}, 6.66666$　　**b.**　irrational: $\sqrt{3}, \pi$

■ **YOUR TURN** Classify the following real numbers as rational or irrational:

$$-\frac{7}{3}, 5.999\overline{9}, 12, 0, -5.27, \sqrt{5}, 2.010010001\ldots$$

Approximations: Rounding and Truncation

When a real number is in decimal form, we often approximate the number by either *rounding off* or *truncating* to a given decimal place. **Truncation** is "cutting off" or eliminating everything beyond a certain point. **Rounding** looks to the right of the specified decimal place and makes a judgment. If the digit to the right is greater than or equal to 5, then the specified digit is rounded up, or increased by 1 unit. If the digit to the right is less than 5, then the specified digit stays the same. In both of these cases, all decimal places to the right of the specified place are removed.

EXAMPLE 2 Approximating Decimals to Two Places

Approximate 17.368204 to two decimal places by

a. truncation　　**b.** rounding

Solution: Approximating to two decimal places implies that the resulting number will only have two digits to the right of the decimal place. The "6" is in the second decimal place.

a. To truncate, eliminate all digits to the right of the 6.　　17.36

b. To round, look to the right of the 6. Because "8" is greater than 5, we round up (add 1 to the 6).　　17.37

■ **YOUR TURN** Approximate 23.02492 to two decimal places by

a. truncation　　**b.** rounding

STUDY TIP

In Examples 2 and 3 we see that rounding and truncation sometimes yield the same approximation, but not always.

EXAMPLE 3 Approximating Decimals to Four Places

Approximate 7.293516 to four decimal places by

a. truncation   **b.** rounding

Solution: The "5" is in the fourth decimal place.

a. To truncate, eliminate all digits to the right of "5."   7.2935

b. To round, look to the right of the "5." Because "1" is less
   than 5, the "5" remains the same.   7.2935

■ **YOUR TURN**   Approximate −2.381865 to four decimal places by

   a. truncation   **b.** rounding

Order of Operations

Addition, subtraction, multiplication, and division are called mathematical operations. When you are handling expressions involving real numbers, it is important to remember the correct *order of operations.* For example, how do we simplify the expression $2 \cdot 5 + 3$? Do we multiply first then add, or add first then multiply? In mathematics, conventional notation implies multiplication first and then addition: $2 \cdot 5 + 3 = 10 + 3 = 13$. Parentheses imply grouping of terms and should always be performed first. If there are nested parentheses, always start with the innermost parentheses and work your way out. Within parentheses, follow the conventional order of operations.

EXAMPLE 4 Using Order of Operations to
Simplify Expressions

Simplify the expression $3[5 \cdot (4 - 1) - 2 \cdot 7]$.

Solution:

Simplify the inner parentheses: $(4 - 1) = 3$. $3[5 \cdot 3 - 2 \cdot 7]$

Inside the parentheses, perform the
multiplication $5 \cdot 3 = 15$ and $2 \cdot 7 = 14$. $3[15 - 14]$

Inside the parentheses, perform subtraction. $3[1]$

Simplify.   3

This approach works well on real numbers when you can perform the operations between the parentheses. In algebra, however, real numbers are often represented by variables. For example, how do we simplify the expression, $3(2x - 5y)$? In this case, we cannot subtract $5y$ from $2x$. Instead, we rely on the basic *properties of real numbers,* or rules of algebra.

Properties of Real Numbers

You probably already know many properties of real numbers. For example, if you add up four numbers, it doesn't matter what order you add them. If you multiply five numbers, it doesn't matter what order you multiply them. If you add 0 to a real number or multiply a real number by 1, the result yields the original real number. Basic properties of real numbers are summarized below.

BASIC PROPERTIES OF REAL NUMBERS

COMMUTATIVE PROPERTY OF ADDITION

The order in which two numbers are added does not affect their sum.
$$x + y = y + x$$

COMMUTATIVE PROPERTY OF MULTIPLICATION

The order in which two numbers are multiplied does not affect their product.
$$xy = yx$$

ASSOCIATIVE PROPERTY OF ADDITION

The order in which three numbers are added does not affect their sum.
$$(x + y) + z = x + (y + z)$$

ASSOCIATIVE PROPERTY OF MULTIPLICATION

The order in which three numbers are multiplied does not affect their product.
$$(xy)z = x(yz)$$

DISTRIBUTIVE PROPERTY

The first term must be multiplied (distributed) to *both* terms within parentheses.
$$x(y + z) = xy + xz$$

ADDITIVE IDENTITY

Adding zero to any number yields the same number.
$$x + 0 = x$$

MULTIPLICATIVE IDENTITY

Multiplying any number by one yields the same number.
$$x(1) = x$$

ADDITIVE INVERSE

A number added to its opposite is zero.
$$x + (-x) = 0$$

MULTIPLICATIVE INVERSE

A number multiplied by its reciprocal is one.
$$x\left(\frac{1}{x}\right) = 1$$

EXAMPLE 5 Using the Distributive Property

Use the distributive property to eliminate the parentheses: $3(x + 5)$.

Solution:

Distribute. $3(x) + 3(5)$

Simplify. $3x + 15$

You also probably know the rules that apply when multiplying negative real numbers. For example, "A negative times a negative is a positive."

NEGATIVE PROPERTIES OF REAL NUMBERS

A negative quantity times a positive quantity is a negative quantity.	$(-x)(y) = -xy$
A negative quantity times a negative quantity is a positive quantity.	$(-x)(-y) = xy$
Minus a negative quantity is a positive quantity.	$-(-x) = x$
A negative sign is distributed throughout parentheses.	$-(x + y) = -x - y$
	$-(x - y) = -x + y$

EXAMPLE 6 Using the Properties of Negative Real Numbers

TECHNOLOGY TIP

Here are the calculator keystrokes for $-5 + 7 - (-2)$.

Scientific calculators:

$\boxed{(-)}\,5\,\boxed{+}\,7\,\boxed{-}\,\boxed{(-)}\,2\,\boxed{=}$

Graphing calculators:

```
-5+7--2
              4
```

Simplify the expressions.

a. $-5 + 7 - (-2)$ **b.** $-(-3)(-4)(-6)$

Solution:

a. Eliminate parentheses. $-5 + 7 + 2$

    Combine the three quantities.   4

b. Group the terms. $[-(-3)][(-4)(-6)]$

    Perform the multiplication inside []. $[3][24]$

    Multiply.   72

EXAMPLE 7 Simplify an Expression Using Properties of Real Numbers

Eliminate the parentheses $-(2x - 3y)$.

COMMON MISTAKE

A common mistake is distributing a negative only to the first term.

 CORRECT   **INCORRECT**

Simplify.   Simplify.

     $-2x + 3y$ $-2x - 3y$

Error is not distributing the negative, $(-)$, through the second term.

CAUTION

Distribute a negative through *all* terms inside the parentheses.

■ **YOUR TURN** Eliminate the parentheses.

  a. $-2(x + 5y)$ **b.** $-(3 - 2b)$

■ **Answer: a.** $-2x - 10y$ **b.** $-3 + 2b$

What is the product of any real number times zero? The answer is zero. This property also leads to the zero product rule, which is the basis for factoring.

ZERO PROPERTY

A number multiplied by zero is zero:

$$x \cdot 0 = 0$$

ZERO PRODUCT RULE

$$\text{If } xy = 0, \text{ then } x = 0 \text{ or } y = 0$$

If the product of two numbers is zero, then one of those numbers has to be zero.

Fractions always seem to intimidate students; in fact most instructors teach students to eliminate fractions in algebraic equations. It is important to realize that you can never divide by zero. Therefore, in the following table of fractional properties we have assumed that no divisors are zero.

FRACTIONAL PROPERTIES (ASSUMING NO DIVISORS ARE ZERO)

Equivalent fractions.	$\dfrac{a}{b} = \dfrac{c}{d}$ if and only if $ad = bc$
Multiplication of fractions.	$\dfrac{a}{b} \cdot \dfrac{c}{d} = \dfrac{ac}{bd}$
Adding fractions with same denominator.	$\dfrac{a}{b} + \dfrac{c}{b} = \dfrac{a + c}{b}$
Subtracting fractions with same denominator.	$\dfrac{a}{b} - \dfrac{c}{b} = \dfrac{a - c}{b}$
Adding fractions with a common denominator.	$\dfrac{a}{b} + \dfrac{c}{d} = \dfrac{ad + bc}{bd}$
Dividing by a fraction is equivalent to multiplying by its reciprocal.	$\dfrac{a}{b} \div \dfrac{c}{d} = \dfrac{a}{b} \cdot \dfrac{d}{c}$

EXAMPLE 8 Performing Operations with Fractions

Perform the indicated operations involving fractions.

a. $\dfrac{2}{3} - \dfrac{1}{4}$ **b.** $\dfrac{2}{3} \div 4$

Solution:

a. Determine the least common denominator. $3 \cdot 4 = 12$

Rewrite fractions using least common denominator. $\dfrac{2(4) - 1(3)}{3(4)}$

Eliminate parentheses. $\dfrac{8-3}{12}$

Add terms in numerator. $\dfrac{5}{12}$

b. Rewrite 4 with an understood 1 in the denominator. $\dfrac{2}{3} \div \dfrac{4}{1}$

Dividing by a fraction is equivalent to multiplying by its reciprocal. $\dfrac{2}{3} \cdot \dfrac{1}{4}$

Multiply numerators and denominators, respectively. $\dfrac{2}{12}$

Reduce the fraction to simplest form. $\dfrac{1}{6}$

It is important to note that the *least* common denominator is not always the product of the two denominators. For example, $\frac{1}{4} - \frac{1}{10}$. The *least* common denominator in this case is 20.

CONCEPT CHECK   What is the difference between simplifying an expression and reducing a fraction?

■ **YOUR TURN**   Perform the indicated operations involving fractions.

a. $\dfrac{3}{5} + \dfrac{1}{2}$ **b.** $\dfrac{1}{5} \div \dfrac{3}{10}$

SECTION 0.1 SUMMARY

In this section, real numbers were defined. Decimals are approximated by either truncating or rounding. Mathematical convention implies the order of operations as parentheses, multiplication/division, and then addition/subtraction. Multiplication and division are done at the same time (whichever appears first) when moving left to right across the problem. Addition and subtraction are handled in a similar manner. The properties of real numbers are used as the basic rules of algebra when dealing with algebraic expressions.

SECTION 0.1 EXERCISES

■ **SKILLS**

Classify the real numbers as rational or irrational.

1. $\dfrac{11}{3}$ **2.** $\dfrac{22}{3}$ **3.** $2.07172737\ldots$ **4.** π

■ **Answer: a.** $\dfrac{11}{10}$ **b.** $\dfrac{2}{3}$

5. 2.7766776677 ...   **6.** 5.222222$\overline{2}$ **7.** $\sqrt{5}$ **8.** $\sqrt{17}$

Approximate the real numbers to three decimal places by (a) rounding (b) truncation.

9. 7.3471   **10.** 9.2549   **11.** 2.9949   **12.** 6.9951

Simplify.

13. $5 + 2 \cdot 3 - 7$ **14.** $2 + 5 \cdot 4 + 3 \cdot 6$ **15.** $2 \cdot (5 + 7 \cdot 4 - 20)$ **16.** $-3 \cdot (2 + 7) + 8 \cdot (7 - 2 \cdot 1)$

17. $-3 - (-6)$ **18.** $-5 + 2 - (-3)$ **19.** $x - (-y) - z$ **20.** $-a + b - (-c)$

21. $-(3x + y)$ **22.** $-(4a - 2b)$ **23.** $\dfrac{-3}{(5)(-1)}$ **24.** $-\dfrac{12}{(-3)(-4)}$

Combine into one expression.

25. $\dfrac{1}{3} + \dfrac{5}{4}$ **26.** $\dfrac{1}{2} - \dfrac{1}{5}$ **27.** $\dfrac{5}{6} - \dfrac{1}{3}$ **28.** $\dfrac{7}{3} - \dfrac{1}{6}$ **29.** $\dfrac{x}{5} + \dfrac{2x}{15}$ **30.** $\dfrac{y}{3} - \dfrac{y}{6}$

Multiply or divide the fractions.

31. $\dfrac{2}{7} \cdot \dfrac{14}{3}$ **32.** $\dfrac{2}{3} \cdot \dfrac{9}{10}$ **33.** $\dfrac{2}{7} \div \dfrac{10}{3}$

34. $\dfrac{4}{5} \div \dfrac{7}{10}$ **35.** $\dfrac{a}{b} \div \dfrac{a^2}{b^3}$ $a > 0, b > 0$ **36.** $\dfrac{3a}{7} \div \dfrac{b}{21}$ $a > 0, b > 0$

 ■ APPLICATIONS

On February 27, 2005, the United States deficit was approximately \$7,724,006,827,525, and at that time, the estimated population was approximately 295,699,283 citizens.

37. Money. Round the deficit to the nearest million.

38. Population. Round the number of citizens to the nearest thousand.

39. Money. If the debt is distributed evenly to all citizens, what is the national debt per citizen? Round your answer to the nearest dollar.

40. Money. If the debt is distributed evenly to all citizens, what is the national debt per citizen? Truncate your answer to the nearest cent.

 ■ CATCH THE MISTAKE

In Exercises 41 and 42, explain the mistake that is made.

41. Round 13.2749 to two decimal places.

Solution:

The "9," to the right of the "4," causes the "4" to round to "5."   13.275

The "5," to the right of the "7," causes the "7" to be rounded to "8."   13.28

13.28 is incorrect. What mistake was made?

42. Simplify the expression: $3(x + 5) - 2(4 + y)$.

Solution:

Eliminate parentheses. $3x + 15 - 8 + y$

Simplify. $3x + 7 + y$

This is incorrect. What mistake was made?

■ **CHALLENGE**

43. T or F: Student-athletes are a subset of the students in the honors program.

44. T or F: The students who are members of fraternities or sororities are a subset of the entire student population.

45. T or F: Every integer is a rational number.

46. T or F: $\dfrac{a}{c} + \dfrac{b}{d} = \dfrac{a + b}{c + d}$.

■ **TECHNOLOGY**

47. Use your calculator to compute $\sqrt{1260}$. Is this a rational or irrational number?

48. Use your calculator to compute $\sqrt{\frac{144}{25}}$. Is this a rational or irrational number?

SECTION 0.2 Exponents and Radicals

Skills Objectives

■ Use properties of exponents to simplify expressions.
■ Use scientific notation to represent real numbers.
■ Use properties of rational exponents.
■ Use radical notation and properties of radicals.
■ Rationalize denominators containing radicals.

Conceptual Objectives

■ Understand that radicals are equivalent to rational exponents.
■ Understand that a radical implies one number (principal root), not two ($\pm$).

Exponents

Exponents are used to represent repeated multiplication. For example, $2^5 = 2 \cdot 2 \cdot 2 \cdot 2 \cdot 2$.

DEFINITION

Let a be a real number and n be a positive integer; then a^n is defined as

$$a^n = \underbrace{a \cdot a \cdot a \cdots a}_{n \text{ factors}} \qquad (n \text{ factors of } a)$$

where n is the **exponent**, or **power**, and a is the **base**: a^n

The expression a^n is stated as "a raised to the nth power" or "a to the nth power." For the special cases of $n = 2$ and $n = 3$, we call a^2 "a squared" and a^3 "a cubed."

EXAMPLE 1 Evaluating Integer Exponents

Evaluate the following expressions.

a. 4^3 **b.** 8^1 **c.** $(-3)^4$ **d.** -3^4

Solution:

a. $4^3 = 4 \cdot 4 \cdot 4 = \boxed{64}$ **b.** $8^1 = \boxed{8}$

c. $(-3)^4 = (-3)(-3)(-3)(-3) = \boxed{81}$

d. $-3^4 = -(3 \cdot 3 \cdot 3 \cdot 3) = \boxed{-81}$

Notice the difference between parts *c* and *d* in Example 1. In mathematics, the implied *order of operations* is (1) <u>P</u>arentheses (2) <u>E</u>xponents (3) <u>M</u>ultiplication (4) <u>D</u>ivision (5) <u>A</u>ddition (6) <u>S</u>ubtraction. The phrase, "<u>P</u>lease <u>E</u>xcuse <u>M</u>y <u>D</u>ear <u>A</u>unt <u>S</u>ally," is often used to remember the correct order. In part *c*, parentheses come first, which brings the negative sign into the base. In part *d*, we can rewrite the negative as an understood multiplication of -1: $-3^4 = (-1)(3)^4$. The exponentiation is performed first followed by multiplication of -1.

The following table lists properties of exponents.

PROPERTIES OF INTEGER EXPONENTS

For *a* and *b* nonzero real numbers and *m* and *n* integers:

1. $a^m \cdot a^n = a^{m+n}$
When multiplying exponentials with the same base, add exponents.

2. $\dfrac{a^m}{a^n} = a^{m-n}$
When dividing exponentials with the same base, subtract the exponents, (numerator − denominator).

3. $(a^m)^n = a^{mn}$
When raising an exponential to a power, multiply exponents.

4. $a^0 = 1$
Any nonzero number raised to zero is 1. Note: 0^0 is not defined.

5. $a^{-n} = \dfrac{1}{a^n} = \left(\dfrac{1}{a}\right)^n$
A negative exponent implies a reciprocal.

6. $(ab)^n = a^n b^n$
A product raised to a power is equal to the product of each factor raised to the power.

7. $\left(\dfrac{a}{b}\right)^n = \dfrac{a^n}{b^n}$
A quotient raised to a power is equal to the quotient of the factors raised to the power.

EXAMPLE 2 Using Properties of Exponents

Simplify the exponential expressions and express in terms of positive exponents.

a. $(3x^2 z^{-4})^{-3}$ **b.** $\dfrac{(x^2 y^{-3})^2}{(x^{-1} y^4)^{-3}}$

Solution:

a. Property 6: $(3x^2 z^{-4})^{-3} = (3)^{-3}(x^2)^{-3}(z^{-4})^{-3}$

Property 3: $= 3^{-3} x^{-6} z^{12}$

TECHNOLOGY TIP
To use the calculator to check the answers, here are the calculator keystrokes for 4^3.
Scientific calculators:

4 $\boxed{x^y}$ 3 $\boxed{=}$

Graphing calculators:

4^3

64

STUDY TIP
It is customary to write expressions with *positive* exponents.

Property 5:
$$= \frac{z^{12}}{3^3 x^6}$$

Simplify.
$$= \frac{z^{12}}{27x^6}$$

b. Property 6:
$$\frac{(x^2 y^{-3})^2}{(x^{-1} y^4)^{-3}} = \frac{x^4 y^{-6}}{x^3 y^{-12}}$$

Property 2:
$$= x^{4-3} y^{-6-(-12)}$$

Simplify.
$$= xy^6$$

■ **YOUR TURN** Simplify the exponential expression and express in terms of positive exponents $\dfrac{(ts^2)^{-3}}{(2t^4 s^3)^{-1}}$.

Scientific Notation

Some quantities are represented by either very large or very small numbers. For example, there are approximately 50 trillion cells in the human body. We write 50 trillion as 50 followed by 12 zeros 50,000,000,000,000. An efficient way of writing such a large number is using **scientific notation**. Notice that 50,000,000,000,000 is **5** followed by **13** zeros, or in scientific notation, 5×10^{13}.

Very small numbers can also be written using scientific notation. For example, in laser communications a pulse width is 2 femtoseconds, or 0.000000000000002 seconds. Notice that if we start with **2.0** and move the decimal point **15** places to the left (adding zeros in between) the result is 0.000000000000002, or in scientific notation, 2×10^{-15}.

> **SCIENTIFIC NOTATION**
>
> **Scientific notation** has the form $\pm c \times 10^n$ where $1 \le c < 10$ and n is an integer. A positive exponent is used for large numbers, and a negative exponent is used for small numbers.

EXAMPLE 3 Expressing a Number in Scientific Notation

Express the following numbers in scientific notation.
a. 3,856,000,000,000,000   **b.** 0.00000275

Solution:

a. This is a large number, so we expect a positive exponent.

Rewrite the number with the implied decimal point.
3,856,000,000,000,000.

Move the decimal point to the *left* 15 places.
3.856×10^{15}

■ **Answer:** $\dfrac{s}{2t^3}$

b. This is a small number, so we expect a negative exponent.

Move the decimal point to the
right 6 places. $\qquad 0.00000275 = \boxed{2.75 \times 10^{-6}}$

YOUR TURN Express the following numbers in scientific notation.
a. 4,520,000,000   **b.** 0.00000043

EXAMPLE 4 Converting from Scientific Notation to Decimals

Write each number as a decimal.

a. 2.869×10^5 **b.** 1.03×10^{-3}

Solution:

a. The exponent is positive, so we expect a large number.

Move the decimal point 5 places to the right
(add zeros in between). $\qquad\qquad \boxed{286,900}$

b. The exponent is negative, so we expect a small number.

Move the decimal point 3 places to the left
(add zeros in between). $\qquad\qquad \boxed{0.00103}$

CONCEPT CHECK Is 3.7×10^{-8} a very large or very small number?

YOUR TURN Write each number as a decimal.
a. 8.1×10^4 **b.** 3.7×10^{-8}

Rational Exponents

Now that we have stated the definitions and properties of integer exponents, let's examine what happens when the exponents are rational numbers. For instance, we know that $4^2 = 16$, but what is $4^{1/2}$?

DEFINITION   **nTH ROOT OF A NUMBER**

For a positive integer $n \geq 2$ with a and b real numbers:

$$\text{if } a^n = b \text{ then } a \text{ is the } \mathbf{nth\ root} \text{ of } b$$

If $n = 2$, the root is called a **square root**. If $n = 3$, the root is called a **cube root**.

■ **Answer: a.** 4.52×10^9 **b.** 4.3×10^{-7} ■ **Answer: a.** 81,000   **b.** 0.000000037

For example,

WORDS	MATH
2 is the 4th root of 16	$2^4 = 16$
3 is the square root of 9	$3^2 = 9$
-3 is the square root of 9	$(-3)^2 = 9$
-2 is the cube root of -8	$(-2)^3 = -8$

Notice in the above example that 9 has two square roots, -3 and 3. For this reason we define the **principal square root** of 9 as the positive root, 3.

DEFINITION   PRINCIPAL nTH ROOT OF A NUMBER

For a positive integer $n \geq 2$ and b a real number:

$$b^{1/n} \text{ is the \textbf{principal } \textit{n}\textbf{th root} of } b$$

1. If n is *even* and b is *positive,* then $b^{1/n}$ represents the nth positive root of b.
2. If n is *even* and b is *negative,* then $b^{1/n}$ does not represent a real number (we will discuss this more in Section 0.7 on complex numbers).
3. If n is *odd,* then $b^{1/n}$ represents the real nth positive root of b. There is only one real root.

For example,

1. $25^{1/2} = 5$ **2.** $(-25)^{1/2}$ is not a real number.   **3.** $(-8)^{1/3} = -2$

A way to put these three examples into words is as follows:

1. Q: What number squared yields 25?   A: 5
2. Q: What number squared yields -25?   A: No real number
3. Q: What number cubed yields -8?   A: -2

In order for the properties of integer exponents to hold for rational exponents we must require the following definition.

DEFINITION   RATIONAL EXPONENTS

For positive integers m and n, and for any real number b:

$$b^{m/n} = (b^{1/n})^m = (b^m)^{1/n}$$

Note: When n is even, b cannot be negative.

TECHNOLOGY TIP

Here are the calculator keystrokes for a. in Example 5.

Scientific calculators:

16 x^y (3 ÷ 2) =

Graphing calculators:

```
16^(3/2)
              64
```

EXAMPLE 5 Simplifying Expressions with Rational Exponents

Simplify each expression.

a. $16^{3/2}$ **b.** $(-8)^{2/3}$

Solution:

a. $16^{3/2} = (16^{1/2})^3 = 4^3 = \boxed{64}$ **b.** $(-8)^{2/3} = [(-8)^{1/3}]^2 = (-2)^2 = \boxed{4}$

■ **YOUR TURN** Simplify $27^{2/3}$.

Radicals

It is sometimes more convenient to use *radical* notation for rational exponents.

$$b^{1/n} = \sqrt[n]{b}$$

The following are examples of *radicals* and their equivalent rational exponents:

$$\sqrt{7} = 7^{1/2} \qquad z^{3/4} = \sqrt[4]{z^3} \qquad 4y^{2/3} = 4\sqrt[3]{y^2}$$

DEFINITION **RADICAL**

For a positive integer $n \geq 2$ and a real number b:

$\sqrt[n]{b}$ is the **principal nth root** of b or the **nth root radical**.

$\sqrt{}$ is called a **radical**, n is called the **index**, and b is called the **radicand**.

Note: If $n = 2$, there is no need to write the 2 because it is implied.

To convert back and forth between radical and fractional notation, we use the following:

$$\sqrt[n]{b^m} = (b^m)^{1/n} = b^{m/n} \qquad \text{or} \qquad b^{m/n} = (b^{1/n})^m = (\sqrt[n]{b})^m$$

All of the properties that were stated for exponents (see the box labeled Properties of Integer Exponents in the beginning of this section) also apply to radicals.

$$\sqrt[n]{ab} = \sqrt[n]{a}\sqrt[n]{b} \qquad \sqrt[n]{\frac{a}{b}} = \frac{\sqrt[n]{a}}{\sqrt[n]{b}} \qquad \sqrt[n]{a^m} = (\sqrt[n]{a})^m$$

For example,

$$\sqrt{81} = 9 \qquad \sqrt[4]{81} = 3 \qquad \sqrt[3]{-8} = -2 \qquad \sqrt{5} \cdot \sqrt{2} = \sqrt{10} \qquad \sqrt[3]{\frac{8}{27}} = \frac{\sqrt[3]{8}}{\sqrt[3]{27}} = \frac{2}{3}$$

Combining Radicals and Rationalizing Denominators

We can add or subtract radicals as long as we obey the following rule.

ADDING OR SUBTRACTING RADICALS

Two or more radicals can be combined as long as they have the same *radicand*. Radicals with the same radicand are called **like terms**. $\sqrt{4} + 3\sqrt{4} = 4\sqrt{4}$

EXAMPLE 6 Combining Radicals

Combine the radicals: $2\sqrt{11} - 3\sqrt{11} + 7\sqrt{11}$.

Solution:

$2\sqrt{11} - 3\sqrt{11} + 7\sqrt{11} = (2 - 3 + 7)\sqrt{11} = \boxed{6\sqrt{11}}$

It is conventional to simplify fractions. For example, $\frac{8}{12}$ is simplified to $\frac{2}{3}$. Similarly, we can use properties of radicals to simplify $\sqrt{12}$.

$$\sqrt{12} = \sqrt{4 \cdot 3} = \sqrt{4}\sqrt{3} = 2\sqrt{3}$$

DEFINITION **SIMPLIFIED FORM OF A RADICAL**

To express a radical in **simplified form,** the following must be satisfied:

- A factor in the radicand is not raised to a power greater than the index.
- The power of the radicand does not share a common factor with the index.
- A denominator does not contain a radical.
- A radical does not contain a fraction.

EXAMPLE 7 Simplifying Radicals

For each of the radicals, explain why it is not simplified and then write the radical in simplified form. Assume $x \geq 0$.

a. $\sqrt{x^3}$ **b.** $\sqrt[8]{x^6}$ **c.** $\dfrac{1}{\sqrt{x}}$ **d.** $\sqrt{\dfrac{3}{5}}$

Solution:

a. The power of the radicand, 3, is greater than the index, 2, which is not simplified.

Write $x^3 = x^2 x$. $\sqrt{x^3} = \sqrt{x^2}\sqrt{x}$

Simplify. $= x\sqrt{x}$

b. The power of the radicand, 6, shares a common factor, 2, with the index, 8.

Write the radical in rational exponent form. $\sqrt[8]{x^6} = x^{6/8}$

Note that $\dfrac{6}{8} = \dfrac{3}{4}$. $= x^{3/4}$

Express using radical notation. $= \sqrt[4]{x^3}$

c. The denominator contains a radical.

Eliminate the radical in the denominator by multiplying the numerator and denominator by $\sqrt{x}$. $\dfrac{1}{\sqrt{x}} = \dfrac{1}{\sqrt{x}} \cdot \dfrac{\sqrt{x}}{\sqrt{x}}$

Simplify. $= \dfrac{\sqrt{x}}{x}$

d. The radical contains a fraction.

Use property of radicals. $\sqrt{\dfrac{3}{5}} = \dfrac{\sqrt{3}}{\sqrt{5}}$

The denominator contains a radical.

Multiply numerator and denominator by $\sqrt{5}$.

$$= \frac{\sqrt{3}}{\sqrt{5}} \cdot \frac{\sqrt{5}}{\sqrt{5}}$$

Simplify.

$$= \frac{\sqrt{15}}{5}$$

Parts *c* and *d* are examples of how to **rationalize a denominator** with a *single radical*. What happens if there are two terms in the denominator? Let's investigate a particular product first. In general,

$$(\sqrt{a} + \sqrt{b})(\sqrt{a} - \sqrt{b}) = (\sqrt{a})^2 - (\sqrt{b})^2 = a - b$$

Notice that the product does not contain a radical.

WORDS	**MATH**
Therefore, to simplify the expression:	$\dfrac{1}{(\sqrt{a} + \sqrt{b})}$
Multiply the numerator and denominator by $(\sqrt{a} - \sqrt{b})$.	$\dfrac{1}{(\sqrt{a} + \sqrt{b})} \cdot \dfrac{(\sqrt{a} - \sqrt{b})}{(\sqrt{a} - \sqrt{b})}$
The denominator now contains no radicals.	$\dfrac{(\sqrt{a} - \sqrt{b})}{(a^2 - b^2)}$

These two factors, $(\sqrt{a} + \sqrt{b})$ and $(\sqrt{a} - \sqrt{b})$, are **conjugates** of each other.

EXAMPLE 8   Rationalizing Denominators with Sums or Differences of Radicals

Write $\dfrac{5}{3 - \sqrt{2}}$ in simplified form.

Solution:

To eliminate the radical in the denominator, multiply the numerator and denominator by the conjugate, $3 + \sqrt{2}$.

$$\frac{5}{(3 - \sqrt{2})} \cdot \frac{(3 + \sqrt{2})}{(3 + \sqrt{2})} = \frac{5(3 + \sqrt{2})}{(3 - \sqrt{2})(3 + \sqrt{2})}$$

The denominator now contains no radicals.

$$= \frac{15 + 5\sqrt{2}}{9 - 2}$$

$$= \frac{15 + 5\sqrt{2}}{7}$$

YOUR TURN   Write the expression $\dfrac{7}{1 - \sqrt{3}}$ in simplified form.

■ **Answer:** $-\dfrac{7(1 + \sqrt{3})}{2}$

| SECTION 0.2 | SUMMARY |

In this section, properties of exponents were discussed. Scientific notation is a convenient way of using exponents to represent either very small or very large numbers. In scientific notation, positive exponents correspond to large numbers and negative exponents correspond to small numbers. When the exponents are rational numbers, it is important to specify the principal root. Radicals are another way of representing rational exponents. It is customary never to leave radicals in a denominator. Rationalization of the denominator is the technique used to eliminate such radicals.

SECTION 0.2 EXERCISES

■ SKILLS

Simplify.

1. $(-27)^{1/3}$ **2.** $8^{2/3}$ **3.** $(-32)^{1/5}$ **4.** $(-1)^{1/3}$ **5.** $1^{5/2}$ **6.** $9^{3/2}$

Simplify using properties of exponents. Express in terms of positive exponents.

7. $x^2 \cdot x^3$ **8.** $x^2 x^{-3}$ **9.** $(x^2)^3$ **10.** $(y^3)^2$ **11.** $(4a)^3$ **12.** $(-2t)^3$

13. $\dfrac{x^5 y^3}{x^7 y}$ **14.** $\dfrac{(2xy)^2}{(-2xy)^3}$ **15.** $\left(\dfrac{b}{2}\right)^{-4}$ **16.** $(9a^{-2}b^3)^{-2}$ **17.** $\dfrac{a^{-2}b^3}{a^4 b^5}$ **18.** $\dfrac{(x^3 y^{-1})^2}{(xy^2)^{-2}}$

Express the given number in scientific notation.

19. 27,600,000   **20.** 144,000,000,000   **21.** 0.0000000567   **22.** 0.00000828

Write the number as a decimal.

23. 4.7×10^7 **24.** 3.9×10^5 **25.** 4.1×10^{-5} **26.** 9.2×10^{-8}

Simplify using properties of rational exponents. Express in terms of positive exponents.

27. $(x^{1/2} y^{2/3})^6$ **28.** $\dfrac{(a^{1/3} b^{1/2})^{-3}}{(a^{-1/2} b^{1/4})^2}$ **29.** $\dfrac{x^{1/2} y^{1/5}}{x^{-2/3} y^{-9/5}}$ **30.** $\dfrac{(2x^{2/3})^3}{(4x^{-1/3})^2}$

Simplify (if possible) the radical expressions.

31. $\sqrt{2} - 5\sqrt{2}$ **32.** $3\sqrt{5} - 2\sqrt{5} + 7\sqrt{5}$ **33.** $\sqrt{3}\sqrt{7}$ **34.** $\sqrt{5}\sqrt{2}$ **35.** $\sqrt{\dfrac{1}{3}}$

36. $\sqrt{\dfrac{2}{5}}$ **37.** $\dfrac{3}{1 - \sqrt{5}}$ **38.** $\dfrac{2}{1 + \sqrt{3}}$ **39.** $\dfrac{1 + \sqrt{2}}{1 - \sqrt{2}}$ **40.** $\dfrac{3 - \sqrt{5}}{3 + \sqrt{5}}$

■ APPLICATIONS

41. Gravity. If a penny is dropped off a building, the time it takes (seconds) to fall d feet is given by $\sqrt{\dfrac{d}{16}}$. If a penny is dropped off a 1280-foot-tall building, how long will it take until it hits the ground?

42. Gravity. If a ball is dropped off a building, the time it takes (seconds) to fall d meters is approximately given by $\sqrt{\dfrac{d}{5}}$. If a ball is dropped off a 600-meter-tall building, how long will it take until it hits the ground?

■ CATCH THE MISTAKE

In Exercises 43–44, explain the mistake that is made.

43. Simplify $(4x^{1/2}y^{1/4})^2$.

Solution:

Use properties of exponents. $4(x^{1/2})^2(y^{1/4})^2$
Simplify. $4xy^{1/2}$

This is incorrect. What mistake was made?

44. Simplify $\dfrac{2}{5 - \sqrt{11}}$.

Solution:

Multiply numerator and denominator by $5 - \sqrt{11}$. $\dfrac{2}{5 - \sqrt{11}} \cdot \dfrac{(5 - \sqrt{11})}{(5 - \sqrt{11})}$

Multiply numerators and denominators. $\dfrac{2(5 - \sqrt{11})}{25 - 11}$

Simplify. $\dfrac{2(5 - \sqrt{11})}{14} = \dfrac{5 - \sqrt{11}}{7}$

■ CHALLENGE

45. T or F: $-2^n = (-2)^n$ $n = $ integer

46. T or F: $\sqrt{a^2 + b^2} = a + b$

47. T or F: $\sqrt{-4} = -2$

48. T or F: $\left((a^2)^2\right)^2 = a^6$

49. Rationalize the denominator and simplify:

$$\frac{4 + \sqrt{5}}{3 + 2\sqrt{5}}$$

50. Rationalize the denominator and simplify:

$$\frac{\sqrt{a + b} - \sqrt{a}}{\sqrt{a + b} + \sqrt{a}}$$

■ TECHNOLOGY

51. Use a calculator to approximate $\sqrt{11}$ to 3 decimal places.

52. Use a calculator to approximate $\sqrt[3]{7}$ to 3 decimal places.

SECTION 0.3 Polynomials: Basic Operations

Skills Objectives

- Add and subtract polynomials.
- Multiply polynomials.
- Multiply binomials using the FOIL method.

Conceptual Objectives

- Understand that binomials are special cases of polynomials.
- Recognize like terms.

Introduction to Polynomials

$$3x^2 - 7x - 1 \qquad 4y^3 - y \qquad 5z$$

are all examples of polynomials because they combine algebraic expressions. A **polynomial** is a single term or the sum of two or more terms containing variables with whole-number exponents. In other words, a polynomial is a combination of terms in the form, ax^k, where x is the variable, a is the **coefficient**, and k is the **degree** of the term.

The **degree** of the polynomial is determined by the highest degree (exponent) of any single term. In the three examples, the polynomials are classified as a *second*-degree polynomial in x, a *third*-degree polynomial in y, and a *first*-degree polynomial in z.

DEFINITION **POLYNOMIAL**

A **polynomial in x** is an algebraic expression of the form

$$a_n x^n + a_{n-1} x^{n-1} + a_{n-2} x^{n-2} + \cdots + a_2 x^2 + a_1 x + a_0$$

where $a_0, a_1, a_2, \ldots, a_n$ are real numbers with $a_n \neq 0$, and n is a nonnegative integer. The polynomial is of **degree** n, a_n is the **leading coefficient**, and a_0 is the **constant term**.

Polynomials with one, two, and three terms are called **monomials**, **binomials**, and **trinomials**, respectively. Polynomials are typically written in **standard form**, decreasing degrees, as shown in the three examples previously.

For example, $3x^2 + 2x + 5$ is a trinomial of degree 2 whose leading coefficient is 3 and constant term is 5. It is important to note that $2x = 2x^1$, so we say that the term $2x$ is degree 1 and the constant term $5 = 5x^0$ has degree zero.

EXAMPLE 1 **Writing Polynomials in Standard Form and Determining Degree**

Write the following polynomials in standard form and state their degree.

a. $4x - 9x^5 + 2$ **b.** $3 - x^2$ **c.** $3x^2 - 8 + 14x^3 - 20x^8 + x$

Solution

a. $-9x^5 + 4x + 2$ Degree 5

b. $-x^2 + 3$ Degree 2

c. $-20x^8 + 14x^3 + 3x^2 + x - 8$ Degree 8

Adding and Subtracting Polynomials

Polynomials are added and subtracted by combining *like terms*. **Like terms** are terms having the same variable and exponent. Like terms can be combined by adding their coefficients, $3\underline{x^2} + 2x + 4\underline{x^2} + 5 = 7\underline{x^2} + 2x + 5$. The $2x$ and 5 could not be combined because they are not like terms.

EXAMPLE 2 **Adding Polynomials**

Simplify the expression $(5x^2 - 2x + 3) + (3x^3 - 4x^2 + 7)$.

Solution

Eliminate parentheses. $5x^2 - 2x + 3 + 3x^3 - 4x^2 + 7$

Identify like terms.

Combine like terms.

Write in standard form.

$5\underline{x^2} - 2x + \underline{\underline{3}} + 3x^3 - 4\underline{x^2} + \underline{\underline{7}}$

$x^2 - 2x + 10 + 3x^3$

$3x^3 + x^2 - 2x + 10$

 YOUR TURN Simplify the expression $(3x^2 + 5x - 2x^5) + (6x^3 - x^2 + 11)$.

EXAMPLE 3 Subtracting Polynomials

Simplify the expression $(3x^3 - 2x + 1) - (x^2 + 5x - 9)$.

COMMON MISTAKE

A common mistake is only distributing the negative sign to the first term in the second polynomial when subtracting.

 CORRECT

Eliminate parentheses.

$3x^3 - 2x + 1 - x^2 - 5x + 9$

Identify like terms.

$3x^3 - 2\underline{x} + \underline{\underline{1}} - x^2 - 5\underline{x} + \underline{\underline{9}}$

Combine like terms.

$3x^3 - x^2 - 7x + 10$

 INCORRECT

Eliminate parentheses.

$3x^3 - 2x + 1 - x^2 + 5x - 9$

$\overset{\nwarrow \quad \nearrow}{\textbf{ERROR}}$

Identify like terms.

$3x^3 - 2\underline{x} + \underline{\underline{1}} - x^2 + 5\underline{x} - \underline{\underline{9}}$

Combine like terms.

$3x^3 - x^2 + 3x - 8$

CAUTION

When you are subtracting polynomials, it is important to distribute the negative through *all* of the terms in the second polynomial.

■ **YOUR TURN** Simplify the expression $(-7x^2 - x + 5) - (2 - x^3 + 3x)$.

Multiplying Polynomials

The product of two monomials is found by using the properties of exponents (Section 0.2). For example,

$$(-5x^3)(9x^2) = (-5)(9)x^{3+2} = -45x^5$$

To multiply a monomial by a polynomial we use the distributive property (Section 0.1).

EXAMPLE 4 Multiplying a Monomial and a Polynomial

Multiply and simplify $5x^2(3x^5 - x^3 + 7x - 4)$.

Solution:

$$5x^2(3x^5 - x^3 + 7x - 4)$$

Distribute the monomial, $5x^2$, throughout the polynomial.

$$5x^2(3x^5) - 5x^2(x^3) + 5x^2(7x) - 5x^2(4)$$

Multiply each set of monomials.

$$15x^7 - 5x^5 + 35x^3 - 20x^2$$

■ **YOUR TURN** Multiply and simplify $-3x^2(4x^2 - 2x + 1)$.

How do we multiply two polynomials if one is not a monomial? For example, how do we find the product of a binomial and a trinomial such as $(2x - 5)(x^2 - 2x + 3)$? Notice that the binomial is a combination of two monomials. Therefore, we treat each monomial, $2x$ and -5, separately and then combine our results.

WORDS	MATH
Multiply each monomial in the binomial by the trinomial.	$(2x - 5)(x^2 - 2x + 3)$ $2x(x^2 - 2x + 3) - 5(x^2 - 2x + 3)$ $2x(x^2 - 2x + 3) - 5(x^2 - 2x + 3)$
Distribute the monomials throughout the trinomial.	$2x^3 - 4x^2 + 6x - 5x^2 + 10x - 15$ $2x^3 - 4x^2 + 6x - 5x^2 + 10x - 15$
Combine like terms.	$2x^3 - 9x^2 + 16x - 15$

The same procedure holds for all polynomials.

EXAMPLE 5 Multiplying Two Polynomials

Multiply the polynomials: $(2x^2 - 3x + 1)(x^2 - 5x + 7)$.

Solution:

Multiply each term of the first trinomial by each term of the second trinomial.

$$2x^2(x^2 - 5x + 7) - 3x(x^2 - 5x + 7) + 1(x^2 - 5x + 7)$$
$$2x^2(x^2 - 5x + 7) - 3x(x^2 - 5x + 7) + 1(x^2 - 5x + 7)$$
$$2x^4 - 10x^3 + 14x^2 - 3x^3 + 15x^2 - 21x + x^2 - 5x + 7$$

Combine like terms.

$$2x^4 - 10x^3 + 14x^2 - 3x^3 + 15x^2 - 21x + x^2 - 5x + 7$$

Simplify.

$$2x^4 - 13x^3 + 30x^2 - 26x + 7$$

■ **Answer:** $-12x^4 + 6x^3 - 3x^2$

■ **YOUR TURN** Multiply $(-x^3 + 2x - 4)(3x^2 - x + 5)$.

Multiplying Binomials Using the FOIL Method

The method used above to multiply polynomials works for all products of polynomials. Let's multiply two binomials, $(5x - 1)(2x + 3)$.

WORDS	**MATH**
Multiply the monomials in each binomial, term by term.	$5x(2x + 3) - 1(2x + 3)$
Distribute each monomial through the binomial.	$5x(2x) + 5x(3) - 1(2x) - 1(3)$
Multiply each set of monomials.	$10x^2 + 15x - 2x - 3$
Combine like terms.	$10x^2 + 13x - 3$

When multiplying two binomials, a more efficient procedure is called the **FOIL method**. **FOIL** stands for First, Outer, Inner, Last. In this procedure we add four products: the product of the first terms, outer terms, inner terms, and last terms. Let's work this problem again using this procedure:

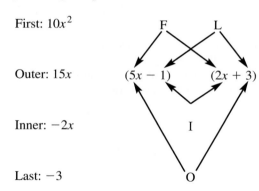

First: $10x^2$

Outer: $15x$

Inner: $-2x$

Last: -3

The inner and outer are like terms, so combining those leads us to:

$$10x^2 + 13x - 3$$

EXAMPLE 6 Multiplying Binomials Using the FOIL Method

Multiply $(3x + 1)(2x - 5)$ using the FOIL method.

Solution:

Multiply the **first** terms.	$(3x)(2x) = 6x^2$
Multiply the **outer** terms.	$(3x)(-5) = -15x$
Multiply the **inner** terms.	$(1)(2x) = 2x$

Multiply the **last** terms. $(1)(-5) = -5$

Add first, outer, inner, last,
and identify like terms. $6x^2 - 15\underline{x} + 2\underline{x} - 5$

☐ Combine like terms. $6x^2 - 13x - 5$

■ **YOUR TURN**   Multiply $(2x - 3)(5x - 2)$.

Recall that a quantity squared is the quantity times itself. For example, $x^2 = x \cdot x$. When a binomial is squared, it is multiplied by itself. For example,

$$(x + 1)^2 = (x + 1)(x + 1)$$
$$= x^2 + x + x + 1$$
$$= x^2 + 2x + 1$$

Similarly,

$$(x - 1)^2 = (x - 1)(x - 1)$$
$$= x^2 - x - x + 1$$
$$= x^2 - 2x + 1$$

It is important to note that the difference in sign, $+$ or $-$, in the binomial corresponds to the difference in sign of the second term of the product. In general, $(a \pm b)^2$ are called **perfect squares**. In addition to squaring binomials, another product that appears frequently produces a **difference of two squares**.

$$(x + 1)(x - 1) = x^2 - 1$$

The three examples above are types of products that appear often, and so it is important for you to know the generalized form:

SPECIAL PRODUCTS OF BINOMIALS

Square of a binomial sum (perfect square):

$$(a + b)^2 = (a + b)(a + b) = a^2 + 2ab + b^2$$

Square of a binomial difference (perfect square):

$$(a - b)^2 = (a - b)(a - b) = a^2 - 2ab + b^2$$

Product of a sum and difference (difference of two squares):

$$(a + b)(a - b) = a^2 - b^2$$

SECTION 0.3 SUMMARY

In this section, polynomials were defined. Polynomials with one, two, and three terms are called monomials, binomials, and trinomials, respectively. Polynomials are added and subtracted by combining like terms. Polynomials are multiplied by distributing the monomials in the first polynomial throughout the second polynomial. In the special case when two binomials are multiplied, the FOIL method can be used to obtain the product.

SECTION 0.3 EXERCISES

 ■ SKILLS

Write the polynomial in standard form, and state the degree of the polynomial.

1. $5x^2 - 2x^3 + 16 - 7x^4$ **2.** $4x + 3 - 6x^3$ **3.** 15 **4.** $y - 2$

Add or subtract the polynomials, gather like terms, and write the simplified expression in standard form.

5. $(2x^2 - x + 7) + (-3x^2 + 6x - 2)$

6. $(3x^2 + 5x + 2) + (2x^2 - 4x - 9)$

7. $(7z^2 - 2) - (5z^2 - 2z + 1)$

8. $(25y^3 - 7y^2 + 9y) - (14y^2 - 7y + 2)$

9. $(2x^2 - 2) - (x + 1) - (x^2 - 5)$

10. $(3x^3 + 1) - (3x^2 - 1) - (5x - 3)$

11. $(4t - t^2 - t^3) - (3t^2 - 2t + 2t^3) + (3t^3 - 1)$

12. $(-z^3 - 2z^2) + (z^2 - 7z + 1) - (4z^3 + 3z^2 - 3z + 2)$

Multiply the polynomials, and write the expressions in standard form.

13. $5xy^2(7xy)$ **14.** $6z(4z^3)$ **15.** $2x^3(1 - x + x^2)$ **16.** $-4z^2(2 + z - z^2)$

17. $(7y - 2y^2)(y - y^2 + 1)$ **18.** $(4 - t^2)(6t + 1 - t^2)$ **19.** $(2x + 1)(3x - 4)$ **20.** $(3z - 1)(4z + 7)$

21. $(x + 2)(x - 2)$ **22.** $(y - 5)(y + 5)$ **23.** $(2x + 3)(2x - 3)$ **24.** $(5y + 1)(5y - 1)$

25. $(t - 2)^2$ **26.** $(t - 3)^2$ **27.** $(z + 2)^2$ **28.** $(z + 3)^2$

29. $y(3y + 4)(2y - 1)$ **30.** $p^2(p + 1)(p - 2)$ **31.** $(x^2 + 1)(x^2 - 1)$ **32.** $(t - 5)^2(t + 5)^2$

33. $(\sqrt{x} - \sqrt{y})(\sqrt{x} + \sqrt{y})$ **34.** $(3 + \sqrt{x})^2$

 ■ APPLICATIONS

35. Profit. Donna decides to sell fabric chord covers for hanging lamps on eBay for $20 a piece. The material for each chord cover costs $9, and advertising on eBay costs $100 a month. Let x be the number of chord covers sold. Write a polynomial representing her monthly profit.

36. Profit. If the revenue associated with selling x units of a product is $R(x) = -x^2 + 100x$ and the cost associated with producing x units of the product is $C(x) = -100x + 7500$, find the polynomial that represents the profit of making and selling x units.

CATCH THE MISTAKE

In Exercises 37–38, explain the mistake that is made.

37. Subtract and simplify: $(2x^2 - 5) - (3x - x^2 + 1)$.

Solution:

Eliminate the parentheses. $2x^2 - 5 - 3x - x^2 + 1$
Collect like terms. $x^2 - 3x - 4$

This is incorrect. What mistake was made?

38. Simplify: $(2 + x)^2$.

Solution:

Write the square of the binomial
as the sum of the squares. $2^2 + x^2$

Simplify. $x^2 + 4$

This is incorrect. What mistake was made?

CHALLENGE

39. T or F: All binomials are polynomials.

40. T or F: The product of two monomials is a binomial.

41. T or F: The sum of two binomials is always a binomial.

42. T or F: $(x - y)^2 = x^2 + y^2$

43. Eliminate the parentheses and write in standard form:
$(7x - 4y^2)^2(7x + 4y^2)^2$.

44. What degree is the *product* of a polynomial of degree n and a polynomial of degree m?

Section 0.4 Factoring Polynomials

Skills Objectives

- Factor out the greatest common factor of a polynomial using the distributive property.
- Factor a trinomial as a product of binomials.
- Factor by grouping.

Conceptual Objectives

- Understand that factoring is using the FOIL method backwards.
- Develop a general strategy for factoring polynomials.

You know how to multiply two binomials, for example,

$$(x + 3)(x + 1) = x^2 + 4x + 3$$

To *factor* a polynomial, you will have to go backwards:

$$x^2 + 4x + 3 = (x + 3)(x + 1)$$

The process of writing a polynomial as a product is called **factoring**. We will use factoring as a means for solving linear and quadratic equations in Chapter 1.

Recall that a real number can be factored into **prime factors** (numbers only divisible by themselves and 1). For instance, $18 = (3)(3)(2)$. A polynomial is **prime** with respect to a particular set of numbers (typically integers or real numbers) if all of the coefficients are from that set of numbers and it cannot be written as a product of two polynomials with coefficients from that same set.

For example, $x^2 - 9$ is *not* prime with respect to the integers because it can be factored into $(x - 3)(x + 3)$. However, $x^2 - 2$ is prime with respect to the integers because it cannot be written as a product of polynomials whose coefficients are integers. Although $x^2 - 2$ is prime with respect to the integers, it is not prime with respect to the real numbers because it can be factored into $(x - \sqrt{2})(x + \sqrt{2})$.

Factoring Polynomials Using the Distributive Property

The simplest type of factoring of polynomials occurs when there is a common factor that can be factored out using the distributive property backwards. For example, $4x^2 - 6x$. Note that $2x$ is common to both terms and it can be factored out as $2x(2x) - 2x(3)$. Using the distributive property backwards, the expression can be written as $2x(2x - 3)$.

EXAMPLE 1 Factoring Polynomials Using the Distributive Property

Factor $3x^2 + 12x$.

Solution:

Identify common factors to all terms.	$3x$
Write each term as a product with $3x$ as a factor.	$3x(x) + 3x(4)$
Using the distributive property backwards, factor out $3x$.	$3x(x + 4)$

$$3x^2 + 12x = 3x(x + 4)$$

■ **YOUR TURN** Factor $4y^3 - 12y$.

Factoring a Trinomial as a Product of Two Binomials

Some trinomials can be factored into a product of two binomials. Recall that the FOIL method (first, outer, inner, last) was used to multiply two binomials. We can use the same procedure to go backwards (factor).

EXAMPLE 2 Factoring Polynomials Using the FOIL Method Backwards

Factor the polynomials into prime factors: $x^2 + 10x + 9$.

Solution:

When using the FOIL method backwards, it is customary to start with the <u>first</u> terms: $x^2 + 10x + 9 = (x \pm \text{?})(x \pm \text{?})$

■ **Answer:** $4y(y^2 - 3)$

At this point, you know that the product of the <u>first</u> terms will yield x^2.

Now let's work on the <u>last</u> terms.

What two numbers when multiplied yield 9? There are several combinations, $(3)(3)$ or $(-3)(-3)$ or $(1)(9)$ or $(-1)(-9)$. Which of these four combinations do we try? Remember that the <u>outer</u> and <u>inner</u> products must combine to yield $10x$. Thus, the only combination we can select is $(1)(9)$.

$$x^2 + 10x + 9 = (x + 9)(x + 1)$$

▮ **YOUR TURN** Factor $x^2 + 22x + 120$.

● **EXAMPLE 3** **Factoring a Trinomial in Which the Lead Coefficient Is Not 1**

Factor $5x^2 + 9x - 2$.

Solution:

Start with the <u>first</u> term. Note that $5x \cdot x = 5x^2$. $(5x \pm \, ?) \, (x \pm \, ?)$

The product of the <u>last</u> terms should yield -2.

Possible combinations are $(-1)(2)$ or $(1)(-2)$ or $(2)(-1)$ or $(-2)(1)$.

Since the <u>outer</u> and <u>inner</u> products must combine to $9x$, the only choice is $(-1)(2)$.

$$5x^2 + 9x - 2 = (5x - 1)(x + 2)$$

▮ **YOUR TURN** Factor $2t^2 + t - 3$.

Recall that in Section 0.3 we consider the special products. Going backwards, we call these perfect squares and difference of two squares.

EXAMPLE 4 **Factoring Trinomials That Are Perfect Squares**

Factor $x^2 + 6x + 9$.

Solution:

Start with the <u>first</u> terms. $(x + \, ?)(x + \, ?)$

$(3)(3), (-3)(-3), (1)(9), (-1)(-9)$ $(x + 3)(x + 3)$

Write the product as a perfect square. $(x + 3)^2$

This polynomial factored into a binomial squared, which we call a *perfect square.*

▮ **Answer:** $(x + 10)(x + 12)$ ▮   **Answer:** $(2t + 3)(t - 1)$

EXAMPLE 5 Factoring Trinomials That Are a Difference of Two Squares

Factor $4x^2 - 9$.

Solution:

Since both the first term, $4x^2 = (2x)^2$, and last term, $9 = (3)^2$, can be written as squares, this polynomial can be classified as a *difference of two squares*.

$$4x^2 - 9 = (2x \pm ?)(2x \pm ?)$$
$$= (2x - 3)(2x + 3)$$

Examples 4 and 5 are examples of polynomials that can be written in a specific type of factored form. The following table summarizes special factored forms.

SPECIAL POLYNOMIAL FACTORED FORMS

$u^2 - v^2 = (u + v)(u - v)$	Difference of two squares
$u^2 + 2uv + v^2 = (u + v)(u + v) = (u + v)^2$	Perfect square
$u^2 - 2uv + v^2 = (u - v)(u - v) = (u - v)^2$	Perfect square
$u^3 + v^3 = (u + v)(u^2 - uv + v^2)$	Sum of two cubes
$u^3 - v^3 = (u - v)(u^2 + uv + v^2)$	Difference of two cubes

■ **YOUR TURN** Factor **a.** $z^2 - 100$ **b.** $9y^2 - 25$.

Factoring by Grouping

Sometimes polynomials with four or more terms can be factored using a strategy called **factoring by grouping**. This method is often overlooked because it is not obvious at first. The strategy is to group the terms that have a common factor.

 EXAMPLE 6 Factoring a Polynomial by Grouping

Factor $x^3 - x^2 + 2x - 2$.

Solution:

Group the terms that have a common factor.	$(x^3 - x^2) + (2x - 2)$
Factor out the common factor in each parentheses.	$x^2(x - 1) + 2(x - 1)$
Use the distributive property.	$(x^2 + 2)(x - 1)$

$$x^3 - x^2 + 2x - 2 = (x^2 + 2)(x - 1)$$

■ **YOUR TURN** Factor $x^3 + x^2 - 3x - 3$.

■ **Answer:** $(x + 1)(x^2 - 3)$ ■ **Answer: a.** $(z - 10)(z + 10)$ **b.** $(3y - 5)(3y + 5)$

SECTION 0.4 SUMMARY

In this section, factoring polynomials was discussed. If all of the terms in a polynomial have a common factor, then a polynomial can be factored using the distributive property. Some trinomials can be factored into a product of two binomials. Special factored forms, such as perfect squares and a difference of squares, arise often. Polynomials with four or more terms sometimes can be factored by grouping.

SECTION 0.4 EXERCISES

 ■ SKILLS

Factor the common term out of the polynomials.

1. $5x + 25$
2. $x^2 + 2x$
3. $4t^2 - 2$
4. $16z^2 - 20z$
5. $3x^3 - 9x^2 + 12x$
6. $14x^4 - 7x^2 + 21x$

Factor as a product of two binomials.

7. $x^2 - 6x + 5$
8. $t^2 - 5t - 6$
9. $y^2 - 2y - 3$
10. $y^2 - 3y - 10$
11. $2y^2 - 5y - 3$
12. $2z^2 - 4z - 6$
13. $3t^2 + 7t + 2$
14. $-6t^2 + t + 2$
15. $x^4 + 2x^2 + 1$
16. $x^6 - 6x^3 + 9$
17. $x^2 - 9$
18. $x^2 - 25$
19. $x^2 + 16$
20. $x^2 + 49$
21. $p^2 + 2pq + q^2$
22. $p^2 - 2pq + q^2$

Factor into a product of a binomial and a trinomial.

23. $t^3 + 27$
24. $z^3 + 64$
25. $y^3 - 64$
26. $x^3 - 1$

Factor into a product of three polynomials.

27. $3x^3 - 5x^2 - 2x$
28. $2y^3 + 3y^2 - 2y$
29. $x^3 - 9x$
30. $w^3 - 25w$

Factor into a product of two binomials.

31. $x^3 - 3x^2 + 2x - 6$
32. $x^5 + 5x^3 - 3x^2 - 15$

 ■ CATCH THE MISTAKE

In Exercises 33–34, explain the mistake that is made.

33. Factor $x^3 - x^2 - 9x + 9$.

    Solution:

    Group terms with common factors. $(x^3 - x^2) + (-9x + 9)$

    Factor out common factors. $x^2(x - 1) - 9(x - 1)$

    Use the distributive property. $(x - 1)(x^2 - 9)$

    Factor $x^2 - 9$. $(x - 1)(x - 3)^2$

    This is incorrect. What mistake was made?

34. Factor $4x^2 + 12x - 40$.

    Solution:

    Factor the trinomial into a product of binomials. $(2x - 4)(2x + 10)$

    Factor out a 2. $2(x - 2)(x + 5)$

    This is incorrect. What mistake was made?

35. T or F: All trinomials can be factored into a product of two binomials.

36. T or F: All polynomials can be factored into prime factors with respect to the real numbers.

37. Factor $a^{2n} - b^{2n}$.

38. Find c such that the trinomial, $x^2 + cx - 14$, can be factored.

SECTION 0.5 Rational Expressions

Skills Objectives	**Conceptual Objectives**

- Simplify rational expressions.
- Add, subtract, multiply, and divide rational expressions.
- Simplify more complicated expressions (complex fractions/mixed quotients).

- Understand why specific real numbers must be eliminated from the domain.

Rational Expressions

Recall that a rational number is a ratio of two integers. Similarly the ratio, or quotient, of two polynomials is a **rational expression**.

$$\text{Rational numbers:} \quad \frac{3}{7}, \quad \frac{5}{9}, \quad \frac{9}{11}$$

$$\text{Rational expressions:} \quad \frac{3x + 2}{x - 5}, \quad \frac{5x^2}{x^2 + 1}, \quad \frac{9}{3x - 2}$$

There is one very important difference between rational numbers and rational expressions. The denominators of rational numbers are never equal to zero. Since the denominators in rational expressions are polynomials, however, sometimes there are values of the variable that will make the denominator zero. In the above rational expressions, $x = 5$ and $x = \frac{2}{3}$ make the first and third expressions' denominators equal to zero. In the second rational expression, there are no real numbers that will correspond to a zero denominator.

The set of real numbers for which an algebraic expression is defined is called the **domain**. Since a rational expression is not defined if its denominator is zero, we must eliminate such values from the domain.

To find the domain of an algebraic expression ask yourself the question, "What can the variable be?" For rational expressions, the answer in general is "any values except those that make the denominator equal to zero."

Let's find the domain of the three rational expressions above:

$\dfrac{3x + 2}{x - 5}$ Q: What can x be?

A: Any number except those that correspond to $x - 5 = 0$.

Therefore $x = 5$ must be eliminated from the domain, or $x \neq 5$.

$\dfrac{5x^2}{x^2 + 1}$ Q: What can x be?

A: Any number except those that correspond to $x^2 + 1 = 0$.

This expression is never equal to zero. Therefore, there are no restrictions on the domain, which implies that x can be any real number .

$\dfrac{9}{3x - 2}$ Q: What can x be?

A: Any number except those that correspond to $3x - 2 = 0$.

Therefore $x = \frac{2}{3}$ must be eliminated from the domain, or $x \neq \frac{2}{3}$.

In this section, we will simplify, multiply, divide, add, and subtract rational expressions. The resulting expressions may not have domain restrictions, but it is important to note that the domain restrictions on the original rational expression still apply.

Simplifying Rational Expressions

Recall that a reduced fraction is written with no common factors.

$$\frac{16}{12} = \frac{4 \cdot 4}{4 \cdot 3} = \frac{4}{4} \cdot \frac{4}{3} = 1 \cdot \frac{4}{3} = \frac{4}{3} \qquad \text{or} \qquad \frac{16}{12} = \frac{\cancel{4} \cdot 4}{\cancel{4} \cdot 3} = \frac{4}{3}$$

REDUCING FRACTIONS

If a, b, and c are real numbers then $\dfrac{ac}{bc} = \dfrac{a}{b}, c \neq 0$.

Similarly, rational fractions are **reduced to lowest form**, or **simplified**, if the numerator and denominator have no common factors other than ± 1. As with real numbers, how readily you write fractions in reduced form depends on your ability to factor.

EXAMPLE 1 Simplify a Rational Expression

Write the expression $\dfrac{x^2 - x - 2}{2x + 2}$ in reduced form.

Solution:

State any domain restrictions. $x \neq -1$

Factor the numerator and denominator. $\dfrac{(x - 2)(x + 1)}{2(x + 1)}$

Cancel the common factor.

$$\frac{(x - 2)(\cancel{x + 1})}{2(\cancel{x + 1})}$$

The reduced, or simplified, expression is written as:

$$\frac{x - 2}{2} \qquad x \neq -1$$

COMMON MISTAKE

A common mistake is eliminating numbers that appear in both the numerator and denominator before factoring.

Simplify, if possible, the rational expression $\dfrac{x + 5}{x^2 + 5}$.

 **CORRECT**

There are no common factors so the rational expression is already in simplified form.

$$\frac{x + 5}{x^2 + 5}$$

 INCORRECT

Cancel the 5s in the numerator and denominator.

$$\frac{x + \cancel{5}}{x^2 + \cancel{5}} \qquad \textbf{ERROR}$$

Simplify. $\dfrac{x}{x^2} = \dfrac{1}{x}$ **INCORRECT**

STUDY TIP

Remember to always factor both the numerator and denominator first, and then cancel common factors.

COMMON MISTAKE

A common mistake is determining domain restrictions from the resulting rational expression, not the original rational expression.

Write the rational expression in simplified form $\dfrac{x}{x^2 + x}$.

 **CORRECT**

Factor the denominator.

$$\frac{x}{x^2 + x} = \frac{x}{x(x + 1)}$$

State domain restrictions.

$$x \neq -1 \text{ and } x \neq 0$$

Cancel common factor, x.

$$\frac{x}{x(x + 1)}$$

Write in simplified form.

$$\frac{1}{x + 1} \qquad x \neq -1, 0$$

 INCORRECT

Factor the denominator.

$$\frac{x}{x^2 + x} = \frac{x}{x(x + 1)}$$

Cancel common factor, x.

$$\frac{x}{x(x + 1)}$$

Simplify. $\dfrac{1}{x + 1}$

State domain restrictions.

$$x \neq -1 \qquad \textbf{ERROR}$$

Write in simplified form.

$$\frac{x}{x^2 + x} = \frac{1}{x + 1} \quad x \neq -1 \text{ \textbf{INCORRECT}}$$

 CAUTION

It is important to note the domain restrictions of the rational expression *before* canceling the common factors.

TECHNOLOGY TIP

Plot the graph of the original expression, $\frac{x^2 - x - 6}{x^2 + x - 2}$.

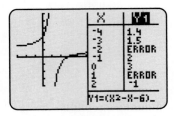

Note that $x = -2$, $x = 1$ are not defined.

Now graph the simplified expression $\frac{x-3}{x-1}$

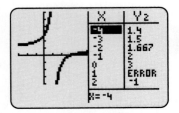

$x = 1$ is not defined. Be careful not to forget that the original domain restrictions ($x \neq -2, 1$) still hold.

EXAMPLE 2 Simplifying a Rational Expression

Write $\dfrac{x^2 - x - 6}{x^2 + x - 2}$ in simplified form.

Solution:

Factor the numerator and denominator. $\qquad \dfrac{(x-3)(x+2)}{(x-1)(x+2)}$

State domain restrictions. $\qquad x \neq -2, 1$

Cancel common factor, $x + 2$. $\qquad \dfrac{(x-3)(\cancel{x+2})}{(x-1)(\cancel{x+2})}$

Write in simplified form. $\qquad \dfrac{x-3}{x-1} \qquad x \neq -2, 1$

■ **YOUR TURN** Write $\dfrac{x^2 + x - 6}{x^2 + 2x - 3}$ in simplified form.

Operations with Rational Expressions

How do we add, subtract, multiply, or divide rational expressions? The same rules for operations with fractions also apply to rational expressions. Let us start with multiplying and dividing rational expressions. Recall that multiplying two fractions involves multiplying numerators and denominators, respectively. And dividing by a fraction is the same as multiplying by its reciprocal.

MULTIPLYING AND DIVIDING FRACTIONS

$$\frac{a}{b} \cdot \frac{c}{d} = \frac{ac}{bd} \qquad b \neq 0, d \neq 0$$

$$\frac{a}{b} \div \frac{c}{d} = \frac{a}{b} \cdot \frac{d}{c} \qquad b \neq 0, d \neq 0, c \neq 0$$

Dividing by a fraction is equivalent to multiplying by its reciprocal.

EXAMPLE 3 Multiplying Rational Expressions

Multiply the rational expressions and write in simplified form.

$$\frac{3x + 1}{4x^2 + 4x} \cdot \frac{x^3 + 3x^2 + 2x}{9x + 3}$$

Solution:

Factor the numerators and denominators.
$$\frac{(3x+1)}{4x(x+1)} \cdot \frac{x(x+1)(x+2)}{3(3x+1)}$$

Identify any restrictions on the domain of the product.
$$x \neq 0,\ x \neq -1,\ x \neq -\frac{1}{3}$$

Cancel common factors in numerator and denominator.
$$\frac{(3\!\!\!\!\diagup x+1)}{4x(x\!\!\!\!\diagup+1)} \cdot \frac{x(x\!\!\!\!\diagup+1)(x+2)}{3(3\!\!\!\!\diagup x+1)}$$

Write in reduced form.
$$\frac{x+2}{12} \qquad x \neq 0,\ x \neq -1,\ x \neq -\frac{1}{3}$$

■ **YOUR TURN** Multiply the rational expressions and write in simplified form.
$$\frac{x^2-x-2}{x^2-x-6} \cdot \frac{x-3}{x+1}$$

EXAMPLE 4 Dividing Rational Expressions

Divide the rational expressions, and write in simplified form.
$$\frac{x^2-4}{x} \div \frac{3x^3-12x}{5x^3}$$

Solution:

Factor numerators and denominators.
$$\frac{(x-2)(x+2)}{x} \div \frac{3x(x-2)(x+2)}{5x^3}$$

Identify domain restrictions of the quotient.
$$x \neq 0$$

Write the quotient as a product.
$$\frac{(x-2)(x+2)}{x} \cdot \frac{5x^3}{3x(x-2)(x+2)}$$

State additional domain restrictions of the product.
$$x \neq -2,\ x \neq 2$$

Cancel common factors.
$$\frac{(x\!\!\!\!\diagup-2)(x\!\!\!\!\diagup+2)}{x} \cdot \frac{5x^3}{3x(x\!\!\!\!\diagup-2)(x\!\!\!\!\diagup+2)}$$

Write in reduced form.
$$\frac{5x}{3} \qquad x \neq -2,\ x \neq 0,\ x \neq 2$$

When combining two or more fractions through addition or subtraction we first find the least common multiple, or least common denominator (LCD). For example,
$$\frac{2}{3} + \frac{1}{6} - \frac{4}{9}$$

■ **Answer:** $\dfrac{x-2}{x+2}$ $x \neq -2, -1, 3$

To find the LCD of these three fractions, factor the denominators into prime factors:

$$3 = 3$$

$$6 = 3 \cdot 2$$
$$9 = 3 \cdot 3$$

$$\frac{2}{3} + \frac{1}{6} - \frac{4}{9} = \frac{12 + 3 - 8}{18} = \frac{7}{18}$$

$$\text{LCD} = 3 \cdot 3 \cdot 2 = 18$$

Similarly, rational expressions are added or subtracted by finding the least common denominator, LCD.

EXAMPLE 5 Subtracting Rational Expressions

Perform the indicated operation and write in simplified form.

$$\frac{5x}{2x - 6} - \frac{(7x - 2)}{x^2 - x - 6}$$

Solution:

Factor numerators and denominators into prime factors.

$$\frac{5x}{2(x - 3)} - \frac{(7x - 2)}{(x - 3)(x + 2)}$$

Identify LCD.

$$2(x - 3)(x + 2)$$

Combine expressions into a single expression.

$$\frac{5x(x + 2) - 2(7x - 2)}{2(x - 3)(x + 2)}$$

Distribute and simplify numerator.

$$\frac{5x^2 - 4x + 4}{2(x - 3)(x + 2)} \qquad x \neq -2, 3$$

■ **YOUR TURN** Perform the indicated operation and write in simplified form.

$$\frac{x}{x - 1} - \frac{(x + 1)}{x^2 - x - 6}$$

Simplifying More Complicated Rational Expressions

If either the numerator or denominator of a rational expression contains a sum or difference of rational expressions, there are two procedures for simplifying these more complicated rational expressions as illustrated in the following example.

■ **Answer:** $\dfrac{x^3 - 2x^2 - 6x + 1}{(x + 1)(x + 2)(x - 3)}$ $\qquad x \neq -2, -1, 3$

EXAMPLE 6 Simplifying a Rational Expression Containing a Rational Expression

Write the rational expression in simplified form.

$$\frac{\dfrac{2}{x} + 1}{1 + \dfrac{1}{x + 1}} \qquad x \neq -2, x \neq -1, x \neq 0$$

Procedure 1:

Add the expressions in both the numerator and denominator.

$$\frac{\dfrac{2 + x}{x}}{\dfrac{(x + 1) + 1}{x + 1}}$$

Simplify.

$$\frac{\dfrac{2 + x}{x}}{\dfrac{x + 2}{x + 1}}$$

Express the quotient as a product.

$$\frac{2 + x}{x} \cdot \frac{x + 1}{x + 2}$$

Cancel common factors.

$$\frac{\cancel{2 + x}}{x} \cdot \frac{x + 1}{\cancel{x + 2}}$$

Write in simplified form.

$$\frac{x + 1}{x} \qquad x \neq -2, x \neq -1, x \neq 0$$

Procedure 2:

Find the LCDs of the numerator and denominator.

$$\frac{\dfrac{2}{x} + 1}{1 + \dfrac{1}{x + 1}}$$

Numerator LCD: x

Denominator LCD: $x + 1$

Combined LCD: $x(x + 1)$

Multiply both numerator and denominator by their combined LCD.

$$\frac{\dfrac{2}{x} + 1}{1 + \dfrac{1}{x + 1}} \cdot \frac{x(x + 1)}{x(x + 1)}$$

Multiply the numerators and denominators, respectively, using the distributive property.

$$\frac{\dfrac{2}{x} \cdot x(x + 1) + 1x(x + 1)}{x(x + 1) + \dfrac{1}{x + 1} \cdot x(x + 1)}$$

Cancel common factors.	$$\dfrac{\dfrac{2}{x} \cdot x(x+1) + 1x(x+1)}{x(x+1) + \dfrac{1}{x+1} \cdot x(x+1)}$$
Simplify.	$$\dfrac{2(x+1) + x(x+1)}{x(x+1) + x}$$
Use the distributive property.	$$\dfrac{2x + 2 + x^2 + x}{x^2 + x + x}$$
Combine like terms.	$$\dfrac{x^2 + 3x + 2}{x^2 + 2x}$$
Factor numerator and denominator.	$$\dfrac{(x+2)(x+1)}{x(x+2)}$$
Cancel common factor.	$$\dfrac{(x+2)(x+1)}{x(x+2)}$$
Write in simplified form.	$$\dfrac{x+1}{x} \qquad x \neq -2, x \neq -1, x \neq 0$$

SECTION 0.5 SUMMARY

In this section, rational expressions, which are quotients of polynomials, were discussed. Simplifying rational expressions involves factoring numerators and denominators and canceling common factors. It is important to note domain restrictions before canceling factors. Rational expressions can be added, subtracted, multiplied, or divided using the same rules for operations on fractions. Least common denominators are found before adding or subtracting rational expressions.

SECTION 0.5 EXERCISES

 ■ SKILLS

State the domain of each rational expression.

1. $\dfrac{3}{x}$ **2.** $\dfrac{3}{x-1}$ **3.** $\dfrac{5x-1}{x+1}$ **4.** $\dfrac{2x}{3-x}$ **5.** $\dfrac{2p^2}{p^2-1}$ **6.** $\dfrac{3t}{t^2-9}$ **7.** $\dfrac{3p-1}{p^2+1}$ **8.** $\dfrac{2t-2}{t^2+4}$

Reduce the rational expression if possible, and state the domain.

9. $\dfrac{(x-3)(x+1)}{2(x+1)}$ **10.** $\dfrac{(2x+1)(x-3)}{3(x-3)}$ **11.** $\dfrac{(5y-1)(y+1)}{25y-5}$ **12.** $\dfrac{(2t-1)(t+2)}{4t+8}$

13. $\dfrac{x^2-4}{x-2}$ **14.** $\dfrac{t^3-t}{t-1}$ **15.** $\dfrac{x+7}{x+7}$ **16.** $\dfrac{x^2+9}{2x+9}$

Multiply the rational expressions and simplify.

17. $\dfrac{2x - 2}{3x} \cdot \dfrac{x^2 + x}{x^2 - 1}$

18. $\dfrac{3x^2 - 12}{x} \cdot \dfrac{x^2 + 5x}{x^2 + 3x - 10}$

19. $\dfrac{t + 2}{3t - 9} \cdot \dfrac{t^2 - 6t + 9}{t^2 + 4t + 4}$

20. $\dfrac{t^2 + 4}{t - 3} \cdot \dfrac{3t}{t + 2}$

Divide the rational expressions and simplify.

21. $\dfrac{3}{x} \div \dfrac{12}{x^2}$

22. $\dfrac{1}{x - 1} \div \dfrac{5}{x^2 - 1}$

23. $\dfrac{2 - p}{p^2 - 1} \div \dfrac{2p - 4}{p + 1}$

24. $\dfrac{36 - n^2}{n^2 - 9} \div \dfrac{n + 6}{n + 3}$

25. $\dfrac{3t^3 - 6t^2 - 9t}{5t - 10} \div \dfrac{6 + 6t}{4t - 8}$

26. $\dfrac{w^2 - w}{w} \div \dfrac{w^3 - w}{5w^3}$

Add or subtract the rational expression and simplify.

27. $\dfrac{3}{x} - \dfrac{2}{5x}$

28. $\dfrac{3}{p - 2} + \dfrac{5p}{p + 1}$

29. $\dfrac{2x + 1}{5x - 1} - \dfrac{3 - 2x}{1 - 5x}$

30. $\dfrac{3y^2}{y + 1} + \dfrac{1 - 2y}{y - 1}$

31. $\dfrac{3x}{x^2 - 4} + \dfrac{3 + x}{x + 2}$

32. $\dfrac{x - 1}{4 - x^2} - \dfrac{x + 1}{2 + x}$

State the domain, and simplify the expressions.

33. $\dfrac{\dfrac{1}{x} - 1}{1 - \dfrac{2}{x}}$

34. $\dfrac{\dfrac{1}{x - 1} + 1}{1 - \dfrac{1}{x + 1}}$

■ CATCH THE MISTAKE

In Exercises 35–36, explain the mistake that is made.

35. Simplify $\dfrac{x^2 + 2x + 1}{x + 1}$.

Solution:

Factor the numerator. $\qquad \dfrac{(x + 1)(x + 1)}{(x + 1)}$

Cancel the common factor, $x + 1$. $\qquad \dfrac{(x + 1)(\cancel{x + 1})}{(\cancel{x + 1})}$

Write in simplified form. $\qquad x + 1$

This is incorrect. What mistake was made?

36. Simplify $\dfrac{x + 1}{x^2 + 2x + 1}$.

Solution:

Cancel the common 1s. $\qquad \dfrac{x + \cancel{1}}{x^2 + 2x + \cancel{1}}$

Factor the denominator. $\qquad \dfrac{x}{x(x + 2)}$

Cancel the common x. $\qquad \dfrac{\cancel{x}}{\cancel{x}(x + 2)}$

Write in simplified form. $\qquad \dfrac{1}{x + 2} \quad x \neq -2$

This is incorrect. What mistake was made?

■ CHALLENGE

37. T or F: $\dfrac{x^2 - 81}{x - 9} = x + 9$

38. T or F: $\dfrac{x - 9}{x^2 - 81} = \dfrac{1}{x + 9} \quad x \neq -9, 9$

39. T or F: When adding or subtracting rational expressions, the LCD is the product of all the denominators.

40. T or F: $\dfrac{x - c}{c - x} = -1$ for all values of x.

41. Perform the operation and write in simplified form (remember to state domain restrictions).

$$\frac{x + a}{x + b} \div \frac{x + c}{x + d}$$

42. Simplify $\dfrac{a^{2n} - b^{2n}}{a^n - b^n}$.

■ TECHNOLOGY

43. Using a graphing technology, plot the expression $y = \dfrac{x + 7}{x + 7}$. Zoom in near $x = -7$. Does this agree with what you found in Exercise 15?

44. Using a graphing technology, plot the expression $y = \dfrac{x^2 - 4}{x - 2}$. Zoom in near $x = 2$. Does this agree with what you found in Exercise 13?

SECTION 0.6 Equations

Skills Objectives

■ Generate equivalent equations.
■ Demonstrate an ability to solve equations.

Conceptual Objectives

■ Understand the difference between expressions and equations.
■ Understand the difference between solution set and no solution.

Introduction

An **expression** can be a combination of terms

$$3x + 2 \qquad 5 - 2y \qquad x + y$$

An **equation** is a statement that says two expressions are equal. For example, the following are all equations in one variable, x

$$x + 7 = 11 \qquad x^2 = 9 \qquad 7 - 3x = 2 - 3x \qquad 4x + 7 = x + 2 + 3x + 5$$

To **solve** an equation means to find all the values of x that make the equation true. The values that make it true are called **solutions**, or **roots**, of the equation. The first of these statements, $x + 7 = 11$, is true when $x = 4$ and false for any other x. We say that $x = 4$ is the solution to the equation.

Sometimes an equation can have more than one solution as in $x^2 = 9$. In this case, there are actually two values of x that make this equation true, $x = -3$, and $x = 3$. Typically, when we have more than one solution we say the **solution set** of the equation is $\{-3, 3\}$. In the third equation, $7 - 3x = 2 - 3x$, there are no values of x that make this statement true. Therefore, we say this equation has **no solution**. And, in the fourth equation, $4x + 7 = x + 2 + 3x + 5$, this statement is true for any choice of x. An equation that is true for any value of the variable, x, is called an **identity**. In this case we say the solution set is the set of all real numbers, denoted $\mathbb{R}$.

Generating Equivalent Equations

Two or more equations that have *exactly* the same solutions are called **equivalent equations**. For example,

$$3x + 7 = 13 \qquad 3x = 6 \qquad x = 2$$

are all equivalent equations because each of them has the solution $x = 2$ and only that solution. Note that $x^2 = 4$ is not equivalent to these three because it has the solution set $\{-2, 2\}$.

Simple equations, such as $3x = 6$, can be solved by asking the question "What value for x makes this statement true?" In searching for the value, the question arises, "What number times 3 is 6?" The answer is 2. However, for more complicated equations such as $x^2 + x = 12$, we use a more structured approach.

The key to solving more complicated equations is to reduce them to simpler equivalent equations. In general, there are five ways to produce equivalent equations. We will discuss how to solve specific types of equations in later chapters, but the following table holds for all equations.

GENERATING EQUIVALENT EQUATIONS

1.	Eliminate parentheses and combine like terms on one or both sides of the equation.	Replace: $\quad 3(x - 6) = 6x - x$ By: $\qquad 3x - 18 = 5x$
2.	Add (or subtract) the same quantity to (from) *both* sides of the equation.	Replace: $\quad 7x + 8 = 29$ $\qquad 7x + 8 - \mathbf{8} = 29 - \mathbf{8}$ By: $\qquad 7x = 21$
3.	Multiply (or divide) both sides of the equation by the same nonzero quantity.	Replace: $\quad 5x = 15$ By: $\qquad x = 3$
4.	Interchange the two sides of the equation.	Replace: $\quad -7 = x$ By: $\qquad x = -7$
5.	If one side of the equation is zero and the other side of the equation can be factored, then we can use the zero product property and set each factor equal to 0.	Replace: $\quad (x - 3)(x + 2) = 0$ By: $\quad x - 3 = 0 \text{ or } x + 2 = 0$

Solving an Equation

To solve equations we generate equivalent equations using the five rules above.

The solution to the equation $7x - 3 = 32$ is also the point of intersection of the graphs of $y_1 = 7x - 3$ and $y_2 = 32$.

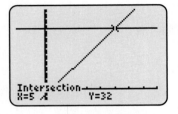

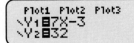

EXAMPLE 1 Solving an Equation

Solve for x: $7x - 3 = 32$.

Solution:

Add 3 to both sides of the equation.

$$7x - 3 = 32$$
$$\underline{+\ 3\ \ +\ 3}$$
$$7x\ \ \ \ = 35$$

Divide both sides by 7.

$$\frac{7x}{7} = \frac{35}{7}$$

Simplify.

$$x = 5$$

Check: Substitute $x = 5$ into $7x - 3 = 32$. $7(5) - 3 = 35 - 3 = 32$ ✓

■ **YOUR TURN** Solve for x: $3x + 15 = 36$.

EXAMPLE 2 Solving an Equation By Factoring

Solve for x: $x^4 - 4x^2 = 0$.

Solution:

Factor out the common x^2 using the distributive property.

$$x^2(x^2 - 4) = 0$$

Use the zero product property, if $a \cdot b = 0$ then $a = 0$ or $b = 0$.

$$x^2 = 0 \text{ or } x^2 - 4 = 0$$

$$x^2 = 0 \text{ or } (x - 2)(x + 2) = 0$$

Solve each of the resulting equations.

$$x = 0 \quad \text{or} \quad x = \pm 2$$

Check:

$x = 0$ $0^4 - 4(0)^2 = 0$ ✓

$x = -2$ $(-2)^4 - 4(-2)^2 = 16 - 16 = 0$ ✓

$x = 2$ $(2)^4 - 4(2)^2 = 16 - 16 = 0$ ✓

■ **YOUR TURN** Solve for x: $3x - 6x^2 = 0$.

Graph $y = x^3 + x^3 - 4x - 4$ and when this graph crosses the x-axis ($y = 0$), the corresponding points are solutions.

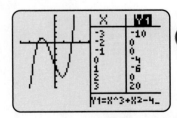

EXAMPLE 3 Solving an Equation Using Factoring By Grouping

Solve for x: $x^3 + x^2 - 4x - 4 = 0$.

Solution:

Group terms with a common factor. $(x^3 + x^2) + (-4x - 4) = 0$

Factor out the common factors. $x^2(x + 1) - 4(x + 1) = 0$

Write as a product of binomials. $(x^2 - 4)(x + 1) = 0$

Use the zero product property. $x^2 - 4 = 0$ or $x + 1 = 0$

$(x - 2)(x + 2) = 0$

Solve each equation for x. $x = \pm 2$ or $x = -1$

■ **YOUR TURN**   Solve for x: $x^3 + 2x^2 - x - 2 = 0$.

SECTION 0.6   SUMMARY

In this section, equations and expressions were defined. A solution, or root, of an equation is the value(s) for the variable that make(s) the equation true. Equivalent equations correspond to equations with the same solution set. Adding or subtracting the same number to both sides or multiplying or dividing by the same number on both sides generates an equivalent equation. Equations were solved by generating equivalent equations and factoring. Factoring makes use of the zero product property to solve equations.

SECTION 0.6   EXERCISES

■ **SKILLS**

Solve each equation.

1. $2x = 10$ **2.** $3y = 27$ **3.** $25z = -125$ **4.** $-8t = 72$

5. $20x = 5$ **6.** $100y = 10$ **7.** $2.5z = -10$ **8.** $3.5t = -21$

9. $2x + 7 = 17$ **10.** $4p + 3 = 47$ **11.** $-2y + 7 = 13$ **12.** $-3z - 10 = 11$

13. $\frac{1}{4}x = -\frac{1}{3}x + 7$ **14.** $\frac{1}{5}y = \frac{1}{4}y - 1$ **15.** $x^2 - 25 = 0$ **16.** $y^2 - 49 = 0$

17. $x^2 + 16 = 0$ **18.** $x^2 + 4 = 0$ **19.** $x^2 = 2x$ **20.** $y^2 = 3y$

21. $x^3 - 2x^2 - 15x = 0$ **22.** $y^3 + 2y^2 - 8y = 0$ **23.** $x^3 - 4x^2 - 4x + 16 = 0$

24. $x^3 - 25x - 2x^2 + 50 = 0$ **25.** $x^3 + x^2 - 2x - 2 = 0$ **26.** $x^3 - x^2 + 2x - 2 = 0$

■ **APPLICATIONS**

For Exercises 27 and 28, the perimeter of a rectangle is related to the length and the width by the equation $P = 2l + 2w$ where P is the perimeter, l is the length, and w is the width.

27. Geometry. The perimeter of a rectangle is 14 inches. If the length is 6 inches, what is the width of the rectangle?

28. Geometry. The perimeter of a rectangle is 20 feet. If the width is 7 feet, what is the length of the rectangle?

■ **Answer:** $x = -2, -1, 1$

For Exercises 29 and 30, the area of a circle is $A_{\text{circle}} = \pi r^2$ and the area of a circular ring, or annulus, is $A_{\text{annulus}} = \pi(r_{\text{outer}}^2 - r_{\text{inner}}^2)$ where r_{outer} is the radius of the outer circle and r_{inner} is the radius of the inner circle.

29. Geometry. If a circle has area 25π square feet, what is the radius?

30. Geometry. If an annulus has an area of 16π square inches and the inner radius is 3 inches, what is the outer radius?

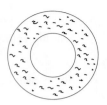

 ■ **CATCH THE MISTAKE**

In Exercises 31–34, explain the mistake that is made.

31. Solve for x: $4x - 10 + x = 6x - 3x - 4$.

Solution:

Combine like terms on each side.

$$(4x + x) - 10 = (6x - 3x) - 4$$
$$5x - 10 = 3x - 4$$

Subtract $3x$ and add 10. $\quad 2x = -4$

Divide by 2. $\quad x = 2$

This is incorrect. What mistakes were made?

32. Solve for x: $x^4 - 9x^2 = 0$.

Solution:

Divide by x^2. $\quad x^2 - 9 = 0$

Add 9 to both sides. $\quad x^2 = 9$

Solve for x. $\quad x = \pm 3$

This is incorrect. What mistake was made?

33. Generate an equivalent equation.

$$x = 1$$
$$2x = 1 + x$$
$$2x - 3 = x - 2$$
$$x^2 + 2x - 3 = x^2 + x - 2$$
$$(x + 3)(x - 1) = (x + 2)(x - 1)$$
$$x + 3 = x + 2$$
$$3 = 2$$

This is incorrect. What mistake was made?

34. Generate an equivalent equation.

$$x = -1$$
$$x^2 + x = x^2 - 1$$
$$x(x + 1) = (x - 1)(x + 1)$$
$$x = x - 1$$
$$0 = -1$$

This is incorrect. What mistake was made?

 ■ **CHALLENGE**

35. T or F: $3x + 2 = 11$ and $x^2 = 9$ are equivalent equations.

36. T or F: $x^4 - x^2 = 0$ and $x^2 - 1 = 0$ are equivalent equations.

37. Solve for x: $ax + b = c \quad a \neq 0$.

38. Solve for x: $ax^4 - 4ax^2 = 0$.

 ■ **TECHNOLOGY**

39. Solve for x and use a calculator to round the solution to 2 decimal places.

$$5.56x - 2.2 = 13.29$$

40. Solve for x and use a calculator to round the solution to 3 decimal places.

$$2.76x + 5.39 = 9.34$$

Skills Objectives	Conceptual Objectives
■ Write radicals of negative numbers as imaginary numbers. ■ Add and subtract complex numbers. ■ Multiply complex numbers. ■ Express quotients of complex numbers in standard form.	■ Understand that real numbers and imaginary numbers are subsets of complex numbers. ■ Understand how to eliminate imaginary numbers in denominators.

For some equations like $x^2 = 1$, the solutions are always real numbers, $x = \pm 1$. However, there are some equations like $x^2 = -1$ that do not have real solutions because the square of a real number cannot be negative. Mathematicians created a new set of numbers so that, when they are squared, the product would be negative. This new set of numbers is called *imaginary* numbers.

DEFINITION **IMAGINARY UNIT, *i***

Imaginary unit is denoted by the letter i and is defined as

$$i = \sqrt{-1}$$

where $i^2 = -1$.

$$i = \sqrt{-1}$$
$$i^2 = -1$$
$$i^3 = i^2 \cdot i = (-1)i = -i$$
$$i^4 = i^2 \cdot i^2 = (-1)(-1) = 1$$

Note that i raised to the fourth power is 1. In simplifying imaginary numbers, we factor out i raised to the largest multiple of 4.

EXAMPLE 1 Simplifying the Imaginary Unit Raised to Powers

Simplify.

a. i^7 **b.** i^{13} **c.** i^{100}

Solution:

a. $i^7 = i^4 \cdot i^3 = (1)(-i) = \boxed{-i}$ **b.** $i^{13} = i^{12} \cdot i = (i^4)^3 \cdot i = 1^3 \cdot i = \boxed{i}$

c. $i^{100} = (i^4)^{25} = 1^{25} = \boxed{1}$

CONCEPT CHECK Concept Check: What is i^{4n} when n is a positive integer?

■ **YOUR TURN** Simplify i^{27}.

EXAMPLE 2 Using Imaginary Numbers to Simplify Radicals

Simplify using imaginary numbers.

a. $\sqrt{-9}$ **b.** $\sqrt{-8}$

Solution:

a. $\sqrt{-9} = \sqrt{9} \cdot \sqrt{-1} = \boxed{3i}$ **b.** $\sqrt{-8} = \sqrt{8} \cdot \sqrt{-1} = 2\sqrt{2} \cdot i = \boxed{2i\sqrt{2}}$

■ **YOUR TURN** Simplify $\sqrt{-144}$.

Complex Numbers

When an imaginary number and a real number are combined, the result is a *complex number*.

DEFINITION   COMPLEX NUMBER

A **complex number** in standard form is defined as

$$a + bi$$

where a and b are real numbers and i is the imaginary unit.

In the definition above, we denote a as the *real part* of the complex number and bi as the *imaginary part* of the complex number. The expressions

$$2 - 3i \quad \text{and} \quad -5 + i$$

are examples of complex numbers in standard form.

DEFINITION   EQUALITY OF COMPLEX NUMBERS

Two complex numbers in **standard form**, $a + bi$ and $c + di$, are **equal** if and only if

$$a = c \quad \text{and} \quad b = d$$

In other words, *two complex numbers are equal if both real parts are equal and both imaginary parts are equal.*

We can treat complex numbers, $a + bi$, similar to binomials, $a + bx$. When adding or subtracting binomials, combine like terms. Similarly, when adding or subtracting complex numbers, real parts are combined with real parts and imaginary parts are combined with imaginary parts.

■ **Answer:** $12i$

EXAMPLE 3 Adding and Subtracting Complex Numbers

Perform the indicated operation and simplify.

a. $(3 - 2i) + (-1 + i)$ **b.** $(2 - i) - (3 - 4i)$

Solution (a):

Eliminate the parentheses.	$3 - 2i - 1 + i$
Group real and imaginary numbers, respectively.	$(3 - 1) + (-2i + i)$
Simplify.	$2 - i$

Solution (b):

Eliminate the parentheses (distribute the negative).	$2 - i - 3 + 4i$
Group real and imaginary numbers, respectively.	$(2 - 3) + (-i + 4i)$
Simplify.	$-1 + 3i$

■ **YOUR TURN** Perform the indicated operation and simplify $(4 + i) - (3 - 5i)$.

Multiplying Complex Numbers

When multiplying complex numbers, you apply all of the methods for multiplying binomials. It is important to remember that $i^2 = -1$.

EXAMPLE 4 Multiplying Complex Numbers

Multiply the complex numbers and express the result in standard form, $a \pm bi$.

a. $(3 - i)(2 + i)$ **b.** $i(-3 + i)$

Solution (a):

Use the FOIL method to multiply.	$3(2) + 3(i) - i(2) - i(i)$
Eliminate parentheses.	$6 + 3i - 2i - i^2$
Substitute $i^2 = -1$.	$6 + 3i - 2i - (-1)$
Group like terms.	$(6 + 1) + (3i - 2i)$
Simplify.	$7 + i$

Solution (b):

Use the distributive property to multiply.	$-3i + i^2$
Substitute $i^2 = -1$.	$-3i - 1$
Write in standard form.	$-1 - 3i$

■ **YOUR TURN** Multiply the complex numbers and express the result in standard form, $a \pm bi$.

$$(4 - 3i)(-1 + 2i)$$

Dividing Complex Numbers

Recall the special product that produces a difference of two squares, $(a + b)(a - b) = a^2 - b^2$. This special product has only first and last terms because the outer and inner terms cancel each other out. If we multiply complex numbers in the same manner, the result is a real number because the imaginary terms cancel each other out.

DEFINITION **COMPLEX CONJUGATE**

The product of a complex number, $a + bi$, and its **complex conjugate**, $a - bi$, is a real number.

$$(a + bi)(a - bi) = a^2 - b^2i^2 = a^2 - b^2(-1) = a^2 + b^2$$

To write a quotient of complex numbers in standard form, $a + bi$, multiply the numerator and denominator by the complex conjugate of the denominator.

EXAMPLE 5 Dividing Complex Numbers

Write the quotient of the complex numbers in standard form: $\dfrac{2 - i}{1 + 3i}$.

Solution:

Multiply numerator and denominator by the complex conjugate of denominator, $1 - 3i$.

$$\left(\frac{2 - i}{1 + 3i}\right)\left(\frac{1 - 3i}{1 - 3i}\right)$$

Multiply numerators and denominators.

$$\frac{(2 - i)(1 - 3i)}{(1 + 3i)(1 - 3i)}$$

Use the FOIL method to multiply the binomials.

$$\frac{2 - 7i + 3i^2}{1 - 9i^2}$$

Substitute $i^2 = -1$.

$$\frac{2 - 7i - 3}{1 + 9}$$

Simplify numerator and denominator.

$$\frac{-1 - 7i}{10}$$

Write in standard form.

$$-\frac{1}{10} - \frac{7}{10}i$$

■ **Answer:** $2 + 11i$

■ **YOUR TURN** Write the complex number in standard form: $\dfrac{3 + 2i}{4 - i}$.

SECTION 0.7 SUMMARY

In this section, imaginary numbers were discussed as a way of representing solutions to equations that have no real solutions, such as $x^2 = -1$. Complex numbers are combinations of both real and imaginary numbers. To add or subtract complex numbers simply add or subtract the real parts and imaginary parts, respectively. Multiplying complex numbers is similar to the FOIL method. It is important to remember that $i^2 = -1$. To write a quotient of complex numbers in standard form, multiply the numerator and denominator by the complex conjugate of the denominator.

SECTION 0.7 EXERCISES

 ■ **SKILLS**

Simplify.

1. i^{15} **2.** i^{99} **3.** i^{40} **4.** i^{18} **5.** $\sqrt{-16}$ **6.** $\sqrt{-100}$ **7.** $\sqrt{-20}$ **8.** $\sqrt{-24}$

Perform indicated operation, simplify, and express in standard form.

9. $(3 - 7i) + (-1 - 2i)$ **10.** $(1 + i) + (9 - 3i)$ **11.** $(4 - 5i) - (2 - 3i)$ **12.** $(-2 + i) - (1 - i)$

13. $(1 - i)(3 + 2i)$ **14.** $(-3 + 2i)(1 - 3i)$ **15.** $(4 + 7i)(4 - 7i)$ **16.** $(-2 + 3i)(-2 - 3i)$

17. $\dfrac{1}{3 - i}$ **18.** $\dfrac{1}{3 + 2i}$ **19.** $\dfrac{4 - 5i}{7 + 2i}$ **20.** $\dfrac{1 - i}{1 + i}$

Write the negative radicals in terms of imaginary numbers and then perform the indicated operation and simplify.

21. $\sqrt{-16} + 2\sqrt{-25} - 3\sqrt{-9}$ **22.** $\sqrt{-8} + \sqrt{-18}$ **23.** $(\sqrt{-4} + 1)(3 - \sqrt{-25})$ **24.** $(3 - \sqrt{-16})(2 - \sqrt{-9})$

 ■ **CATCH THE MISTAKE**

In Exercises 25–28, explain the mistake that is made.

25. Simplify $\sqrt{-4} \cdot \sqrt{-4}$.

   Solution:

   Multiply the two radicals. $\sqrt{(-4)(-4)}$

   Simplify. $\sqrt{16}$

   Evaluate the radical. 4

   This is incorrect. What mistake was made?

26. Simplify i^{103}.

   Solution:

   $$i^{103} = (i^{100})(i^3) = (-1)(-i) = i$$

   This is incorrect. What mistake was made?

■ **Answer:** $\dfrac{10}{17} + \dfrac{11}{17}i$

27. Write the quotient in standard form: $\dfrac{2}{4 - i}$.

Solution:

Multiply numerator and denominator by $4 - i$.

$$\dfrac{2}{4 - i} \cdot \dfrac{4 - i}{4 - i}$$

Multiply numerator using distributive property and denominator using FOIL method.

$$\dfrac{8 - 2i}{16 - 1}$$

Simplify.

$$\dfrac{8 - 2i}{15}$$

Write in standard form.

$$\dfrac{8}{15} - \dfrac{2}{15}i$$

This is incorrect. What mistake was made?

28. Write the product in standard form: $(2 - 3i)(5 + 4i)$.

Solution:

Use the FOIL method to multiply the complex numbers.

$$10 - 7i - 12i^2$$

Simplify.

$$-2 - 7i$$

This is incorrect. What mistake was made?

■ **CHALLENGE**

29. T or F: The product is a real number: $(a + bi)(a - bi)$.

30. T or F: Imaginary numbers are a subset of the complex numbers.

31. T or F: Real numbers are a subset of the complex numbers.

32. T or F: There is no complex number that equals its conjugate.

33. Simplify: $i^{26} - i^{15} + i^{17} - i^{1000}$.

34. Solve: $x^4 + 2x^2 + 1 = 0$.

Q: Kaitlyn is working on homework problems for a genetics class. She knows that if two parents are carriers of a recessive trait, then an offspring of those parents has a one in four chance of actually having that trait. One such recessive trait is the negative Rh factor of a blood type. Kaitlyn knows that two parents who are Rh positive but are carriers of a recessive Rh negative gene will produce a child with an Rh negative blood type $\frac{1}{4}$ of the time. Thus, she calculates that the probability of these same parents having three children with Rh negative blood is $\left(\frac{1}{4}\right)^3$, which she calculates to be approximately 0.125 or 12%. However, she learns that she is incorrect and that the probability is actually about 0.016 or 1.6%. Assuming that her probability formula is correct, how did she make an error?

A: Kaitlyn's error occurred in applying exponent rules when she raised $\frac{1}{4}$ to the 3rd power. Her calculation was as follows:

$$\left(\frac{1}{4}\right)^3 = \left(\frac{1}{2^2}\right)^3 = (2^{-2})^3 = 2^{-3}$$

Her work was correct until her last step. When a power is raised to a power, the exponents should be multiplied. Thus, $(2^{-2})^3 = 2^{-6} \approx 0.016$.

Kaitlyn calculates that the probability that the same two parents would produce five Rh negative children is $\left(\frac{1}{4}\right)^5 = \left(\frac{1}{2^2}\right)^5$. Convert this value to a percentage.

TYING IT ALL TOGETHER

Rafael is planning to paint his house, and is trying to estimate how much paint he will need to buy and how much it will cost. His house is 40 feet wide, 50 feet long, and 10 feet tall to the base of the roof. On the two sides of the house, the roof rises to a peak which is 8 feet tall from the base of the roof. The volume of the house (including the attic) is 28,000 ft^3 and the area of the roof is about 2154 ft^2. The paint he is purchasing costs \$25 per gallon and each gallon covers 350 ft^2. If h = height of the house (to the base of the roof), l = length of the house, w = width of the house, and r = height from the base to the peak of the roof, write a formula to calculate the amount (in gallons) of the paint needed in terms of these variables if he plans to apply two coats of paint. Approximately how much will the paint cost? Assume that he is not painting the trim, and that he is not subtracting out area for the doors and windows since he is just trying to estimate the cost of the paint.

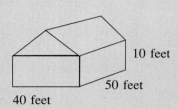

10 feet

50 feet

40 feet

CHAPTER O REVIEW

SECTION	TOPIC	PAGES	REVIEW EXERCISES	KEY CONCEPTS
0.1	Real numbers	4–10	1–10	
	Rational/irrational numbers	4–5		Rational: $\frac{a}{b}$ or decimal that repeats or terminates Irrational: nonrepeating decimal
	Approximating decimals	5–6	1 and 2	Rounding: Use digit to the right Digit $<$ 5 stay the same Digit $\geq$ 5 round up Truncating: Eliminate to the right
	Properties of real numbers	6–10	3–10	Distributive Property: $x(y + z) = xy + xz$ Zero product property: If $xy = 0$ then $x = 0$ or $y = 0$
0.2	Exponents and radicals	12–20	11–20	$a^n = \underbrace{a \cdot a \cdot a \cdot \cdots \cdot a}_{n \text{ factors}}$
	Properties of exponents	12–14	11–14, 17	$a^m \cdot a^n = a^{m+n} \quad \dfrac{a^m}{a^n} = a^{m-n} \quad a^0 = 1$ $(a^m)^n = a^{mn} \quad a^{-n} = \dfrac{1}{a^n} = \left[\dfrac{1}{a}\right]^n$
	Scientific notation	14–15	15 and 16	$\pm c \times 10^n$ $\quad c$ is a real number $1 \leq c < 10$. n is an integer. n positive—large number n negative—small number
	Radicals	15–18	18 and 19	$\sqrt[n]{b}$ is the **principal nth root** of b.
	Rationalizing denominators	19–20	20	Never leave radicals in denominators. $(\sqrt{a} + \sqrt{b})(\sqrt{a} - \sqrt{b}) = a - b$ (real)
0.3	Polynomials	21–27	21–30	
	Add and subtract polynomials	22–23	21 and 22	Combine like terms: Monomials of the same degree.
	Multiply polynomials	23–25	23 and 24	Distributive property
	Multiply two binomials using the FOIL method	25–26	25–30	$(x + a)(x + b) = x^2 + (ax + bx) + ab$ Special products: $(a + b)^2 = (a + b)(a + b) = a^2 + 2ab + b^2$ $(a - b)^2 = (a - b)(a - b) = a^2 - 2ab + b^2$ $(a + b)(a - b) = a^2 - b^2$
0.4	Factoring polynomials	28–32	31–40	
	Greatest common factor	29	31 and 32	Factor out using distributive property.
	Factor a trinomial into a product of binomials	29–31	33–38	Use FOIL backwards.
	Factoring by grouping	31	39 and 40	Group terms with common factors. Applies to polynomials with 4 or more terms.

Section	Topic	Pages	Review Exercises	Key Concepts
0.5	Rational expressions	33–40	41–50	Rational expression is a $\dfrac{\text{polynomial}}{\text{polynomial}}$.
	Simplify rational expressions	34–36	41–46	Use properties of rational numbers. Note domain restrictions when denominator is equal to zero.
	Multiply and divide rational expressions	36–37	47 and 48	Use properties of rational numbers.
	Add and subtract rational expressions	37–38	49 and 50	Least common denominator
	Simplify more complicated rational expressions	38–40	50	Two strategies: **1.** Write sum/difference in numerator/denominator as a rational expression. **2.** Multiply by LCD of the numerator and denominator.
0.6	Equations	42–45	51–60	
	Solution set	42–43		Values of variable that make equation true
	Equivalent equations	43		Two equations are equivalent if they have the exact same solution set.
	Solving equations	43–44	51–60	Generate equivalent equations. If a number is added, subtracted, multiplied, or divided on one side, it must also be done on the other side of the equation.
	Solving equations by factoring	44–45	54–60	If polynomial = 0, factor the polynomial and use the zero product property.
0.7	Complex numbers	47–51	61–70	
	i	47–48	61–64	$i = \sqrt{-1},\ i^2 = -1,\ i^3 = -i,\ i^4 = 1$
	Complex number	48		$a + bi$, a and b are real numbers.
	Adding and subtracting complex numbers	49	65 and 66	Combine real parts with real parts and imaginary parts with imaginary parts.
	Multiplying complex numbers	49	65 and 66	Use FOIL method and $i^2 = -1$ to simplify.
	Writing the quotient of complex numbers in standard form	50	67–70	If $a + bi$ is in denominator, then multiply numerator and denominator by $a - bi$. The result is a real number in denominator.

CHAPTER 0 REVIEW EXERCISES

0.1 Real Numbers

Approximate to 2 decimal places by (a) rounding and (b) truncating.

1. 5.21597

2. 7.3623

Simplify.

3. $7 - 2 \cdot 5 + 4 \cdot 3 - 5$

4. $-2(5 + 3) + 7(3 - 2 \cdot 5)$

5. $-\dfrac{16}{(-2)(-4)}$

6. $-3(x - y) + 4(3x - 2y)$

Perform the indicated operation and simplify.

7. $\dfrac{x}{4} - \dfrac{x}{3}$

8. $\dfrac{y}{3} + \dfrac{y}{5} - \dfrac{y}{6}$

9. $\dfrac{12}{7} \cdot \dfrac{21}{4}$

10. $\dfrac{a^2}{b^3} \div \dfrac{2a}{b^2}$

0.2 Exponents and Radicals

Simplify using properties of exponents.

11. $(-2z)^3$

12. $81^{1/4}$

13. $(-64)^{1/3}$

14. $\dfrac{(2x^2y^3)^2}{(4xy)^3}$

15. Express 0.00000215 in scientific notation.

16. Express 7.2×10^9 as a real number.

Simplify.

17. $\dfrac{(3x^{2/3})^2}{(4x^{1/3})^2}$

18. $\sqrt{20}$

19. $(2 + \sqrt{5})(1 - \sqrt{5})$

20. $\dfrac{1}{2 - \sqrt{3}}$

0.3 Polynomials

Perform the indicated operation and write the results in standard form.

21. $(14z^2 + 2) + (3z - 4)$

22. $(27y^2 - 6y + 2) - (y^2 + 3y - 7)$

23. $5xy^2(3x - 4y)$

24. $-2st^2(-t + s - 2st)$

25. $(x - 7)(x + 9)$

26. $(2x + 1)(3x - 2)$

27. $(2x - 3)^2$

28. $(5x - 7)(5x + 7)$

29. $(x^2 + 1)^2$

30. $(1 - x^2)^2$

0.4 Factoring Polynomials

Factor out the common factor.

31. $14x^2y^2 - 10xy^3$

32. $30x^4 - 20x^3 + 10x^2$

Factor the trinomial into a product of two binomials.

33. $2x^2 + 9x - 5$

34. $6x^2 - 19x - 7$

35. $16x^2 - 25$

36. $9x^2 - 30x + 25$

Factor into a product of three polynomials.

37. $2x^3 + 4x^2 - 30x$

38. $6x^3 - 5x^2 + x$

Factor into a product of two binomials by grouping.

39. $x^3 + x^2 - 2x - 2$

40. $2x^3 - x^2 + 6x - 3$

0.5 Rational Expressions

State the domain of the rational expression.

41. $\dfrac{4x^2 - 3}{x^2 - 9}$

42. $\dfrac{1}{x^2 + 1}$

Simplify.

43. $\dfrac{x^2 - 4}{x - 2}$

44. $\dfrac{x - 5}{x - 5}$

45. $\dfrac{t^2 + t - 6}{t^2 - t - 2}$

46. $\dfrac{z^3 - z}{z^2 + z}$

Perform the indicated operation and simplify.

47. $\dfrac{x^2 + 3x - 10}{x^2 + 2x - 3} \cdot \dfrac{x^2 + x - 2}{x^2 + x - 6}$

48. $\dfrac{x^2 - x - 2}{x^3 + 3x^2} \div \dfrac{x + 1}{x^2 + 2x}$

49. $\dfrac{1}{x + 1} - \dfrac{1}{x + 3}$

50. $\dfrac{1}{x} - \dfrac{1}{x + 1} + \dfrac{1}{x + 2}$

0.6 Equations

Solve the equation.

51. $3(z + 2) - 1 = 4z + 10$

52. $6x + 6 = 8x + 3$

53. $\dfrac{1}{5}y - \dfrac{1}{3}y = -2$

54. $y^2 + 100 = 0$

55. $x^2 - 144 = 0$

56. $x^2 = 5x$

57. $x^2 - 6x + 8 = 0$

58. $y^3 - 4y = 0$

59. $x^3 - x^2 - 4x + 4 = 0$

60. $x^3 + x^2 + 3x + 3 = 0$

0.7 Complex Numbers

Simplify.

61. $\sqrt{-169}$

62. $\sqrt{-32}$

63. i^{19}

64. i^9

65. $(\sqrt{-4} + 2)(3 - \sqrt{-9})$

66. $(\sqrt{-36} + 1)(1 + \sqrt{-25})$

Express the quotient in standard form.

67. $\dfrac{1}{2 - i}$

68. $\dfrac{1}{3 + i}$

69. $\dfrac{6 - 5i}{3 - 2i}$

70. $\dfrac{7 + 2i}{4 + 5i}$

CHAPTER 0 PRACTICE TEST

Simplify.

1. $\sqrt{16}$ 2. $\sqrt{18}$ 3. $\sqrt[5]{-32}$

4. $\sqrt{-12}$ 5. i^{17} 6. $\dfrac{(x^2y^{-3}z^{-1})^{-2}}{(x^{-1}y^2z^3)^{1/2}}$

Perform the indicated operation and simplify.

7. $(3y^2 - 5y + 7) - (y^2 + 7y - 13)$

8. $(2x - 3)(5x + 7)$

Factor.

9. $2x^2 - x - 1$ 10. $6y^2 - y - 1$

11. $2t^3 - t^2 - 3t$ 12. $2r^3 - \dfrac{r}{2}$

Divide the rational expressions and state the domain.

13. $\dfrac{x - 1}{2x - 1} \div \dfrac{x^2 - 1}{1 - 2x}$ 14. $\dfrac{1 - t}{3t + 1} \div \dfrac{t^2 - 2t + 1}{7t + 21t^2}$

Solve the equations.

15. $3x - 2 = 25$ 16. $3x^2 - 3x = 0$

Write the resulting expression in standard form.

17. $(1 - 3i)(7 - 5i)$

18. $\dfrac{2 - 11i}{4 + i}$

19. Factor $x^{4n} - y^{4n}$.

20. Rationalize the denominator. $\dfrac{7 - 2\sqrt{3}}{4 - 5\sqrt{3}}$

Equations and Inequalities

UCF Athletics Association, Inc.

UCF Golden Knights

A tight end can run the 100 yard dash in 12 seconds. A defensive back can do it in 10 seconds. The tight end catches a pass at his own 20 yard line with the defensive back at the 15 yard line. If no other players are nearby, at what yard line will the defensive back catch up to the tight end? A defensive coordinator needs to know this so he can position his defensive backs against particular receivers. Many football fans can guess that the tackle will be made about midfield, but there is actually a mathematical equation that will determine that the tackle will be made at the tight end's own 45 yard line. This is just one example of how the real world can be modeled with mathematical equations.

In this chapter you will solve linear and quadratic equations and then solve application problems that involve such equations. You will also solve inequalities, both linear and polynomial, and absolute value equations and inequalities.

Equations and Inequalities

Equations

- Linear
- Quadratic
- Radical
- Absolute Value

Applications

- **Equations**
  - Linear
  - Quadratic
  - Absolute Value
- **Inequalities**
  - Linear
  - Polynomial
  - Rational
  - Absolute Value

Inequalities

- Linear
- Polynomial
- Rational
- Absolute Value

CHAPTER OBJECTIVES

- Classify equations.
- Solve linear, quadratic, radical, and absolute value equations.
- Solve linear, quadratic, rational, and absolute value inequalities.
- Solve application problems involving equations and inequalities.
- Eliminate solutions that are extraneous or not in the domain of the variable.

NAVIGATION THROUGH SUPPLEMENTS

DIGITAL VIDEO SERIES #1

STUDENT SOLUTIONS MANUAL CHAPTER 1

BOOK COMPANION SITE
www.wiley.com/college/young

- Identify different types of solutions and equations.
- Solve linear equations in one variable.
- Solve equations involving fractional expressions.

- Relate types of equations to types of solutions.
- Eliminate solutions that are not in the domain of the variable.

Types of Equations and Solutions

An **equation** in x is a statement that equates two algebraic expressions. An example of an equation follows.

$$3x - 5 = 1 + x$$

To **solve** an equation for x means to find all values for x that make the statement true. These values are called **solutions**, or **roots**. For example, we say that $x = 3$ is a solution to the above equation because when $x = 3$ the statement $3x - 5 = 1 + x$ is true. We say that $x = 3$ **satisfies** the equation $3x - 5 = 1 + x$.

$$3(3) - 5 = 1 + (3)$$
$$9 - 5 = 1 + 3$$
$$4 = 4$$

The set of all such solutions is the **solution set**. For example, the solution set of the equation $3x - 5 = 1 + x$ is $\{3\}$.

Equations can be classified according to the types of solutions they have. An equation that is true for *every* real number in the domain of the variable is called an **identity**. For example,

$$2x + 1 + 4 = x + x + 5$$

is an identity because it is a true statement for any real values of x. Another example of an identity is

$$\frac{x}{5x^2} = \frac{1}{5x} \qquad x \neq 0$$

because this statement is true for all nonzero real numbers. An equation that is true for only *some* of the real numbers in the domain of the variable is called a **conditional equation**. An example of a conditional equation is

$$x^2 = 4$$

since the only real values of x that make the statement true are $x = -2$ and $x = 2$. We say the solution set of the equation $x^2 = 4$ is $\{-2, 2\}$. An **inconsistent equation** or **contradiction** is an equation where there are *no* real values of x that make the statement true. An example of an inconsistent equation is

$$x + 5 = x - 2$$

since no real values of x make this statement true.

The following table summarizes the types of equations and solutions.

TYPE OF EQUATION	SOLUTION SET (REAL VALUES)
Identity	All real numbers
Conditional	Some real numbers
Contradiction	No solution

Solving Linear Equations in One Variable

You probably already know how to solve simple linear equations. The goal is to solve for the variable (most of the time we use x but any letter can be used to represent a variable). How do we solve for x? We gather x's to one side of the equals sign and constants to the other side of the equals sign.

EXAMPLE 1 Solving a Linear Equation

Solve the equation $3x + 4 = 16$.

Solution:

Subtract 4 from both sides of the equation.

$$3x + 4 = 16$$
$$\underline{-4 \quad -4}$$
$$3x \quad = 12$$

Divide both sides by 3.

$$\frac{3x}{3} = \frac{12}{3}$$

The solution is $x = 4$.

$$x = 4$$

The solution set is $\{4\}$.

■ **YOUR TURN**   Solve the equation $2x + 3 = 9$.

You are solving linear equations. What is a linear equation?

DEFINITION   LINEAR EQUATION

A **linear equation** in one variable, x, can be written in the form

$$ax + b = 0$$

where a and b are real numbers and $a \neq 0$.

What makes this equation linear is that x is raised to the first power. We can also classify a linear equation as a **first degree** equation.

EQUATION	DEGREE	GENERAL NAME
$x - 7 = 0$	First	Linear
$x^2 - 9 = 0$	Second	Quadratic
$x^3 - 8 = 0$	Third	Cubic

TECHNOLOGY TIP

Use a graphing utility to display graphs of $y_1 = 3x + 4$ and $y_2 = 16$.

The x-coordinate of the point of intersection is the solution to the equation $3x + 4 = 16$.

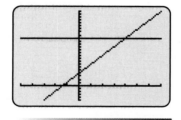

■ **Answer:** The solution is $x = 3$. The solution set is $\{3\}$.

In this section we work with linear equations. In a later section, quadratic equations will be discussed and in a later chapter, higher degree equations will be covered.

EXAMPLE 2 Solving a Linear Equation

Solve the equation $5x - (7x - 4) - 2 = 5 - (3x + 2)$.

Solution:

Write the equation.
$$5x - (7x - 4) - 2 = 5 - (3x + 2)$$

Eliminate the parentheses.
$$5x - (7x - 4) - 2 = 5 - (3x + 2)$$

Don't forget to distribute the negative 1 through *both* terms inside the parentheses.
$$5x - 7x + 4 - 2 = 5 - 3x - 2$$

Combine x terms on left, constants on right.
$$-2x + 2 = 3 - 3x$$

Add $3x$ to both sides.
$$\underline{+3x \qquad\qquad +3x}$$
$$x + 2 = 3$$

Subtract 2 from both sides.
$$\underline{-2 \; -2}$$
$$\boxed{x = 1}$$

Check to verify that $x = 1$ is a solution to the original equation.

$$5 \cdot 1 - (7 \cdot 1 - 4) - 2 = 5 - (3 \cdot 1 + 2)$$
$$5 - (7 - 4) - 2 = 5 - (3 + 2)$$
$$5 - (3) - 2 = 5 - (5)$$
$$0 = 0$$

Since the solution, $x = 1$, checks, the solution set is $\{1\}$.

YOUR TURN Solve the equation $4(x - 1) - 2 = x - 3(x - 2)$.

Equations Involving Fractional Expressions

To solve an equation involving fractional expressions, find the least common denominator (LCD) of all terms and multiply the entire equation by the LCD. We will first review how to find the LCD.

Recall (Least Common Denominator)

To add the fractions $\frac{1}{2} + \frac{1}{6} + \frac{2}{5}$, we must first get a common denominator. Some people are taught to find the lowest number that 2, 6, and 5 all divide evenly into. Others prefer a more systematic approach in terms of prime factors.

Answer: The solution is $x = 2$.

<u>Prime Factors</u>

$$2 = 2$$

$$6 = 2 \cdot 3$$

$$5 = \qquad 5$$

$$\text{LCD} = 2 \cdot 3 \cdot 5 = 30$$

EXAMPLE 3 Solving a Linear Equation Involving Fractions

Solve the equation $\frac{1}{2}p - 5 = \frac{3}{4}p$.

Solution:

Write the equation.

$$\frac{1}{2}p - 5 = \frac{3}{4}p$$

Multiply each term in the equation by the LCD, 4.

$$(4)\frac{1}{2}p - (4)5 = (4)\frac{3}{4}p$$

Now, it is clear that this is a linear equation.

$$2p - 20 = 3p$$

Subtract $2p$ from both sides.

$$\underline{-2p \qquad\qquad -2p}$$

$$-20 = p$$

$$p = -20$$

Since $p = -20$ satisfies the original equation, the solution set is $\{-20\}$.

CONCEPT CHECK What is the LCD of $\frac{1}{4}, \frac{1}{12}, 3$?

YOUR TURN Solve the equation $\frac{1}{4}m = \frac{1}{12}m - 3$.

TECHNOLOGY TIP

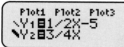

Use a graphing utility to display graphs of $y_1 = \frac{1}{2}p - 5$ and $y_2 = \frac{3}{4}p$. When entering the equations into the graphing utility, use x in place of p.

```
Plot1  Plot2  Plot3
\Y1◻1/2X-5
\Y2◻3/4X
```

The x-coordinate of the point of intersection is the solution to the equation.

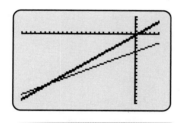

EXAMPLE 4 Solving an Equation with the Variable in the Denominator

Solve the equation $\dfrac{2}{3x} + \dfrac{1}{2} = \dfrac{4}{x} + \dfrac{4}{3}$.

Solution:

We can let x be anything except 0.

$$\frac{2}{3x} + \frac{1}{2} = \frac{4}{x} + \frac{4}{3} \qquad x \neq 0$$

Multiply *each term* by the LCD, $6x$.

$$6x\left(\frac{2}{3x}\right) + 6x\left(\frac{1}{2}\right) = 6x\left(\frac{4}{x}\right) + 6x\left(\frac{4}{3}\right)$$

Simplify both sides.

$$4 + 3x = 24 + 8x$$

Subtract $3x$.

$$\underline{-3x \qquad\qquad -3x}$$

$$4 = 24 + 5x$$

Subtract 24.

$$\underline{-24 \quad -24}$$

$$-20 = 5x$$

Answer: The solution is $m = -18$. The solution set is $\{-18\}$.

Divide by 5. $-4 = x$

$$x = -4$$

Since $x = -4$ satisfies the original equation, the solution set is $\{-4\}$.

■ **YOUR TURN** Solve the equation $\dfrac{3}{y} + 2 = \dfrac{7}{2y}$.

When a variable is in the denominator of a fraction, then the LCD will contain the variable. Often this results in an extraneous solution. An **extraneous solution** is one that does not satisfy the original equation.

EXAMPLE 5 An Equation with an Extraneous Solution

Solve the equation $\dfrac{3x}{x - 1} + 2 = \dfrac{3}{x - 1}$.

Solution:

We can let x be any real number except 1. $\qquad \dfrac{3x}{x - 1} + 2 = \dfrac{3}{x - 1} \quad x \neq 1$

Eliminate the fractions by multiplying each term by the LCD, $x - 1$.
$$\dfrac{3x}{x - 1} \cdot (x - 1) + 2 \cdot (x - 1) = \dfrac{3}{x - 1} \cdot (x - 1)$$

Simplify both sides.
$$\dfrac{3x}{x - 1} \cdot (x - 1) + 2 \cdot (x - 1) = \dfrac{3}{x - 1} \cdot (x - 1)$$
$$3x + 2(x - 1) = 3$$

Eliminate parentheses. $\qquad 3x + 2x - 2 = 3$

Combine x terms on left. $\qquad 5x - 2 = 3$

Add 2 to both sides. $\qquad 5x = 5$

Divide both sides by 5. $\qquad x = 1$

Students may think $x = 1$ is the solution. However, the original equation had the restriction $x \neq 1$. Therefore, $x = 1$ is an extraneous solution and must be eliminated as a possible solution. Thus, the equation $\dfrac{3x}{x - 1} + 2 = \dfrac{3}{x - 1}$ has no solution .

CAUTION

When a variable is in the denominator of a fraction, then the LCD will contain the variable. Often this results in an extraneous solution.

■ **YOUR TURN** Solve the equation $\dfrac{2}{x} + \dfrac{1}{x + 1} = -\dfrac{1}{x(x + 1)}$.

In addition to multiplying by the LCD to eliminate fractions, in some problems, we can also cross multiply to eliminate them. This method can only be used when we have a fraction equal to a fraction.

Recall (Cross Multiplication)

When two fractions are equal, we can cross multiply to also generate an equivalent relationship.

$$\frac{2}{3} = \frac{4}{6}$$

Cross multiplying yields another equivalent relationship, $2 \cdot 6 = 3 \cdot 4$ or $12 = 12$.

EXAMPLE 6 Solving an Equation Using Cross Multiplication

Solve the equation $\dfrac{2}{x-3} = \dfrac{-3}{2-x}$.

Solution:

What values make *either* denominator equal to zero? The values $x = 2$ and $x = 3$ must be excluded from possible solutions to the equation.

We can let x be any real value
except 2 or 3. $\qquad \dfrac{2}{x-3} = \dfrac{-3}{2-x} \qquad x \neq 2, x \neq 3$

Cross multiply. $\qquad 2(2-x) = -3(x-3)$

Eliminate parentheses. $\qquad 4 - 2x = -3x + 9$

Collect x terms on left,
constants on right. $\qquad \boxed{x = 5}$

Since $x = 5$ satisfies the original equation, the solution set is $\{5\}$.

> **YOUR TURN** Solve the equation $\dfrac{-4}{x+8} = \dfrac{3}{x-6}$.

EXAMPLE 7 Solving an Equation Using Cross Multiplication

Solve the equation $\qquad \dfrac{x + 1 - \dfrac{2}{x}}{1 - \dfrac{1}{x}} = x + 2.$

Solution:

There are two numbers that must be excluded as possible solutions ($x = 0$ and $x = 1$) because they result in a denominator equal to 0.

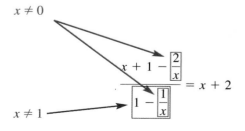

We can let x can be any real number except 0 or 1.

$$\frac{x + 1 - \dfrac{2}{x}}{1 - \dfrac{1}{x}} = x + 2 \qquad x \neq 0, x \neq 1$$

Rewrite right side with the implied 1 in denominator.

$$\frac{x + 1 - \dfrac{2}{x}}{1 - \dfrac{1}{x}} = \frac{x + 2}{1}$$

Cross multiply.

$$x + 1 - \frac{2}{x} = (x + 2)\left(1 - \frac{1}{x}\right)$$

Use the FOIL method to multiply the quantities on the right side.

$$x + 1 - \frac{2}{x} = x - 1 + 2 - \frac{2}{x}$$

Add constants on right.

$$x + 1 - \frac{2}{x} = x + 1 - \frac{2}{x}$$

Multiply each term by LCD, x.

$$x^2 + x - 2 = x^2 + x - 2$$

What do you notice about the left and right sides? Since they are the same, we call this an identity. Any x value makes this equation true except for the values we eliminated at the beginning. Therefore we say that the solution is **all real numbers except 0 and 1.** The solution set is $\{\mathbb{R} \mid x \neq 0 \text{ or } x \neq 1\}$, where the symbol $\mathbb{R}$ represents all real numbers.

SECTION 1.1   ## SUMMARY

In this section you have practiced solving linear equations. Useful techniques are cross multiplication and eliminating fractions by multiplying through by the least common denominator (LCD).

Steps for Solving a Linear Equation
Step 1: Determine any values that make denominators zero.
Step 2: Eliminate any fractions by cross multiplying or by multiplying the entire equation by the LCD.
Step 3: Eliminate parentheses and simplify.
Step 4: Solve by collecting like terms (variables on one side, constants on other).
Step 5: Eliminate any solutions that were noted in Step 1.

SECTION 1.1 EXERCISES

■ SKILLS

Solve for the given equation.

1. $3x - 5 = 7$ **2.** $4p + 5 = 9$ **3.** $9m - 7 = 11$ **4.** $2x + 4 = 5$

5. $5t + 11 = 18$ **6.** $7x + 4 = 21 + 24x$ **7.** $3x - 5 = 25 + 6x$ **8.** $5x + 10 = 25 + 2x$

9. $20n - 30 = 20 - 5n$ **10.** $14c + 15 = 43 + 7c$ **11.** $2(x - 1) + 3 = x - 3(x + 1)$

12. $4(y + 6) - 8 = 2y - 4(y + 2)$ **13.** $5p + 6(p + 7) = 3(p + 2)$ **14.** $3(z + 5) - 5 = 4z + 7(z - 2)$

15. $2a - 9(a + 6) = 6(a + 3) - 4a$

16. $25 - [2 + 5y - 3(y + 2)] = -3(2y - 5) - [5(y - 1) - 3y + 3]$

17. $32 - [4 + 6x - 5(x + 4)] = 4(3x + 4) - [6(3x - 4) + 7 - 4x]$

18. $12 - [3 + 4m - 6(3m - 2)] = -7(2m - 8) - 3[(m - 2) + 3m - 5]$

19. $20 - 4[c - 3 - 6(2c + 3)] = 5(3c - 2) - [2(7c - 8) - 4c + 7]$

20. $46 - [7 - 8y + 9(6y - 2)] = -7(4y - 7) - 2[6(2y - 3) - 4 + 6y]$

21. $\dfrac{1}{5}m = \dfrac{1}{60}m + 1$ **22.** $\dfrac{1}{12}z = \dfrac{1}{24}z + 3$ **23.** $\dfrac{x}{7} = \dfrac{2x}{63} + 4$ **24.** $\dfrac{a}{11} = \dfrac{a}{22} + 9$

25. $\dfrac{1}{3}p = 3 - \dfrac{1}{24}p$ **26.** $\dfrac{3x}{5} - x = \dfrac{x}{10} - \dfrac{5}{2}$ **27.** $\dfrac{5y}{3} - 2y = \dfrac{2y}{84} + \dfrac{5}{7}$ **28.** $2m - \dfrac{5m}{8} = \dfrac{3m}{72} + \dfrac{4}{3}$

29. $p + \dfrac{p}{4} = \dfrac{5}{2}$ **30.** $\dfrac{c}{4} - 2c = \dfrac{5}{4} - \dfrac{c}{2}$

In Exercises 31–48, specify any values that must be excluded from the solution set and then solve the equation.

31. $\dfrac{4}{y} - 5 = \dfrac{5}{2y}$ **32.** $\dfrac{4}{x} + 10 = \dfrac{2}{3x}$ **33.** $7 - \dfrac{1}{6x} = \dfrac{10}{3x}$

34. $\dfrac{7}{6t} = 2 + \dfrac{5}{3t}$ **35.** $\dfrac{2}{a} - 4 = \dfrac{4}{3a}$ **36.** $\dfrac{1}{x} + \dfrac{1}{x - 1} = \dfrac{1}{x(x - 1)}$

37. $\dfrac{1}{n} + \dfrac{1}{n + 1} = \dfrac{-1}{n(n + 1)}$ **38.** $\dfrac{1}{c - 2} + \dfrac{1}{c} = \dfrac{2}{c(c - 2)}$ **39.** $\dfrac{3}{a} - \dfrac{2}{a + 3} = \dfrac{9}{a(a + 3)}$

40. $\dfrac{5}{m} + \dfrac{3}{m - 2} = \dfrac{6}{m(m - 2)}$ **41.** $\dfrac{n - 5}{6n - 6} = \dfrac{1}{9} - \dfrac{n - 3}{4n - 4}$ **42.** $\dfrac{-2}{x + 4} = -\dfrac{3}{x + 6}$

43. $\dfrac{2}{5x + 1} = \dfrac{1}{2x - 1}$ **44.** $\dfrac{p^2 - 1}{2p + 1} = \dfrac{p}{2}$ **45.** $\dfrac{t - 1}{1 - t} = \dfrac{3}{2}$

46. $\dfrac{s^2 - 1}{s + 1} = s - 1$ **47.** $\dfrac{1 - \dfrac{1}{x}}{1 + \dfrac{1}{x}} = 1$ **48.** $\dfrac{t + \dfrac{1}{t}}{\dfrac{1}{t} - 1} = 1$

 ■ APPLICATIONS

49. Temperature. To calculate temperature in degrees Fahrenheit we use the formula $F = \dfrac{9}{5}C + 32$, where F is degrees Fahrenheit and C is degrees Celsius. Find the formula to convert from Fahrenheit to Celsius.

50. Geometry. The perimeter, P, of a rectangle is related to the length, L, and width, W, of the rectangle through the equation: $P = 2L + 2W$. Determine the width in terms of the perimeter and length.

51. Speed of Light. The frequency, f, of an optical signal in hertz (Hz) is related to the wavelength, λ, in meters (m) of

a laser through the equation $f = \dfrac{c}{\lambda}$, where c is the speed of light in a vacuum and is typically taken to be $c = 3.0 \times 10^8$ meters per second (m/s). What values must be eliminated from the wavelengths?

52. Optics. For objects placed near a lens, an image forms at a distinct position determined by the object distance. The position of the image is found using the thin lens equation:

$$\frac{1}{f} = \frac{1}{s_o} + \frac{1}{s_i},$$

where s_o is the distance of the object from the lens, s_i is the distance the image forms from the lens, and f is the focal length of the lens. Solve for the object distance, s_o, in terms of the focal length and image distance.

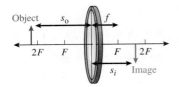

■ CATCH THE MISTAKE

In Exercises 53–56, explain the mistake that is made.

53. Solve the equation $4x + 3 = 6x - 7$.

Solution:

Subtract $4x$ and add 7 to the equation. $3 = 6x$
Divide by 3. $x = 2$

54. Solve the equation $3(x + 1) + 2 = x - 3(x - 1)$.

Solution:
$$3x + 3 + 2 = x - 3x - 3$$
$$3x + 5 = -2x - 3$$
$$5x = -8$$
$$x = -\frac{8}{5}$$

55. Solve the equation $\dfrac{4}{p} - 3 = \dfrac{2}{5p}$.

Solution:

Cross multiply.
$$(p - 3)2 = 4(5p)$$
$$2p - 6 = 20p$$
$$-6 = 18p$$
$$y = -\frac{6}{18}$$
$$y = -\frac{1}{3}$$

56. Solve the equation $\dfrac{1}{x} + \dfrac{1}{x - 1} = \dfrac{1}{x(x - 1)}$.

Solution:

Multiply by LCD, $x(x - 1)$.

$$\frac{x(x - 1)}{x} + \frac{x(x - 1)}{x - 1} = \frac{x(x - 1)}{x(x - 1)}$$

Simplify.
$$(x - 1) + x = 1$$
$$x - 1 + x = 1$$
$$2x = 2$$
$$x = 1$$

■ CHALLENGE

57. T or F: The solution to the equation $x = \dfrac{1}{\dfrac{1}{x}}$ is all real numbers.

58. T or F: The solution to the equation $\dfrac{1}{(x - 1)(x + 2)} = \dfrac{1}{x^2 + x - 2}$ is all real numbers.

59. T or F: $x = -1$ is a solution to the equation $\dfrac{x^2 - 1}{x - 1} = x + 1$.

60. T or F: $x = 1$ is a solution to the equation $\dfrac{x^2 - 1}{x - 1} = x + 1$.

61. Solve the equation for x: $\dfrac{b + c}{x + a} = \dfrac{b - c}{x - a}$. Are there any restrictions given that $a \neq 0, c \neq 0$?

62. Solve the equation for y: $\dfrac{1}{y - a} + \dfrac{1}{y + a} = \dfrac{2}{y - 1}$. Does y have any restrictions?

63. Solve the equation for x in terms of y:
$$y = \frac{a}{1 + \dfrac{b}{x + c}}$$

64. Find the number a for which $y = 2$ is a solution of the equation $y - a = y + 5 - 3ay$.

■ **TECHNOLOGY**

In Exercises 65–70, plot both sides of each equation in the same viewing rectangle, and classify each equation as an identity, a conditional equation, or a contradiction (inconsistent equation).

65. $3(x + 2) - 5x = 3x - 4$

66. $-5(x - 1) - 7 = 10 - 9x$

67. $2x + 6 = 4x - 2x + 8 - 2$

68. $10 - 20x = 10x - 30x + 20 - 10$

69. $\dfrac{x(x - 1)}{x^2} = 1$

70. $\dfrac{2x(x + 3)}{x^2} = 2$

SECTION 1.2 Applications Involving Linear Equations

Skills Objectives

- Develop mathematical models to solve application or real life problems.
- Use common formulas to solve application problems.
  - Solve geometry problems.
  - Solve interest problems.
  - Solve mixture problems.
  - Solve distance-rate-time problems.

Conceptual Objectives

- Understand the mathematical modeling process.
- Judge solutions to applied problems using common sense and intuition.

Solving Application Problems Using Mathematical Models

In this section we will use algebra to solve problems that occur in our day-to-day lives. You typically will read the problem in words, develop a mathematical model for the problem, solve that problem, and write the answer in words.

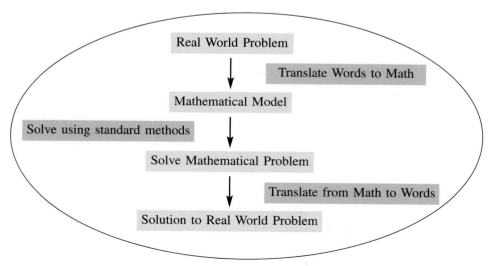

Although there is no *specific formula* for solving all word problems, there is a universal *procedure* for approaching them.

PROCEDURE FOR SOLVING WORD PROBLEMS

Step 1: Identify the question. Read the problem *one* time and note what you are asked to find.
Step 2: Make notes. Read until you can note something (an amount, a picture, anything). Continue reading and making notes until you have read the problem a second time.
Step 3: Set up an equation.
Step 4: Solve the equation.
Step 5: Check the solution. Run the solution past the "common sense department" using estimation.

EXAMPLE 1   How Long Was the Trip?

During a camping trip in North Bay, Ontario, a couple went one-third of the way by boat, 10 miles by foot, and one-sixth of the way by horse. How long was the trip?

Solution:

STEP 1 **Identify the question.**
How many miles was the trip?

STEP 2 **Make notes.**

Read	Write
... one-third of the way by boat	BOAT—$\frac{1}{3}$ of the trip
... 10 miles by foot	FOOT—10 miles
... and one-sixth of the way by horse	HORSE—$\frac{1}{6}$ of the trip

STEP 3 **Set up an equation.**
The total distance of the trip is the sum of all the distances by boat, foot, and horse.

Distance by boat + Distance by foot + Distance by horse = Total distance of trip

Distance of total trip in miles $= x$

Distance by boat $= \frac{1}{3}x$

Distance by foot $= 10$ miles

Distance by horse $= \frac{1}{6}x$

$$\underbrace{\frac{1}{3}x}_{\text{boat}} + \underbrace{10}_{\text{foot}} + \underbrace{\frac{1}{6}x}_{\text{horse}} = \underbrace{x}_{\text{total}}$$

STEP 4 **Solve the equation.** $\frac{1}{3}x + 10 + \frac{1}{6}x = x$

Collect x terms on right. $10 = x - \frac{1}{3}x - \frac{1}{6}x$

Simplify right side. $10 = \frac{1}{2}x$

The trip was 20 miles. $\boxed{x = 20}$

STEP 5 **Check the solution.**

Estimate: The boating distance, $\frac{1}{3}$ of 20 miles, is approximately 7 miles; the riding distance on horse, $\frac{1}{6}$ of 20 miles, is approximately 3 miles. Adding these two distances to the 10 miles by foot gives a trip distance of 20 miles.

YOUR TURN   A family arrives at the Walt Disney World parking lot. To get from their car in the parking lot to the gate at the Magic Kingdom they walk $\frac{1}{4}$ mile, take a tram for $\frac{1}{3}$ of their total distance, and take a monorail for $\frac{1}{2}$ of their total distance. How far is it from their car to the gate of Magic Kingdom?

EXAMPLE 2 Find the Numbers

Find three consecutive even integers so that the sum of the three numbers is 2 more than twice the third.

Solution:

STEP 1 **Identify the question.**

What are the three consecutive even integers?

STEP 2 **Make notes.**

Examples of three consecutive even integers are 14, 16, 18 or 20, 22, 24 or 2, 4, 6. Let n represent the first even integer. The next consecutive even integer is $n + 2$ and the next consecutive even integer after that is $n + 4$.

$n = $ 1st integer
$n + 2 = $ 2nd consecutive even integer
$n + 4 = $ 3rd consecutive even integer

Read	**Write**
... sum of the first three	$n + (n + 2) + (n + 4)$
... is	$=$
... two more than	$+2$
... twice the third	$2(n + 4)$

STEP 3 **Set up an equation.**

$$\underbrace{n + (n + 2) + (n + 4)}_{\text{Sum of the first three}} \underbrace{=}_{\text{is}} \underbrace{2}_{\text{2 more than}} + \underbrace{2(n + 4)}_{\text{twice the third}}$$

STEP 4 **Solve the equation.** $n + (n + 2) + (n + 4) = 2 + 2(n + 4)$

Eliminate parentheses. $n + n + 2 + n + 4 = 2 + 2n + 8$

Simplify both sides. $3n + 6 = 2n + 10$

Collect n terms on left and constants on right. $n = 4$

The three consecutive even integers are 4, 6, and 8 .

STEP 5 **Check the solution.**

Substitute the solution into the problem to see if it makes sense. The sum of the three integers $(4 + 6 + 8)$ is 18. Twice the third is 16. Since 2 more than twice the third is 18, the solution checks.

■ **YOUR TURN**   Find a number such that 12 less than $\frac{3}{5}$ the number is $\frac{2}{3}$ the number.

Geometry Problems

Some problems require geometric formulas in order to be solved. The following geometric formulas may be useful.

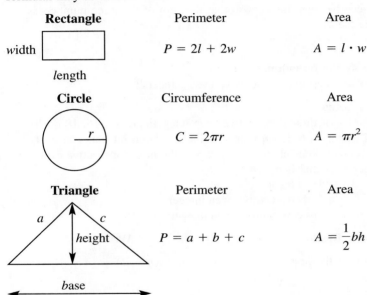

Rectangle	Perimeter	Area
width	$P = 2l + 2w$	$A = l \cdot w$

Circle	Circumference	Area
r	$C = 2\pi r$	$A = \pi r^2$

Triangle	Perimeter	Area
a c height base	$P = a + b + c$	$A = \dfrac{1}{2}bh$

EXAMPLE 3 Geometry

A rectangle 24 meters long has the same area as a square, with 12 meter sides. What are the dimensions of the rectangle?

Solution:

STEP 1 **Identify the question.**

What are the dimensions (length and width) of the rectangle?

STEP 2 **Make notes.**

Read	Write/Draw
A rectangle 24 meters long	w ☐ $l = 24$

... has the same area

... as a square that is
12 meters long

rectangle area $= l \cdot w = 24w$

area of square $= 12 \times 12 = 144$

12 m $\boxed{144 \text{ m}^2}$

12 m

STEP 3 **Set up an equation.**
The area of the rectangle is
equal to the area of the square.

Substitute in known quantities.

rectangle area = square area

$$24w = 144$$

STEP 4 **Solve the equation.**

$$24w = 144$$

Divide by 24.

$$w = \frac{144}{24} = 6$$

The dimensions of the rectangle are 24 meters long and 6 meters wide .

6 m $\boxed{}$

24 m

STEP 5 **Check the solution.**
A 24 m by 6 m rectangle has an area of 144 m^2.

CONCEPT CHECK   How do you find the perimeter of a triangle?

■ **YOUR TURN**   Find the perimeter of a triangle if one side is 12 inches,
another side is 2 inches more than $\frac{1}{2}$ the perimeter, and the third side is $\frac{1}{4}$ the
perimeter.

Interest Problems

In financial planning we deal with problems involving money. In particular, we discuss
interest. **Interest** is money paid for the use of money. The total amount borrowed is
called the **principal**. The principal can be the price of our new car and we pay the bank
interest for loaning us money. The principal can also be the amount we keep in a CD or
money market account that the bank uses and pays us interest. Typically interest rate,
expressed as a percentage, is the amount charged for the use of the principal for a given
time, usually in years.

DEFINITION   SIMPLE INTEREST

If a principal of P dollars is borrowed for a period of t years at an annual interest rate r (expressed in decimal form), the interest I charged is

$$I = Prt$$

This formula is classified as **simple interest**.

EXAMPLE 4 Simple Interest

Through a summer job Morgan is able to save $2,500. If she puts that money into a 6-month certificate of deposit (CD) that pays a simple interest rate of 3% a year, how much money will she have in her CD at the end of the 6 months?

Solution:

STEP 1 **Identify the question.**
How much money does Morgan have after 6 months?

STEP 2 **Make notes.**
The principal is $2,500.

The annual interest rate is 3%, which in decimal form is 0.03.

The time the money spends accruing interest is 6 months, or $\frac{1}{2}$ of a year.

STEP 3 **Set up an equation.**
Write the simple interest
rate formula. $I = Prt$

Label the known quantities. $P = 2500$, $r = 0.03$, and $t = 0.5$

STEP 4 **Solve the equation.** $I = Prt$

$$I = (2500)(0.03)(0.5) = 37.5$$

The interest paid on the CD is $37.50. Adding this to the principal gives a total of

$$\$2{,}500 + \$37.50 = \boxed{\$2{,}537.50}$$

STEP 5 **Check the solution.**
This answer agrees with our intuition. Had we made a mistake such as moving one decimal to the right, then the interest would have been $375, which would have seemed larger than we expected.

EXAMPLE 5 Multiple Investments

Theresa earns a full athletic scholarship for college so her parents have given her the $20,000 they had saved to pay for her college tuition. She decides to invest that money with an overall goal of earning 11% interest. She wants to put some of the money in a low risk investment that has been earning 8% a year and the rest of the money in a medium risk investment that typically earns 12% a year. How much money should she put in each investment to reach her goal?

Solution:

STEP 1 **Identify the question.**
How much money is invested in each (8% and 12%) account?

STEP 2 **Make notes.**

Read	Write/Draw

Theresa has $20,000 to invest.

If part is invested at 8% and the rest at 12%, how much should be invested at each rate to yield 11% on the total amount invested?

$20,000

Some at 8%   Some at 12%

$20,000 at 11%

STEP 3 **Set up an equation.**

If we let x represent the amount Theresa puts into the 8% investment, how much of the $20,000 is left for her to put in the 12% investment?

Amount in the 8% investment $= x$

Amount in the 12% investment $= 20{,}000 - x$

Simple Interest Formula: $I = Prt$

Investment	Principal	Rate	Time (yr)	Interest
8% Account	x	0.08	1	$0.08x$
12% Account	$20{,}000 - x$	0.12	1	$0.12(20{,}000 - x)$
Total	$20{,}000$	0.11	1	$0.11(20{,}000)$

Adding the interest earned in the 8% investment to the interest earned in the 12% investment should earn an average of 11% on the total investment.

$$0.08x + 0.12(20{,}000 - x) = 0.11(20{,}000)$$

STEP 4 **Solve the equation.**

We first eliminate the decimals by multiplying the entire equation by 100.

$$8x + 12(20{,}000 - x) = 11(20{,}000)$$

Eliminate the parentheses.

$$8x + 240{,}000 - 12x = 220{,}000$$

Collect x terms on the left, constants on the right.

$$-4x = -20{,}000$$

Divide by -4.

$$x = 5000$$

Calculate amount at 12%.

$$20{,}000 - 5{,}000 = 15{,}000$$

Theresa should invest $5,000 at 8% and $15,000 at 12% to reach her goal .

STEP 5 **Check the solution.**

If money is invested at 8% and 12% with a goal of averaging 11%, our intuition tells us that more should be invested at 12% than 8%, which is what we found. The exact check is as follows:

$$0.08(5000) + 0.12(15{,}000) = 0.11(20{,}000)$$

$$400 + 1800 = 2200$$

$$2200 = 2200$$

> ■ **YOUR TURN** You win $24,000 and you decide to invest the money in two different investments: one paying 18% and the other paying 12%. A year later you collectively have $27,480. How much did you originally invest in each account?

Mixture Problems

Mixtures occur every day. The different candies that sell for different prices make up a movie snack. New blends of coffees are developed by coffee connoisseurs. Chemists mix different concentrations of acids in their labs. Whenever two or more quantities are combined the result is a **mixture**.

Our choices at the gas pumps are typically 87, 89, or 93 octane. We would expect the relationship between price and octane to be linear, but it is not. 89 octane is significantly overpriced. Therefore, if your car requires 89 octane, it would be more cost effective to mix 87 and 93 octane.

EXAMPLE 6 Mixture Problem

The manual for your new car suggests using gasoline that is 89 octane. In order to save money, you decide to use some 87 octane and some 93 octane in combination with the 89 octane currently in your tank in order to have an approximate 89 octane mixture. Assuming you have 1 gallon of 89 octane remaining in your tank (your tank capacity is 16 gallons), how many gallons of 87 and 93 octane should be used to fill up your tank to achieve a mixture of 89 octane?

Solution:

STEP 1 **Identify the question.**
How many gallons of 87 octane and how many gallons of 93 octane should be used?

STEP 2 **Make notes.**

Read

Assuming you have one gallon of 89 octane remaining in your tank (your tank capacity is 16 gallons), how many gallons of 87 and 93 octane should you add?

Write/Draw

89 octane + 87 octane + 93 octane = 89 octane

[1 gallon] [? gallons] [? gallons] [16 gallons]

STEP 3 **Set up an equation.**

x = gallons of 87 octane gasoline added at the pump

$15 - x$ = gallons of 93 octane gasoline added at the pump

1 = gallons of 89 octane gasoline already in the tank

$0.89(1) + 0.87x + 0.93(15 - x) = 0.89(16)$

STEP 4 **Solving the equation.**

$$0.89(1) + 0.87x + 0.93(15 - x) = 0.89(16)$$

Eliminate parentheses. $0.89 + 0.87x + 13.95 - 0.93x = 14.24$

Collect x terms on left side. $-0.06x + 14.84 = 14.24$

Subtract 14.84 from both sides
of the equation. $-0.06x = -0.6$

Divide both sides by -0.06. $x = 10$

Calculate the amount of 93 octane. $15 - 10 = 5$

Add 10 gallons of 87 octane and 5 gallons of 93 octane .

STEP 5 **Check the solution.**
Estimate: Our intuition tells us that if the desired mixture is
89 octane then we should add approximately 1 part 93 octane and
2 parts 87 octane. The solution we found, 10 gallons of 87 octane and
5 gallons of 93 octane, agrees with this.

■ **YOUR TURN**   For a certain experiment, a student requires 100 ml
of a solution that is 11% HCl (hydrochloric acid). The storeroom has only
solutions that are 5% HCl and 15% HCl. How many milliliters of each
available solution should be mixed to get 100 ml of 11% HCl?

Distance-Rate-Time Problems

The next example deals with distance, rate, and time. On a road trip you see a sign that
says your destination is 90 miles away and your speedometer reads 60 miles per hour.
Dividing 90 miles by 60 miles per hour allows you to estimate that your arrival will be
in 1.5 hours.

If the rate, or speed, is assumed to be constant, then the equation that relates distance
(d), rate (r), and time (t), is given by $d = r \cdot t$. In the above driving example,

$$d = 90 \text{ miles} \qquad r = 60\frac{\text{miles}}{\text{hour}}$$

Substituting these into $d = r \cdot t$ we arrive at

$$90 \text{ miles} = \left[60\frac{\text{miles}}{\text{hour}}\right] \cdot t$$

Solving for t, we get $t = \dfrac{90 \text{ miles}}{60\dfrac{\text{miles}}{\text{hour}}} = 1.5 \text{ hours}$

When attending the 2002 Ryder Cup, it took 8 hours to go from Orlando to London and 9.5 hours for the return flight. The reason for the difference in time is the jet stream (constant wind speed). On the flight from Orlando to London there was a tailwind, and on the flight from London to Orlando there was a headwind. For example, a plane that flies 600 mph in still air actually flies 800 mph relative to the ground (ground speed) when it has a 200 mph tailwind, and that plane flies 400 mph relative to the ground when it has a 200 mph headwind.

EXAMPLE 7 Distance-Rate-Time

Because of the jet stream, it takes 8 hours to fly from Orlando to London and 9.5 hours to return. If an airplane averages 550 mph in still air, what is the average rate of the wind blowing in the direction from Orlando to London?

Solution:

STEP 1 **Identify the question.**
What is the rate that the wind is blowing in mph?

STEP 2 **Make notes.**

Read	Write/Draw
It takes 8 hours to fly from Orlando to London and 9.5 hours to return.	8 hours Orlando → London 9.5 hours
If the airplane averages 550 mph in still air...	550 mph + wind Orlando → London 550 mph − wind

STEP 3 **Set up an equation.**
The formula relating distance, rate, and time is $d = r \cdot t$. The distance, d, of each flight is the same. On the Orlando to London flight the time is 8 hours due to an increased speed from a tailwind. On the London to Orlando flight the time is 9.5 hours and the speed is decreased due to the headwind. Let w represent the wind speed.

Orlando to London: $d = (550 + w)8$

London to Orlando: $d = (550 - w)9.5$

These distances are the same, so set them equal to each other:

$$(550 + w)8 = (550 - w)9.5$$

STEP 4 **Solve the equation.**

Eliminate parentheses. $4400 + 8w = 5225 - 9.5w$

Collect w terms on left, constants on right. $17.5w = 825$

Divide by 17.5. $w = 47.1429 \approx 47$

The wind is blowing approximately 47 mph in the direction from Orlando to London.

STEP 5 **Check the solution.**

Estimate: Going from Orlando to London the tailwind is approximately 47 mph, which added to the plane's 550 mph speed yields a ground speed of 597 mph. The Orlando to London route took 8 hours. The distance of that flight is (597 mph)(8 hr), which is 4776 miles. The return trip experienced a headwind of approximately 47 mph, so subtracting the 47 from 550 gives an average speed of 503 mph. That route took 9.5 hours, so the distance of the London to Orlando flight was (503 mph)(9.5 hr), which is 4778.5 miles. Note that 4776 and 4778.5 miles are approximately equal.

■ **YOUR TURN** A boat traveling up the St. John's River can make the trip from Sanford to Deland in 2 hours, but the return trip takes 3 hours. The river runs at a rate of 5 mph. What is the average speed of the boat in still water?

SECTION 1.2 SUMMARY

In this section application or real world problems are solved. We used the five-step procedure to solve all problems. Some problems require development of a mathematical model, while others rely on common formulas.

SECTION 1.2 EXERCISES

■ **APPLICATIONS**

1. **Money.** Donna uses a 10% off coupon at her local nursery. After buying azaleas, bougainvillea, and bags of potting soil, her checkout price before tax is $217.95. How much would she have paid without the coupon?

2. **Money.** The original price of a pair of binoculars is $74. The sale price is $51.80. How much was the markdown?

3. **Money.** Jeff, Tom, and Chelsea order a large pizza. They decide to split the cost according to how much they will eat. Tom pays $5.16, Chelsea eats $\frac{1}{8}$ of the pizza, and Jeff eats $\frac{1}{2}$ the pizza. How much did the pizza cost?

4. **Budget.** A couple decides to analyze their monthly spending habits. They break down expenses into four categories: monthly bills, groceries, investments, and miscellaneous. The monthly bills are 50% of their take-home pay, and they invest 20% of their take-home pay.

They spend $560 on groceries and 23% goes to miscellaneous. How much is their take-home pay a month?

5. **Money.** A builder of tract homes reduced the price of a model by 15%. If the new price is $125,000, what was its original price? How much can be saved by purchasing the model?

6. **Money.** A college bookstore marks up the price it pays the publisher for a book by 25%. If the selling price of a book is $79, how much did the bookstore pay for the book?

7. **Puzzle.** Angela is on her way from Jersey City into New York City for dinner. She walks 1 mile to the train station, takes the train $\frac{3}{4}$ of the way, and takes a taxi $\frac{1}{6}$ of the way to the restaurant. How far does Angela live from the restaurant?

8. **Puzzle.** An employee at Kennedy Space Center lives in Daytona Beach and works in the VAB building. She

carpools to work with a colleague. On the days that her colleague drives the car pool, she drives 7 miles to the park and rides with her colleague to the KSC headquarters building, and then takes the KSC shuttle from the headquarters building to the VAB. The drive from the park and ride to the headquarters building is $\frac{5}{6}$ of her total trip and the shuttle ride is $\frac{1}{20}$ of her total trip. How many miles does she travel from her house to the VAB on days when her colleague drives?

9. Puzzle. A typical college student spends $\frac{1}{3}$ of her waking time in class, $\frac{1}{5}$ of her waking time eating, $\frac{1}{10}$ of her waking time working out, 3 hours studying, and 2.5 hours doing other things. How many hours of sleep does the typical college student get?

10. Diet. A particular 1550-calories-per-day diet suggests eating breakfast, lunch, dinner, and two snacks. Dinner is twice the calories of breakfast. Lunch is 100 calories more than breakfast. The two snacks are 100 and 150 calories. How many calories are each meal?

11. Budget. A company has a total of $20,000 allocated for monthly costs. Fixed costs are $15,000 per month and variable costs are $18.50 per unit. How many units can be manufactured a month?

12. Budget. A woman decides to start a small business making monogrammed cocktail napkins. She can set aside $1,870 for monthly costs. Fixed costs are $1,329.50 per month and variable costs are $3.70 per set of napkins. How many sets of napkins can she afford to make per month?

13. Numbers. Find a number such that 10 less than $\frac{2}{3}$ the number is $\frac{1}{4}$ the number.

14. Numbers. Find a positive number such that 10 times the number is 16 more than twice the number.

15. Numbers. Find two consecutive even integers such that 4 times the smaller number is two more than 3 times the larger number.

16. Numbers. Find three consecutive integers such that the sum of the three is equal to 2 times the sum of the first two integers.

17. Geometry. Find the perimeter of a triangle if one side is 11 inches, another side is $\frac{1}{5}$ the perimeter, and the third side is $\frac{1}{4}$ the perimeter.

18. Geometry. Find the dimensions of a rectangle whose length is one foot longer than twice its width and whose perimeter is 20 feet.

19. Geometry. An NFL playing field is a rectangle. The length of the field (excluding the end zones) is 40 more yards than twice the width. The perimeter of the playing field is 260 yards. What are the dimensions of the field in yards?

20. Geometry. The length of a rectangle is 2 more than 3 times the width and the perimeter is 28 inches. What are the dimensions of the rectangle?

21. Geometry. Consider two circles, a smaller one and a larger one. If the larger one has a radius that is 3 feet larger than the smaller circle and the ratio of the circumferences is 2:1, what are the radii of the two circles?

22. Geometry. The perimeter of a semicircle is doubled when the radius is increased by 1. Find the radius of the semicircle.

23. Geometry. A man wants to remove a tall pine tree from his yard. To save money he wants to do it himself, but before he goes to Home Depot he needs to know how tall of an extension ladder he needs to purchase. He measures the shadow of the tree to be 225 feet long. At the same time he measures the shadow of a 4 foot stick to be 3 feet. Approximately how tall is the pine tree?

24. Geometry. The same man in Exercise 23 realizes he also wants to remove a dead oak tree. Later in the day he measures the shadow of the oak tree to be 880 feet long and the 4 foot stick now has a shadow of 10 feet. Approximately how tall is the oak tree?

25. Alligators. It is common to see alligators in ponds, lakes, and rivers in Florida. The ratio of head size (back of the head to the end of the snout) to the full body length of an alligator is typically constant. If a $3\frac{1}{2}$ foot alligator has a head length of 6 inches, how long would you expect an alligator to be whose head length is 9 inches?

26. Snakes. In the African rainforest there is a snake called a Gaboon Viper. The fang size of this snake is proportional to the length of the snake. A 3 foot snake typically has 2 inch fangs. If a herpetologist finds Gaboon Viper fangs that are 2.6 inches long, how big of a snake would she expect to find?

27. Money. Ashley has $120,000 to invest and decides to put some in a CD that earns 4% interest per year and the rest in a low risk stock that earns 7%. How much did she invest in each to earn $7,800 interest in the first year?

28. Money. You inherit $13,000 and you decide to invest the money in two different investments: one paying 10% and the other paying 14%. A year later your investments are worth $14,580. How much did you originally invest in each account?

29. Money. Wendy was awarded a volleyball scholarship to the University of Michigan, so on graduation her parents gave her the money they had saved for her college tuition, $14,000. She opted to invest some money in a privately held company that pays 10% per year and evenly split the remaining money between a money market account yielding 2% and a high risk stock that yielded 40%. At the end of the first year she had $16,610 total. How much did she invest in each of the three?

30. Money. A high school student was able to save $5,000 by working a part-time job every summer. He invested half the money in a money market account and half the money in a stock that paid three times as much interest as the money market account. After a year he earned $150 in interest. What were the interest rates of the money market account and the stock?

31. Budget. When landscaping their yard, a couple budgeted $4,200. The irrigation system is $2,400 and the sod is $1,500. The rest they will spend on trees and shrubs. Trees each cost $32 and shrubs each cost $4. Combined they plant a total of 33 trees and shrubs. How many of each did they plant in their yard?

32. Budget. At the deli Jennifer bought spicy turkey and provolone cheese. The turkey costs $6.32 per pound and the cheese costs $4.27 per pound. In total, she bought 3.2 pounds and the price was $17.56. How many pounds of each did she buy?

33. Chemistry. For a certain experiment, a student requires 100 ml of a solution that is 8% HCl (hydrochloric acid). The storeroom has only solutions that are 5% HCl and 15% HCl. How many milliliters of each available solution should be mixed to get 100 ml of 8% HCl?

34. Overhead. A professor is awarded two research grants, each having different overhead rates. The research project conducted on campus has a rate of 42.5% overhead and the project conducted in the field, off campus, has a rate of 26% overhead. If she was awarded $1,170,000 total for the two awards with an average overhead rate of 39%, how much was the research project on campus and how much was the research project off campus?

35. Food. On the way to the movies a family picks up a custom-made bag of candies. The parents like caramels ($1.50/lb) and the children like gummy bears ($2.00/lb). Combined they bought a 1.25 lb bag of candies that cost $2.50. How much of each candy did they buy?

36. Food. Mr. Parkinson, who is on a low-sodium diet, is craving chicken noodle soup. A serving size, 120 ml, of regular chicken noodle soup contains approximately 890 mg of sodium. A low-sodium version of the soup contains 450 mg of sodium per serving. If he mixes the two soups to yield a soup that contains 582 mg of sodium, what percentage of the mixture was the regular version and what percentage was the low-sodium version?

37. Communications. The speed of light is approximately 3.0×10^8 meters per second (670,616,629 mph). The distance from Earth to Mars varies because of the orbits of the planets around the sun. On average Mars is 100 million miles from Earth. If we use laser communication systems, what will the delay be between Houston and NASA astronauts on the red planet?

38. Sound. The speed of sound is approximately 760 mph in air. If a gun is fired $\frac{1}{2}$ mile away, how long will it take the sound to reach you?

39. Boating. A motorboat can maintain a constant speed of 16 miles per hour relative to the water. The boat makes a trip upstream to a marina in 20 minutes. The return trip takes 15 minutes. What is the speed of the current?

40. Flying. A Cessna 175 can average 130 mph. If a trip takes 2 hours one way and the return takes 1 hour and 15 minutes, find the wind speed assuming it is constant.

41. Exercise. A jogger and a walker cover the same distance. The jogger finishes in 40 minutes. The walker takes an hour. How fast is each exerciser moving if the jogger runs 2 mph faster than the walker?

42. Travel. A high school student in Seattle, Washington, decides to attend the University of Central Florida. On the way to UCF he took a southern route. After graduation he returned to Seattle via a northern trip. On both trips he had the same average speed. If the southern trek took 45 hours, and the northern trek took 50 hours, and the northern trek was 300 miles longer, how long was each trip?

43. Work. Christopher can paint the interior of his house in 15 hours. If he hires Cynthia to help him, they can do the same job together in 9 hours. If he lets Cynthia work alone, how long will it take her to paint the interior of his house?

44. Work. Joshua can deliver his newspapers in 30 minutes. It takes Amber 20 minutes to do the same route. How long would it take them to deliver the newspapers if they worked together?

45. Sound. A major chord in music is composed of notes whose frequencies are in the ratio 4:5:6. If the first note of a chord has a frequency of 264 hertz (middle C on the piano), find the frequencies of the other two notes. (*Hint:* Set up two proportions using 4:5 and 4:6.)

46. Sound. A minor chord in music is composed of notes whose frequencies are in the ratio 10:12:15. If the first note of a minor chord is A, with a frequency of 220 hertz, what are the frequencies of the other two notes?

47. Grades. Danielle's test scores are 86, 80, 84, and 90. The final exam will count as $\frac{2}{3}$ of the final grade. What score does Danielle need on the final in order to earn a B, which requires an average score of 80? What does she need to earn an A, which requires an average of 90?

48. Grades. Sam's final exam will count as two tests. His test scores are 80, 83, 71, 61, and 95. What score does Sam need on the final in order to have an average score of 80?

49. Sports. In Superbowl XXXVII, the Tampa Bay Buccaneers scored a total of 48 points. All of their points came from field goals and touchdowns. Field goals are worth 3 points and each touchdown was worth 7 points (Martin Gramatica was successful in every extra point attempt). They scored a total of 8 times. How many field goals and touchdowns were scored, respectively?

50. Sports. A tight end can run the 100 yard dash in 12 seconds. A defensive back can do it in 10 seconds. The tight end catches a pass at his own 20 yard line with the defensive back at the 15 yard line. If no other players are nearby, at what yard line will the defensive back catch up to the tight end?

51. Seesaw. How do two children of different weights balance on a seesaw? The heavier child sits closer to the center and the lighter child sits further away. When the product of the weight of the child and the distance from the center is equal on both sides, the seesaw should be horizontal to the ground. Suppose Max weighs 42 lbs and Maria weighs 60 lbs. If Max sits 5 feet from the center, how far should Maria sit from the center in order to balance the seesaw horizontal to the ground?

52. Seesaw. Suppose Martin, who weighs 33 lbs, sits on the side with Max. If their average distance to the center is 4 feet, how far should Maria sit from the center in order to balance the seesaw horizontal to the ground?

53. Seesaw. If a seesaw has an adjustable bench, then the board can slide along the fulcrum. Maria and Max in Exercise 51 decide to sit on the very edge of the board on each side. Where should the fulcrum be placed along the board in order to balance the seesaw horizontally to the ground? Give answer in terms of distance from each child's end.

54. Seesaw. Add Martin (Exercise 52) to Max's side and recalculate Exercise 53.

For Exercises 55–58, refer to this lens law.

The position of the image is found using the thin lens equation:

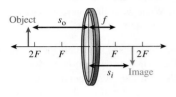

$$\frac{1}{f} = \frac{1}{s_o} + \frac{1}{s_i},$$

where s_o is the distance of the object from the lens, s_i is the distance the image forms from the lens, and f is the focal length of the lens.

55. Optics. If the focal length of a lens is 3 cm and the image distance is 5 cm from the lens, what is the distance from the object to the lens?

56. Optics. If the focal length of the lens is 8 cm and the image distance is 2 cm from the lens, what is the distance from the object to the lens?

57. Optics. The focal length of a lens is 1.2 cm. If the image distance from the lens is 1 cm more than the distance from the object to the lens, find the object distance.

58. Optics. The focal length of a lens is 8 cm. If the image distance from the lens is half the distance from the object to the lens, find the object distance.

■ TECHNOLOGY

59. Suppose you bought a house for $132,500 and sold it 3 years later for $168,190. Plot these points using a graphing utility. Assuming a linear relationship, how much could you have sold the house for had you waited 2 additional years?

60. Suppose you bought a house for $132,500 and sold it 3 years later for $168,190. Plot these points using a graphing utility. Assuming a linear relationship how much could you have sold the house for had you sold it 1 year after buying it?

61. A golf club membership has two options. Option A is a $300 monthly fee plus $15 cart fee every time you play. Option B has a $150 monthly fee and a $42 fee every time you play. Write a mathematical model for monthly costs for each plan and graph both in the same viewing rectangle using a graphing utility. Explain when Option A is the better deal and when Option B is the better deal.

62. A phone provider offers two calling plans. Plan A has a $30 monthly charge and a $.10 per minute charge on every call. Plan B has a $50 monthly charge and a $.03 per minute charge on every call. Explain when Plan A is the better deal and when Plan B is the better deal.

SECTION 1.3 Quadratic Equations

Skills Objectives

- Use factoring to solve quadratic equations.
- Use the square root method to solve quadratic equations.
- Complete the square to solve quadratic equations.
- Use the quadratic formula to solve quadratic equations.
- Solve applied problems that result in quadratic equations.

Conceptual Objectives

- Classify equations as quadratic.
- Choose appropriate methods for solving quadratic equations.
- Interpret different types of solution sets (real, imaginary, repeated roots).
- Derive the quadratic formula.

Examples of *quadratic equations,* also called second-degree equations, are

$$x^2 + 3 = 7 \qquad 5x^2 + 4x - 7 = 0 \qquad x^2 - 3 = 0$$

DEFINITION   QUADRATIC EQUATION

A **quadratic equation** in x is an equation that can be written in the **standard form**

$$ax^2 + bx + c = 0$$

where a, b, and c are real numbers and $a \neq 0$.

Factoring

The factoring method uses the **zero product property**:

WORDS	MATH
If a product is zero, then at least one of its factors has to be zero.	If $a \cdot b = 0$, then $a = 0$ or $b = 0$.

Consider $(x - 3)(x + 2) = 0$. The zero product property says that $x - 3 = 0$ or $x + 2 = 0$, which leads to $x = -2$ or $x = 3$. The solution set is $\{-2, 3\}$.

When a quadratic equation is written in standard form, $ax^2 + bx + c = 0$, it may be possible to factor the left side of the equation as a product of two first-degree polynomials. We use the zero product property and set each linear factor equal to zero. We solve the resulting two linear equations to obtain the solutions of the quadratic equation.

EXAMPLE 1 Solving a Quadratic Equation by Factoring

Solve the equation $x^2 - 6x - 16 = 0$.

Solution:

The quadratic equation is already in standard form.

$$x^2 - 6x - 16 = 0$$

TECHNOLOGY TIP

The graph of $y = x^2 - 6x - 16$ is shown. The x-intercepts are the solutions to the equation $x^2 - 6x - 16 = 0$.

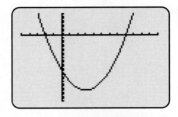

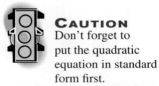

Factor the left side into a product of two linear factors.

$$(x - 8)(x + 2) = 0$$

If a product equals zero, one of its factors has to be zero.

$$x - 8 = 0 \text{ or } x + 2 = 0$$

Solve both linear equations.

$$x = 8 \text{ or } x = -2$$

The solution set is $\{-2, 8\}$.

▪ YOUR TURN Solve the quadratic equation $x^2 + x - 20 = 0$ by factoring.

EXAMPLE 2 Solving a Quadratic Equation by Factoring

Solve the equation $x^2 - 6x + 5 = -4$.

CAUTION
Don't forget to put the quadratic equation in standard form first.

COMMON MISTAKE

A common mistake is to forget to put the equation in standard form first and then use the zero product property incorrectly.

✓ CORRECT

Write the original equation.

$$x^2 - 6x + 5 = -4$$

Write the equation in standard form by adding 4 to both sides.

$$x^2 - 6x + 9 = 0$$

Factor the left side.

$$(x - 3)(x - 3) = 0$$

Use the zero product property and set each factor equal to zero.

$$x - 3 = 0 \text{ or } x - 3 = 0$$

Solve each linear equation.

$$x = 3$$

Don't forget to put the quadratic equation in standard form first.

✗ INCORRECT

Write the original equation.

$$x^2 - 6x + 5 = -4$$

Factor the left side.

$$(x - 5)(x - 1) = -4$$

The **error** occurs here.

$$x - 5 = -4 \text{ or } x - 1 = -4$$

The error above leads to an **incorrect answer**.

$$x = 1 \text{ or } x = -3$$

NOTE: The equation had one solution, or root, which is 3. Since the linear factors were the same, or repeated, we say that 3 is a **double root**, or **repeated root**.

▪ **Answer:** The solution set is $\{4, -5\}$.

■ **YOUR TURN** Solve the quadratic equation $9p^2 = 24p - 16$ by factoring.

EXAMPLE 3 Solving a Quadratic Equation by Factoring

Solve the equation $2x^2 = 3x$.

 **COMMON MISTAKE**

The common mistake here is dividing both sides by x, which is not allowed because x might be 0.

 CORRECT

Write the equation in standard form by subtracting $3x$.

$$2x^2 - 3x = 0$$

Factor the left side.

$$x(2x - 3) = 0$$

Use the zero product property and set each factor equal to zero.

$$x = 0 \text{ or } 2x - 3 = 0$$

Solve each linear equation.

$$x = 0 \text{ or } x = \frac{3}{2}$$

The solution set is $\left\{0, \dfrac{3}{2}\right\}$.

INCORRECT

Write the original equation.

$$2x^2 = 3x$$

The **error** occurs here when both sides are divided by x.

$$2x = 3$$

This is **incorrect** because one solution, $x = 0$, is missing.

$$x = \frac{3}{2}$$

The root $x = 0$ is lost when the original quadratic equation is divided by x. Remember to put the equation in standard form first and then factor.

 CAUTION

Never divide by a variable (it may be zero). Always bring terms to one side and factor.

Square Root Method

The square root of 16, $\sqrt{16}$, is 4, *not* ± 4. In the review (Chapter 0) the **principal square root** was discussed. The solutions to $x^2 = 16$, however, are $x = -4$ and $x = 4$. Let us now investigate quadratic equations that do not have a first-degree term. They have the form

$$ax^2 + c = 0 \qquad a \neq 0.$$

The method we use to solve such equations uses the square root property.

■ **Answer:** The solution set is $\left\{\dfrac{4}{3}\right\}$, which is a double root.

> **SQUARE ROOT PROPERTY**
>
> | **WORDS** | **MATH** |
> |---|---|
> | If an expression squared is equal to a constant, then that expression is equal to the positive or negative square root of the constant. | If $u^2 = A$, then $u = \pm\sqrt{A}$. |

EXAMPLE 4 Using the Square Root Property

Solve the equation $3x^2 - 27 = 0$.

Solution:

This quadratic is missing the first-degree term. $\quad 3x^2 = 27$

Divide both sides by 3. $\qquad\qquad\qquad x^2 = 9$

Apply the square root property. $\qquad\quad x = \pm\sqrt{9} = \pm 3$

The solution set is $\{-3, 3\}$.

If we alter Example 4 by changing the negative sign to positive, we see in Example 5 that we get imaginary roots (as opposed to real roots), which we discussed in Chapter 0.

EXAMPLE 5 Using the Square Root Property

Solve the equation $3x^2 + 27 = 0$.

Solution:

Subtract 27 from both sides. $\qquad 3x^2 = -27$

Divide by 3. $\qquad\qquad\qquad\qquad x^2 = -9$

Apply the square root property. $\qquad x = \pm\sqrt{-9}$

Simplify. $\qquad\qquad\qquad\qquad x = \pm i\sqrt{9} = \pm 3i$

The solution set is $\{-3i, 3i\}$.

■ **YOUR TURN** Solve the equations $y^2 - 147 = 0$ and $v^2 + 64 = 0$.

EXAMPLE 6 Using the Square Root Property

Solve the equation $(x - 2)^2 = 16$.

Solution:

We use the square root property.

If an expression squared is 16, then the expression equals $\pm\sqrt{16}$. $\qquad (x - 2) = \pm\sqrt{16}$

■ **Answer:** $y = \pm 7\sqrt{3}$ $\quad v = \pm 8i$

Separate into two equations.

$$x - 2 = \sqrt{16} \quad \text{or} \quad x - 2 = -\sqrt{16}$$
$$x - 2 = 4 \qquad\qquad x - 2 = -4$$
$$x = 6 \qquad\qquad x = -2$$

The solution set is $\{-2, 6\}$.

It is acceptable notation to keep the equations together.

$$(x - 2) = \pm\sqrt{16}$$
$$x - 2 = \pm 4$$
$$x = 2 \pm 4$$
$$x = -2, 6$$

Completing the Square

Factoring and the square root method are two efficient, quick procedures for solving many quadratic equations. Some equations, however, cannot be solved directly by these methods. We will see that $y^2 - 10y - 3 = 0$ cannot. A more general procedure we will develop to solve this kind of equation is called **completing the square**. The idea behind completing the square is to transform any standard quadratic equation $ax^2 + bx + c = 0$ into the form $(x + A)^2 = B$, where A and B are constants and the left side, $(x + A)^2$, is called a **perfect square**. This last equation can easily be solved by the square root method. How do we transform the first equation into the second equation?

Note that the above-mentioned example, $y^2 - 10y - 3 = 0$, is not easily factored, nor does it contain a perfect square. Let us transform this equation into a form that does contain a perfect square.

WORDS	**MATH**
Write the original equation.	$y^2 - 10y - 3 = 0$
Add 3 to both sides.	$y^2 - 10y = 3$
Add 25 to both sides.	$y^2 - 10y + 25 = 3 + 25$
The left side can be written as a perfect square.	$(y - 5)^2 = 28$
Use the square root method.	$y - 5 = \pm\sqrt{28}$
Add 5 to both sides.	$y = 5 \pm 2\sqrt{7}$

Why did we **add 25 to both sides?** Recall that $(x + c)^2 = x^2 + 2xc + c^2$. Using this product we can write $(x + \frac{b}{2})^2 = x^2 + bx + (\frac{b}{2})^2$. The coefficient of the first-degree term, b, is *halved* (cut in half) and then *squared*. In our example, $y^2 - 10y - 3 = 0$, the coefficient of the first-degree term is -10.

$$\left(\frac{-10}{2}\right)^2 = (-5)^2 = 25$$

COMPLETING THE SQUARE PROCEDURE

WORDS	**MATH**
Express the quadratic equation in the following form.	$x^2 + bx = c$

Cut b in half; square that quantity; and add the result to both sides.	$x^2 + bx + \left(\dfrac{b}{2}\right)^2 = c + \left(\dfrac{b}{2}\right)^2$
Write the left side of the equation as a perfect square.	$\left(x + \dfrac{b}{2}\right)^2 = c + \left(\dfrac{b}{2}\right)^2$
Solve using the square root method.	

EXAMPLE 7 Completing the Square

Solve the quadratic equation $x^2 + 8x - 3 = 0$ by completing the square.

Solution:

Add 3 to both sides.	$x^2 + 8x = 3$
Add $\left(\dfrac{1}{2} \cdot 8\right)^2$ to both sides.	$x^2 + 8x + 4^2 = 3 + 4^2$
Write the left side as a perfect square and simplify the right side.	$(x + 4)^2 = 19$
Use the square root method to solve.	$x + 4 = \pm\sqrt{19}$
Subtract 4 from both sides.	$x = -4 \pm\sqrt{19}$

In Example 7, the lead coefficient (the coefficient of x^2 term) is 1. When the lead coefficient is not 1, always start by first dividing the equation by that lead coefficient.

STUDY TIP

When the lead coefficient is not 1, always start by first dividing the equation by that lead coefficient.

EXAMPLE 8 Completing the Square When the Lead Coefficient Is Not 1

Solve the equation $3x^2 - 12x + 13 = 0$ by completing the square.

Solution:

Make the leading coefficient 1 by dividing by 3.	$x^2 - 4x + \dfrac{13}{3} = 0$
Collect variables to one side of the equation and constants to the other side.	$x^2 - 4x = -\dfrac{13}{3}$
Add $\left(\dfrac{-4}{2}\right)^2$ to both sides.	$x^2 - 4x + 4 = -\dfrac{13}{3} + 4$
Write the left side of the equation as a perfect square and simplify the right side.	$(x - 2)^2 = -\dfrac{1}{3}$
Solve using the square root method.	$x - 2 = \pm\sqrt{-\dfrac{1}{3}}$
Simplify.	$x = 2 \pm i\sqrt{\dfrac{1}{3}}$
Rationalize the denominator (Chapter 0).	$x = 2 \pm \dfrac{i\sqrt{3}}{3}$

CONCEPT CHECK What do you have to do to the lead coefficient of a quadratic equation before completing the square?

■ **YOUR TURN** Solve the equation $2x^2 - 4x + 3 = 0$ by completing the square.

Quadratic Formula

Let us now consider the most general quadratic equation:

$$ax^2 + bx + c = 0, \text{ where } a \neq 0$$

We can solve this equation by completing the square.

WORDS	MATH
Make the lead coefficient 1 by dividing the equation by a.	$x^2 + \dfrac{b}{a}x + \dfrac{c}{a} = 0$
Subtract $\dfrac{c}{a}$ from both sides.	$x^2 + \dfrac{b}{a}x = -\dfrac{c}{a}$
Square half of $\dfrac{b}{a}$ and add the result, $\left(\dfrac{b}{2a}\right)^2$, to both sides.	$x^2 + \dfrac{b}{a}x + \left(\dfrac{b}{2a}\right)^2 = \left(\dfrac{b}{2a}\right)^2 - \dfrac{c}{a}$
Write the left side of the equation as a perfect square and the right side as a single fraction.	$\left(x + \dfrac{b}{2a}\right)^2 = \dfrac{b^2 - 4ac}{4a^2}$
Solve using the square root method.	$x + \dfrac{b}{2a} = \pm\sqrt{\dfrac{b^2 - 4ac}{4a^2}}$
Subtract $\left(\dfrac{b}{2a}\right)$ from both sides and simplify the radical.	$x = -\dfrac{b}{2a} \pm \dfrac{\sqrt{b^2 - 4ac}}{2a}$
Write as a single fraction.	$x = \dfrac{-b \pm \sqrt{b^2 - 4ac}}{2a}$

We have derived the **quadratic formula**.

QUADRATIC FORMULA

If $ax^2 + bx + c = 0$, $a \neq 0$, then $x = \dfrac{-b \pm \sqrt{b^2 - 4ac}}{2a}$.

STUDY TIP

$x = \dfrac{-b \pm \sqrt{b^2 - 4ac}}{2a}$.

"negative b plus or minus the square root of b squared minus $4ac$ all over $2a$."

■ **Answer:** $x = 1 \pm \dfrac{i\sqrt{2}}{2}$

We read this formula as *negative b plus or minus the square root of b squared minus 4ac all over 2a.* The quadratic formula should be memorized and used when simpler methods (factoring and the square root method) fail.

TYPES OF SOLUTIONS

The term inside the radical, $b^2 - 4ac$, is called the **discriminant**. The discriminant gives important information about the corresponding solutions or roots of $ax^2 + bx + c = 0$.

$b^2 - 4ac$	Solutions (roots)
Positive	Two distinct real roots
0	One real root (a double or repeated root)
Negative	Two complex roots, complex conjugates

EXAMPLE 9 Quadratic Formula, Two Distinct Real Roots Appear

Use the quadratic formula to solve the quadratic equation $x^2 - 4x - 1 = 0$.

Solution:

For this problem $a = 1$, $b = -4$, and $c = -1$.

The discriminant is positive, $b^2 - 4ac = 20 > 0$; therefore expect two distinct real roots.

Write the quadratic formula.

$$x = \frac{-b \pm \sqrt{b^2 - 4ac}}{2a}$$

Use parentheses to avoid a minus sign getting lost.

$$x = \frac{-(\ ) \pm \sqrt{(\ )^2 - 4(\ )(\ )}}{2(\ )}$$

Substitute values for a, b, and c into the parentheses.

$$x = \frac{-(-4) \pm \sqrt{(-4)^2 - 4(1)(-1)}}{2(1)}$$

Simplify.

$$x = \frac{4 \pm \sqrt{16 + 4}}{2} = \frac{4 \pm \sqrt{20}}{2} = \frac{4 \pm 2\sqrt{5}}{2} = 2 \pm \sqrt{5}$$

The solution set $\{2 - \sqrt{5}, 2 + \sqrt{5}\}$ contains two distinct real numbers.

EXAMPLE 10 Quadratic Formula, Two Imaginary Roots Appear

Use the quadratic formula to solve the quadratic equation $x^2 + 8 = 4x$.

Solution:

Write this equation in standard form $x^2 - 4x + 8 = 0$ in order to identify $a = 1$, $b = -4$, and $c = 8$.

The discriminant is negative, $b^2 - 4ac = -16 < 0$; therefore expect two imaginary roots.

Write the quadratic formula.

$$x = \frac{-b \pm \sqrt{b^2 - 4ac}}{2a}$$

Use parentheses to avoid a minus sign getting lost.

$$x = \frac{-(\ ) \pm \sqrt{(\ )^2 - 4(\ )(\ )}}{2(\ )}$$

Substitute the values for a, b, c into the parentheses.

$$x = \frac{-(-4) \pm \sqrt{(-4)^2 - 4(1)(8)}}{2(1)}$$

Simplify.

$$x = \frac{4 \pm \sqrt{16 - 32}}{2} = \frac{4 \pm \sqrt{-16}}{2} = \frac{4 \pm 4i}{2} = 2 \pm 2i$$

The solution set $\{2 - 2i, 2 + 2i\}$ contains two complex numbers. Note that they are complex conjugates of each other.

Applications Involving Quadratic Equations

In Section 1.2 we developed a procedure for solving applied problems. The procedure is the same for applications involving quadratic equations. The only difference is that the mathematical equations will be quadratic, as opposed to linear as in Section 1.1.

EXAMPLE 11   Stock Value

From 1999 to 2001 the price of Abercrombie & Fitch's (ANF) stock is approximately given by $P = 0.2t^2 - 5.6t + 50.2$, where P is the price of stock in dollars, t is in months, and $t = 1$ corresponds to January 1999. When was the value of the stock worth $30?

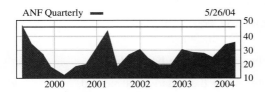

Solution:

STEP 1 **Identify the question.**
When is the price of the stock equal to $30?

STEP 2 **Make notes.**
Stock price:

$$P = 0.2t^2 - 5.6t + 50.2$$
$$P = 30$$

STEP 3 **Set up an equation.**

$$30 = 0.2t^2 - 5.6t + 50.2$$

STEP 4 **Solve the equation.**
Subtract 30 from both sides.

$$0.2t^2 - 5.6t + 20.2 = 0$$

Solve using the quadratic formula.

$$t = \frac{-(-5.6) \pm \sqrt{(-5.6)^2 - 4(0.2)(20.2)}}{2(0.2)}$$

Simplify.

$$t \approx \frac{5.6 \pm 3.9}{0.4} = 4.25, 23.75$$

Rounding these two numbers we find that $t \approx 4$ and $t \approx 24$. Since $t = 1$ corresponds to January 1999, these two solutions correspond to April 1999 and December 2000 .

STEP 5 Check the solution.
Look at the figure. The horizontal axis represents the year (2000 corresponds to January 2000), and the vertical axis represents the stock price. Estimating when the stock price is approximately $30, we find April 1999 and December 2000.

EXAMPLE 12 Pythagorean Theorem

Hitachi makes a 60 inch HDTV that has a 60 inch diagonal. If the width of the screen is approximately 52 inches, what is the approximate height of the screen?

Solution:

STEP 1 Identify the question.
What is the approximate height of the HDTV screen?

STEP 2 Make notes.

STEP 3 Set up an equation.

Recall the Pythagorean theorem. $\quad a^2 + b^2 = c^2$

Substitute in known values. $\quad h^2 + 52^2 = 60^2$

STEP 4 Solve the equation.

Simplify constants. $\quad h^2 + 2704 = 3600$

Subtract 2704 from both sides. $\quad h^2 = 896$

Solve using the square root method. $\quad h = \pm \sqrt{896} \approx \pm 30$

The height cannot be negative, so the negative value is eliminated and the height is approximately 30 inches .

STEP 5 Check the solution. $\quad 30^2 + 52^2 \overset{?}{=} 60^2$

$$900 + 2704 \overset{?}{=} 3600$$

$$3604 \approx 3600 \checkmark$$

SECTION 1.3 **SUMMARY**

In this section you have learned and practiced solving quadratic equations using four primary methods: factoring, the square root method, completing the square, and the quadratic formula. Factoring and the square root method are the quickest and easiest but cannot always be used. The quadratic formula and completing the square works for all quadratic equations and can yield three types of solutions: two distinct real roots, one real root (repeated), or two complex roots (conjugates of each other).

SECTION 1.3 EXERCISES

 ■ **SKILLS**

In Exercises 1–16, solve by factoring.

1. $x^2 - 5x + 6 = 0$ **2.** $3x^2 + 10x - 8 = 0$ **3.** $x^2 = 12 - x$ **4.** $11x = 2x^2 + 12$

5. $16x^2 + 8x = -1$ **6.** $v^2 + 7v + 6 = 0$ **7.** $8y^2 = 16y$ **8.** $3A^2 = -12A$

9. $9p^2 = 12p - 4$ **10.** $x^2 - 9 = 0$ **11.** $x(x + 4) = 12$ **12.** $3t^2 - 48 = 0$

13. $x + \dfrac{12}{x} = 7$ **14.** $x - \dfrac{10}{x} = -3$ **15.** $\dfrac{4(x - 2)}{x - 3} + \dfrac{3}{x} = \dfrac{-3}{x(x - 3)}$ **16.** $\dfrac{5}{y + 4} = 4 + \dfrac{3}{y - 2}$

In Exercises 17–28, solve using the square root method.

17. $p^2 - 8 = 0$ **18.** $y^2 - 72 = 0$ **19.** $x^2 + 9 = 0$ **20.** $v^2 + 16 = 0$

21. $(x - 3)^2 = 36$ **22.** $(x - 1)^2 = 25$ **23.** $(2x + 3)^2 = -4$ **24.** $(4x - 1)^2 = -16$

25. $(5x - 2)^2 = 27$ **26.** $(3x + 8)^2 = 12$ **27.** $(1 - x)^2 = 9$ **28.** $(1 - x)^2 = -9$

In Exercises 29–38, what number should be added to complete the square of each expression?

29. $x^2 + 6x$ **30.** $x^2 - 8x$ **31.** $x^2 - 12x$ **32.** $x^2 + 20x$ **33.** $x^2 - \dfrac{1}{2}x$

34. $x^2 - \dfrac{1}{3}x$ **35.** $x^2 + \dfrac{2}{5}x$ **36.** $x^2 + \dfrac{4}{5}x$ **37.** $x^2 - 2.4x$ **38.** $x^2 + 1.6x$

In Exercises 39–50, solve by completing the square.

39. $x^2 + 2x = 3$ **40.** $y^2 + 8y - 2 = 0$ **41.** $t^2 - 6t = -5$ **42.** $x^2 + 10x = -21$

43. $y^2 - 4y + 3 = 0$ **44.** $x^2 - 7x + 12 = 0$ **45.** $2p^2 + 8p = -3$ **46.** $2x^2 - 4x + 3 = 0$

47. $2x^2 - 7x + 3 = 0$ **48.** $3x^2 - 5x - 10 = 0$ **49.** $\dfrac{x^2}{2} - 2x = \dfrac{1}{4}$ **50.** $\dfrac{t^2}{3} + \dfrac{2t}{3} + \dfrac{5}{6} = 0$

In Exercises 51–62, solve using the quadratic formula.

51. $t^2 + 3t - 1 = 0$ **52.** $t^2 + 2t = 1$ **53.** $s^2 + s + 1 = 0$ **54.** $2s^2 + 5s = -2$

55. $3x^2 - 3x - 4 = 0$ **56.** $4x^2 - 2x = 7$ **57.** $x^2 - 2x + 17 = 0$ **58.** $4m^2 + 7m + 8 = 0$

59. $5x^2 + 7x = 3$ **60.** $3x^2 + 5x = -11$ **61.** $-0.5x^2 + 1.2x - 3.7 = 0$ **62.** $-0.3x^2 - 1.4x + 2.1 = 0$

In Exercises 63–68, determine if the discriminant is positive, negative, or zero and indicate the number and type of root to expect. Do not solve.

63. $x^2 - 22x + 121 = 0$ **64.** $x^2 - 28x + 196 = 0$ **65.** $2y^2 - 30y + 68 = 0$

66. $-3y^2 + 27y + 66 = 0$ **67.** $9x^2 - 7x + 8 = 0$ **68.** $-3x^2 + 5x - 7 = 0$

In Exercises 69–82, solve using any method.

69. $v^2 - 8v = 20$ **70.** $v^2 - 8v = -20$ **71.** $t^2 + 5t - 6 = 0$ **72.** $t^2 + 5t + 6 = 0$

73. $(x + 3)^2 = 16$ **74.** $(x + 3)^2 = -16$ **75.** $8w^2 + 2w + 21 = 0$ **76.** $8w^2 + 2w - 21 = 0$

77. $3p^2 - 9p + 1 = 0$ **78.** $3p^2 - 9p - 1 = 0$ **79.** $\frac{2}{3}t^2 - \frac{4}{3}t = \frac{1}{5}$ **80.** $\frac{1}{2}x^2 + \frac{2}{3}x = \frac{2}{5}$

81. $x^2 - 0.1x = 0.12$ **82.** $y^2 - 0.5y = -0.06$

In Exercises 83–86, solve for the indicated variable in terms of other variables.

83. Solve $s = \frac{1}{2}gt^2$ for t.   **84.** Solve $A = P(1 + r)^2$ for r.   **85.** Solve $a^2 + b^2 = c^2$ for c.   **86.** Solve $P = EI - RI^2$ for I.

■ APPLICATIONS

87. Exercise. A jogger and a walker both cover a distance of 5 miles. The runner is traveling 1.5 times faster than the walker and finishes in 25 minutes less time. How fast is each going?

88. Boating. A speedboat takes 1 hour longer to go 24 miles up a river than to return. If the boat cruises at 10 miles per hour in still water, what is the rate of the current?

89. Numbers. Find two consecutive numbers such that their sum is 35 and their product is 306.

90. Numbers. Find two consecutive odd integers such that their sum is 24 and their product is 143.

91. Geometry. If a circle's radius is increased by 2 meters, the area is increased by 24π square meters. What are the radii of the two circles?

92. Geometry. A rectangle has an area of 31.5 square meters. If the length is 2 more than twice the width, find the dimensions of the rectangle.

93. Geometry. A triangle has a height that is 2 more than 3 times the base and the area is 60 square units. Find the base and height.

94. Geometry. A square's side is increased by 3 yards, which corresponds to an increase in the area by 69 square yards. How many yards is the side of the initial square?

95. Stock Value. From June 2003 until April 2004 JetBlue airlines stock (JBLU) was approximately worth $P = -4t^2 + 80t - 360$ where P denotes the price of the stock in dollars and t corresponds to months with $t = 1$ corresponding to January 2003. During what months was the stock equal to $24?

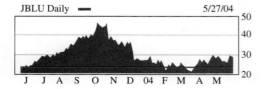

JBLU Daily —   5/27/04

J J A S O N D 04 F M A M

96. Stock Value. From November 2003 until March 2004 Wal-Mart Stock (WMT) was approximately worth

$P = 2t^2 - 12t + 70$ where P denotes the price of the stock in dollars and t corresponds to months with $t = 1$ corresponding to November 2003. During what months was the stock equal to $60?

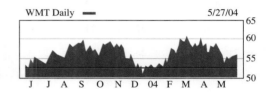

WMT Daily —   5/27/04

J J A S O N D 04 F M A M

97. Gravity. If a person drops a water balloon off the rooftop of a 100 foot building, the height of the water balloon is given by the equation $h = -16t^2 + 100$ where t is in seconds. When will the water balloon hit the ground?

98. Gravity. If the person in Exercise 97 throws the water balloon downward with a speed of 5 feet per second, the height of the water balloon is given by the equation $h = -16t^2 - 5t + 100$ where t is in seconds. When will the water balloon hit the ground?

99. Gardening. A square garden has an area 900 square feet. If a sprinkler (with a circular pattern) is placed in the center of the garden, what is the minimum radius of spray the sprinkler would need in order to water all of the garden?

100. Sports. A baseball diamond is a square. The distance from base to base is 90 feet. What is the distance from home plate to second base?

101. Volume. A flat square piece of cardboard is used to construct an open box. Cutting a 1 foot by 1 foot square off of each corner and folding up the edges will yield an open box (assuming these edges are taped together). If the desired volume of the box is 9 cubic feet, what are the dimensions of the original square piece of cardboard?

102. Volume. A rectangular piece of cardboard whose length is twice its width is used to construct an open box. Cutting a 1 foot by 1 foot square off of each corner and folding up the edges will yield an open box. If the desired volume is 12 cubic feet, what are the dimensions of the original rectangular piece of cardboard?

103. Gardening. A landscaper has planted a rectangular garden that measures 8 feet by 5 feet. He has ordered 1 yard (27 cubic feet) of stones for a border along the outside of the garden. If the border needs to be 4 inches deep and he wants to use all of the stones, how wide should the border be?

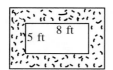

104. Gardening. A gardener has planted a semicircular rose garden with a radius of 6 feet, and 2 yards of mulch (1 yard = 27 cubic feet) is being delivered. Assuming she uses all of the mulch, how deep will the layer of mulch be?

105. Work. Lindsay and Kimmie, working together, can balance the financials for the Kappa Kappa Gamma sorority in 6 days. Lindsay by herself can complete the job in 5 days less than Kimmie. How long will it take Lindsay to complete the job by herself?

106. Work. When Jack cleans the house it takes him 4 hours. When Ryan cleans the house it takes him 6 hours. How long would it take both of them if they worked together?

■ **CATCH THE MISTAKE**

In Exercises 107–110, explain the mistake that is made.

107.
$$t^2 - 5t - 6 = 0$$
$$(t - 3)(t - 2) = 0$$
$$t = 2, 3$$

108.
$$(2y - 3)^2 = 25$$
$$2y - 3 = 5$$
$$2y = 8$$
$$y = \frac{5}{4}$$
$$y = 4$$

109.
$$16a^2 + 9 = 0$$
$$16a^2 = -9$$
$$a^2 = -\frac{9}{16}$$
$$a = \pm\sqrt{\frac{9}{16}}$$
$$a = \pm\frac{3}{4}$$

110.
$$2x^2 - 4x - 3 = 0$$
$$2x^2 - 4x = 3$$
$$2(x^2 - 2x) = 3$$
$$2(x^2 - 2x + 1) = 3 + 1$$
$$2(x - 1)^2 = 4$$
$$(x - 1)^2 = 2$$
$$x - 1 = \pm\sqrt{2}$$
$$x = 1 \pm\sqrt{2}$$

■ **CHALLENGE**

111. T or F: The equation $(3x + 1)^2 = 16$ has the same solution set as the equation $3x + 1 = 4$.

112. T or F: The quadratic equation $ax^2 + bx + c = 0$ can only be solved by the square root method if $b = 0$.

113. T or F: If a complex number is a root of a quadratic equation, then its complex conjugate is the other root.

114. T or F: A repeated root can be either real or imaginary.

115. Show that the sum of the roots of a quadratic equation is equal to $-\frac{b}{a}$.

116. Show that the product of the roots of a quadratic equation is equal to $\frac{c}{a}$.

117. Write a quadratic equation in general form whose solution set is $\{3 + \sqrt{5}, 3 - \sqrt{5}\}$.

118. Write a quadratic equation in general form whose solution set is $\{2 - i, 2 + i\}$.

119. Find a quadratic equation whose two distinct real roots are the negatives of the two distinct real roots of the equation $ax^2 + bx + c = 0$.

120. Find a quadratic equation whose two distinct real roots are the reciprocals of the two distinct real roots of the equation $ax^2 + bx + c = 0$.

■ **TECHNOLOGY**

121. Solve the equation $x^2 - x = 2$ by first writing in standard form and then factoring. Now plot both sides of the equation in the same viewing screen ($y_1 = x^2 - x$

and $y_2 = 2$). At what x values do these two graphs intersect? Do those points agree with the solution set you found?

122. Solve the equation $x^2 - 2x = -2$ by first writing in standard form and then using the quadratic formula. Now plot both sides of the equation in the same viewing screen ($y_1 = x^2 - 2x$ and $y_2 = -2$). Do these graphs intersect? Does this agree with the solution set you found?

123. Solve the equation $x^2 - 3.4x = -2.89$ by first writing in standard form and then using the quadratic formula. Now plot both sides of the equation in the same viewing screen ($y_1 = x^2 - 3.4x$ and $y_2 = -2.89$). Do these graphs intersect and if so, at what point(s)? Does this agree with the solution set you found?

SECTION 1.4 Radical Equations; Equations Quadratic in Form

Skills Objectives

- Solve radical equations.
- Use u-substitution to solve equations that can be transformed into quadratic form.

Conceptual Objectives

- Transform a difficult problem into a simpler problem, solve the simpler problem, and transform back.
- Recognize the need to check solutions when the transformation process may produce extraneous solutions.

Radical Equations

Radical equations are equations where the variable is inside a radical (that is, in a square root, cube root, or higher root). Examples of radical equations follow.

$$\sqrt[3]{x - 3} = 2 \qquad \sqrt{2x + 3} = x \qquad \sqrt{x + 2} + \sqrt{7x + 2} = 6$$

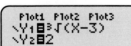

TECHNOLOGY TIP
Use a graphing utility to display graphs of $y_1 = \sqrt[3]{x - 3}$ and $y_2 = 2$.

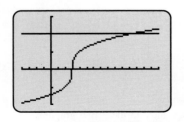

The x-coordinate of the point of intersection is the solution to the equation $\sqrt[3]{x - 3} = 2$.

Until now your experience has been with linear and quadratic equations. Often you can transform a radical equation into a simple linear or quadratic equation. Sometimes the transformation process yields **extraneous solutions**, or apparent solutions that may solve the transformed problem but are not solutions of the original radical equation. Therefore, it is very important to always check your answers.

EXAMPLE 1 Solving an Equation Involving a Radical

Solve the equation $\sqrt[3]{x - 3} = 2$.

Solution:

The goal is to solve for x.

Write the radical in terms of fractional exponents. $\qquad (x - 3)^{1/3} = 2$

Cube both sides of the equation. $\qquad \left[(x - 3)^{1/3}\right]^3 = 2^3$

Simplify. $\qquad x - 3 = 8$

Solve the resulting linear equation. $\qquad \boxed{x = 11}$

Check by substituting $x = 11$ into the original equation and making sure the statement is true.

■ **YOUR TURN**   Solve the equation $\sqrt{3p + 4} = 5$.

EXAMPLE 2 Solving an Equation Involving a Radical

Solve the equation $\sqrt{2x + 3} = x$.

Solution:

The goal is to solve for x.

Square both sides of the equation. $2x + 3 = x^2$

Write quadratic equation in standard form. $x^2 - 2x - 3 = 0$

Factor. $(x - 3)(x + 1) = 0$

Use the zero product property. $x = 3$ or $x = -1$

Check these values to see if they *both* make the original equation statement true.

$x = 3$: $\sqrt{2(3) + 3} = 3 \Rightarrow \sqrt{6 + 3} = 3 \Rightarrow \sqrt{9} = 3 \Rightarrow 3 = 3$ (True)

$x = -1$: $\sqrt{2(-1) + 3} = -1 \Rightarrow \sqrt{-2 + 3} = -1 \Rightarrow \sqrt{1} = -1 \Rightarrow 1 \neq -1$ (False)

Solution: $x = 3$

STUDY TIP

Extraneous solutions are common in radical equations so check your answers.

What happened in Example 2? When we transformed the radical equation into a quadratic equation, we created an **extraneous solution**, $x = -1$, a solution that appears to solve the original equation but does not. When solving radical equations, you must take the time to check answers to avoid including extraneous solutions in your solution set.

■ **YOUR TURN**   Solve the equation $\sqrt{12 + t} = t$.

■ **YOUR TURN**   Solve the equation $\sqrt{2x + 6} = x + 3$.

Examples 1 and 2 contained only one radical each. We transformed the radical equation into a linear (Example 1) or quadratic (Example 2) equation with one step. The next example contains two radicals. Our technique will be to isolate one radical on one side of the equation with the other radical on the other side of the equation.

EXAMPLE 3 Solving an Equation with More than One Radical

Solve the equation $\sqrt{x + 2} + \sqrt{7x + 2} = 6$.

Solution:

Subtract $\sqrt{x + 2}$ from both sides. $\sqrt{7x + 2} = 6 - \sqrt{x + 2}$

Square both sides. $(\sqrt{7x + 2})^2 = (6 - \sqrt{x + 2})^2$

Simplify. $7x + 2 = (6 - \sqrt{x + 2})(6 - \sqrt{x + 2})$

■ **Answer:** $p = 7$ ■ **Answer:** $t = 4$ ■ **Answer:** $x = -1$ and $x = -3$

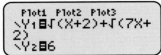
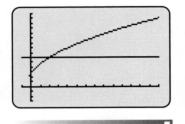
Use the FOIL method to multiply the right side.

$$7x + 2 = 36 - 12\sqrt{x+2} + (x+2)$$

We are still left with a radical.

Isolate the radical on the left side.

$$12\sqrt{x+2} = 36 + x + 2 - 7x - 2$$

Combine like terms on the right side.

$$12\sqrt{x+2} = 36 - 6x$$

Divide by 6.

$$2\sqrt{x+2} = 6 - x$$

Square both sides.

$$4(x+2) = (6-x)^2$$

Simplify.

$$4x + 8 = 36 - 12x + x^2$$

Rewrite quadratic equation in standard form.

$$x^2 - 16x + 28 = 0$$

Factor.

$$(x-14)(x-2) = 0$$

Solve.

$$x = 14 \text{ and } x = 2$$

The apparent solutions are 2 and 14. Note that $x = 14$ does not satisfy the original equation. Therefore it is extraneous. The solution is $\boxed{x = 2}$.

PROCEDURE FOR SOLVING RADICAL EQUATIONS

1. Isolate a radical on one side.
2. Raise both sides of the equation to the power that will eliminate this radical, and simplify the equation.
3. If there remains a radical, repeat Steps 1 and 2.
4. Solve the linear or quadratic equation.
5. Check the solutions and eliminate any extraneous solutions.

Note: It does not matter which radical is isolated first.

Equations Quadratic in Form: U-Substitution

Equations that are higher order or that have fractional powers often can be transformed into a quadratic equation by introducing a u-substitution. When this is the case we say that equations are **quadratic in form**. In the table below the two original equations are quadratic in form because they can be transformed into a quadratic equation given the correct substitution.

ORIGINAL EQUATION	SUBSTITUTION	NEW EQUATION
$x^4 - 3x^2 - 4 = 0$	$u = x^2$	$u^2 - 3u - 4 = 0$
$t^{2/3} + 2t^{1/3} + 1 = 0$	$u = t^{1/3}$	$u^2 + 2u + 1 = 0$

For example, the equation $x^4 - 3x^2 - 4 = 0$ is a fourth-degree equation in x. How did we know that $u = x^2$ would transform the original equation into a quadratic equation?

If we rewrite the original equation as $(x^2)^2 - 3(x^2) - 4 = 0$, the expression in parentheses is the u-substitution.

Let us introduce the substitution $u = x^2$—note that squaring both sides implies $u^2 = x^4$. We then replace x^2 in the original equation with u, and x^4 in the original equation with u^2, which leads to a quadratic equation in u: $u^2 - 3u - 4 = 0$.

WORDS	MATH
Solve for x.	$x^4 - 3x^2 - 4 = 0$
Introduce u-substitution.	$u = x^2$ [Note that $u^2 = x^4$.]
Write quadratic equation in u.	$u^2 - 3u - 4 = 0$
Factor.	$(u - 4)(u + 1) = 0$
Solve for u.	$u = 4$ or $u = -1$
Transform back to x, $u = x^2$	$x^2 = 4$ or $x^2 = -1$
Solve for x.	$x = \pm 2$ or $x = \pm i$

It is important to correctly determine the appropriate substitution in order to arrive at an equation quadratic in form. For example, $t^{2/3} + 2t^{1/3} + 1 = 0$ is an original equation given in the above table. If we rewrite this equation as $(t^{1/3})^2 + 2(t^{1/3}) + 1 = 0$, then it becomes apparent the correct substitution is $u = t^{1/3}$, which transforms the equation in t into a quadratic equation in u: $u^2 + 2u + 1 = 0$.

PROCEDURE FOR SOLVING EQUATIONS QUADRATIC IN FORM

1. Identify the substitution.
2. Transform the equation into a quadratic equation.
3. Solve the quadratic equation.
4. Use the substitution to rewrite the solution in terms of the original variable.
5. Solve the resulting equation.
6. Check the solutions in the original equation.

EXAMPLE 4 Solving an Equation Quadratic in Form

Find the solutions to the equation $(2x + 2)^2 - 6(2x + 2) + 8 = 0$.

Solution:

Rewrite original equation.	$(2x + 2)^2 - 6(2x + 2) + 8 = 0$
Determine u-substitution.	$u = 2x + 2$
The original equation in x becomes a quadratic equation in u.	$u^2 - 6u + 8 = 0$
Factor.	$(u - 4)(u - 2) = 0$
Solve for u.	$u = 4$ or $u = 2$

The most common mistake is forgetting to transform back to x.

The original transformation is $u = 2x + 2$ and we have two solutions for u, $u = 4$ or $u = 2$, which yield two equations for x.

CAUTION

The most common mistake is forgetting to transform back to x.

$$2x + 2 = 4 \qquad\qquad 2x + 2 = 2$$
$$2x = 2 \qquad\qquad\qquad 2x = 0$$
$$x = 1 \qquad\qquad\qquad x = 0$$

The solutions to the original equation are $x = 1$ or $x = 0$.

Check them and you will find that both solutions make the equation true.

CONCEPT CHECK What is an appropriate u-substitution for $(6 - s)^2 - 7(6 - s) + 12 = 0$?

■ **YOUR TURN** Solve the equation $(6 - s)^2 - 7(6 - s) + 12 = 0$.

EXAMPLE 5 Solving an Equation Quadratic in Form with Fractional Exponents

Find the solutions to the equation $x^{2/3} - 3x^{1/3} - 10 = 0$.

Solution:

Rewrite the original equation. $\qquad\qquad (x^{1/3})^2 - 3x^{1/3} - 10 = 0$

Identify substitution as $u = x^{1/3}$. $\qquad\qquad u^2 - 3u - 10 = 0$

Factor. $\qquad\qquad\qquad\qquad\qquad (u - 5)(u + 2) = 0$

Solve for u. $\qquad\qquad u = 5 \qquad$ or $\qquad u = -2$

Let $u = x^{1/3}$ again. $\qquad\qquad x^{1/3} = 5 \qquad\qquad x^{1/3} = -2$

Cube both sides of the equations. $\quad (x^{1/3})^3 = (5)^3 \qquad (x^{1/3})^3 = (-2)^3$

Simplify. $\qquad\qquad\qquad\qquad x = 125 \qquad\qquad x = -8$

The solution set is $\{-8, 125\}$, which a check will confirm.

■ **YOUR TURN** Find the solution to the equation $2t - 5t^{1/2} - 3 = 0$.

SECTION 1.4 **SUMMARY**

In this section we discussed equations involving radicals and equations quadratic in form.

Things to Remember

1. Check the solutions in order to eliminate extraneous solutions.

2. Don't forget to transform back to the original variable.

■ **Answer:** $s = 3$ and $s = 2$ ■ **Answer:** $t = 9$

SECTION 1.4 EXERCISES

■ SKILLS

In Exercises 1–30, solve the radical equation for the given variable.

1. $\sqrt{t - 5} = 2$

2. $\sqrt{2t - 7} = 3$

3. $(4p - 7)^{1/2} = 5$

4. $11 = (21 - p)^{1/2}$

5. $\sqrt{u + 1} = -4$

6. $-\sqrt{3 - 2u} = 9$

7. $\sqrt[3]{5x + 2} = 3$

8. $\sqrt[3]{1 - x} = -2$

9. $(4y + 1)^{1/3} = -1$

10. $(5x - 1)^{1/3} = 4$

11. $\sqrt{12 + x} = x$

12. $x = \sqrt{56 - x}$

13. $y = 5\sqrt{y}$

14. $\sqrt{y} = \dfrac{y}{4}$

15. $s = 3\sqrt{s - 2}$

16. $-2s = \sqrt{3 - s}$

17. $\sqrt{2x + 6} = x + 3$

18. $\sqrt{8 - 2x} = 2x - 2$

19. $\sqrt{1 - 3x} = x + 1$

20. $\sqrt{2 - x} = x - 2$

21. $\sqrt{x^2 - 4} = x - 1$

22. $\sqrt{25 - x^2} = x + 1$

23. $\sqrt{x^2 - 2x - 5} = x + 1$

24. $\sqrt{2x^2 - 8x + 1} = x - 3$

25. $\sqrt{2x - 1} - \sqrt{x - 1} = 1$

26. $\sqrt{8 - x} = 2 + \sqrt{2x + 3}$

27. $\sqrt{3x - 5} = 7 - \sqrt{x + 2}$

28. $\sqrt{x + 5} = 1 + \sqrt{x - 2}$

29. $\sqrt{2 + \sqrt{x}} = \sqrt{x}$

30. $\sqrt{2 - \sqrt{x}} = \sqrt{x}$

In Exercises 31–56, solve the equations by introducing a substitution that transforms these equations to quadratic form.

31. $x^{2/3} + 2x^{1/3} = 0$

32. $x^{1/2} - 2x^{1/4} = 0$

33. $x^4 - 3x^2 + 2 = 0$

34. $x^4 - 8x^2 + 16 = 0$

35. $2x^4 + 7x^2 + 6 = 0$

36. $x^8 - 17x^4 + 16 = 0$

37. $(2x + 1)^2 + 5(2x + 1) + 4 = 0$

38. $(x - 3)^2 + 6(x - 3) + 8 = 0$

39. $4(t - 1)^2 - 9(t - 1) = -2$

40. $2(1 - y)^2 + 5(1 - y) - 12 = 0$

41. $x^{-8} - 17x^{-4} + 16 = 0$

42. $2u^{-2} + 5u^{-1} - 12 = 0$

43. $3y^{-2} + y^{-1} - 4 = 0$

44. $5a^{-2} + 11a^{-1} + 2 = 0$

45. $z^{2/5} - 2z^{1/5} + 1 = 0$

46. $2x^{1/4} + x^{1/2} - 1 = 0$

47. $6t^{-2/3} - t^{-1/3} - 1 = 0$

48. $t^{-2/3} - t^{-1/3} - 6 = 0$

49. $3 = \dfrac{1}{(x + 1)^2} + \dfrac{2}{(x + 1)}$

50. $\dfrac{1}{(x + 1)^2} + \dfrac{4}{(x + 1)} + 4 = 0$

51. $\left(\dfrac{1}{2x - 1}\right)^2 + \left(\dfrac{1}{2x - 1}\right) - 12 = 0$

52. $\dfrac{5}{(2x + 1)^2} - \dfrac{3}{(2x + 1)} = 2$

53. $u^{4/3} - 5u^{2/3} = -4$

54. $u^{4/3} + 5u^{2/3} = -4$

55. $t = \sqrt[4]{t^2 + 6}$

56. $u = \sqrt[4]{-2u^2 - 1}$

■ APPLICATIONS

57. Insurance. Cost for health insurance with a private policy is given by $C = \sqrt{10 + a}$, where C is the cost per day and a is the insured's age in years. Health insurance for a 6 year old, $a = 6$, is $4 a day (or $1,460 per year). At what age would someone be paying $9 a day (or $3,285 per year)?

58. Insurance. Cost for life insurance is given by $C = \sqrt{5a + 1}$, where C is the cost per day and a is the insured's age in years. Life insurance for a newborn, $a = 0$, is $1 a day (or $365 per year). At what age would someone be paying $20 a day (or $7,300 per year)?

59. Stock Value. The stock price of MGI Pharmaceutical (MOGN) from March 2004 to June 2004 can be approximately modeled by the equation $P = 5\sqrt{t^2 + 1} + 50$, where P is the price of the stock in dollars and t is the month with $t = 0$ corresponding to March 2004. Assuming this trend continues, when would the stock be worth $85?

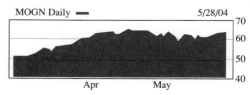

60. Grades. The average combined math and verbal SAT score of incoming freshmen at a university is given by the equation $S = 1000 + 10\sqrt{2t}$, where t is in years and $t = 0$ corresponds to 1990. What year will the incoming class have an average SAT score of 1230?

61. Sound. A person buys a house with an old well but does not know how deep the well is. To get an estimate he decides to drop a rock at the opening of the well and count how long it takes until he hears the *splash*. The total elapsed time, T given by $T = t_1 + t_2$, is the sum of the time it takes for the rock to reach the water, t_1, and

the time it takes for the sound of the splash to travel to the top of the well, t_2. The time (seconds) that it takes for the rock to reach the water is given by $t_1 = \sqrt{d}/4$, where d is the depth of the well in feet. Since the speed of sound is 1100 ft/s, the time (seconds) it takes for the sound to reach the top of the well is $t_2 = d/1100$. If the splash is heard after 3 seconds, how deep is the well?

62. Sound. If the owner of the house in Exercise 61 forgot to account for the speed of sound, what would he have calculated the depth of the well to be?

■ CATCH THE MISTAKE

In Exercises 63–66, explain the mistake that is made.

63. Solve the equation: $\sqrt{3t + 1} = -4$.

Solution:
$$3t + 1 = 16$$
$$3t = 15$$
$$t = 5$$

64. Solve the equation: $x = \sqrt{x + 2}$.

Solution:
$$x^2 = x + 2$$
$$x^2 - x - 2 = 0$$
$$(x - 2)(x + 1) = 0$$
$$x = -1, x = 2$$

65. Solve the equation: $x^{2/3} - x^{1/3} - 20 = 0$.

Solution:
$$u = x^{1/3}$$
$$u^2 - u - 20 = 0$$
$$(u - 5)(u + 4) = 0$$
$$x = 5, x = -4$$

66. Solve the equation: $x^4 - 2x^2 = 3$.

Solution:
$$x^4 - 2x^2 - 3 = 0$$
$$u = x^2$$
$$u^2 - 2u - 3 = 0$$
$$(u - 3)(u + 1) = 0$$
$$u = -1, u = 3$$
$$u = x^2$$
$$x^2 = -1, x^2 = 3$$
$$x = \pm 1, x = \pm 3$$

■ CHALLENGE

67. T or F: The equation $(2x - 1)^6 + 4(2x - 1)^3 + 3 = 0$ is quadratic in form.

68. T or F: The equation $t^{25} + 2t^5 + 1 = 0$ is quadratic in form.

69. T or F: If two solutions are found and one does not check then the other does not check.

70. T or F: Squaring both sides of $\sqrt{x + 2} + \sqrt{x} = \sqrt{x + 5}$ leads to $x + 2 + x = x + 5$.

71. Solve the equation $3x^2 + 2x = \sqrt{3x^2 + 2x}$ *without squaring* both sides.

72. Solve the equation $3x^{7/12} - x^{5/6} - 2x^{1/3} = 0$.

73. Solve the equation $\sqrt{x + 6} + \sqrt{11 + x} = 5\sqrt{3 + x}$.

74. Solve the equation $\sqrt[4]{2x\sqrt[3]{x\sqrt{x}}} = 2$.

75. Solve the equation $\sqrt{x-3} = 4 - \sqrt{x+2}$. Plot both sides of the equation in the same viewing screen, $y_1 = \sqrt{x-3}$ and $y_2 = 4 - \sqrt{x+2}$, and zoom in on the x-coordinate of the point of intersection. It should agree with your solution.

76. Solve the equation $2\sqrt{x+1} = 1 + \sqrt{3-x}$. Plot both sides of the equation in the same viewing screen, $y_1 = 2\sqrt{x+1}$ and $y_2 = 1 + \sqrt{3-x}$, and zoom in on the x-coordinate of the points of intersection. It should agree with your solution set.

77. Solve the equation $-4 = \sqrt{x+3}$. Plot both sides of the equation in the same viewing screen, $y_1 = -4$ and $y_2 = \sqrt{x+3}$. Does the graph agree or disagree with your solution?

78. Solve the equation $x^{1/4} = -4x^{1/2} + 21$. Plot both sides of the equation in the same viewing screen, $y_1 = x^{1/4}$ and $y_2 = -4x^{1/2} + 21$. Does the point(s) of intersection agree with your solution?

SECTION **1.5** Linear Inequalities

Skills Objectives

- Use interval notation.
- Solve linear inequalities and express solutions using interval notation and graphs.
- Solve application problems involving linear inequalities.

Conceptual Objectives

- Apply intersection and union concepts.
- Compare and contrast equations and inequalities.
- Understand that solution sets to inequalities are intervals, not discrete numbers.

An example of a linear equation is $3x - 2 = 7$. If the equal sign is changed to an inequality symbol, as in $3x - 2 \le 7$, then we have a **linear inequality**. One difference between a linear equation and a linear inequality is that the equation has only one solution, or value of x, that makes the statement true, whereas the inequality has a range or continuum of numbers that make the statement true. For example, the inequality $x \le 4$ denotes all real numbers x that are less than or equal to 4.

Inequality Symbols

Four inequality symbols are used.

SYMBOL	IN WORDS
$<$	Less than
$>$	Greater than
$\le$	Less than **or** equal to
$\ge$	Greater than **or** equal to

We call $<$ and $>$ **strict inequalities**.

For any two real numbers a and b, one of three things *must* be true:

$$a < b \qquad \text{or} \qquad a = b \qquad \text{or} \qquad a > b$$

This property is called the **trichotomy property** of real numbers.

If x is less than 5, $x < 5$, and x is greater than or equal to -2, $x \geq -2$, then we can represent this as a **double** (or **combined**) **inequality**, $-2 \leq x < 5$, which means that x is greater than or equal to -2 and less than 5.

Graphing Inequalities and Interval Notation

We will express solutions to inequalities four ways: as an inequality, solution set, interval, and graph. The following are ways of expressing all real numbers greater than or equal to a and less than b.

- Inequality notation: $a \leq x < b$
- Solution set: $\{x \mid a \leq x < b\}$
- Interval notation: $[a, b)$
- Graph/number line: or

In this example, a is referred to as the **left endpoint** and b is referred to as the **right endpoint**. If it is a strict inequality, $<$ or $>$, then the graph and interval notation use *parentheses*. If it includes an endpoint, $\geq$ or $\leq$, then the graph and interval notation use *brackets*. Number lines are drawn with either open/closed circles or brackets/parentheses. In this text the brackets/parentheses notation will be used.

Intervals are classified as follows.

Open (,)   Closed [,]   Half open (,] or [,)

X IS...	INEQUALITY	SET NOTATION	INTERVAL	GRAPH
greater than a and less than b	$a < x < b$	$\{x \mid a < x < b\}$	(a, b)	
greater than or equal to a and less than b	$a \leq x < b$	$\{x \mid a \leq x < b\}$	$[a, b)$	
greater than a and less than or equal to b	$a < x \leq b$	$\{x \mid a < x \leq b\}$	$(a, b]$	
greater than or equal to a and less than or equal to b	$a \leq x \leq b$	$\{x \mid a \leq x \leq b\}$	$[a, b]$	
less than a	$x < a$	$\{x \mid x < a\}$	$(-\infty, a)$	
less than or equal to a	$x \leq a$	$\{x \mid x \leq a\}$	$(-\infty, a]$	
greater than b	$x > b$	$\{x \mid x > b\}$	(b, ∞)	
greater than or equal to b	$x \geq b$	$\{x \mid x \geq b\}$	$[b, \infty)$	
all real numbers	$\mathbb{R}$	$\mathbb{R}$	$(-\infty, \infty)$	

Note

1. *Infinity*, ∞, is not a number. It is a symbol which means continuing indefinitely to the right on the number line. Similarly, *negative infinity*, $-\infty$,

means continuing indefinitely to the left on the number line. Since both are unbounded we use a parenthesis, never a bracket.

2. In interval notation, the <u>smaller</u> number is always written to the <u>left</u>. Write the inequality in interval notation: $-1 \leq x < 3$

 **CORRECT:** $[-1, 3)$  **INCORRECT:** $(3, -1]$

EXAMPLE 1 Expressing Inequalities Using Interval Notation and a Graph

Express the following as an inequality, interval, and graph.

a. x is greater than -3
b. x is less than or equal to 5
c. x is greater than or equal to -1 and less than 4
d. x is greater than or equal to 0 and less than or equal to 4

Solution:

Inequality	Interval	Graph
a. $x > -3$	$(-3, \infty)$	
b. $x \leq 5$	$(-\infty, 5]$	
c. $-1 \leq x < 4$	$[-1, 4)$	
d. $0 \leq x \leq 4$	$[0, 4]$	

Since the solutions to inequalities are sets of real numbers, it is useful to discuss two operations on sets called **intersection** and **union**.

DEFINITION   UNION AND INTERSECTION

The **union** of sets A and B, denoted $A \cup B$, is the set formed by combining all the elements in A with all the elements in B.

$$A \cup B = \{x \mid x \text{ is in } A \textbf{ or } B\}$$

The **intersection** of sets A and B, denoted $A \cap B$, is the set formed by the elements that are in both A and B.

$$A \cap B = \{x \mid x \text{ is in } A \textbf{ and } B\}$$

The notation $\{x \mid x \text{ is in } A \text{ and/or } B\}$ is read "all x such that x is in A and/or B." The vertical line represents "such that."

As an example of intersection and union, consider the following sets of people.

$$A = \{\text{Austin, Brittany, Jonathan}\} \qquad B = \{\text{Anthony, Brittany, Elise}\}$$

$$A \cap B = \{\text{Brittany}\} \quad A \cup B = \{\text{Anthony, Austin, Brittany, Elise, Jonathan}\}$$

EXAMPLE 2 Determining Unions and Intersections: Intervals and Graphs

If $A = [-3, 2]$ and $B = (1, 7)$, determine $A \cup B$ and $A \cap B$. Write these sets in interval notation, and graph.

Solution:

Set	Interval notation	Graph
A	$[-3, 2]$	
B	$(1, 7)$	
$A \cup B$	$[-3, 7)$	
$A \cap B$	$(1, 2]$	

■ **YOUR TURN** If $C = [-3, 3)$ and $D = (0, 5]$, find $C \cup D$ and $C \cap D$. Express the intersection and union in interval notation, and graph.

Solving Linear Inequalities

As mentioned at the beginning of this section, if we were to solve the equation $3x - 2 = 7$, we would add 2 to both sides, divide by 3, and find that $x = 3$ is the solution, the *only* value that makes the equation true. If we were to solve the linear inequality $3x - 2 \leq 7$, we would follow the same procedure: add 2 to both sides, divide by 3, and find that $x \leq 3$, which is a *range or interval* of numbers that make the inequality true.

In solving linear inequalities we follow the same procedures that we used in solving linear equations with one general exception: if you multiply or divide an inequality by a negative number, then you must change the direction of the inequality sign. For example, if $-2x < -10$, then the solution set includes real numbers such as $x = 6$ and $x = 7$. Note that real numbers such as $x = -6$ and $x = -7$ are not included in the solution set. Therefore, when this inequality is divided by -2, the sign must also be reversed: $x > 5$. If $a < b$, then $ac < bc$ if $c > 0$ and $ac > bc$ if $c < 0$.

The most common mistake that occurs when solving an inequality is forgetting to change direction of, or reverse, the inequality symbol when the inequality is multiplied or divided by a negative number.

CAUTION

If you multiply or divide an inequality by a negative number, then you must change the direction of the inequality sign.

INEQUALITY PROPERTIES

PROCEDURES THAT DO NOT CHANGE THE INEQUALITY SIGN

1. Simplify by eliminating parentheses and collecting like terms.	$3(x - 6) < 6x - x$ $3x - 18 < 5x$
2. Adding or subtracting the same quantity on both sides.	$7x + 8 \geq 29$ $7x \geq 21$

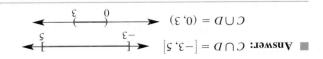

■ **Answer:** $C \cup D = [-3, 5]$

$C \cap D = (0, 3)$

3. Multiplying or dividing by the same *positive* quantity.

$5x \le 15$
$x \le 3$

PROCEDURES THAT CHANGE (REVERSE) THE INEQUALITY SIGN

1. Interchanging the two sides of the inequality.

$x \le 4$ is equivalent to $4 \ge x$

2. Multiplying or dividing by the same *negative* quantity.

$-5x \le 15$ is equivalent to $x \ge -3$

EXAMPLE 3 Solving a Linear Inequality

Solve and graph the inequality $5 - 3x < 23$.

Solution:

Write original inequality. $5 - 3x < 23$

Subtract 5 from both sides. $-3x < 18$

Divide both sides by -3 and reverse the inequality.

$$\frac{-3x}{-3} > \frac{18}{-3}$$

Simplify. $x > -6$

Solution set: $\{x \mid x > -6\}$ Interval notation: $(-6, \infty)$ Graph:

CONCEPT CHECK   What must happen if an inequality is divided by a negative number?

■ **YOUR TURN**   Solve the inequality $5 \le 3 - 2x$. Express solution in set and interval notation, and graph.

EXAMPLE 4 Solving Linear Inequalities with Fractions

Solve the inequality $\dfrac{5x}{3} \le \dfrac{4 + 3x}{2}$.

COMMON MISTAKE

A common mistake is using cross multiplication to solve inequalities.

CORRECT

Eliminate the fractions by multiplying by the LCD, 6.

$$6\left(\frac{5x}{3}\right) \le 6\left(\frac{4 + 3x}{2}\right)$$

INCORRECT

Cross multiply. $\dfrac{5x}{3} \le \dfrac{4 + 3x}{2}$

Error is cross multiplying.

CAUTION

Cross multiplication cannot be used in solving inequalities.

Simplify.
$$10x \leq 3(4 + 3x)$$

Eliminate the parentheses.
$$10x \leq 12 + 9x$$

Subtract $9x$ from both sides.
$$x \leq 12$$

Simplify.
$$3(4 + 3x) \leq 2(5x) \quad \textbf{INCORRECT}$$

Eliminate parentheses.
$$12 + 9x \leq 10x \quad \textbf{INCORRECT}$$

Subtract $9x$ from both sides.
$$12 \leq x \quad \textbf{INCORRECT}$$

Check: Select a number that lies in the solution set, $x \leq 12$, such as $x = 0$. Show that $x = 0$ satisfies the original inequality:
$$\frac{5(0)}{3} \leq \frac{4 + 3(0)}{2}$$
$$0 \leq 2$$

It is important to remember that cross multiplication cannot be used in solving inequalities.

EXAMPLE 5 Solving a Double Linear Inequality

Solve the inequality $-2 < 3x + 4 \leq 16$.

Solution:

This double inequality can be written as two inequalities. $-2 < \overbrace{3x + 4} \leq 16$

Inequalities that must both be satisfied. $-2 < 3x + 4 \qquad 3x + 4 \leq 16$

Subtract 4 from both sides of each inequality. $-6 < 3x \qquad 3x \leq 12$

Divide each inequality by 3. $-2 < x \qquad x \leq 4$

Combining these two inequalities gives us $-2 < x \leq 4$ in inequality notation; in interval notation we have $(-2, \infty) \cap (-\infty, 4]$ or $(-2, 4]$.

Notice that the steps we took in solving these inequalities individually were identical. This leads us to a **shortcut method** in which we solve them together:

Write the combined inequality. $-2 < 3x + 4 \leq 16$

Subtract 4 from each part. $-6 < 3x \leq 12$

Divide each part by 3. $-2 < x \leq 4$

Interval notation: $(-2, 4]$

For the remainder of this section we will use the shortcut method for solving inequalities.

EXAMPLE 6 Solving a Double Linear Inequality

Solve the inequality $1 \leq \dfrac{-2 - 3x}{7} < 4$. Express the solution set in interval notation, and graph.

Solution:

Write the original double inequality. $1 \leq \dfrac{-2 - 3x}{7} < 4$

Multiply each part by 7.	$7 \leq -2 - 3x < 28$
Add 2 to each part.	$9 \leq -3x < 30$
Divide by each part by -3 and reverse the signs.	$-3 \geq x > -10$
Write in standard form.	$-10 < x \leq -3$

Interval notation: $(-10, -3]$ Graph: ◄─(──────┤+─►
                                              -10 $-3\ 0$

Applications Involving Linear Inequalities

EXAMPLE 7 Temperature Ranges

New York City on average has yearly temperature ranges from 23 degrees Fahrenheit to 86 degrees Fahrenheit. What is the range in degrees Celsius given that the conversion relation is $F = 32 + \dfrac{9}{5}C$?

Solution:

Temperature ranges from 23 °F to 86 °F.	$23 \leq F \leq 86$
Replace F with Celsius conversion.	$23 \leq 32 + \dfrac{9}{5}C \leq 86$
Subtract 32 from all three parts.	$-9 \leq \dfrac{9}{5}C \leq 54$
Multiply all three parts by $\dfrac{5}{9}$.	$-5 \leq C \leq 30$

New York City has an average yearly temperature range from $-5\ °C$ to $30\ °C$.

EXAMPLE 8 Comparative Shopping

Two car rental companies have advertised weekly specials on full-size cars. Hertz is advertising a $80 rental fee plus an additional $.10 per mile. Thrifty is advertising $60 and $.20 per mile. How many miles must you drive for the rental car from Hertz to be the better deal?

Solution:

Let x = number of miles driven during the week.

Write cost for Hertz rental.	$80 + 0.1x$
Write cost for Thrifty rental.	$60 + 0.2x$
Write inequality if Hertz is less than Thrifty.	$80 + 0.1x < 60 + 0.2x$
Subtract $0.1x$ from both sides.	$80 < 60 + 0.1x$
Subtract 60 from both sides.	$20 < 0.1x$
Divide both sides by 0.1.	$200 < x$

You must drive more than   200 miles   for Hertz to be the better deal.

SECTION 1.5 **SUMMARY**

In this section we discussed linear inequalities. Their solutions are solution sets that can be expressed four ways:

1. Inequality notation $\quad a < x \leq b$
2. Set notation $\quad\quad \{x \mid a < x \leq b\}$
3. Interval notation $\quad (a, b]$
4. Graph (number line)
$\quad\quad\quad\quad\quad\quad\quad\quad\quad\quad a \quad\quad\quad b$

Linear inequalities are solved using the same procedures as linear equations with one exception: **when you multiply or divide by a negative number you must reverse the inequality sign.**

NOTE: Cross multiplication cannot be used with inequalities.

SECTION 1.5 **EXERCISES**

■ **SKILLS**

In Exercises 1–8, rewrite in interval notation and graph.

1. $x \geq 3$ **2.** $x < -2$ **3.** $-2 \leq x < 3$ **4.** $-4 \leq x \leq -1$

5. $-3 < x \leq 5$ **6.** $0 < x < 6$ **7.** $0 \leq x \leq 0$ **8.** $-7 \leq x \leq -7$

In Exercises 9–16, rewrite in set notation.

9. $[0, 2)$ **10.** $(0, 3]$ **11.** $(-7, -2)$ **12.** $[-3, 2]$

13. $(-\infty, 6]$ **14.** $(5, \infty)$ **15.** $(-\infty, \infty)$ **16.** $[4, 4]$

In Exercises 17–20, write in inequality and interval notation.

17.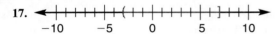

18.

19.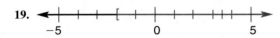

20.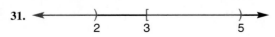

In Exercises 21–30, graph the indicated set and write as a single interval if possible.

21. $(-5, 2] \cup (-1, 3)$ **22.** $(2, 7) \cup [-5, 3)$ **23.** $(-\infty, 1] \cap [-1, \infty)$ **24.** $(-\infty, -5) \cap (-\infty, 7]$

25. $[-5, 2) \cap [-1, 3]$ **26.** $(-\infty, -3] \cup [-3, \infty)$ **27.** $(-\infty, -3] \cup [3, \infty)$ **28.** $(-2, 2) \cap [-3, 1]$

29. $(-\infty, \infty) \cap (-3, 2]$ **30.** $(-\infty, \infty) \cup (-4, 7)$

In Exercises 31–34, write in interval notation.

31.

32.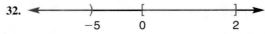

33.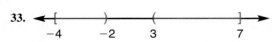

34.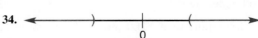

In Exercises 35–60, solve and express solution in interval notation.

35. $x - 3 < 7$

36. $x + 4 > 9$

37. $3x - 2 \leq 4$

38. $3x + 7 \geq -8$

39. $-5p \geq 10$

40. $-4u < 12$

41. $3 - 2x \leq 7$

42. $4 - 3x > -17$

43. $3(t + 1) > 2t$

44. $2(y + 5) \leq 3(y - 4)$

45. $7 - 2(1 - x) > 5 + 3(x - 2)$

46. $4 - 3(2 + x) < 5$

47. $\dfrac{x + 2}{3} - 2 \geq \dfrac{x}{2}$

48. $\dfrac{t - 5}{3} \leq -4$

49. $\dfrac{2}{3}y - \dfrac{1}{2}(5 - y) < \dfrac{5y}{3} - (2 + y)$

50. $\dfrac{s}{2} - \dfrac{(s - 3)}{3} > \dfrac{s}{4} - \dfrac{1}{12}$

51. $-2 < x + 3 < 5$

52. $1 < x + 6 < 12$

53. $-8 \leq 4 + 2x < 8$

54. $0 < 2 + x \leq 5$

55. $-3 < 1 - x \leq 9$

56. $3 \leq -2 - 5x \leq 13$

57. $0 < 2 - \dfrac{1}{3}y < 4$

58. $3 < \dfrac{1}{2}A - 3 < 7$

59. $\dfrac{1}{2} \leq \dfrac{1 + y}{3} \leq \dfrac{3}{4}$

60. $-1 < \dfrac{2 - z}{4} \leq \dfrac{1}{5}$

 ■ **APPLICATIONS**

61. Weight. A healthy weight range for a woman is given by the following formula:

> 110 lb for the first 5 feet (tall)
> 2–6 lb per inch for every inch above 5 feet

Write an inequality representing a healthy weight, w, for a 5 foot 9 inch woman.

62. Weight. NASA has more stringent weight allowances for their astronauts. Write an inequality representing allowable weight for a female 5 foot 9 inch mission specialist given 105 lb for the first 5 feet, and 1–5 lb/inch for every additional inch.

63. Profit. A seamstress decides to open a dress shop. Her fixed costs are $4,000 per month and it costs her $20 to make each dress. If the price of each dress is $100, how many dresses does she have to sell per month to make a profit?

64. Profit. Labrador retrievers that compete in field trials typically cost $2,000 at birth. Professional trainers charge $400 to $1,000 per month to train the dogs. If the dog is a champion by age 2, it sells for $30,000. What is the range of profit for a champion at age 2?

65. Cell Phones. A cell phone company charges $50 for an 800-minute monthly plan, plus an additional $0.22/minute for every minute over 800. If a customer's bill ranged from a low of $67.16 to a high of $96.86 over a 6-month period, what were the most minutes used in a single month? What were the least?

66. Grades. In your general biology class, your first three test scores are 67, 77, and 84. What is the lowest score you can get on the fourth test to earn a B for the course? Assume each test is of equal weight and a B is any score greater than or equal to 80.

67. Profit. Typical markup on new cars is 15%–30%. If the sticker price is $27,999, write an inequality that gives the range of the invoice price (what the dealer paid the manufacturer for the car).

68. Temperature. A digital thermometer has an error of 1.3%. If normal body temperature is assumed to be 98.6 °F, write an inequality representing a range of readings the thermometer could display and over which your temperature would be considered normal.

69. Lasers. A circular laser beam with a radius, r_T, is transmitted from one tower to another tower. If the received beam radius, r_R, fluctuates 10% from the transmitted beam radius due to atmospheric turbulence, write an inequality representing the received beam radius.

70. Communications. Communication systems are often evaluated based on their signal-to-noise ratio (SNR), which is the ratio of the average power of received signal, S, to average power of noise, N, in the system. If the SNR is required to be at least 2 at all times, write an inequality representing the received signal power if the noise can fluctuate 10%.

71. Real Estate. The Aguileras are listing their house with a real estate agent. They are trying to determine a listing price, L, for the house. Their realtor advises them that most buyers traditionally offer a buying price, B, that is 85%–95% of the listing price. Write an inequality that relates the buying price to the listing price.

72. Humidity. The National Oceanic and Atmospheric Administration (NOAA) has stations on buoys in the oceans to measure atmosphere and ocean characteristics such as temperature, humidity, and wind. The humidity sensors have an error of 5%. Write an inequality relating the humidity measurement, h_m, and the true humidity, h_t.

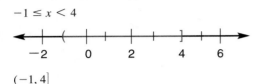

■ CATCH THE MISTAKE

In Exercises 73–76, explain the mistake that is made.

73. Rewrite in interval notation.

$-1 \le x < 4$

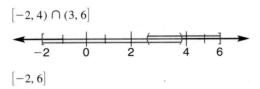

$(-1, 4]$

74. Graph the indicated set and write as a single interval if possible.

$[-2, 4) \cap (3, 6]$

$[-2, 6]$

75. Solve the inequality $2 - 3p \le -4$ and express the solution in interval notation.

Solution: $2 - 3p \le -4$

$-3p \le -6$

$p \le 2$

$(-\infty, 2]$

76. Solve the inequality $3 - 2x \le 7$ and express the solution in interval notation.

Solution: $3 - 2x \le 7$

$-2x \le 4$

$x \ge -2$

$(-\infty, -2]$

■ CHALLENGE

In Exercises 77–80, select any that could be true.

  a. $m > 0$ and $n > 0$ **c.** $m > 0$ and $n < 0$

  b. $m < 0$ and $n < 0$ **d.** $m < 0$ and $n > 0$

77. If $mn > 0$ **79.** $\dfrac{m}{n} > 0$

78. $mn < 0$ **80.** $\dfrac{m}{n} < 0$

In Exercises 81–82, select any that could be true.

  a. $n = 0$ **b.** $n > 0$ **c.** $n < 0$

81. $m + n < m - n$ **82.** $m + n \ge m - n$

■ TECHNOLOGY

83. a. Solve the inequality $2.7x + 3.1 < 9.4x - 2.5$.
   b. Graph each side of the inequality in the same viewing screen. Find the range of x values when the graph of the left side lies *below* the graph of the right side.
   c. Do (a) and (b) agree?

84. a. Solve the inequality $-0.5x + 2.7 > 4.1x - 3.6$.
   b. Graph each side of the inequality in the same viewing screen. Find the range of x values when the graph of the left side lies *above* the graph of the right side.
   c. Do (a) and (b) agree?

85. a. Solve the inequality $x - 3 < 2x - 1 < x + 4$.
   b. Graph all three expressions of the inequality in the same viewing screen. Find the range of x values when the graph of the middle expression lies above the graph of the left side and below the graph of the right side.
   c. Do (a) and (b) agree?

86. a. Solve the inequality $x + 3 < x + 5$.
   b. Graph each side of the inequality in the same viewing screen. Find the range of x values when the graph of the left side lies *below* the graph of the right side.
   c. Do (a) and (b) agree?

Skills Objectives	Conceptual Objectives
■ Solve polynomial inequalities. ■ Solve rational inequalities.	■ Determine domain restrictions on rational inequalities. ■ Understand critical points.

Polynomial Inequalities

In this section we will focus primarily on quadratic inequalities, but the procedures outlined are also valid for higher degree polynomial inequalities. An example of a quadratic inequality is $x^2 + x - 2 < 0$. This statement is true when the polynomial on the left side is negative. For any value of x, a polynomial is either positive, negative, or zero. A polynomial must pass through zero before it changes from positive to negative or from negative to positive. **Critical points** are the values of x that make the polynomial equal to zero. These zeros, or critical points, divide the real number line into **test intervals** where the polynomial is either positive or negative. For example, if we set the above polynomial equal to zero and solve:

$$x^2 + x - 2 = 0$$

$$(x + 2)(x - 1) = 0$$

$$x = -2 \quad \text{or} \quad x = 1$$

we find that $x = -2$ and $x = 1$ are the critical points. These critical points divide the real number line into three test intervals: $(-\infty, -2)$, $(-2, 1)$, and $(1, \infty)$.

Since the polynomial is equal to zero at $x = -2$ and $x = 1$, the value of the polynomial in each of these three intervals is either positive or negative. We select one real number that lies in each of the three intervals and test to see if the value of the polynomial at each point is either positive or negative. In this example we select the real numbers: $x = -3$, $x = 0$, and $x = 2$. The polynomial is written as the product $(x + 2)(x - 1)$; therefore we simply look for the sign in each set of parentheses.

$$(x + 2)(x - 1)$$

In this example, the statement $x^2 + x - 2 < 0$ is true when the polynomial (in factored form), $(x + 2)(x - 1)$, is negative. The interval $(-2, 1)$ is when the polynomial is negative. Thus, the solution to the inequality $x^2 + x - 2 < 0$ is $(-2, 1)$. To check the solution, select any number in the interval and substitute it into the original inequality to make sure it makes the statement true. The value $x = -1$ lies in the interval $(-2, 1)$. On substituting into the original inequality we find that $x = -1$ satisfies the inequality $(-1)^2 + (-1) - 2 = -2 < 0$.

PROCEDURE FOR SOLVING POLYNOMIAL INEQUALITIES

Step 1: Write the inequality in *standard form.*
Step 2: Factor the polynomial.
Step 3: Identify critical points.
Step 4: Draw a number line with critical points labeled.
Step 5: Select a real number in each interval and check sign of product.
Step 6: Identify which intervals make the inequality true.
Step 7: Write the solution in interval notation.

STUDY TIP

If the original polynomial is < 0 then the interval(s) that yield *negative* numbers should be selected. If the original polynomial is > 0 then the interval(s) that yield *positive* numbers should be selected.

Note: Be careful on Step 7. If the original polynomial is < 0, then the interval(s) that yield negative numbers should be selected. If the original polynomial is > 0, then the interval(s) that yield positive numbers should be selected.

EXAMPLE 1 Solving a Quadratic Inequality

Solve the inequality $x^2 - x > 12$.

Solution:

Write the inequality in standard form. $\qquad x^2 - x - 12 > 0$

Factor the left side. $\qquad (x + 3)(x - 4) > 0$

Identify the critical points. $\qquad (x + 3)(x - 4) = 0$

$\qquad x = -3 \text{ and } x = 4$

Draw a number line with critical points labeled.

Test each interval. $\qquad (x + 3)(x - 4)$

Positive intervals make this inequality true. $\qquad (-\infty, -3) \text{ and } (4, \infty)$

Write the solution in interval notation. $\qquad (-\infty, -3) \cup (4, \infty)$

The inequality in Example 1, $x^2 - x > 12$, is a *strict* inequality and so we use parentheses when we express the solution in interval notation $(-\infty, -3) \cup (4, \infty)$. It is important to note that if we change the inequality sign from $>$ to $\geq$, then the critical points $x = -3$ and $x = 4$ also make the inequality true. Therefore the solution to $x^2 - x \geq 12$ is $(-\infty, -3] \cup [4, \infty)$.

YOUR TURN Solve the inequality $x^2 - 5x \leq 6$ and express the solution in interval notation.

EXAMPLE 2 Solving a Quadratic Inequality

Solve the inequality $x^2 \leq 4$.

COMMON MISTAKE

Do not take the square root of both sides. You must write the inequality in standard form and factor.

 **CORRECT**

Write the inequality in standard form.
$$x^2 - 4 \leq 0$$

Factor.
$$(x - 2)(x + 2) \leq 0$$

Identify critical points.
$$(x - 2)(x + 2) = 0$$
$$x = 2 \text{ and } x = -2$$

Draw number line and test intervals.

$$(x - 2)(x + 2)$$

Negative intervals make the inequality true.

$$(-2, 2)$$

The endpoints, $x = -2$ and $x = 2$, satisfy the inequality so they are included in the solution and

expressed with brackets. $[-2, 2]$

INCORRECT

Write the inequality.
$$x^2 \leq 4$$

Take the square root of both sides.
$$x \leq 2 \qquad \textbf{ERROR}$$

INCORRECT solution:
$$(-\infty, 2]$$

When solving quadratic inequalities, you must first write in standard form and then factor to identify critical points.

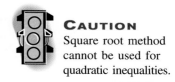 **CAUTION**

Square root method cannot be used for quadratic inequalities.

Not all inequalities have a solution. For example, $x^2 < 0$ has no real solution. Any real number squared is always nonnegative, so there are no real values that when squared will yield a negative number. The critical point is $x = 0$, which divides the real number line into two intervals: $(-\infty, 0)$ and $(0, \infty)$. Both of these intervals, however, are positive, so there are no intervals that satisfy the inequality. We say that this inequality has no real solution.

EXAMPLE 3 Solving a Quadratic Inequality

Solve the inequality $x^2 > -5x$.

COMMON MISTAKE

A common mistake is to divide by x because if x is negative the sign must be reversed. Always start by writing the inequality in standard form and factoring to determine critical points.

CAUTION
Do not divide inequalities by a variable.

 **CORRECT**

Write inequality in standard form.
$$x^2 + 5x > 0$$

Factor.
$$x(x + 5) > 0$$

Identify critical points.
$$x = 0, x = -5$$

Draw number line and test intervals.

$x(x + 5)$

$(-)(-)=(+) \quad (-)(+)=(-) \quad (+)(+)=(+)$

$-6 \quad -5 \quad -1 \quad 0 \quad 1$

Positive intervals satisfy the inequality.

$$(-\infty, -5) \text{ and } (0, \infty)$$

Express the solution in interval notation.

$$(-\infty, -5) \cup (0, \infty)$$

 INCORRECT

Write the original inequality.
$$x^2 > -5x$$

Divide both sides by x.
$$x > -5 \quad \textbf{ERROR}$$

INCORRECT solution:
$$(-5, \infty)$$

Dividing by x is the mistake. If x is negative, the inequality sign must be reversed. And what if x is zero?

The solution to the inequality $x^2 > -5x$ is $(-\infty, -5) \cup (0, \infty)$.

STUDY TIP
When solving quadratic inequalities you must first write in standard form and then factor to identify critical points.

When solving quadratic polynomials, once the inequality is written in standard form and factored, then a product is either positive or negative. If a product is positive, $ab > 0$, then either both factors are positive, $a > 0$ and $b > 0$, or both factors are negative, $a < 0$ and $b < 0$. Similarly, if a product is negative, $ab < 0$, then one factor is positive and the other factor is negative, $a > 0$ and $b < 0$ or $a < 0$ and $b > 0$. The critical points are $a = 0$ and $b = 0$. Next we will solve rational inequalities, which are quotients (as opposed to products), but we will treat them in the same manner with critical points and test intervals.

Rational Inequalities

We use a similar procedure to solve rational inequalities, such as $\dfrac{x - 3}{x^2 - 4} \geq 0$ as we used to solve polynomial inequalities with one exception. You must eliminate values for x that make the denominator equal to 0. In this example, we must eliminate $x \neq -2$ and $x \neq 2$, because they make the denominator, $x^2 - 4$, equal to zero.

We will proceed with a similar procedure involving critical points and test intervals that was outlined for polynomial inequalities. However, in rational inequalities, the **critical points** are any values that make *either* the numerator *or* denominator equal to zero.

EXAMPLE 4 Solving a Rational Inequality

Solve the inequality $\dfrac{x - 3}{x^2 - 4} \geq 0$.

Solution:

Factor denominator.

$$\frac{(x - 3)}{(x - 2)(x + 2)} \geq 0$$

Identify critical points.

$$x = -2, x = 2, x = 3$$

Draw a number line and label critical points.

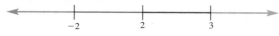

Test intervals.

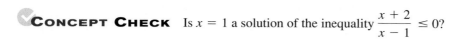

Positive intervals satisfy this inequality. $(-2, 2)$ and $(3, \infty)$

Since this inequality is greater than or equal to, we include $x = 3$ in our solution because it satisfies the inequality. However, the points $x = -2$ and $x = 2$ are not included in the solution because they make the denominator zero.

The solution is $(-2, 2) \cup [3, \infty)$.

CONCEPT CHECK   Is $x = 1$ a solution of the inequality $\dfrac{x + 2}{x - 1} \leq 0$?

YOUR TURN   Solve the inequality $\dfrac{x + 2}{x - 1} \leq 0$.

EXAMPLE 5 Solving a Rational Inequality

Solve the inequality $\dfrac{x}{x + 2} < 3$.

COMMON MISTAKE

Do not cross multiply.

 **CORRECT**

Subtract 3 from both sides.

$$\frac{x}{x + 2} - 3 < 0$$

 **INCORRECT**

Do not cross multiply.

$$x < 3(x + 2) \quad \textbf{ERROR}$$

 CAUTION

Rational inequalities cannot be solved by cross multiplication.

Write as a single rational function.

$$\frac{x - 3(x + 2)}{x + 2} < 0$$

Eliminate parentheses.

$$\frac{x - 3x - 6}{x + 2} < 0$$

Simplify numerator.

$$\frac{-2x - 6}{x + 2} < 0$$

Factor numerator.

$$\frac{-2(x + 3)}{x + 2} < 0$$

Identify the critical points.

$$x = -3 \text{ and } x = -2$$

Draw number line and test intervals.

$$\frac{-2(x + 3)}{x + 2} < 0$$

$$\frac{(-)(-)}{(-)} = (-) \qquad \frac{(-)(+)}{(-)} = (+) \qquad \frac{(-)(+)}{(+)} = (-)$$

$$\begin{array}{ccccc} \,|\, & \,|\, & \,|\, & \,|\, & \,|\, \\ -4 & -3 & -2.5 & -2 & 0 \end{array}$$

Negative intervals satisfy the inequality. $(-\infty, -3)$ and $(-2, \infty)$.

Note that $x = -2$ is not included in the solution because it makes the denominator zero and $x = -3$ is not included because it does not satisfy the strict inequality.

The solution is:

$$(-\infty, -3) \cup (-2, \infty)$$

Eliminate parentheses.

$$x < 3x + 6$$

Collect terms.

$$-2x < 6$$

Divide by -2, and reverse sign.

$$x > -3 \quad \textbf{INCORRECT}$$

INCORRECT solution:

$$(-3, \infty)$$

This solution is incorrect because the sign of $(x + 2)$ can be positive or negative depending on the value of x. It is not appropriate to multiply both sides of an inequality by a variable expression.

Applications

EXAMPLE 6 Stock Prices

From November 2003 until March 2004 Wal-Mart stock (WMT) was worth approximately $P = 2t^2 - 12t + 70$ where P denotes the price of the stock in dollars and t corresponds to months. November 2003 represents $t = 1$,

December 2003 represents $t = 2$, January 2004 represents $t = 3$, etc. During what months was the stock value at least \$54?

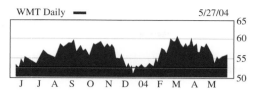

Solution:

Set price greater than or equal to 54.	$2t^2 - 12t + 70 \geq 54 \quad 1 \leq t \leq 5$
Write in standard form.	$2t^2 - 12t + 16 \geq 0$
Divide by 2.	$(t^2 - 6t + 8) \geq 0$
Factor.	$(t - 4)(t - 2) \geq 0$
Identify the critical points.	$t = 4$ and $t = 2$
Test intervals.	$(t - 4)(t - 2) \geq 0$

$$(-)(-) = (+) \quad (-)(+) = (-) \quad (+)(+) = (+)$$

$$\begin{array}{ccccc} 1 & 2 & 3 & 4 & 5 \end{array}$$

Positive intervals satisfy the inequality. $[1, 2]$ and $[4, 5]$

The Wal-Mart stock price was at least \$54 during November 2003, December 2003, February 2004, and March 2004.

SECTION 1.6   SUMMARY

In this section we have discussed polynomial and rational inequalities. In solving these we follow the same procedure:

1. Write in standard form—zero on one side.
2. Factor the polynomial; if a rational function, factor the numerator and denominator.
3. Determine the critical points.
4. Draw a number line labeling critical points and intervals.
5. Test intervals to determine if positive or negative.
6. Select intervals according to sign of inequality.
7. Write the solution in interval notation.

SECTION 1.6 EXERCISES

■ SKILLS

In Exercises 1–22, solve the polynomial inequality and express the solution set in interval notation.

1. $x^2 - 3x - 10 \geq 0$
2. $x^2 + 2x - 3 < 0$
3. $p^2 + 4p < -3$
4. $p^2 - 2p \geq 15$

5. $2t^2 - 3 \leq t$
6. $3t^2 \geq -5t + 2$
7. $5v - 1 > 6v^2$
8. $12t^2 < 37t + 10$

9. $2s^2 - 5s \geq 3$

10. $8s + 12 \leq -s^2$

11. $y^2 + 2y \geq 4$

12. $y^2 + 3y \leq 1$

13. $u^2 \geq 3u$

14. $u^2 \leq -4u$

15. $-2x \leq -x^2$

16. $-3x \leq x^2$

17. $x^2 > 9$

18. $x^2 \geq 16$

19. $z^2 > -16$

20. $z^2 \geq -2$

21. $y^2 < -4$

22. $y^2 \leq -25$

In Exercises 23–40, solve the rational inequality and graph the solution on the real number line.

23. $-\dfrac{3}{x} \leq 0$

24. $\dfrac{3}{x} \leq 0$

25. $\dfrac{y}{y + 3} > 0$

26. $\dfrac{y}{2 - y} \leq 0$

27. $\dfrac{s + 1}{4 - s^2} \geq 0$

28. $\dfrac{s + 5}{4 - s^2} \leq 0$

29. $\dfrac{2u^2 + u}{3} < 1$

30. $\dfrac{u^2 - 3u}{3} \geq 6$

31. $\dfrac{3t^2}{t + 2} \geq 5t$

32. $\dfrac{-2t - t^2}{4 - t} \geq t$

33. $\dfrac{3p - 2p^2}{4 - p^2} < \dfrac{3 + p}{2 - p}$

34. $-\dfrac{7p}{p^2 - 100} \leq \dfrac{p + 2}{p + 10}$

35. $\dfrac{x^2}{5 + x^2} < 0$

36. $\dfrac{x^2}{5 + x^2} \leq 0$

37. $\dfrac{x^2 + 10}{x^2 + 16} > 0$

38. $-\dfrac{x^2 + 2}{x^2 + 4} < 0$

39. $\dfrac{v^2 - 9}{v - 3} \geq 0$

40. $\dfrac{v^2 - 1}{v + 1} \leq 0$

■ APPLICATIONS

41. Profit. A Web-based embroidery company makes monogrammed napkins. The profit associated with producing x orders of napkins is governed by the equation

$$P(x) = -x^2 + 130x - 3000$$

Determine the range of orders the company should accept in order to make a profit.

42. Profit. Repeat Exercise 41 using $P(x) = x^2 - 130x + 3600$.

43. Car Value. The term "upside down" on car payments refers to owing more than a car is worth. Assume you buy a new car and finance 100% over 5 years. The difference between the value of the car and what is owed on the car is governed by the expression $\frac{t}{t - 3}$ where t is age (in years) of the car. Determine the time period when the car is worth more than you owe ($\frac{t}{t - 3} > 0$). When do you owe more than it's worth ($\frac{t}{t - 3} < 0$)?

44. Car Value. Repeat Exercise 43 using the expression $-\dfrac{2 - t}{4 - t}$.

45. Bullet Speed. A .22 caliber gun fires a bullet at a speed of 1200 feet per second. If a .22 caliber is fired straight upward into the sky, the height of the bullet in feet is given by the equation $h = -16t^2 + 1200t$ where t is the time in seconds with $t = 0$ corresponding to the instant the gun is fired. How long is the bullet in the air?

46. Bullet Speed. A .38 caliber gun fires a bullet at a speed of 600 feet per second. If a .38 caliber gun is fired straight upward into the sky, the height of the bullet in feet is given by the equation $h = -16t^2 + 600t$. How many seconds is the bullet in the air?

47. Geometry. A rectangular area is fenced in with 100 feet of fence. If the minimum area enclosed is to be 600 square feet, what is the range of feet allowed for the length of the rectangle?

48. Stock Value. From June 2003 until April 2004 JetBlue airlines stock (JBLU) was approximately worth $P = -4t^2 + 80t - 360$ where P denotes the price of the stock in dollars and t corresponds to months with $t = 1$ corresponding to January 2003. During what months was the stock value at least $36?

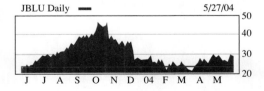

49. Real Estate. A person is selling a piece of land that she advertises as 400 acres ($+/-7$ acres) for $1.36 million. If you pay that price, what is the range of dollars per acre you have paid? Round to the nearest dollar.

50. Real Estate. A person is selling a piece of land that she advertises as 1000 acres ($+/-10$ acres) for $1 million. If you pay that price, what is the range of dollars per acre you have paid? Round to the nearest dollar.

■ **CATCH THE MISTAKE**

In Exercises 51–54, explain the mistake that is made.

51. Solve the inequality $3x < x^2$.

Solution:

Divide by x.	$3 < x$
Write solution in interval notation.	$(3, \infty)$

52. Solve the inequality $u^2 < 25$.

Solution:

Take the square root of both sides.	$u < -5$
Write the solution in interval notation.	$(-\infty, -5)$

53. Solve the inequality $\dfrac{x^2 - 4}{x + 2} > 0$.

Solution:

Factor the numerator and denominator.	$\dfrac{(x - 2)(x + 2)}{(x + 2)} > 0$
Cancel the $(x + 2)$ common factor.	$x - 2 > 0$
Solve.	$x > 2$

54. Solve the inequality $\dfrac{x + 4}{x} < -\dfrac{1}{3}$.

Solution:

Cross multiply.	$3(x + 4) < -1(x)$
Eliminate parentheses.	$3x + 12 < -x$
Combine like terms.	$4x < -12$
Divide both sides by 4.	$x < -3$

■ **CHALLENGE**

55. T or F: If $x < a^2$ then the solution is $(-\infty, a)$.

56. T or F: If $x \geq a^2$ then the solution is $[a, \infty)$.

In Exercises 57–60, solve for x given that a and b are both real positive numbers.

57. $-x^2 \leq a^2$

58. $\dfrac{x^2 - b^2}{x + b} < 0$

59. $\dfrac{x^2 + a^2}{x^2 + b^2} \geq 0$

60. $\dfrac{a}{x^2} < -b$

■ **TECHNOLOGY**

In Exercises 61–64, plot the left side and the right side of each inequality in the same screen and use the zoom feature to determine the range of values for which the inequality is true.

61. $1.4x^2 - 7.2x + 5.3 > -8.6x + 3.7$

62. $17x^2 + 50x - 19 < 9x^2 + 2$

63. $11x^2 < 8x + 16$

64. $0.1x + 7.3 > 0.3x^2 - 4.1$

SECTION 1.7 Absolute Value in Equations and Inequalities

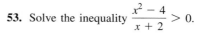

Skills Objectives

- Solve absolute value equations.
- Solve absolute value inequalities and express solution sets using interval notation.

Conceptual Objectives

- Understand absolute value in terms of distance on the number line.
- Compare and contrast absolute value equations and absolute value inequalities.

Absolute Value

The **absolute value** of a real number can be interpreted algebraically and graphically. Algebraically, the absolute value of 5 is 5, $|5| = 5$, and the absolute value of -5 is 5, $|-5| = 5$. Graphically, the absolute value of a real number is the distance between the real number and the origin. The distance on the real number line from 0 to either -5 or 5 is 5.

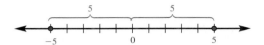

The absolute value of a real number is never negative. When $a = -5$, this definition says $|-5| = -(-5) = 5$.

Absolute value can be used to define the distance between two points on the real number line.

EXAMPLE 1 **Finding the Distance between Two Points on a Number Line**

Find the distance between -4 and 3 on the real number line.

Solution:

The distance between -4 and 3 is given by the absolute value of the difference

$$|-4 - 3| = |-7| = 7$$

Note that if we reversed the numbers the result is the same,

$$|3 - (-4)| = |7| = 7$$

We check this by counting the units between -4 and 3 on the number line.

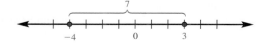

Equations Involving Absolute Value

When absolute value is involved in algebraic equations, we interpret the definition of absolute value as follows.

> **DEFINITION**   **ABSOLUTE VALUE EQUATION**
>
> If $|x| = a$ then $x = -a$ or $x = a$ where $a \geq 0$.

In words, "if the absolute value of a number is a, then that number equals $-a$ or a." For example, the equation $|x| = 7$ is true if $x = -7$ or $x = 7$. We say the equation $|x| = 7$ has the solution set $\{-7, 7\}$.

EXAMPLE 2 Solving an Absolute Value Equation

Solve the equation $|x - 3| = 8$ algebraically and graphically.

Solution:

Using the absolute value equation definition, we see that if the absolute value of an expression is 8, then that expression is either -8 or 8. This leads us to two equations:

$$x - 3 = -8 \quad \text{or} \quad x - 3 = 8$$
$$x = -5 \qquad\qquad x = 11$$

The solution set is $\{-5, 11\}$.

Graph: The absolute value equation $|x - 3| = 8$ is interpreted as "what numbers are 8 units away from 3 on the number line?" We find that 8 units to the right of 3 is 11 and 8 units to the left of 3 is -5.

TECHNOLOGY TIP
Use a graphing utility to display graphs of $y_1 = |x - 3|$ and $y_2 = 8$. The x-coordinates of any point of intersection is the solution to $|x - 3| = 8$.

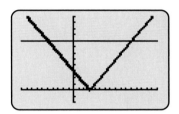

> **YOUR TURN** Solve the equation $|x + 5| = 7$.

EXAMPLE 3 Solving an Absolute Value Equation

Solve the equation $|1 - 3x| = 7$.

Solution:

If the absolute value of an expression is 7, then that expression is -7 or 7.

$$1 - 3x = -7 \quad \text{or} \quad 1 - 3x = 7$$
$$-3x = -8 \qquad\qquad -3x = 6$$
$$x = \frac{8}{3} \qquad\qquad\quad x = -2$$

The solution set is $\left\{ -2, \dfrac{8}{3} \right\}$.

■ **Answer:** $x = 2$ or $x = -12$ The solution set is $\{-12, 2\}$.

■ **YOUR TURN** Solve the equation $|1 + 2x| = 5$.

EXAMPLE 4 Finding That an Absolute Value Equation Has No Solution

Solve the equation $|1 - 3x| = -7$.

Solution: No solution .

The absolute value of an expression is never negative. Therefore no values of x make this equation true.

EXAMPLE 5 Solving a Quadratic Absolute Value Equation

Solve the equation $|5 - x^2| = 1$.

Solution:

If the absolute value of an expression is 1, that expression is either -1 or 1, which leads to two equations.

$$
\begin{array}{ll}
5 - x^2 = -1 & \quad \text{or} \quad 5 - x^2 = 1 \\
-x^2 = -6 & \qquad\qquad -x^2 = -4 \\
x^2 = 6 & \qquad\qquad x^2 = 4 \\
x = \pm\sqrt{6} & \qquad\qquad x = \pm\sqrt{4} = \pm 2
\end{array}
$$

Check to verify that all four values satisfy the absolute value equation.

The solution set is $\{\pm 2, \pm\sqrt{6}\}$.

■ **YOUR TURN** Solve the equation $|7 - x^2| = 2$.

Inequalities Involving Absolute Value

To solve the inequality, $|x| < 3$, look for all real numbers that make this statement true. Some numbers that make it true are $-2, -1, 0, 1$, and 2. Some numbers that make it false are $-7, -5, -3, 3$, and 4. If we interpret this inequality as distance, we ask *what numbers are less than 3 units from the origin?* We can represent the solution in the following ways.

Inequality notation: $-3 < x < 3$

Interval notation: $(-3, 3)$

Graph:

■ **Answer:** $x = -3$ or $x = 2$ The solution set is $\{-3, 2\}$.

■ **Answer:** $x = \pm\sqrt{5}$ or $x = \pm 3$ The solution set is $\{\pm\sqrt{5}, \pm 3\}$.

Similarly, to solve the inequality, $x \geq 3$, look for all real numbers that make the statement true. If we interpret this inequality as a distance, we ask *what numbers are at least 3 units from the origin?* We can represent the solution in the following three ways.

Inequality notation: $x \leq -3$ or $x \geq 3$

Interval notation: $(-\infty, -3] \cup [3, \infty)$

Graph: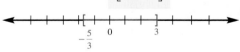

This discussion leads us to the following equivalence relations.

PROPERTIES OF ABSOLUTE VALUE INEQUALITIES

1. $|x| < a$ is equivalent to $-a < x < a$

2. $|x| \leq a$ is equivalent to $-a \leq x \leq a$

3. $|x| > a$ is equivalent to $x < -a$ or $x > a$

4. $|x| \geq a$ is equivalent to $x \leq -a$ or $x \geq a$

It is important to realize that in the above four properties the variable x can be any algebraic expression.

EXAMPLE 6 Solving an Inequality Involving an Absolute Value

Solve the inequality $|3x - 2| \leq 7$.

Solution:

We apply property (2) and squeeze the absolute value expression between -7 and 7. $-7 \leq 3x - 2 \leq 7$

Add 2 to all three parts. $-5 \leq 3x \leq 9$

Divide all three parts by 3. $-\dfrac{5}{3} \leq x \leq 3$

The solution in interval notation is $\left[-\dfrac{5}{3}, 3 \right]$.

Graph:

YOUR TURN   Solve the inequality $|2x + 1| < 11$.

STUDY TIP

Less than inequalities can be written as a single statement.

Greater than inequalities must be written as two statements.

It is often helpful to note that for absolute value inequalities

- *less than* inequalities can be written as a single statement (see Example 6).
- *greater than* inequalities must be written as two statements (see Example 7).

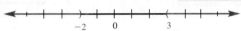**EXAMPLE 7 Solving an Inequality Involving an Absolute Value**

Solve the inequality $|1 - 2x| > 5$.

Solution:

Apply property (3).	$1 - 2x < -5$ or	$1 - 2x > 5$
Subtract 1 from all expressions.	$-2x < -6$	$-2x > 4$
Divide by -2 and reverse the sign.	$x > 3$	$x < -2$
Express the solution in interval notation.		$(-\infty, -2) \cup (3, \infty)$

Graph:

$$\longleftarrow \;|\;\;|\;\;|\;\;)\;\;|\;\;|\;\;|\;\;|\;\;(\;\;|\;\;|\; \longrightarrow$$
$$\quad\quad -2 \quad\; 0 \quad\quad\;\; 3$$

CONCEPT CHECK Write the absolute value inequality $|5 - 2x| \geq 1$ as two inequalities.

YOUR TURN Solve the inequality $|5 - 2x| \geq 1$.

SECTION 1.7 SUMMARY

In this section we solved equations and inequalities involving absolute values by writing the equations or inequalities in terms of two equations or inequalities.

Equations

$|x| = A$ is equivalent to $x = -A$ or $x = A$

Inequalities

$|x| < A$ is equivalent to $-A < x < A$

$|x| > A$ is equivalent to $x < -A$ or $x > A$

SECTION 1.7 EXERCISES

 ■ **SKILLS**

In Exercises 1–22, solve the equation.

1. $|x| = 3$ **2.** $|x| = 2$ **3.** $|x| = -4$ **4.** $|x| = -2$

5. $|t + 3| = 2$ **6.** $|t - 3| = 2$ **7.** $|4 - y| = 1$ **8.** $|2 - y| = 11$

9. $|3x| = 9$ **10.** $|5x| = 50$ **11.** $|2x + 7| = 9$ **12.** $|7 - 2x| = 9$

■ **Answer:** Inequality notation: $x \leq 2$ or $x \geq 3$ Interval notation: $(-\infty, 2] \cup [3, \infty)$

13. $|4.7 - 2.1x| = 3.3$ **14.** $|5.2x + 3.7| = 2.4$ **15.** $\left|\dfrac{2}{3}x - \dfrac{4}{7}\right| = \dfrac{5}{3}$ **16.** $\left|\dfrac{1}{2}x + \dfrac{3}{4}\right| = \dfrac{1}{16}$

17. $|4 - x^2| = 1$ **18.** $|7 - x^2| = 3$ **19.** $|x^2 + 1| = 5$ **20.** $|x^2 - 1| = 5$

21. $|x - a| = b$, where a and b are both real, positive numbers   **22.** $|x - a| = -b$, where a and b are both real, positive numbers

In Exercises 23–26, write an absolute value equation that describes the statement.

23. Any number whose distance from $x = 5$ on the number line is 3.

24. Any number whose distance from $x = -5$ on the number line is 3.

25. Any number whose distance from $x = c$ on the number line is 2.

26. Any number whose distance from $x = -3c$ on the number line is 1.

In Exercises 27–46, solve the inequality and express the solution in interval notation and graph.

27. $|x| < 7$ **28.** $|y| < 9$ **29.** $|y| \geq 5$ **30.** $|x| \geq 2$

31. $|x + 3| < 7$ **32.** $|x + 2| \leq 4$ **33.** $|x - 4| > 2$ **34.** $|x - 1| < 3$

35. $|4 - x| \leq 1$ **36.** $|1 - y| < 3$ **37.** $|2x| > -3$ **38.** $|2x| < -3$

39. $|4 - 3x| \geq 0$ **40.** $|4 - 3x| \geq 1$ **41.** $|1 - 2x| < \dfrac{1}{2}$ **42.** $\left|\dfrac{2 - 3x}{5}\right| \geq \dfrac{2}{5}$

43. $|2.6x + 5.4| < 1.8$ **44.** $|3.7 - 5.5x| > 4.3$ **45.** $|x^2 - 1| \leq 8$ **46.** $|x^2 + 4| \geq 29$

In Exercises 47–52, write an inequality that describes the statement.

47. Any real numbers less than 7 units from 2.

48. Any real numbers more than 3 units from -2.

49. Any real numbers at least $\frac{1}{2}$ unit from $\frac{3}{2}$.

50. Any real number no more than $\frac{5}{3}$ units from $\frac{11}{3}$.

51. Any real numbers no more than 2 units from a.

52. Any real number at least a units from -3.

 ■ APPLICATIONS

53. Temperature. If the average temperature in Hawaii is 83 degrees Fahrenheit plus or minus 15 degrees, write an absolute value inequality representing the temperature in Hawaii.

54. Temperature. If the average temperature of a human is 97.8 (± 1.2) degrees Fahrenheit, write an absolute value inequality describing normal human body temperature.

55. Sports. Two women tee off the green of a par-3 hole on a golf course. They are playing "closest to the pin." If the

first woman tees off and lands exactly 4 feet from the hole, write an inequality that describes where the second woman must land in order to win the hole. What equation would suggest a tie? Let $d =$ the distance from where the second woman lands to the tee.

56. Electric Frequencies. A bandpass filter in electronics allows certain frequencies within a range (or band) to pass through to the receiver and eliminates all other frequencies. Write an absolute value inequality that allows any frequency, f, within 15 Hz of the carrier frequency, f_c, to pass.

 ■ CATCH THE MISTAKE

In Exercises 57–60, explain the mistake that is made.

57. Solve the absolute value equation $|x - 3| = 7$.

Solution:

Eliminate absolute value sign.	$x - 3 = 7$
Add 3 to both sides.	$x = 10$
Check.	$|10 - 3| = 7$

58. Solve the inequality $|x - 3| < 7$.

Solution:

Eliminate absolute value signs.	$x - 3 < -7$ or	$x - 3 > 7$
Add 3 to both sides.	$x < -4$	$x > 10$

The solution is $(-\infty, -4) \cup (10, \infty)$.

59. Solve the inequality $|5 - 2x| \leq 1$.

Solution:

Eliminate absolute values. $-1 \leq 5 - 2x \leq 1$

Subtract 5. $-1 \leq -2x \leq 1$

Divide by -2 and reverse the sign. $\dfrac{1}{2} \geq x \geq -\dfrac{1}{2}$

Write the solution in interval notation. $\left[-\dfrac{1}{2}, \dfrac{1}{2} \right]$

60. Solve the equation $|5 - 2x| = -1$.

Solution:

$5 - 2x = -1$ or $5 - 2x = 1$

$-2x = -6$ $-2x = -4$

$x = 3$ $x = 2$

The solution is $\{2, 3\}$.

■ CHALLENGE

In Exercises 61–64, assuming a and b are real positive numbers, solve the equation or inequality and express the solution in interval notation.

61. $|x - a| < b$ **62.** $|a - x| > b$

63. $|x| \geq -a$ **64.** $|x| \leq -b$

65. T or F: $-|m| \leq m \leq |m|$

66. T or F: $|n|^2 = n^2$

67. T or F: $|m + n| = |m| + |n|$ is true only when m and n are both nonnegative.

68. For what values of x does the absolute value equation $|x - 7| = x - 7$ hold?

69. For what values of x does the absolute value equation $|x + 1| = 4 + |x - 2|$ hold?

70. Solve the inequality $|3x^2 - 7x + 2| > 8$.

■ TECHNOLOGY

71. Graph $y_1 = |x - 7|$ and $y_2 = x - 7$ in the same screen. Do the x values where these two graphs coincide agree with your result in Exercise 68?

72. Graph $y_1 = |x + 1|$ and $y_2 = |x - 2| + 4$ in the same screen. Do the x values where these two graphs coincide agree with your result in Exercise 69?

73. Graph $y_1 = |3x^2 - 7x + 2|$ and $y_2 = 8$ in the same screen. Do the x values where y_1 lies above y_2 agree with your result in Exercise 70?

74. Solve the inequality $|2.7x^2 - 7.9x + 5| \leq |5.3x^2 - 9.2|$ by graphing both sides of the inequality and identify which x values make this statement true.

Q: To earn some extra money on the weekends, Jake started pressure washing homes. It takes him about 8 hours to pressure wash the outside of an average-sized house. Since he had a busy weekend planned and needed to pressure wash a house, he hired his friend Jen to work on the job with him. From experience Jake knew that Jen could pressure wash an average-sized house in about 6 hours. Since Jen did not own a pressure washer, Jake rented one for the 7 hours he anticipated it would take them working together to finish the job. They finished the job in just under 3½ hours. How were they able to work so much more quickly than Jake had calculated?

A: The average time it would take each person to do the job individually (which is what was calculated) would not be the amount of time it would take the two people working together to do the same job. If the calculation had been correct, that would mean that Jake caused Jen to work more slowly than if she had worked alone.

To calculate the amount of time it would take for both people to do the job together use the fact that (amount of work completed by first person in 1 hour) + (amount of work completed by second person in 1 hour) = (amount of work completed together in 1 hour).

Jake is considering hiring another friend who can pressure wash an average-sized house in 9 hours. If all three of them work together on the house, how long would it take to complete the job? If Jake decides to pay his friends $15 an hour to help him and it costs $5 an hour to rent the pressure washers for his friends (Jake owns the pressure washer he uses), would Jake profit by hiring both friends if he earns $150 for pressure washing the house?

TYING IT ALL TOGETHER

You have been doing well in your College Algebra class throughout the semester. Now that final exam week is approaching, you would like to determine what you need to earn on the final exam in order to earn an A for the course. There were four tests worth 20 points each (although the lowest test grade was dropped), five quizzes worth 10 points each, and five homework assignments worth 20 points each. Your test average is 95%, quiz average is 86%, and your homework average is 85%. Your individual grades are listed below:

TESTS	QUIZZES	HOMEWORK
18	8	17
19	9	18
16	9	16
20	7	19
	10	15

If 50% of your grade is based on your test average, 10% is based on your quiz average, 15% is based on your homework average, and the remaining 25% is based on your final exam score, in what range of grades must you score on your final exam in order to earn an A for the course (an A is 89.5% or higher)? If you have already earned two bonus points added to your course average, then in what range of grades must your final exam score fall in order to earn an A for the course?

SECTION	TOPIC	PAGES	REVIEW EXERCISES	KEY CONCEPTS
1.1	Linear equations	60–66	1–22	$ax + b = 0$
	Solve linear equations in one variable	61–63	1–12	Isolate xs on one side and constants on the other side.
	Solve linear equations with domain restrictions	63–66	13–20	Any values that make the denominator equal to 0 must be eliminated as possible solutions.
1.2	Applications involving linear equations	69–79	23–34	5-step procedure $d = rt, I = Prt$
1.3	Quadratic equations	83–92	35–62	$ax^2 + bx + c = 0$
	Factoring	83–85	35–38	If $(x - h)(x - k) = 0$ then $x = h$ or $x = k$.
	Square root method	85–87	39–42	If $u^2 = A$, then $u = \pm\sqrt{A}$.
	Completing the square	87–89	43–46	Half of b; square that quantity; add the result to both sides.
	Quadratic formula	89–91	47–50	$x = \dfrac{-b \pm \sqrt{b^2 - 4ac}}{2a}$
	Applications	91–92	57–62	
1.4	Radical equations; equations quadratic in form	96–100	63–82	Transform into linear or quadratic equations.
	Solving radical equations	96–98	63–74	Check solutions to avoid extraneous solutions.
	Solving equations quadratic in form	98–100	75–82	Use a u-substitution to write the equation quadratic in form.
1.5	Linear inequalities	103–110	83–106	Range of solutions.
	Inequality, interval notation, and graphs	103–106	83–96	$a < x < b$ is equivalent to (a, b). $x \le a$ is equivalent to $(-\infty, a]$. $x > a$ is equivalent to (a, ∞).
	Solving linear inequalities	106–110	97–106	If an inequality is multiplied or divided by a *negative* number, the sign must be reversed.
1.6	Polynomial and rational inequalities	113–119	107–116	Critical points divide the number line into test intervals.
	Polynomial inequalities	113–116	107–112	Critical points are values that make the polynomial equal to 0.
	Rational inequalities	116–118	113–116	Critical points are values that make either the numerator or denominator equal to 0.
1.7	Absolute value in equations and inequalities	121–126	117–130	$|b - a|$ is the distance between the points a and b on the number line.
	Absolute value in equations	123–124	117–120	If $|x| = a$ then $x = -a$ or $x = a$.
	Absolute value in inequalities	124–126	121–128	$|x| \le a$ is equivalent to $-a \le x \le a$. $|x| > a$ is equivalent to $x < -a$ or $x > a$.
	Applications involving absolute values	127	129–130	

CHAPTER 1 REVIEW EXERCISES

1.1 Linear Equations

Solve for the variable.

1. $7x - 4 = 12$

2. $13d + 12 = 7d + 6$

3. $20p + 14 = 6 - 5p$

4. $4(x - 7) - 4 = 4$

5. $3(x + 7) - 2 = 4(x - 2)$

6. $7c + 3(c - 5) = 2(c + 3) - 14$

7. $14 - [-3(y - 4) + 9] = [4(2y + 3) - 6] + 4$

8. $[6 - 4x + 2(x - 7)] - 52 = 3(2x - 4) + 6[3(2x - 3) + 6]$

9. $\dfrac{12}{b} - 3 = \dfrac{6}{b} + 4$

10. $\dfrac{g}{3} + g = \dfrac{7}{9}$

11. $\dfrac{13x}{7} - x = \dfrac{x}{4} - \dfrac{3}{14}$

12. $5b + \dfrac{b}{6} = \dfrac{b}{3} - \dfrac{29}{6}$

Specify any values that must be excluded from the solution set and then solve.

13. $\dfrac{1}{x} - 4 = 3(x - 7) + 5$

14. $\dfrac{4}{x + 1} - \dfrac{8}{x - 1} = 3$

15. $\dfrac{2}{t + 4} - \dfrac{7}{t} = \dfrac{6}{t(t + 4)}$

16. $\dfrac{3}{2x - 7} = \dfrac{-2}{3x + 1}$

17. $\dfrac{3}{2x} - \dfrac{6}{x} = 9$

18. $\dfrac{3 - \dfrac{5}{m}}{2 + \dfrac{5}{m}} = 1$

19. $7x - (2 - 4x) = 3[-6 + (4 - 2x + 7)] + 12$

20. $\dfrac{x}{5} - \dfrac{x - 3}{15} = -6$

In Exercises 21–22, solve for the specified variable.

21. Solve for x in terms of y: $3x - 2[(y + 4)3 - 7] = y - 2x + 6(x - 3)$.

22. If $y = \dfrac{x + 3}{1 + 2x}$ find $\dfrac{y + 2}{1 - 2y}$ in terms of x.

1.2 Applications Involving Linear Equations

23. **Travel.** Maria is on her way from Orlando to Tampa for a rock concert. She walks two miles to the bus station, takes the bus $\frac{1}{4}$ of the way, and a taxi $\frac{1}{3}$ of the way to the stadium. How far does Maria live from the stadium?

24. **Diet.** A particular 2000 calorie per day diet suggests eating breakfast, lunch, dinner, and four snacks. Each snack is $\frac{1}{4}$ the calories of lunch. Lunch has 100 calories less than dinner. Dinner has 1.5 times as many calories as breakfast. How many calories are each meal and snack?

25. **Numbers.** Find a number such that 12 more than $\frac{1}{4}$ the number is $\frac{1}{3}$ the number.

26. **Numbers.** Find four consecutive odd integers such that the sum of the four is equal to three more than three times the fourth integer.

27. **Geometry.** The length of a rectangle is 1 more than 2 times the width and the perimeter is 20 inches. What are the dimensions of the rectangle?

28. **Geometry.** Find the perimeter of a triangle if one side is 10 inches, another side is $\frac{1}{3}$ of the perimeter, and the third side is $\frac{1}{6}$ of the perimeter.

29. **Money.** You win $25,000 and you decide to invest the money in two different investments: one paying 20% and the other paying 8%. A year later you collectively have $27,600. How much did you originally invest in each account?

30. **Money.** A college student on summer vacation was able to make $5,000 by working a full-time job every summer. He invested half the money in a mutual fund and half the money in a stock that yielded four times as much interest as the mutual fund. After a year he earned $250 in interest. What were the interest rates of the mutual fund and the stock?

31. **Chemistry.** For an experiment, a student requires 150 ml of a solution that is 8% NaCl (sodium chloride). The storeroom has only solutions that are 10% NaCl and 5% NaCl. How many milliliters of each available solution should be mixed to get 150 ml of 8% NaCl?

32. **Chemistry.** A mixture containing 8% salt is to be mixed with 4 ounces of a mixture that is 20% salt, in order to obtain a solution that is 12% salt. How much of the first solution must be used?

33. **Grades.** Going into the College Algebra exam, which will count as two tests, Danny has test scores of 95, 82, 90, and 77. If his final exam is higher than his lowest test score, then it will count for the final exam and replace the lowest test score. What score does Danny need on the final in order to have an average score of at least 90?

34. **Car Value.** A car salesperson reduced the price of a model car by 20%. If the new price is $25,000, what was its original price? How much can be saved by purchasing the model?

1.3 Quadratic Equations

Solve by factoring.

35. $b^2 = 4b + 21$

36. $x(x - 3) = 54$

37. $4x^2 = 8x$

38. $6y^2 - 7y - 5 = 0$

Solve by the square root method.

39. $q^2 - 169 = 0$

40. $c^2 + 36 = 0$

41. $(2x - 4)^2 = -64$

42. $(d + 7)^2 - 4 = 0$

Solve by completing the square.

43. $x^2 - 4x - 12 = 0$

44. $2x^2 - 5x - 7 = 0$

45. $\dfrac{x^2}{2} = 4 + \dfrac{x}{2}$

46. $8m = m^2 + 15$

Solve by the quadratic formula.

47. $3t^2 - 4t = 7$

48. $4x^2 + 5x + 7 = 0$

49. $8f^2 - \dfrac{1}{3}f = \dfrac{7}{6}$

50. $x^2 = -6x + 6$

Solve by any method.

51. $5q - 3q^2 - 3 = 0$

52. $(x - 7)^2 = -12$

53. $2x^2 - 3x - 5 = 0$

54. $(g - 2)(g + 5) = -7$

55. $7x^2 = -19x + 6$

56. $7 = 2(b^2 + 1)$

Solve for the indicated variable.

57. $S = \pi r^2 h$ for r

58. $V = \dfrac{\pi r^3 h}{3}$ for r

59. $h = vt - 16t^2$ for v

60. $A = 2\pi r^2 + 2\pi rh$ for h

61. Find the base and height of a triangle with an area of 2 square feet if its base is 3 feet longer than its height.

62. A man is standing on top of a building 500 feet tall. If he drops a penny off the roof, the height of the penny is given by $h = -16t^2 + 500$ where t is in seconds. Determine how many seconds it takes until the penny hits the ground.

1.4 Radical Equations; Equations Quadratic in Form

Solve the radical equation for the given variable.

63. $\sqrt[3]{2x - 4} = 2$

64. $\sqrt{x - 2} = -4$

65. $(2x - 7)^{1/5} = 3$

66. $x = \sqrt{7x - 10}$

67. $x - 4 = \sqrt{x^2 + 5x + 6}$

68. $\sqrt{2x - 7} = \sqrt{x + 3}$

69. $\sqrt{x + 3} = 2 - \sqrt{3x + 2}$

70. $4 + \sqrt{x - 3} = \sqrt{x - 5}$

71. $x - 2 = \sqrt{49 - x^2}$

72. $\sqrt{2x - 5} - \sqrt{x + 2} = 3$

73. $-x = \sqrt{3 - x}$

74. $\sqrt{15 + 2\sqrt{x - 4}} + \sqrt{x} = 5$

Solve the equation by introducing a substitution that transforms these equations to quadratic form.

75. $-28 = (3x - 2)^2 - 11(3x - 2)$

76. $x^4 - 6x^2 + 9 = 0$

77. $\left(\dfrac{x}{1 - x}\right)^2 = 15 - 2\left(\dfrac{x}{1 - x}\right)$

78. $3(x - 4)^4 - 11(x - 4)^2 - 20 = 0$

79. $(3x^{1/3} + 2x^{2/3}) = 5$

80. $2x^{2/3} - 3x^{1/3} - 5 = 0$

81. $x^4 + 5x^2 = 36$

82. $3 - 4x^{-1/2} + x^{-1} = 0$

1.5 Linear Inequalities

Rewrite using interval notation.

83. $x \leq -4$

84. $-1 < x \leq 7$

85. $2 \leq x \leq 6$

86. $x > -1$

Rewrite using inequality notation.

87. $(-6, \infty)$

88. $(-\infty, 0]$

89. $[-3, 7]$

90. $(-5, 2]$

Express each interval using inequality and interval notation.

91.

92.

Graph the indicated set and write as a single interval if possible.

93. $(4, 6] \cup [5, \infty)$

94. $(-\infty, -3] \cup [-7, 2]$

95. $(3, 12] \cap [8, \infty)$

96. $(-\infty, -2) \cap [-2, 9)$

Solve and graph.

97. $2x < 5 - x$ **98.** $6x + 4 \leq 2$

99. $4(x - 1) > 2x - 7$ **100.** $\dfrac{x + 3}{3} \geq 6$

101. $6 < 2 + x \leq 11$ **102.** $-6 \leq 1 - 4(x + 2) \leq 16$

103. $\dfrac{2}{3} \leq \dfrac{1 + x}{6} \leq \dfrac{3}{4}$ **104.** $\dfrac{x}{3} + \dfrac{x + 4}{9} > \dfrac{x}{6} - \dfrac{1}{3}$

Applications.

105. Grades. In your algebra class your first four exam grades are 72, 65, 69, and 70. What is the lowest score you can get on the fifth exam to earn a C for the course? Assume each exam is equal in weight and a C is any score greater than or equal to 70.

106. Profit. A tailor decided to open a men's custom suit business. His fixed costs are $8,500 per month and it costs him $50 for the materials to make each suit. If the price he charges per suit is $300, how many suits does he have to tailor per month to make a profit?

1.6 Polynomial and Rational Inequalities

Solve the polynomial inequality and express the solution set using interval notation.

107. $x^2 \leq 36$ **108.** $6x^2 - 7x < 20$

109. $4x \leq x^2$ **110.** $-x^2 \geq 9x + 14$

111. $-x^2 < -7x$ **112.** $x^2 < -4$

Solve the rational inequality and express the solution set using interval notation.

113. $\dfrac{x}{x - 3} < 0$ **114.** $\dfrac{x - 1}{x - 4} > 0$

115. $\dfrac{x^2 - 3x}{3} \geq 18$ **116.** $\dfrac{x^2 - 49}{x - 7} \geq 0$

1.7 Absolute Value in Equations and Inequalities

Solve the equation.

117. $|x - 3| = -4$ **118.** $|2 + x| = 5$

119. $|3x - 4| = 1.1$ **120.** $|x^2 - 6| = 3$

Solve the inequality and express the solution using interval notation.

121. $|x| < 4$ **122.** $|x - 3| < 6$

123. $|x + 4| > 7$ **124.** $|-7 + y| \leq 4$

125. $|2x| > 6$ **126.** $\left| \dfrac{4 + 2x}{3} \right| \geq \dfrac{1}{7}$

127. $|2 + 5x| \geq 0$ **128.** $|1 - 2x| \leq 4$

Applications.

129. Temperature. If the average temperature in Phoenix is 85 degrees Fahrenheit plus or minus 10 degrees, write an inequality representing the average temperature, T, in Phoenix.

130. Blood Alcohol Level. If a person registers a 0.08 blood alcohol level, he will be issued a DUI ticket in the state of Florida. If the test is accurate within 0.007, write a linear inequality representing an actual blood alcohol level that will not be issued a ticket.

Solve the equation.

1. $-2(z - 1) + 3 = -3z + 3(z - 1)$

2. $\dfrac{x^2 - 81}{x + 9} = x - 9$

3. $6x^2 - 13x = 8$

4. $2x^{2/3} + 3x^{1/3} - 2 = 0$

5. $\sqrt{3y - 2} = 3 - \sqrt{3y + 1}$

Solve the inequality and express the solution in interval notation.

6. $7 - 5x > -18$

7. $4.2 \le 1.7 - 2.5x \le 7.3$

8. $3p^2 \ge p + 4$

9. $\left| \dfrac{2}{3} - \dfrac{1}{5}x \right| \ge \dfrac{1}{10}$

10. $|x^2 - 6| \le -3$

11. **Puzzle.** A piling supporting a bridge sits so that $\frac{1}{4}$ of the piling is in the sand, 150 feet is in the water, and $\frac{3}{5}$ of the piling is in the air. What is the total height of the piling?

12. **Real Estate.** As a realtor you earn 7% of the sale price. The owners of a house you have listed at $150,000 will entertain offers within 10% of the list price. Write an inequality that models the commission you could make on this sale.

13. **Cell Phones.** A cell phone company charges $49 for a 600-minute monthly plan plus an additional $0.17/minute for every minute over 600. If a customer's bill ranged from a low of $53.59 to a high of $69.74 over a 6-month period, write an inequality expressing the number of monthly minutes used over the 6-month period.

14. **Television.** Television and film formats are classified as ratios of width to height. Traditional televisions have a 4:3 ratio (1.33:1), and movies are typically made in widescreen format with a 21:9 ratio (2.35:1). If you own a traditional 25 inch television (20 inch × 15 inch screen)

and you play a widescreen DVD on it, there will be black bars above and below the image. What are the dimensions of the movie and of the black bars?

Andy Washnik

PRACTICE TEST

Graphs

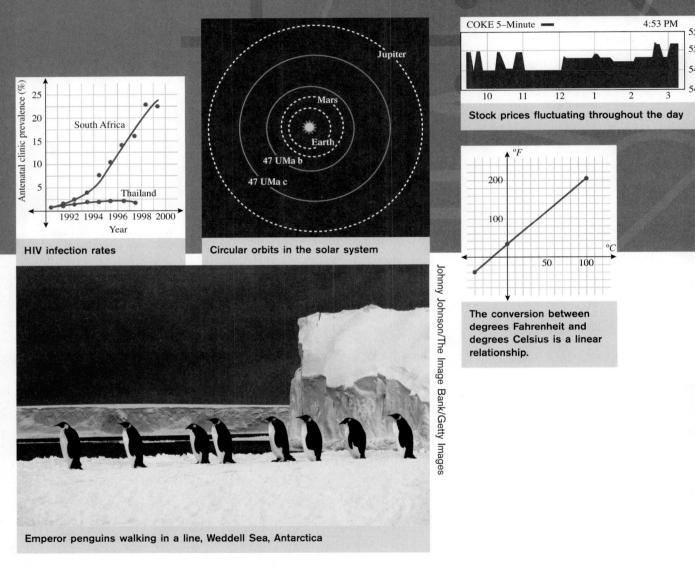

Antenatal clinic prevalence (%)

South Africa

Thailand

1992 1994 1996 1998 2000
Year

HIV infection rates

Jupiter

Mars

Earth

47 UMa b

47 UMa c

Circular orbits in the solar system

COKE 5–Minute ▬ 4:53 PM

55.2

55.0

54.7

54.5

10 11 12 1 2 3

Stock prices fluctuating throughout the day

°F

200

100

50 100 °C

The conversion between degrees Fahrenheit and degrees Celsius is a linear relationship.

Johnny Johnson/The Image Bank/Getty Images

Emperor penguins walking in a line, Weddell Sea, Antarctica

Graphs are used in many ways. There is only one temperature that yields the same number in degrees Celsius and degrees Fahrenheit. Do you know what it is? The penguins are a clue.

In this chapter, we will review the Cartesian plane. We'll calculate the distance and midpoint between two points. We will then plot graphs of equations. We focus on graphing two types of equations: lines and circles.

Graphs

Cartesian Plane

- Coordinates/Axes
- Distance/Midpoint

Graphing Equations

- Point Plotting
- Symmetry

Common Graphs

- Lines
- Circles

⊙ CHAPTER OBJECTIVES

- Plot points in the Cartesian plane and algebraically determine the distance and the midpoint between two points.
- Graph equations by point plotting.
- Determine if a graph has symmetry.
- Graph lines and circles.
- Determine equations of lines and circles.

NAVIGATION THROUGH SUPPLEMENTS

DIGITAL VIDEO SERIES #2

STUDENT SOLUTIONS MANUAL CHAPTER 2

BOOK COMPANION SITE
www.wiley.com/college/young

Skills Objectives

- Plot points in the Cartesian plane.
- Calculate the distance between two points.
- Calculate the midpoint between two points.

Conceptual Objectives

- Expand the concept of a one-dimensional number line to a two-dimensional plane.

Cartesian Plane

HIV infection rates, circular orbits, stock prices, and temperature conversions are all examples of relationships between two quantities that can be expressed in a two-dimensional graph, which is called a **plane**.

The **axes** are two perpendicular real number lines in the plane that intersect at their origins. Typically, the horizontal axis is called the **x-axis** and the vertical axis is denoted as the **y-axis**. The point where the axes intersect is called the **origin**. The axes divide the plane into four **quadrants**, numbered by roman numerals and ordered counterclockwise.

Each point in the plane is called an **ordered pair**, denoted (x, y). The first number of the ordered pair indicates the position in the horizontal direction and is often called the x-coordinate or **abscissa**. The second number indicates the position in the vertical direction and is often called the y-coordinate or **ordinate**. The origin is denoted $(0, 0)$. Examples of other coordinates are given on the graph below:

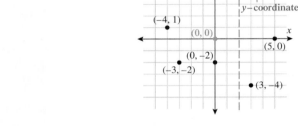

The point $(2, 4)$ lies in quadrant I. To **plot** this point, start at the origin $(0, 0)$ and move two units to the right and four units up.

All points in quadrant I have positive coordinates and all points in quadrant III have negative coordinates. Quadrant II has negative x-coordinates and positive y-coordinates; quadrant IV has positive x-coordinates and negative y-coordinates.

This representation is called the **rectangular coordinate system** or **Cartesian coordinate system**, named after the French mathematician René Descartes.

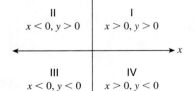

EXAMPLE 1 Plotting Points in a Cartesian Plane

a. Plot and label the points $(-1, -4)$, $(2, 2)$, $(-2, 3)$, $(2, -3)$, $(0, 5)$, and $(-3, 0)$ in the Cartesian plane.

b. List the points and corresponding quadrant or axis in a table.

Solution:

a.

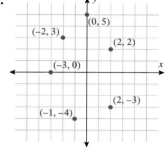

b.

Point	Quadrant
$(2, 2)$	I
$(-2, 3)$	II
$(-1, -4)$	III
$(2, -3)$	IV
$(0, 5)$	y-axis
$(-3, 0)$	x-axis

Distance Between Two Points

Suppose you want to find the distance between any two points in the plane. In the previous graph, to find the distance between the points $(2, -3)$ and $(2, 2)$, count the units between the two points. The distance is 5. What if the two points do not lie along a horizontal or vertical line? Example 2 uses the Pythagorean theorem to help find the distance between any two points.

EXAMPLE 2 Finding the Distance Between Two Points

Find the distance between the points $(-2, -1)$ and $(1, 3)$.

Solution:

STEP 1 Plot and label the two points in the Cartesian plane and draw a segment indicating the distance, d, between the two points.

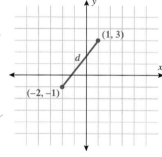

STEP 2 Form a right triangle by connecting the points to a third point, $(1, -1)$.

STEP 3 Calculate the length of the horizontal segment. $3 = |1 - (-2)|$
Calculate the length of the vertical segment. $4 = |3 - (-1)|$

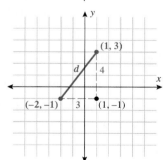

STEP 4 Use the Pythagorean theorem to calculate the distance, d.

$$d^2 = 3^2 + 4^2$$
$$d^2 = 9 + 16 = 25$$
$$d = 5$$

WORDS

MATH

For any two points,
(x_1, y_1) and (x_2, y_2).

The distance along the horizontal
segment is the absolute value.

$|x_2 - x_1|$

The distance along the vertical
segment is the absolute value.

$|y_2 - y_1|$

Use the Pythagorean theorem to
calculate the distance, d.

$d^2 = |x_2 - x_1|^2 + |y_2 - y_1|^2$

$|a|^2 = a^2$ for all real numbers a.

$d^2 = (x_2 - x_1)^2 + (y_2 - y_1)^2$

Use the square root property.

$d = \pm\sqrt{(x_2 - x_1)^2 + (y_2 - y_1)^2}$

Distance can only be positive.

$d = \sqrt{(x_2 - x_1)^2 + (y_2 - y_1)^2}$

DISTANCE FORMULA

The **distance**, d, between two points $P_1 = (x_1, y_1)$ and $P_2 = (x_2, y_2)$ is given by

$$d = \sqrt{(x_2 - x_1)^2 + (y_2 - y_1)^2}$$

The distance between two points is the square root of the distance between the x-coordinates squared plus the distance between the y-coordinates squared.

Note: It does not matter which point is taken to be the first point or the second point.

EXAMPLE 3 **Using the Distance Formula to Find the Distance Between Two Points**

Find the distance between $(-3, 7)$ and $(5, -2)$.

Solution:

Write the distance formula.

$d = \sqrt{[x_2 - x_1]^2 + [y_2 - y_1]^2}$

Substitute $(x_1, y_1) = (-3, 7)$
and $(x_2, y_2) = (5, -2)$.

$d = \sqrt{[5 - (-3)]^2 + [-2 - 7]^2}$

Simplify.

$d = \sqrt{[5 + 3]^2 + [-2 - 7]^2}$

$d = \sqrt{8^2 + (-9)^2} = \sqrt{64 + 81} = \sqrt{145}$

Solve for d.

$d = \sqrt{145}$

YOUR TURN Find the distance between $(4, -5)$ and $(-3, -2)$.

Midpoint of a Segment Joining Two Points

The **midpoint**, (x, y), of a segment connecting two points (x_1, y_1) and (x_2, y_2) is defined as the point that lies on the segment which has the same distance, d, from both points.

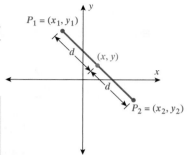

In other words, the midpoint of a segment lies halfway between the given endpoints. The coordinates of the midpoint are found by *averaging* the x-coordinates and averaging the y-coordinates.

> **MIDPOINT FORMULA**
>
> The **midpoint**, (x, y), between two points (x_1, y_1) and (x_2, y_2) is given by
>
> $$(x, y) = \left(\frac{x_1 + x_2}{2}, \frac{y_1 + y_2}{2} \right)$$
>
> *The midpoint can be found by averaging the x-coordinates and averaging the y-coordinates.*

 EXAMPLE 4 Finding the Midpoint of a Segment

Find the midpoint of the segment joining points $(2, 6)$ and $(-4, -2)$.

Solution:

Write the midpoint formula.

$$(x, y) = \left(\frac{x_1 + x_2}{2}, \frac{y_1 + y_2}{2} \right)$$

Substitute $(x_1, y_1) = (2, 6)$ and $(x_2, y_2) = (-4, -2)$.

$$(x, y) = \left(\frac{2 + (-4)}{2}, \frac{6 + (-2)}{2} \right)$$

Simplify.

$$(x, y) = (-1, 2)$$

One way to check your answer is to plot the points and midpoint to make sure your answer looks reasonable.

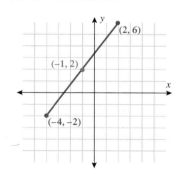

■ **Answer:** $d = \sqrt{58}$

CONCEPT CHECK What quadrant would you expect the midpoint of the segment joining the points $(3, -4)$ and $(5, 8)$ to lie in?

■ **YOUR TURN** Compute the midpoint of the segment joining points $(3, -4)$ and $(5, 8)$.

SECTION 2.1 SUMMARY

In this section we discussed point plotting in a plane. Formulas for the distance and midpoint between two points were developed.

Cartesian Plane

- Plotting coordinates: (x, y)
- Quadrants: I, II, III, and IV

Distance Between Two Points

$$d = \sqrt{(x_2 - x_1)^2 + (y_2 - y_1)^2}$$

Midpoint of Segment Joining Two Points

$$\text{Midpoint} = (x, y) = \left(\frac{x_1 + x_2}{2}, \frac{y_1 + y_2}{2} \right)$$

SECTION 2.1 EXERCISES

■ **SKILLS**

In Exercises 1–6, give the coordinates for each point labeled.

1. Point A
2. Point B
3. Point C
4. Point D
5. Point E
6. Point F

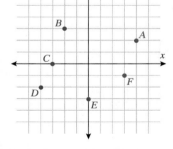

In Exercises 7 and 8, plot each point in the Cartesian plane and indicate in which quadrant or on which axis the point lies.

7. A: $(-2, 3)$ B: $(1, 4)$ C: $(-3, -3)$ D: $(5, -1)$ E: $(0, -2)$ F: $(4, 0)$

8. A: $(-1, 2)$ B: $(1, 3)$ C: $(-4, -1)$ D: $(3, -2)$ E: $(0, 5)$ F: $(-3, 0)$

■ **Answer:** Midpoint $= (4, 2)$

9. Plot the points $(-3, 1), (-3, 4), (-3, -2), (-3, 0), (-3, -4)$. Describe the line containing points of the form $(-3, y)$.

10. Plot the points $(-1, 2), (-3, 2), (0, 2), (3, 2), (5, 2)$. Describe the line containing points of the form $(x, 2)$.

In Exercises 11–24, calculate the distance and midpoint between the segment joining the given points.

11. $(1, 3)$ and $(5, 3)$

12. $(-2, 4)$ and $(-2, -4)$

13. $(-1, 4)$ and $(3, 0)$

14. $(-3, -1)$ and $(1, 3)$

15. $(-10, 8)$ and $(-7, -1)$

16. $(-2, 12)$ and $(7, 15)$

17. $(-\frac{1}{2}, \frac{1}{3})$ and $(\frac{7}{2}, \frac{10}{3})$

18. $(\frac{1}{5}, \frac{7}{3})$ and $(\frac{9}{5}, -\frac{2}{3})$

19. $(-\frac{2}{3}, -\frac{1}{5})$ and $(\frac{1}{4}, \frac{1}{3})$

20. $(\frac{7}{5}, \frac{1}{9})$ and $(\frac{1}{2}, -\frac{7}{3})$

21. $(-1.5, 3.2)$ and $(2.1, 4.7)$

22. $(-1.2, -2.5)$ and $(3.7, 4.6)$

23. $(-14.2, 15.1)$ and $(16.3, -17.5)$

24. $(1.1, 2.2)$ and $(3.3, 4.4)$

In Exercises 25 and 26, calculate (to two decimal places) the perimeter of the triangle with the following vertices:

25. Points A, B, and C

26. Points C, D, and E

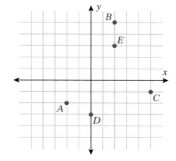

In Exercises 27–30, determine if the triangle with the given vertices is a right triangle, isosceles triangle, neither, or both. (Recall that a right triangle satisfies the Pythagorean theorem and an isosceles triangle has at least two sides of equal length.)

27. $(0, -3), (3, -3)$, and $(3, 5)$

28. $(0, 2), (-2, -2)$, and $(2, -2)$

29. $(1, 1), (3, -1)$, and $(-2, -4)$

30. $(-3, 3), (3, 3)$, and $(-3, -3)$

 ■ **APPLICATIONS**

31. Cell Phones. A cellular phone company currently has three towers: one in Tampa, one in Orlando, and one in Gainesville to serve the central Florida region. If Orlando is 80 miles east of Tampa and Gainesville is 100 miles north of Tampa, what is the distance from Orlando to Gainesville?

32. Cell Phones. The same cellular phone company in Exercise 31 has decided to add additional towers "halfway" between each city. How many miles from Tampa is each "halfway" tower?

33. Travel. A retired couple who lives in Columbia, South Carolina, decides to take their motor home and visit two children who live in Atlanta and in Savannah, Georgia. Savannah is 160 miles south of Columbia and Atlanta is 215 miles west of Columbia. How far apart do the children live from each other?

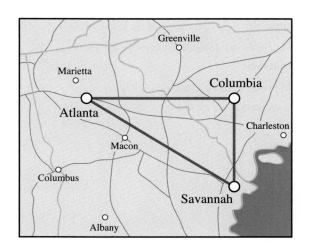

34. Sports. In the 1984 Orange Bowl, Doug Flutie, the 5 foot 9 inch quarterback for Boston College, shocked the world as he threw a "hail Mary" pass that was caught in the end zone with no time left on the clock, defeating the Miami Hurricanes 47–45. Although the record books have it listed as a 48 yard pass, what was the actual distance the ball was thrown?

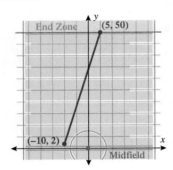

35. NASCAR Revenue. Action Performance Inc., the leading seller of NASCAR merchandise, recorded $260 million in revenue in 2002 and $400 million in revenue in 2004. Calculate the midpoint to estimate the revenue Action Performance Inc. recorded in 2003.

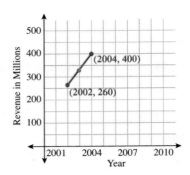

36. Ticket Price. In 1993 a Miami Dolphins average ticket price was $28 and in 2001 the average price was $56. Find the midpoint of the segment joining these two points to estimate the ticket price in 1997.

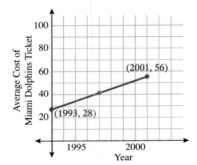

■ **CATCH THE MISTAKE**

In Exercises 37–40, explain the mistake that is made.

37. Calculate the distance between $(2, 7)$ and $(9, 10)$.

Solution:

Write the distance formula. $d = \sqrt{(x_2 - x_1)^2 + (y_2 - y_1)^2}$

Substitute $(2, 7)$ and $(9, 10)$. $d = \sqrt{(7 - 2)^2 + (10 - 9)^2}$

Simplify. $d = \sqrt{(5)^2 + (1)^2} = \sqrt{26}$

38. Calculate the distance between $(-2, 1)$ and $(3, -7)$.

Solution:

Write the distance formula. $d = \sqrt{(x_2 - x_1)^2 + (y_2 - y_1)^2}$

Substitute $(-2, 1)$ and $(3, -7)$. $d = \sqrt{(3 - 2)^2 + (-7 - 1)^2}$

Simplify. $d = \sqrt{(1)^2 + (-8)^2} = \sqrt{65}$

39. Compute the midpoint between the points $(-3, 4)$ and $(7, 9)$.

Solution:

Write the midpoint formula. $(x, y) = \left(\dfrac{x_1 + x_2}{2}, \dfrac{y_1 + y_2}{2} \right)$

Substitute $(-3, 4)$ and $(7, 9)$. $(x, y) = \left(\dfrac{-3 + 4}{2}, \dfrac{7 + 9}{2} \right)$

Simplify. $(x, y) = \left(\dfrac{1}{2}, \dfrac{16}{2} \right) = \left(\dfrac{1}{2}, 4 \right)$

40. Compute the midpoint between the points $(-1, -2)$ and $(-3, -4)$.

Solution:

Write the midpoint formula. $(x, y) = \left(\dfrac{x_1 - x_2}{2}, \dfrac{y_1 - y_2}{2} \right)$

Substitute $(-1, -2)$ and $(-3, -4)$. $(x, y) = \left(\dfrac{-1 - (-3)}{2}, \dfrac{-2 - (-4)}{2} \right)$

Simplify. $(x, y) = (1, 1)$

■ CHALLENGE

41. T or F: The distance from the origin to the point (a, b) is $d = \sqrt{a^2 + b^2}$.

42. T or F: The midpoint between the origin and the point (a, a) is $(\frac{a}{2}, \frac{a}{2})$.

43. T or F: The midpoint of any segment joining two points in quadrant I also lies in quadrant I.

44. T or F: The midpoint of any segment joining a point in quadrant I to a point in quadrant 3 also lies in either quadrant I or quadrant III.

45. Calculate the length and the midpoint for the segment joining the points (a, b) and (b, a).

46. Calculate the length and the midpoint for the segment joining the points (a, b) and $(-a, -b)$.

47. Assume two points, (x_1, y_1) and (x_2, y_2), are connected by a segment. Prove that the distance from the midpoint of the segment to either of the two points is the same.

48. Prove that the diagonals of a parallelogram in the figure intersect at their midpoints.

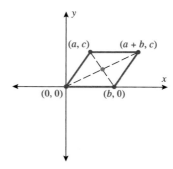

■ TECHNOLOGY

In Exercises 49–52, calculate the distance between the two points. Use a graphing utility to graph the segment joining the two points and find the length of the segment.

49. $(-2.3, 4.1)$ and $(3.7, 6.2)$

50. $(-4.9, -3.2)$ and $(5.2, 3.4)$

51. $(1.1, 2.2)$ and $(3.3, 4.4)$

52. $(-1.3, 7.2)$ and $(2.3, -4.5)$

SECTION 2.2 Graphing Equations: Point Plotting and Symmetry

Skills Objectives

■ Plot equations by plotting points.
■ Conduct a test for symmetry about x-axis, y-axis, and origin.

Conceptual Objectives

■ Relate symmetry graphically and algebraically.

We will learn how to graph equations in this section using point plotting. Later we will learn other techniques. All equations can be graphed by plotting points. However, as we discuss graphing aids in Chapter 3, you will see that other techniques can be more efficient.

Point Plotting

An equation in two variables, such as $y = x^2$, has an infinite number of ordered pairs which are its solutions. For example, $(0, 0)$ is a solution to $y = x^2$ because when $x = 0$ and $y = 0$ the equation is true. Two other solutions are $(-1, 1)$ and $(1, 1)$.

The **graph of an equation** in two variables, x and y, consists of all the points in the xy-plane whose coordinates (x, y) satisfy the equation. A procedure for plotting the graphs of equations is outlined below and is illustrated with the above example, $y = x^2$.

WORDS

MATH

Step 1: In a table, list several pairs of coordinates that make the equation true.

x	$y = x^2$	(x, y)
0	0	$(0, 0)$
-1	1	$(-1, 1)$
1	1	$(1, 1)$
-2	4	$(-2, 4)$
2	4	$(2, 4)$

Step 2: Plot these points on a graph and connect the points with a smooth curve. Use arrows to indicate that the graph continues.

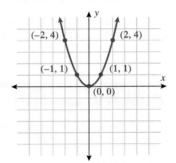

EXAMPLE 1 **Graphing an Equation of a Line by Plotting Points**

Graph the equation $y = 2x - 1$.

Solution:

STEP 1 In a table, list several pairs of coordinates that make the equation true.

x	$y = 2x - 1$	(x, y)
0	-1	$(0, -1)$
-1	-3	$(-1, -3)$
1	1	$(1, 1)$
-2	-5	$(-2, -5)$
2	3	$(2, 3)$

STEP 2 Plot these points on a graph and connect the points, resulting in a line. Arrows indicate that the graph continues.

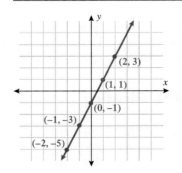

CONCEPT CHECK How many points are necessary to draw a straight line?

YOUR TURN The graph of the equation $y = -x + 1$ is a line. Graph the line.

EXAMPLE 2 Graphing an Equation of a Parabola by Plotting Points

Graph the equation $y = x^2 - 5$.

Solution:

STEP 1 In a table, list several pairs of coordinates that make the equation true.

x	$y = x^2 - 5$	(x, y)
0	-5	$(0, -5)$
-1	-4	$(-1, -4)$
1	-4	$(1, -4)$
-2	-1	$(-2, -1)$
2	-1	$(2, -1)$
-3	4	$(-3, 4)$
3	4	$(3, 4)$

STEP 2 Plot these points on a graph and connect the points with a smooth curve, indicating with arrows that the curve continues.

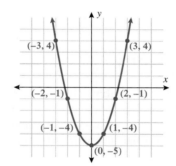

This graph is called a *parabola* and will be discussed in further detail in Chapter 5.

YOUR TURN Graph $x = y^2 - 1$.

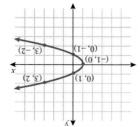

Answer:

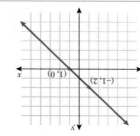

Answer:

EXAMPLE 3 Graphing an Equation by Plotting Points

Graph the equation $y = x^3$.

Solution:

STEP 1 In a table, list several pairs of coordinates in a table that satisfy the equation.

x	$y = x^3$	(x, y)
0	0	(0, 0)
-1	-1	$(-1, -1)$
1	1	(1, 1)
-2	-8	$(-2, -8)$
2	8	(2, 8)

STEP 2 Plot these points on a graph and connect the points with a smooth curve, indicating with arrows that the curve continues in both the positive and negative directions.

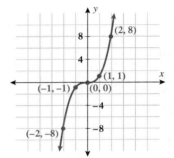

Symmetry

The word **symmetry** conveys balance. Suppose you have two pictures to hang on a wall. If you space them equally apart on the wall, then you prefer a symmetric décor. This is an example of symmetry about a line. The word (water) written below is identical if you rotate the word 180 degrees. This is an example of symmetry about a point.

Symmetric graphs have the characteristic that their mirror image can be obtained about a reference, typically a line or a point.

In Example 2, the points $(-2, -1)$ and $(2, -1)$ both lie on the graph as do the points $(-1, -4)$ and $(1, -4)$. Notice that the graph on the right side of the y-axis is a mirror image of the part of the graph to the left of the *y*-axis. This graph illustrates *symmetry* with respect to the *y-axis* (the line $x = 0$).

In the *Your Turn* following Example 2, the points $(0, 1)$ and $(0, -1)$ both lie on the graph as do the points $(3, 2)$ and $(3, -2)$. Notice that the part of the graph above the *x*-axis is a mirror image of the part of the graph below the *x*-axis. This graph illustrates *symmetry* with respect to the *x-axis* (the line $y = 0$).

In Example 3, the points $(-1, -1)$ and $(1, 1)$ both lie on the graph. Notice that rotating this graph 180 degrees results in an identical graph. This is an example of *symmetry* with respect to the *origin* $(0, 0)$.

Symmetry aids in graphing by giving information "for free." For example, if a graph is symmetric about the *y*-axis, then once the graph to the right of the *y*-axis is found, the left side of the graph is the mirror image of that. If a graph is symmetric about the origin, then once the graph is known in quadrant I the graph in quadrant III is found by rotating the known graph 180 degrees.

It would be beneficial to know if a graph of an equation is symmetric about a line or point before the equation is plotted. Although a graph can be symmetric about any line or point, we will only discuss symmetry about the *x*-axis, *y*-axis, or origin. These types of symmetry and the algebraic procedure for testing for symmetry are outlined in the box below.

Types and Tests for Symmetry

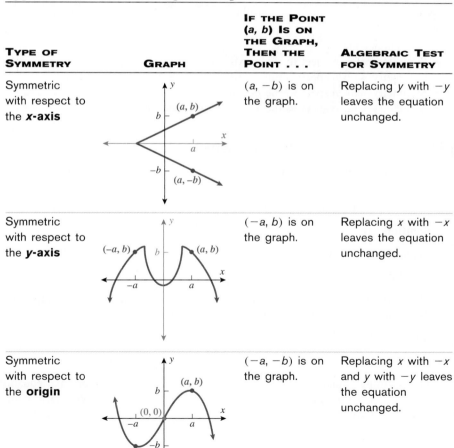

TYPE OF SYMMETRY	GRAPH	IF THE POINT (*a, b*) IS ON THE GRAPH, THEN THE POINT . . .	ALGEBRAIC TEST FOR SYMMETRY
Symmetric with respect to the **x-axis**		$(a, -b)$ is on the graph.	Replacing *y* with $-y$ leaves the equation unchanged.
Symmetric with respect to the **y-axis**		$(-a, b)$ is on the graph.	Replacing *x* with $-x$ leaves the equation unchanged.
Symmetric with respect to the **origin**		$(-a, -b)$ is on the graph.	Replacing *x* with $-x$ and *y* with $-y$ leaves the equation unchanged.

STUDY TIP

Symmetry gives us information about the graph for "free."

TECHNOLOGY TIP
To enter the graph of $y^2 = x^3$, solve for y first. The graphs of $y_1 = \sqrt{x^3}$ and $y_2 = -\sqrt{x^3}$ are shown.

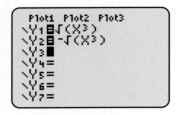

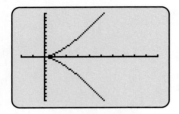

EXAMPLE 4 **Testing for Symmetry with Respect to the Axes**

Test the equation $y^2 = x^3$ for symmetry with respect to the axes.

Solution:

Test for symmetry with respect to the x-axis.

Replace y with $-y$. $(-y)^2 = x^3$

Simplify. $y^2 = x^3$

The resulting equation is the same as the original equation, $y^2 = x^3$. Therefore, $y^2 = x^3$ is **symmetric with respect to the x-axis**.

Test for symmetry with respect to the y-axis.

Replace x with $-x$. $y^2 = (-x)^3$

Simplify. $y^2 = -x^3$

The resulting equation, $y^2 = -x^3$, is not the same as the original equation, $y^2 = x^3$. Therefore, $y^2 = x^3$ is **not** symmetric with respect to the y-axis.

When testing for symmetry about the x-axis, y-axis, or origin, there are *five* possibilities:

- No symmetry
- Symmetry with respect to the x-axis
- Symmetry with respect to the y-axis
- Symmetry with respect to the origin
- Symmetry with respect to the x-axis, y-axis, and origin

EXAMPLE 5 **Testing for Symmetry**

Determine what type of symmetry (if any) the graphs of the equations exhibit:

a. $y = x^2 + 1$ **b.** $y = x^3 + 1$

Solution (a):

Replace x with $-x$. $y = (-x)^2 + 1$

Simplify. $y = x^2 + 1$

The resulting equation is equivalent to the original equation, so the graph of the equation $y = x^2 + 1$ is symmetric with respect to the y-axis.

Replace y with $-y$. $(-y) = x^2 + 1$

Simplify. $y = -x^2 - 1$

The resulting equation, $y = -x^2 - 1$, is not equivalent to the original equation, $y = x^2 + 1$, so the graph of the equation, $y = x^2 + 1$, is not symmetric with respect to the x-axis.

Replace x with $-x$ and y with $-y$. $(-y) = (-x)^2 + 1$

Simplify. $-y = x^2 + 1$

                                                     $y = -x^2 - 1$

The resulting equation, $y = -x^2 - 1$, is not equivalent to the original equation, $y = x^2 + 1$, so the graph of the equation, $y = x^2 + 1$, is not symmetric with respect to the origin.

The equation $y = x^2 + 1$ is **symmetric with respect to the y-axis** .

Solution (b):

Replace x with $-x$. $\qquad\qquad\qquad y = (-x)^3 + 1$

Simplify. $\qquad\qquad\qquad\qquad\qquad y = -x^3 + 1$

The resulting equation, $y = -x^3 + 1$, is not equivalent to the original equation, $y = x^3 + 1$. Therefore, the graph of the equation, $y = x^3 + 1$, is not symmetric with respect to the y-axis.

Replace y with $-y$. $\qquad\qquad\qquad (-y) = x^3 + 1$

Simplify. $\qquad\qquad\qquad\qquad\qquad y = -x^3 - 1$

The resulting equation, $y = -x^3 - 1$, is not equivalent to the original equation, $y = x^3 + 1$. Therefore, the graph of the equation, $y = x^3 + 1$, is not symmetric with respect to the x-axis.

Replace x with $-x$ and y with $-y$. $\qquad (-y) = (-x)^3 + 1$

Simplify. $\qquad\qquad\qquad\qquad\qquad -y = -x^3 + 1$

$$y = x^3 - 1$$

The resulting equation, $y = x^3 - 1$, is not equivalent to the original equation, $y = x^3 + 1$. Therefore, the graph of the equation, $y = x^3 + 1$, is not symmetric with respect to the origin.

The equation $y = x^3 + 1$ exhibits **no symmetry** .

■ **YOUR TURN** Determine the symmetry (if any) for $x = y^2 - 1$.

Using Symmetry as a Graphing Aid

How can we use symmetry to assist us in graphing? Look back at Example 2, $y = x^2 - 5$. We selected seven x-coordinates and solved the equation to find the corresponding y-coordinates. If we had known that this graph was symmetric with respect to the y-axis, then we would have only had to find the solutions to the positive x-coordinates, since we get the negative x-coordinates for free. For example, we found the point $(1, -4)$ to be a solution to the equation. The rules of symmetry tell us that $(-1, -4)$ is also on the graph.

EXAMPLE 6 Using Symmetry as a Graphing Aid

For the equation, $x^2 + y^2 = 25$, use symmetry to help you graph the equation using the point plotting technique.

Solution:

Test for symmetry with respect to the y-axis.

Replace x with $-x$. $\qquad$ $(-x)^2 + y^2 = 25$

Simplify. $\qquad$ $x^2 + y^2 = 25$

The resulting equation is equivalent to the original, so the graph of $x^2 + y^2 = 25$ is symmetric with respect to the y-axis.

Test for symmetry with respect to the x-axis.

Replace y with $-y$. $\qquad$ $x^2 + (-y)^2 = 25$

Simplify. $\qquad$ $x^2 + y^2 = 25$

The resulting equation is equivalent to the original, so the graph of $x^2 + y^2 = 25$ is symmetric with respect to the x-axis.

Test for symmetry with respect to the origin.

Replace x with $-x$ and y with $-y$. $\qquad$ $(-x)^2 + (-y)^2 = 25$

Simplify. $\qquad$ $x^2 + y^2 = 25$

The resulting equation is equivalent to the original, so the graph of $x^2 + y^2 = 25$ is symmetric with respect to the origin.

Since the graph is symmetric with respect to the y-axis, x-axis, and origin, we need to determine solutions to the equation on only the positive x- and y-axes and in quadrant I because of the following symmetries:

- Symmetry with respect to the y-axis gives the solutions in quadrant II.
- Symmetry with respect to the origin gives the solutions in quadrant III.
- Symmetry with respect to the x-axis yields solutions in quadrant IV.

Solutions to $x^2 + y^2 = 25$.

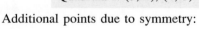

Axes: $(0, 5)$ and $(5, 0)$

Quadrant I: $(3, 4), (4, 3)$

Additional points due to symmetry:

Axes: $(0, -5)$ and $(-5, 0)$

Quadrant II: $(-3, 4), (-4, 3)$

Quadrant III: $(-3, -4), (-4, -3)$

Quadrant IV: $(3, -4), (4, -3)$

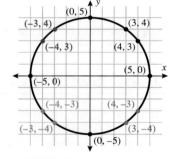

Connecting the points with a smooth curve yields a **circle**. We will discuss circles in more detail in Section 2.4.

TECHNOLOGY TIP

To enter the graph of $x^2 + y^2 = 25$, solve for y first. The graphs of
$y_1 = \sqrt{25 - x^2}$ and
$y_2 = -\sqrt{25 - x^2}$ are shown.

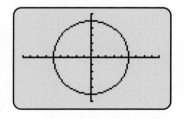

SECTION 2.2 **SUMMARY**

In this section you graphed equations using a point plotting technique, tested equations for symmetry, and used symmetry about the x-axis, y-axis, and origin as a graphing aid.

SECTION 2.2 EXERCISES

■ **SKILLS**

In Exercises 1–4, determine whether each point lies on the graph of the equation.

1. $y = x^2 - 2x + 1$ **a.** $(-1, 4)$ **b.** $(0, -1)$ **2.** $y = x^3 - 1$ **a.** $(-1, 0)$ **b.** $(-2, -9)$

3. $y = \sqrt{x} + 2$ **a.** $(7, 3)$ **b.** $(-6, 4)$ **4.** $y = 2 + |3 - x|$ **a.** $(9, -4)$ **b.** $(-2, 7)$

In Exercises 5–8, complete the table and use the table to sketch a graph of the equation.

5.

x	$y = 2 + x$	(x, y)
-2		
0		
1		

6.

x	$y = 3x - 1$	(x, y)
-1		
0		
2		

7.

x	$y = x^2 - x$	(x, y)
-1		
0		
$\frac{1}{2}$		
1		
2		

8.

x	$y = -\sqrt{x + 2}$	(x, y)
-2		
-1		
2		
7		

In Exercises 9–14, graph the equation by plotting points.

9. $y = -3x + 2$ **10.** $y = x^2 - x - 2$ **11.** $x = y^2 - 1$

12. $x = |y + 1| + 2$ **13.** $y = \frac{1}{2}x - \frac{3}{2}$ **14.** $y = 0.5|x - 1|$

In Exercises 15–20, match the graph with the corresponding symmetry.

a. No symmetry   **b.** Symmetry with respect to the x-axis   **c.** Symmetry with respect to the y-axis
d. Symmetry with respect to the origin   **e.** Symmetry with respect to the x-axis, y-axis, and origin

15.

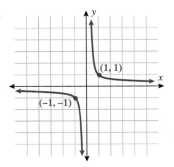

16.

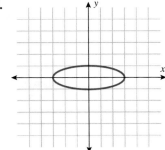

17.

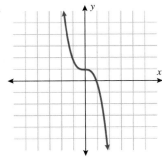

18. 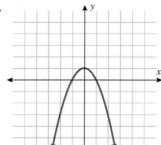   **19.** 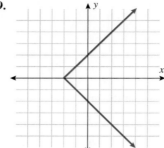   **20.**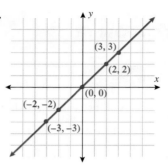

In Exercises 21–26, a point that lies on a graph is given along with that graph's symmetry. State the other points that must also lie on the graph.

POINT ON A GRAPH	THE GRAPH IS SYMMETRIC ABOUT THE	POINT ON A GRAPH	THE GRAPH IS SYMMETRIC ABOUT THE
21. $(-1, 3)$	x-axis	**22.** $(-2, 4)$	y-axis
23. $(7, -10)$	origin	**24.** $(-1, -1)$	origin
25. $(3, -2)$	x-axis, y-axis, and origin	**26.** $(-1, 7)$	x-axis, y-axis, and origin

In Exercises 27–40, use algebraic tests to determine whether the equation's graph is symmetric with respect to the x-axis, y-axis, or origin.

27. $x = y^2 + 4$

28. $x = 2y^2 + 3$

29. $y = x^3 + x$

30. $y = x^5 + 1$

31. $x = |y|$

32. $x = |y| - 2$

33. $x^2 - y^2 = 100$

34. $x^2 + 2y^2 = 30$

35. $y = x^{2/3}$

36. $x = y^{2/3}$

37. $x^2 + y^3 = 1$

38. $y = \sqrt{1 + x^2}$

39. $y = \dfrac{2}{x}$

40. $xy = 1$

In Exercises 41–52, use symmetry to help you graph the given equations.

41. $y = x$

42. $y = x^2 - 1$

43. $y = \dfrac{x^3}{2}$

44. $x = y^2 + 1$

45. $y = \dfrac{1}{x}$

46. $xy = -1$

47. $y = |x|$

48. $|x| = |y|$

49. $x^2 + y^2 = 16$

50. $\dfrac{x^2}{4} + \dfrac{y^2}{9} = 1$

51. $x^2 - y^2 = 16$

52. $x^2 - \dfrac{y^2}{25} = 1$

 ■ APPLICATIONS

53. Bomb. A particular bomb has a destruction pattern governed by the equation $x^2 + y^2 = 9$ where the xy-plane represents a town with the origin $(0, 0)$ as the town center. Use symmetry as a graphing aid to draw a picture of the destruction area.

54. Sprinkler. A sprinkler will water grass in the shape of $x^2 + \frac{y^2}{9} = 1$. Use symmetry to draw the watered area assuming the sprinkler is located at the origin.

55. Signals. The received power of an electromagnetic signal is a fraction of the power transmitted. The relationship is given by

$$P_{\text{received}} = P_{\text{transmitted}} \cdot \dfrac{1}{R^2}$$

where R is the distance that the signal has traveled in meters. Plot the points showing the percentage of

transmitted power that is received for $R = 100$ m, 1 km, and 10,000 km.

56. Signals. The wavelength, λ, and the frequency, f, of a signal are related by the equation

$$f = \dfrac{c}{\lambda}$$

where c is the speed of light in a vacuum, $c = 3.0 \times 10^8$ meters per second. For the values, $\lambda = 0.001, \lambda = 1$, and $\lambda = 100$ mm, plot the points corresponding to frequency, f. What do you notice about the relationship between frequency and wavelength? Note that the frequency will have units Hz = 1/seconds.

■ CATCH THE MISTAKE

In Exercises 57–60, explain the mistake that is made.

57. Graph the equation $y = x^2 + 1$.

Solution:

x	$y = x^2 + 1$	(x, y)
0	1	(0, 1)
1	2	(1, 2)

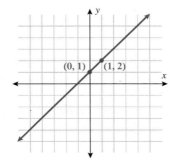

58. Test $y = -x^2$ for symmetry with respect to the y-axis.

Solution:

Replace x with $-x$. $y = -(-x)^2$

Simplify. $y = x^2$

The resulting equation, $y = x^2$, is not equivalent to the original equation. $y = -x^2$ is not symmetric with respect to the y-axis.

59. Test $x = |y|$ for symmetry with respect to the y-axis.

Solution:

Replace y with $-y$. $x = |-y|$

Simplify. $x = |y|$

The resulting equation is equivalent to the original equation. $x = |y|$ is symmetric with respect to the y-axis.

60. Use symmetry to help you graph $x^2 = y - 1$.

Solution:

Replace x with $-x$. $(-x)^2 = y - 1$

Simplify. $x^2 = y - 1$

$x^2 = y - 1$ is symmetric with respect to the x-axis.

Determine points that lie on the graph in quadrant I.

y	$x^2 = y - 1$	(x, y)
1	0	(0, 1)
2	1	(1, 2)
5	2	(2, 5)

Symmetry with respect to the x-axis implies that $(0, -1)$, $(1, -2)$, and $(2, -5)$ are also points that lie on the graph.

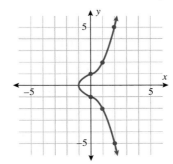

■ CHALLENGE

61. T or F: If the point (a, b) lies on a graph that is symmetric about the x-axis, then the point $(-a, b)$ also must lie on the graph.

62. T or F: If the point (a, b) lies on a graph that is symmetric about the y-axis, then the point $(-a, b)$ also must lie on the graph.

63. T or F: If the point $(a, -b)$ lies on a graph that is symmetric about the x-axis, y-axis, and origin, then the points (a, b), $(-a, -b)$ and $(-a, b)$ must also lie on the graph.

64. Determine if the graph of $y = \dfrac{ax^2 + b}{cx^3}$ has any symmetry, where a, b, and c are real numbers.

■ TECHNOLOGY

In Exercises 65–68, graph the equation using a graphing utility and state if there is any symmetry.

65. $y = 16.7x^4 - 3.3x^2 + 7.1$ **66.** $y = 0.4x^5 + 8.2x^3 - 1.3x$ **67.** $2.3x^2 = 5.5\,|y|$ **68.** $3.2x^2 - 5.1y^2 = 1.3$

What is the shortest path between two points? The answer is a *straight line*. In this section we will discuss characteristics of lines such as slope and intercepts. We will also discuss types of lines such as horizontal, vertical, falling, and rising, and recognize relationships between lines such as perpendicular and parallel. At the end of this section you should be able to find the equation of a line given two specific pieces of information about the line.

Graphing a Line

First-degree equations in two variables, such as

$$y = -2x + 4, \qquad 3x + y = 6, \qquad y = 2, \qquad \text{and} \qquad x = -3,$$

have graphs that are straight lines. One way of writing an equation of a straight line is called *standard form*.

EQUATION OF A STRAIGHT LINE: STANDARD FORM

If A, B, and C are constants and x and y are variables, then the equation $Ax + By = C$ is in **standard form** and its graph is a straight line.
Note: A or B (but not both) can be zero.

The equation $2x - y = -2$ is a first-degree equation in standard form so its graph is a straight line. To graph this line, list two solutions in a table, plot those points, and use a straight edge to draw the **line.**

x	y	(x, y)
-2	-2	$(-2, -2)$
1	4	$(1, 4)$

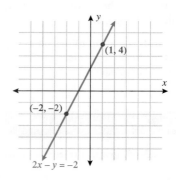

Intercepts

The point where the line crosses, or intersects, the x-axis is called the **x-intercept**. The point where the line crosses, or intersects, the y-axis is called the **y-intercept**. By inspecting the graph of the previous line, the x-**intercept** is $(-1, 0)$ and the y-**intercept** is $(0, 2)$.

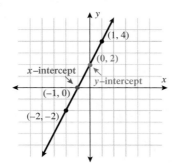

Since the x-axis corresponds to $y = 0$, the algebraic way of finding the x-intercept is to set $y = 0$ in the equation and solve for x. If we set $y = 0$ in the equation $2x - y = -2$, we find that $2x = -2$ or $x = -1$. Therefore, the coordinates of the x-**intercept** are $(-1, 0)$, which is what we found by inspecting the graph above.

Similarly, the y-axis corresponds to $x = 0$. Therefore, the algebraic procedure for finding the y-intercept is to set $x = 0$ in the equation and solve for y. If we set $x = 0$ in the equation $2x - y = -2$, we find that $-y = -2$ or $y = 2$. Therefore, the coordinates of the y-**intercept** are $(0, 2)$, which is what we found by inspecting the graph above.

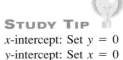

STUDY TIP

x-intercept: Set $y = 0$
y-intercept: Set $x = 0$

DETERMINING INTERCEPTS

	COORDINATES	AXIS CROSSED	ALGEBRAIC METHOD
x-intercept:	$(a, 0)$	x-axis	Set $y = 0$ and solve for x.
y-intercept:	$(0, b)$	y-axis	Set $x = 0$ and solve for y.

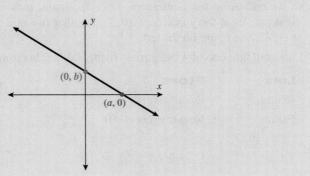

Slope

If the graph of $2x - y = -2$ represented an incline that you were about to walk on, would you classify that incline as steep? In the language of mathematics, we use the word **slope** as a measure of steepness. Slope is the ratio of the change in y over the change in x. An easy way to remember this is *rise over run.*

SLOPE OF A LINE

A line passing through two points (x_1, y_1) and (x_2, y_2) has slope, m, given by the formula

$$m = \frac{y_2 - y_1}{x_2 - x_1}, \text{ where } x_1 \neq x_2 \quad m = \frac{\text{rise}}{\text{run}} = \frac{\text{vertical change}}{\text{horizontal change}}$$

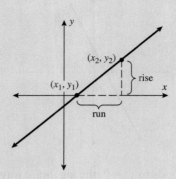

Let's find the slope of our graph of $2x - y = -2$. We'll let $(x_1, y_1) = (-2, -2)$ and $(x_2, y_2) = (1, 4)$ in the slope formula:

$$m = \frac{y_2 - y_1}{x_2 - x_1} = \frac{(4 - (-2))}{(1 - (-2))} = \frac{6}{3} = 2$$

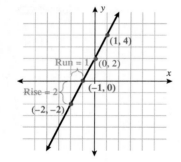

Notice that if we had chosen the two intercepts instead, $(x_1, y_1) = (0, 2)$ and $(x_2, y_2) = (-1, 0)$, we still would have found the slope to be $m = 2$.

When interpreting slope, always read the graph from *left to right*. Since we have determined the slope to be 2, or $\frac{2}{1}$, we can interpret this as rising 2 units and running (to the right) 1 unit. If we start at the point $(-2, -2)$ and move 2 units up and 1 unit to the right, we end up at the x-intercept, $(-1, 0)$. Again, moving 2 units up and 1 unit to the right puts us at the y-intercept, $(0, 2)$. Another rise of 2 and run of 1 takes us to the point $(1, 4)$. See figure on the left.

Lines fall into one of 4 categories: rising, falling, horizontal, or vertical.

LINE	SLOPE
Rising	Positive $(m > 0)$
Falling	Negative $(m < 0)$
Horizontal	Zero $(m = 0)$ hence $y = b$
Vertical	Undefined $\quad x = a$

The slope of a horizontal line is 0 because the y coordinates of any two points are the same. The change in y in the slope formula's numerator is 0, hence $m = 0$.

The slope of a vertical line is undefined because the x-coordinates of any two points are the same. The change in x in the slope formula's denominator is zero; hence m is undefined.

Slope–Intercept Form

As mentioned earlier, the standard form for an equation of a line is $Ax + By = C$. A more convenient way to write an equation of a line is in slope–intercept form because it identifies the slope and the y-intercept.

> **EQUATION OF A STRAIGHT LINE: SLOPE–INTERCEPT FORM**
>
> The slope–intercept form for the equation of a line is
> $$y = mx + b$$
> Its graph has slope m and y-intercept $(0, b)$.

For example, $2x - y = -2$ is in **standard form**. To write this equation in **slope–intercept form**, we isolate the y variable:
$$y = 2x + 2$$

Recall that the **slope** of this line is **2** and the **y-intercept** is $(0, 2)$.

We began this section by starting with an equation, plotting its corresponding line, and defining its slope and intercepts by looking at the graph. We could have proceeded another way by starting with the equation of the line in slope–intercept form, noting its slope and intercepts, and drawing its graph.

Equation of the line in slope–intercept form: $y = 2x + 2$

The slope is 2 and the y-intercept is $(0, 2)$. Therefore, if we start at the y-intercept and go up 2 and over 1, we find the point $(1, 4)$. We can label that point and draw the line.

The following examples will reinforce the concepts of slope and intercepts.

EXAMPLE 1   Determining *x*- and *y*-intercepts

Determine the x- and y-intercepts (if they exist) for the lines given by the following equations:

a. $2x + 4y = 10$ **b.** $x = -2$

Sometimes it is easier to plot the line and graphically determine the intercepts, and other times it is more convenient to apply the algebraic approach. In this example the algebraic approach will be used for the first equation and the graphing approach will be used for the second equation.

Solution (a): $2x + 4y = 10$

To find the x-intercept, set $y = 0$. $2x + 4(0) = 10$

Solve for x. $2x = 10$
$$x = 5$$

The x-intercept is $(5, 0)$.

To find the y-intercept, set $x = 0$. $2(0) + 4y = 10$

Solve for y. $4y = 10$

$$y = \frac{10}{4} = \frac{5}{2}$$

The y-intercept is $\left(0, \dfrac{5}{2}\right)$.

Solution (b): $x = -2$

This vertical line consists of all points $(-2, y)$.

The graph shows that the x-intercept is $(-2, 0)$.

We also find that the line never crosses the y-axis, so the y-intercept does not exist.

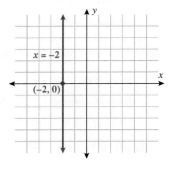

CONCEPT CHECK Does a horizontal line have any x-intercepts or y-intercepts?

■ **YOUR TURN** Determine the x- and y-intercepts (if they exist) for the lines given by the following equations:

a. $3x - y = 2$ **b.** $y = 5$

EXAMPLE 2 Graph, Classify the Line, and Determine the Slope

Sketch a line through each pair of points, classify the line as rising, falling, vertical, or horizontal, and determine its slope.

a. $(-1, -3)$ and $(1, 1)$ **b.** $(3, 1)$ and $(-3, 3)$

c. $(-1, -2)$ and $(3, -2)$ **d.** $(1, -4)$ and $(1, 3)$

Solution (a): $(-1, -3)$ and $(1, 1)$

This line is rising, so its slope is positive.

$$m = \frac{1 - (-3)}{1 - (-1)} = \frac{4}{2} = \frac{2}{1} = 2$$

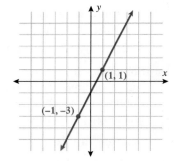

Solution (b): $(3, 1)$ and $(-3, 3)$

This line is falling, so its slope is negative.

$$m = \frac{1 - 3}{3 - (-3)} = \frac{-2}{6} = -\frac{1}{3}$$

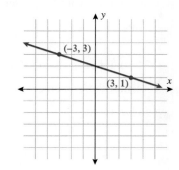

■ **Answer: a.** x-intercept $\left(\frac{2}{3}, 0\right)$ **b.** x-intercept does not exist

y-intercept $(0, -2)$ y-intercept $(0, 5)$

Solution (c): $(-1, -2)$ and $(3, -2)$

This is a horizontal line, so its slope is zero.

$$m = \frac{-2 - (-2)}{3 - (-1)} = \frac{0}{4} = 0$$

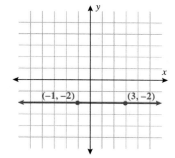

Solution (d): $(1, -4)$ and $(1, 3)$

This is a vertical line, so its slope is undefined.

$$m = \frac{3 - (-4)}{1 - 1} = \frac{7}{0}, \text{ which is undefined.}$$

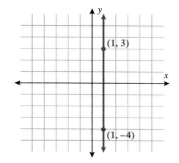

YOUR TURN For each pair of points classify the line that passes through them as rising, falling, vertical, or horizontal, and determine its slope. Do not graph.

a. $(2, 0)$ and $(1, 5)$ **b.** $(-2, -3)$ and $(2, 5)$
c. $(-3, -1)$ and $(-3, 4)$ **d.** $(-1, 2)$ and $(3, 2)$

COMMON MISTAKE

The most common mistake in calculating slope is writing the coordinates in the wrong order, which results in the slope being opposite in sign.
Find the slope of the line containing the two points $(1, 2)$ and $(3, 4)$.

 CORRECT

Label the points.
$(x_1, y_1) = (1, 2)$ $(x_2, y_2) = (3, 4)$

Write the slope formula.
$$m = \frac{y_2 - y_1}{x_2 - x_1}$$

 INCORRECT

Label the points.
$(x_1, y_1) = (1, 2)$ $(x_2, y_2) = (3, 4)$

Apply the slope formula (**ERROR**).
$$m = \frac{4 - 2}{1 - 3}$$

 CAUTION
Interchanging the coordinates can result in a sign error in the slope.

Answer: a. $m = -5$, falling   **b.** $m = 2$, rising   **c.** slope is undefined, vertical   **d.** $m = 0$, horizontal

Substitute the coordinates.

$$m = \frac{4 - 2}{3 - 1}$$

Simplify. $m = \frac{2}{2} = 1$

$$m = 1$$

The calculated slope is **INCORRECT** by a negative sign.

$$m = \frac{2}{-2} = -1$$

Finding Equations of Lines

Instead of starting with equations of lines and characterizing them, let us now start with particular features of a line and derive its governing equation. Suppose that you are given the y-intercept and the slope of a line. Using the slope–intercept form of an equation of a line, $y = mx + b$, you could find its equation.

 EXAMPLE 3 Using Slope–Intercept Form to Find the Equation of a Line

Find the equation of a line that has slope $\frac{2}{3}$ and y-intercept $(0, 1)$.

Solution:

Write the slope–intercept form of an equation of a line. $y = mx + b$

Label the slope. $m = \frac{2}{3}$

Label the y-intercept. $b = 1$

The equation of the line in slope–intercept form is $y = \frac{2}{3}x + 1$.

■ YOUR TURN Find the equation of the line that has slope $-\frac{3}{2}$ and y-intercept $(0, 2)$.

Suppose now, that the two pieces of information you are given about an equation are its slope and one point that lies on it. You still have enough information to write an equation of the line. Recall the formula for slope:

$$m = \frac{y_2 - y_1}{x_2 - x_1}, \qquad \text{where } x_2 \neq x_1$$

We are given the slope, m, and we know a particular point that lies on the line, (x_1, y_1). All other points that lie on the line we refer to as (x, y).

Substituting these values into the slope formula gives us

$$m = \frac{y - y_1}{x - x_1}$$

■ Answer: $y = -\frac{3}{2}x + 2$

Cross multiplying yields

$$y - y_1 = m(x - x_1)$$

This is called the *point–slope form* of an equation of a line.

EQUATION OF A STRAIGHT LINE: POINT–SLOPE FORM

The point–slope form for the equation of a line is

$$y - y_1 = m(x - x_1)$$

Its graph passes through the point (x_1, y_1), and its slope is m.

Note: This formula does not hold for vertical lines since their slope is undefined.

EXAMPLE 4   Using Point–Slope Form to Find the Equation of the Line

Find the equation of the line that has slope $-\frac{1}{2}$ and passes through the point $(-1, 2)$.

Solution:

Write the point–slope form of an equation of a line. $y - y_1 = m(x - x_1)$

Substitute the values $m = -\dfrac{1}{2}$ and $(x_1, y_1) = (-1, 2)$. $y - 2 = -\dfrac{1}{2}(x - (-1))$

Eliminate parentheses. $y - 2 = -\dfrac{1}{2}x - \dfrac{1}{2}$

Isolate y. $y = -\dfrac{1}{2}x + \dfrac{3}{2}$

Note that we did not have to use the point–slope formula in Example 4. We could have started with the slope–intercept form, $y = mx + b$, and substituted $m = -\frac{1}{2}$, giving us $y = -\frac{1}{2}x + b$. We still need the y-intercept. We do know, however, that the line passes through $(-1, 2)$, which means that $x = -1$ and $y = 2$ is a solution to the equation. We will substitute these values into the equation, $2 = -\frac{1}{2}(-1) + b$, and solve for b, $b = \frac{3}{2}$. Therefore, the equation of the line is $y = -\frac{1}{2}x + \frac{3}{2}$, which is the same as the equation we found in Example 2 using the point–slope formula.

STUDY TIP

This alternative method always works when given the slope and a point lying on the line, which eliminates the need to memorize the point–slope formula.

YOUR TURN   Derive the equation of the line that has slope $\frac{1}{4}$ and passes through the point $(1, -\frac{1}{2})$. Give the answer in slope–intercept form.

Suppose the slope of a line is not given at all. Instead, two points that lie on the line are given. If we know two points that lie on the line, then we can calculate the slope. Then, using the slope and *either* of the two points, the equation of the line can be derived.

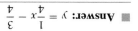 **Answer:** $y = \dfrac{1}{4}x - \dfrac{3}{4}$

EXAMPLE 5 Finding the Equation of a Line Given Two Points

Find the equation of the line that passes through the points $(-2, -1)$ and $(3, 2)$.

Solution:

STEP 1 **Write the equation of a line.**

$$y = mx + b$$

STEP 2 **Calculate the slope.**

$$m = \frac{y_2 - y_1}{x_2 - x_1}$$

Substitute $(x_1, y_1) = (-2, -1)$ and $(x_2, y_2) = (3, 2)$.

$$m = \frac{2 - (-1)}{3 - (-2)} = \frac{3}{5}$$

STEP 3 **Substitute $\frac{3}{5}$ for the slope.**

$$y = \frac{3}{5}x + b$$

STEP 4 **Let $(x, y) = (3, 2)$.**

(Either point satisfies the equation.)

$$2 = \frac{3}{5}(3) + b$$

STEP 5 **Solve for b.**

$$b = \frac{1}{5}$$

Write the equation in slope–intercept form $\quad y = \frac{3}{5}x + \frac{1}{5}$.

■ **YOUR TURN** Find the equation of the line that passes through the points $(-1, 3)$ and $(2, -4)$.

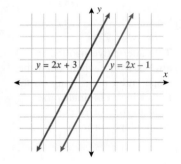

Parallel and Perpendicular Lines

Two nonintersecting lines in a plane are *parallel*. How can we tell whether the two lines in the graph on the left are parallel? Parallel lines must have the same steepness. It turns out that parallel lines must have the same slopes. The two lines shown on the left are parallel because they have the same slope, 2.

> **PARALLEL LINES**
>
> Two lines in a plane are **parallel** if and only if their slopes are equal.

In other words, if two lines in a plane are parallel, then their slopes are equal, and if the slopes of two lines in a plane are equal, then the lines are parallel.

WORDS	MATH
Lines L_1 and L_2 are parallel.	$L_1 \| L_2$
Two parallel lines have the same slope.	$m_1 = m_2$

■ **Answer:** $y = -\frac{7}{3}x + \frac{2}{3}$ ■

EXAMPLE 6 Determining If Two Lines Are Parallel

Determine if the lines $-x + 3y = -3$ and $y = \frac{1}{3}x - 6$ are parallel.

Solution:

Write the first line in slope–intercept form. $-x + 3y = -3$

  Add x to both sides. $3y = x - 3$

  Divide by 3. $y = \dfrac{1}{3}x - 1$

Compare the two lines. $y = \dfrac{1}{3}x - 1$ and $y = \dfrac{1}{3}x - 6$

Both lines have the same slope, $\frac{1}{3}$. Thus,   the two lines are parallel .

EXAMPLE 7 Finding an Equation of a Parallel Line

Find the equation of the line that passes through the point $(1, 1)$ and is parallel to the line $y = 3x + 1$.

Solution:

Write the slope–intercept equation of a line. $y = mx + b$

Parallel lines have equal slope. $m = 3$

Substitute the slope into the equation of the line. $y = 3x + b$

Since the line passes through $(1, 1)$, this point
must satisfy the equation. $1 = 3 + b$

Solve for b. $b = -2$

The equation of the line is $y = 3x - 2$.

■ **YOUR TURN**   Find the equation of the line parallel to $y = 2x - 1$ that passes through the point $(-1, 3)$.

Two *perpendicular* lines form a right angle at their point of intersection. Notice the slopes of the two perpendicular lines in the figure to the right. They are $-\frac{1}{2}$ and 2, negative reciprocals of each other. It turns out that almost all perpendicular lines share this property. Horizontal ($m = 0$) and vertical (m undefined) lines do not share this property.

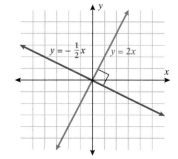

PERPENDICULAR LINES

Except for vertical and horizontal lines, two lines in a plane are **perpendicular** if and only if their slopes are negative reciprocals of each other.

In other words, if two lines in a plane are perpendicular, their slopes are negative reciprocals, and if the slopes of two lines in a plane are negative reciprocals, then the lines are perpendicular, provided their slopes are defined and are not zero.

■ **Answer:** $y = 2x + 5$

WORDS	MATH
Lines L_1 and L_2 are perpendicular.	$L_1 \perp L_2$
Two perpendicular lines have negative reciprocal slopes. For example, if a line has slope equal to 3, then a line perpendicular to it has slope $-\frac{1}{3}$.	$m_1 = -\dfrac{1}{m_2}$ $m_1 \neq 0, m_2 \neq 0$

EXAMPLE 8 Finding an Equation of a Line That Is Perpendicular to Another Line

Write in slope–intercept form the equation of a line perpendicular to the line with the equation $-x + 4y = -12$.

Solution:

STEP 1 Rewrite the given equation in slope–intercept form.

$$-x + 4y = -12$$
$$4y = x - 12$$
$$y = \frac{1}{4}x - 3$$

STEP 2 Identify the slope of the given line.

$$m_2 = \frac{1}{4}$$

STEP 3 Find the slope of a perpendicular line, which is the negative reciprocal of $\frac{1}{4}$.

$$m_1 = -\frac{1}{m_2} = -\frac{1}{\frac{1}{4}} = -4$$

STEP 4 Write the slope–intercept form for the equation of a perpendicular line with slope -4.

$$y = -4x + b$$

EXAMPLE 9 Finding an Equation of a Particular Perpendicular Line

Find the equation of the line that passes through the point $(3, 0)$ and is perpendicular to the line $y = 3x + 1$.

STEP 1 **Write the equation of a line in slope–intercept form.**

$$y = mx + b$$

STEP 2 **Determine the slope of the desired line.**
Slope of the given line is 3, and perpendicular lines have negative reciprocal slopes.

$$m = -\frac{1}{3}$$

STEP 3 **Substitute the slope into the equation of the line.**

$$y = -\frac{1}{3}x + b$$

STEP 4 **Find b.**
Since the desired line passes through $(3, 0)$, this point must satisfy the equation.

$$0 = -\frac{1}{3}(3) + b$$
$$0 = -1 + b$$

Solve for b.

$$b = 1$$

The equation of the line is $y = -\frac{1}{3}x + 1$.

CONCEPT CHECK   All lines perpendicular to the line $y = -\frac{1}{2}x + 4$ have what slope?

■ **YOUR TURN**   Find the equation of the line that passes through the point $(1, -5)$ and is perpendicular to the line $y = -\frac{1}{2}x + 4$.

Applications Involving Linear Equations

Slope is the ratio of the change in y over the change in x. In applications slope can often be interpreted as the **rate of change**, as illustrated in the next example.

EXAMPLE 10   Slope as a Rate of Change

The average age of a person when they first get married has been increasing over the last several decades. In 1970 the average age was 20 and in 1990 it was 25 years old, and in 2010 it is expected that the average age will be 30 years old at the time of one's first marriage. Find the slope of the line passing through these points. Describe what that slope represents.

Solution:

If we let x represent the year and y represent the age, then two points that lie on the line are $(1970, 20)$ and $(2010, 30)$.

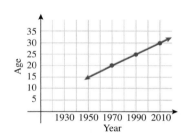

Write the slope formula.

$$m = \frac{\text{change in } y}{\text{change in } x}$$

Substitute the points into the slope formula.

$$m = \frac{30 - 20}{2010 - 1970} = \frac{10}{40} = \frac{1}{4}$$

The slope is $\frac{1}{4}$ and can be interpreted as the rate of change of the average age when a person is first married.   Every 4 years the average age at the first marriage is 1 year older .

In Example 10 we chose to use the points $(1970, 20)$ and $(2010, 30)$. We could have also used the point $(1990, 25)$ with either of the other points.

EXAMPLE 11   Service Charges

Suppose that your two neighbors both use the same electrician. One neighbor had a 2-hour job, which cost her $100 and another neighbor had a 3-hour job that cost him $130. Assuming a linear equation governs the service charge of this electrician, what will your cost be for a 5-hour job?

■ **Answer:** $y = 2x - 7$

Solution:

STEP 1 **Identify the question.**

Determine the linear equation for this electrician's service charge and calculate the charge for a 5-hour job.

STEP 2 **Make notes.**

3-hour job costs $130, and 2-hour job costs $100

STEP 3 **Set up an equation.**

Let x equal the number of hours and y equal the service charge in dollars.

Linear equation: $y = mx + b$

Two points that must satisfy this equation are (3, 130) and (2, 100).

STEP 4 **Solve the equation.**

Calculate the slope. $m = \dfrac{130 - 100}{3 - 2} = \dfrac{30}{1} = 30$

Substitute slope into linear equation. $y = 30x + b$

Either point must satisfy
the equation. $100 = 30(2) + b$

We'll use (2, 100). $100 = 60 + b$

$b = 40$

The service charge, y, is given by $y = 30x + 40$.

Substitute $x = 5$ into this equation
for a 5-hour job. $y = 30(5) + 40 = 190$

The 5-hour job will cost $190 .

STEP 5 **Check the solution.**

The service charge, $y = 30x + 40$, can be interpreted as a $40 charge for coming to your home and a $30 per hour fee for the job. A 5-hour job would cost $190. Additionally, a 5-hour job should cost less than the sum of a 2-hour job and a 3-hour job ($100 + $130 = $230) since the $40 fee is only charged once.

■ **YOUR TURN** You decide to hire a tutor that some of your friends recommended. The tutor comes to your home, so she charges a flat fee per session and then an hourly rate. One friend prefers 2-hour sessions, and the charge is $60 per session. Another friend has 5-hour sessions that cost $105 per session. How much should you be charged for a 3-hour session?

SECTION 2.3 SUMMARY

In this section we graphed lines by plotting points and connecting with a straight edge and then using slope and intercepts. We then were able to find the equation of a line given any two pieces of information (slope and a point, or two points). The most convenient way of expressing the equation of a line is in slope–intercept form: $y = mx + b$. Parallel lines have the same slope, and perpendicular lines have slopes that are negative reciprocals. Application problems were solved using linear equations.

SECTION 2.3 EXERCISES

■ **SKILLS**

In Exercises 1–8, find the slope of the line that passes through the given points.

1. $(1, 3)$ and $(2, 6)$ **2.** $(2, 1)$ and $(4, 9)$ **3.** $(-2, 5)$ and $(2, -3)$ **4.** $(-1, -4)$ and $(4, 6)$

5. $(-7, 9)$ and $(3, -10)$ **6.** $(11, -3)$ and $(-2, 6)$ **7.** $(0.2, -1.7)$ and $(3.1, 5.2)$ **8.** $\left(\dfrac{1}{2}, \dfrac{3}{5}\right)$ and $\left(-\dfrac{3}{4}, \dfrac{7}{5}\right)$

For each graph in Exercises 9–14, identify (by inspection) the x- and y-intercepts and slope if they exist, and classify the line as rising, falling, horizontal, or vertical.

9.

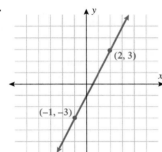

10.

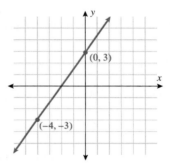

11.

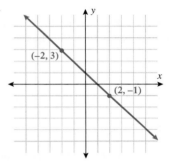

12.

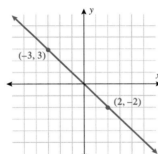

13.

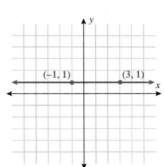

14.

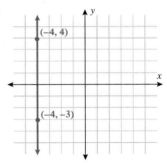

In Exercises 15–28, find the x- and y-intercepts and slope if they exist and graph the corresponding line.

15. $y = 2x - 3$ **16.** $y = -3x + 2$ **17.** $y = -\dfrac{1}{2}x + 2$ **18.** $y = \dfrac{1}{3}x - 1$

19. $2x - 3y = 4$ **20.** $-x + y = -1$ **21.** $\dfrac{1}{2}x + \dfrac{1}{2}y = -1$ **22.** $\dfrac{1}{3}x - \dfrac{1}{4}y = \dfrac{1}{12}$

23. $x = -1$ **24.** $y = -3$ **25.** $y = 1.5$ **26.** $x = -7.5$

27. $x = -\dfrac{7}{2}$ **28.** $y = \dfrac{5}{3}$

In Exercises 29–36, write the equation of the line given the slope and intercept.

29. Slope: $m = 2$ **30.** Slope: $m = -2$ **31.** Slope: $m = -\dfrac{1}{3}$ **32.** Slope: $m = \dfrac{1}{2}$

     y-intercept: $(0, 3)$ y-intercept: $(0, 1)$ y-intercept: $(0, 0)$ y-intercept: $(0, -3)$

33. Slope: $m = 0$

y-intercept: $(0, 2)$

34. Slope: $m = 0$

y-intercept: $(0, -1.5)$

35. Slope: undefined

x-intercept: $\left(\dfrac{3}{2}, 0\right)$

36. Slope: undefined

x-intercept: $(-3.5, 0)$

In Exercises 37–46, write an equation of the line in slope–intercept form given the slope and a point that lies on the line.

37. Slope: $m = 5$
$(-1, -3)$

38. Slope: $m = 2$
$(1, -1)$

39. Slope: $m = -3$
$(-2, 2)$

40. Slope: $m = -1$
$(3, -4)$

41. Slope: $m = \dfrac{3}{4}$

$(1, -1)$

42. Slope: $m = -\dfrac{1}{7}$

$(-5, 3)$

43. Slope: $m = 0$

$(-2, 4)$

44. Slope: $m = 0$

$(3, -3)$

45. Slope: undefined
$(-1, 4)$

46. Slope: undefined
$(4, -1)$

In Exercises 47–58, write the equation of the line that passes through the given points. Express the equation in slope–intercept form or in the form $x = 0$ or $y = b$.

47. $(-2, -1)$ and $(3, 2)$

48. $(-4, -3)$ and $(5, 1)$

49. $(-1, 4)$ and $(2, -5)$

50. $(-2, 3)$ and $(2, -3)$

51. $\left(\dfrac{1}{2}, \dfrac{3}{4}\right)$ and $\left(\dfrac{3}{2}, \dfrac{9}{4}\right)$

52. $\left(-\dfrac{2}{3}, -\dfrac{1}{2}\right)$ and $\left(\dfrac{7}{3}, \dfrac{1}{2}\right)$

53. $(3, 5)$ and $(3, -7)$

54. $(-5, -2)$ and $(-5, 4)$

55. $(3, 7)$ and $(9, 7)$

56. $(-2, -1)$ and $(3, -1)$

57. $(0, 6)$ and $(-5, 0)$

58. $(0, -3)$ and $(0, 2)$

In Exercises 59–68, find the equation of the line that passes through the given point and also satisfies the additional piece of information. Express your answer in slope–intercept form.

59. $(-3, 1)$; parallel to the line $y = 2x - 1$

60. $(1, 3)$; parallel to the line $y = -x + 2$

61. $(0, 0)$; perpendicular to the line $2x + 3y = 12$

62. $(0, 6)$; perpendicular to the line $x - y = 7$

63. $(3, 5)$; parallel to the x-axis

64. $(3, 5)$; parallel to the y-axis

65. $(-1, 2)$; perpendicular to the y-axis

66. $(-1, 2)$; perpendicular to the x-axis

67. $(-2, -7)$; parallel to the line $\dfrac{1}{2}x - \dfrac{1}{3}y = 5$

68. $(1, 4)$; perpendicular to the line $-\dfrac{2}{3}x + \dfrac{3}{2}y = -2$

 ■ **APPLICATIONS**

69. Budget. The cost of having your bathroom remodeled is the combination of material costs and labor costs. The materials (tile, grout, toilet, fixtures, etc.) cost is $1,200 and the labor cost is $25 per hour. Write an equation that models the total cost, C, of having your bathroom remodeled as a function of hours, h. How much will the job cost if the workman estimates 32 hours?

70. Budget. The cost of a one day car rental is the sum of the rental fee, $50, plus $.39 per mile. Write an equation that models the total cost associated with the car rental.

71. Budget. The monthly costs associated with driving a new Honda Accord is the monthly loan payment plus $25 every time you fill up with gasoline. If you fill up 5 times in a month, your total monthly cost is $500. How much is your loan payment?

72. Budget. The monthly costs associated with driving a Ford Explorer are the monthly loan payment plus the cost of filling up your tank with gasoline. If you fill up 3 times in a month, your total monthly cost is $520. If you fill up 5 times in a month, your total monthly cost is $600. How much is your monthly loan, and how much does it cost every time you fill up with gasoline?

73. Temperature. The National Oceanic and Atmospheric Administration (NOAA) has an online conversion chart that relates degrees Fahrenheit, °F, to degrees Celsius, °C. 77° F is equivalent to 25° C, and 68° F is equivalent to 20° C. Assuming the relationship is linear, write the equation relating degrees Celsius, C, to degrees Fahrenheit, F. What temperature is the same in both degrees Celsius and degrees Fahrenheit?

74. Temperature. According to NOAA, a "standard day" is 15 °C at sea level, and every 500 feet elevation above sea level corresponds to a 1 °C temperature drop. Assuming the relationship between temperature and elevation is linear, write an equation that models this relationship. What is the expected temperature at 2500 feet on a "standard day?"

75. Height. The average height of a man has increased over the last century.

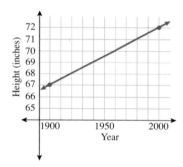

What is the rate of change in inches per year of the average height of men?

76. Height. The average height of a woman has increased over the last century.

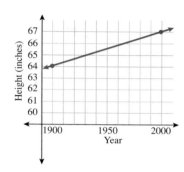

What is the rate of change in inches per year of the average height of women?

77. Weight. The average weight of a baby born in 1900 was 6 pounds 4 ounces. In 2000, the average weight of a newborn was 6 pounds 10 ounces. What is the rate of change of baby weight in ounces per year? What do we expect babies to weigh at birth in 2040?

78. Sports. The fastest a man could run a mile in 1906 was 4 minutes and 30 seconds. In 1957, Don Bowden became the first American to break the 4-minute mile. Calculate the rate of change in mile speed per year.

■ **CATCH THE MISTAKE**

In Exercises 79–82, explain the mistake that is made.

79. Find the x- and y-intercepts of the line with equation $2x - 3y = 6$.

Solution:

x-intercept: set $x = 0$ and solve for y. $\qquad -3y = 6$
$\qquad\qquad\qquad\qquad\qquad\qquad\qquad\qquad\quad y = -2$

The x-intercept is $(0, -2)$.

y-intercept: set $y = 0$ and solve for x. $\qquad 2x = 6$
$\qquad\qquad\qquad\qquad\qquad\qquad\qquad\qquad\quad x = 3$

The y-intercept is $(3, 0)$.

80. Find the slope of the line that passes through the points $(-2, 3)$ and $(4, 1)$.

Solution:

Write the slope formula. $\qquad m = \dfrac{y_2 - y_1}{x_2 - x_1}$

Substitute $(-2, 3)$ and $(4, 1)$. $\quad m = \dfrac{1 - 3}{-2 - 4} = \dfrac{-2}{-6} = \dfrac{1}{3}$

81. Find the slope of the line that passes through the points $(-3, 4)$ and $(-3, 7)$.

Solution:

Write the slope formula.

$$m = \frac{y_2 - y_1}{x_2 - x_1}$$

Substitute $(-3, 4)$ and $(-3, 7)$.

$$m = \frac{-3 - (-3)}{4 - 7} = 0$$

82. Given the slope, classify the line as rising, falling, horizontal, or vertical.

a. $m = 0$ **b.** m undefined

c. $m = 2$ **d.** $m = -1$

Solution:

a. vertical line   **b.** horizontal line

c. rising   **d.** falling

 **CHALLENGE**

83. T or F: A line can have at most one x-intercept.

84. T or F: A line must have at least one y-intercept.

85. T or F: If the slopes of two lines are $-\frac{1}{5}$ and 5, then the lines are parallel.

86. T or F: If the slopes of two lines are -1 and 1, then the lines are perpendicular.

87. If a line has slope zero, describe a line that is perpendicular to it.

88. If a line has no slope, describe a line that is parallel to it.

89. Find an equation of a line that passes through the point $(-B, A + 1)$ and is parallel to the line $Ax + By = C$.

90. Find an equation of a line that passes through the point $(A, B + 1)$ and is perpendicular to the line $Ax + By = C$.

 TECHNOLOGY

For Exercises 91–94, determine whether the lines are parallel, perpendicular, or neither, and then graph both lines in the same viewing screen using a graphing utility to confirm your answer.

91. $y_1 = 17x + 22$

$y_2 = -\frac{1}{17}x - 13$

92. $y_1 = 0.35x + 2.7$

$y_2 = 0.35x - 1.2$

93. $y_1 = 0.25x + 3.3$

$y_2 = -4x + 2$

94. $y_1 = \frac{1}{2}x + 5$

$y_2 = 2x - 3$

SECTION 2.4 Circles

Skills Objectives

■ Identify the center and radius of a circle from the standard equation.
■ Graph a circle.
■ Transform equations of circles to the standard form by completing the square.

Conceptual Objectives

■ Understand algebraic and graphical representations of circles.

Standard Equation of a Circle

Most people understand the geometry of a circle. The goal in this section is to develop the equation of a circle.

DEFINITION   CIRCLE

A **circle** is the set of all points that are a fixed distance from a point, the **center**. The center, C, is typically denoted (h, k) and the fixed distance, or **radius**, r.

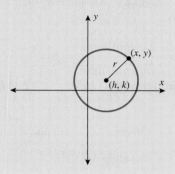

What is the equation of a circle? We'll use the distance formula from Section 2.1.

$$\text{Distance formula: } d = \sqrt{(x_2 - x_1)^2 + (y_2 - y_1)^2}$$

The distance between the center (h, k) and any point (x, y) on the circle is the radius, r. Substitute these values

$$d = r \qquad\qquad (x_1, y_1) = (h, k) \qquad\qquad (x_2, y_2) = (x, y)$$

into the distance formula

$$r = \sqrt{(x - h)^2 + (y - k)^2}$$

Square both sides.

$$(x - h)^2 + (y - k)^2 = r^2$$

All circles can be written in standard form, which makes it easy to identify the center and radius.

EQUATION OF A CIRCLE

The standard form of the equation of a **circle** with **radius** r and **center** (h, k) is

$$(x - h)^2 + (y - k)^2 = r^2$$

For the special case of a circle with center at the origin $(0, 0)$, the equation simplifies to

$$x^2 + y^2 = r^2$$

UNIT CIRCLE

A circle with radius 1 and center $(0, 0)$ is called the **unit circle**:

$$x^2 + y^2 = 1$$

Note that if $x^2 + y^2 = 0$, the radius is 0, so the "circle" is just a point.

TECHNOLOGY TIP

To enter the graph of
$(x - 2)^2 + (y + 1)^2 = 4$, solve
for y first. The graphs of
$y_1 = \sqrt{4 - (x - 2)^2} - 1$ and
$y_2 = -\sqrt{4 - (x - 2)^2} - 1$
are shown.

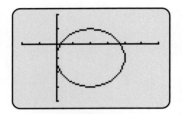

EXAMPLE 1 Finding the Center and Radius of a Circle

Identify the center and radius of the given circle and graph:

$$(x - 2)^2 + (y + 1)^2 = 4$$

Solution:

Rewrite this equation in standard form. $[x - 2]^2 + [y - (-1)]^2 = 2^2$

Identify h, k, and r by comparing this
equation to the standard form of a circle. $(x - h)^2 + (y - k)^2 = r^2$

$$h = 2, k = -1, \text{ and } r = 2$$

$$Center = (2, -1) \text{ and } r = 2$$

To draw the circle, label the center $(2, -1)$.

Label four additional points 2 units
(the radius) away from the center:

$$(4, -1), (0, -1), (2, 1), (2, -3)$$

Note that the easiest such 4 points to get are
those obtained by going out from the center
both horizontally and vertically.

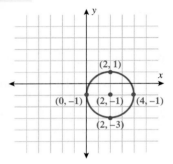

Connect those four points with a smooth curve.

■ YOUR TURN   Identify the center and radius of the given circle; graph:

$$(x + 1)^2 + (y + 2)^2 = 9$$

EXAMPLE 2 Graphing a Circle: Fractions and Radicals

Identify the center and radius of the given circle and sketch its graph.

$$\left(x - \frac{1}{2}\right)^2 + \left(y + \frac{1}{3}\right)^2 = 20$$

Solution:

In order to rewrite this equation in standard form, we must first compute the radius.
If $r^2 = 20$ then $r = \sqrt{20} = 2\sqrt{5}$.

Write the equation in standard form. $(x - h)^2 + (y - k)^2 = r^2$

$$\left(x - \frac{1}{2}\right)^2 + \left(y - \left(-\frac{1}{3}\right)\right)^2 = (2\sqrt{5})^2$$

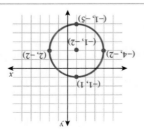

■ Answer: center $= (-1, -2)$
radius $= 3$

Identify the center and radius.

$$Center = \left(\frac{1}{2}, -\frac{1}{3}\right) \qquad r = 2\sqrt{5}$$

To graph the circle, we'll use decimal approximations of the fractions and radicals: $(0.5, -0.3)$ for the center and 4.5 for the radius. Four points on the circle that are 4.5 units from the center are $(-4, -0.3)$, $(5, -0.3)$, $(0.5, 4.2)$, and $(0.5, -4.8)$. Connect them with a smooth curve.

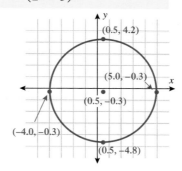

EXAMPLE 3 Determining the Equation of a Circle Given the Center and Radius

Find the equation of a circle with radius 5 and center $(-2, 3)$. Graph the circle.

Solution:

Substitute $(h, k) = (-2, 3)$ and $r = 5$ into the standard equation of a circle.

$$(x - (-2))^2 + (y - 3)^2 = 5^2$$

Simplify.

$$(x + 2)^2 + (y - 3)^2 = 25$$

To graph the circle, plot the center, $(-2, 3)$, and plot four points 5 units away from the center:

$(-7, 3)$, $(3, 3)$, $(-2, -2)$, and $(-2, 8)$

Connect them with a smooth curve.

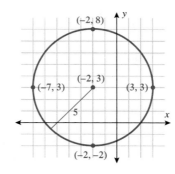

YOUR TURN Find the equation of a circle with radius 3 and center $(0, 1)$; graph.

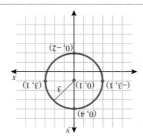

Answer: $x^2 + (y - 1)^2 = 9$

Let's change the look of the equation given in Example 1.

In Example 1 the equation of the circle was given as: $(x - 2)^2 + (y + 1)^2 = 4$

Eliminate the parentheses.

$$x^2 - 4x + 4 + y^2 + 2y + 1 = 4$$

Group like terms and subtract 4 from both sides.

$$x^2 - 4x + y^2 + 2y + 1 = 0$$

$$x^2 + y^2 - 4x + 2y + 1 = 0$$

We have written the *general form* of the equation of the circle in Example 1.

> The general form of the equation of a circle is
> $$x^2 + y^2 + ax + by + c = 0$$

Transforming Equations of Circles to the Standard Form by Completing the Square

If the equation of a circle is given in general form, it must be rewritten in standard form in order to identify its center and radius. To transform equations of circles from general to standard form, complete the square on both the x and y variables. We first discussed completing the square in Section 1.3.

 EXAMPLE 4 Finding the Center and Radius of a Circle by Completing the Square

Find the center and radius of the circle with equation:

$$x^2 - 8x + y^2 + 20y + 107 = 0$$

Solution:

Our goal is to transform this equation into standard form:

$$(x - h)^2 + (y - k)^2 = r^2$$

STEP 1 Group x and y terms, respectively, on the left side of the equation; move constants to the right side. $(x^2 - 8x) + (y^2 + 20y) = -107$

STEP 2 Complete the square on both the x and y expressions. $(x^2 - 8x + \_) + (y^2 + 20y + \_) = -107$

Add $\left(-\dfrac{8}{2}\right)^2 = 16$ and $\left(\dfrac{20}{2}\right)^2 = 100$ to both sides.

$$(x^2 - 8x + 16) + (y^2 + 20y + 100) = -107 + 16 + 100$$

Factor the perfect squares on the left and simplify the right. $(x - 4)^2 + (y + 10)^2 = 9$

STEP 3 Write in standard form. $(x - 4)^2 + (y - (-10))^2 = 3^2$

The center is $(4, -10)$ and the radius is 3 .

■ **YOUR TURN** Find the center and radius of the circle with equation:
$$x^2 + y^2 + 4x - 6y - 12 = 0$$

COMMON MISTAKE

A common mistake is forgetting to add *both* constants to the right side of the equation. Identify the center and radius of the circle with equation:
$$x^2 + y^2 + 16x + 8y + 44 = 0$$

 **CORRECT**

$$x^2 + y^2 + 16x + 8y + 44 = 0$$

$$(x^2 + 16x) + (y^2 + 8y) = -44$$

$$(x^2 + 16x + \_) + (y^2 + 8y + \_) = -44$$

$$(x^2 + 16x + 64) + (y^2 + 8y + 16)$$
$$= -44 + 64 + 16$$

$$(x^2 + 8)^2 + (y + 4)^2 = 36$$

center $= (-8, -4)$ radius $= 6$

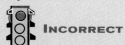 **INCORRECT**

$$x^2 + y^2 + 16x + 8y + 44 = 0$$

$$(x^2 + 16x) + (y^2 + 8y) = -44$$

$$(x^2 + 16x + \_) + (y^2 + 8y + \_) = -44$$

$$(x^2 + 16x + 64) + (y^2 + 8y + 16)$$
$$= -44 + 64$$

ERROR: Don't forget to add 16 to the right.

$$(x^2 + 8)^2 + (y + 4)^2 = 20 \text{ (Incorrect)}$$

center $= (-8, -4)$ radius $= 2\sqrt{5}$

(Radius is incorrect; the correct radius is 6.)

CAUTION
Don't forget to add *both* constants to each side of the equation when completing the square for x and y.

Finding the Equation of a Circle Given Its Center and a Point

Suppose you are given a point that lies on a circle and its center. Can you find the equation of the circle?

EXAMPLE 5 Finding the Equation of a Circle Given Its Center and One Point

The point $(10, -4)$ lies on a circle centered at $(7, -8)$. Find the equation of the circle in general form.

Solution:

This circle is centered at $(7, -8)$, so its standard equation is $(x - 7)^2 + (y + 8)^2 = r^2$.

Since the point $(10, -4)$ lies on the circle, it must satisfy the equation of the circle.

Substitute $(x, y) = (10, -4)$.

$$(10 - 7)^2 + (-4 + 8)^2 = r^2$$

$$r = \sqrt{3^2 + 4^2} = \sqrt{9 + 16} = \sqrt{25} = 5$$

■ **Answer:** center $= (-2, 3)$ radius $= 5$

Substitute $r = 5$ into the standard equation.	$(x - 7)^2 + (y + 8)^2 = 5^2$
Eliminate parentheses and simplify.	$x^2 - 14x + 49 + y^2 + 16y + 64 = 25$
Write in general form.	$x^2 + y^2 - 14x + 16y + 88 = 0$

SECTION 2.4 SUMMARY

In this section we have learned how to identify and graph circles by identifying the center, (h, k), and the radius, r, when the equation is in standard form: $(x - h)^2 + (y - k)^2 = r^2$. When the equation is given in general form, $x^2 + y^2 + ax + by + c = 0$, we first transform the equation into standard form by completing the square.

SECTION 2.4 EXERCISES

 ■ SKILLS

In Exercises 1–16, write the equation of the circle in standard form.

1. Center $= (-3, -4)$
   $r = 10$

2. Center $= (-1, -2)$
   $r = 4$

3. Center $= (5, 7)$
   $r = 9$

4. Center $= (2, 8)$
   $r = 6$

5. Center $= (-11, 12)$
   $r = 13$

6. Center $= (6, -7)$
   $r = 8$

7. Center $= (0, 0)$
   $r = 2$

8. Center $= (0, 0)$
   $r = 3$

9. Center $= (0, 0)$
   $r = \sqrt{2}$

10. Center $= (-1, 2)$
   $r = \sqrt{7}$

11. Center $= (5, -3)$
   $r = 2\sqrt{3}$

12. Center $= (-4, -1)$
   $r = 3\sqrt{5}$

13. Center $= \left(\dfrac{2}{3}, -\dfrac{3}{5}\right)$
   $r = \dfrac{1}{4}$

14. Center $= \left(-\dfrac{1}{3}, -\dfrac{2}{7}\right)$
   $r = \dfrac{2}{5}$

15. Center $= (1.3, 2.7)$
   $r = 3.2$

16. Center $= (-3.1, 4.2)$
   $r = 5.5$

In Exercises 17–28, find the center and radius of the circle with the given equations.

17. $(x - 1)^2 + (y - 3)^2 = 25$

18. $(x + 1)^2 + (y + 3)^2 = 11$

19. $(x - 2)^2 + (y + 5)^2 = 49$

20. $(x + 3)^2 + (y - 7)^2 = 81$

21. $(x - 4)^2 + (y - 9)^2 = 20$

22. $(x + 1)^2 + (y + 2)^2 = 8$

23. $\left(x - \dfrac{2}{5}\right)^2 + \left(y - \dfrac{1}{7}\right)^2 = \dfrac{4}{9}$

24. $\left(x - \dfrac{1}{2}\right)^2 + \left(y - \dfrac{1}{3}\right)^2 = \dfrac{9}{25}$

25. $(x - 1.5)^2 + (y + 2.7)^2 = 1.69$

26. $(x + 3.1)^2 + (y - 7.4)^2 = 56.25$

27. $x^2 + y^2 - 50 = 0$

28. $x^2 + y^2 - 8 = 0$

In Exercises 29–38, transform the equation into standard form by completing the square and state the center and radius of each circle.

29. $x^2 + y^2 + 6x + 8y - 75 = 0$

30. $x^2 + y^2 + 2x + 4y - 9 = 0$

31. $x^2 + y^2 - 10x - 14y - 7 = 0$

32. $x^2 + y^2 - 4x - 16y + 32 = 0$

33. $x^2 + y^2 - 10x + 6y + 22 = 0$

34. $x^2 + y^2 + 8x + 2y - 28 = 0$

35. $x^2 + y^2 - x + y + \dfrac{1}{4} = 0$

36. $x^2 + y^2 - \dfrac{x}{2} - \dfrac{3y}{2} + \dfrac{3}{8} = 0$

37. $x^2 + y^2 - 2.6x - 5.4y - 1.26 = 0$

38. $x^2 + y^2 - 6.2x - 8.4y - 3 = 0$

In Exercises 39–42, find the equation of each circle.

39. Centered at $(-1, -2)$ and passing through the point $(1, 0)$.

40. Centered at $(4, 9)$ and passing through the point $(2, 5)$.

41. Centered at $(-2, 3)$ and passing through the point $(3, 7)$.

42. Centered at $(1, 1)$ and passing through the point $(-8, -5)$.

 ■ **APPLICATIONS**

43. Cell Phones. If a cellular phone tower has a reception radius of 100 miles and you live 95 miles north and 33 miles east of the tower, can you use your cell phone while at home?

44. Cell Phones. Repeat Exercise 43 assuming you live 45 miles south and 87 miles west of town.

45. Backyard. A couple and their dog moved into a new house that does not have a fenced-in backyard. The backyard is square with dimensions 100 ft × 100 ft. If they put a stake in the center of the backyard with a leash, write the equation of the circle that will map out the dog's outer perimeter.

46. Backyard. Repeat Exercise 45 assuming the couple wants to put in a pool and a garden. What coordinates represent the center of the circle? What is the radius?

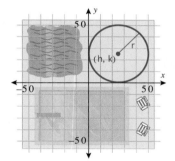

47. Campus. A university designs its campus with a master plan of two concentric circles. All of the academic buildings are within the inner circle (so that students can get between classes in less than 10 minutes) and an outer circle that contains all the dormitories, the Greek park, cafeterias, the gymnasium, and intramural fields.

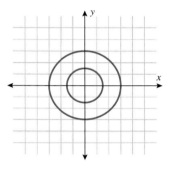

Assuming the center of campus is the origin, write an equation for the inner circle if the diameter is 3000 feet.

48. Campus. Repeat Exercise 47 for the outer circle with a diameter of 6000 feet.

 ■ **CATCH THE MISTAKE**

In Exercises 49–52, explain the mistake that is made.

49. Identify the center and radius of the circle with equation $(x - 4)^2 + (y + 3)^2 = 25$.

Solution: center = $(4, 3)$ and radius = 5

50. Identify the center and radius of the circle with equation $(x - 2)^2 + (y + 3)^2 = 2$.

Solution: center = $(2, -3)$ and radius = 2

51. Graph the solution to the equation $(x - 1)^2 + (y + 2)^2 = -16$.

Solution: center = $(1, -2)$ and radius = 4

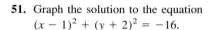
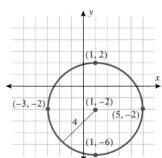

52. Find the center and radius of the circle with the given equation $x^2 + y^2 - 6x + 4y - 3 = 0$.

Solution:

Group like terms. $(x^2 - 6x) + (y^2 + 4y) = 3$

Complete the square.

$$(x^2 - 6x + 9) + (y^2 + 4y + 4) = 12$$

$$(x - 3)^2 + (y + 2)^2 = (2\sqrt{3})^2$$

The center is $(3, -2)$ and the radius is $2\sqrt{3}$.

 ■ **CHALLENGE**

53. T or F: The equation whose graph is depicted has infinitely many solutions.

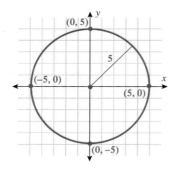

54. T or F: The equation $(x - 7)^2 + (y + 15)^2 = -64$ has no solution.

55. T or F: The equation $(x - 2)^2 + (y + 5)^2 = -20$ has no solution.

56. T or F: The equation $(x - 1)^2 + (y + 3)^2 = 0$ has only one solution.

57. Describe the graph (if it exists) of $x^2 + y^2 + 10x - 6y + 34 = 0$.

58. Describe the graph (if it exists) of $x^2 + y^2 - 4x + 6y + 49 = 0$.

59. Find the equation of a circle that has a diameter with endpoints $(5, 2)$ and $(1, -6)$.

60. Find the equation of a circle that has a diameter with endpoints $(3, 0)$ and $(-1, -4)$.

61. For the equation $x^2 + y^2 + ax + by + c = 0$, specify conditions on a, b, and c so that the graph is a single point.

62. For the equation, $x^2 + y^2 + ax + by + c = 0$, specify conditions on a, b, and c that correspond to no graph.

■ **TECHNOLOGY**

In Exercises 63–66, use a graphing utility to graph each equation given. Does this agree with the answer you gave in the challenge section?

63. $(x - 2)^2 + (y + 5)^2 = -20$ (See Exercise 55 for comparison.)

64. $(x - 1)^2 + (y + 3)^2 = 0$ (See Exercise 56 for comparison.)

65. $x^2 + y^2 + 10x - 6y + 34 = 0$ (See Exercise 57 for comparison.)

66. $x^2 + y^2 - 4x + 6y + 49 = 0$ (See Exercise 58 for comparison.)

Q: Jack lives in central Florida, which experienced three hurricanes within a 6-week period in 2004. Unfortunately, his roof (which rises to a single peak at the center from front to back) was badly damaged and needs to be reshingled. Before his roofer arrives, Jack would like to have an idea of the cost of a new roof. He learns that roofers measure roofs in terms of "squares." Each square is a 10 foot by 10 foot area that is 100 square feet, and the roofer will charge $240 a square to install a three-tab shingled roof. Jack knows that his house (including his garage) is 1920 square feet, so he estimates that a new roof will cost $4,608. However, the roofer's estimate is $5,538 (rounded to the nearest dollar). Is the roofer giving Jack a fair price?

A: Since Jack's roof is not flat, the square footage of his house and the square footage of his roof are not equal. The roofer's estimate is correct.

For every 12 feet of horizontal distance the roof rises 8 feet. The house itself measures 48 feet from front to back and 40 feet from side to side. To calculate the total square footage of the roof, calculate the square footage of the roof from base to peak (one side of the roof) and double it. To find the distance from the base to the peak of the roof divide the gable into two right triangles and use the given lengths, the slope of the roof, and the Pythagorean theorem. Jack's roof is $640\sqrt{13}$ square feet, which is $6.4\sqrt{13}$ squares.

Jack's neighbor Alan also needs a new roof. His roof also rises to a single peak at the center from front to back. If Alan's house is 40 feet from front to back, 55 feet across, and the roof has an area of $4.4\sqrt{29}$ squares, find the slope of Alan's roof.

TYING IT ALL TOGETHER

An air traffic controller working on a naval aircraft carrier sends out a patrol aircraft to scout for ships in the area. The patrol aircraft uses radar to detect objects within a certain radius of itself. This area within the field of the radar is thus a circle with the patrol aircraft at the center. However, since the patrol aircraft is scouting on behalf of the aircraft carrier, an *xy*-coordinate plane is used with the aircraft carrier at the origin.

The patrol aircraft finds two ships on the outer boundary of its radar field: one is $125\sqrt{2}$ miles east and $125\sqrt{2}$ miles north (in other words, directly northeast) of the carrier, while the other ship is $75\sqrt{2}$ miles south and $75\sqrt{2}$ miles west (directly southwest) of the carrier. It also detects two other ships: one is 150 miles east and 40 miles north of the carrier, while the other ship is 70 miles east and $4\sqrt{2}$ miles south of the carrier. What are the coordinates of the patrol aircraft at the moment it detects the four ships? Write the equation in standard form for the outer boundary of radar coverage. Would a ship 90 miles east and 150 miles south of the carrier be visible on the radar of the patrol aircraft?

CHAPTER 2 REVIEW EXERCISES

2.1 Basic Tools: Cartesian Plane, Distance, and Midpoint

Plot each point and indicate which quadrant the point lies in.

1. $(-4, 2)$ **2.** $(4, 7)$ **3.** $(-1, -6)$ **4.** $(2, -1)$

Calculate the distance between the two points.

5. $(-2, 0)$ and $(4, 3)$ **6.** $(1, 4)$ and $(4, 4)$

7. $(-4, -6)$ and $(2, 7)$ **8.** $\left(\dfrac{1}{4}, \dfrac{1}{12}\right)$ and $\left(\dfrac{1}{3}, -\dfrac{7}{3}\right)$

Calculate the midpoint of the segment joining the two points.

9. $(2, 4)$ and $(3, 8)$ **10.** $(-2, 6)$ and $(5, 7)$

11. $(2.3, 3.4)$ and $(5.4, 7.2)$ **12.** $(-a, 2)$ and $(a, 4)$

Applications.

13. Sports. A quarterback drops back to pass. At the point $(-5, -20)$ he throws the ball to his wide receiver located at $(10, 30)$. Find the distance the ball has traveled. Assume the width of the football field is $[-15, 15]$ and the length is $[-50, 50]$.

14. Sports. Suppose in the above exercise a defender was midway between the quarterback and the receiver. At what point was the defender located when the ball was thrown over his head?

2.2 Graphing Equations: Point Plotting and Symmetry

Use algebraic tests to determine symmetry with respect to the x-axis, y-axis, or origin.

15. $x^2 + y^3 = 4$ **16.** $y = x^2 - 2$

17. $xy = 4$ **18.** $y^2 = 5 + x$

Use symmetry as a graphing aid and point plot the given equations.

19. $y = x^2 - 3$ **20.** $y = |x| - 4$ **21.** $y = \sqrt[3]{x}$

22. $x = y^2 - 2$ **23.** $y = x\sqrt{9 - x^2}$ **24.** $x^2 + y^2 = 36$

Applications.

25. Track. A track around a high school football field is in the shape of the graph of $8x^2 + y^2 = 8$. Graph using symmetry and by plotting points.

26. Highways. A "bypass" around a town follows the graph of $y = x^3 + 2$ where the origin is the center of town. Graph the equation.

2.3 Straight Lines

Find the x- and y-intercepts and the slope of each line if they exist and graph.

27. $y = 4x - 5$ **28.** $y = \dfrac{-3}{4}x - 3$ **29.** $x + y = 4$

30. $x = -4$ **31.** $y = 2$ **32.** $\dfrac{-1}{2}x - \dfrac{1}{2}y = 3$

Write the equation of the line given the slope and the intercepts.

33. Slope: $m = 4$
y-intercept: $(0, -3)$

34. Slope: $m = 0$
y-intercept: $(0, 4)$

35. Slope: m is undefined
x-intercept: $(-3, 0)$

36. Slope: $m = \dfrac{-2}{3}$
y-intercept: $\left(0, \dfrac{3}{4}\right)$

Write an equation of the line given the slope and a point that lies on the line.

37. $m = -2$ $(-3, 4)$ **38.** $m = \dfrac{3}{4}$ $(2, 16)$

39. $m = 0$ $(-4, 6)$ **40.** m is undefined $(2, -5)$

Write the equation of the line that passes through the given points. Express the equation in slope–intercept form or in the form of $x = a$ or $y = b$.

41. $(-4, -2)$ $(2, 3)$ **42.** $(-1, 4)$ $(-2, 5)$

43. $\left(-\dfrac{3}{4}, \dfrac{1}{2}\right)$ $\left(-\dfrac{7}{4}, \dfrac{5}{2}\right)$ **44.** $(3, -2)$ $(-9, 2)$

Find the equation of the line that passes through the given point and also satisfies the additional piece of information.

45. $(-2, -1)$ parallel to the line $2x - 3y = 6$

46. $(5, 6)$ perpendicular to the line $5x - 3y = 0$

47. $\left(-\dfrac{3}{4}, \dfrac{5}{2}\right)$ perpendicular to the line $\dfrac{2}{3}x - \dfrac{1}{2}y = 12$

48. $(a + 2, b - 1)$ parallel to the line $Ax + By = C$

Applications.

49. Grades. Before a student starts a GRE prep class, he or she must take a pretest and then a posttest after the completion of the course. Two students' results were

Pretest	Posttest
1020	1324
950	1240

Give a linear equation to represent the given data. Graph the linear equation.

50. **Budget.** The cost of having the air conditioner in your car repaired is the combination of material costs and labor costs. The materials (tubing, coolant, etc.) are $250 and the labor costs $38 per hour. Write an equation that models the total cost, C, of having your air conditioner repaired as a function of hours, t. Graph this equation with t as the horizontal axis and C representing the vertical axis. How much will the job cost if the mechanic works 1.5 hours?

2.4 Circles

Write the equation of the circle in standard form.

51. center = $(-2, 3)$
    $r = 6$

52. center = $(-6, -8)$
    $r = 3\sqrt{6}$

53. center = $\left(\dfrac{3}{4}, \dfrac{5}{2}\right)$
    $r = \dfrac{2}{5}$

54. center = $(1.2, -2.4)$
    $r = 3.6$

Find the center and the radius of the circle given by the equation.

55. $(x + 2)^2 + (y + 3)^2 = 81$

56. $(x - 4)^2 + (y + 2)^2 = 32$

57. $\left(x + \dfrac{3}{4}\right)^2 + \left(y - \dfrac{1}{2}\right)^2 = \dfrac{16}{36}$

58. $x^2 + y^2 + 4x - 2y = 0$

59. $x^2 + y^2 + 2y - 4x + 11 = 0$

60. $3x^2 + 3y^2 - 6x - 7 = 0$

61. $9x^2 + 9y^2 - 6x + 12y - 76 = 0$

62. $x^2 + y^2 + 3.2x - 6.6y - 2.4 = 0$

63. Find the equation of a circle centered at $(2, 7)$ and passing through $(3, 6)$.

64. Find the equation of a circle that has the diameter with endpoints $(-2, -1)$ and $(5, 5)$.

1. Find the distance between the points $(-7, -3)$ and $(2, -2)$.

2. Find the midpoint between $(-3, 5)$ and $(5, -1)$.

3. Determine the length and the midpoint of a segment that joins the points $(-2, 4)$ and $(3, 6)$.

4. **Research Triangle.** The Research Triangle in North Carolina was established as a collaborative research center among Duke University (Durham, NC), North Carolina State University (Raleigh, NC), and the University of North Carolina (Chapel Hill, NC).

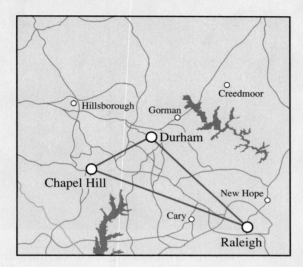

Durham is 10 miles north and 8 miles east of Chapel Hill, and Raleigh is 28 miles east and 15 miles south of Chapel Hill. What is the perimeter of the research triangle? Round your answer to the nearest mile.

5. Determine the two values for y so that the point $(3, y)$ is 5 units away from the point $(6, 5)$.

6. If the point $(3, -4)$ is on a graph that is symmetric with respect to the y-axis, what point must also be on the graph?

7. If the point $(1, -1)$ is on the graph that is symmetric with respect to the x-axis, y-axis, and origin, what other points also must lie on the graph?

In Exercises 8–12, find the equation of the line that is characterized by the given information. Graph the line.

8. Slope $= 4$; y-intercept $(0, 3)$

9. Passes through the points $(-3, 2)$ and $(4, 9)$

10. Parallel to the line $y = 4x + 3$ and passes through the point $(1, 7)$

11. Perpendicular to the line $2x - 4y = 5$ and passes through the point $(1, 1)$

12. x-intercept $(3, 0)$; y-intercept $(0, 6)$

In Exercises 13 and 14, write the equation of the line that corresponds to the graph.

13.

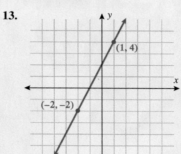

14.

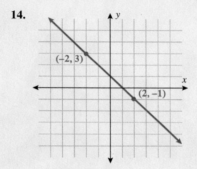

15. Write the equation of a circle that has center $= (6, -7)$ and radius $r = 8$.

16. Determine the center and radius of the circle $x^2 + y^2 - 10x + 6y + 22 = 0$.

17. Find the equation of the circle that is centered at $(4, 9)$ and passes through the point $(2, 5)$.

18. **Solar System.** Earth is approximately 93 million miles from the sun. Approximating Earth's orbit around the sun, write an equation governing Earth's path around the sun. Locate the sun at the origin.

19. How many real solutions does $x^2 + y^2 + 6x + 4y + 22 = 0$ have?

20. Describe the graph of $x^2 + y^2 - 2x - 8y = -17$.

PRACTICE TEST

185

CHAPTER 3

Functions and Their Graphs

Corbis Images

There is a sales rack of clothes at a department store. The original price of a skirt was $90, but it has been discounted 20%. As a preferred shopper, you get an automatic additional 10% off the sale price at the register. How much will you pay for this skirt?

Twenty percent off of $90 is a savings of $18. Therefore, the sale price is $72. You then take an additional 10% off of the sale price. Ten percent off of $72 is an additional savings of $7.20. The checkout price of the skirt is now $64.80. Naïve shoppers often are lured into thinking that 20% and 10% equates to 30% off the original price. But they end up paying more than they expected because the additional 10% is off the sale (not original) price. Most experienced sale shoppers can figure out the checkout price before they get to the register—they have mastered *composite functions.*

A composition of *functions* can be thought of as a function of a function. One function takes an input (original price) and maps it to an output (sale price), and then another function takes that output as its input (sale price) and maps that to an output (checkout price). This will be discussed in further detail in Section 3.4.

In this chapter you will find that people use functions every day: converting from degrees Celsius to degrees Fahrenheit, the cost of a wedding, DNA testing in forensic science, stock values for day traders, and the sale price of a skirt. We will take examples you are familiar with and develop a more complete, thorough understanding of functions. At the end of this chapter, you should be able to determine if a relationship is a function, determine if functions are one-to-one, graph functions, and perform operations and composition of functions.

Functions

Graphs of Functions

- Common Functions
- Graphing Aids

Types of Functions

- Even
- Odd
- One-to-One
- Inverse

Operations on Functions

- Addition/Subtraction
- Multiplication/Division
- Composition

CHAPTER OBJECTIVES

- Determine if a relation is a function.
- Use function notation.
- Perform operations on functions (addition, subtraction, multiplication, and division).
- Perform composition of functions.
- Determine domain restrictions of functions.
- Find inverse functions.
- Graph functions.

NAVIGATION THROUGH SUPPLEMENTS

DIGITAL VIDEO SERIES #3

STUDENT SOLUTIONS MANUAL CHAPTER 3

BOOK COMPANION SITE
www.wiley.com/college/young

SECTION 3.1 Functions

Skills Objectives

- Determine if a relation is a function.
- Use the vertical line test for functions defined as equations.
- Use function notation.
- Find the value of a function.
- Determine the domain of a function.

Conceptual Objectives

- Understand the difference between discrete functions (sets) and continuous functions (equations).
- Think of function notation as a placeholder or mapping.

Introduction to Functions

What do the following relations have in common?

- Every person has a blood type.
- Every real number has an absolute value.
- Every person has a DNA sequence.
- Every working household phone in the United States has a 10-digit phone number.

They all describe a particular correspondence between two groups. A **relation** is a correspondence between two sets. A relation is a **function** when each element in one group or set corresponds to *only* one element in another group or set. The first set (or input set) is called the **domain**, and the corresponding second set (or output set) is called the **range**. If each member of the domain has *exactly one* corresponding member in the range, then that relation is called a function. All of the above examples are functions.

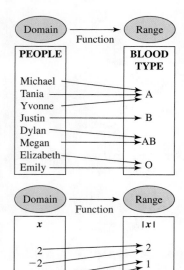

In the blood-type example on the left, each person can have only *one* blood type. For instance, Michael cannot be both type A and type O. Because each person in the domain corresponds to a single blood type in the range, this relation is a function. Notice that two elements in the domain may correspond to the same element in the range. In this example, both Elizabeth and Emily have type O blood.

One way of determining if this example constitutes a function is to ask the question: If I know the person's name, do I know his or her blood type? And the answer is yes. Before DNA testing revolutionized the forensic sciences, blood types were used to narrow down the list of suspects in investigations. Because no two people have the same DNA, forensic scientists now can match DNA to a specific suspect. The DNA example also constitutes a function because every person corresponds to a unique DNA structure.

In the absolute value example on the left, each number in the domain corresponds to a single value. For instance, the absolute value of -1 is 1.

Notice that two numbers in the domain may correspond to the same number in the range. For example, both -1 and 1 in the domain correspond to 1 in the range. The key to a relation being classified as a function is that each element in the domain has *one and only one* element in the range that it maps to. If an element in the domain corresponds to two or more elements in the range, then the relationship is *not* a function.

In the following examples, elements in the domain map to more than one element in the range. Hence the two examples are not functions.

At a university, four primary sports typically overlap in the late fall: football, volleyball, soccer, and basketball. The start times are 1 P.M., 3 P.M., and 7 P.M. The 1 P.M. start time corresponds to one and only one event: football. However, the 3 P.M. start time corresponds to *both* volleyball and soccer. Because 3 P.M. corresponds to *both* volleyball and soccer, this is not a unique relation and, hence, is *not* a function.

If the temperature is measured throughout the day, there can be more than one time during the day when the same temperature is recorded. If we know the temperature, do we know the time of day? No, because each temperature does not correspond to one and only one time.

In the example on the right, although 85 °F corresponds to only 11 A.M. and 91 °F corresponds to only 4 P.M., 87 °F corresponds to two different times (2 P.M. and 6 P.M.). Since one element in the domain corresponds to two elements in the range, this is *not* a function.

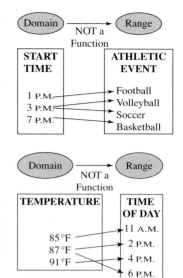

FUNCTION

A **function** is a relation that maps each element in the domain to *one and only one* element in the range.

If we let x represent an element in the domain and y represent an element in the range, then we can think of a function as a set of ordered pairs (x, y). Because a function maps each element in the domain to one and only one element in the range, then functions are ordered pairs that have no common x values. For example, here are three sets of ordered pairs. Which sets represent a function and why?

- $\{(1, 2), (3, 4), (5, 6), (7, 8)\}$ Function
- $\{(1, 2), (3, 2), (5, 6), (7, 6)\}$ Function
- $\{(1, 2), (1, 3), (5, 6), (5, 7)\}$ Not a function

In the first set, all of the pairs are distinct with no first elements repeated. Therefore, it is a function. In the second set, each first element corresponds to one and only one second element. Although two first elements map to the same second element, this still represents a function. The last set does not represent a function because there are repeated first elements, indicating multiple mappings for each element in the domain.

EXAMPLE 1 Classifying Relationships as Functions

Classify the following relationships as functions or not functions and justify your answer.

a. $\{(-3, 4), (2, 4), (3, 5), (6, 4)\}$ **b.** $\{(-3, 4), (2, 4), (3, 5), (2, 2)\}$

c. Domain = set of all items for sale in a grocery store; Range = price

Solution:

a. Function: No x value is repeated. Therefore, each x maps to one and only one y.

b. Not a function: $x = 2$ maps to *both* $y = 2$ and $y = 4$.

c. Function: Each item has one and only one price. It is still a function even though some items have the same price (or elements in the domain map to the same element in the range).

All of the examples we have discussed thus far are finite, discrete sets. They represent a finite number of distinct pairs of (x, y). For instance, for a set consisting of certain people, we have their corresponding blood type: (Michael, A), (Tania, A) (Yvonne, A), (Justin, B), (Dylan, AB), (Megan, AB), (Elizabeth, O), and (Emily, O).

Using ordered pairs of numbers, we described the set $\{(1, 2), (3, 4), (5, 6), (7, 8)\}$ as a function. If we plot this set of ordered pairs, we have 4 distinct points on the graph.

Domain	Range
x	y
1	2
3	4
5	6
7	8

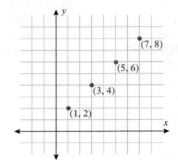

There are only four elements in the domain and four elements in the range.

Notice that these points all lie on the line, $y = x + 1$. What if we wanted other points that lie on this line? We could use this equation to map x values to corresponding y values. Some additional points are $(0, 1)$, $(-2, -1)$, and $(8, 9)$.

Function: $y = x + 1$.

Domain	Range
x	y
1	2
3	4
5	6
7	8
-2	-1
0	1
8	9

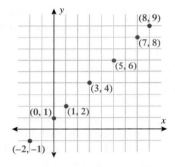

We now have seven distinct pairs of numbers. What if we want *all* of the points that lie on this line? The equation, $y = x + 1$, describes the correspondence between the x and y values. To describe the mapping from all real values of x to the corresponding real values of y, we no longer use a discrete set of points but rather indicate an infinite set of points, or continuum, described by the equation: $y = x + 1$.

Function: $y = x + 1$.

Domain	Range
x	y
$\mathbb{R}$	$\mathbb{R}$

The symbol $\mathbb{R}$ denotes all real numbers.

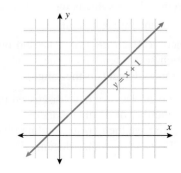

This function maps an infinite set of real numbers to a corresponding infinite set of real numbers and forms a continuum of ordered pairs as opposed to discrete points.

Functions Defined by Equations

Displaying a function as a table or a set of ordered pairs only works if the domain and range are finite sets. If the domain and/or range are infinite sets of real numbers, then functions must be defined using equations.

Let's start with the equation $y = x^2 - 3x$, where x can be any real number. This equation assigns each x *one and only one* corresponding value for y.

x	$y = x^2 - 3x$	y
1	$y = (1)^2 - 3(1)$	-2
5	$y = (5)^2 - 3(5)$	10
$-\dfrac{2}{3}$	$y = \left(-\dfrac{2}{3}\right)^2 - 3\left(-\dfrac{2}{3}\right)$	$\dfrac{22}{9}$
1.2	$y = (1.2)^2 - 3(1.2)$	-2.16

Since the variable y *depends* on what value of x is selected, we denote y as the **dependent variable**. The variable x can be any number in the domain; therefore we denote x as the **independent variable**. Although it is customary to use the variables x and y, any variables can be used in a function relation. The variable in the domain is the independent variable, and the corresponding variable in the range is the dependent variable.

Words that are synonymous with *domain* are *input* or *independent variable*. Words that are synonymous with *range* are *output* or *dependent variable*. Typically, the variable x is used to denote the domain or input, and the variable y is often used to denote the range or output. The coordinates (x, y) represent a corresponding pair of elements: one in the domain and one in the range.

Function

Domain	Range
Input	Output
x	y
Independent variable	Dependent variable

Although functions are defined by equations, it is important to recognize that *not all equations are functions*. The requirement for an equation to define a function is that each element in the domain corresponds to only one element in the range.

CAUTION
Not all equations are functions.

EQUATIONS THAT ARE FUNCTIONS	EQUATIONS THAT ARE NOT FUNCTIONS
$y = x^2$	$x = y^2$
$y = \lvert x \rvert$	$x^2 + y^2 = 1$
$y = x^3$	$x = \lvert y \rvert$

In the "equations that are functions," every x corresponds to only one y. Some points that correspond to these functions are

$$y = x^2: \quad (-1, 1) \ (0, 0) \ (1, 1)$$
$$y = |x|: \quad (-1, 1) \ (0, 0) \ (1, 1)$$
$$y = x^3: \quad (-1, -1) \ (0, 0) \ (1, 1)$$

The fact that $x = -1$ and $x = 1$ both correspond to $y = 1$ in the first two examples does not violate the definition of a function.

In the "equations that are NOT functions," some x values correspond to *more than one y* value. Some points that correspond to these equations are

$$x = y^2: \quad (1, -1) \ (0, 0) \ (1, 1) \qquad x = 1 \text{ maps to } \textbf{both } y = -1 \text{ and } y = 1$$
$$x^2 + y^2 = 1: \quad (0, -1) \ (0, 1) \ (-1, 0) \ (1, 0) \qquad x = 0 \text{ maps to } \textbf{both } y = -1 \text{ and } y = 1$$
$$x = |y|: \quad (1, -1) \ (0, 0) \ (1, 1) \qquad x = 1 \text{ maps to } \textbf{both } y = -1 \text{ and } y = 1$$

Let's look at the graphs of the three **functions:**

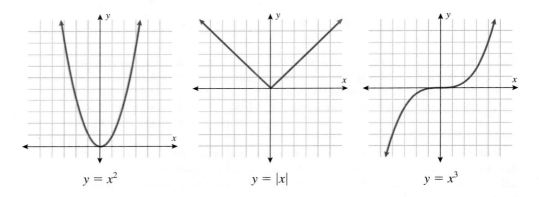

$$y = x^2 \qquad\qquad y = |x| \qquad\qquad y = x^3$$

Let's take any value for x, say $x = a$. This corresponds to a vertical line. A function maps one x to only one y; therefore there should be only one point of intersection with any vertical line. We see in the three graphs of functions above that if a vertical line is drawn at any value of x on any of the three graphs, the vertical line only intersects the graph in one place. Look at the graphs of the three equations that are **not functions**.

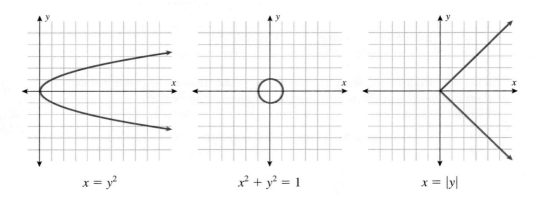

$$x = y^2 \qquad\qquad x^2 + y^2 = 1 \qquad\qquad x = |y|$$

If a vertical line is drawn at $x = \frac{1}{2}$ in any of the three graphs, that vertical line will intersect each of these graphs at two points. Thus, there are two y values that correspond to each x in the domain, which is why these equations do not define functions.

VERTICAL LINE TEST

Given a graph of an equation, if any vertical line that can be drawn intersects the graph at no more than one point, the equation defines a function. This test is called the **vertical line** test.

EXAMPLE 2 Using the Vertical Line Test

Use the vertical line test to determine if the graphs of equations define functions.

a.

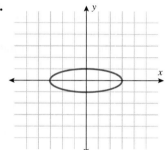

b.

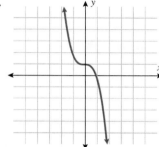

Solution:

Apply the vertical line test.

a.

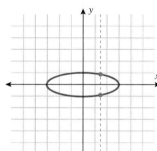

b.

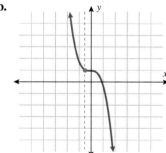

a. Because the vertical line intersects the graph of the equation at two points, this equation is *not a function.*

b. Because the vertical line intersects the graph of the equation at *one and only one* point for any x-value, this equation *is a function.*

■ **YOUR TURN**   Determine if the equation $(x - 3)^2 + (y + 2)^2 = 16$ is a function.

■ **Answer:** The graph of the equation is a circle which does not pass the vertical line test. Therefore, the equation does not define a function.

Function Notation

It was mentioned earlier that functions can be defined by equations such as $y = x^2 - 3x$, where y is the dependent variable. We often use an equivalent **function notation** instead:

$$f(x) = x^2 - 3x$$

The symbol $f(x)$ is read "f evaluated at x" or "f of x" and represents the value in the range that the function corresponds to, given a specified value of x in the domain. For instance, $f(1)$ represents the value of the function when $x = 1$.

$$f(x) = x^2 - 3x$$
$$f(1) = (1)^2 - 3(1)$$
$$f(1) = 1 - 3 = -2$$

STUDY TIP

When using function notation, it helps to think of the independent variable as a placeholder.

Therefore, the value $x = 1$ in the domain corresponds to the value $f(1) = -2$ in the range. A graphical interpretation of this correspondence is that the point $(1, -2)$ lies on the graph of the function, f. When using function notation, it helps to think of the independent variable as a placeholder. For instance, the function above, $f(x) = x^2 - 3x$, can be thought of as:

$$f(\ ) = (\ )^2 - 3(\ )$$

It is important to note that $f(x)$ **does not** represent f "times" x.

In the above example, we have used x to represent the independent variable. Because $y = f(x)$, we say that y and $f(x)$ both represent the dependent variable. We say that "f maps x into $f(x)$." In other words, the function, f, maps the independent variable, x in the domain, to a dependent variable, $f(x)$ in the range.

Although f is the most logical letter to represent a function, other letters can be used to denote a function. The most common are F, G, or g, but any letter can be used to represent either the independent or dependent variable.

We state that "G maps t into $G(t)$." In other words, the function, G, maps the independent variable, t in the domain, to a dependent variable, $G(t)$ in the range.

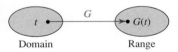

EXAMPLE 3 Evaluating Functions: Interpreting Mapping

Given the function $f(x) = 2x^3 - 3x^2 + 4$ find $f(-1)$.

Solution:

The independent variable, x, is a placeholder. $f(\ ) = 2(\ )^3 - 3(\ )^2 + 6$

To find $f(-1)$, substitute $x = -1$
into the function. $f(-1) = 2(-1)^3 - 3(-1)^2 + 6$

Evaluate the right side. $f(-1) = -2 - 3 + 6$

Simplify. $f(-1) = 1$

We can interpret $f(-1) = 1$ as "the function $f(x) = 2x^3 - 3x^2 + 6$ maps -1 in the domain to 1 in the range.

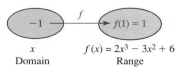

The independent variable is also referred to as the **argument** of a function. In the case of the function $f(x) = x^2 - 3x$, we say that x is the argument of f. The function $f(x) = x^2 - 3x$ can be thought of as $f(\ ) = (\ )^2 - 3(\ )$, because the argument is a placeholder. A way to describe this function is "the argument squared minus 3 times the argument." *Any* expression can be substituted in for the argument.

$$f(1) = (1)^2 - 3(1)$$

$$f(x + 1) = (x + 1)^2 - 3(x + 1)$$

$$f(-x) = (-x)^2 - 3(-x)$$

 EXAMPLE 4 Evaluating Functions with Variable Arguments

Evaluate $f(x + 1)$, given that $f(x) = x^2 - 3x$.

COMMON MISTAKE

A common misunderstanding is to interpret the notation $f(x + 1)$ as a sum: $f(x + 1) \neq f(x) + f(1)$.

 CORRECT

Write original function.

$$f(x) = x^2 - 3x$$

Replace the argument, x, with a placeholder.

$$f(\ ) = (\ )^2 - 3(\ )$$

Substitute $x + 1$ for the argument.

$$f(x + 1) = (x + 1)^2 - 3(x + 1)$$

Eliminate parentheses.

$$f(x + 1) = x^2 + 2x + 1 - 3x - 3$$

Combine like terms.

$$f(x + 1) = x^2 - x - 2$$

 INCORRECT

The **ERROR** is interpreting as a sum.

$$f(x + 1) = f(x) + f(1)$$

Substituting $f(x) = x^2 - 3x$ and $f(1) = -2$ into the right side leads to the **INCORRECT** answer.

$$f(x + 1) = x^2 - 3x - 2$$

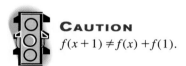 **CAUTION**

$f(x + 1) \neq f(x) + f(1)$.

YOUR TURN Evaluate $g(x - 1)$ given that $g(x) = x^2 - 2x + 3$.

EXAMPLE 5 Evaluating Functions: Sums

For the given function $H(x) = x^2 + 2x$ evaluate:

a. $H(x + 1)$ **b.** $H(x) + H(1)$

Solution (a):

Write the function H in placeholder notation.

$$H(\ ) = (\ )^2 + 2(\ )$$

Substitute $x + 1$ in for the argument of H.

$$H(x + 1) = (x + 1)^2 + 2(x + 1)$$

Answer: $g(x - 1) = x^2 - 4x + 6$

TECHNOLOGY TIP

Use a graphing utility to display graphs of $y_1 = H(x + 1) = (x + 1)^2 + 2(x + 1)$ and $y_2 = H(x) + H(1) = x^2 + 2x + 3$ in a $[-6, 3]$ by $[-5, 10]$ viewing rectangle.

The graphs are not the same.
$H(x + 1) \neq H(x) + H(1)$

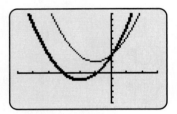

Eliminate the parentheses on the right side.

$$H(x + 1) = x^2 + 2x + 1 + 2x + 2$$

Combine like terms on the right side. $H(x + 1) = x^2 + 4x + 3$

Solution (b):

Write $H(x)$. $H(x) = x^2 + 2x$

Evaluate H at $x = 1$. $H(1) = (1)^2 + 2(1) = 3$

Evaluate the sum $H(x) + H(1)$. $H(x) + H(1) = x^2 + 2x + 3$

$$H(x) + H(1) = x^2 + 2x + 3$$

Note: Comparing the results of part a and part b, we see that $H(x + 1) \neq H(x) + H(1)$.

EXAMPLE 6 Evaluating Functions: Negatives

For the given function $G(t) = t^2 - t$ evaluate:

a. $G(-t)$ **b.** $-G(t)$

Solution (a):

Write the function G in placeholder notation. $G(\ ) = (\ )^2 - (\ )$

Substitute $-t$ in for the argument of G. $G(-t) = (-t)^2 - (-t)$

Eliminate the parentheses on the right side. $G(-t) = t^2 + t$

Solution (b):

Write $G(t)$. $G(t) = t^2 - t$

Multiply both sides by -1. $-G(t) = -(t^2 - t)$

Eliminate parentheses on right side. $-G(t) = -t^2 + t$

Note: Comparing the results of part a and part b, we see that $G(-t) \neq -G(t)$.

EXAMPLE 7 Evaluating Functions: Quotients

For the given function $F(x) = 3x + 5$ evaluate:

a. $F\left(\dfrac{1}{2}\right)$ **b.** $\dfrac{F(1)}{F(2)}$

Solution (a):

Write F in placeholder notation. $F(\ ) = 3(\ ) + 5$

Replace the argument with $\frac{1}{2}$. $F\left(\dfrac{1}{2}\right) = 3\left(\dfrac{1}{2}\right) + 5$

Simplify the right side. $F\left(\dfrac{1}{2}\right) = \dfrac{13}{2}$

Solution (b):

Evaluate $F(1)$. $F(1) = 3(1) + 5 = 8$

Evaluate $F(2)$. $F(2) = 3(2) + 5 = 11$

Divide $F(1)$ by $F(2)$.

$$\frac{F(1)}{F(2)} = \frac{8}{11}$$

CAUTION

$$f\left(\frac{a}{b}\right) \neq \frac{f(a)}{f(b)}$$

Note: Comparing the results of part a and part b, we see that $F\left(\dfrac{1}{2}\right) \neq \dfrac{F(1)}{F(2)}$.

YOUR TURN   Given the function, $G(t) = 3t - 4$, evaluate:

a. $G(t - 2)$ **b.** $\dfrac{G(1)}{G(3)}$

Domain of a Function

The word **domain** has been used to represent the input or independent variables that are mapped to a range via the function correspondence. What elements are allowed in the domain? Let's investigate three different functions and ask the question: Are there any restrictions on what elements are allowed in the domain?

$$f(x) = x^3 - 4x^2 + 17 \qquad g(x) = \sqrt{x} \qquad h(x) = \frac{1}{x}$$

The first function, $f(x) = x^3 - 4x^2 + 17$, maps each value in the domain to one and only one corresponding value in the range. Any real number can be cubed, squared, and combined with other constants, so there are no restrictions on what numbers can be "input" into this function. Therefore, we say the domain of the function f is all real numbers, $\mathbb{R}$.

The second function, $g(x) = \sqrt{x}$, maps numbers from the domain to the range by taking the square root. Are there any real numbers on which the square root is taken that yield something other than a real number? In other words, what can x be? The square root of a positive real number yields some other positive real number. The square root of zero is zero. The square root of a negative real number, however, is an imaginary number. Therefore, in order to have real numbers in the range, we must restrict the domain to nonnegative real numbers, $[0, \infty)$.

The third function, $h(x) = \frac{1}{x}$, maps quantities in the domain to unique quantities in the range by taking the reciprocal of the input to yield the output. We again ask the question: What can x be? The variable x can be any real number except zero. Therefore, we must restrict the domain to all real numbers except zero, which can be written as *all real numbers except 0, $x \neq 0$*, or in interval notation as $(-\infty, 0) \cup (0, \infty)$.

EXAMPLE 8   Determining the Domain of a Function

State the domain of the given functions.

a. $F(x) = \dfrac{3}{x^2 - 25}$ **b.** $H(x) = \sqrt{9 - 2x}$ **c.** $G(t) = |4 - t|$

Solution (a):

Write the original equation.

$$F(x) = \frac{3}{x^2 - 25}$$

Determine any restrictions on the values of x.

$$x^2 - 25 = 0$$

Answer: a. $G(t - 2) = 3t - 10$ **b.** $\dfrac{G(1)}{G(3)} = -\dfrac{1}{5}$

Solve the restriction equation.

$$x^2 = 25 \text{ or } x = \pm\sqrt{25} = \pm 5$$

State the domain restrictions.

$$x \neq \pm 5$$

Write domain in interval notation.

$$(-\infty, -5) \cup (-5, 5) \cup (5, \infty)$$

The domain is restricted to all real numbers except -5 and 5.

Solution (b):

Write the original equation.

$$H(x) = \sqrt{9 - 2x}$$

Determine any restrictions on the values of x.

$$9 - 2x \geq 0$$

Solve the restriction equation.

$$9 \geq 2x$$

State the domain restrictions.

$$\frac{9}{2} \geq x \text{ or } x \leq \frac{9}{2}$$

Write domain in interval notation.

$$\left(-\infty, \frac{9}{2}\right]$$

The domain is restricted to all real numbers less than or equal to $\frac{9}{2}$.

Solution (c):

Write the original equation.

$$G(t) = |4 - t|$$

Determine any restrictions on the values of t.

no restrictions

State the domain.

$$\mathbb{R}$$

Write domain in interval notation.

$$(-\infty, \infty)$$

The domain is the set of all real numbers.

TECHNOLOGY TIP

Graph of $H(x) = \sqrt{9 - 2x}$.

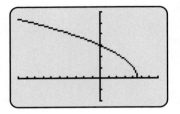

Notice that $H(x)$ is defined for all x less than or equal to $\frac{9}{2}$.

CONCEPT CHECK Is the function $f(x) = \sqrt{x - 3}$ defined for $x = 0$?

Is the function $g(x) = \dfrac{1}{x^2 - 4}$ defined for $x = 2$?

YOUR TURN State the domain of the given functions.

a. $f(x) = \sqrt{x - 3}$ **b.** $g(x) = \dfrac{1}{x^2 - 4}$

Applications

Functions that are used in applications often have restrictions on the domains due to physical constraints. For example, the volume of a square cube is given by the function $V(x) = x^3$ where x is the length of a side. The function $f(x) = x^3$ has no restrictions on x, and therefore the domain is the set of all real numbers. However, the volume of a box has the restriction that the length of a side can never be negative or zero.

■ **Answer: a.** The domain is the set of all real numbers greater than or equal to 3, which can be written as $x \geq 3$ or $[3, \infty)$.

b. The domain is the set of all real numbers except -2 and 2, which can be written as $x \neq \pm 2$ or $(-\infty, -2) \cup (-2, 2) \cup (2, \infty)$.

EXAMPLE 9 Price of Gasoline

Following the capture of Saddam Hussain in Iraq in 2003, gas prices in the United States escalated and then finally returned to their precapture prices. Over a 6-month period the average price of a gallon of 87 octane gasoline was given by the function $C(x) = -0.05x^2 + 0.3x + 1.7$ where C is the cost function and x represents the number of months after the capture.

a. Determine the domain of the cost function.

b. What was the average price of gas per gallon 3 months after the capture?

Solution (a):

Since the cost function $C(x) = -0.05x^2 + 0.3x + 1.7$ modeled the price of gas only for 6 months after the capture, the domain is $0 \le x \le 6$ or $[0, 6]$.

Solution (b):

Write the cost function. $\qquad C(x) = -0.05x^2 + 0.3x + 1.7 \quad 0 \le x \le 6$

Find the value of the function
when $x = 3$. $\qquad C(3) = -0.05(3)^2 + 0.3(3) + 1.7$

Simplify. $\qquad C(3) = 2.15$

The average price per gallon 3 months after the capture was $2.15 .

EXAMPLE 10 The Dimensions of a Pool

Express the volume of a 30 foot $\times$ 10 foot rectangular swimming pool as a function of its depth.

Solution:

The volume of any rectangular box is $V = lwh$ where V is the volume, l is the length, w is the width, and h is the height. In this example the length is 30 feet, the width is 10 feet, and the height is the depth, d.

Write the volume as a function of depth, d. $\qquad V(d) = (30)(10)d$

Simplify. $\qquad V(d) = 300d$

Determine any restrictions on the domain. $\qquad d > 0$

SECTION 3.1 SUMMARY

Determining if a relation is a function can be done both algebraically and graphically.

- Algebraically: Each element in the domain maps to one and only one element in the range. It is still a function if two elements in the domain map to the same element in the range.

- Graphically: Vertical line test—If a vertical line is drawn anywhere on a graph and intersects the graph of an equation in more than one point, then the graph does not represent a function. If they intersect in at most one point, then it is a function.

When evaluating functions the independent variable acts like a placeholder.

To determine the domain of a function, ask the question: Are there any restrictions on what the input can be?

SECTION 3.1 EXERCISES

■ **SKILLS**

In Exercises 1–24, determine whether each relation is a function. Assume that the coordinate pair (x, y) represents the independent variable x and the dependent variable y.

1.

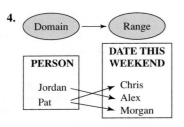

2.

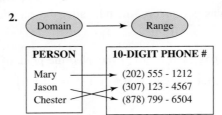

3.

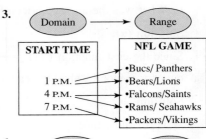

4.

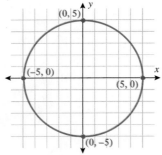

5.

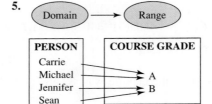

6.

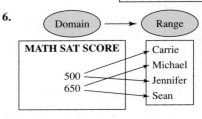

7. $\{(0, -3), (0, 3), (-3, 0), (3, 0)\}$

8. $\{(2, -2), (2, 2), (5, -5), (5, 5)\}$

9. $\{(0, 0), (9, -3), (4, -2), (4, 2), (9, 3)\}$

10. $\{(0, 0), (-1, -1), (-2, -8), (1, 1), (2, 8)\}$

11. $\{(0, 1), (1, 0), (2, 1), (-2, 1), (5, 4), (-3, 4)\}$

12. $\{(0, 1), (1, 1), (2, 1), (3, 1)\}$

13. $x^2 + y^2 = 9$

14. $x = |y|$

15. $x = y^2$

16. $y = x^3$

17. $y = |x - 1|$

18. $y = 3$

19.

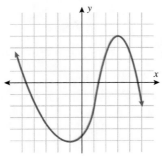

20.

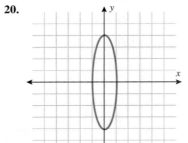

21.

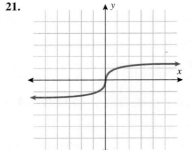

22.

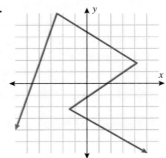

23.

24.

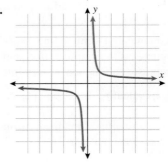

In Exercises 25–48, evaluate the given quantities using the following four functions.

$$f(x) = 2x - 3 \qquad F(t) = 4 - t^2 \qquad g(t) = 5 + t \qquad G(x) = x^2 + 2x - 7$$

25. $f(-2)$

26. $G(-3)$

27. $g(1)$

28. $F(-1)$

29. $f(-2) + g(1)$

30. $G(-3) - F(-1)$

31. $3f(-2) - 2g(1)$

32. $2F(-1) - 2G(-3)$

33. $\dfrac{f(-2)}{g(1)}$

34. $\dfrac{G(-3)}{F(-1)}$

35. $\dfrac{f(0) - f(-2)}{g(1)}$

36. $\dfrac{G(0) - G(-3)}{F(-1)}$

37. $f(x + 1) - f(x - 1)$

38. $F(t + 1) - F(t - 1)$

39. $g(x + a) - f(x + a)$

40. $G(x + b) + F(b)$

41. $\dfrac{f(x + h) - f(x)}{h}$

42. $\dfrac{F(t + h) - F(t)}{h}$

43. $\dfrac{g(t + h) - g(t)}{h}$

44. $\dfrac{G(x + h) - G(x)}{h}$

45. $\dfrac{f(-2 + h) - f(-2)}{h}$

46. $\dfrac{F(-1 + h) - F(-1)}{h}$

47. $\dfrac{g(1 + h) - g(1)}{h}$

48. $\dfrac{G(-3 + h) - G(-3)}{h}$

In Exercises 49–66, find the domain of the given function. Express the domain in interval notation.

49. $f(x) = 2x - 5$

50. $f(x) = -2x - 5$

51. $g(t) = t^2 + 3t$

52. $h(x) = 3x^4 - 1$

53. $P(x) = \dfrac{x + 5}{x - 5}$

54. $Q(t) = \dfrac{2 - t^2}{t + 3}$

55. $T(x) = \dfrac{2}{x^2 - 4}$

56. $R(x) = \dfrac{1}{x^2 - 1}$

57. $F(x) = \dfrac{1}{x^2 + 1}$

58. $G(t) = \dfrac{2}{t^2 + 4}$

59. $q(x) = \sqrt{7 - x}$

60. $k(t) = \sqrt{t - 7}$

61. $f(x) = \sqrt{2x + 5}$

62. $g(x) = \sqrt{5 - 2x}$

63. $G(t) = \sqrt{t^2 - 4}$

64. $F(x) = \sqrt{x^2 - 25}$

65. $F(x) = \dfrac{1}{\sqrt{x - 3}}$

66. $G(x) = \dfrac{2}{\sqrt{5 - x}}$

 ■ APPLICATIONS

67. Budget. The cost associated with a catered wedding reception is $45 per person for a reception for more than 75 people. Write the cost of the reception in terms of the number of guests and state any domain restrictions.

68. Budget. The cost of a local home phone plan is $35 for basic service and $.10 per minute for any domestic long-distance calls. Write the cost of monthly phone service in terms of the number of monthly long-distance minutes and state any domain restrictions.

69. Temperature. The average temperature in Tampa, Florida, in the springtime is given by the function $T(x) = -0.7x^2 + 16.8x - 10.8$ where T is the temperature in degrees Fahrenheit and x is the time of day in military time and is restricted to $6 \le x \le 18$ (sunrise to sunset). What is the temperature at 6 A.M.? What is the temperature at noon?

70. Firecrackers. A firecracker is launched straight up, and its height is a function of time, $h(t) = -16t^2 + 128t$ where h is the height in feet and t is the time in seconds with $t = 0$ corresponding to the instant it launches. What is the height 4 seconds after launch? What is the domain of this function?

71. Collector Card. The price of a signed Alex Rodriguez card is a function of how many are for sale. When he was traded from the Texas Rangers to the New York Yankees, the going rate for a signed card on eBay in 2004 was $P(x) = 10 + \sqrt{400,000 - 100x}$ where x represents the number of signed cards for sale. What was the value of the card when there were 10 signed cards for sale? What was the value of the card when there were 100 signed cards for sale?

72. Collector Card. In Exercise 71, what was the lowest price on eBay, and how many cards were available then? What was the highest price on eBay, and how many cards were available then?

73. Volume. An open box is constructed from a square 10 inch piece of cardboard by cutting squares of length x out of each corner and folding the sides up. Express the volume of the box as a function of x and state the domain.

74. Volume. A cylindrical water basin will be built to harvest rainwater. The basin is limited in that the largest radius it can have is 10 feet. Write a function representing the volume of water, V, as a function of height, h. How many additional gallons of water will be collected if you increase the height by 2 feet? (Hint: 1 cubic foot = 7.48 gallons.)

■ CATCH THE MISTAKE

In Exercises 75–80, explain the mistake that is made.

75. Determine if the relationship is a function.

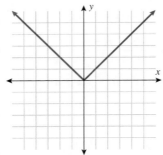

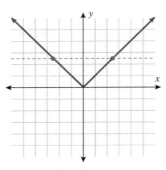

Solution: Apply the horizontal line test.

Because the horizontal line intersects the graph in two places, this is not a function.

76. Given the function, $H(x) = 3x - 2$, evaluate the quantity $H(3) - H(-1)$.

Solution: $H(3) - H(-1) = H(3) + H(1)$

$$H(3) = 7 \text{ and } H(1) = 1$$

$$H(3) + H(1) = 7 + 1 = 8$$

77. Given the function, $f(x) = x^2 - x$, evaluate the quantity $f(x + 1)$.

Solution: $f(x + 1) = f(x) + f(1)$

$$f(x) = x^2 - x \text{ and } f(1) = 0$$

$$f(x + 1) = x^2 - x$$

78. Determine the domain of the function $g(t) = \sqrt{3 - t}$, and express in interval notation.

Solution:

What can t be? Any nonnegative real number.

$$3 - t > 0$$

$$3 > t \text{ or } t < 3$$

Domain: $(-\infty, 3)$

79. Given the function, $G(x) = x^2$, evaluate

$$\frac{G(-1 + h) - G(-1)}{h}.$$

Solution:

$$\frac{G(-1 + h) - G(-1)}{h} = \frac{G(-1) + G(h) - G(-1)}{h} = \frac{G(h)}{h}$$

$$G(x) = x^2 \Rightarrow G(\ ) = (\ )^2 \Rightarrow G(h) = (h)^2$$

$$\frac{G(h)}{h} = \frac{h^2}{h} = h$$

80. Given the function, $f(x) = |x - A| - 1$ and $f(1) = -1$, find A.

Solution:

Since $f(1) = -1$, the point $(-1, 1)$ must satisfy the function. $-1 = |-1 - A| - 1$

Add 1 to both sides of the equation. $|-1 - A| = 0$

The absolute value of zero is zero so there is no need for the absolute value signs.

$$-1 - A = 0 \Rightarrow A + 1 = 0 \Rightarrow A = -1$$

■ CHALLENGE

81. T or F: If a vertical line does not intersect the graph of an equation, then that equation does not represent a function.

82. T or F: If a horizontal line intersects a graph of an equation more than once, the equation does not represent a function.

83. T or F: If $f(-a) = f(a)$, then f does not represent a function.

84. T or F: If $f(-a) = f(a)$, then f may or may not represent a function.

85. If $f(x) = Ax^2 - 3x$ and $f(1) = -1$, find A.

86. If $g(x) = \dfrac{1}{b - x}$, and $g(3)$ is undefined, find b.

87. If $F(x) = \dfrac{C - x}{D - x}$, $F(-2)$ is undefined and $F(-1) = 4$, find C and D.

88. Construct a function that is undefined at $x = 5$ and the point $(1, -1)$ lies on the graph of the function.

■ **TECHNOLOGY**

89. Using a graphing utility, graph the temperature function in Exercise 69. What time of day is it the warmest? What is the temperature? Looking at this function, explain why this model for Tampa, Florida, is only valid from sunrise to sunset (6 A.M. to 6 P.M. [1800]).

90. Using a graphing utility, graph the height of the firecracker in Exercise 70. How long after liftoff is the firecracker airborne? What is the maximum height that the firecracker attains? Explain why this height model is only valid for the first 8 seconds.

91. Using a graphing utility, graph the price function in Exercise 71. What are the lowest and highest prices of the cards? Does this agree with what you found in Exercise 72?

92. The makers of malted milk balls are considering increasing the size of the spherical treats. The thin chocolate coating on a malted milk ball can be approximated by the surface area, $S(r) = 4\pi r^2$. If the radius is increased 3 mm, what is the resulting increase in required chocolate for the thin outer coating?

SECTION 3.2 Graphs of Functions: Common Functions and Piecewise-Defined Functions

Skills Objectives

- Recognize and graph common functions.
- Classify functions as even, odd, or neither.
  - Algebraically
  - By graphing
- Determine if intervals are increasing, decreasing, or constant.
- Graph piecewise-defined functions.
- Solve application problems involving piecewise functions.

Conceptual Objectives

- Identify common functions.
- Develop and graph piecewise-defined functions.
  - Identify and graph points of discontinuity.
  - State the domain and range.

Common Functions

In Section 2.2, point-plotting techniques were discussed, and it was mentioned that we would discuss more efficient ways of graphing functions later in this chapter. There are nine main functions that we will discuss that will constitute a "library" of functions that you should commit to memory. We will draw on this library of functions in the next section when graphing aids are discussed. Several of these functions have been shown previously in this chapter, but now we will classify them specifically by name and discuss properties that each function exhibits.

In Section 2.3 we discussed equations and graphs of lines. All lines (with the exception of vertical lines) pass the vertical line test, and hence are classified as functions. Instead of the traditional notation of a line, $y = mx + b$, we use function notation and classify a function whose graph is a *line* as a *linear* function.

LINEAR FUNCTION

$$f(x) = mx + b \qquad m \text{ and } b \text{ are real numbers.}$$

The domain of a linear function, $f(x) = mx + b$, is the set of all real numbers, $\mathbb{R}$. The graph of this function has slope m and y-intercept $(0, b)$, Two specific linear functions are the constant function (when $m = 0$) and the identity function (when $b = 0$ and $m = 1$).

When the slope is zero, $m = 0$, the function reduces to the special case of a constant function.

CONSTANT FUNCTION

$$f(x) = b \qquad\qquad b \text{ is any real number.}$$

The graph of a constant function, $f(x) = b$, is a horizontal line. The y-intercept is the point $(0, b)$. The domain of a constant function is the set of all real numbers, $\mathbb{R}$. The range, however, is a single value, b. In other words, all x values correspond to a single y value.

Points that lie on the graph of a constant function, $f(x) = b$, are

$(-5, b)$

$(-1, b)$

$(0, b)$

$(2, b)$

$(4, b)$

$\dots (x, b)$

Another specific example of a linear function is when the slope is one, $m = 1$, and the y-intercept is zero, $b = 0$. This special case is called the identity function.

IDENTITY FUNCTION

$$f(x) = x$$

The graph of the identity function has the following properties: It passes through the origin, and every point that lies on the line has equal x- and y-coordinates. Both the domain and the range of the identity function are the set of all real numbers, $\mathbb{R}$.

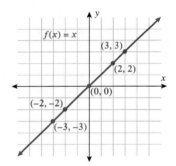

A function that squares the input is called the square function.

SQUARE FUNCTION

$$f(x) = x^2$$

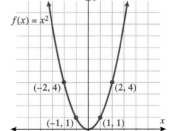

The graph of the square function is called a parabola and will be discussed in further detail in Chapters 4 and 8. The domain of the square function is the set of all real numbers, $\mathbb{R}$. Because squaring a real number always yields a positive number or zero, the range of the square function is the set of all nonnegative numbers. Notice that the intercept is the origin and the square function is symmetric about the y-axis. This graph is contained in quadrants I and II.

A function that cubes the input is called the cube function.

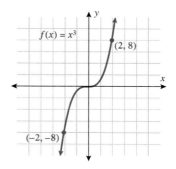

CUBE FUNCTION

$$f(x) = x^3$$

The domain of the cube function is the set of all real numbers, $\mathbb{R}$. Because cubing a negative number yields a negative number, the range of the cube function is also the set of all real numbers, $\mathbb{R}$. Notice that the intercept is the origin and the cube function is symmetric about the origin. This graph extends only into quadrants I and III.

The next two functions have similar names to the previous two functions: square root and cube root. When a function takes the square root of the input or the cube root of the input, the function is called the square root function or the cube root function, respectively. In the following discussion we restrict ourselves to real numbers.

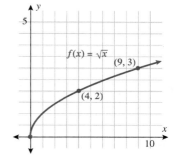

SQUARE ROOT FUNCTION

$$f(x) = \sqrt{x} \quad \text{or} \quad f(x) = x^{1/2}$$

What values for x will yield real values for the function? All negative numbers must be eliminated from the domain. Therefore, we state the domain to be all nonnegative real numbers. The output of the function will be all real numbers greater than or equal to zero. The graph of this function will be contained in quadrant I.

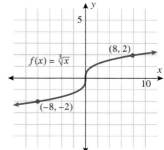

CUBE ROOT FUNCTION

$$f(x) = \sqrt[3]{x} \quad \text{or} \quad f(x) = x^{1/3}$$

The cube root function does not have any restrictions on the domain and, hence, the range. In fact, the domain and range both are the set of all real numbers, $\mathbb{R}$. This graph is contained in quadrants I and III and passes through the origin. This function is symmetric about the origin.

In Section 1.7, absolute value equations and inequalities were discussed. Now we shift our focus to the graph of the absolute value function.

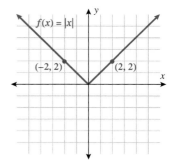

ABSOLUTE VALUE FUNCTION

$$f(x) = |x|$$

Some points that are on the graph of the absolute value function are $(-1, 1)$, $(0, 0)$, and $(1, 1)$. The domain of the absolute value function is the set of all real numbers, $\mathbb{R}$, yet the range is the set of nonnegative real numbers. The graph of this function is symmetric with respect to the y-axis and is contained in quadrants I and II.

A function that takes the reciprocal of the input is called the reciprocal function.

RECIPROCAL FUNCTION

$$f(x) = \frac{1}{x} \qquad x \neq 0$$

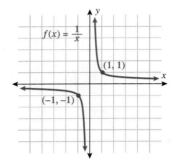

$f(x) = \dfrac{1}{x}$

$(1, 1)$

$(-1, -1)$

The only restriction on the domain of the reciprocal function is that $x \neq 0$. Hence, we say the domain is the set of all real numbers excluding zero. The range is the set of all real numbers except zero. This function is symmetric with respect to the origin and is contained in quadrants I and III.

Even and Odd Functions

Of the nine functions discussed above, several have similar properties of symmetry. The constant function, square function, and absolute value function are all symmetric with respect to the y-axis. The identity function, cube function, cube root function, and reciprocal function are all symmetric with respect to the origin. The term **even** is used to describe functions that are symmetric with respect to the y-axis, or vertical axis, and the term **odd** is used to describe functions that are symmetric with respect to the origin. Recall from Section 2.2 that symmetry can be determined both graphically and algebraically. The box below summarizes the graphic and algebraic characteristics of even and odd functions.

TECHNOLOGY TIP

Graph $f(x) = |x|$.

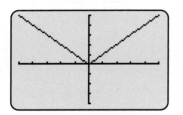

EVEN AND ODD FUNCTIONS

FUNCTION	SYMMETRIC WITH RESPECT TO	ON REPLACING x WITH $-x$
Even	y-axis or vertical axis	$f(-x) = f(x)$
Odd	origin	$f(-x) = -f(x)$

The algebraic method for determining symmetry with respect to the y-axis, or vertical axis, is to substitute in $-x$ for x. If it results in an equivalent equation, the function is symmetric with respect to the y-axis. Some examples of even functions are $f(x) = b$, $f(x) = x^2$, $f(x) = x^4$, and $f(x) = |x|$. In any of these equations, if $-x$ is substituted for x, the result is the same, $f(-x) = f(x)$. Also note that, with the exception of the absolute value function, these examples are all even-degree polynomial equations. One common mistake is assuming the constant function is odd if the constant is odd or even if the constant is even. All constant functions are even functions.

The algebraic method for determining symmetry with respect to the origin is to substitute $-x$ for x. If the result is the negative of the original function, $f(-x) = -f(x)$, then the function is symmetric with respect to the origin and, hence, classified as an odd function. Examples of odd functions are $f(x) = x$, $f(x) = x^3$, $f(x) = x^5$, and $f(x) = x^{1/3}$. In any of these functions, if $-x$ is substituted in for x, the result is the negative of the original function. Notice that with the exception of the cube root function, these equations are odd-degree polynomials.

Be careful, though, because functions that are combinations of even- and odd-degree polynomials can turn out to be neither even nor odd, as we will see in the next example.

EXAMPLE 1 Determining If a Function Is Even, Odd, or Neither

Determine if the functions are even, odd, or neither.

a. $f(x) = x^2 - 3$ **b.** $g(x) = x^5 + x^3$ **c.** $h(x) = x^2 - x$

Solution (a):

Original function. $\qquad\qquad\qquad\qquad f(x) = x^2 - 3$

Replace x with $-x$. $\qquad\qquad\qquad\quad f(-x) = (-x)^2 - 3$

Simplify.

$$f(-x) = x^2 - 3 = f(x)$$

Because $f(-x) = f(x)$, we say that $f(x)$ is an *even* function .

Solution (b):

Original function:

$$g(x) = x^5 + x^3$$

Replace x with $-x$.

$$g(-x) = (-x)^5 + (-x)^3$$

Simplify.

$$g(-x) = -x^5 - x^3 = -(x^5 + x^3) = -g(x)$$

Because $g(-x) = -g(x)$, we say that $g(x)$ is an *odd* function .

Solution (c):

Original function:

$$h(x) = x^2 - x$$

Replace x with $-x$.

$$h(-x) = (-x)^2 - (-x)$$

Simplify.

$$h(-x) = x^2 + x$$

$h(-x)$ is neither $-h(x)$ nor $h(x)$; therefore the function $h(x)$ is neither even nor odd .

In parts a, b, and c, we classified these functions as either even, odd, or neither, using the algebraic test. Look back at them now and reflect on whether these classifications agree with your intuition. In part (a), we combined two functions: the square function and constant function. Both of these functions are even, and adding even functions yields another even function. In part (b), we combined two odd functions: the fifth-power function and the cube function. Both of these functions are odd, and adding two odd functions yields another odd function. In part (c), we combined two functions: the square function and the identity function. The square function is even, and the identity function is odd. In this part, combining an even function with an odd function yields a function that is neither even nor odd and, hence, has no symmetry with respect to the vertical axis or the origin.

■ **YOUR TURN**   Classify the functions as even, odd, or neither.

  a. $f(x) = |x| + 4$ **b.** $f(x) = x^3 - 1$

Increasing and Decreasing Functions

Functions can be described as increasing, decreasing, or constant. There is no fourth option. Look at the figure on the right. If we start at the left side of the graph and trace the red curve with our pen, we see that the function values (values in the vertical direction) are decreasing until arriving at the point $(-2, 2)$. Then, the function values increase until arriving at the point $(-1, 1)$. The value then remains constant $(y = 1)$ between the points $(-1, 1)$ and $(0, 1)$. Proceeding beyond the point $(0, 1)$, the function values decrease again until the point $(2, -2)$. Beyond the point $(2, -2)$ the function

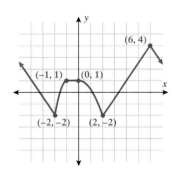

■ **Answer: a.** even   **b.** neither.

values increase again until the point (6, 4). Finally, the function values decrease and continue to do so.

When specifying intervals of increasing, decreasing, and constant function values, the intervals are classified according to the x-coordinate. For instance, in this graph, we say the function is increasing when x is between $x = -2$ and $x = -1$ and again when x is between $x = 2$ and $x = 6$. The graph is classified as decreasing when x is less than $x = -2$ and again when x is between $x = 0$ and $x = 2$ and again when x is greater than $x = 6$. The graph is classified as constant when x is between $x = -1$ and $x = 0$. In interval notation, this is summarized as

DECREASING	INCREASING	CONSTANT
$(-\infty, -2) \cup (0, 2) \cup (6, \infty)$	$(-2, -1) \cup (2, 6)$	$(-1, 0)$

An algebraic test for increasing, decreasing, or constant is to compare the value, $f(x)$, of the function for particular points in intervals.

INCREASING, DECREASING, AND CONSTANT INTERVALS

If $x_1 < x_2$ then

Increasing interval:	$f(x_1) < f(x_2)$
Decreasing interval:	$f(x_1) > f(x_2)$
Constant interval:	$f(x_1) = f(x_2)$

Let's now classify our library of functions according to their intervals of increasing, decreasing, or constant. The constant function stays constant for all x. Therefore, the interval that is constant is $(-\infty, \infty)$, which accounts for all real numbers—there are no intervals of increasing or decreasing. The identity function, cube function, and cube root function always increase from left to right on the graph. Therefore, we classify those intervals of increasing as $(-\infty, \infty)$, and, hence, those functions have no intervals of decreasing or constant values. The square root function always increases. Since it is only defined for nonnegative numbers, we say the square root function has an increasing interval of $(0, \infty)$. The square and absolute value functions, however, both decrease until reaching the origin, then increase. Therefore, we classify those functions as having no intervals of constant value, and the interval, $(-\infty, 0)$, is decreasing and the interval, $(0, \infty)$, is increasing.

Piecewise-Defined Functions

Most of the functions that we have seen in this text are functions defined by polynomials. Sometimes, the need arises to define functions in terms of *pieces*. For example, most plumbers charge a flat fee for a house call and then an additional hourly rate for the job. For instance, if a particular plumber charges $100 to drive out to your house and work for 1 hour and then an additional $25 an hour for every additional hour he/she works on your job, we would define this function in pieces. If we let h be the number of hours worked, then the charge is defined as

$$\text{Plumbing charge} = \begin{cases} 100 & h \leq 1 \\ 100 + 25(h - 1) & h > 1 \end{cases}$$

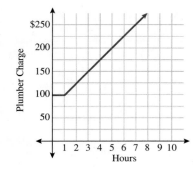

If we were to graph this function, we would see that there is 1 hour that is constant and after that the function continually increases.

Another piecewise-defined function is the absolute value function. The absolute value function can be thought of as two pieces: the line $y = -x$ (when x is negative) and the line $y = x$ (when x is positive). We start by graphing these two lines on the same graph.

The absolute value function behaves like the line $y = -x$ when x is negative (erase the blue graph in quadrant IV) and like the line $y = x$ when x is positive (erase the purple graph in quadrant III).

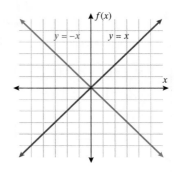

Absolute value piecewise-defined function

$$f(x) = |x| = \begin{cases} -x & x < 0 \\ x & x \geq 0 \end{cases}$$

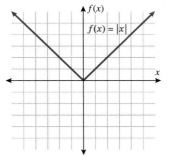

The next example is a piecewise-defined function given in terms of functions in our "library of functions." Because the function is defined in terms of pieces of other functions, we draw the graph of each individual function and, then, for each function darken the piece corresponding to its part of the domain. This is like the procedure above for the absolute value function.

EXAMPLE 2 Graphing Piecewise-Defined Functions

Graph the piecewise-defined function, and state the domain, range, and intervals when the function is increasing, decreasing, or constant.

$$G(x) = \begin{cases} x^2 & x < -1 \\ 1 & -1 \leq x \leq 1 \\ x & x > 1 \end{cases}$$

Solution:

Graph each of the functions on the same plane.

Square function:

$f(x) = x^2$

Constant function:

$f(x) = 1$

Identity function:

$f(x) = x$

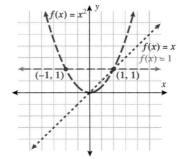

The points to focus on in particular are the x values where the pieces change over, that is, $x = -1$ and $x = 1$.

Let's now investigate each piece. When $x < -1$, this function is defined by the square function, $f(x) = x^2$, so darken that particular function to the left of $x = -1$. When $-1 \leq x \leq 1$, the function is defined by the constant function, $f(x) = 1$, so darken that particular function between the x values of -1

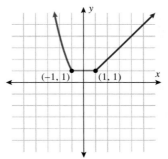

TECHNOLOGY TIP
Graph the piecewise function
using a graphing utility.

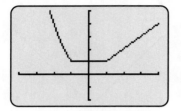

and 1. When $x > 1$, the function is defined by the identity function, $f(x) = x$, so darken that function to the right of $x = 1$. Erase everything that is not darkened, and the resulting graph of the piecewise-defined function is given below.

This function is defined for all real values x, so the domain of this function is the set of all real numbers. The values that this function yields in the vertical direction are all real numbers greater than or equal to 1. Hence, the range of this function is $[1, \infty)$. The intervals of increasing, decreasing, and constant are as follows:

$$\text{Decreasing: } (-\infty, -1)$$

$$\text{Constant: } (-1, 1)$$

$$\text{Increasing: } (1, \infty)$$

Notice that the points $x = -1$ and $x = 1$ are not classified as increasing, decreasing, or constant. Those two points represent the transition between increasing, decreasing, and constant.

The term **continuous** implies that there are no holes or jumps and the graph can be drawn without picking up your pencil. A function that does have holes or jumps and cannot be drawn in one motion without picking up your pencil is classified as **discontinuous**, and the points where the holes or jumps occur are called *points of discontinuity*.

The previous example illustrates a *continuous* piecewise-defined function. At the $x = -1$ junction, the square function and constant function both pass through the point $(-1, 1)$. At the $x = 1$ junction, the constant function and the identity function both pass through the point $(1, 1)$. Since the graph of this piecewise function has no holes or jumps, we classify it as a continuous function.

The next example illustrates a *discontinuous* piecewise-defined function.

EXAMPLE 3 Graphing a Discontinuous Piecewise-Defined Function

Graph the piecewise-defined function, and state the intervals where the function is increasing, decreasing, or constant, along with the domain and range.

$$f(x) = \begin{cases} 1 - x & x < 0 \\ x & 0 \le x < 2 \\ -1 & x > 2 \end{cases}$$

Solution:

Graph these functions on the same plane:

Linear function:

$f(x) = 1 - x$

Identity function:

$f(x) = x$

Constant function:

$f(x) = -1$

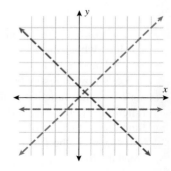

Unlike Example 2, at the junction points, $x = 0$ and $x = 2$, the functions do not match up. There will be a jump at those values of x. Particular attention must be given to the inequality. The function is defined by the **identity function** at the point, $x = 0$, and notice that the function is *not defined* at the point $x = 2$. The use of open circles denotes up to but not including the point, and the use of filled-in circles denotes the value corresponding to that point.

Let's darken the piecewise function on the graph. For all values less than zero, $x < 0$, the function is defined by the **linear function**. Note the use of the open circle at zero, indicating up to, but not including, that point. For values, $0 \le x < 2$, the function is defined by the **identity function**.

Because the left endpoint includes the value at zero, since the right endpoint of this interval does not include the point, a filled-in circle is used at $x = 0$, and an open circle is used at $x = 2$. For all values greater than 2, $x > 2$, the function is defined by the **constant function**. Because this interval does not include the point, $x = 2$, an open circle is used.

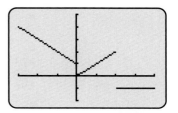

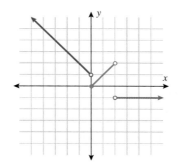

What are the intervals when the function is increasing, decreasing, or constant? Remember that the intervals correspond to the x-values.

Decreasing: $(-\infty, 0)$ Increasing: $(0, 2)$ Constant: $(2, \infty)$

What is the domain of this function? Is the function defined for all values of x? No, it is defined for all values except $x = 2$. The domain is expressed in interval notation below.

Domain: $(-\infty, 2) \cup (2, \infty)$

What is the range of this function? The output in the vertical direction takes on the y-values $y \ge 0$ and the additional single value $y = -1$. Note that this function does not assign the values between 0 and -1 or any numbers less than -1. The range is given in interval notation below.

Range: $[-1, -1] \cup [0, \infty)$

It was mentioned earlier that a discontinuous function has a graph that exhibits holes or jumps. In Example 3, the point $x = 0$ corresponds to a jump, because you would have to pick up your pencil to continue drawing the graph, but the function is defined at $x = 0$. The point $x = 2$ corresponds to both a hole and a jump. The hole indicates that the function is not defined at that point, and there is still a jump because the identity function and the constant function do not approach the same y value at $x = 2$.

■ **YOUR TURN** 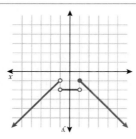 Graph the piecewise-defined function, and state the intervals where the function is increasing, decreasing, or constant, along with the domain and range.

$$f(x) = \begin{cases} -x & x \le -1 \\ 2 & -1 < x < 1 \\ x & x > 1 \end{cases}$$

SECTION 3.2 SUMMARY

Common Functions

- Linear function
- Constant function
- Identity function
- Square function
- Cube function
- Square root function
- Cube root function
- Absolute value function
- Reciprocal function

Even and Odd Functions

- Even function: Symmetry with respect to the y-axis and $f(-x) = f(x)$
- Odd function: Symmetry with respect to the origin and $f(-x) = -f(x)$

Intervals When a Function is Increasing, Decreasing, or Constant

Piecewise-Defined Functions

- Continuous: Draw without picking up your pencil.
- Discontinuous
  - Have holes, a point where the function is not defined.
  - Have jumps, a point where you must pick up your pencil to continue drawing a graph.

SECTION 3.2 EXERCISES

 ■ **SKILLS**

In Exercises 1–24, determine if the function is even, odd, or neither.

1. $G(x) = x + 4$

2. $h(x) = 3 - x$

3. $f(x) = 3x^2 + 1$

4. $F(x) = x^4 + 2x^2$

5. $g(t) = 5t^3 - 3t$

6. $f(x) = 3x^5 + 4x^3$

7. $h(x) = x^2 + 2x$

8. $G(x) = 2x^4 - 3x^3$

■ **Answer:** Increasing: $(1, \infty)$
Decreasing: $(-\infty, -1)$
Constant: $(-1, 1)$
Domain: $(-\infty, 1) \cup (1, \infty)$
Range: $[1, \infty)$

9. $h(x) = x^{1/3} - x$ **10.** $g(x) = x^{-1} + x$

11. $f(x) = |x| + 5$ **12.** $f(x) = |x| + x^2$

13. $f(x) = |x|$ **14.** $f(x) = |x^3|$

15. $G(t) = |t - 3|$ **16.** $g(t) = |t + 2|$

17. $G(t) = \sqrt{t - 3}$ **18.** $f(x) = \sqrt{2 - x}$

19. $g(x) = \sqrt{x^2 + x}$ **20.** $f(x) = \sqrt{x^2 + 2}$

21. $h(x) = \dfrac{1}{x} + 3$ **22.** $h(x) = \dfrac{1}{x} - 2x$

23.

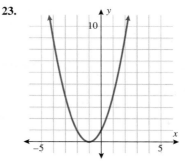

24.

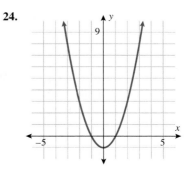

In Exercises 25–48, graph the piecewise-defined functions. State the domain and range in interval notation. Determine the intervals when the function is increasing, decreasing, or constant.

25. $f(x) = \begin{cases} x & x < 2 \\ 2 & x \geq 2 \end{cases}$

26. $f(x) = \begin{cases} -x & x < -1 \\ -1 & x \geq -1 \end{cases}$

27. $f(x) = \begin{cases} 1 & x < -1 \\ x^2 & x \geq -1 \end{cases}$

28. $f(x) = \begin{cases} x^2 & x < 2 \\ 4 & x \geq 2 \end{cases}$

29. $f(x) = \begin{cases} x & x < 0 \\ x^2 & x \geq 0 \end{cases}$

30. $f(x) = \begin{cases} -x & x \leq 0 \\ x^2 & x > 0 \end{cases}$

31. $f(x) = \begin{cases} -x + 2 & x < 1 \\ x^2 & x \geq 1 \end{cases}$

32. $f(x) = \begin{cases} 2 + x & x \leq -1 \\ x^2 & x > -1 \end{cases}$

33. $G(x) = \begin{cases} -1 & x < -1 \\ x & -1 \leq x \leq 3 \\ 3 & x > 3 \end{cases}$

34. $G(x) = \begin{cases} -1 & x < -1 \\ x & -1 < x < 3 \\ 3 & x > 3 \end{cases}$

35. $G(t) = \begin{cases} 1 & t < 1 \\ t^2 & 1 \leq t \leq 2 \\ 4 & t > 2 \end{cases}$

36. $G(t) = \begin{cases} 1 & t < 1 \\ t^2 & 1 < t < 2 \\ 4 & t > 2 \end{cases}$

37. $f(x) = \begin{cases} -x - 1 & x < -2 \\ x + 1 & -2 < x < 1 \\ -x + 1 & x \geq 1 \end{cases}$

38. $f(x) = \begin{cases} -x - 1 & x \leq -2 \\ x + 1 & -2 < x < 1 \\ -x + 1 & x > 1 \end{cases}$

39. $G(x) = \begin{cases} 0 & x < 0 \\ \sqrt{x} & x \geq 0 \end{cases}$

40. $G(x) = \begin{cases} 1 & x < 1 \\ \sqrt[3]{x} & x > 1 \end{cases}$

41. $G(x) = \begin{cases} 0 & x = 0 \\ \dfrac{1}{x} & x \neq 0 \end{cases}$

42. $G(x) = \begin{cases} 0 & x = 0 \\ -\dfrac{1}{x} & x \neq 0 \end{cases}$

43. $G(x) = \begin{cases} -\sqrt[3]{x} & x \leq -1 \\ x & -1 < x < 1 \\ -\sqrt{x} & x > 1 \end{cases}$

44. $G(x) = \begin{cases} \sqrt[3]{x} & x < -1 \\ x & -1 \leq x < 1 \\ \sqrt{x} & x > 1 \end{cases}$

45. $f(x) = \begin{cases} x + 3 & x \leq -2 \\ |x| & -2 < x < 2 \\ x^2 & x \geq 2 \end{cases}$

46. $f(x) = \begin{cases} |x| & x < -1 \\ 1 & -1 < x < 1 \\ |x| & x > 1 \end{cases}$

47. $f(x) = \begin{cases} x & x \leq -1 \\ x^3 & -1 < x < 1 \\ x^2 & x > 1 \end{cases}$

48. $f(x) = \begin{cases} x^2 & x \leq -1 \\ x^3 & -1 < x < 1 \\ x & x \geq 1 \end{cases}$

■ APPLICATIONS

In Exercises 49–58, formulate a piecewise-defined function that models the real-world scenario.

49. Budget. The Kappa Kappa Gamma sorority decides to order custom-made T-shirts for their *Kappa Krush* mixer with the Sigma Alpha Epsilon fraternity. If they order 50 or fewer T-shirts, the cost is $10 per shirt. If they order more than 50 but 100 or fewer T-shirts, the cost is $9 per shirt. If they order more than 100, the cost is $8 per shirt.

Find the cost function, $C(x)$, as a function of the number of T-shirts, x, ordered.

50. Budget. The marching band at a university is ordering some additional uniforms to replace existing uniforms that are worn out. If they order 50 or fewer, the cost is $176.12

per uniform. If they order more than 50, but fewer than 100, the cost is $159.73 per uniform. Find the cost function, $C(x)$, as a function of the number of new uniforms ordered, x.

51. **Budget.** The Richmond rowing club is planning to enter the *Head of the Charles* race in Boston and is trying to figure out how much money to raise. The entry fee is $250 per boat for the first 10 boats, and $175 for each additional boat. Find the cost function, $C(x)$, as a function of the number of boats they enter, x.

52. **Phones.** A phone company charges $.39 per minute for the first 10 minutes of an international long-distance phone call and $.12 per minute every minute after that. Find the cost function, $C(x)$, as a function of the length of the phone call in minutes, x.

53. **Wedding.** A young couple is planning their wedding reception at a yacht club. The yacht club charges a flat rate of $1,000 to reserve the dining room for a private party. The cost of food is $35 per person for the first 100 people and $25 per person for every additional person beyond the first 100. Write the cost function, $C(x)$, as a function of the number of people, x, attending the reception.

54. **Budget.** An irrigation company gives you an estimate for an eight zone sprinkler system. The parts are $1,400, and the labor is $25 per hour. Write a function, $C(x)$, that determines the cost of a new sprinkler system if you choose this irrigation company.

55. **Royalties.** A famous author negotiates with her publisher the monies she will receive for her next suspense novel. She will receive $50,000 up front and a 15% royalty rate on the first 100,000 books sold, and 20% on any books sold beyond that. If the book sells for $20 and royalties are based on the selling price, write a royalties function, $R(x)$, as a function of total number of books sold, x.

56. **Royalties.** Rework Exercise 55 if the author receives $35,000 up front, 15% for the first 100,000 books sold, and 25% on any books sold beyond that.

57. **Profit.** A group of artists is trying to decide if they will make a profit if they set up a Web-based business to market and sell stained glass that they make. The costs associated with this business are $100 per month for the website and $700 per month for the studio they rent. The materials for each work in stained glass cost $35 each, and the artists charge $100 for each one they sell. Write the monthly profit as a function of the number of stained glass units they sell.

58. **Profit.** Philip decides to host a shrimp boil at his house as a fundraiser for his daughter's AAU basketball team. He orders gulf shrimp to be flown in from New Orleans. The shrimp costs $5 per pound. The shipping costs $30. If he charges $10 per person, write a function, $F(x)$, that represents either his loss or profit as a function of the number of people that attend, x. Assume each person will eat 1 pound of shrimp.

 ■ **CATCH THE MISTAKE**

In Exercises 59–62, explain the mistake that is made.

59. Graph the piecewise-defined function. State the domain and range.

$$f(x) = \begin{cases} -x & x < 0 \\ x & x > 0 \end{cases}$$

Solution:

Draw the graphs of $f(x) = -x$ and $f(x) = x$

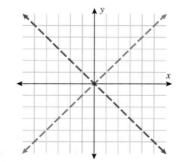

Darken the function, $f(x) = -x$, when $x < 0$ and the function $f(x) = x$, when $x > 0$. This gives us the familiar absolute value graph.

Domain: $(-\infty, \infty)$ or $\mathbb{R}$

Range: $[0, \infty)$

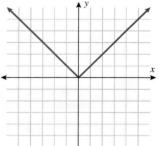

60. Graph the piecewise-defined function. State the domain and range.

$$f(x) = \begin{cases} -x & x \le 1 \\ x & x > 1 \end{cases}$$

Solution:

Draw the graphs of $f(x) = -x$ and $f(x) = x$.

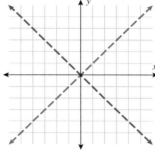

Darken the function, $f(x) = -x$, when $x < 1$ and the function $f(x) = x$, when $x > 1$.

The resulting graph is as shown.

Domain: $(-\infty, \infty)$ or $\mathbb{R}$

Range: $(-1, \infty)$

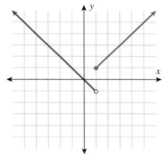

61. The cost of airport Internet access is $15 for the first 30 minutes and $1 per minute for each additional minute. Write a function describing the cost of the service as a function of minutes used on-line.

Solution: $C(x) = \begin{cases} 15 & x \le 30 \\ 15 + x & x > 30 \end{cases}$

62. Most money market accounts pay a higher interest with a higher principal. If the credit union is offering 2% on accounts with less than $10,000 and 4% on the additional money over $10,000, write the interest function, $I(x)$, that represents the interest earned on an account as a function of dollars in the account.

Solution: $I(x) = \begin{cases} 0.02x & x \le 10,000 \\ 0.02(10,000) + 0.04x & x > 10,000 \end{cases}$

■ CHALLENGE

63. T or F: The identity function is a special case of the linear function.

64. T or F: The constant function is a special case of the linear function.

65. T or F: If an odd function has an interval when the function is increasing, then it has to also have an interval when the function is decreasing.

66. T or F: If an even function has an interval when the function is increasing, then it has to also have an interval when the function is decreasing.

In Exercises 67 and 68, for a and b real numbers, can the function given ever be a continuous function? If so, specify the value for a and b that would make it so.

67. $f(x) = \begin{cases} ax & x \le 2 \\ bx^2 & x > 2 \end{cases}$

68. $f(x) = \begin{cases} -\dfrac{1}{x} & x < a \\ \dfrac{1}{x} & x \ge a \end{cases}$

■ TECHNOLOGY

69. In trigonometry you will learn about the sine function, $\sin x$. Using a graphing utility, plot the function $f(x) = \sin x$. It should look like the graph below. Is the sine function even, odd, or neither?

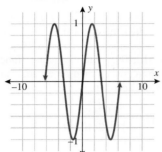

70. In trigonometry you will learn about the cosine function $\cos x$. Using a graphing utility, plot the function $f(x) = \cos x$. It should look like the graph below. Is the cosine function even, odd, or neither?

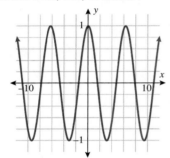

71. In trigonometry you will learn about the tangent function, $\tan x$. Using a graphing utility, plot the function $f(x) = \tan x$. If you restrict the values of x so that $-\frac{\pi}{2} < x < \frac{\pi}{2}$ the graph should resemble the graph below. Is the tangent function even, odd, or neither?

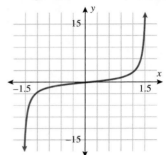

72. Plot the function $f(x) = \dfrac{\sin x}{\cos x}$. What function is this?

SECTION 3.3 Graphing Techniques: Transformations

Skills Objectives	Conceptual Objectives
■ Graph horizontal and vertical shifting of common functions. ■ Graph reflections of common functions about the x-axis or y-axis. ■ Graph expansion and contraction of common functions.	■ Identify the common functions by their graphs. ■ Use multiple transformations of common functions to obtain graphs of functions.

The focus of the previous section was to learn the graphs that correspond to particular functions such as identity, square, cube, square root, cube root, absolute value, and reciprocal. Therefore, at this point, you should be able to recognize and generate the graphs of $y = x$, $y = x^2$, $y = x^3$, $y = \sqrt{x}$, $y = \sqrt[3]{x}$, $y = |x|$, and $y = \frac{1}{x}$. In this section, we will discuss how to graph functions that look almost like these functions. For instance, a common function may be shifted (horizontally or vertically), reflected, or stretched (or compressed). Collectively, these techniques are called **transformations**.

Horizontal and Vertical Shifting

Let's take the absolute value function as an example. The graph of $f(x) = |x|$ was given in the last section. Now look at two examples that are much like this function: $g(x) = |x| + 2$ and $h(x) = |x - 1|$. Graphing these functions by point plotting yields

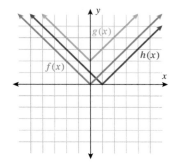

x	$f(x)$
-2	2
-1	1
0	0
1	1
2	2

x	$g(x)$
-2	4
-1	3
0	2
1	3
2	4

x	$h(x)$
-2	3
-1	2
0	1
1	0
2	1

Instead of point plotting the function, $g(x) = |x| + 2$, we could have started with the function, $f(x) = |x|$, and shifted the entire graph *up* 2 units. Similarly, we could have generated the graph of the function, $h(x) = |x - 1|$, by shifting the function $f(x) = |x|$ to the *right* 1 unit. In both cases, the base or starting function is $f(x) = |x|$. Why did we go up for $g(x)$ and to the right for $h(x)$?

Notice that we could rewrite the functions $g(x)$ and $h(x)$ in terms of $f(x)$:

$$g(x) = |x| + 2 \quad = f(x) + 2$$
$$h(x) = |x - 1| \quad = f(x - 1)$$

In the case of $g(x)$, the shift ($+2$) occurs *outside* the function. Therefore, the output for $g(x)$ is 2 more than the typical output for $f(x)$. Because the output corresponds to the vertical axis, this results in a shift *upward* of two units. In general, shifts that occur outside the function correspond to a vertical shift that corresponds to the sign of the shift. For instance, had the function been $G(x) = |x| - 2$, this graph would have started with the function $f(x)$ and shifted down 2 units.

In the case of $h(x)$, the shift occurs *inside* the function. Notice that the point $(0, 0)$ that lies on $f(x)$ was shifted to the point $(1, 0)$ on the function $h(x)$. The y value remained the same, but the x value shifted to the right 1 unit. Similarly, the points $(-1, 1)$ and $(1, 1)$ were shifted to the points $(0, 1)$ and $(2, 1)$, respectively. In general, shifts that occur inside the function correspond to a horizontal shift that corresponds to opposite the sign. In this case, $h(x) = |x - 1|$, shifted the function $f(x)$ to the right 1 unit. If instead, we had the function, $H(x) = |x + 1|$, this graph would have started with the function $f(x)$ and shifted to the left 1 unit.

VERTICAL SHIFTS

Assuming c is a positive constant, shifts **outside** the function correspond to a **vertical** shift that goes **with the sign**.

To GRAPH	SHIFT THE FUNCTION $f(x)$
$y = f(x) + c$	c units upward
$y = f(x) - c$	c units downward

HORIZONTAL SHIFTS

Assuming c is a positive constant, shifts **inside** the function correspond to a **horizontal** shift that goes **opposite the sign**.

TO GRAPH	SHIFT THE FUNCTION $f(x)$
$y = f(x + c)$	c units to the left
$y = f(x - c)$	c units to the right

EXAMPLE 1 Plotting Horizontal and Vertical Translations

Plot the given functions using graph-shifting techniques:

a. $g(x) = x^2 - 1$ **b.** $H(x) = (x + 1)^2$

Solution: In both cases, the function to start with is $f(x) = x^2$.

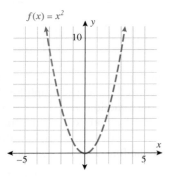

a. $g(x) = x^2 - 1$ can be rewritten as $g(x) = f(x) - 1$.

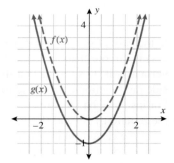

1. The shift (1 unit) occurs *outside* of the function. Therefore, we expect a vertical shift that goes *with* the sign.

2. Since the sign is *negative,* this corresponds to a *downward* shift.

3. Shifting the function $f(x) = x^2$ down 1 unit yields the graph of $g(x) = x^2 - 1$.

b. $H(x) = (x + 1)^2$ can be rewritten as $H(x) = f(x + 1)$.

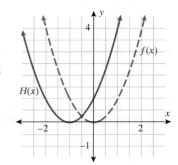

1. The shift (1 unit) occurs *inside* of the function. Therefore, we expect a horizontal shift that goes *opposite* the sign.

2. Since the sign is *positive,* this corresponds to a *left* shift.

3. Shifting the function $f(x) = x^2$ to the left 1 unit yields the graph of $H(x) = (x + 1)^2$.

CONCEPT CHECK What does the graph of $y = x^2$ look like?

■ **YOUR TURN** Plot the given functions using graph-shifting techniques.

a. $g(x) = x^2 + 1$ **b.** $H(x) = (x - 1)^2$

EXAMPLE 2 Horizontal and Vertical Shifts, and Changes in Domain and Range

Graph the functions using graph-shifting techniques, and state the domain and range of each function.

a. $g(x) = \sqrt{x + 1}$ **b.** $G(x) = \sqrt{x} - 2$

Solution:

In both cases the function to start with is $f(x) = \sqrt{x}$.

   Domain: $[0, \infty)$

   Range: $[0, \infty)$

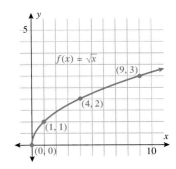

a.

  1. $g(x) = \sqrt{x + 1}$ can be rewritten as $g(x) = f(x + 1)$.

  2. The shift (1 unit) is *inside* the function, which corresponds to a *horizontal* shift *opposite the sign*.

  3. Shifting $f(x) = \sqrt{x}$ to the *left* 1 unit yields the graph of $g(x) = \sqrt{x + 1}$. Notice that the point $(0, 0)$, which lies on $f(x)$, gets shifted to the point $(-1, 0)$ on the graph of $g(x)$.

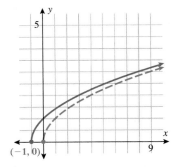

Although the original function, $f(x) = \sqrt{x}$, had an implicit restriction on the domain: $[0, \infty)$, the function, $g(x) = \sqrt{x + 1}$, has the implicit restriction that

---

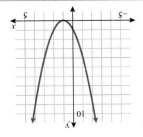

b.

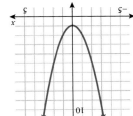

■ **Answer: a.**

$x \geq -1$. We see that the output or range of $g(x)$ is the same as the output of the original function, $f(x)$.

Domain: $[-1, \infty)$ Range: $[0, \infty)$

b.

1. $G(x) = \sqrt{x} - 2$ can be rewritten as $G(x) = f(x) - 2$.

2. The shift (2 units) is *outside* the function, which corresponds to a *vertical* shift *with the sign*.

3. $G(x) = \sqrt{x} - 2$ is found by shifting $f(x) = \sqrt{x}$ to the *down* 2 units. Notice that the point $(0, 0)$, which lies on $f(x)$, gets shifted to the point $(0, -2)$ on the graph of $G(x)$.

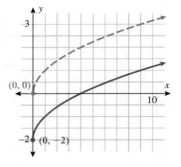

The original function, $f(x) = \sqrt{x}$, has an implicit restriction on the domain: $[0, \infty)$. The function, $G(x) = \sqrt{x} - 2$, also has the implicit restriction that $x \geq 0$. The output or range of $G(x)$ is always 2 units less than the output of the original function, $f(x)$.

Domain: $[0, \infty)$ Range: $[-2, \infty)$

■ YOUR TURN Plot the graph of the functions using graph-shifting techniques.

a. $G(x) = x^3 + 1$ **b.** $h(x) = (x + 2)^3$

The previous examples have involved graphing functions by shifting a known function either in the horizontal or vertical direction. Let us now look at combinations of horizontal and vertical translations.

EXAMPLE 3 Combination of Horizontal and Vertical Translations

Graph the function: $F(x) = (x + 1)^2 - 2$.

Solution:

The base function is $y = x^2$.

1. The shift 1 unit is *inside* the function, so it represents a *horizontal* shift *opposite the sign*.

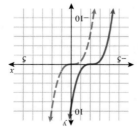

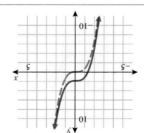

■ Answer: a. $G(x) = x^3 + 1$ is $y = x^3$ shifted up 1.   **b.** $h(x) = (x + 2)^3$ is $y = x^3$ shifted to the left 2.

2. The -2 shift is *outside* the function, which represents a *vertical* shift *with the sign*.

3. Therefore, we shift the graph of $y = x^2$ to the left 1 unit and down 2 units. For instance, the point $(0, 0)$ on $y = x^2$ shifts to the point $(-1, -2)$ on $F(x) = (x + 1)^2 - 2$.

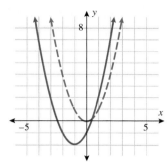

■ **YOUR TURN** Graph the function $f(x) = |x - 2| + 1$.

All of the previous transformation examples involve starting with a common function and shifting the function in either the horizontal or vertical direction (or a combination of both). Now, let's investigate *reflections* of functions about the *x*-axis or *y*-axis.

Reflection About the Axes

Plot $f(x) = x^2$ and $g(x) = -x^2$ on the same graph. Start by first listing points that are on each of the graphs and then connecting the points with smooth curves.

x	$f(x)$
-2	4
-1	1
0	0
1	1
2	4

x	$g(x)$
-2	-4
-1	-1
0	0
1	-1
2	-4

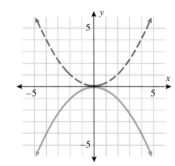

Notice that if the graph of $f(x) = x^2$ is reflected around the *x*-axis, the result is the graph of $g(x) = -x^2$. Also note the function $g(x)$ can be written as the negative of the function $f(x)$, $g(x) = -f(x)$. In general, **reflection about the *x*-axis** is produced by multiplying a function by -1.

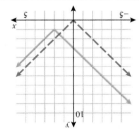

$f(x) = |x|$

■ **Answer:** $f(x) = |x - 2| + 1$

Let's now investigate reflection about the y-axis. Plot the functions $f(x) = \sqrt{x}$ and $g(x) = \sqrt{-x}$ on the same graph. Start by listing points that are on each of the graphs and then connecting the points with smooth curves.

x	$f(x)$
0	0
1	1
4	2
9	3

x	$g(x)$
-9	3
-4	2
-1	1
0	0

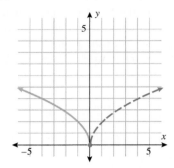

Notice that if the graph of $f(x) = \sqrt{x}$ is reflected around the y-axis, the result is the graph of $g(x) = \sqrt{-x}$. Also note that the function $g(x)$ can be written as $g(x) = f(-x)$. In general, **reflection about the y-axis** is produced by replacing x with $-x$ in the function.

REFLECTION ABOUT THE AXES

The graph of $-f(x)$ is obtained by reflecting the function $f(x)$ about the x-axis.

The graph of $f(-x)$ is obtained by rotating the function $f(x)$ about the y-axis.

Combinations of Shifting and Reflection Techniques

When multiple steps are involved, some students find it easier to use function notation, $f(x)$, and others find it easier to use y in place of $f(x)$. In Example 4, function notation is used, and more detailed explanations are given. In Example 5 the y notation is used, and a step-by-step approach is described. Either way of describing the transformations is acceptable.

EXAMPLE 4 Graph Using Combinations of Translations and Reflections Relying on Function Notation

Graph the function $G(x) = -\sqrt{x + 1}$.

Solution:

1. Start with the square root function $f(x) = \sqrt{x}$.
2. The shift 1 unit is *inside* the function, which corresponds to a *horizontal* shift *opposite the sign*.
3. Shift the graph of $f(x) = \sqrt{x}$ to the left 1 unit to yield $f(x + 1) = \sqrt{x + 1}$.
4. The negative outside the function $G(x) = -\sqrt{x + 1}$ corresponds to a reflection of $f(x + 1) = \sqrt{x + 1}$ about the x-axis.

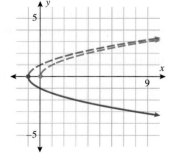

The question always arises: How do you know in which order to do the transformations? In Example 4, the horizontal shift was performed first. The reflection about the x-axis was performed second. Had we reflected about the x-axis and then shifted to the left instead, the same graph would have resulted.

EXAMPLE 5 **Graph Using Combinations of Translations and Reflections Relying on y Notation**

Graph the function $y = \sqrt{2 - x} + 1$.

Solution:

1. Start with the square
   root function. $y = \sqrt{x}$

2. Shift $y = \sqrt{x}$ 2 units
   to the left. $y = \sqrt{x + 2}$

3. Replacement of x with
   $-x$ results in reflection
   about the y-axis. $y = \sqrt{-x + 2}$

4. Shift 1 unit upward. $y = \sqrt{2 - x} + 1$

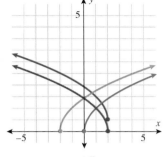

The graph of the function
$y = \sqrt{2 - x} + 1$ is:

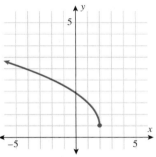

What graph would have resulted in Example 5 had we performed the shifts and reflection in another order? The same graph. For instance, if we start with the function $y = \sqrt{x}$, then vertically shift up 1 unit, the result is the function $y = \sqrt{x} + 1$. Then that function is reflected about the y-axis, resulting in the function $y = \sqrt{-x} + 1$. Shifting that function to the left 2 leads to the graph of $y = \sqrt{-x + 2} + 1$ or $y = \sqrt{2 - x} + 1$. This is exactly what we found before.

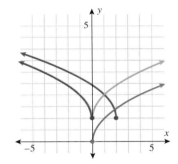

■ **YOUR TURN** Use shifts and reflection to graph the function $f(x) = -\sqrt{x - 1} + 2$. State the domain and range of $f(x)$.

Expansion and Contraction

In addition to shifts and reflections, another graphing aid is expansion or contraction. An expansion or contraction of a graph occurs when the function is multiplied by a positive

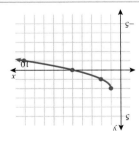

■ **Answer:** Domain: $[1, \infty)$
Range: $(-\infty, 2]$

constant. For example, plot the function $f(x) = x^2$ on the same graph with the functions $g(x) = 2f(x) = 2x^2$ and $h(x) = \frac{1}{2}f(x) = \frac{1}{2}x^2$. Depending on whether or not the constant is larger than 1 or smaller than 1 will determine if it corresponds to a stretch or compression in the vertical direction.

x	$f(x)$
-2	4
-1	1
0	0
1	1
2	4

x	$g(x)$
-2	8
-1	2
0	0
1	2
2	8

x	$h(x)$
-2	2
-1	$\frac{1}{2}$
0	0
1	$\frac{1}{2}$
2	2

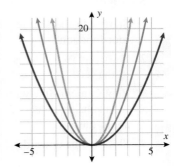

Notice that when the function is multiplied by 2, the result is a graph expanded (stretched) in the vertical direction. When the function is multiplied by $\frac{1}{2}$, the result is a graph that is contracted (compressed) in the vertical direction. Let the function $f(x) = x^2$ represent this year's profit for your company. If profit doubles next year, $g(x) = 2f(x) = 2x^2$, you have *expanded* your profit. If the profit is cut in half next year, $h(x) = \frac{1}{2}f(x) = \frac{1}{2}x^2$, you have *contracted* your profit.

EXPANSION AND CONTRACTION OF GRAPHS

The graph of $cf(x)$ is found by

    Expanding the graph of $f(x)$ if $c > 1$

    Contracting the graph of $f(x)$ if $0 < c < 1$

EXAMPLE 6 Expanding and Contracting Graphs

Graph the function $h(x) = \dfrac{1}{4}x^3$.

Solution:

1. Start with the cube function. $f(x) = x^3$

2. *Contraction* is expected because $\frac{1}{4}$ is less than 1. $h(x) = \dfrac{1}{4}x^3$

3. Determine a few points that lie on the graph of h. $(0, 0)\ (2, 2)\ (-2, -2)$

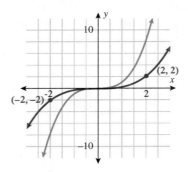

■ **YOUR TURN** Graph the function $g(x) = 4x^3$.

Combining Graphing Aids

Combinations of shifting, reflecting, and expansion or contraction can be used to generate graphs from our library of common functions.

EXAMPLE 7 Combination of Graphing Aids

Use the appropriate graphing aids to plot the function $f(x) = -2(x - 3)^2$.

1. Start with $y = x^2$.   **2.** Shift to the right 3 units $y = (x - 3)^2$.

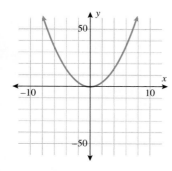   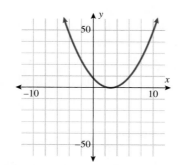

3. Reflect about the x-axis.   **4.** Expand in vertical direction.

  $y = -(x - 3)^2$ $y = -2(x - 3)^2$

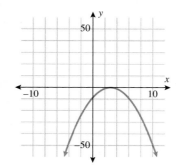   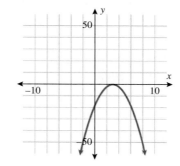

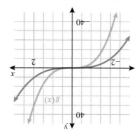

■ **Answer:** *Expansion of the graph* $f(x) = x^3$

SECTION 3.3 **SUMMARY**

In this section, we discussed three types of graphing aids that transform the graph of a known function into a graph of a similar function using techniques such as

- Horizontal and vertical shifts
- Reflection about the x-axis or y-axis
- Expansion and contraction in the vertical direction

Vertical Shifts

To graph	Shift the function $f(x)$
$y = f(x) + c$	c units upward
$y = f(x) - c$	c units downward

Horizontal Shifts

To graph	Shift the function $f(x)$
$y = f(x + c)$	c units to the left
$y = f(x - c)$	c units to the right

Reflection about the Axes

To graph	Reflect the function $f(x)$ about the
$-f(x)$	x-axis
$f(-x)$	y-axis

Expansion and Contraction

To graph	The graph of $f(x)$ is
$cf(x)$ if $c > 1$	expanded in the vertical direction
$cf(x)$ if $0 < c < 1$	contracted in the vertical direction

A series or combination of these techniques can be used. *They can be used in any order.*

SECTION 3.3 EXERCISES

■ **SKILLS**

In Exercises 1–12, match the function to the graph.

a.

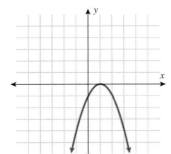

b.

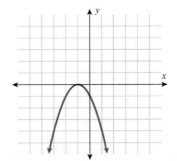

c.

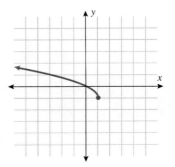

d.

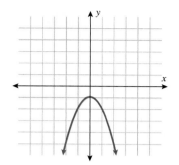

e.

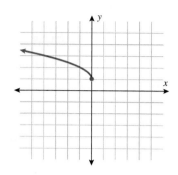

f.

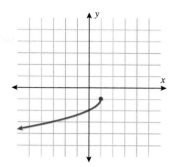

g.

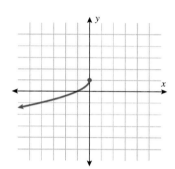

h.

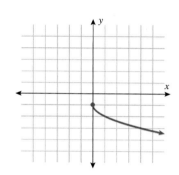

i.

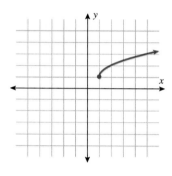

j.

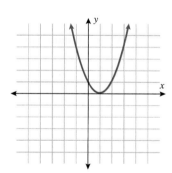

k.

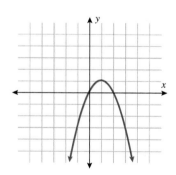

l.

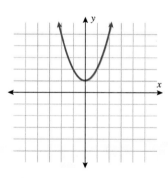

1. $f(x) = x^2 + 1$

2. $f(x) = (x - 1)^2$

3. $f(x) = -(1 - x)^2$

4. $f(x) = -x^2 - 1$

5. $f(x) = -(x + 1)^2$

6. $f(x) = -(1 - x)^2 + 1$

7. $f(x) = \sqrt{x - 1} + 1$

8. $f(x) = -\sqrt{x} - 1$

9. $f(x) = \sqrt{1 - x} - 1$

10. $f(x) = \sqrt{-x} + 1$

11. $f(x) = -\sqrt{-x} + 1$

12. $f(x) = -\sqrt{1 - x} - 1$

In Exercises 13–18, write the function whose graph is the graph of $y = |x|$ but is transformed accordingly.

13. Shifted up 3

14. Shifted to the left 4

15. Reflected about the y-axis

16. Reflected about the x-axis

17. Expanded by a factor of 3

18. Contracted by a factor of 3

In Exercises 19–24, write the function whose graph is the graph of $y = x^3$ but is transformed accordingly.

19. Shifted down 4

20. Shifted to the right 3

21. Shifted up 3 and to the left 1

22. Reflected about the x-axis

23. Reflected about the y-axis

24. Reflected about both the x-axis and the y-axis

In Exercises 25–32, write the function that results from the transformation.

25. The function $f(x) = x^2$ that is shifted to the right 1 and up 2 and reflected about the x-axis.

26. The function $f(x) = \sqrt[3]{x}$ that is shifted up 10 and reflected about the x-axis.

27. The function $f(x) = \sqrt{x}$ that is reflected about the y-axis and shifted to the right 1 and down 2.

28. The function $f(x) = |x|$ that is reflected about the y-axis and shifted up 3 and to the right 2.

29. The function $f(x) = \frac{1}{x}$ that is shifted to the right 2 and up 5 and reflected about the x-axis.

30. The function $f(x) = x^2 + 3x - 1$ that is shifted to the left 2 and reflected about the y-axis.

31. The function $f(x) = x^5 + x$ that is reflected about the y-axis and shifted up 2.

32. The function $f(x) = x^4 + x^3$ that is reflected about the x-axis and shifted to the right 2 and down 3.

In Exercises 33–58, graph the function using graphing aids.

33. $y = x^2 - 2$

34. $y = x^2 + 3$

35. $y = (x + 1)^2$

36. $y = (x - 2)^2$

37. $y = (x - 3)^2 + 2$

38. $y = (x + 2)^2 + 1$

39. $y = -(1 - x)^2$

40. $y = -(x + 2)^2$

41. $y = |-x|$

42. $y = -|x|$

43. $y = -|x + 2| - 1$

44. $y = |1 - x| + 2$

45. $y = 2x^2 + 1$

46. $y = 2|x| + 1$

47. $y = -\sqrt{x - 2}$

48. $y = \sqrt{2 - x}$

49. $y = -\sqrt{2 + x} - 1$

50. $y = \sqrt{2 - x} + 3$

51. $y = \sqrt[3]{x - 1} + 2$

52. $y = \sqrt[3]{x + 2} - 1$

53. $y = \dfrac{1}{x + 3} + 2$

54. $y = \dfrac{1}{3 - x}$

55. $y = 2 - \dfrac{1}{x + 2}$

56. $y = 2 - \dfrac{1}{1 - x}$

57. $y = 5\sqrt{-x}$

58. $y = -\dfrac{1}{5}\sqrt{x}$

In Exercises 59–64, transform the function into the form $f(x) = c(x - h)^2 + k$, where c, k, and h are constants, by completing the square. Use graph-shifting techniques to graph the function.

59. $y = x^2 - 6x + 11$

60. $f(x) = x^2 + 2x - 2$

61. $f(x) = -x^2 - 2x$

62. $f(x) = -x^2 + 6x - 7$

63. $f(x) = 2x^2 - 8x + 3$

64. $f(x) = 3x^2 - 6x + 5$

 ■ **APPLICATIONS**

65. Salary. A manager hires an employee at a rate of $10 per hour. Write the function that describes the current salary of the employee as a function of number of hours worked per week, x. After a year, the manager decides to award the employee a raise equivalent to paying him for an additional 5 hours per week. Write a function that describes the salary of the employee after the raise.

66. Profit. The profit associated with St. Augustine sod in Florida is typically $P(x) = -x^2 + 14,000x - 48,700,000$ where x is the number of pallets sold per year in a normal year. In rainy years Sod King gives away 10 free pallets per year. Write the function that describes the profit of x pallets of sod in rainy years.

67. Taxes. Every year in the United States each working American typically pays in taxes a percentage of his or her earnings (minus the standard deduction). Karen's 2005 taxes were calculated based on the formula $T(x) = 0.22(x - 6500)$. That year the standard deduction was $6,500 and her tax bracket paid 22%. Write the function that will determine her 2006 taxes assuming she receives a raise that places her in a 33% bracket.

68. Medication. The amount of medication that an infant requires is typically a function of the baby's weight. The number of milliliters of an antiseizure medication, A, is given by $A(x) = \sqrt{x} + 2$ where x is the weight of the infant in ounces. In emergencies there is often not enough time to weigh the infant so nurses have to estimate the baby's weight. What is the function that is the actual amount the infant is given if his weight is overestimated by 3 ounces?

■ CATCH THE MISTAKE

In Exercises 69–72, explain the mistake that is made.

69. Describe a procedure for graphing the function $f(x) = \sqrt{x - 3} + 2$.

Solution:

a. Start with the function $f(x) = \sqrt{x}$.

b. Shift the function to the left 3.

c. Shift the function up 2.

70. Describe a procedure for graphing the function $f(x) = -\sqrt{x + 2} - 3$.

Solution:

a. Start with the function $f(x) = \sqrt{x}$.

b. Shift the function to the left 2.

c. Reflect the function about the y-axis.

d. Shift the function down 3.

71. Describe a procedure for graphing the function $f(x) = |3 - x| + 1$.

Solution:

a. Start with the function $f(x) = |x|$.

b. Reflect about the y-axis.

c. Shift to the left 3.

d. Shift up 1.

72. Describe a procedure for graphing the function $f(x) = -2x^2 + 1$.

Solution:

a. Start with the function $f(x) = x^2$.

b. Reflect graph about the y-axis.

c. Shift up 1 unit.

d. Expand in the vertical direction by a factor of 2.

■ CHALLENGE

73. T or F: The graph of $y = |-x|$ is the same as the graph of $y = |x|$.

74. T or F: The graph of $y = \sqrt{-x}$ is the same as the graph of $y = \sqrt{x}$.

75. T or F: If the graph of an odd function is reflected around the x-axis and then the y-axis, the result is the graph of the original odd function.

76. T or F: If the graph of $y = \frac{1}{x}$ is reflected around the x-axis, it produces the same graph as if it had been reflected about the y-axis.

77. The point (a, b) lies on the graph of the function $y = f(x)$. What point is guaranteed to lie on the graph of $f(x - 3) + 2$?

78. The point (a, b) lies on the graph of the function $y = f(x)$. What point is guaranteed to lie on the graph of $-f(-x) + 1$?

■ TECHNOLOGY

79. Use a graphing utility to graph

a. $y = x^2 - 2$ and $y = |x^2 - 2|$

b. $y = x^3 + 1$ and $y = |x^3 + 1|$

What is the relationship between $f(x)$ and $|f(x)|$?

80. Use a graphing utility to graph

a. $y = x^2 - 2$ and $y = |x|^2 - 2$

b. $y = x^3 + 1$ and $y = |x|^3 + 1$

What is the relationship between $f(x)$ and $f(|x|)$?

81. Use a graphing utility to graph

a. $y = \sqrt{x}$ and $y = \sqrt{0.1x}$ **b.** $y = \sqrt{x}$ and $y = \sqrt{10x}$

What is the relationship between $f(x)$ and $f(ax)$ assuming a is positive?

82. Use a graphing utility to graph

a. $y = \sqrt{x}$ and $y = 0.1\sqrt{x}$ **b.** $y = \sqrt{x}$ and $y = 10\sqrt{x}$

What is the relationship between $f(x)$ and $af(x)$ assuming a is positive?

SECTION 3.4 Operations on Functions and Composition of Functions

Skills Objectives

- Add, subtract, multiply, and divide functions.
- Evaluate composite functions.
- Determine domain of functions resulting from operations and composite functions.

Conceptual Objectives

- Understand domain restrictions when dividing functions.
- Determine domain restrictions during composition of functions.

Two different functions can be combined using mathematical operations such as addition, subtraction, multiplication, and division. Also, there is an operation on functions called composition, which can be thought of as a function of a function. When we combine functions, we do so algebraically. Special attention must be paid to the domain and range of the combined functions.

Operations on Functions

Consider the two functions $f(x) = x^2 + 2x - 3$ and $g(x) = x + 1$. The domain of both of these functions is the set of all real numbers. Therefore, we can add, subtract, or multiply these functions for any real number x.

$$\text{Addition: } f(x) + g(x) = x^2 + 2x - 3 + x + 1 = x^2 + 3x - 2$$

The result is in fact a new function, which we denote:

$$(f + g)(x) = x^2 + 3x - 2 \quad \text{This is the \textbf{sum function}.}$$

$$\text{Subtraction: } f(x) - g(x) = x^2 + 2x - 3 - (x + 1) = x^2 + x - 4$$

The result is in fact a new function, which we denote:

$$(f - g)(x) = x^2 + x - 4 \quad \text{This is the \textbf{difference function}.}$$

$$\text{Multiplication: } f(x) \cdot g(x) = (x^2 + 2x - 3)(x + 1) = x^3 + 3x^2 - x - 3$$

The result is in fact a new function, which we denote:

$$(f \cdot g)(x) = x^3 + 3x^2 - x - 3 \quad \text{This is the \textbf{product function}.}$$

Although both f and g are defined for all real numbers x, we must restrict x such that $x \neq -1$ to form the quotient f/g.

$$\text{Division: } \frac{f(x)}{g(x)} = \frac{x^2 + 2x - 3}{x + 1}, \quad x \neq -1$$

The result is in fact a new function, which we denote:

$$\left(\frac{f}{g}\right)(x) = \frac{x^2 + 2x - 3}{x + 1}, \quad x \neq -1 \quad \text{This is called the } \textbf{quotient function}.$$

The new function that arises from adding, subtracting, multiplying, or dividing two functions is defined for all real values that both f and g are defined for. The exception is the quotient function, which eliminates values based on what numbers make the value of the denominator equal to zero.

The previous examples involved polynomials, whose domain is all real numbers. Adding, subtracting, and multiplying polynomials result in other polynomials, which have domains of all real numbers. Let's now investigate operations applied to functions that have a restricted domain.

The domain of the sum function, difference function, or product function is the *intersection* of the individual domains of the two functions. The quotient function has a similar domain in that it is the intersection of the two domains. But any values that make the denominator zero must also be eliminated.

FUNCTION	NOTATION	DOMAIN
Sum	$(f + g)(x) = f(x) + g(x)$	{domain of f} $\cap$ {domain of g}
Difference	$(f - g)(x) = f(x) - g(x)$	{domain of f} $\cap$ {domain of g}
Product	$(f \cdot g)(x) = f(x) \cdot g(x)$	{domain of f} $\cap$ {domain of g}
Quotient	$\left(\dfrac{f}{g}\right)(x) = \dfrac{f(x)}{g(x)}$	{domain of f} $\cap$ {domain of g} $\cap$ {$g(x) \neq 0$}

We can think of this in the following way. Any number that is in the domain of *both* of the functions is in the domain of the combined function. The exception to this is the quotient function, which also eliminates values that make the denominator equal to zero.

EXAMPLE 1 Operations on Functions: Determining Domains of New Functions

For the functions $f(x) = \sqrt{x - 1}$ and $g(x) = \sqrt{4 - x}$, determine the sum function, difference function, product function, and quotient function. State the domain of these four new functions.

Solution:

Sum function: $\qquad\qquad f(x) + g(x) = \sqrt{x - 1} + \sqrt{4 - x}$

Difference function: $\qquad f(x) - g(x) = \sqrt{x - 1} - \sqrt{4 - x}$

Product function: $\qquad\quad f(x) \cdot g(x) = \sqrt{x - 1} \cdot \sqrt{4 - x}$

$$\qquad\qquad\qquad\qquad\quad = \sqrt{(x - 1)(4 - x)} = \sqrt{-x^2 + 5x - 4}$$

Quotient function: $\qquad \dfrac{f(x)}{g(x)} = \dfrac{\sqrt{x - 1}}{\sqrt{4 - x}} = \sqrt{\dfrac{x - 1}{4 - x}}$

The domain of the square root function is determined by setting the argument under the radical greater than or equal to zero.

$$\text{Domain of } f(x): \quad [1, \infty)$$

$$\text{Domain of } g(x): \quad (-\infty, 4]$$

Domain of the sum, difference, and product functions is

$$[1, \infty) \cap (-\infty, 4] = [1, 4]$$

The quotient function has the additional constraint that the denominator cannot be zero. This implies that $x \neq 4$, so the domain of the quotient function is $[1, 4)$.

CONCEPT CHECK Are there any restrictions on the domain of the functions $f(x) = \sqrt{x + 3}$ and $g = \sqrt{1 - x}$?

YOUR TURN Given the function $f(x) = \sqrt{x + 3}$ and $g = \sqrt{1 - x}$ find $(f + g)(x)$ and state its domain.

EXAMPLE 2 Quotient Function and Domain Restrictions

Given the functions $F(x) = \sqrt{x}$ and $G(x) = |x - 3|$, find the quotient function, $\left(\frac{F}{G}\right)(x)$, and state its domain.

Solution:

The quotient function is written as

$$\left(\frac{F}{G}\right)(x) = \frac{F(x)}{G(x)} = \frac{\sqrt{x}}{|x - 3|}$$

Domain of $F(x)$: $[0, \infty)$ Domain of $G(x)$: $(-\infty, \infty)$

The real numbers that are in both the domain for $F(x)$ and the domain for $G(x)$ are represented by the intersection: $[0, \infty) \cap (-\infty, \infty) = [0, \infty)$. Also, the denominator of the quotient function is equal to zero when $x = 3$, so we must eliminate this value from the domain.

$$\text{Domain of } \left(\frac{F}{G}\right)(x): \quad [0, 3) \cup (3, \infty)$$

YOUR TURN For the functions given in Example 2, determine the quotient function $\left(\frac{G}{F}\right)(x)$, and state its domain.

■ Answer: $(f + g)(x) = \sqrt{x + 3} + \sqrt{1 - x}$ domain: $[-3, 1]$

■ Answer: $\left(\frac{G}{F}\right)(x) = \frac{G(x)}{F(x)} = \frac{|x - 3|}{\sqrt{x}}$ domain: $(0, \infty)$

Composition of Functions

Recall that a function maps every element in the domain to one and only one corresponding element in the range as shown in the figure on the right.

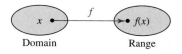

Suppose there is a sales rack of clothes in the juniors department at Macy's. Let x correspond to the original price of each item on the rack. These clothes have recently been marked down 20%. Therefore, the function, $f(x) = 0.80x$, represents the current sale price of each item. You have been invited to a special sale that lets you take 10% off the current sale price and an additional \$5 off every item at checkout. The function, $g(f(x)) = 0.90f(x) - 5$, determines the checkout price. Note that the input of the function g is the output of the function f as shown in the figure below.

Domain g Range g

x $f(x) = 0.80x$ $f(x)$ $g(x) = 0.90f(x) - 5$ $g(f(x))$

Domain f Range f

Original price   Sale price 20% off original price   Additional 10% off sale price and \$5 off at checkout

This is an example of a **composition of functions**, when the output of one function is the input of another function. It is commonly referred to as a function of a function.

An algebraic example of this is the function $y = \sqrt{x^2 - 2}$. Suppose we let $g(x) = x^2 - 2$ and $f(x) = \sqrt{x}$. Recall that the independent variable in function notation is a placeholder. Since $f(\ ) = \sqrt{(\ )}$, then $f(g(x)) = \sqrt{(g(x))}$. Substituting the expression for $g(x)$, we find $f(g(x)) = \sqrt{x^2 - 2}$. The function $y = \sqrt{x^2 - 2}$ is said to be a composite function, $y = f(g(x))$.

Note that the domain of $g(x)$ is the set of all real numbers, and the domain of $f(x)$ is the set of all nonnegative numbers. The domain of a composite function is the set of all x such that $g(x)$ is in the domain of f. For instance, in the composite function $y = f(g(x))$, we know that the allowable inputs into f are all numbers greater than or equal to zero. Therefore, we restrict the outputs of $g(x) \geq 0$ and find the corresponding x values. Those x values are the only allowable inputs and constitute the domain of the composite function $y = f(g(x))$.

The symbol that represents composition of functions is a small open circle, $(f \circ g)(x) = f(g(x))$ and is read aloud as "f of g." It is important not to confuse this with the multiplication sign, $(f \cdot g)(x) = f(x)g(x)$.

CAUTION
$f \circ g \neq f \cdot g$

COMPOSITE FUNCTIONS

Given two functions f and g, the composite function is denoted by $f \circ g$ and defined by

$$(f \circ g)(x) = f(g(x))$$

The domain of $f \circ g$ is the set of all numbers x in the domain of g where $g(x)$ is in the domain of f.

It is important to realize that there are two "filters" that allow certain values of x into the domain. The first filter is $g(x)$. If x is not in the domain of $g(x)$, it cannot be in the domain of $(f \circ g)(x) = f(g(x))$. Of those values for x that are in the domain of $g(x)$, only some pass through, because we restrict the output of $g(x)$ to values that are allowable as input into f. This adds an additional filter.

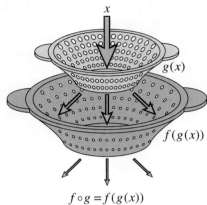

STUDY TIP

The domain of $f \circ g$ is always a subset of the domain of g, and the range of $f \circ g$ is always a subset of the range of f.

The domain of $f \circ g$ is always a subset of the domain of g, and the range of $f \circ g$ is always a subset of the range of f.

EXAMPLE 3 Finding a Composite Function

Given the functions $f(x) = x^2 + 1$ and $g(x) = x - 3$, find the composite function, $f \circ g$.

Solution:

Write $f(x)$ using placeholder notation. $f(\ ) = (\ )^2 + 1$

Substitute $g(x) = x - 3$ into f. $f(g(x)) = (x - 3)^2 + 1$

Eliminate parentheses on right side. $f(g(x)) = x^2 - 6x + 10$

$$(f \circ g)(x) = f(g(x)) = x^2 - 6x + 10$$

■ **YOUR TURN** Given the functions in Example 3, find $g \circ f$.

EXAMPLE 4 Determining the Domain of a Composite Function

Given the functions $f(x) = \dfrac{1}{x - 1}$ and $g(x) = \dfrac{1}{x}$, determine the composite function, $f \circ g$, and state its domain.

Solution:

Write $f(x)$ using placeholder notation. $f(\ ) = \dfrac{1}{(\ ) - 1}$

Substitute $g(x) = \dfrac{1}{x}$ into f. $f(g(x)) = \dfrac{1}{\dfrac{1}{x} - 1}$

■ **Answer:** $g \circ f = g(f(x)) = x^2 - 2$

Multiply right side by $\dfrac{x}{x}$.

$$f(g(x)) = \dfrac{1}{\dfrac{1}{x} - 1} \cdot \dfrac{x}{x} = \dfrac{x}{1 - x}$$

$$(f \circ g) = f(g(x)) = \dfrac{x}{1 - x}$$

What is the domain of $(f \circ g)(x) = f(g(x))$? By inspecting the final result of $f(g(x))$, we see that the denominator is zero when $x = 1$. Therefore, $x \neq 1$. Are there any other values for x that are not allowed? The function $g(x)$ has the domain $x \neq 0$, therefore we must also exclude zero.

The domain of $(f \circ g)(x) = f(g(x))$ is $x \neq 0$ and $x \neq 1$ or in interval notation

$$(-\infty, 0) \cup (0, 1) \cup (1, \infty)$$

The domain of the composite function cannot always be determined by examining the final form of $f \circ g$.

CAUTION
The domain of the composite function cannot always be determined by examining the final form of $f \circ g$.

■ **YOUR TURN** For the functions f and g given in Example 4, determine the composite function $g \circ f$ and state its domain.

EXAMPLE 5 Evaluating a Composite Function

Given the functions $f(x) = x^2 - 7$ and $g(x) = 5 - x^2$, evaluate

a. $f(g(1))$ **b.** $f(g(-2))$ **c.** $g(f(3))$ **d.** $g(f(-4))$

Solution:

One way of evaluating these composite functions is to calculate the two individual composites in terms of x: $f(g(x))$ and $g(f(x))$. Once those functions are known, the values can be substituted for x and evaluated.

Another way of proceeding is as follows:

a. Write desired quantity. $f(g(1))$

  Find the value of the inner function, g. $g(1) = 5 - 1^2 = 4$

  Substitute $g(1) = 4$ into f. $f(g(1)) = f(4)$

  Evaluate $f(4)$. $f(4) = 4^2 - 7 = 9$

$$f(g(1)) = 9$$

b. Write desired quantity. $f(g(-2))$

  Find the value of the inner function, g. $g(-2) = 5 - (-2)^2 = 1$

  Substitute $g(-2) = 1$ into f. $f(g(-2)) = f(1)$

  Evaluate $f(1)$. $f(1) = 1^2 - 7 = -6$

$$f(g(-2)) = -6$$

■ **Answer:** $(g \circ f)(x) = g(f(x)) = x - 1$
Domain of $g \circ f$ is $x \neq 1$ or in interval notation $(-\infty, 1) \cup (1, \infty)$

c. Write desired quantity. $\qquad$ $g(f(3))$

Find the value of the inner function, f. $\qquad$ $f(3) = 3^2 - 7 = 2$

Substitute $f(3) = 2$ into g. $\qquad$ $g(f(3)) = g(2)$

Evaluate $g(2)$. $\qquad$ $g(2) = 5 - 2^2 = 1$

$$g(f(3)) = 1$$

d. Write desired quantity. $\qquad$ $g(f(-4))$

Find the value of the inner function, f. $\qquad$ $f(-4) = (-4)^2 - 7 = 9$

Substitute $f(-4) = 9$ into g. $\qquad$ $g(f(-4)) = g(9)$

Evaluate $g(9)$. $\qquad$ $g(9) = 5 - 9^2 = -76$

$$g(f(-4)) = -76$$

■ **YOUR TURN** Given the functions $f(x) = x^3 - 3$ and $g(x) = 1 + x^3$, evaluate $f(g(1))$ and $g(f(1))$.

Application Problems

Recall the first example in this section regarding the clothes that are on sale. Often, real-world applications are modeled with composite functions. In the clothes example, x is the original price of each item. The first function maps the input (original price) to an output (sale price). The second function maps the input (sale price) to an output (check-out price). The next example is another real-world application of composite functions.

Three temperature scales are commonly used

■ The degree Celsius (°C) scale

  ■ This scale was devised by dividing the range between the freezing and boiling of pure water at sea level into 100 equal parts. This scale is used in science and is one of the standards of the "metric" (SI) system of measurements.

  ■ The **degrees Celsius** range from 0 °C (freezing) to 100 °C (boiling).

■ The Kelvin (K) temperature scale

  ■ This scale extends the Celsius scale down to absolute zero, a hypothetical temperature at which there is a complete absence of heat energy.

  ■ Temperatures on this scale are called **kelvins**, *not* degrees kelvin, and kelvin is not capitalized. The symbol for kelvin is K.

■ The degree Fahrenheit (°F) scale

  ■ This scale evolved over time and is still widely used mainly in the United States, although Celsius is the preferred "metric" scale.

  ■ With respect to pure water at sea level, the **degrees Fahrenheit** correspond to 32 °F (freezing) and 212 °F (boiling).

The equations that relate these temperature scales are

$$F = \frac{9}{5}C + 32 \qquad C = K - 273.15$$

■ **Answer:** $f(g(1)) = 5$ and $g(f(1)) = -7$

EXAMPLE 6 Applications Involving Composite Functions

Determine degrees Fahrenheit as a function of kelvins.

Solution:

Degrees Fahrenheit is a function of degrees Celsius.

$$F = \frac{9}{5}C + 32$$

Now substitute $C = K - 273.15$ into the equation for F.

$$F = \frac{9}{5}(K - 273.15) + 32$$

Simplify.

$$F = \frac{9}{5}K - 491.67 + 32$$

$$F = \frac{9}{5}K - 459.67$$

SECTION 3.4 SUMMARY

Operations on Functions

FUNCTION	NOTATION
Sum	$(f + g)(x) = f(x) + g(x)$
Difference	$(f - g)(x) = f(x) - g(x)$
Product	$(f \cdot g)(x) = f(x) \cdot g(x)$
Quotient	$\left(\dfrac{f}{g}\right)(x) = \dfrac{f(x)}{g(x)} \qquad g(x) \neq 0$

The domain of the sum, difference, and product functions is the intersection, or common, domain shared by both $f(x)$ and $g(x)$. The domain of the quotient function is also the intersection of the domain shared by both $f(x)$ and $g(x)$ with an additional restriction that $g(x) \neq 0$.

Composition of Functions

$$(f \circ g)(x) = f(g(x))$$

The domain restrictions cannot be determined only by inspecting the final form of $f(g(x))$. The domain of the composite function is a subset of the domain of $g(x)$. Values for x must be eliminated if their corresponding values, $g(x)$, are not in the domain of f.

SECTION 3.4 EXERCISES

 **SKILLS**

In Exercises 1–10, given the functions f and g, find $f + g$, $f - g$, $f \cdot g$, and $\dfrac{f}{g}$, and state the domain of each.

1. $f(x) = 2x + 1$
   $g(x) = 1 - x$

2. $f(x) = 3x + 2$
   $g(x) = 2x - 4$

3. $f(x) = 2x^2 - x$
   $g(x) = x^2 - 4$

4. $f(x) = 3x + 2$
   $g(x) = x^2 - 25$

5. $f(x) = \dfrac{1}{x}$
   $g(x) = x$

6. $f(x) = \dfrac{2x + 3}{x - 4}$

$g(x) = \dfrac{x - 4}{3x + 2}$

7. $f(x) = \sqrt{x}$

$g(x) = 2\sqrt{x}$

8. $f(x) = \sqrt{x - 1}$

$g(x) = 2x^2$

9. $f(x) = \sqrt{4 - x}$

$g(x) = \sqrt{x + 3}$

10. $f(x) = \sqrt{1 - 2x}$

$g(x) = \dfrac{1}{x}$

In Exercises 11–20, for the given functions f and g, find the composite functions $f \circ g$ and $g \circ f$, and state their domains.

11. $f(x) = 2x + 1$

$g(x) = x^2 - 3$

12. $f(x) = x^2 - 1$

$g(x) = 2 - x$

13. $f(x) = \dfrac{1}{x - 1}$

$g(x) = x + 2$

14. $f(x) = \dfrac{2}{x - 3}$

$g(x) = 2 + x$

15. $f(x) = |x|$

$g(x) = \dfrac{1}{x - 1}$

16. $f(x) = |x - 1|$

$g(x) = \dfrac{1}{x}$

17. $f(x) = \sqrt{x - 1}$

$g(x) = x + 5$

18. $f(x) = \sqrt{2 - x}$

$g(x) = x^2 + 2$

19. $f(x) = x^3 + 4$

$g(x) = (x - 4)^{1/3}$

20. $f(x) = \sqrt[3]{x^2 - 1}$

$g(x) = x^{2/3} + 1$

In Exercises 21–28, evaluate the functions for the specified values, if possible.

$$f(x) = x^2 + 10 \qquad g(x) = \sqrt{x - 1}$$

21. $(f + g)(2)$

22. $(f - g)(5)$

23. $(f \cdot g)(4)$

24. $\left(\dfrac{f}{g}\right)(2)$

25. $f(g(2))$

26. $g(f(4))$

27. $f(g(0))$

28. $g(f(0))$

In Exercises 29–34, evaluate $f(g(1))$ and $g(f(2))$, if possible.

29. $f(x) = \dfrac{1}{x} \quad g(x) = 2x + 1$

30. $f(x) = x^2 + 1 \quad g(x) = \dfrac{1}{2 - x}$

31. $f(x) = \sqrt{1 - x} \quad g(x) = x^2 + 2$

32. $f(x) = \sqrt{3 - x} \quad g(x) = x^2 + 1$

33. $f(x) = \dfrac{1}{|x - 1|} \quad g(x) = x + 3$

34. $f(x) = \dfrac{1}{x} \quad g(x) = |2x - 3|$

In Exercises 35–40, show that $f(g(x)) = x$ and $g(f(x)) = x$.

35. $f(x) = 2x + 1 \quad g(x) = \dfrac{x - 1}{2}$

36. $f(x) = \dfrac{x - 2}{3} \quad g(x) = 3x + 2$

37. $f(x) = \sqrt{x - 1} \quad g(x) = x^2 + 1 \quad$ for $x \geq 1$

38. $f(x) = 2 - x^2 \quad g(x) = \sqrt{2 - x} \quad$ for $x \leq 2$

39. $f(x) = \dfrac{1}{x} \quad g(x) = \dfrac{1}{x} \quad$ for $x \neq 0$

40. $f(x) = (5 - x)^{1/3} \quad g(x) = 5 - x^3$

In Exercises 41–46, write the function as a composite of two functions f and g. (More than one answer is correct.)

41. $f(g(x)) = 2(3x - 1)^2 + 5(3x - 1)$

42. $f(g(x)) = \dfrac{1}{1 + x^2}$

43. $f(g(x)) = \dfrac{2}{|x - 3|}$

44. $f(g(x)) = \sqrt{1 - x^2}$

45. $f(g(x)) = \dfrac{3}{\sqrt{x + 1} - 2}$

46. $f(g(x)) = \dfrac{\sqrt{x}}{3\sqrt{x} + 2}$

■ **APPLICATIONS**

Exercises 47 and 48 depend on the relationship between degrees Fahrenheit, degrees Celsius, and kelvins:

$$F = \tfrac{9}{5}C + 32 \quad C = K - 273.15$$

47. Temperature. Write a composite function that converts degrees kelvins into Fahrenheit.

48. Temperature. Convert the following degrees Fahrenheit to kelvins: 32 °F and 212 °F.

49. Dog Run. Suppose you want to build a *square* fenced-in area for your dog. Fence is purchased in linear feet.

   a. Write a composite function that determines the area of your dog pen as a function of how many linear feet are purchased.

   b. If you purchase 100 linear feet, what is the area of your dog pen?

   c. If you purchase 200 linear feet, what is the area of your dog pen?

50. Dog Run. Suppose you want to build a *circular* fenced-in area for your dog. Fence is purchased in linear feet.

   a. Write a composite function that determines the area of your dog pen as a function of how many linear feet are purchased.

   b. If you purchase 100 linear feet, what is the area of your dog pen?

   c. If you purchase 200 linear feet, what is the area of your dog pen?

51. Market Price. Typical supply and demand relationships state that as the number of units for sale increases, the market price decreases. Assume the market price, p, and the number of units for sale, x, are related by the demand equation:

$$p = 3000 - \frac{1}{2}x$$

Assume the cost, $C(x)$, of producing x items is governed by the equation

$$C(x) = 2000 + 10x$$

And the revenue, $R(x)$, generated by selling x units is governed by

$$R(x) = 100x$$

   a. Write the cost as a function of price, p.

   b. Write the revenue as a function of price, p.

   c. Write the profit as a function of price, p.

52. Market Price. Typical supply and demand relationships state that as the number of units for sale increases, the market price decreases. Assume the market price, p, and the number of units for sale, x, are related by the demand equation:

$$p = 10{,}000 - \frac{1}{4}x$$

Assume the cost, $C(x)$, of producing x items is governed by the equation

$$C(x) = 30000 + 5x$$

And the revenue, $R(x)$, generated by selling x units is governed by

$$R(x) = 1000x$$

   a. Write the cost as a function of price, p.

   b. Write the revenue as a function of price, p.

   c. Write the profit as a function of price, p.

53. Oil Spill. An oil spill makes a circular pattern around a ship. If the radius, in feet, grows as a function of time, in hours: $r(t) = 150\sqrt{t}$. Find the area of the spill as a function of time.

54. Pool Volume. A 20 foot $\times$ 10 foot rectangular pool has been built. If 50 cubic feet of water is pumped into the pool per hour, write the water level height (feet) as a function of time (hours).

55. Fireworks. A family is watching a fireworks display. If they are 2 miles from where the fireworks are being launched and the fireworks travel vertically, what is the distance between the family and the fireworks as a function of height above ground?

56. Real Estate. A couple are about to put their house up for sale. They bought the house for $172,000 a few years ago, and when they list it with a realtor they will pay 6% commission. Write a function that represents the amount of money they will make on their home as a function of asking price, p.

■ CATCH THE MISTAKE

In Exercises 57–60, for the functions $f(x) = x + 2$ and $g(x) = x^2 - 4$, find the composite function and state its domain. Explain the mistake made in each problem.

57. $\dfrac{g}{f}$

   Solution: $\dfrac{g(x)}{f(x)} = \dfrac{x^2 - 4}{x + 2}$

$$= \dfrac{(x - 2)(x + 2)}{(x + 2)}$$

$$= x - 2$$

Domain: All real numbers

58. $\dfrac{f}{g}$

   Solution: $\dfrac{f(x)}{g(x)} = \dfrac{x + 2}{x^2 - 4}$

$$= \dfrac{x + 2}{(x - 2)(x + 2)} = \dfrac{1}{x - 2}$$

$$= \dfrac{1}{x - 2}$$

Domain: All real numbers except $x = 2$

59. $f \circ g$

Solution: $f \circ g = f(x)g(x)$

$$= (x + 2)(x^2 - 4)$$
$$= x^3 + 2x^2 - 4x - 8$$

Domain: All real numbers

60. Given the function $f(x) = x^2 + 7$ and $g(x) = \sqrt{x - 3}$, find $f \circ g$, and state the domain.

Solution: $f \circ g = f(g(x)) = (\sqrt{x - 3})^2 + 7$

$$= f(g(x)) = x - 3 + 7$$
$$= x - 4$$

Domain: All real numbers

■ CHALLENGE

61. T or F: When adding, subtracting, multiplying, or dividing two functions, the domain of the resulting function is the union of the domains of the individual functions.

62. T or F: For any functions f and g, $f(g(x)) = g(f(x))$ for all values of x that are in the domain of both f and g.

63. T or F: For any functions f and g, $(f \circ g)(x)$ exists for all values of x that are in the domain of $g(x)$ provided the range of g is a subset of the domain of f.

64. T or F: The domain of a composite function can be found by inspection, without knowledge of the domain of the individual functions.

65. For the functions $f(x) = x + a$ and $g(x) = \dfrac{1}{x - a}$, find $g \circ f$ and state its domain.

66. For the functions $f(x) = ax^2 + bx + c$ and $g(x) = \dfrac{1}{x - c}$, find $g \circ f$ and state its domain.

67. For the functions $f(x) = \sqrt{x + a}$ and $g(x) = x^2 - a$, find $g \circ f$ and state its domain.

68. For the functions $f(x) = \dfrac{1}{x^a}$ and $g(x) = \dfrac{1}{x^b}$, find $g \circ f$ and state its domain.

■ TECHNOLOGY

69. Using a graphing utility plot $y_1 = \sqrt{x + 7}$ and $y_2 = \sqrt{9 - x}$. Plot $y_3 = y_1 + y_2$. Explain what you see for y_3.

70. Using a graphing utility plot $y_1 = \sqrt{1 - x}$, $y_2 = x^2 + 2$, and $y_3 = y_1^2 + 2$. If y_1 represents a function f and y_2 represents a function g, then y_3 represents the composite function $g \circ f$. The graph of y_3 is only defined for the domain of $g \circ f$.

SECTION 3.5 One-to-One Functions and Inverse Functions

Skills Objectives

■ Determine if a function is a one-to-one function
  ■ Algebraically
  ■ Graphically
■ Find the inverse of a function.
■ Graph the inverse function given the graph of the function.

Conceptual Objectives

■ Visualize the relationships between domain and range of a function and the domain and range of its inverse.
■ Understand why functions and their inverses are symmetric about $y = x$.

Every human being has a blood type, and every human being has a DNA sequence. These are examples of functions when a person is the input and the output is blood type or DNA sequence. These relationships are classified as functions because each person can have one and only one blood type or DNA strand. The difference between these functions is that many people have the same blood type, but DNA is unique to each individual. Could we map backwards? For instance, if you know the blood type, do you know specifically which person it came from? No. But, if you know the DNA sequence, you know exactly which person it corresponds to. When a function has a one-to-one correspondence, like the DNA example, then mapping backwards is possible. The map back is called the *inverse function.*

One-to-One Functions

In Section 3.1, we defined a function as a relationship that maps an input (domain) to one and only one output (range). Algebraically, each value for x can only correspond to a single value for y. Recall the square, identity, absolute value, and reciprocal functions from our library of functions in Section 3.3 (See graphs on the right.)

All of the graphs of these functions satisfy the vertical line test. Although the square function and the absolute value function map each value of x to one and only one value for y, these two functions map two values of x to the same value for y. For example $(-1, 1)$ and $(1, 1)$ lie on both graphs. The identity and reciprocal functions, on the other hand, map each x to a single value for y, and no two x values map to the same y value. These two functions are examples of one-to-one functions.

ONE-TO-ONE FUNCTION

A function $f(x)$ is **one-to-one** if every two distinct values for x in the domain, $x_1 \neq x_2$, correspond to two distinct values of the function, $f(x_1) \neq f(x_2)$.

In other words, no two inputs map to the same output. Just as there is a graphical test for functions, the vertical line test, there is a graphical test for one-to-one functions, the *horizontal line test.* Note that a horizontal line can be drawn on the square and absolute value functions so that it intersects the graph of each function at two points. The identity and reciprocal functions, however, will only intersect a horizontal line in at most one point. This leads us to the horizontal line test for one-to-one functions.

HORIZONTAL LINE TEST

If every horizontal line intersects the graph of a function in at most one point, then the function is classified as a one-to-one function.

EXAMPLE 1 Determine If a Function Defined as a Set of Points Is One-to-One

For each of the three relationships, determine if the relationship is a function. If it is a function, determine if it is a one-to-one function.

$$f = \{(0, 0), (1, 1), (1, -1)\}$$

$$g = \{(-1, 1), (0, 0), (1, 1)\}$$

$$h = \{(-1, -1), (0, 0), (1, 1)\}$$

Square function: $f(x) = x^2$

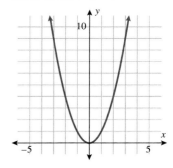

Identity function: $f(x) = x$

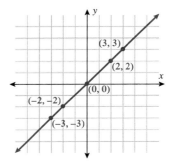

Absolute value function: $f(x) = |x|$

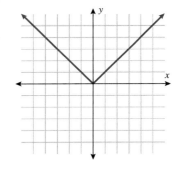

Reciprocal function: $f(x) = 1/x$

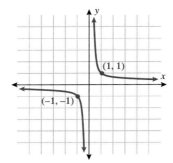

Solution:

	f			g			h	
Domain		Range	Domain		Range	Domain		Range
0	$\longrightarrow$	0	0	$\longrightarrow$	0	−1	$\longrightarrow$	−1
1		−1	−1	$\longrightarrow$	1	0	$\longrightarrow$	0
		1	1			1	$\longrightarrow$	1

f is not a function.

g is a function, but not one-to-one.

h is a one-to-one function.

EXAMPLE 2 **Determine If a Function Defined by an Equation Is One-to-One**

For each of the three relationships, determine if the relationship is a function. If it is a function, determine if it is a one-to-one function. Assume that x is the independent variable and y is the dependent variable.

$$x = y^2 \qquad y = x^2 \qquad y = x^3$$

Solution:

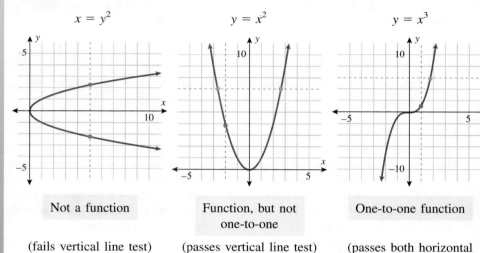

$x = y^2$

$y = x^2$

$y = x^3$

Not a function

Function, but not one-to-one

One-to-one function

(fails vertical line test)

(passes vertical line test) (fails horizontal line test)

(passes both horizontal and vertical line tests)

CONCEPT CHECK Does a parabola pass the horizontal line test?

YOUR TURN Determine if the functions are one to one.

a. $f(x) = x + 2$ **b.** $f(x) = x^2 + 1$

Inverse Functions

If a function is one-to-one, then the function maps each x to one and only one y, and no two x values map to the same y value. This implies that there is a one-to-one correspondence between the input (domain) and output (range) of a one-to-one function $f(x)$. In the special case of a one-to-one function, it would be possible to map from the output (range of f) back to the input (domain of f), and this mapping would also be a function. The function that maps the output back to the input of a function, f, is called the **inverse function** and is denoted $f^{-1}(x)$.

A one-to-one function, f, maps every x in the domain to a unique and distinct corresponding y in the range. Therefore, the inverse, f^{-1}, maps every y back to a unique and distinct x.

The function notation $f(x) = y$ and $f^{-1}(y) = x$ indicates that if the point (x, y) satisfies the function, then the point (y, x) satisfies the inverse function.

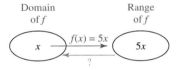

For example, the function $h(x) = \{(-1, 0), (1, 2), (3, 4)\}$

$$h = \{(-1, 0)\ (1, 2)\ (3, 4)\}$$

Domain	Range
-1 ⇌ 0	
1 ⇌ 2	
3 ⇌ 4	
Range	Domain

h is a one-to-one function

$$h^{-1} = \{(0, -1), (2, 1), (4, 3)\}$$

The inverse function undoes whatever the function does. For example, if $f(x) = 5x$, then the function f maps any value x in the domain to a value $5x$ in the range.

If we want to map backwards or undo the $5x$, we develop a function called the inverse function that takes $5x$ as input and maps back to x as output.

The inverse function is $f^{-1}(x) = \frac{1}{5}x$. Notice that if we input $5x$ into the inverse function, the output is x: $f^{-1}(5x) = \frac{1}{5}(5x) = x$.

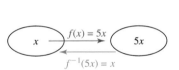

DEFINITION   INVERSE FUNCTIONS

If f and g denote two one-to-one functions such that

$$f(g(x)) = x \text{ for every } x \text{ in the domain of } g$$

and

$$g(f(x)) = x \text{ for every } x \text{ in the domain of } f,$$

then the function g is the **inverse** of the function f. The function g is denoted by f^{-1} (read "f-inverse").

Note: f^{-1} is used to denote the inverse of f. The -1 is not used as an exponent and, therefore, does not represent the reciprocal of f, $\frac{1}{f}$.

CAUTION

$$f^{-1} \neq \frac{1}{f}$$

Two properties hold true relating one-to-one functions to their inverses: (1) the range of the function is the domain of the inverse, and the range of the inverse is the domain of the function and (2) the composite function that results with a function and its inverse (and vice versa) yields x.

Domain of f = range of f^{-1} and range of f = domain of f^{-1}

$$f^{-1}(f(x)) = x \quad \text{and} \quad f(f^{-1}(x)) = x$$

EXAMPLE 3 Verifying Inverse Functions

Verify that $f^{-1}(x) = \frac{1}{2}x - 2$ is the inverse of $f(x) = 2x + 4$.

Solution:

Show that $f^{-1}(f(x)) = x$ and $f(f^{-1}(x)) = x$.

Write f^{-1} using placeholder notation.
$$f^{-1}(\ ) = \frac{1}{2}(\ ) - 2$$

Substitute $f(x) = 2x + 4$ into f^{-1}.
$$f^{-1}(f(x)) = \frac{1}{2}(2x + 4) - 2$$

Simplify.
$$f^{-1}(f(x)) = x + 2 - 2 = x$$
$$f^{-1}(f(x)) = x$$

Write f using placeholder notation.
$$f(\ ) = 2(\ ) + 4$$

Substitute $f^{-1}(x) = \frac{1}{2}x - 2$ into f.
$$f(f^{-1}(x)) = 2\left(\frac{1}{2}x - 2\right) + 4$$

Simplify.
$$f(f^{-1}(x)) = x - 4 + 4 = x$$
$$f(f^{-1}(x)) = x$$

Note the relationship between the domain and range of f and f^{-1}.

	Domain	Range
$f(x) = 2x + 4$	$\mathbb{R}$	$\mathbb{R}$
$f^{-1}(x) = \frac{1}{2}x - 2$	$\mathbb{R}$	$\mathbb{R}$

EXAMPLE 4 Verifying Inverse Functions with Domain Restrictions

Verify that $f^{-1}(x) = x^2, x \geq 0$ is the inverse of $f(x) = \sqrt{x}$.

Solution:

Show that $f^{-1}(f(x)) = x$ and $f(f^{-1}(x)) = x$.

Write f^{-1} using placeholder notation.
$$f^{-1}(\ ) = (\ )^2$$

Substitute $f(x) = \sqrt{x}$ into f^{-1}.
$$f^{-1}(f(x)) = (\sqrt{x})^2 = x$$
$$f^{-1}(f(x)) = x \text{ for } x \geq 0$$

Write f using placeholder notation.
$$f(\ ) = \sqrt{(\ )}$$

Substitute $f^{-1}(x) = x^2, x \geq 0$ into f.
$$f(f^{-1}(x)) = \sqrt{x^2} = x, \ x \geq 0$$
$$f(f^{-1}(x)) = x \text{ for } x \geq 0$$

	Domain	Range
$f(x) = \sqrt{x}$	$[0, \infty)$	$[0, \infty)$
$f^{-1}(x) = x^2, x \geq 0$	$[0, \infty)$	$[0, \infty)$

Graphical Interpretation of Inverse Functions

In Example 3, we showed that $f^{-1}(x) = \frac{1}{2}x - 2$ is the inverse of $f(x) = 2x + 4$. Let's now investigate the graphs that correspond to the function f and its inverse f^{-1}.

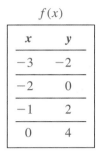

$f(x)$

x	y
-3	-2
-2	0
-1	2
0	4

$f^{-1}(x)$

x	y
-2	-3
0	-2
2	-1
4	0

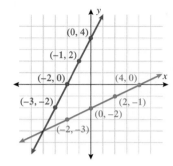

Notice that the point $(-3, -2)$ lies on the function and the point $(-2, -3)$ lies on the inverse. In fact, every point (a, b) that lies on the function corresponds to a point (b, a) that lies on the inverse.

Draw the line $y = x$ on the graph. In general, the point (b, a) on the inverse, $f^{-1}(x)$, is the reflection (about $y = x$) of the point (a, b) on the function, $f(x)$.

In general, if the point (a, b) is on the graph of a function, then the point (b, a) is on the graph of its inverse.

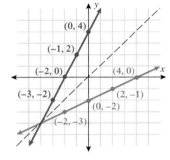

 EXAMPLE 5 Graphing the Inverse Function

Given the graph of the function, $f(x)$, plot the graph of its inverse, $f^{-1}(x)$.

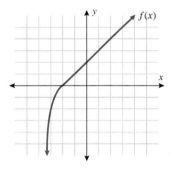

Solution:

Because the points $(-3, -2)$, $(-2, 0)$, $(0, 2)$, and $(2, 4)$ lie on the graph of f, then the points $(-2, -3)$, $(0, -2)$, $(2, 0)$ and $(4, 2)$ lie on the graph of f^{-1}.

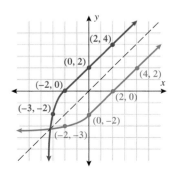

CONCEPT CHECK If the point (2, 0) lies on the graph of a one-to-one function, what point lies on the graph of its inverse?

■ **YOUR TURN** Given the graph of a function, f, plot the inverse function.

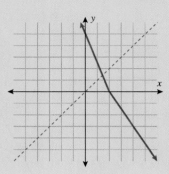

We have developed the definition of an inverse function, and properties of inverses. At this point you should be able to verify if two functions are inverses of one another. Let's turn our attention to another problem: How do you find the inverse of a function?

Finding the Inverse Function

Recall that if the point (a, b) lies on the graph of a function, then the point (b, a) lies on the graph of the inverse function. The symmetry about the line $y = x$ tells us that the roles of x and y interchange. Therefore, if we start with every point (x, y) that lies on the graph of a function, every point (y, x) lies on the graph of its inverse. Algebraically, this corresponds to interchanging x and y. Finding the inverse of a finite set of ordered pairs is easy: Simply interchange the x and y coordinates. Earlier, we found that if $h(x) = \{(-1, 0),(1, 2),(3, 4)\}$ then $h^{-1}(x) = \{(0, -1),(2, 1),(4, 3)\}$. But how do we find the inverse of a function defined by an equation?

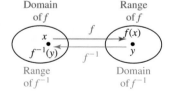

Recall the mapping relationship if f is a one-to-one function. This relationship implies that $f(x) = y$ and $f^{-1}(y) = x$. Let's use these two identities in finding the inverse. Consider the function defined by $f(x) = 3x - 1$. To find f^{-1}, we let $f(x) = y$, which yields $y = 3x - 1$. Solve for the variable, x: $x = \frac{1}{3}y + \frac{1}{3}$.

Recall that $f^{-1}(y) = x$, so we have found the inverse to be $f^{-1}(y) = \frac{1}{3}y + \frac{1}{3}$. It is customary to write the independent variable as x, so we write the inverse as

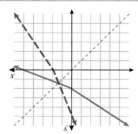

$f^{-1}(x) = \frac{1}{3}x + \frac{1}{3}$. Now that we have found the inverse, let's confirm that the property $f^{-1}(f(x)) = x$ and $f(f^{-1}(x)) = x$, holds.

$$f(f^{-1}(x)) = 3\left(\frac{1}{3}x + \frac{1}{3}\right) - 1 = x + 1 - 1 = x$$

$$f^{-1}(f(x)) = \frac{1}{3}(3x - 1) + \frac{1}{3} = x - \frac{1}{3} + \frac{1}{3} = x$$

Finding inverses is often summarized with a seven-step approach.

FINDING THE INVERSE OF A FUNCTION

Step 1: Verify that $f(x)$ is a one-to-one function.
Step 2: Let $y = f(x)$.
Step 3: Interchange x and y.
Step 4: Solve for y.
Step 5: Let $y = f^{-1}(x)$.
Step 6: Note any domain restrictions on $f^{-1}(x)$.
Step 7: Check.

EXAMPLE 6 The Inverse of a Square Root Function Is a Square Function

Find the inverse of the function $f(x) = \sqrt{x + 2}$.

Solution:

STEP 1 $f(x)$ is a one-to-one function because it passes the horizontal line test.

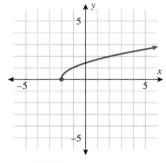

STEP 2   Let $y = f(x)$. $y = \sqrt{x + 2}$

STEP 3   Interchange x and y. $x = \sqrt{y + 2}$

STEP 4   Solve for y.

Square both sides of the equation. $x^2 = y + 2$

Subtract 2 from both sides. $x^2 - 2 = y$ or $y = x^2 - 2$

STEP 5   Let $y = f^{-1}(x)$. $f^{-1}(x) = x^2 - 2$

STEP 6   Note any domain restrictions.

f: Domain: $[-2, \infty)$ Range: $[0, \infty)$

f^{-1}:   Domain: $[0, \infty)$ Range: $[-2, \infty)$

The inverse of $f(x) = \sqrt{x + 2}$ is $f^{-1}(x) = x^2 - 2$ for $x \geq 0$.

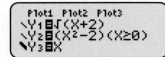

TECHNOLOGY TIP

Using a graphing utility, plot
$y_1 = f(x) = \sqrt{x + 2}$,
$y_2 = f^{-1}(x) = x^2 - 2$, and
$y_3 = x$.

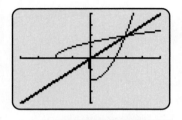

```
Plot1 Plot2 Plot3
\Y1◘√(X+2)
\Y2◘(X²-2)(X≥0)
\Y3◘X
```

Note that the function $f(x)$ and its inverse $f^{-1}(x)$ are symmetric about the line $y = x$.

STEP 7 Check.

$f^{-1}(f(x)) = x$ for all x in the domain of f.

$$f^{-1}(f(x)) = (\sqrt{x + 2})^2 - 2$$
$$= x + 2 - 2 \quad \text{for } x \geq -2$$
$$= x$$

$f(f^{-1}(x)) = x$ for all x in the domain of f^{-1}.

$$f(f^{-1}(x)) = \sqrt{(x^2 - 2) + 2}$$
$$= \sqrt{x^2} \quad \text{for } x \geq 0$$
$$= x$$

Note that the function $f(x) = \sqrt{x + 2}$ and its inverse $f^{-1}(x) = x^2 - 2$ for $x \geq 0$ are symmetric about the line $y = x$.

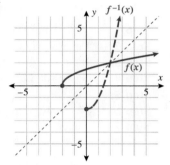

YOUR TURN Find the inverse of the given function. State the domain and range of the inverse function.

a. $f(x) = 7x - 3$ **b.** $g(x) = \sqrt{x - 1}$

EXAMPLE 7 When the Inverse of a Function Does Not Exist

Find the inverse of the function $f(x) = |x|$.

Solution:

The function $f(x) = |x|$ fails the horizontal line test and therefore is not a one-to-one function. Because f is not a one-to-one function, its inverse does not exist.

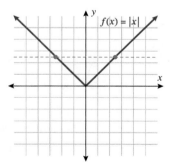

Answer: a. $f^{-1}(x) = \dfrac{x + 3}{7}$ domain: $(-\infty, \infty)$ range: $(-\infty, \infty)$

b. $g^{-1}(x) = x^2 + 1$ domain: $[0, \infty)$ range: $[1, \infty)$

EXAMPLE 8 Finding the Inverse of a Function

The function $f(x) = \dfrac{2}{x+3}$, $x \neq -3$ is a one-to-one function. Find its inverse.

Solution:

STEP 1 The problem states that $f(x)$ is a one-to-one function.

STEP 2 Let $y = f(x)$.

$$y = \frac{2}{x+3}$$

STEP 3 Interchange x and y.

$$x = \frac{2}{y+3}$$

STEP 4 Solve for y.

Multiply equation by $(y + 3)$. $x(y+3) = 2$

Eliminate parentheses. $xy + 3x = 2$

Subtract $3x$ from both sides. $xy = -3x + 2$

Divide equation by x. $y = \dfrac{-3x+2}{x} = -3 + \dfrac{2}{x}$

STEP 5 Let $y = f^{-1}(x)$. $f^{-1}(x) = -3 + \dfrac{2}{x}$

STEP 6 Note any domain restrictions on $f^{-1}(x)$. $x \neq 0$

The inverse of the function $f(x) = \dfrac{2}{x+3}$, $x \neq -3$ is $f^{-1}(x) = -3 + \dfrac{2}{x}, x \neq 0$.

STEP 7 Check.

$$f^{-1}(f(x)) = -3 + \frac{2}{\left(\dfrac{2}{x+3}\right)} = -3 + (x+3) = x \qquad x \neq -3$$

$$f(f^{-1}(x)) = \frac{2}{\left(-3 + \dfrac{2}{x}\right) + 3} = \frac{2}{\left(\dfrac{2}{x}\right)} = x \qquad x \neq 0$$

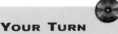

■ **YOUR TURN**   The function $f(x) = \dfrac{4}{x-1}$, $x \neq 1$ is a one-to-one function. Find its inverse.

■ **Answer:** $f^{-1}(x) = 1 + \dfrac{4}{x}$ $x \neq 0$

SECTION 3.5 **SUMMARY**

One-to-One Functions

- Algebraic test: No two *x* values map to the same *y* value.
- Graph test: Horizontal line test

Properties of Inverses

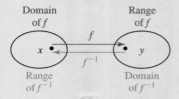

1. If *f* is one-to-one, then f^{-1} exists.
2. Domain of f^{-1} = range of *f*.
   Range of f^{-1} = domain of *f*.
3. $f^{-1}(f(x)) = x$ (for all *x* in domain of $f(x)$) and $f(f^{-1}(y)) = y$ (for all *y* in domain of $f^{-1}(y)$).
4. *f* and f^{-1} are symmetric with respect to the line $y = x$.

There is a seven-step procedure for finding the inverse of a function.

SECTION 3.5 EXERCISES

■ SKILLS

In Exercises 1–20, determine whether the given function is a one-to-one function.

1.

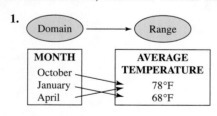

MONTH	AVERAGE TEMPERATURE
October	
January	78°F
April	68°F

2.

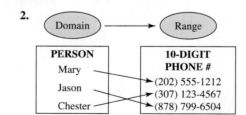

PERSON	10-DIGIT PHONE #
Mary	(202) 555-1212
Jason	(307) 123-4567
Chester	(878) 799-6504

3.

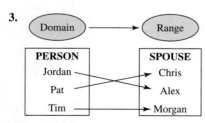

PERSON	SPOUSE
Jordan	Chris
Pat	Alex
Tim	Morgan

4.

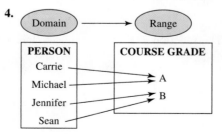

PERSON	COURSE GRADE
Carrie	
Michael	A
Jennifer	B
Sean	

5. $\{(0, 1),(1, 2),(2, 3),(3, 4)\}$

6. $\{(0, -2),(2, 0),(5, 3),(-5, -7)\}$

7. $\{(0, 0),(9, -3),(4, -2),(4, 2),(9, 3)\}$

8. $\{(0, 1),(1, 1),(2, 1),(3, 1)\}$

9. $\{(0, 1),(1, 0),(2, 1),(-2, 1),(5, 4),(-3, 4)\}$

10. $\{(0, 0),(-1, -1),(-2, -8),(1, 1),(2, 8)\}$

11.

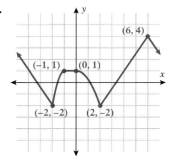

12.

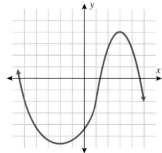

13.

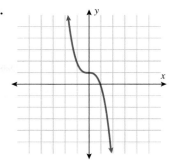

14.

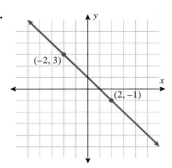

15.

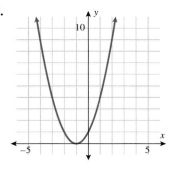

16.

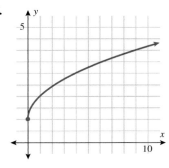

17. $f(x) = |x - 3|$ **18.** $f(x) = (x - 2)^2 + 1$ **19.** $f(x) = \dfrac{1}{x - 1}$ **20.** $f(x) = \sqrt[3]{x}$

In Exercises 21–30, verify that the function, $f^{-1}(x)$, is the inverse of $f(x)$ by showing that $f(f^{-1}(x)) = x$ and $f^{-1}(f(x)) = x$. Graph $f(x)$ and $f^{-1}(x)$ on the same axes to show the symmetry about the line $y = x$.

21. $f(x) = 2x + 1$ $f^{-1}(x) = \dfrac{x - 1}{2}$

22. $f(x) = \dfrac{x - 2}{3}$ $f^{-1}(x) = 3x + 2$

23. $f(x) = \sqrt{x - 1}$ $x \geq 1$ $f^{-1}(x) = x^2 + 1$ $x \geq 0$

24. $f(x) = 2 - x^2$ $x \geq 0$ $f^{-1}(x) = \sqrt{2 - x}$ $x \leq 2$

25. $f(x) = \dfrac{1}{x}$ $f^{-1}(x) = \dfrac{1}{x}$ $x \neq 0$

26. $f(x) = (5 - x)^{1/3}$ $f^{-1}(x) = 5 - x^3$

27. $f(x) = \dfrac{1}{2x + 6}$ $x \neq -3$ $f^{-1}(x) = \dfrac{1}{2x} - 3$ $x \neq 0$

28. $f(x) = \dfrac{3}{4 - x}$ $x \neq 4$ $f^{-1}(x) = 4 - \dfrac{3}{x}$ $x \neq 0$

29. $f(x) = \dfrac{x + 3}{x + 4}$ $x \neq -4$ $f^{-1}(x) = \dfrac{3 - 4x}{x - 1}$ $x \neq 1$

30. $f(x) = \dfrac{x - 5}{3 - x}$ $x \neq 3$ $f^{-1}(x) = \dfrac{3x + 5}{x + 1}$ $x \neq -1$

In Exercises 31–38, a graph of a one-to-one function is given; plot its inverse.

31.

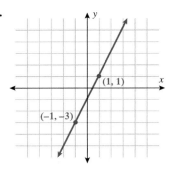

32.

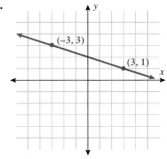

33.

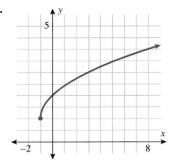

34.

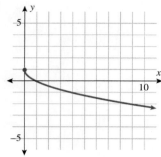

35.

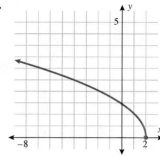

36.

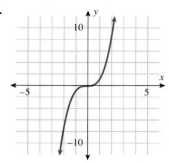

37.

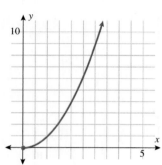

38.

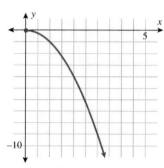

In Exercises 39–56, the function *f* is one-to-one. Find its inverse, and check your answer. State the domain and range of both *f* and f^{-1}.

39. $f(x) = x - 1$

40. $f(x) = 7x$

41. $f(x) = -3x + 2$

42. $f(x) = 2x + 3$

43. $f(x) = x^3 + 1$

44. $f(x) = x^3 - 1$

45. $f(x) = \sqrt{x - 3}$

46. $f(x) = \sqrt{3 - x}$

47. $f(x) = x^2 - 1, \, x \geq 0$

48. $f(x) = -x^2 + 1, \, x \geq 0$

49. $f(x) = (x + 2)^2 - 3, \, x \geq -2$

50. $f(x) = (x - 3)^2 - 2, \, x \geq 3$

51. $f(x) = \dfrac{2}{x}$

52. $f(x) = -\dfrac{3}{x}$

53. $f(x) = \dfrac{2}{3 - x}$

54. $f(x) = \dfrac{7}{x + 2}$

55. $f(x) = \dfrac{7x + 1}{5 - x}$

56. $f(x) = \dfrac{2x + 5}{7 + x}$

In Exercises 57–60, graph the piecewise-defined function to determine if it is a one-to-one function. If it is a one-to-one function, find its inverse.

57. $G(x) = \begin{cases} 0 & x < 0 \\ \sqrt{x} & x \geq 0 \end{cases}$

58. $G(x) = \begin{cases} \dfrac{1}{x} & x < 0 \\ \sqrt{x} & x \geq 0 \end{cases}$

59. $f(x) = \begin{cases} x & x \leq -1 \\ x^3 & -1 < x < 1 \\ x & x \geq 1 \end{cases}$

60. $f(x) = \begin{cases} x + 3 & x \leq -2 \\ |x| & -2 < x < 2 \\ x^2 & x \geq 2 \end{cases}$

■ APPLICATIONS

61. Temperature. The equation used to convert from degrees Celsius, *x*, to degrees Fahrenheit is $f(x) = \frac{9}{5}x + 32$. Determine the inverse function, $f^{-1}(x)$. What does the inverse function represent?

62. Temperature. The equation used to convert from degrees Fahrenheit, *x*, to degrees Celsius is $C(x) = \frac{5}{9}(x - 32)$.

Determine the inverse function, $C^{-1}(x)$. What does the inverse function represent?

63. Budget. The Richmond rowing club is planning to enter the Head of the Charles race in Boston and is trying to figure out how much money to raise. The entry fee is $250 per boat for the first 10 boats, and $175 for each additional boat. Find the cost function, $C(x)$, as a function of the number of boats they enter, x. Find the inverse function that will yield how many boats they can enter as a function of how much money they raise.

64. Phones. A phone company charges $.39 per minute for the first 10 minutes of a long-distance phone call and $.12 per minute every minute after that. Find the cost function, $C(x)$, as a function of the length of the phone call in minutes, x. Suppose you buy a "prepaid" phone card that is planned for a single call. Find the inverse function that determines how many minutes you can talk as a function of how much you prepaid.

65. Salary. A student is working at Target making $7 per hour and the weekly number of hours worked per week, x, varies. If Target withholds 25% of his earnings for taxes and social security, write a function, $E(x)$, that expresses the student's take-home pay each week. Find the inverse function, $E^{-1}(x)$. What does the inverse function tell you?

66. Salary. A grocery store pays you $8 per hour for the first 40 hours per week and time and a half for overtime. Write a piecewise-defined function that represents your weekly earnings, $E(x)$, as a function of number of hours worked, x. Find the inverse function, $E^{-1}(x)$. What does the inverse function tell you?

■ **CATCH THE MISTAKE**

In Exercises 67–70, explain the mistake that is made.

67. Is $x = y^2$ a one-to-one function?

Solution:

Yes, this graph represents a one-to-one function because it passes the horizontal line test.

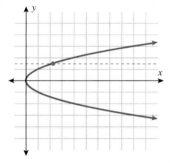

68. A linear one-to-one function is graphed below. Draw its inverse.

Solution:

Note that the points $(3, 3)$ and $(0, -4)$ lie on the graph of the function.

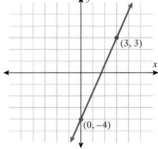

By symmetry, the points $(-3, -3)$ and $(0, 4)$ lie on the graph of the inverse.

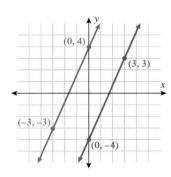

69. Given the function, $f(x) = x^2$, find the inverse function, $f^{-1}(x)$.

Solution:

Step 1: Let $y = f(x)$.	$y = x^2$
Step 2: Solve for x.	$x = \sqrt{y}$
Step 3: Interchange x and y.	$y = \sqrt{x}$
Step 4: Let $y = f^{-1}(x)$.	$f^{-1}(x) = \sqrt{x}$

Check: $f(f^{-1}(x)) = (\sqrt{x})^2 = x$ and $f^{-1}(f(x)) = \sqrt{x^2} = x$.

The inverse of $f(x) = x^2$ is $f^{-1}(x) = \sqrt{x}$.

70. Given the function, $f(x) = \sqrt{x - 2}$, find the inverse function, $f^{-1}(x)$, and state the domain restrictions on $f^{-1}(x)$.

Solution:

Step 1: Let $y = f(x)$.	$y = \sqrt{x - 2}$
Step 2: Interchange x and y.	$x = \sqrt{y - 2}$
Step 3: Solve for y.	$y = x^2 + 2$
Step 4: Let $f^{-1}(x) = y$.	$f^{-1}(x) = x^2 + 2$
Step 5: Domain restrictions $f(x) = \sqrt{x - 2}$ has the domain restriction that $x \geq 2$.	

The inverse of $f(x) = \sqrt{x - 2}$ is $f^{-1}(x) = x^2 + 2$.

The domain of $f^{-1}(x)$ is $x \geq 2$.

71. T or F: Every even function is a one-to-one function.

72. T or F: Every odd function is a one-to-one function.

73. T or F: It is not possible for $f = f^{-1}$.

74. T or F: A function f has an inverse. If the function lies in quadrant II, then its inverse lies in quadrant IV.

75. If $(0, b)$ is the y-intercept of a one-to-one function f, what is the x-intercept of the inverse, f^{-1}?

76. The unit circle is not a function. If we restrict ourselves to the semicircle that lies in quadrants I and II, the

graph represents a function, but it is not a one-to-one function. If we further restrict ourselves to the quarter circle lying in quadrant I, the graph does represent a one-to-one function. Determine the equations of both the one-to-one function and its inverse. State the domain restrictions of both.

77. Under what conditions is the linear function $f(x) = mx + b$ a one-to-one function?

78. Assuming the conditions found in Exercise 77 are met, determine the inverse of the linear function.

■ TECHNOLOGY

In Exercises 79–82, graph the following functions and determine if they are one-to-one.

79. $f(x) = |4 - x^2|$

80. $f(x) = \dfrac{3}{x^3 + 2}$

81. $f(x) = x^{1/3} - x^5$

82. $f(x) = \dfrac{1}{x^{1/2}}$

In Exercises 83 and 84, graph the functions f and g and the line $y = x$ in the same screen. Are the two function inverses of each other?

83. $f(x) = \sqrt{3x - 5}$ $g(x) = \dfrac{x^2}{3} + \dfrac{5}{3}$

84. $f(x) = (x - 7)^{1/3} + 2$ $g(x) = x^3 - 6x^2 + 12x - 1$

Q: Beth's company offers a product whose profit can be modeled by the function $P(x) = 40x - \dfrac{1}{200}x^2 - 10{,}000$ where x is the number of units sold. The company's profits increase until 4000 units have been sold. At this point, the demand for the product decreases and the profits begin to decrease. Beth is doing some research into the relationship between her company's profit and number of units sold. She hires an analyst to find a formula to give her the number of units sold given the profit. The analyst tells her that this is not possible unless some restrictions are made. Why is there a problem?

A: Beth assumed that the profit function had an inverse. However, in this case profit is a quadratic function, so it is not one-to-one and does not have an inverse.

The analyst can still help Beth by writing two different functions for x given the profit P. If she restricts the domain of $P(x)$ so that the resulting functions are one-to-one and include all of the profit values from the range, then an inverse for each of these functions can be found. One inverse function will correspond to the values of x as the profit is increasing, and the other inverse function will correspond to values of x as the profit is decreasing.

Find the rules for the each of these functions.

TYING IT ALL TOGETHER

Janie has been a residential real estate agent for 3 years. She receives a monthly base salary of $2,000 and works an average of 50 hours per week. Her commission rate is 6% of all monthly sales up to and including $1,000,000 and 8% of the portion of monthly sales that exceed $1,000,000. Her sales last month were $724,625 and were $1,212,900 for the month previous to that. Write a function $I(x)$ for Janie's monthly income where x represents her monthly sales.

Next year, Janie will receive a $100 a month raise to her base salary. However, her commission rate will change so that she will earn 6% commission on monthly sales up to and including $1,200,000 and 9% on the portion of her monthly sales that exceed $1,200,000. Her average monthly sales for the past 12 months have been $915,500. Write a function $P(x)$ for Janie's projected monthly income next year where x represents her monthly sales.

SECTION	TOPIC	PAGES	REVIEW EXERCISES	KEY CONCEPTS
3.1	Functions	188–199	1–26	
	Determine if a relationship is a function	188–193	1–8	Each x corresponds to one and only one y.
	Vertical line test	193	9 and 10	A vertical line can only intersect a function in at most one point.
	Evaluate a function	194–197	11–18	Placeholder notation
	Domain of a function	197–199	19–24	Are there any restrictions on x?
3.2	Graphs of functions: Common functions and piecewise-defined functions	203–212	27–40	
	Common functions	203–206		$f(x) = mx + b,\ f(x) = x,\ f(x) = x^2,$ $f(x) = x^3,\ f(x) = \sqrt{x},\ f(x) = \sqrt[3]{x},$ $f(x) = |x|,\ f(x) = \frac{1}{x}$
	Even functions	206–207	27–34	$f(-x) = f(x)$ Symmetry about y-axis
	Odd functions	206–207	27–34	$f(-x) = -f(x)$ Symmetry about origin
	Piecewise-defined functions	208–212	35–38	Points of discontinuity
	Applications	209	39 and 40	
3.3	Graphing techniques: Transformations	216–226	41–54	
	Horizontal shifts	217–221	41–54	$y = f(x + c)$ c units to the left $y = f(x - c)$ c units to the right
	Vertical shifts	217–221	41–54	$y = f(x) + c$ c units upward $y = f(x) - c$ c units downward
	Reflection about the x-axis	221–222	41–54	$y = -f(x)$
	Reflection about the y-axis	221–222	41–54	$y = f(-x)$
	Expansion	223–225	41–54	$y = cf(x)$ if $c > 1$ Expanded in the vertical direction
	Contraction	223–225	41–54	$y = cf(x)$ if $0 < c < 1$ Contracted in the vertical direction
3.4	Operations on functions and composition of functions	230–237	55–74	
	Adding, subtracting, and multiplying functions	230–232	55–60	$(f + g)(x) = f(x) + g(x)$ $(f - g)(x) = f(x) - g(x)$ $(f \cdot g)(x) = f(x) \cdot g(x)$ Domain of the resulting function is the intersection of the individual domains.

Section	Topic	Pages	Review Exercises	Key Concepts
	Dividing functions	230–232	55–60	$\left(\dfrac{f}{g}\right)(x) = \dfrac{f(x)}{g(x)} \qquad g(x) \neq 0$ Domain of quotient is intersection of domains of f and g, and any points when $g(x) = 0$ must be eliminated.
	Composition of functions	233–237	61–72	$(f \circ g)(x) = f(g(x))$ The domain of the composite function is a subset of the domain of $g(x)$. Values for x must be eliminated if their corresponding values $g(x)$ are not in the domain of f.
3.5	One-to-one functions and inverse functions	240–250	75–92	
	Determine if a function is one-to-one	241–242	75–80	No two x values map to the same y value.
	Horizontal line test	241–242	79 and 80	A horizontal line may only intersect a one-to-one function in at most one point.
	Properties of inverses	242–246	81–84	Only one-to-one functions have inverses. $f^{-1}(f(x)) = x$ and $f(f^{-1}(x)) = x$. Domain of f = range of f^{-1}. Range of f = domain of f^{-1}. The graph of a function and its inverse are symmetric about the line $y = x$. If the point (a, b) lies on the graph of a function, then the point (b, a) lies on the graph of its inverse.
	Find an inverse of a function	246–249	85–90	Seven-step procedure

3.1 Functions

Determine whether each relationship is a function.

1.

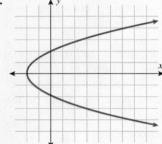

2. $\{(1, 2),(3, 4),(2, 4),(3, 7)\}$

3. $\{(-2, 3),(1, -3),(0, 4),(2, 6)\}$

4. $\{(4, 7),(2, 6),(3, 8),(1, 7)\}$ **5.** $x^2 + y^2 = 36$

6. $x = 4$ **7.** $y = |x + 2|$ **8.** $y = \sqrt{x}$

9.

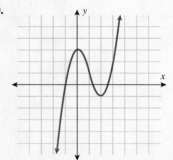

10.

Evaluate the given quantities using the following three functions.

$$f(x) = 4x - 7 \quad F(t) = t^2 + 4t - 3 \quad g(x) = |x^2 + 2x + 4|$$

11. $f(3)$ **12.** $F(4)$ **13.** $f(-7) \cdot g(3)$

14. $\dfrac{F(0)}{g(0)}$ **15.** $\dfrac{f(2) - F(2)}{g(0)}$ **16.** $f(3 + h)$

17. $\dfrac{f(3 + h) - f(3)}{h}$ **18.** $\dfrac{F(t + h) - F(t)}{h}$

Find the domain of the given function. Express the domain in interval notation.

19. $f(x) = -3x - 4$ **20.** $g(x) = x^2 - 2x + 6$

21. $h(x) = \dfrac{1}{x + 4}$ **22.** $F(x) = \dfrac{7}{x^2 + 3}$

23. $G(x) = \sqrt{x - 4}$ **24.** $H(x) = \dfrac{1}{\sqrt{2x - 6}}$

Challenge.

25. If $f(x) = \dfrac{D}{x^2 - 16}$, $f(4)$ and $f(-4)$ are undefined and $f(5) = 2$, find D.

26. Construct a function that is undefined at $x = -3$ and $x = 2$, and the point $(0, -4)$ lies on the graph of the function.

3.2 Graphs of Functions: Common Functions and Piecewise-Defined Functions

Determine if the function is even, odd, or neither.

27. $f(x) = 2x - 7$ **28.** $g(x) = 7x^5 + 4x^3 - 2x$

29. $h(x) = x^3 - 7x$ **30.** $f(x) = x^4 + 3x^2$

31. $f(x) = x^{1/4} + x$ **32.** $f(x) = \sqrt{x + 4}$

33. $f(x) = \dfrac{1}{x^3} + 3x$ **34.** $f(x) = \dfrac{1}{x^2} + 3x^4 + |x|$

Graph the piecewise-defined function. State the domain and range in interval notation.

35. $F(x) = \begin{cases} x^2 & x < 0 \\ 2 & x \geq 0 \end{cases}$

36. $f(x) = \begin{cases} -2x - 3 & x \leq 0 \\ 4 & 0 < x \leq 1 \\ x^2 + 4 & x > 1 \end{cases}$

37. $f(x) = \begin{cases} x^2 & x \leq 0 \\ -\sqrt{x} & 0 < x \leq 1 \\ |x + 2| & x > 1 \end{cases}$

38. $F(x) = \begin{cases} x^2 & x < 0 \\ x^3 & 0 < x < 1 \\ -|x| - 1 & x \geq 1 \end{cases}$

Applications.

39. Tutoring Costs. A tutoring company charges $25 for the first hour of tutoring and $10.50 for every 30-minute period after that. Find the cost function, $C(x)$, as a function of the length of the tutoring session. Let $x =$ number of 30-minute periods.

40. Salary. An employee who makes \$30 per hour also earns time and a half for overtime (any hours worked above the normal 40-hour work week). Write a function, $E(x)$, that describes her weekly earnings as a function of number of hours worked, x.

3.3 Graphing Techniques: Transformations

Graph the following functions using graphing aids.

41. $y = -(x - 2)^2 + 4$

42. $y = |-x + 5| - 7$

43. $y = \sqrt[3]{x - 3} + 2$

44. $y = \dfrac{1}{x - 2} - 4$

45. $y = \dfrac{-1}{2}x^3$

46. $y = 2x^2 + 3$

Write the function whose graph is the graph of $y = \sqrt{x}$ but is transformed accordingly, and state the domain of the resulting function.

47. shifted to left 3 units

48. shifted down 4 units

49. shifted to the right 2 units and up 3 units

50. reflected about the y-axis

51. expanded by a factor of 5 and shifted down 6 units

52. compressed by a factor of 2 and shifted up 3 units

Transform the function into the form $f(x) = c(x - h)^2 + k$ by completing the square and graph the resulting function using transformations.

53. $y = x^2 + 4x - 8$

54. $y = 2x^2 + 6x - 5$

3.4 Operations on Functions and Composition of Functions

Given the functions g and h, find $g + h$, $g - h$, $g \cdot h$, $\dfrac{g}{h}$, and state the domain.

55. $g(x) = -3x - 4$
$h(x) = x - 3$

56. $g(x) = 2x + 3$
$h(x) = x^2 + 6$

57. $g(x) = \dfrac{1}{x^2}$
$h(x) = \sqrt{x}$

58. $g(x) = \dfrac{x + 3}{2x - 4}$
$h(x) = \dfrac{3x - 1}{x - 2}$

59. $g(x) = \sqrt{x - 4}$
$h(x) = \sqrt{2x + 1}$

60. $g(x) = x^2 - 4$
$h(x) = x + 2$

For the given functions f and g, find the composite functions $f \circ g$ and $g \circ f$, and state the domains.

61. $f(x) = 3x - 4$
$g(x) = 2x + 1$

62. $f(x) = x^3 + 2x - 1$
$g(x) = x + 3$

63. $f(x) = \dfrac{2}{x + 3}$
$g(x) = \dfrac{1}{4 - x}$

64. $f(x) = \sqrt{2x^2 - 5}$
$g(x) = \sqrt{x + 6}$

Evaluate $f(g(3))$ and $g(f(-1))$, if possible.

65. $f(x) = 4x^2 - 3x + 2$
$g(x) = 6x - 3$

66. $f(x) = \sqrt{4 - x}$
$g(x) = x^2 + 5$

67. $f(x) = \dfrac{x}{|2x - 3|}$
$g(x) = |5x + 2|$

68. $f(x) = \dfrac{1}{x - 1}$
$g(x) = x^2 - 1$

Write the function as a composite of two functions f and g.

69. $h(x) = 3(x - 2)^2 + 4(x - 2) + 7$

70. $h(x) = \dfrac{\sqrt[3]{x}}{1 - \sqrt[3]{x}}$

71. $h(x) = \dfrac{1}{\sqrt{x^2 + 7}}$

72. $h(x) = \sqrt{|3x + 4|}$

Applications.

73. Rain. A rain drop hitting a lake makes a circular ripple. If the radius, in inches, grows as a function of time, in minutes: $r(t) = 25\sqrt{t + 2}$, find the area of the ripple as a function of time.

74. Geometry. Let the area of a rectangle be given by: $42 = l \cdot w$ and let the perimeter be $36 = 2 \cdot l + 2 \cdot w$. Express the perimeter in terms of w.

3.5 One-to-One Functions and Inverse Functions

Determine whether the given function is a one-to-one function.

75.

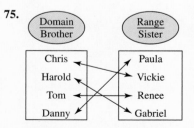

76. $\{(2, 3), (-1, 2), (3, 3), (-3, -4), (-2, 1)\}$

77. $\{(-2, 0), (4, 5), (3, 7)\}$

78. $\{(-8, -6), (-4, 2), (0, 3), (2, -8), (7, 4)\}$

79. $y = \sqrt{x}$

80. $y = x^2$

Verify that the function $f^{-1}(x)$ is the inverse of $f(x)$ by showing that $f(f^{-1}(x)) = x$. Graph $f(x)$ and $f^{-1}(x)$ on the same graph and show the symmetry about the line $y = x$.

81. $f(x) = 3x + 4 \qquad f^{-1}(x) = \dfrac{x - 4}{3}$

82. $f(x) = \dfrac{1}{4x - 7} \qquad f^{-1}(x) = \dfrac{1 + 7x}{4x}$

83. $f(x) = \sqrt{x + 4} \qquad f^{-1}(x) = x^2 - 4 \qquad x \geq 0$

84. $f(x) = \dfrac{x + 2}{x - 7} \qquad f^{-1}(x) = \dfrac{7x + 2}{x - 1}$

The function f is one-to-one. Find its inverse and check your answer. State the domain and range of both f and f^{-1}.

85. $f(x) = 2x + 1$

86. $f(x) = x^5 + 2$

87. $f(x) = \sqrt{x + 4}$

88. $f(x) = (x + 4)^2 + 3 \quad x \geq -4$

89. $f(x) = \dfrac{x + 6}{x + 3}$

90. $f(x) = 2\sqrt[3]{x - 5} - 8$

Applications.

91. **Salary.** A pharmaceutical salesperson makes $22,000 base salary a year plus 8% of the total products sold. Write a function $S(x)$ that represents her yearly salary as a function of the total dollars worth of products sold, x. Find $S^{-1}(x)$. What does this inverse function tell you?

92. **Volume.** Express the volume of a rectangular box, V, that has a square base of length s and is 3 feet high as a function of the square length. Find V^{-1}. If a certain volume is desired what does the inverse tell you?

Assuming x represents the independent variable and y represents the dependent variable, classify the relationships in Exercises 1–3 as

   a. not a function   **b.** function, but not one-to-one

   c. one-to-one function

1. $f(x) = |2x + 3|$ **2.** $x = y^2 + 2$ **3.** $y = \sqrt[3]{x + 1}$

In Exercises 4–7, use $f(x) = \sqrt{x - 2}$ and $g(x) = x^2 + 11$, and determine the desired quantity or expression. In the case of an expression, state the domain.

4. $f(11) - 2g(-1)$ **5.** $\left(\dfrac{f}{g}\right)(x)$

6. $\left(\dfrac{g}{f}\right)(x)$ **7.** $g(f(x))$

In Exercises 8–10, determine if the function is odd, even, or neither.

8. $f(x) = |x| - x^2$

9. $f(x) = 9x^3 + 5x - 3$

10. $f(x) = \dfrac{2}{x}$

In Exercises 11–14, graph the functions. State the domain and range of each function.

11. $f(x) = -\sqrt{x - 3} + 2$

12. $f(x) = -2(x - 1)^2$

13. $f(x) = \dfrac{1}{x - 2} + 3$

14. $f(x) = \begin{cases} -x & x < -1 \\ 1 & -1 < x < 2 \\ x^2 & x \geq 2 \end{cases}$

In Exercises 15–18, given the function f, find the inverse if it exists. State the domain and range of both f and f^{-1}.

15. $f(x) = \sqrt{x - 5}$ **16.** $f(x) = x^2 + 5$

17. $f(x) = \dfrac{2x + 1}{5 - x}$ **18.** $f(x) = \begin{cases} -x & x \leq 0 \\ -x^2 & x > 0 \end{cases}$

19. What domain restriction can be made so that $f(x) = x^2$ has an inverse?

20. If the point $(-2, 5)$ lies on the graph of a function, what point lies on the graph of its inverse function?

21. Discount. Suppose a suit has been marked down 40% off the original price. An advertisement in the newspaper has an "additional 30% off the sale price" coupon. Write a function that determines the "checkout" price of the suit.

22. Temperature. Degrees Fahrenheit (F), degrees Celsius (C), and kelvins (K) are related by the two equations: $F = \frac{9}{5}C + 32$ and $K = C + 273.15$. Write a function whose input is kelvins and output is degrees Fahrenheit.

23. Circles. If a quarter circle is drawn by tracing the unit circle in quadrant III, what does the inverse of that function look like? Where is it located?

24. Sprinkler. A sprinkler head malfunctions at midfield of an NFL football field. The puddle of water forms a circular pattern around the sprinkler head with a radius in yards that grows as a function of time, in hours: $r(t) = 10\sqrt{t}$. When will the puddle reach the sidelines? (A football field is 30 yards from sideline to sideline.)

25. Internet. The cost of airport Internet access is $15 for the first 30 minutes and $1 per minute for each minute after that. Write a function describing the cost of the service as a function of minutes used.

CHAPTER 4

Polynomial and Rational Functions

Ray Guy

Indoor football stadiums are designed so that punters will not kick the roof. One of the greatest NFL punters of all time was Ray Guy, who played 14 seasons from 1973 to 1986. "In the 1976 Pro Bowl, one of his punts hit the giant TV screen hanging from the rafters in the Louisiana Superdome. Not only did Guy punt high and far—'hang time' came into the NFL lexicon during his tenure—once he even had an opponent take a ball he punted and test it for helium!" (www.prokicker.com; Ray Guy Fact Sheet)

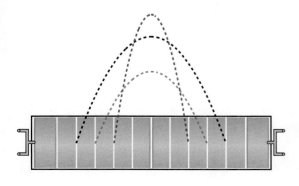

The path that punts typically follow is called a *parabola* and is classified as a *quadratic function.* The distance of a punt is measured in the horizontal direction. The yard line where the punt is kicked from and the yard line where the punt either hits the field or is caught are the zeros of the quadratic function. Zeros are the points where the function value is equal to zero.

Polynomial and Rational Functions

Dividing Polynomials

- Long Division
- Synthetic Division

Zeros of Polynomials

- Factoring Polynomials
- Possible Roots
  - Rational Zero Test
  - Descartes's Rule of Signs
- Remainder Theorem
- Factor Theorem

Graphs of Polynomials

- Finding Real Zeros
- Parabolas
  - Quadratic Functions
- Other Polynomial Functions
  - Finding Intercepts
  - Test Points in Intervals
  - End Behavior

Rational Functions

- Ratios of Polynomial Functions
- Domain Restrictions
- Asymptotes
- Graphs

CHAPTER OBJECTIVES

- Divide polynomials.
- Factor polynomials that are third-degree or higher.
- Graph polynomial functions.
- Graph rational functions.

NAVIGATION THROUGH SUPPLEMENTS

DIGITAL VIDEO SERIES #4

STUDENT SOLUTIONS MANUAL CHAPTER 4

BOOK COMPANION SITE
www.wiley.com/college/young

SECTION 4.1 Quadratic Functions

Skills Objectives

- Graph a quadratic function in standard form.
- Write a quadratic function in standard form.
- Find the equation of a parabola.
- Solve application problems that involve quadratic functions.

Conceptual Objectives

- Recognize characteristics of parabolas:
  - whether the parabola opens up or down
  - if the vertex is a maximum or minimum
  - the axis of symmetry

In Chapter 3 we studied functions in general. In this chapter we will learn about a special group of functions called polynomial functions. Polynomial functions are simple functions, and often, more complicated functions are approximated by polynomial functions. Polynomial functions are used to model many real-world applications such as the stock market, football punts, business costs, revenues, and profits, and the flight path of NASA's "vomit comet." Let's first start by defining a polynomial function.

DEFINITION   **POLYNOMIAL FUNCTION**

Let n be a nonnegative integer, and let $a_n, a_{n-1}, ..., a_2, a_1, a_0$ be real numbers with $a_n \neq 0$. The function

$$f(x) = a_n x^n + a_{n-1} x^{n-1} + ... + a_2 x^2 + a_1 x + a_0$$

is called a **polynomial function of x with degree n**. The coefficient a_n is called the **leading coefficient**.

Polynomials of particular degrees have special names. In Chapter 3, the library of functions included the constant function, $f(x) = b$, which is a horizontal line; the linear function, $f(x) = mx + b$, which is a line with slope m and y-intercept $(0, b)$; the square function $f(x) = x^2$; and the cube function $f(x) = x^3$. These are all special cases of a polynomial function.

Here are other examples of polynomials of a particular degree and their names:

POLYNOMIAL	DEGREE	SPECIAL NAME
$f(x) = 3$	0	Constant function
$f(x) = -2x + 1$	1	Linear function
$f(x) = 7x^2 - 5x + 19$	2	Quadratic function
$f(x) = 4x^3 + 2x - 7$	3	Cubic function

The leading coefficients of these functions are 3, -2, 7, and 4, respectively. In Section 2.3, we discussed graphs of linear functions, which are first-degree polynomial functions. In this section, we will discuss graphs of quadratic functions, which are second-degree polynomial functions.

Graph of a Quadratic Function

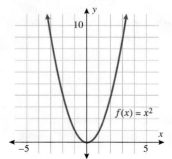

$f(x) = x^2$

In Section 3.2, the library of functions that we compiled included the square function, $f(x) = x^2$, whose graph is a parabola. See graph on the left.

In Section 3.3, we worked with graphing functions using transformation techniques such as $F(x) = (x + 1)^2 - 2$, which can be graphed by starting with the square function $y = x^2$ and shifting 1 unit to the left and 2 units down to arrive at $F(x) = (x + 1)^2 - 2$. See the graph on the right.

Note that if we eliminated the parentheses in $F(x) = (x + 1)^2 - 2$

$$F(x) = x^2 + 2x + 1 - 2$$

$$= x^2 + 2x - 1$$

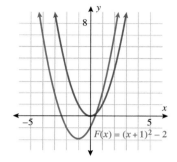

the result is a function defined by a second-degree polynomial (a polynomial with a square as the highest term), which is also called a **quadratic function**.

DEFINITION   QUADRATIC FUNCTION

Let a, b, and c be real numbers with $a \neq 0$. The function

$$f(x) = ax^2 + bx + c$$

is called a **quadratic function**.

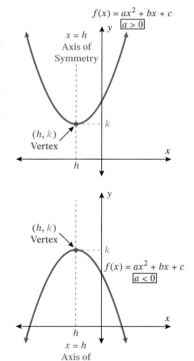

The graph of all quadratic functions is a **parabola**. Parabolas are shaped like cups or the letter U. If the leading coefficient, a, is *positive,* then the parabola opens *upward.* If the leading coefficient, a, is *negative,* then the parabola opens *downward.* The **vertex** (or turning point) is the minimum point, or low point, on the graph if the parabola opens upward or the maximum point, or high point, on the graph if the parabola opens downward. The vertical line that intersects the parabola at the vertex is called the **axis of symmetry**.

The axis of symmetry is the line $x = h$, and the vertex is located at the point (h, k) for some values h and k as shown in the two figures on the right.

At the start of this section, the function $F(x) = x^2 + 2x - 1$, which can also be written as $F(x) = (x + 1)^2 - 2$, was shown through graph-shifting techniques to have a graph of a parabola. Looking at the graph on top of the margin, we see that the parabola opens upward ($a = 1 > 0$), has a vertex at the point $(-1, -2)$, and an axis of symmetry of $x = -1$.

Graphing Quadratic Functions in Standard Form

In general, writing a quadratic function in the form,

$$f(x) = a(x - h)^2 + k$$

allows the vertex (h, k) and the axis of symmetry $x = h$ to be determined by inspection. In other words, you can "eyeball" a parabola. This form is a convenient way to express a quadratic function in order to quickly determine its corresponding graph. Hence, this form is called *standard form.*

QUADRATIC FUNCTION: STANDARD FORM

The quadratic function

$$f(x) = a(x - h)^2 + k$$

is in **standard form**. The graph of f is a parabola whose vertex is the point (h, k). The parabola is symmetric with respect to the line $x = h$. If $a > 0$, the parabola opens upward. If $a < 0$, the parabola opens downward.

Recall that graphing linear equations requires finding two points on the line, or a point and the slope of the line. However, simply knowing two points that lie on the graph no longer suffices for quadratic equations in general. Which leads us to the question: Is there a systematic way to graph quadratic equations? Yes, here is a general step-by-step procedure for graphing parabolas with the quadratic function in standard form.

PROCEDURE FOR GRAPHING PARABOLAS

To graph $f(x) = a(x - h)^2 + k$

Step 1: Determine whether the parabola is opening upward or downward.

$$a > 0 \quad \text{upward}$$
$$a < 0 \quad \text{downward}$$

Step 2: Determine the vertex (h, k).
Step 3: Find the y-intercept (by setting $x = 0$).
Step 4: Find any x-intercepts (by setting $f(x) = 0$ and solving for x).
Step 5: Plot the vertex and intercepts and connect with a smooth curve.

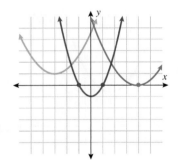

Note that in Step 4 it says to "find any x-intercepts." Parabolas opening up or down will always have a y-intercept. However, they can have **one**, **two**, or **no** x-intercepts. The figure on the left illustrates this for parabolas opening up, and the same can be said about parabolas opening down.

EXAMPLE 1 Graphing a Quadratic Function Given in Standard Form

Graph the quadratic function $f(x) = (x - 3)^2 - 1$.

Solution:

STEP 1 The parabola opens up. $a = 1$, so $a > 0$

STEP 2 Determine the vertex. $(h, k) = (3, -1)$

STEP 3 Find the y-intercept. $f(0) = (-3)^2 - 1 = 8$
    $(0, 8)$ is the y-intercept

STEP 4 Find any x-intercepts. $f(x) = (x - 3)^2 - 1 = 0$
    $(x - 3)^2 = 1$

Use square root method. $x - 3 = \pm 1$

Solve. $x = 2$ or $x = 4$
    $(2, 0)$ and $(4, 0)$ are x-intercepts

Graphing Quadratic Functions in General Form

A quadratic function is often written in one of two forms:

■   Standard form: $f(x) = a(x - h)^2 + k$
■   General form: $f(x) = ax^2 + bx + c$

When the quadratic function is expressed in standard form, the graph is easily obtained by identifying the vertex (h, k) and the intercepts and drawing a smooth curve that either opens upward or downward depending on the sign of a.

Typically, quadratic functions are expressed in general form and a graph is the ultimate goal, so we must first express the quadratic function in standard form. The technique that is used to transform a quadratic function from general form to standard form was introduced in Section 1.3 and is called *completing the square*.

EXAMPLE 3 Graph a Quadratic Function Given in General Form

Graph the quadratic function $f(x) = x^2 - 6x + 4$.

Solution:

Express the function in standard form by completing the square.

Write original function.	$f(x) = x^2 - 6x + 4$
Group the variable terms together.	$(x^2 - 6x) + 4$

Complete the square.

Half of -6 is -3; -3 squared is 9.

Add and subtract 9 within the parentheses.	$(x^2 - 6x + 9 - 9) + 4$
Write the -9 outside the parentheses.	$(x^2 - 6x + 9) - 9 + 4$
Write the expression inside parentheses as a perfect square and simplify.	$(x - 3)^2 - 5$

Now that once the quadratic function is written in standard form, $f(x) = (x - 3)^2 - 5$, we follow our step-by-step procedure for graphing a quadratic function in standard form.

STEP 1 The parabola opens upward. $a = 1$, so $a > 0$

STEP 2 Determine the vertex. $(h, k) = (3, -5)$

STEP 3 Find the y-intercept.
$$f(0) = (0)^2 - 6(0) + 4 = 4$$
$(0, 4)$ is the y-intercept

STEP 4 Find any x-intercepts.
$$f(x) = 0$$
$$f(x) = (x - 3)^2 - 5 = 0$$
$$(x - 3)^2 = 5$$
$$x - 3 = \pm\sqrt{5}$$
$$x = 3 \pm\sqrt{5}$$

$(3 + \sqrt{5}, 0)$ and $(3 - \sqrt{5}, 0)$ are the x-intercepts.

STUDY TIP

Although either form (standard or general) can be used to find the intercepts, it is often more convenient to use the general form when finding the y-intercept and the standard form when finding the x-intercept.

STEP 5 Plot the vertex and intercepts $(3, -5)$, $(0, 4)$, $(3 + \sqrt{5}, 0)$, and $(3 - \sqrt{5}, 0)$ and connect with a smooth parabolic curve.

Note: $3 + \sqrt{5} \approx 5.24$ and $3 - \sqrt{5} \approx 0.76$.

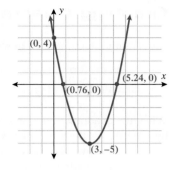

■ **YOUR TURN** Graph the quadratic function $f(x) = x^2 - 8x + 14$.

EXAMPLE 4 Graph a Quadratic Function Whose Equation Is in General Form with a Negative Leading Coefficient

Graph the quadratic function $f(x) = -3x^2 + 6x + 2$.

Solution:

Express the function in standard form by completing the square.

Write original function. $\qquad f(x) = -3x^2 + 6x + 2$

Group the variable terms together. $\qquad (-3x^2 + 6x) + 2$

Factor -3 is common. $\qquad -3(x^2 - 2x) + 2$

Add and subtract 1 inside the parentheses to create a perfect square. $\qquad -3(x^2 - 2x + 1 - 1) + 2$

Regroup terms. $\qquad -3(x^2 - 2x + 1) - 3(-1) + 2$

Write the expression inside parentheses as a perfect square and simplify. $\qquad -3(x - 1)^2 + 5$

Now that once the quadratic function is written in standard form, $f(x) = -3(x - 1)^2 + 5$, we follow our step-by-step procedure for graphing a quadratic function in standard form.

STEP 1 The parabola is opening downward. $\qquad a = -3$ therefore $a < 0$

STEP 2 Determine the vertex. $\qquad (h, k) = (1, 5)$

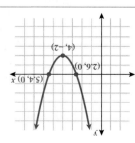

■ **Answer:**

STEP 3 Find the *y*-intercept using
the general form.

$$f(0) = -3(0)^2 + 6(0) + 2 = 2$$
$(0, 2)$ is the *y*-intercept

STEP 4 Find any *x*-intercepts using
the standard form.

$$f(x) = -3(x - 1)^2 + 5 = 0$$
$$-3(x - 1)^2 = -5$$
$$(x - 1)^2 = \frac{5}{3}$$
$$x - 1 = \pm\sqrt{\frac{5}{3}}$$
$$x = 1 \pm\sqrt{\frac{5}{3}} = 1 \pm\frac{\sqrt{15}}{3}$$

The *x*-intercepts are $\left(1 + \dfrac{\sqrt{15}}{3}, 0\right)$ and $\left(1 - \dfrac{\sqrt{15}}{3}, 0\right)$.

STEP 5 Plot the vertex and intercepts
$(1, 5)$, $(0, 2)$, $\left(1 + \dfrac{\sqrt{15}}{3}, 0\right)$, and
$\left(1 - \dfrac{\sqrt{15}}{3}, 0\right)$ and connect with a
smooth curve.

Note: $1 + \dfrac{\sqrt{15}}{3} \approx 2.3$ and
$1 - \dfrac{\sqrt{15}}{3} \approx -0.3$.

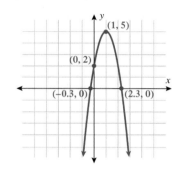

CONCEPT CHECK   For the function $f(x) = -2x^2 - 4x + 1$ would you
use the standard form or the general form to find the *y*-intercept?

■ **YOUR TURN**   Graph the quadratic function $f(x) = -2x^2 - 4x + 1$.

Find the Equation of a Parabola

It is important to understand that the equation, $y = x^2$, is equivalent to the quadratic
function $f(x) = x^2$. Both have the same parabolic graph. Thus far, we have been given
the equation and then asked to find characteristics (vertex and intercepts) in order to

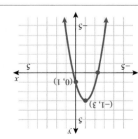

■ **Answer:**

graph. We now turn our attention to the problem of determining the equation, or function, given certain characteristics.

Suppose that we know the maximum height (vertex) of a punted football. Knowing where the ball was kicked from (an x-intercept) gives us enough information to determine the equation of the parabola. Once the equation is known, other information such as the other x-intercept (where the ball will land) can be determined.

EXAMPLE 5 Finding the Equation of a Parabola with an x-Intercept

Find the equation of a parabola whose vertex is $(3, 4)$ and passes through the point $(1, 0)$. Express the equation in both standard and general forms.

Solution:

STEP 1 Write the standard form of the equation of a parabola. $\qquad f(x) = a(x - h)^2 + k$

STEP 2 Substitute the vertex $(h, k) = (3, 4)$. $\qquad f(x) = a(x - 3)^2 + 4$

STEP 3 Use the point $(1, 0)$ to find a.

The point $(1, 0)$ implies $f(1) = 0$. $\qquad f(1) = a(1 - 3)^2 + 4 = 0$

Solve for a. $\qquad\qquad\qquad a(1 - 3)^2 + 4 = 0$
$$a(-2)^2 + 4 = 0$$
$$4a + 4 = 0$$
$$a = -1$$

STEP 4 Write both forms of the equation of a parabola.

Standard form: $\quad f(x) = -(x - 3)^2 + 4$

General form: $\quad f(x) = -x^2 + 6x - 5$

▪ **YOUR TURN** Find the standard form of the equation of a parabola whose vertex is $(-3, 5)$ and passes through the point $(2, 0)$.

EXAMPLE 6 Find the Equation of a Parabola without an x-Intercept

Find the standard form and the general form of the equation of a parabola whose vertex is $(3, 1)$ and passes through the point $(0, 4)$.

Solution:

STEP 1 Write the standard form of the equation of a parabola. $\qquad f(x) = a(x - h)^2 + k$

STEP 2 Substitute the vertex $(h, k) = (3, 1)$. $\qquad f(x) = a(x - 3)^2 + 1$

▪ **Answer: Standard form:** $f(x) = -\dfrac{1}{5}(x + 3)^2 + 5$

STEP 3 Use the point $(0, 4)$ to find a.

The point $(0, 4)$ implies $f(0) = 4$. $f(0) = a(0 - 3)^2 + 1 = 4$

Solve for a. $a(0 - 3)^2 + 1 = 4$

$9a + 1 = 4$

$9a = 3$

$a = \dfrac{1}{3}$

STEP 4 Write both forms of the equation of a parabola.

Standard form: $f(x) = \dfrac{1}{3}(x - 3)^2 + 1$

General form: $f(x) = \dfrac{1}{3}x^2 - 2x + 4$

■ **YOUR TURN**   Find the equation of a parabola whose vertex is $(-1, 3)$ and passes through the point $(0, 5)$. Express the equation in both standard and general form.

Note that in Example 6 and in this Your Turn there are no x-intercepts. Instead, we used the y-intercept to find the coefficient a. It is not necessary to be given intercepts. Once the vertex is known, the lead coefficient, a, can be found with any point that lies on the parabola.

Application Problems That Involve Quadratic Functions

What is the minimum distance a driver has to maintain between her car and the car in front as a function of speed? How many units produced will yield a maximum profit for a company? Given the number of linear feet of fence, what rectangular dimensions will yield a maximum fenced-in area? If a particular stock price has been shown to follow a quadratic trend, when will the stock achieve a maximum value? If a gun is fired into the air, where will the bullet land? These are all problems that can be solved using quadratic functions. Because the vertex of a parabola represents either the minimum or maximum value of the quadratic function, in application problems, it often suffices simply to find the vertex.

Recall in Section 1.3 that the general quadratic equation can be solved by completing the square, which leads to the quadratic formula. Therefore, we can always use the quadratic formula to solve quadratic equations, as opposed to completing the square. Similarly, the general quadratic function $f(x) = ax^2 + bx + c$ can be expressed in standard form, $f(x) = a(x - h)^2 + k$, by completing the square. It can be shown that the standard form of the general quadratic function is $f(x) = a\left(x + \dfrac{b}{2a}\right)^2 + c - \dfrac{b^2}{4a}$.

Thus, the vertex occurs when $x = -\dfrac{b}{2a}$.

■ **Answer:** Standard form: $f(x) = 2(x + 1)^2 + 3$ General form: $f(x) = 2x^2 + 4x + 5$

VERTEX OF A PARABOLA

For the graph of a quadratic function, $f(x) = ax^2 + bx + c$, the **vertex** of the parabola occurs at the point $\left(-\dfrac{b}{2a}, f\left(-\dfrac{b}{2a}\right)\right)$.

In the beginning of this section, the function $F(x) = x^2 + 2x - 1$, which can also be written as $F(x) = (x + 1)^2 - 2$, was shown to have a vertex at the point $(-1, -2)$. Suppose this function represents profit (or loss) and all we care about is the "worst case." This would be when the profit is at the lowest point (vertex of a parabola opening up). Instead of rewriting the function in standard form through completing the square, we use the vertex formula above.

$$F(x) = x^2 + 2x - 1 \qquad \text{Coefficients: } a = 1, \ b = 2, \ c = -1$$

The vertex occurs at the point $x = -\dfrac{b}{2a} = -\dfrac{2}{2(1)} = -1$.

The value of the function at $x = -1$ is $f(-1) = (-1)^2 + 2(-1) - 1 = 1 - 2 - 1 = -2$.

Therefore the vertex is located at the point $(-1, -2)$.

EXAMPLE 7 Find the Minimum Cost of Manufacturing a Toy

A toy company that produces handmade baby dolls has daily production costs of

$$C(x) = 2000 - 15x + 0.05x^2$$

where C is the total cost in dollars and x is the number of baby dolls produced. How many dolls can be produced each day while keeping costs to a minimum?

Solution:

The graph of the quadratic function is a parabola.

STEP 1 Rewrite the quadratic function
in general form. $\qquad\qquad\qquad C(x) = 0.05x^2 - 15x + 2000$

STEP 2 The parabola opens upward,
because a is positive. $\qquad\qquad a = 0.05 > 0$

STEP 3 Because the parabola opens
upward, the vertex of the parabola
is a minimum.

STEP 4 Find the x-coordinate of the vertex. $\qquad x = -\dfrac{b}{2a} = -\dfrac{(-15)}{2(0.05)} = 150$

The toymaker keeps costs to a minimum when 150 baby dolls are produced each day.

CONCEPT CHECK When a parabola opens downward, is the vertex a maximum or a minimum?

■ **YOUR TURN**   The revenue associated with manufacturing vitamins is
$$R(x) = 500x - 0.001x^2$$
where R is the revenue in dollars and x is the number of bottles of vitamins sold. Determine how many bottles of vitamins should be sold to maximize the revenue.

EXAMPLE 8 **Find the Dimensions That Yield a Maximum Area**

You have just bought a puppy and want to fence in an area in the backyard for her to roam. You buy 100 linear feet of fence from Home Depot and have decided to make a rectangular fenced-in area using the back of your house as a side. Determine the dimensions of the rectangular pen that will maximize the area for your puppy to roam.

Solution:

Recall the five-step procedure for solving application problems presented in Section 1.2.

STEP 1 **Identify the question.**
Find the dimensions of the rectangular pen.

STEP 2 **Make notes.**

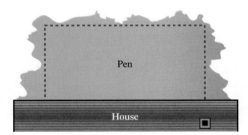

STEP 3 **Set up an equation.**
If we let x represent the length of one side of the rectangle, then the opposite side is also length x. Because there are 100 feet of fence, the remaining fence left for the side opposite the house is $100 - 2x$.

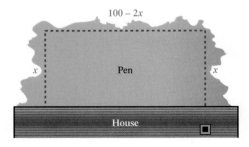

The area of a rectangle is equal to length times width:
$$A(x) = x(100 - 2x)$$

■ **Answer:** 250,000 bottles

TECHNOLOGY TIP

Use a graphing utility to graph the Area function

$y_1 = -2x^2 + 100x$

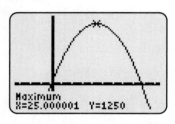

The maximum occurs when $x = 25$.

STEP 4 **Solve the equation.**

$$A(x) = x(100 - 2x) = -2x^2 + 100x$$

Find the maximum of the parabola that corresponds to the quadratic equation for area: $A(x) = -2x^2 + 100x$.

$a = -2$ and $b = 100$; therefore, the maximum is at

$$x = -\frac{b}{2a} = -\frac{100}{2(-2)} = 25$$

Replacing x with 25 in our original diagram:

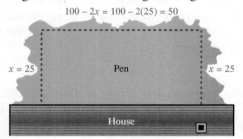

The dimensions of the rectangle are 25 feet by 50 feet .

STEP 5 **Check the solution.**

Two sides are 25 feet and one side is 50 feet, and together they account for all 100 feet of fence.

■ **YOUR TURN** Suppose you have 200 linear feet of fence to enclose a rectangular garden. Determine the dimensions of the rectangle that will yield the greatest area.

SECTION 4.1 **SUMMARY**

All quadratic functions $f(x) = ax^2 + bx + c$ have graphs that are parabolas:

- If $a > 0$, the parabola opens upward.
- If $a < 0$, the parabola opens downward.
- The vertex is at the point $(h, k) = \left(-\dfrac{b}{2a}, f\left(-\dfrac{b}{2a} \right) \right)$.
- When the quadratic function is given in general form, completing the square can be used to rewrite the function in standard form.

SECTION 4.1 **EXERCISES**

 **SKILLS**

In Exercises 1–4, match the quadratic function with the graph.

1. $f(x) = 3(x + 2)^2 - 5$ **2.** $f(x) = 2(x - 1)^2 + 3$ **3.** $f(x) = -\dfrac{1}{2}(x + 3)^2 + 2$ **4.** $f(x) = -\dfrac{1}{3}(x - 2)^2 + 3$

■ **Answer:** 50 feet by 50 feet

a.

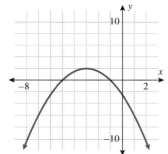

b.

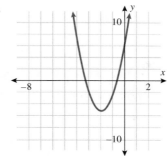

c.

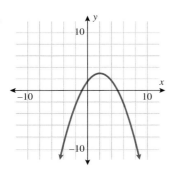

d.

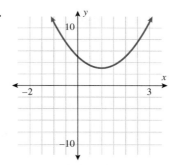

In Exercises 5–8, match the quadratic function with the graph.

5. $f(x) = 3x^2 + 5x - 2$ **6.** $f(x) = 3x^2 - x - 2$ **7.** $f(x) = -x^2 + 2x - 1$ **8.** $f(x) = -2x^2 - x + 3$

a.

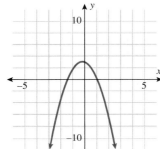

b.

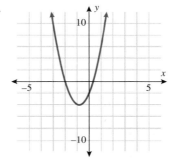

c.

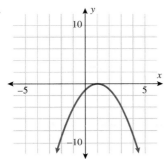

d.

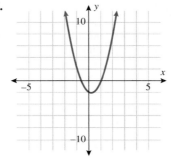

In Exercises 9–22, graph the quadratic function given in standard form.

9. $f(x) = (x + 1)^2 - 2$

10. $f(x) = (x + 2)^2 - 1$

11. $f(x) = (x - 2)^2 - 3$

12. $f(x) = (x - 4)^2 + 2$

13. $f(x) = -(x - 3)^2 + 9$

14. $f(x) = -(x - 5)^2 - 4$

15. $f(x) = -(x + 1)^2 - 3$

16. $f(x) = -(x - 2)^2 + 6$

17. $f(x) = 2(x - 2)^2 + 2$

18. $f(x) = -3(x + 2)^2 - 15$ **19.** $f(x) = \left(x - \dfrac{1}{3}\right)^2 + \dfrac{1}{9}$ **20.** $f(x) = \left(x + \dfrac{1}{4}\right)^2 - \dfrac{1}{2}$

21. $f(x) = -0.5(x - 0.25)^2 + 0.75$ **22.** $f(x) = -0.2(x + 0.6)^2 + 0.8$

In Exercises 23–32, rewrite the quadratic function in standard form by completing the square.

23. $f(x) = x^2 + 6x - 3$ **24.** $f(x) = x^2 + 8x + 2$ **25.** $f(x) = -x^2 - 10x + 3$ **26.** $f(x) = -x^2 - 12x + 6$

27. $f(x) = 2x^2 + 8x - 2$ **28.** $f(x) = 3x^2 - 9x + 11$ **29.** $f(x) = -4x^2 + 16x - 7$ **30.** $f(x) = -5x^2 + 100x - 36$

31. $f(x) = \dfrac{1}{2}x^2 - 4x + 3$ **32.** $f(x) = -\dfrac{1}{3}x^2 + 6x + 4$

In Exercises 33–40, graph the quadratic function.

33. $f(x) = x^2 + 6x - 7$ **34.** $f(x) = x^2 - 3x + 10$ **35.** $f(x) = x^2 + 2x - 15$ **36.** $f(x) = -x^2 + 3x + 4$

37. $f(x) = -2x^2 - 12x - 16$ **38.** $f(x) = -3x^2 + 12x - 12$ **39.** $f(x) = \dfrac{1}{2}x^2 - \dfrac{1}{2}$ **40.** $f(x) = -\dfrac{1}{3}x^2 + \dfrac{4}{3}$

In Exercises 41–48, find the vertex of the parabola associated with each quadratic function.

41. $f(x) = 33x^2 - 2x + 15$ **42.** $f(x) = 17x^2 + 4x - 3$ **43.** $f(x) = \dfrac{1}{2}x^2 - 7x + 5$

44. $f(x) = -\dfrac{1}{3}x^2 + \dfrac{2}{5}x + 4$ **45.** $f(x) = -0.002x^2 - 0.3x + 1.7$ **46.** $f(x) = 0.05x^2 + 2.5x - 1.5$

47. $f(x) = -\dfrac{2}{5}x^2 + \dfrac{3}{7}x + 2$ **48.** $f(x) = -\dfrac{1}{7}x^2 - \dfrac{2}{3}x + \dfrac{1}{9}$

In Exercises 49–58, find the quadratic function that has the given vertex and goes through the given point.

49. vertex: $(-1, 4)$ point: $(0, 2)$ **50.** vertex: $(2, -3)$ point: $(0, 1)$ **51.** vertex: $(2, 5)$ point: $(3, 0)$

52. vertex: $(1, 3)$ point: $(-2, 0)$ **53.** vertex: $(-1, -3)$ point: $(-4, 2)$ **54.** vertex: $(0, -2)$ point: $(3, 10)$

55. vertex: $\left(\dfrac{1}{2}, -\dfrac{3}{4}\right)$ point: $\left(\dfrac{3}{4}, 0\right)$ **56.** vertex: $\left(-\dfrac{5}{6}, \dfrac{2}{3}\right)$ point: $(0, 0)$ **57.** vertex: $(2.5, -3.5)$ point: $(4.5, 1.5)$

58. vertex: $(1.8, 2.7)$ point: $(-2.2, 7.5)$

■ APPLICATIONS

59. Sports. Assume that the punter for the Tampa Bay Buccaneers on average punts a ball that follows a parabolic path given by the equation $P(x) = -\frac{1}{2}x^2 + 20x$. P is the height above ground, and x is the horizontal distance in yards. If the ball is punted from his own 20 yard line, what yard line should it be caught at and what is the maximum height that his punt will attain?

60. Sports. Assume that the punter for the Carolina Panthers on average punts a ball that follows a parabolic path given by the equation, $P(x) = -\frac{2}{15}x^2 + 8x$. P is the height above ground, and x is the horizontal distance in yards. If the ball is punted from his own 20 yard line, what yard line should it be caught at and what is the maximum height that his punt will attain?

61. Ranching. A rancher has 10,000 linear feet of fencing and wants to enclose a rectangular field and then divide it into two equal pastures with an internal fence parallel to one of the rectangular sides. What is the maximum area of each pasture? Round to the nearest square foot.

62. Ranching. A rancher has 30,000 linear feet of fencing and wants to enclose a rectangular field and then divide it into four equal pastures with three internal fences parallel to one of the rectangular sides. What is the maximum area of each pasture?

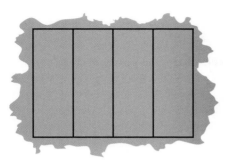

63. Gravity. A person standing near the edge of a cliff 100 feet high throws a rock upward with an initial speed of 32 feet per second. The height of the rock above the lake at the bottom of the cliff is a function of time and is described by

$$h(t) = -16t^2 + 32t + 100$$

a. How many seconds will it take until the rock reaches its maximum height? What is that height?

b. At what time will the rock hit the water?

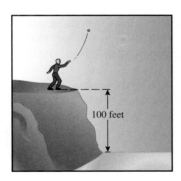

100 feet

64. Gravity. A person holds a pistol straight upward and fires. The initial velocity of most bullets is around 1200 feet/second. The height of the bullet is a function of time and is described by

$$h(t) = -16t^2 + 1200t$$

How long, after the gun is fired, does the person have to get out of the way of the bullet falling from the sky?

65. Zero Gravity. As part of their training, astronauts ride the "vomit comet," NASA's reduced gravity KC 135A aircraft that performs parabolic flights to simulate weightlessness. The plane starts at an altitude of 20,000 feet and makes a steep climb at 52° with the horizon for 20–25 seconds and then dives at that same angle back down, repeatedly.

NASA's "Vomit Comet"

The equation governing the altitude of the flight is

$$A(x) = -0.0003x^2 + 9.3x - 46,075$$

where $A(x)$ is altitude and x is horizontal distance in feet. What is the maximum altitude the plane attains? Over what horizontal distance is the entire maneuver performed?

66. Mileage. Gas mileage (miles per gallon, mpg) can be approximated by a quadratic function of speed. For a particular automobile, assume the vertex occurs when the speed is 50 mph (the mpg will be 30).

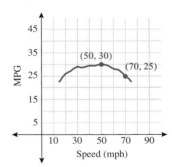

a. Write a quadratic function that models this relationship, assuming 70 mph corresponds to 25 mpg.

b. What gas mileage would you expect for this car driving 90 mph?

For Exercises 67 and 68, use the following information:

One function of particular interest in economics is the **profit function.** We denote this function by $P(x)$. It is defined to be the difference between revenue, $R(x)$, and cost, $C(x)$, so that:

$$P(x) = R(x) - C(x)$$

The total revenue received from the sale of x goods at price p is given by

$$R(x) = px$$

The total cost function relates the cost of production to the level of output, x. This includes both fixed costs, C_f, and variable costs, C_v (costs per unit produced). The total cost in producing x goods is given by

$$C(x) = C_f + C_v x$$

Thus, the profit function is

$$P(x) = px - C_f - C_v x$$

Assume fixed costs are \$1,000, and variable costs per unit are \$20 and the demand function is

$$p = 100 - x$$

67. Profit. How many units should the company produce to break even?

68. Profit. What is the maximum profit?

■ CATCH THE MISTAKE

In Exercises 69–72, explain the mistake that is made. There may be a single mistake or there may be more than one mistake.

69. Plot the quadratic function: $f(x) = (x + 3)^2 - 1$.

Solution:

Step 1: The parabola opens upward, because $a = 1 > 0$.
Step 2: The vertex is $(3, -1)$.
Step 3: The y-intercept is $(0, 8)$.
Step 4: The x-intercepts are $(2, 0)$ and $(4, 0)$.
Step 5: Plot the vertex and intercepts, and connect with a smooth curve.

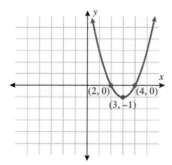

70. Determine the vertex of the quadratic function:
$f(x) = -2x^2 + 6x - 18$.

Solution:

Step 1: The vertex is given by $(h, k) = \left(\dfrac{b}{2a}, f\left(\dfrac{b}{2a}\right)\right)$.

In this case, $a = -2$ $b = 6$.

Step 2: The x-coordinate of the vertex:

$$x = \frac{6}{2(-2)} = -\frac{6}{4} = -\frac{3}{2}$$

Step 3: The y-coordinate of the vertex:

$$f\left(-\frac{3}{2}\right) = -2\left(-\frac{3}{2}\right)^2 + 6\left(-\frac{3}{2}\right) - 18$$

$$= -2\left(\frac{9}{4}\right) - \frac{18}{2} - 18$$

$$= -\frac{9}{2} - 9 - 18$$

$$= -\frac{63}{2}$$

71. Rewrite the quadratic function in standard form.
$$f(x) = -x^2 + 2x + 3$$

Step 1: Group variables together. $\quad (-x^2 + 2x) + 3$
Step 2: Factor out a negative. $\quad -(x^2 + 2x) + 3$
Step 3: Add and subtract 1 inside the parentheses. $\quad -(x^2 + 2x + 1 - 1) + 3$
Step 4: Factor out the -1. $\quad -(x^2 + 2x + 1) + 1 + 3$
Step 5: Simplify. $\quad -(x + 1)^2 + 4$

72. Find the quadratic function whose vertex is $(2, -3)$ and passes through the point $(9, 0)$.

Step 1: Write the quadratic equation in standard form. $\quad f(x) = a(x - h)^2 + k$
Step 2: Substitute $(h, k) = (2, -3)$. $\quad f(x) = a(x - 2)^2 - 3$
Step 3: Substitute the point $(9, 0)$ and solve for a. $\quad f(0) = a(0 - 2)^2 - 3 = 9$

$$4a - 3 = 9$$
$$4a = 12$$
$$a = 3$$

The quadratic function sought: $f(x) = 3(x - 2)^2 - 3$.

73. T or F: A quadratic function must have a y-intercept.

74. T or F: A quadratic function must have an x-intercept.

75. T or F: A quadratic function may have more than one y-intercept.

76. T or F: A quadratic function may have more than one x-intercept.

77. For the general quadratic equation, $f(x) = ax^2 + bx + c$,

show that the vertex is $(h, k) = \left(-\dfrac{b}{2a}, f\left(-\dfrac{b}{2a}\right)\right)$.

78. Given the standard equation: $f(x) = a(x - h)^2 + k$, determine the x- and y-intercepts in terms of a, h, and k.

79. A rancher has 1000 feet of fence to enclose a pasture.
   a. Determine the maximum area if a rectangular fence is used.
   b. Determine the maximum area if a circular fence is used.

80. A 600 room hotel in Orlando is filled to capacity every night when the rate is \$90 per night. For every \$5 increase in the rate, 10 fewer rooms are filled. How much should the hotel charge to produce the maximum income? What is the maximum income?

■ TECHNOLOGY

81. On a graphing calculator, plot the quadratic function:
   $f(x) = -.002x^2 + 5.7x - 23$
   a. Identify the vertex of this parabola.
   b. Identify the y-intercept.
   c. Identify the x-intercepts (if any).
   d. What is the axis of symmetry?

82. Determine the quadratic function whose vertex is $(-0.5, 1.7)$ and passes through the point $(0, 4)$.
   a. Write the quadratic function in general form.
   b. Use a graphing calculator to plot this quadratic function.
   c. Zoom in on the vertex and y-intercept. Do they agree with the given values?

SECTION 4.2 Polynomial Functions of Higher Degree

Skills Objectives

- Identify polynomials and their degree.
- Graph polynomial functions using transformations.
- Identify zeros of a polynomial function and their multiplicity.
- Determine the end behavior of a polynomial function.

Conceptual Objectives

- Understand that zeros of polynomial functions can sometimes correspond to x-intercepts, but not always.

Polynomial functions model many real-world applications. For example, the number of active duty military in the United States, the number of new cases in a spread of an epidemic, and stock prices, can be modeled with polynomial functions.

Identifying Polynomials and Their Degree

What is a polynomial function?

> **DEFINITION** **POLYNOMIAL FUNCTION**
>
> Let n be a nonnegative integer and let $a_n, a_{n-1}, ..., a_2, a_1, a_0$ be real numbers with $a_n \neq 0$. The function
>
> $$f(x) = a_n x^n + a_{n-1} x^{n-1} + ... + a_2 x^2 + a_1 x + a_0$$
>
> is called a **polynomial function of x with degree n**. The coefficient a_n is called the leading coefficient.

EXAMPLE 1 **Identifying Polynomials and Their Degree**

For each of the functions given, determine if the function is a polynomial function. If it is a polynomial function, then state the degree of the polynomial. If it is not a polynomial function, justify your answer.

a. $f(x) = 3 - 2x^5$ **b.** $F(x) = \sqrt{x} + 1$ **c.** $g(x) = 2$

d. $h(x) = 3x^2 - 2x + 5$ **e.** $H(x) = 4x^5(2x - 3)^2$

Solution:

a. $f(x)$ is a polynomial function of degree 5.

b. $F(x)$ is not a polynomial function. The variable x is raised to the $\frac{1}{2}$ power, which is not an integer.

c. $g(x)$ is a polynomial function of degree zero, also known as a constant function. Note that $g(x) = 2$ can also be written as $g(x) = 2x^0$.

d. $h(x)$ is a polynomial function of degree 2. A polynomial function of degree 2 is called a quadratic function.

e. $H(x)$ is a polynomial function of degree 7.

> ■ **YOUR TURN** For each of the functions given, determine if the function is a polynomial function. If it is a polynomial function, then state the degree of the polynomial. If it is not a polynomial function, justify your answer.
>
> **a.** $f(x) = \dfrac{1}{x} + 2$ **b.** $g(x) = 3x^8(x - 2)^2(x + 1)^3$

Graphs of Polynomial Functions

Whenever we discussed a particular polynomial, we have depicted it, too. The graph of a constant function (degree 0) is a horizontal line. The graph of a general linear function (degree 1) is a slanted line. The graph of a quadratic function (degree 2) is a parabola. These functions are summarized in the box below.

POLYNOMIAL	DEGREE	SPECIAL NAME	GRAPH
$f(x) = c$	0	Constant function	Horizontal line
$f(x) = mx + b$	1	Linear function	Line Slope $= m$ y-intercept $(0, b)$
$f(x) = ax^2 + bx + c$	2	Quadratic function	Parabola Opening upward if $a > 0$. Opening downward if $a < 0$.

How do we graph polynomial functions that are degree 3 or higher, and why do we care? Polynomial functions model real-world applications as mentioned earlier. One example is the percentage of fat in our bodies as we age. When a baby comes home from the hospital, it usually experiences weight loss. Then typically there is an increase in the percent body fat when the baby is nursing. When infants start to walk, the increase in exercise is associated with a drop in the percentage of fat. Growth spurts in children are examples of the percent body fat increases and decreases. Later in life, our metabolism slows down, and typically the percent body fat increases. We will model this with a higher degree polynomial.

Polynomial functions are considered simple functions. Graphs of all polynomial functions are both *continuous* and *smooth*. **Continuous** means that the entire graph can be drawn without picking up your pencil (the graph has no jumps or holes). **Smooth** means there are no sharp corners.

EXAMPLE 2 Classifying Functions as Continuous and Smooth

Determine if each function is continuous. If it is, also determine if it is smooth.

a. $g(x) = \begin{cases} 1 & x > 0 \\ -1 & x \leq 0 \end{cases}$ **b.** $f(x) = |x|$ **c.** $f(x) = x^3$

Solution:

a. Recall from Section 3.2 the graph of a piecewise-defined function.

This function is *not continuous* (discontinuous) because there is a jump at $x = 0$.

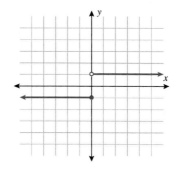

b. Recall from Section 3.2 the graph of the absolute value function.

This function is *continuous* because it can be drawn without picking up the pencil. This function is *not smooth* because of the sharp corner at the point (0, 0).

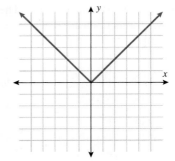

c. Recall the graph of a cubic function from the library of functions in Section 3.2.

This polynomial (cubic) function is both *continuous* and *smooth,* since there are no holes or jumps and no sharp corners.

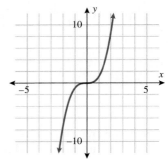

■ **YOUR TURN** Determine if each function is continuous. If it is, also determine if it is smooth.

a.

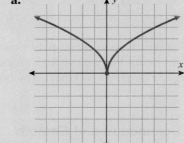

b.

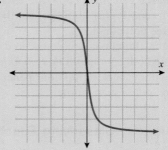

c.

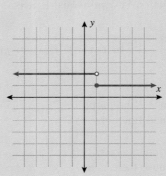

All polynomial functions have graphs that are both continuous and smooth. Recall from Chapter 3 that graphs of functions can be drawn by hand using graphing aids such as

■ **Answer: a.** Continuous but not smooth **b.** Continuous and smooth **c.** Discontinuous

intercepts and symmetry. The graphs of polynomial functions can be graphed using these aids. Let's start with the simplest type of polynomial functions called **power functions**. Power functions are monomial functions of the form, $f(x) = x^n$, where n is an integer greater than zero.

DEFINITION   POWER FUNCTION

Let n be a positive integer and the coefficient $a \neq 0$ be a real number. The function

$$f(x) = ax^n$$

is called a **power function of degree n**.

$n = 1$ When $n = 1$, the power function simplifies to a linear function whose graph is a straight line with slope a and y-intercept $(0, 0)$, which implies that the line passes through the origin.

$n = 2$ When $n = 2$, the power function simplifies to a quadratic function whose graph is a parabola that opens up if a is positive and down if a is negative. Note that the vertex of this parabola is at the origin.

$n = 3$ When $n = 3$, the power function simplifies to a cubic function whose graph we first learned in the library of functions in Chapter 3. Note that this function passes through the origin.

$n = 4$ When $n = 4$, the power function has a graph very similar to the graph of a parabola that opens up if a is positive and down if a is negative. The vertex is at the origin. When comparing a fourth-degree monomial to a second-degree monomial, the graph of x^4 is closer to the x-axis when $-1 < x < 1$ and closer to the y-axis otherwise.

$n = 5$ When $n = 5$, the power function has a graph very similar to the graph of a cubic function. The graph also passes through the origin. When comparing a fifth-degree monomial to a third-degree monomial, the graph of the fifth-degree monomial is closer to the x-axis when $-1 < x < 1$ and closer to the y-axis otherwise.

In the figures below, we have graphed these 5 functions for the special case $a = 1$.

$y = x$	$y = x^2$ and $y = x^4$	$y = x^3$ and $y = x^5$

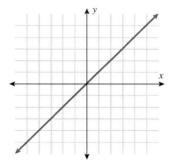
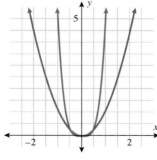
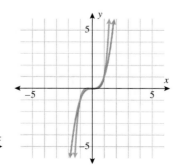

All even power functions have similar characteristics to a quadratic function (parabola), and all odd ($n > 1$) power functions have similar characteristics to a cubic function. For example, all even functions are symmetric with respect to the y-axis, whereas all odd functions are symmetric with respect to the origin. This table summarizes their characteristics.

CHARACTERISTICS OF POWER FUNCTIONS: $f(x) = x^n$

	n even	*n* odd
Symmetry	y-axis	origin
Domain	$(-\infty, \infty)$	$(-\infty, \infty)$
Range	$[0, \infty)$	$(-\infty, \infty)$
Points That Lie on Graph	$(-1, 1)$ $(0, 0)$, and $(1, 1)$	$(-1, -1)$ $(0, 0)$, and $(1, 1)$

Graphing Polynomial Functions Using Transformations of Power Functions

We now have the tools to graph polynomial functions that are transformations of power functions. We will use the power functions summarized above, $f(x) = x^n$, combined with our graphing techniques such as horizontal and vertical shifting, reflection, and compression.

EXAMPLE 3 Graphing Transformations of Power Functions

Graph the function $f(x) = (x - 1)^3$.

Solution:

STEP 1 Start with the graph of $y = x^3$:

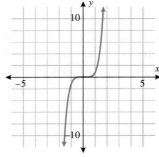

STEP 2 Shift $y = x^3$ to the right 1 unit, which yields the graph of $f(x) = (x - 1)^3$:

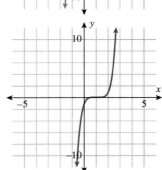

CONCEPT CHECK   Just by scanning the function $f(x) = 1 - x^4$ can you tell whether the "parabola-like" graph opens up or down?

■ **YOUR TURN**   Graph the function $f(x) = 1 - x^4$.

Real Zeros of a Polynomial Function

How do we graph general polynomial functions of degree greater than or equal to 3 if they cannot be written as transformations of power functions? We start by identifying the x-intercepts of the polynomial function. Recall that we determine the x-intercepts by setting the function equal to *zero* and solving for x. Therefore, an alternative name for x-intercept is *zero*. In our experience, when a quadratic function is equal to zero the first step is to factor the quadratic expression into linear factors and then set each factor equal to zero. Therefore, there are four equivalent relationships that are summarized in the following box.

REAL ZEROS OF POLYNOMIAL FUNCTIONS

If $f(x)$ is a polynomial function and a is a real number, then the following statements are equivalent.

Relationship 1: $x = a$ is a **solution**, or **root**, of the equation $f(x) = 0$.
Relationship 2: $(a, 0)$ is an **x-intercept** of the graph of $f(x)$.
Relationship 3: $x = a$ is a **zero** of the function $f(x)$.
Relationship 4: $(x - a)$ is a **factor** of $f(x)$.

Let's use a simple polynomial function to illustrate these four relationships. We'll focus on the quadratic function $f(x) = x^2 - 1$. The graph of this function is a parabola that opens upward and has as its vertex the point $(0, -1)$.

Illustration of Relationship 1

Set the function equal to zero. $f(x) = x^2 - 1 = 0$
Factor. $(x - 1)(x + 1) = 0$
Solve. $x = 1$ or $x = -1$

$x = -1$ and $x = 1$ are solutions, or roots, of the equation $x^2 - 1 = 0$.

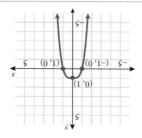

■ **Answer:** $f(x) = 1 - x^4$

Illustration of Relationship 2

The graph of $f(x) = x^2 - 1$ can be obtained by shifting the parabola $y = x^2$ down 1 unit.

The x-intercepts are $(-1, 0)$ and $(1, 0)$.

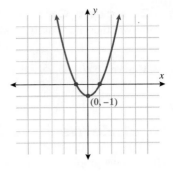

Illustration of Relationship 3

The graph of $f(x) = x^2 - 1$ can be obtained by shifting the parabola $y = x^2$ down 1 unit.

The value of the function at $x = -1$ and $x = 1$ is 0.

$$f(-1) = 0 \text{ and } f(1) = 0$$

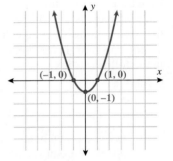

Illustration of Relationship 4

$$f(x) = x^2 - 1$$
$$= (x - 1)(x + 1)$$

$x = 1$ is a zero of the polynomial, so $(x - 1)$ is a factor.
$x = -1$ is a zero of the polynomial, so $(x + 1)$ is a factor.

We have a good reason for wanting to know the x-intercepts, or zeros. When the value of a continuous function transitions from negative to positive and vice versa, it must pass through zero. Therefore, the zeros break the domain into intervals that allow us to determine if the function is above or below the x-axis in those intervals. Keep in mind, though, that just because there is a zero does not imply that the function will change signs—as you will see in the examples of graphing general polynomial functions.

EXAMPLE 4 Identify the Zeros of a Polynomial Function

Find the zeros of the polynomial function $f(x) = x^3 + x^2 - 2x$.

Solution:

Set the function equal to zero. $\qquad x^3 + x^2 - 2x = 0$

Factor out an x common to all three terms. $\qquad x(x^2 + x - 2) = 0$

Factor the quadratic expression inside the parentheses. $\qquad x(x + 2)(x - 1) = 0$

Apply the zero property. $x = 0$ or $(x + 2) = 0$ or $(x - 1) = 0$

Solve. $x = 0$ or $x = -2$ or $x = 1$

The zeros are -2, 0, and 1.

> **YOUR TURN** Find the zeros of the polynomial function
> $f(x) = x^3 - 7x^2 + 12x$.

TECHNOLOGY TIP

Graph $y_1 = x^3 + x^2 - 2x$.

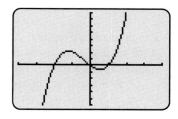

The x-intercepts are the zeros of the function, -2, 0, and 1.

Recall that when we were factoring a quadratic equation, if the factor was raised to a power greater than 1, the corresponding root, or zero, was repeated. For example, the quadratic equation $x^2 - 2x + 1 = 0$ when factored is written as $(x - 1)^2 = 0$. The solution, or root, in this case is $x = 1$, and we say that it is a **repeated** root. Similarly, when determining zeros of higher order polynomial functions, if a factor is repeated, we say that the zero is a repeated, or **multiple**, zero of the function. The number of times that a zero repeats is called its *multiplicity*.

DEFINITION | **MULTIPLICITY OF A ZERO**

If $(x - a)^n$ is a factor of a polynomial f, then a is called a **zero of multiplicity n** of f.

EXAMPLE 5 Finding the Multiplicities of Zeros of a Polynomial Function

Find the zeros, and state their multiplicities, of the polynomial function

$$g(x) = (x - 1)^2 \left(x + \frac{3}{5}\right)^7 (x + 5).$$

Solution: 1 is a zero of multiplicity 2.

$-\frac{3}{5}$ is a zero of multiplicity 7.

-5 is a zero of multiplicity 1.

Note: Adding the multiplicities yields the degree of the polynomial. The polynomial $g(x)$ is degree 10, since $2 + 7 + 1 = 10$.

✓**CONCEPT CHECK** What is the degree of the polynomial $h(x) = x^2(x - 2)^3(x + \frac{1}{2})^5$?

> **YOUR TURN** For the polynomial $h(x)$, determine the zeros and state their multiplicity.
>
> $$h(x) = x^2(x - 2)^3\left(x + \frac{1}{2}\right)^5$$

> ■ **Answer:** The zeros are 0, 3, and 4.

> ■ **Answer:** 0 is a zero of multiplicity 2. 2 is a zero of multiplicity 3. $-\frac{1}{2}$ is a zero of multiplicity 5.

EXAMPLE 6 Finding a Polynomial from Its Zeros

Find a polynomial of degree 7 whose zeros are:

-2 (multiplicity 2)   0 (multiplicity 4)   1 (multiplicity 1)

Solution:

If $x = a$ is a zero, then $(x - a)$ is a factor.	$f(x) = (x + 2)^2(x - 0)^4(x - 1)^1$
Simplify.	$x^4(x + 2)^2(x - 1)$
Square the binomial.	$x^4(x^2 + 4x + 4)(x - 1)$
Multiply the two polynomials.	$x^4(x^3 + 3x^2 - 4)$
Distribute x^4.	$x^7 + 3x^6 - 4x^4$

Graphing General Polynomial Functions

Let's develop a strategy for sketching an approximate graph of any polynomial function. First, we determine the x- and y-intercepts. Then we use the x-intercepts, or zeros, to divide the domain into intervals where the polynomial is positive or negative so that we can find points in those intervals to assist in sketching a smooth and continuous graph. Note: It is not always possible to find x-intercepts. Sometimes there are no x-intercepts.

STUDY TIP

It is not always possible to find x-intercepts. Sometimes there are no x-intercepts.

EXAMPLE 7 Using a Strategy for Sketching the Graph of a Polynomial Function

Sketch the graph of $f(x) = (x + 2)(x - 1)^2$.

Solution:

STEP 1 Find the y-intercept. $f(0) = (2)(-1)^2 = 2$
(Let $x = 0$.) $(0, 2)$ is the y-intercept

STEP 2 Find any x-intercepts. $f(x) = (x + 2)(x - 1)^2 = 0$
(Set $f(x) = 0$.) $x = -2$ or $x = 1$
$(-2, 0)$ and $(1, 0)$ are the x-intercepts

STEP 3 Plot the intercepts.

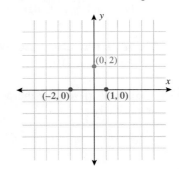

STEP 4 Divide the x-axis into intervals*: $(-\infty, -2)$, $(-2, 1)$, and $(1, \infty)$

*The x-intercepts ($x = -2$ and $x = 1$) divide the x-axis into three intervals similar to critical points when we studied inequalities in Section 1.5.

STEP 5 **Select a number in each interval and test each interval.**

The function $f(x)$ either *crosses* the x-axis at an x-intercept or *touches* the x-axis at an x-intercept. Therefore, we need to check each of these intervals to determine if the function is positive (above the x-axis) or negative (below the x-axis). We do so by selecting numbers in the intervals and determining the value of the function at those points.

Interval	$(-\infty, -2)$	$(-2, 1)$	$(1, \infty)$
Number Selected in Interval	-3	-1	2
Value of Function	$f(-3) = -16$	$f(-1) = 4$	$f(2) = 4$
Point on Graph	$(-3, -16)$	$(-1, 4)$	$(2, 4)$
Interval Relation to x-Axis	Below x-axis	Above x-axis	Above x-axis

From the table, we find three additional points on the graph: $(-3, -16)$, $(-1, 4)$, and $(2, 4)$. The point $x = -2$ is an intercept where the function *crosses* the x-axis, because it is below the x-axis to the left of -2 and above the x-axis to the right of -2. The point $x = 1$ is an intercept where the function *touches* the x-axis, because it is above the x-axis on both sides of $x = 1$. Connecting these points with a smooth curve yields the graph.

STEP 6 **Sketch a plot of the function.**

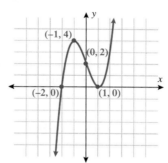

YOUR TURN Sketch the graph of $f(x) = x^3(x + 3)^2$.

In Example 7, we found that the function crosses through the x-axis at the point $x = -2$. Note that the -2 is a zero of multiplicity 1. We also found that the function touches the x-axis at the point $x = 1$. Note that 1 is a zero of multiplicity 2. In general, zeros with even multiplicity correspond to intercepts where the function touches the x-axis and zeros with odd multiplicity correspond to intercepts where the function crosses the x-axis.

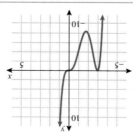

■ **Answer:**

MULTIPLICITY OF A ZERO AND RELATION TO THE GRAPH OF A POLYNOMIAL

If a is a zero of $f(x)$, then:

Multiplicity of a	$f(x)$ on either side of $x = a$	Graph of Function at the Intercept
Even	Does not change sign	Touches x-axis at $x = a$
Odd	Changes sign	Crosses x-axis at $x = a$

STUDY TIP

If f is a polynomial of degree n, then the graph of f has at most $n - 1$ turning points.

Also in Example 7, we know that somewhere in the interval $(-2, 1)$ the function must reach a maximum and then turn back toward the x-axis, because both $x = -2$ and $x = 1$ correspond to x-intercepts. The point $x = -1$ does correspond to a local maximum point on the graph, and we refer to $x = -1$ as a *turning point*. In general, if f is a polynomial of degree n, then the graph of f has at most $n - 1$ turning points.

Intercepts and turning points assist us in sketching graphs of polynomial functions. Another piece of information that will assist us in graphing polynomial functions is knowledge of the *end behavior*. All polynomials eventually rise or fall without bound as x gets large in either the positive $(x \to \infty)$ or negative $(x \to -\infty)$ direction. The *lead term,* or highest degree monomial within the polynomial, dominates the *end behavior.*

END BEHAVIOR

As x gets large in either the positive $(x \to \infty)$ or negative $(x \to -\infty)$ direction, the graph of the polynomial

$$f(x) = a_n x^n + a_{n-1} x^{n-1} + \ldots + a_2 x^2 + a_1 x + a_0$$

has the same behavior as the power function

$$y = a_n x^n$$

Power functions behave much like a quadratic function (parabola) for even-degree polynomial functions and a cubic function for odd-degree polynomial functions. There are only four possibilities because the lead coefficient can be positive or negative.

$$y = a_n x^n$$

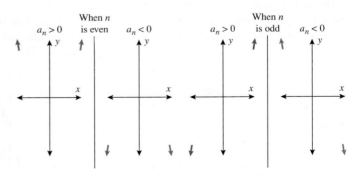

$x \to -\infty$	$x \to \infty$	n	a_n
Function *rises*	Function *rises*	Even	Positive
Function *falls*	Function *falls*	Even	Negative
Function *rises*	Function *falls*	Odd	Negative
Function *falls*	Function *rises*	Odd	Positive

EXAMPLE 8 Graphing a Polynomial Function

Sketch a graph of the polynomial function $f(x) = 2x^4 - 8x^2$.

Solution:

STEP 1 Determine the y-intercept. $f(0) = 0$
($x = 0$)
                                                     The y-intercept is $(0, 0)$.

STEP 2 Find the zeros of the polynomial. $f(x) = 2x^4 - 8x^2$

Factor out common $2x^2$. $2x^2(x^2 - 4)$

Factor quadratic binomial. $2x^2(x - 2)(x + 2)$

Set $f(x) = 0$. $2x^2(x - 2)(x + 2) = 0$

0 is a zero of multiplicity 2. Expect the graph to *touch* the x-axis.

2 is a zero of multiplicity 1. Expect the graph to *cross* the x-axis.

-2 is a zero of multiplicity 1. Expect the graph to *cross* the x-axis.

STEP 3 Determine end behavior. $f(x) = 2x^4 - 8x^2$ behaves
like $y = 2x^4$.

$y = 2x^4$ is even degree and the lead coefficient is positive, so the graph rises without bound as x gets large in either the positive or negative direction.

STEP 4 Sketch the intercepts and
end behavior.

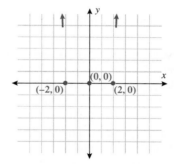

STEP 5 Test intervals.

The x-intercepts (zeros) divide the x-axis into four intervals.

Interval	$(-\infty, -2)$	$(-2, 0)$	$(0, 2)$	$(2, \infty)$
Number Selected in Interval	-3	-1	1	3
Value of Function	$f(-3) = 90$	$f(-1) = -6$	$f(1) = -6$	$f(3) = 90$
Point on Graph	$(-3, 90)$	$(-1, -6)$	$(1, -6)$	$(3, 90)$
Interval Relation to x-Axis	Above	Below	Below	Above

STEP 6 Sketch final graph.

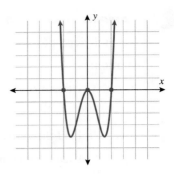

Note the symmetry about the y-axis. This function is an even function, $f(-x) = f(x)$.

CONCEPT CHECK Is the function $f(x) = x^5 - 4x^3$ even, odd, or neither? Based on this, is there any symmetry?

■ **YOUR TURN** Sketch a graph of the polynomial function $f(x) = x^5 - 4x^3$.

SECTION 4.2 SUMMARY

In general, graphing polynomials can be done one of two ways:

- Monomial
  - Use graph-shifting techniques with power functions.
- General Polynomial
  - Identify intercepts, if possible.
  - Determine the zeros and their multiplicity, and ascertain if the graph crosses or touches the x-axis at each zero.
  - x-Intercepts (zeros) divide the x-axis into intervals. Test points in intervals to determine if the graph is above or below the x-axis.
  - Determine the end behavior by investigating the end behavior of the highest degree monomial.
  - Sketch with a smooth curve.

■ **Answer:**

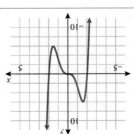

SECTION 4.2 EXERCISES

■ **SKILLS**

In Exercises 1–10, determine which functions are polynomials, and for those, state their degree.

1. $f(x) = -3x^2 + 15x - 7$ **2.** $f(x) = 2x^5 - x^2 + 13$ **3.** $g(x) = (x + 2)^3\left(x - \dfrac{3}{5}\right)^2$ **4.** $g(x) = x^4(x - 1)^2(x + 2.5)^3$

5. $h(x) = \sqrt{x} + 1$ **6.** $h(x) = (x - 1)^{1/2} + 5x$ **7.** $F(x) = x^{1/3} + 7x^2 - 2$ **8.** $F(x) = 3x^2 + 7x - \dfrac{2}{3x}$

9. $G(x) = \dfrac{x + 1}{x^2}$ **10.** $H(x) = \dfrac{x^2 + 1}{2}$

In Exercises 11–18, match the polynomial function with its graph.

11. $f(x) = -3x + 1$ **12.** $f(x) = -3x^2 - x$ **13.** $f(x) = x^2 + x$ **14.** $f(x) = -2x^3 + 4x^2 - 6x$

15. $f(x) = x^3 - x^2$ **16.** $f(x) = 2x^4 - 18x^2$ **17.** $f(x) = -x^4 + 5x^3$ **18.** $f(x) = x^5 - 5x^3 + 4x$

a.

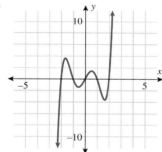

b.

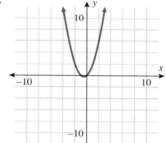

c.

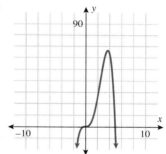

d.

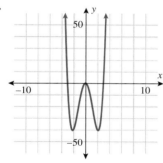

e.

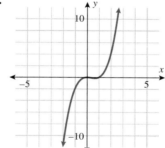

f.

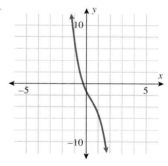

g.

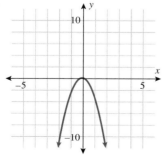

h.

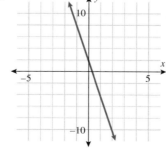

In Exercises 19–26, graph each function by transforming a power function, $y = x^n$.

19. $f(x) = -x^5$

20. $f(x) = x^4$

21. $f(x) = (x - 2)^4$

22. $f(x) = (x + 2)^5$

23. $f(x) = x^5 + 3$

24. $f(x) = -x^4 - 3$

25. $f(x) = 3 - (x + 1)^4$

26. $f(x) = (x - 3)^5 - 2$

In Exercises 27–38, find all the real zeros (and state their multiplicity) of each polynomial function.

27. $f(x) = 2(x - 3)(x + 4)^3$

28. $f(x) = -3(x + 2)^3(x - 1)^2$

29. $f(x) = 4x^2(x - 7)^2(x + 4)$

30. $f(x) = 5x^3(x + 1)^4(x - 6)$

31. $f(x) = 4x^2(x - 1)^2(x^2 + 4)$

32. $f(x) = 4x^2(x^2 - 1)(x^2 + 9)$

33. $f(x) = 8x^3 + 6x^2 - 27x$

34. $f(x) = 2x^4 + 5x^3 - 3x^2$

35. $f(x) = -2.7x^3 - 8.1x^2$

36. $f(x) = 1.2x^6 - 4.6x^4$

37. $f(x) = \frac{1}{3}x^6 + \frac{2}{5}x^4$

38. $f(x) = \frac{2}{7}x^5 - \frac{3}{4}x^4 + \frac{1}{2}x^3$

In Exercises 39–52, find a polynomial of minimum degree (there are many) that have the given zeros.

39. $-3, 0, 1, 2$

40. $-2, 0, 2$

41. $-5, -3, 0, 2, 6$

42. $0, 1, 3, 5, 10$

43. $-\frac{1}{2}, \frac{2}{3}, \frac{3}{4}$

44. $-\frac{3}{4}, -\frac{1}{3}, 0, \frac{1}{2}$

45. $1 - \sqrt{2}, 1 + \sqrt{2}$

46. $1 - \sqrt{3}, 1 + \sqrt{3}$

47. -2 (multiplicity 3), 0 (multiplicity 2)

48. -4 (multiplicity 2), 5 (multiplicity 3)

49. -3 (multiplicity 2), 7 (multiplicity 5)

50. 0 (multiplicity 1), 10 (multiplicity 3)

51. $-\sqrt{3}$ (multiplicity 2), -1 (multiplicity 1), 0 (multiplicity 2), $\sqrt{3}$ (multiplicity 2)

52. $-\sqrt{5}$ (multiplicity 2), 0 (multiplicity 1), 1 (multiplicity 2), $\sqrt{5}$ (multiplicity 2)

In Exercises 53–64, for each polynomial function given: (a) list each real zero and its multiplicity; (b) determine whether the graph touches or crosses at each x-intercept; (c) find the y-intercept and a few points on the graph; (d) determine the end behavior; and (e) sketch the graph.

53. $f(x) = -x^2 - 6x - 9$

54. $f(x) = x^2 + 4x + 4$

55. $f(x) = (x - 2)^3$

56. $f(x) = -(x + 3)^3$

57. $f(x) = -x^4 - 3x^3$

58. $f(x) = x^5 - x^3$

59. $f(x) = 12x^6 - 36x^5 - 48x^4$

60. $f(x) = 7x^5 - 14x^4 - 21x^3$

61. $f(x) = -(x + 2)^2(x - 1)^2$

62. $f(x) = (x - 2)^3(x + 1)^3$

63. $f(x) = x^2(x - 2)^3(x + 3)^2$

64. $f(x) = -x^3(x - 4)^2(x + 2)^2$

 ■ APPLICATIONS

65. Weight. Jennifer has joined a gym to lose weight and feel better. Jennifer still likes to cheat a little and will enjoy the occasional bad meal with an ice cream dream dessert and then miss the gym for a couple of days. Given below is Jennifer's weight for a period of 8 months. Her weight can be modeled as a polynomial. What is the lowest degree polynomial that can represent Jennifer's weight?

Month	Weight
1	169
2	158
3	150
4	161
5	154
6	159
7	148
8	153

66. Stock Value. A day trader checks the stock price of Coca-Cola during a 4-hour period (given below). The price of Coca-Cola stock during this 4-hour period can be modeled as a polynomial function. What is the lowest degree polynomial that can represent the Coca-Cola stock price?

Period Watching Stock Market	Price
1	$53.00
2	$56.00
3	$52.70
4	$51.50

67. Gym Membership. During the month of January, memberships at local gyms increase because people make New Year's resolutions. The membership can be modeled as a polynomial equation. Give the graph of the polynomial equation. Find any turning points that exist. Find any zeros that exist in the polynomial, too. Let y represent the number of hundreds of memberships sold and x represent the month.

$$y = x(x - 3)^2 \qquad x \geq 0$$

68. Stock Value. The stock prices for Coca-Cola during a 4-hour period on another day yield the following results. If a third-degree polynomial models this stock, do you expect the stock to go up or down in the fifth period?

Period Watching Stock Market	Price
1	$52.80
2	$53.00
3	$56.00
4	$52.70

69. Stock Value. The price of Tommy Hilfiger stock during a 4-hour period is given below. If a fourth-degree polynomial models this stock, do you expect the stock to go up or down in the fifth period?

Period Watching Stock Market	Price
1	$15.10
2	$14.76
3	$14.85
4	$15.50

70. Stock Value. Assuming Tommy Hilfiger stock can be represented by a third-degree polynomial, do you expect the price to go up or down in the fifth period?

Period Watching Stock Market	Price
1	$15.10
2	$14.76
3	$14.85
4	$14.70

■ **CATCH THE MISTAKE**

In Exercises 71–74, explain the mistake that is made.

71. Find a fourth-degree polynomial function with zeros -2, -1, 3, 4.

Solution: $f(x) = (x - 2)(x - 1)(x + 3)(x + 4)$

72. Determine the end behavior of the polynomial function $f(x) = x(x - 2)^3$.

Solution:

This polynomial has similar end behavior to the graph of $y = x^3$.

End behavior falls to the left and rises to the right.

73. Graph the polynomial function $f(x) = (x - 1)^2(x + 2)^3$.

Solution:

The zeros are -2 and 1, and therefore, the x-intercepts are $(-2, 0)$ and $(1, 0)$.

The y-intercept is $(0, 8)$.

Plotting these points and connecting with a smooth curve yield:

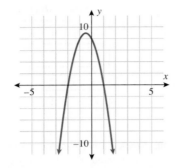

This graph is incorrect. What did we forget to do?

74. Graph the polynomial function $f(x) = (x + 1)^2(x - 1)^2$.

Solution:

The zeros are -1 and 1, so the x-intercepts are $(-1, 0)$ and $(1, 0)$.

The y-intercept is $(0, 1)$.

Plotting these points and connecting with a smooth curve yield:

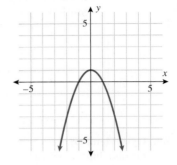

This graph is incorrect. What did we forget to do?

■ **CHALLENGE**

75. T or F: The graph of a polynomial function might not have any y-intercepts.

76. T or F: The graph of a polynomial function might not have any x-intercepts.

77. T or F: The domain of all polynomial functions is $(-\infty, \infty)$.

78. T or F: The range of all polynomial functions is $(-\infty, \infty)$.

79. What is the maximum number of zeros that a polynomial of degree n can have?

80. Find a seventh-degree polynomial that has the following graph characteristics: The graph touches at $x = -1$, and the graph crosses at $x = 3$. Plot this polynomial function.

81. Determine the zeros of the polynomial
$f(x) = x^3 + (b - a)x^2 - abx$.

82. Graph the function $f(x) = x^2(x - a)^2(x - b)^2$ for the positive real numbers a, b, where $b > a$.

■ **TECHNOLOGY**

In Exercises 83 and 84, use a graphing calculator or computer to graph each polynomial. From that graph, estimate the x-intercepts (if any). Set the function equal to zero, and solve for the zeros of the polynomial. Compare the zeros with the x-intercepts.

83. $f(x) = x^4 + 2x^2 + 1$

84. $f(x) = 1.1x^3 - 2.4x^2 + 5.2x$

For each polynomial in Exercises 85 and 86, determine the power function that has similar end behavior. Plot this power function and the polynomial. Do they have similar end behavior?

85. $f(x) = -2x^5 - 5x^4 - 3x^3$

86. $f(x) = x^4 - 6x^2 + 9$

Skills Objectives	Conceptual Objectives
▪ Divide polynomials using long division. ▪ Divide polynomials using synthetic division. ▪ Check the results using the division algorithm.	▪ Extend long division of real numbers to polynomials. ▪ Understand *when* synthetic division can be used.

To divide polynomials, we rely on the technique we use for dividing real numbers. For example, if you were asked to divide 346 by 5, the following long division method would be used.

5 goes into 34 how many times? **6** $5\overline{\smash{\big)}\,346}$

6 times 5 is **30**. $\overset{69}{5\overline{\smash{\big)}\,346}}$

Subtract **30**. Bring down the 6. $-30\downarrow$

5 goes into 46 how many times? 9 46

9 times 5 is 45. Subtract 45. -45

Remainder is 1. 1

This solution can be written two ways.

$$346 \div 5 = 69 \quad R1 \quad \text{or} \quad 346 \div 5 = 69 + \frac{1}{5}$$

In this example the **dividend** is 346, the **divisor** is 5, the **quotient** is 69, and the **remainder** is 1. A similar technique is used when dividing polynomials.

Long Division of Polynomials

Let's start with an example whose answer we already know. We know that a quadratic expression can be factored into the product of two linear factors: $x^2 + 4x - 5 = (x + 5)(x - 1)$. Therefore, if we divide both sides of the equation by $(x - 1)$, we get:

$$\frac{x^2 + 4x - 5}{x - 1} = x + 5$$

We can state this as $x^2 + 4x - 5$ divided by $x - 1$ is equal to $x + 5$. Let's use long division to confirm this statement.

$$x - 1\overline{\smash{\big)}\,x^2 + 4x - 5}$$

Note that although this is standard division notation, the dividend and the divisor are both polynomials that consist of multiple terms. The *lead* terms of each algebraic expression will guide us.

WORDS	MATH

x times what quantity gives x^2? x

$$x - 1 \overline{)x^2 + 4x - 5}$$

$$\begin{array}{r} x \phantom{{}+4x-5} \\ x - 1 \overline{)x^2 + 4x - 5} \end{array}$$

$$\begin{array}{r} x \phantom{{}+4x-5} \\ x - 1 \overline{)x^2 + 4x - 5} \\ x^2 - x \phantom{{}-5} \end{array}$$

Multiply: $x(x - 1) = x^2 - x$.

(*Draw the line, change the sign, and add.*)
Subtract $-(x^2 - x) = -x^2 + x$

$$\begin{array}{r} x \phantom{{}+4x-5} \\ x - 1 \overline{)x^2 + 4x - 5} \\ \underline{-x^2 + x} \phantom{{}-5} \\ 5x - 5 \end{array}$$

Bring down the -5.

x times what quantity is $5x$? 5

$$\begin{array}{r} x + 5 \phantom{{}} \\ x - 1 \overline{)x^2 + 4x - 5} \\ \underline{-x^2 + x} \phantom{{}-5} \\ 5x - 5 \end{array}$$

Multiply: $5(x - 1) = 5x - 5$.

$$5x - 5$$

$$\begin{array}{r} x + 5 \phantom{{}} \\ x - 1 \overline{)x^2 + 4x - 5} \\ \underline{-x^2 + x} \phantom{{}-5} \\ 5x - 5 \end{array}$$

Subtract: $-(5x - 5) = -5x + 5$.

$$\begin{array}{r} \underline{-5x + 5} \\ 0 \end{array}$$

As expected, the remainder $= 0$. By long division we have shown that

$$\frac{x^2 + 4x - 5}{x - 1} = x + 5$$

Check: Multiplying the equation by $x - 1$ yields $x^2 + 4x - 5 = (x + 5)(x - 1)$, which we knew to be true.

EXAMPLE 1 Dividing Polynomials Using Long Division, Zero Remainder

Divide $2x^3 - 9x^2 + 7x + 6$ by $2x + 1$.

Solution:

$$\begin{array}{r} x^2 - 5x + 6 \phantom{{}} \\ 2x + 1 \overline{)2x^3 - 9x^2 + 7x + 6} \end{array}$$

Multiply: $x^2(2x + 1)$.

$$\underline{-(2x^3 + x^2)} \phantom{{}}$$

Subtract. Bring down the $7x$.

$$-10x^2 + 7x$$

Multiply: $-5x(2x + 1)$.

$$\underline{-(-10x^2 - 5x)}$$

Subtract. Bring down the 6.

$$12x + 6$$

Multiply: $6(2x + 1)$.

$$\underline{-(12x + 6)}$$

Subtract.

$$0$$

Quotient: $\boxed{x^2 - 5x + 6}$

☐ Check: $(2x + 1)(x^2 - 5x + 6) = 2x^3 - 9x^2 + 7x + 6$.

Why are we interested in dividing polynomials? Because it helps us find zeros of polynomials. In Example 1, using long division, we found that $2x^3 - 9x^2 + 7x + 6 = (2x + 1)(x^2 - 5x + 6)$. Factoring the quadratic expression $x^2 - 5x + 6 = (x - 3)(x - 2)$ enables us to write the cubic polynomial as a product of three linear factors:

$$2x^3 - 9x^2 + 7x + 6 = (2x + 1)(x - 3)(x - 2)$$

We can now identify the zeros of this cubic polynomial as $x = -\frac{1}{2}$, $x = 3$, and $x = 2$.

■ **YOUR TURN** Divide $4x^3 + 13x^2 - 2x - 15$ by $4x + 5$.

In Example 1 and in the Your Turn the remainder was 0. Sometimes there is a nonzero remainder. For example, if 19 is divided by 7 we get

$$
\begin{array}{r}
2 \\
7{\overline{\smash{\big)}\,19}} \\
\underline{14} \\
5
\end{array}
$$

We write this result as: $\dfrac{19}{7} = 2 + \dfrac{5}{7}$

EXAMPLE 2 Dividing Polynomials Using Long Division, Nonzero Remainder

Divide $6x^2 - x - 2$ by $x + 1$.

Solution:

$$
\begin{array}{r}
6x - 7 \\
x + 1{\overline{\smash{\big)}\,6x^2 - x - 2}} \\
\underline{-(6x^2 + 6x)} \\
-7x - 2 \\
\underline{-(-7x - 7)} \\
+5
\end{array}
$$

Multiply: $6x\,(x + 1)$.
Subtract.
Multiply: $-7(x + 1)$.
Remainder.

$$\underset{\text{Divisor}}{\underbrace{\frac{\overset{\text{Dividend}}{6x^2 - x - 2}}{x + 1}}} = \overset{\text{Quotient}}{6x - 7} + \overset{\text{Remainder}}{\frac{5}{x + 1}}$$

Check:

Multiply by $x + 1$.
$$
\begin{aligned}
6x^2 - x - 2 &= (6x - 7)(x + 1) + 5 \\
&= 6x^2 - x - 7 + 5 \\
&= 6x^2 - x - 2
\end{aligned}
$$

■ **Answer:** $x^2 + 2x - 3$, remainder 0

■ **YOUR TURN**   Divide $2x^3 + x^2 - 4x - 3$ by $x - 1$.

In general, when a polynomial is divided by another polynomial, we express the result in the following form

$$\frac{P(x)}{d(x)} = Q(x) + \frac{r(x)}{d(x)}$$

where $P(x)$ is the **dividend**, $d(x) \neq 0$ is the divisor, $Q(x)$ is the quotient, and $r(x)$ is the remainder. Multiplying this equation by the divisor, $d(x)$, leads us to the division algorithm.

THE DIVISION ALGORITHM

If $P(x)$ and $d(x)$ are polynomials with $d(x) \neq 0$, and the degree of $P(x)$ is greater than or equal to the degree of $d(x)$, then unique polynomials $Q(x)$ and $r(x)$ exist such that

$$P(x) = d(x) \cdot Q(x) + r(x)$$

If the remainder $r(x) = 0$, then we say that $d(x)$ divides evenly into $P(x)$ and that $d(x)$ and $Q(x)$ are factors of $P(x)$.

EXAMPLE 3 Long Division of Polynomials with "Missing" Terms

Divide $x^3 - 8$ by $x - 2$.

Solution:

Insert $0x^2 + 0x$ for placeholders.

Multiply: $x^2(x - 2) = x^3 - 2x^2$.

Subtract and bring down $0x$.

Multiply: $2x(x - 2) = 2x^2 - 4x$.

Subtract and bring down -8.

Multiply: $4(x - 2) = 4x - 8$.

Subtract and get remainder 0.

$$
\begin{array}{r}
x^2 + 2x + 4 \\
x - 2 \overline{)\; x^3 + 0x^2 + 0x - 8} \\
\underline{-(x^3 - 2x^2)} \\
2x^2 + 0x \\
\underline{-(2x^2 - 4x)} \\
4x - 8 \\
\underline{-(4x - 8)} \\
0
\end{array}
$$

Since the remainder is 0, $x - 2$ divides *evenly* into $x^3 - 8$.

$$\frac{x^3 - 8}{x - 2} = x^2 + 2x + 4, \qquad x \neq 2$$

Check: $x^3 - 8 = (x^2 + 2x + 4)(x - 2)$

$$= x^3 + 2x^2 + 4x - 2x^2 - 4x - 8$$

$$= x^3 - 8$$

■ **Answer:** $2x^2 + 3x - 1\ R\ -4$ or $2x^2 + 3x - 1 - \dfrac{4}{x - 1}$

■ **YOUR TURN** Divide $x^3 - 1$ by $x - 1$.

EXAMPLE 4 Long Division of Polynomials

Divide $3x^4 + 2x^3 + x^2 + 4$ by $x^2 - 2x + 1$.

Solution:

Insert $0x$ as a placeholder.

Multiply: $3x^2(x^2 - 2x + 1)$

Subtract and bring down $0x$.

Multiply: $8x(x^2 - 2x + 1)$

Subtract and bring down 4.

Multiply: $14(x^2 - 2x + 1)$

Subtract and get remainder $20x - 10$.

$$
\begin{array}{r}
3x^2 + 8x + 14 \\
x^2 - 2x + 1\overline{)3x^4 + 2x^3 + x^2 + 0x + 4} \\
-(3x^4 - 6x^3 + 3x^2) \\
\hline
8x^3 - 2x^2 + 0x \\
-(8x^3 - 16x^2 + 8x) \\
\hline
14x^2 - 8x + 4 \\
-(14x^2 - 28x + 14) \\
\hline
20x - 10
\end{array}
$$

$$\frac{3x^4 + 2x^3 + x^2 + 4}{x^2 - 2x + 1} = 3x^2 + 8x + 14 + \frac{20x - 10}{x^2 - 2x + 1}$$

 CONCEPT CHECK Before dividing $2x^5 + 3x^2 + 12$ by $x^3 - 3x - 4$, are any zero placeholders needed?

■ **YOUR TURN** Divide $2x^5 + 3x^2 + 12$ by $x^3 - 3x - 4$.

Look back at Examples 1 through 4. Notice the following:

■ Degree of $Q(x) = $ degree of $P(x) - $ degree of $d(x)$
■ Degree of $r(x) < $ degree of $d(x)$

Example 1: A cubic polynomial divided by a linear polynomial yields a quotient that is a quadratic polynomial.

Example 2: A quadratic polynomial divided by a linear polynomial yields a quotient that is a linear polynomial.

Example 3: A cubic polynomial divided by a linear polynomial yields a quotient that is a quadratic polynomial.

Example 4: A fourth degree polynomial divided by a quadratic polynomial yields a quotient that is a quadratic polynomial.

Synthetic Division of Polynomials

In the special case when the divisor is a linear factor of the form $x - a$ or $x + a$, there is another, more efficient way to divide polynomials. This method is called **synthetic**

■ **Answer:** $\dfrac{2x^5 + 3x^2 + 12}{x^3 - 3x - 4} = 2x^2 + 6 + \dfrac{11x^2 + 18x + 36}{x^3 - 3x - 4}$

■ **Answer:** $\dfrac{x^3 - 1}{x - 1} = x^2 + x + 1, \; x \neq 1$

division. It is called synthetic because it is a contrived shorthand way of dividing a polynomial by a linear factor. A detailed step-by-step procedure is given below for synthetic division. Let's divide $x^4 - x^3 - 2x + 2$ by $x + 1$ using synthetic division.

STEP 1 Write the division in synthetic form.

Coefficients of Dividend

a. List the coefficients of the dividend. Remember to use 0 for a placeholder.

$$-1 \ \big| \ 1 \quad -1 \quad 0 \quad -2 \quad 2$$

Divisor

b. The divisor is $x + 1$, so -1 is used.

If $(x - a)$ is divisor, then a is the number used in synthetic division.

STEP 2 *Bring down* the first term (1) in the dividend.

$$-1 \ \big| \ 1 \quad -1 \quad 0 \quad -2 \quad 2$$

Bring down the 1

$$1$$

STEP 3 *Multiply* the divisor (-1) times the (1), and place the product up and to the right.

$$-1 \ \big| \ 1 \quad -1 \quad 0 \quad -2 \quad 2$$

-1

Multiply $(-1)(1) = -1$

1

STEP 4 *Add* each column.

$$-1 \ \big| \ 1 \quad -1 \quad 0 \quad -2 \quad 2$$

-1 ADD

$$1 \quad -2$$

STEP 5 Repeat Steps 3 and 4 until all columns are filled.

$$-1 \ \big| \ 1 \quad -1 \quad 0 \quad -2 \quad 2$$

$-1 \quad 2 \quad -2 \quad 4$

$$1 \quad -2 \quad 2 \quad -4 \quad 6$$

STEP 6 Identify the quotient and the remainder.

$$-1 \ \big| \ 1 \quad -1 \quad 0 \quad -2 \quad 2$$

$-1 \quad 2 \quad -2 \quad 4$

$$1 \quad -2 \quad 2 \quad -4 \quad 6$$

Quotient Coefficients · Remainder

$$x^3 - 2x^2 + 2x - 4$$

Let's compare dividing $x^4 - x^3 - 2x + 2$ by $x + 1$ using both long division and synthetic division.

Long Division

$$
\begin{array}{r}
x^3 - 2x^2 + 2x - 4 \\
x + 1 \overline{)\ x^4 - x^3 + 0x^2 - 2x + 2} \\
\underline{x^4 + x^3} \\
-2x^3 + 0x^2 \\
\underline{-(-2x^3 - 2x^2)} \\
2x^2 - 2x \\
\underline{-(2x^2 + 2x)} \\
-4x + 2 \\
\underline{-(-4x - 4)} \\
+6
\end{array}
$$

Synthetic Division

$$-1 \ \big| \ 1 \quad -1 \quad 0 \quad -2 \quad 2$$

$-1 \quad 2 \quad -2 \quad 4$

$$1 \quad -2 \quad 2 \quad -4 \quad \boxed{6}$$

$$x^3 - 2x^2 + 2x \ - 4$$

Both long division and synthetic division yield the same answer:

$$\frac{x^4 - x^3 - 2x + 2}{x + 1} = x^3 - 2x^2 + 2x - 4 + \frac{6}{x + 1}$$

 EXAMPLE 5 Synthetic Division

Use synthetic division to divide $3x^5 - 2x^3 + x^2 - 7$ by $x + 2$.

Solution:

STEP 1 Write the division in synthetic form.

    a. List the coefficients of the dividend. Remember to use 0 for a placeholder.

$$-2 \,\big|\, 3 \quad 0 \quad -2 \quad 1 \quad 0 \quad -7$$

    b. The divisor is $x + 2$, so -2 is used.

STEP 2 Perform the synthetic division steps.

$$
\begin{array}{r}
-2 \,\big|\, \begin{array}{rrrrrr} 3 & 0 & -2 & 1 & 0 & -7 \\ & -6 & 12 & -20 & 38 & -76 \\ \hline 3 & -6 & 10 & -19 & 38 & -83 \end{array}
\end{array}
$$

STEP 3 Identify the quotient and remainder.

$$
\begin{array}{r}
-2 \,\big|\, \begin{array}{rrrrrr} 3 & 0 & -2 & 1 & 0 & -7 \\ & -6 & 12 & -20 & 38 & -76 \\ \hline 3 & -6 & 10 & -19 & 38 & \boxed{-83} \end{array}
\end{array}
$$

$$3x^4 - 6x^3 + 10x^2 - 19x + 38$$

Solution:

$$\frac{3x^5 - 2x^3 + x^2 - 7}{x + 2} = 3x^4 - 6x^3 + 10x^2 - 19x + 38 - \frac{83}{x + 2}$$

 CAUTION

Synthetic division can only be used when the divisor is of the form $x - a$ or $x + a$.

■ **YOUR TURN**   Use synthetic division to divide $2x^3 - x + 3$ by $x - 1$.

Synthetic division can only be used when the divisor is of the form $x - a$ or $x + a$.

SECTION 4.3 SUMMARY

Division of Polynomials

- Long division can always be used.
- Synthetic division can be used only if the divisor is $(x - a)$ or $(x + a)$.

Expressing Results

- $\dfrac{\text{Dividend}}{\text{Divisor}} = \text{quotient} + \dfrac{\text{remainder}}{\text{divisor}}$
- Dividend = (quotient)(divisor) + remainder

When Remainder Is Zero

- Dividend = (quotient)(divisor)
- Quotient and divisor are factors of the dividend

■ **Answer:** $\dfrac{2x^3 - x + 3}{x - 1} = 2x^2 + 2x + 1 + \dfrac{4}{x - 1}$

SECTION 4.3 EXERCISES

 ■ SKILLS

Use long division to divide the polynomials. If you use a calculator, do not round off. Use exact values. Express the answer in the form $Q(x) =$ $r(x) =$.

1. $(4x^2 - 9) \div (2x + 3)$

2. $(x^2 - 4) \div (x + 4)$

3. $(2x^2 + 5x - 3) \div (x + 3)$

4. $(2x^2 + 5x - 3) \div (x - 3)$

5. $(3x^2 - 13x - 10) \div (x + 5)$

6. $(3x^2 - 13x - 10) \div (x - 5)$

7. $(11x + 20x^2 + 12x^3 + 2) \div (3x + 2)$

8. $(12x^3 + 2 + 11x + 20x^2) \div (2x + 1)$

9. $(4x^3 - 2x + 7) \div (2x + 1)$

10. $(6x^4 - 2x^2 + 5) \div (-3x + 2)$

11. $(4x^3 - 12x^2 - x + 3) \div \left(x - \dfrac{1}{2}\right)$

12. $(12x^3 + 1 + 7x + 16x^2) \div \left(x + \dfrac{1}{3}\right)$

13. $(-2x^5 + 3x^4 - 2x^2) \div (x^3 - 3x^2 + 1)$

14. $(-9x^6 + 7x^4 - 2x^3 + 5) \div (3x^4 - 2x + 1)$

15. $\dfrac{x^4 - 1}{x^2 - 1}$

16. $\dfrac{x^4 - 9}{x^2 + 3}$

17. $\dfrac{40 - 22x + 7x^3 + 6x^4}{6x^2 + x - 2}$

18. $\dfrac{-13x^2 + 4x^4 + 9}{4x^2 - 9}$

19. $\dfrac{-3x^4 + 7x^3 - 2x + 1}{x - 0.6}$

20. $\dfrac{2x^5 - 4x^3 + 3x^2 + 5}{x - 0.9}$

21. $(x^4 + 0.8x^3 - 0.26x^2 - 0.168x + 0.0441) \div (x^2 + 1.4x + 0.49)$

22. $(x^5 + 2.8x^4 + 1.34x^3 - 0.688x^2 - 0.2919x + 0.0882) \div (x^2 - 0.6x + 0.09)$

Use synthetic division to divide the polynomial by the linear factor. Indicate the quotient, $Q(x)$, and the remainder, $r(x)$.

23. $(3x^2 + 7x + 2) \div (x + 2)$

24. $(2x^2 + 7x - 15) \div (x + 5)$

25. $(7x^2 - 3x + 5) \div (x + 1)$

26. $(4x^2 + x + 1) \div (x - 2)$

27. $(3x^2 + 4x - x^4 - 2x^3 - 4) \div (x + 2)$

28. $(3x^2 - 4 + x^3) \div (x - 1)$

29. $(x^4 + 1) \div (x + 1)$

30. $(x^4 + 9) \div (x + 3)$

31. $(x^4 - 16) \div (x + 2)$

32. $(x^4 - 81) \div (x - 3)$

33. $(2x^3 - 5x^2 - x + 1) \div \left(x + \dfrac{1}{2}\right)$

34. $(3x^3 - 8x^2 + 1) \div \left(x + \dfrac{1}{3}\right)$

35. $(2x^4 - 3x^3 + 7x^2 - 4) \div \left(x - \dfrac{2}{3}\right)$

36. $(3x^4 + x^3 + 2x - 3) \div \left(x - \dfrac{3}{4}\right)$

37. $(2x^4 + 9x^3 - 9x^2 - 81x - 81) \div (x + 1.5)$

38. $(5x^3 - x^2 + 6x + 8) \div (x + 0.8)$

39. $\dfrac{x^7 - 8x^4 + 3x^2 + 1}{x - 1}$

40. $\dfrac{x^6 + 4x^5 - 2x^3 + 7}{x + 1}$

41. $(x^6 - 49x^4 - 25x^2 + 1225) \div (x - \sqrt{5})$

42. $(x^6 - 4x^4 - 9x^2 + 36) \div (x - \sqrt{3})$

Divide the polynomials using either long division or synthetic division.

43. $(6x^2 - 23x + 7) \div (3x - 1)$

44. $(6x^2 + x - 2) \div (2x - 1)$

45. $(x^3 - x^2 - 9x + 9) \div (x - 1)$

46. $(x^3 + 2x^2 - 6x - 12) \div (x + 2)$

47. $(x^5 + 4x^3 + 2x^2 - 1) \div (x - 2)$

48. $(x^4 - x^2 + 3x - 10) \div (x + 5)$

49. $(x^4 - 25) \div (x^2 - 1)$

50. $(x^3 - 8) \div (x^2 - 2)$

51. $(x^7 - 1) \div (x - 1)$

52. $(x^6 - 27) \div (x - 3)$

■ **APPLICATIONS**

53. Geometry. The area of a rectangle is $6x^4 + 4x^3 - x^2 - 2x - 1$ square feet. If the length of the rectangle is $2x^2 - 1$ feet, what is the width of the rectangle?

54. Geometry. If the rectangle in Exercise 53 is the base of a rectangular box with volume $18x^5 + 18x^4 + x^3 - 7x^2 - 5x - 1$ cubic feet, what is the height of the box?

55. Travel. If a car travels a distance of $x^3 + 60x^2 + x + 60$ miles at an average speed of $x + 60$ miles per hour, how long did the trip take?

56. Sports. If a quarterback throws a ball $-x^2 - 5x + 50$ yards in $(5 - x)$ seconds, how fast is the football traveling?

 ■ **CATCH THE MISTAKE**

In Exercises 57–60, explain the mistake that is made.

57. Divide $x^3 - 4x^2 + x + 6$ by $x^2 + x + 1$.

Solution:

$$
\begin{array}{r}
x - 3 \\
x^2 + x + 1 \overline{)x^3 - 4x^2 + x + 6} \\
\underline{x^3 + x^2 + x} \\
-3x^2 + 2x + 6 \\
\underline{-3x^2 - 3x - 3} \\
-x + 3
\end{array}
$$

58. Divide $x^4 - 3x^2 + 5x + 2$ by $x - 2$.

Solution:

$$
\begin{array}{r|rrrr}
-2 & 1 & -3 & 5 & 2 \\
& & -2 & 10 & -30 \\
\hline
& 1 & -5 & 15 & \boxed{-28}
\end{array}
$$

$$x^2 - 5x + 15$$

59. Divide $x^3 - 3x^2 + 4x - 12$ by $x - 3$.

Solution:

$$
\begin{array}{r|rrrr}
-3 & 1 & -3 & 4 & -12 \\
& & -3 & 18 & -66 \\
\hline
& 1 & -6 & 22 & \boxed{-78}
\end{array}
$$

$$x^2 - 6x + 22$$

60. Divide $x^3 + 3x^2 - 2x + 1$ by $x^2 + 1$.

Solution:

$$
\begin{array}{r|rrrr}
-1 & 1 & 3 & -2 & 1 \\
& & -1 & -2 & 4 \\
\hline
& 1 & 2 & -4 & \boxed{5}
\end{array}
$$

$$x^2 - 2x - 4$$

 ■ **CHALLENGE**

61. T or F: A fifth-degree polynomial divided by a third-degree polynomial will yield a quadratic quotient.

62. T or F: A third-degree polynomial divided by a linear polynomial will yield a linear quotient.

63. T or F: Synthetic division can be used whenever the degree of the dividend is exactly one more than the degree of the divisor.

64. T or F: When the remainder is zero, the divisor is a factor of the dividend.

65. Is $(x + b)$ a factor of $x^3 + (2b - a)x^2 + (b^2 - 2ab)x - ab^2$?

66. Is $(x + b)$ a factor of $x^4 + (b^2 - a^2)x^2 - a^2b^2$?

67. $(x - 1)$ is a factor of $x^3 - 2x^2 - 5x + 6$. Find the other two factors and write the cubic in terms of a product of three linear factors.

68. Divide $x^{3n} + 5x^{2n} + 8x^n + 4$ by $x^n + 1$.

■ **TECHNOLOGY**

69. Plot $\dfrac{2x^3 - x^2 + 10x - 5}{x^2 + 5}$. What type of function is it? Perform this division using long division, and confirm that the graph corresponds to the quotient.

70. Plot $\dfrac{x^3 - 3x^2 + 4x - 12}{x - 3}$. What type of function is it? Perform this division using synthetic division, and confirm that the graph corresponds to the quotient.

71. Plot $\dfrac{x^4 + 2x^3 - x - 2}{x + 2}$. What type of function is it? Perform this division using synthetic division, and confirm that the graph corresponds to the quotient.

72. Plot $\dfrac{x^5 - 9x^4 + 18x^3 + 2x^2 - 5x - 3}{x^4 - 6x^3 + 2x + 1}$. What type of function is it? Perform this division using long division, and confirm that the graph corresponds to the quotient.

SECTION 4.4 Properties and Tests of Zeros of Polynomial Functions

Skills Objectives

- Apply the remainder theorem to evaluate a polynomial.
- Rely on the factor theorem to factor a polynomial.
- Rely on the rational zero theorem to determine possible zeros.
- Apply Descartes's rule of signs to determine the possible combinations of positive or negative zeros.

Conceptual Objectives

- Extend the domain of polynomial functions to complex numbers.
- Understand how the fundamental theorem of algebra guarantees at least one zero.
- Understand why complex zeros occur in conjugate pairs.

STUDY TIP

The zeros of a polynomial can be complex numbers. Only when the zeros are real numbers do we interpret zeros as *x*-intercepts.

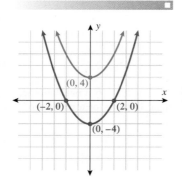

In this chapter we have assumed the coefficients of polynomial functions are real numbers. The domain and range of polynomial functions thus far have been the set of all real numbers. Now, we consider a more general case. In this section, the coefficients of the polynomial function and the domain of the polynomial function are *complex numbers*. Note that the set of real numbers is a subset of the complex numbers. (Choose the imaginary part to be zero.)

It is important to note, however, that when we are discussing *graphs* of polynomial functions, we will restrict the domain to the set of real numbers.

A *zero* of a polynomial $P(x)$ is the *solution* or *root* of the equation $P(x) = 0$. The *zeros of a polynomial can be complex numbers.* However, since the *xy*-plane represents real numbers, only when the zeros are real numbers do we interpret zeros as *x*-intercepts.

We can illustrate the relationship between real and complex zeros of polynomial functions and their graphs with two similar examples. Let's take the two quadratic functions $f(x) = x^2 - 4$ and $g(x) = x^2 + 4$. The graphs of these two functions are parabolas that open upward with $f(x)$ shifted down 4 units and $g(x)$ shifted up 4 units as shown on the left.

Setting each function equal to zero and solving for *x*, we find that the zeros for $f(x)$ are -2 and 2 and the zeros for $g(x)$ are $-2i$ and $2i$. Notice that the *x*-intercepts for $f(x)$ are -2 and 2 and $g(x)$ has no *x*-intercepts.

The Remainder Theorem and the Factor Theorem

The remainder obtained in division of polynomials is related to the evaluation of the polynomial. This relationship is given by the *remainder theorem.*

REMAINDER THEOREM

If a polynomial $P(x)$ is divided by $x - a$, then the remainder is $r = P(a)$.

The remainder theorem tells you that polynomial division can be used to evaluate a polynomial function at a particular point.

EXAMPLE 1 Two Methods for Evaluating Polynomials

Let $P(x) = 4x^5 - 3x^4 + 2x^3 - 7x^2 + 9x - 5$ evaluate $P(2)$ by

a. Evaluating $P(2)$ directly

b. Using the remainder theorem and synthetic division

Solution:

a. $P(2) = 4(2)^5 - 3(2)^4 + 2(2)^3 - 7(2)^2 + 9(2) - 5$
$ = 4(32) - 3(16) + 2(8) - 7(4) + 9(2) - 5$
$ = 128 - 48 + 16 - 28 + 18 - 5$
$ = \boxed{81}$

b.
$$
\begin{array}{r|rrrrrr}
2 & 4 & -3 & 2 & -7 & 9 & -5 \\
  &   & 8 & 10 & 24 & 34 & 86 \\
\hline
  & 4 & 5 & 12 & 17 & 43 & \boxed{81}
\end{array}
$$

> ▪ **YOUR TURN**   Let $P(x) = -x^3 + 2x^2 - 5x + 2$ and evaluate $P(-2)$ using the remainder theorem and synthetic division.

Recall that when a polynomial is divided by $x - a$, if the remainder is zero, we say that $x - a$ is a factor of the polynomial. And through the remainder theorem, we now know that the remainder is related to evaluation of the polynomial at the point $x = a$. We are then led to the *factor theorem.*

FACTOR THEOREM

If $P(a) = 0$, then $x - a$ is a factor of $P(x)$. Conversely, if $x - a$ is a factor of $P(x)$, then $P(a) = 0$.

▪ **Answer:** $P(-2) = 28$

TECHNOLOGY TIP

Note that $P(-2) = 0$, $P(3) = 0$, and $P(1) = 0$. Graphing $P(x) = x^3 - 2x^2 - 5x + 6$

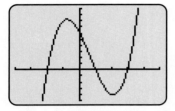

```
Plot1 Plot2 Plot3
\Y1◼X^3-2X²-5X+6
\Y2=◼
```

The three zeros of the function are $x = -2$, $x = 1$, $x = 3$.

EXAMPLE 2 Using the Factor Theorem to Factor a Polynomial

Determine if $x + 2$ is a factor of $P(x) = x^3 - 2x^2 - 5x + 6$. If so, factor $P(x)$ completely.

Solution:

By the factor theorem, $x + 2$ is a factor of $P(x) = x^3 - 2x^2 - 5x + 6$ if $P(-2) = 0$. By the remainder theorem, if we divide $P(x) = x^3 - 2x^2 - 5x + 6$ by $x + 2$, then the remainder is equal to $P(-2)$.

STEP 1 Divide $P(x) = x^3 - 2x^2 - 5x + 6$ by $x + 2$ using synthetic division.

$$
\begin{array}{r|rrrr}
-2 & 1 & -2 & -5 & 6 \\
   &   & -2 & 8 & -6 \\
\hline
   & 1 & -4 & 3 & \boxed{0} \\
\end{array}
$$
$$\underbrace{}_{x^2 - 4x + 3}$$

Since the remainder is zero, $P(-2) = 0$, and $x + 2$ is a factor of $P(x) = x^3 - 2x^2 - 5x + 6$.

STEP 2 Write $P(x)$ as a product. $P(x) = (x + 2)(x^2 - 4x + 3)$

STEP 3 Factor the quadratic polynomial. $P(x) = (x + 2)(x - 3)(x - 1)$

✓ **CONCEPT CHECK** Given $P(x) = x^3 - 4x^2 - 7x + 10$, find $P(1)$.

■ **YOUR TURN** Determine if $x - 1$ is a factor of $P(x) = x^3 - 4x^2 - 7x + 10$. If so, factor $P(x)$ completely.

EXAMPLE 3 Using the Factor Theorem to Factor a Polynomial

Determine if $x - 3$ and $x + 2$ are factors of $P(x) = x^4 - 13x^2 + 36$. If so, factor $P(x)$ completely.

Solution:

STEP 1 Divide $P(x) = x^4 - 13x^2 + 36$ by $x - 3$ using synthetic division.

$$
\begin{array}{r|rrrrr}
3 & 1 & 0 & -13 & 0 & 36 \\
  &   & 3 & 9 & -12 & -36 \\
\hline
  & 1 & 3 & -4 & -12 & \boxed{0} \\
\end{array}
$$
$$\underbrace{}_{x^3 + 3x^2 - 4x - 12}$$

Because the remainder is 0, $x - 3$ is a factor, and we can write the polynomial as:

$$P(x) = (x - 3)(x^3 + 3x^2 - 4x - 12)$$

STEP 2 Divide the remaining cubic polynomial, $(x^3 + 3x^2 - 4x - 12)$, by $x + 2$ using synthetic division.

$$
\begin{array}{r|rrrr}
-2 & 1 & 3 & -4 & -12 \\
   &   & -2 & -2 & 12 \\
\hline
   & 1 & 1 & -6 & \boxed{0} \\
\end{array}
$$
$$\underbrace{}_{x^2 + x - 6}$$

■ **Answer:** $P(x) = (x - 5)(x - 1)(x + 2)$

Because the remainder is 0, $x + 2$ is a factor, and we can now write the polynomial as:

$$P(x) = (x - 3)(x + 2)(x^2 + x - 6)$$

STEP 3 Factor the quadratic polynomial, $x^2 + x - 6 = (x + 3)(x - 2)$.

STEP 4 Write $P(x)$ as a product of linear factors:

$$P(x) = (x - 3)(x - 2)(x + 2)(x + 3)$$

■ **YOUR TURN** Determine if $x - 3$ and $x + 2$ are factors of $P(x) = x^4 - x^3 - 7x^2 + x + 6$. If so, factor $P(x)$ completely.

The Search for Zeros

All of the examples thus far have started with a polynomial. We have been given one or more zeros (or linear factors), and we used polynomial (or synthetic) division to divide the polynomial by the given linear factor. This operation results in another polynomial (typically quadratic) that we can then factor. Now we will not be given any zeros to start. Instead, we will have to develop methods to search for them.

First, we will determine how many zeros and what types (real or complex) of zeros to expect. We will develop a list of possible zeros, and then test those to determine the actual zeros.

We will use the *fundamental theorem of algebra* and the *n zeros theorem* to determine how many zeros to expect. Then, if the degree of the polynomial is even or odd, we will determine possible combinations of real and imaginary zeros. We will use the *rational zero theorem* to determine a list of possible zeros. We will use synthetic division to test those possibilities until we find a remainder that is zero. When the remainder is zero, the quotient will be another polynomial (one degree lower than the original), and the process starts again until we can get the polynomial into a product of linear factors.

The Fundamental Theorem of Algebra

What are the minimum and maximum number of zeros a polynomial can have? Every polynomial has *at least one zero* (provided the degree is greater than zero). The largest number of zeros a polynomial can have is equal to the degree of the polynomial.

THE FUNDAMENTAL THEOREM OF ALGEBRA

Every polynomial $P(x)$ of degree $n > 0$ has *at least one zero* in the complex number system.

n **ZEROS THEOREM**

Every polynomial $P(x)$ of degree $n > 0$ can be expressed as the product of n linear factors. Hence, $P(x)$ has exactly n zeros, not necessarily distinct.

STUDY TIP

The largest number of zeros a polynomial can have is equal to the degree of the polynomial.

■ **Answer:** $P(x) = (x - 3)(x + 2)(x - 1)(x + 1)$

These two theorems are illustrated with five polynomials below.

a. The **first**-degree polynomial $f(x) = x + 3$ has exactly **one** zero $x = -3$.
b. The **second**-degree polynomial $f(x) = x^2 + 10x + 25 = (x + 5)(x + 5)$ has exactly **two** zeros $x = -5$ and $x = -5$. It is customary to write this as a single zero of multiplicity 2 or refer to it as a repeated root.
c. The **third**-degree polynomial $f(x) = x^3 + 16x = x(x^2 + 16) = x(x + 4i)(x - 4i)$ has exactly **three** zeros $x = 0$, $x = -4i$, and $x = 4i$.
d. The **fourth**-degree polynomial $f(x) = x^4 - 1 = (x^2 - 1)(x^2 + 1) = (x - 1)(x + 1)(x - i)(x + i)$ has exactly **four** zeros $x = 1$, $x = -1$, $x = i$, and $x = -i$.
e. The **fifth**-degree polynomial $f(x) = x^5 = x \cdot x \cdot x \cdot x \cdot x$ has exactly **five** zeros $x = 0$, which has multiplicity **five**.

The fundamental theorem of algebra and the n zeros theorem only tell you that the zeros *exist*—not how to find them. We must rely on other techniques to determine the zeros.

Imaginary Zeros

STUDY TIP

If we restrict the coefficients of a polynomial to real numbers, complex zeros always come in conjugate pairs.

Often, at a grocery store or a drugstore, we see signs for special offers—"two for the price of one." This is true when zeros are complex. If a zero is a complex number, then another zero will always be the complex conjugate. Look at the third-degree polynomial in the above illustration, part c, where two of the zeros were $x = -4i$ and $x = 4i$, and in part d, where two of the zeros were $x = i$, and $x = -i$. In general, if we restrict the coefficients of a polynomial to real numbers, complex zeros always come in conjugate pairs.

IMAGINARY ZEROS THEOREM

If a polynomial $P(x)$ has real coefficients, and if $a + bi$ is a zero of $P(x)$, then $a - bi$ is also a zero of $P(x)$.

EXAMPLE 4 Zeros That Appear as Complex Conjugates

Find the zeros of the polynomial $P(x) = x^2 - 4x + 13$.

Solution:

Set the polynomial equal to zero.
$$P(x) = x^2 - 4x + 13 = 0$$

Use the quadratic formula to solve for x.
$$x = \frac{-(-4) \pm \sqrt{(-4)^2 - 4(1)(13)}}{2(1)}$$

Simplify.
$$x = 2 \pm 3i$$

The zeros are the complex conjugates $x = 2 - 3i$ and $x = 2 + 3i$.

Check: This is a *second-degree* polynomial, so we expect *two* zeros.

In Example 4, we solved a quadratic equation to find the two zeros that are complex conjugates of one another. In the next example, we use the imaginary zeros theorem to assist us in factoring a higher degree polynomial.

EXAMPLE 5 Factoring a Polynomial with Imaginary Zeros

Factor the polynomial $P(x) = x^4 - x^3 - 5x^2 - x - 6$ given that i is a zero of $P(x)$.

Since $P(x)$ is a *fourth*-degree polynomial we expect *four* zeros. The goal in this problem is to write $P(x)$ as a product of four linear factors: $P(x) = (x - a)(x - b)(x - c)(x - d)$ where a, b, c, and d are complex numbers and represent the zeros of the polynomial.

Solution:

STEP 1 Write known zeros and linear factors.
Since i is a zero, we know that $-i$ is a zero. $x = i$ and $x = -i$
We now know two linear factors of $P(x)$. $(x - i)$ and $(x + i)$

STEP 2 Write $P(x)$ as a product of four factors.
$$P(x) = (x - i)(x + i)(x - c)(x - d)$$

STEP 3 Multiply the known two factors.
$$(x + i)(x - i) = x^2 - i^2 = x^2 - (-1) = x^2 + 1$$

STEP 4 Rewrite polynomial.
$$P(x) = (x^2 + 1)(x - c)(x - d)$$

STEP 5 Divide both sides of the equation by $x^2 + 1$.
$$\frac{P(x)}{x^2 + 1} = (x - c)(x - d)$$

STEP 6 Divide $P(x) = x^4 - x^3 - 5x^2 - x - 6$ by $x^2 + 1$ using long division.

$$
\begin{array}{r}
x^2 - x - 6 \\
x^2 + 0x + 1 \overline{)x^4 - x^3 - 5x^2 - x - 6} \\
\underline{-(x^4 + 0x^3 + x^2)} \\
-x^3 - 6x^2 - x \\
\underline{-(-x^3 + 0x^2 - x)} \\
-6x^2 + 0x - 6 \\
\underline{-(-6x^2 + 0x - 6)} \\
0
\end{array}
$$

STEP 7 Factor the quotient.
$$x^2 - x - 6 = (x - 3)(x + 2)$$

STEP 8 Write $P(x)$ as a product of four linear factors.
$$P(x) = (x - i)(x + i)(x - 3)(x + 2)$$

Check: $P(x)$ is a *fourth*-degree polynomial and we found *four* zeros, two of which are complex conjugates.

CONCEPT CHECK If $x - 2i$ is a factor of $P(x) = x^4 - 3x^3 + 6x^2 - 12x + 8$, what other factor do we know?

■ **YOUR TURN** Factor the polynomial $P(x) = x^4 - 3x^3 + 6x^2 - 12x + 8$ given that $x - 2i$ is a factor.

■ **Answer:** $P(x) = (x - 2i)(x + 2i)(x - 1)(x - 2)$ Note: The zeros of $P(x)$ are 1, 2, 2i, and $-2i$.

EXAMPLE 6 Factoring a Polynomial with Complex Zeros

Factor the polynomial $P(x) = x^4 - 2x^3 + x^2 + 2x - 2$, given that $1 + i$ is a zero of $P(x)$.

Since $P(x)$ is a *fourth*-degree polynomial, we expect *four* zeros. The goal in this problem is to write $P(x)$ as a product of four linear factors:
$P(x) = (x - a)(x - b)(x - c)(x - d)$ where a, b, c, and d are complex numbers and represent the zeros of the polynomial.

Solution:

STEP 1 Write known zeros and linear factors.
Since $1 + i$ is a zero, we know
that $1 - i$ is a zero. $\qquad\qquad x = 1 + i$ and $x = 1 - i$

We now know two linear factors
of $P(x)$. $\qquad\qquad \left[x - (1 + i)\right]$ and $\left[x - (1 - i)\right]$

STEP 2 Write $P(x)$ as a product
of four factors. $\qquad P(x) = \left[x - (1 + i)\right]\left[x - (1 - i)\right]\left[x - c\right]\left[x - d\right]$

STEP 3 Multiply $\left[x - (1 + i)\right]\left[x - (1 - i)\right]$.

First group the real parts
together in each bracket. $\qquad\qquad \left[(x - 1) - i\right]\left[(x - 1) + i\right]$

Use the special product
$(a - b)(a + b) = a^2 - b^2$,
where a is $(x - 1)$ and b is i. $\qquad (x - 1)^2 - i^2$
$\qquad\qquad\qquad (x^2 - 2x + 1) - (-1)$
$\qquad\qquad\qquad x^2 - 2x + 2$

STEP 4 Rewrite the polynomial. $\qquad P(x) = (x^2 - 2x + 2)(x - c)(x - d)$

STEP 5 Divide both sides of the equation by $x^2 - 2x + 2$, and substitute in the original polynomial $P(x) = x^4 - 2x^3 + x^2 + 2x - 2$.

$$\frac{x^4 - 2x^3 + x^2 + 2x - 2}{x^2 - 2x + 2} = (x - c)(x - d)$$

STEP 6 Divide the left side of
the equation using
long division. $\qquad \dfrac{x^4 - 2x^3 + x^2 + 2x - 2}{x^2 - 2x + 2} = x^2 - 1$

STEP 7 Factor $x^2 - 1$. $\qquad\qquad\qquad\qquad (x - 1)(x + 1)$

STEP 8 Write $P(x)$ as a product of four linear factors.

$$P(x) = \left[x - (1 + i)\right]\left[x - (1 - i)\right]\left[x - 1\right]\left[x + 1\right]$$

■ **YOUR TURN** Factor the polynomial $P(x) = x^4 - 2x^2 + 16x - 15$ given that $1 + 2i$ is a factor.

■ **Answer:** $P(x) = \left[x - (1 + 2i)\right]\left[x - (1 - 2i)\right]\left[x - 1\right]\left[x + 3\right]$ Note: The zeros of $P(x)$ are $1, -3, 1 + 2i$, and $1 - 2i$.

Because an n-degree polynomial has exactly n zeros and since complex zeros always come in conjugate pairs, if the degree of the polynomial is **odd**, there is guaranteed to be **at least one zero that is a real number**. If the degree of the polynomial is even, there is no guarantee that a zero will be real—all the zeros could be complex.

EXAMPLE 7 Find Possible Combinations of Real and Imaginary Zeros

List the possible combinations of real and imaginary zeros for the given polynomials.

a. $17x^5 + 2x^4 - 3x^3 + x^2 - 5$ **b.** $5x^4 + 2x^3 - x + 2$

Solution:

a. Since this is a *fifth*-degree polynomial, there are *five* zeros. Because complex zeros come in conjugate pairs, the table describes the possible five zeros.

Real Zeros	Imaginary Zeros
1	4
3	2
5	0

b. Because this is a *fourth*-degree polynomial, there are *four* zeros. Since complex zeros come in conjugate pairs, the table describes the possible four zeros.

Real Zeros	Imaginary Zeros
0	4
2	2
4	0

■ **YOUR TURN** List the possible combinations of real and imaginary zeros for

$$P(x) = x^6 - 7x^5 + 8x^3 - 2x + 1$$

The Rational Zero Theorem and Descartes's Rule of Signs

When the coefficients of a polynomial are integers, then the *rational zero theorem* gives us a list of possible rational roots. We can use trial and error to test them all to find the actual zeros. When the coefficients of a polynomial are real, *Descartes's rule of signs* tells us the possible combinations of *positive* real zeros and *negative* real zeros. Using

■ **Answer:**

Real Zeros	Imaginary Zeros
6	0
4	2
2	4
0	6

Descartes's rule of signs will assist us in narrowing down the large list of possible zeros generated through the rational zero theorem to a (hopefully) shorter list of possible zeros.

> **THE RATIONAL ZERO THEOREM**
>
> If the polynomial function $P(x) = a_n x^n + a_{n-1} x^{n-1} + \ldots + a_2 x^2 + a_1 x + a_0$ has *integer* coefficients, then every rational zero of $P(x)$ has the form:
>
> $$\text{Rational zero} = \frac{\text{factors of } a_0}{\text{factors of } a_n} = \frac{\text{factors of constant term}}{\text{factors of leading coefficient}}$$

To use this theorem, simply list all combinations of factors of both the constant term, a_0, and the leading coefficient term, a_n, and take all combinations of ratios. This procedure is illustrated in Example 8. Notice that when the leading coefficient is 1, then the possible rational zeros will simply be the possible factors of the constant term.

EXAMPLE 8 Using the Rational Zero Theorem

Use the rational zero theorem to determine possible rational zeros for the polynomial $P(x) = x^4 - x^3 - 5x^2 - x - 6$. Test each one to find all rational zeros.

Solution:

STEP 1 List factors of the constant
and leading coefficient terms. $a_0 = -6 \qquad \pm 1, \pm 2, \pm 3, \pm 6$

$a_n = 1 \qquad \pm 1$

STEP 2 List possible rational zeros $\dfrac{a_0}{a_n}$. $\pm 1, \pm 2, \pm 3, \pm 6$

There are three ways to test if any of these are zeros: Substitute these values into the polynomial to see which ones yield zero or use either polynomial division or synthetic division to divide the polynomial by these zeros and look for a zero remainder.

STEP 3 Test possible zeros by looking for zero remainders.

1 is not a zero: $P(1) = (1)^4 - (1)^3 - 5(1)^2 - (1) - 6 = -12$

-1 is not a zero: $P(-1) = (-1)^4 - (-1)^3 - 5(-1)^2 - (-1) - 6 = -8$

We could continue testing with direct substitution but let us now use synthetic division as an alternative.

2 is not a zero:

$$
\begin{array}{r|rrrr}
2 & 1 & -1 & -5 & -1 & -6 \\
  &   & 2 & 2 & -6 & -14 \\
\hline
  & 1 & 1 & -3 & -7 & \boxed{-20}
\end{array}
$$

-2 *is* a zero:

$$
\begin{array}{r|rrrr}
-2 & 1 & -1 & -5 & -1 & -6 \\
   &   & -2 & 6 & -2 & 6 \\
\hline
   & 1 & -3 & 1 & -3 & \boxed{0}
\end{array}
$$

Since -2 is a zero, then $x + 2$ is a factor of $P(x)$, and the remaining quotient is $x^3 - 3x^2 + x - 3$. Therefore, if there are any other real roots remaining, we can now use the simpler $x^3 - 3x^2 + x - 3$ for the dividend.

3 is a zero:

$$
\begin{array}{r|rrrr}
3 & 1 & -3 & 1 & -3 \\
  &   & 3 & 0 & 3 \\
\hline
  & 1 & 0 & 1 & \boxed{0}
\end{array}
$$

We now know that -2 and 3 are confirmed zeros. If we continue testing, we will find that the other possible zeros fail. This is a fourth-degree polynomial, and we have found two rational real zeros. Therefore, the other two zeros must be either irrational real zeros or a complex conjugate pair.

YOUR TURN Determine the possible rational zeros of the polynomial $P(x) = x^4 + 2x^3 - 2x^2 + 2x - 3$, and determine all real zeros.

For polynomials with coefficients that are real numbers, there is a test—Descartes's rule of signs—that determines the possible combinations of positive real zeros and negative real zeros through variations of sign. A *variation in sign* is when consecutive coefficients show a sign change.

Sign Change
− to +

$$P(x) = 2x^6 - 5x^5 - 3x^4 + 2x^3 - x^2 - x - 1$$

Sign Change   **Sign Change**
+ to −   + to −

This polynomial experiences three sign changes or variations in sign.

DESCARTES'S RULE OF SIGNS

If the polynomial function $P(x) = a_n x^n + a_{n-1} x^{n-1} + \ldots + a_2 x^2 + a_1 x + a_0$ has real coefficients and $a_0 \neq 0$, then:

- The number of **positive** real zeros of the polynomial is either equal to the number of variations of sign of $P(x)$ or less than that number by an even integer.
- The number of **negative** real zeros of the polynomial is either equal to the number of variations of sign of $P(-x)$ or less than that number by an even integer.

EXAMPLE 9 Using Descartes's Rule of Signs

Determine the possible combinations of zeros for $P(x) = x^3 - 2x^2 - 5x + 6$.

Solution:

STEP 1 Determine the number of variations of sign in $P(x)$.

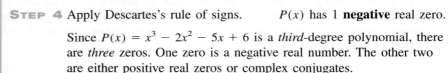

$P(x)$ has **2** variations in sign:

STEP 2 Apply Descartes's rule of signs. $P(x)$ has *either* 2 or 0 **positive** real zeros.

STEP 3 Determine the number of variations of sign in $P(-x)$.

$(-x)^3 - 2(-x)^2 - 5(-x) + 6$

$-x^3 - 2x^2 + 5x + 6$

$P(-x)$ has **1** variation in sign.

$$P(x) = -x^3 - 2x^2 + 5x + 6$$

STEP 4 Apply Descartes's rule of signs. $P(x)$ has 1 **negative** real zero.

Since $P(x) = x^3 - 2x^2 - 5x + 6$ is a *third*-degree polynomial, there are *three* zeros. One zero is a negative real number. The other two are either positive real zeros or complex conjugates.

STEP 5 Illustrate possible zero combinations with a table.

Positive Real Zeros	Negative Real Zeros	Imaginary Zeros
0	1	2
2	1	0

Check: Now look back at Example 2 and see that in fact there was 1 negative real zero and 2 positive real zeros.

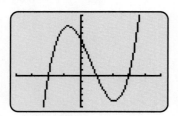

TECHNOLOGY TIP
Graph $P(x) = x^3 - 2x^2 - 5x + 6$.

There is 1 negative real zero and 2 positive real zeros.

EXAMPLE 10 Using Descartes's Rule of Signs to Find Possible Combinations of Zeros

Determine the possible combinations of zeros for $P(x) = x^4 - 2x^3 + x^2 + 2x - 2$.

Solution:

STEP 1 $P(x)$ has **3** variations in sign.

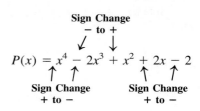

STEP 2 Apply Descartes's
rule of signs. $P(x)$ has *either* **3** or **1 positive** real zeros.

STEP 3 Find $P(-x)$. $(-x)^4 - 2(-x)^3 + (-x)^2 + 2(-x) - 2$
                                   $x^4 + 2x^3 + x^2 - 2x - 2$

STEP 4 $P(-x)$ has **1** variation
in sign. $P(-x) = x^4 + 2x^3 + x^2 - 2x - 2$

Sign Change
+ to −

STEP 5 Apply Descartes's rule of signs. $P(x)$ has **1 negative** real zero.

Since $P(x) = x^4 - 2x^3 + x^2 + 2x - 2$ is a *fourth*-degree polynomial,
there are *four* zeros. One zero is a negative real number. The other
three are either positive real zeros, or one positive real zero and two
complex conjugate zeros.

STEP 6 Summarize possible zero combinations in a table.

Positive Real Zeros	Negative Real Zeros	Imaginary Zeros
1	1	2
3	1	0

TECHNOLOGY TIP

Graph $P(x) = x^4 - 2x^3 + x^2 + 2x - 2$.

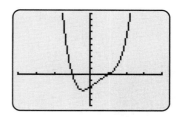

There is 1 negative real zero and
1 positive real zero. So, $P(x)$ has
2 complex conjugate zeros.

CONCEPT CHECK How many zeros does the polynomial function
$P(x) = x^4 + 2x^3 + x^2 + 8x - 12$ have?

■ **YOUR TURN** Determine the possible combinations of zeros for
$P(x) = x^4 + 2x^3 + x^2 + 8x - 12$.

Factoring Polynomials

Now let's draw on the tests discussed in this section to help us in finding all zeros of a
polynomial. Doing so will enable us to factor polynomials.

EXAMPLE 11 Factoring a Polynomial

Factor the polynomial $P(x) = x^5 + 2x^4 - x - 2$.

Solution:

STEP 1 Determine variations in sign.

$P(x)$ has 1 sign change. $P(x) = x^5 + 2x^4 - x - 2$

$P(-x)$ has 2 sign changes. $P(-x) = -x^5 + 2x^4 + x - 2$

Positive Real Zeros	Negative Real Zeros	Imaginary Zeros
1	1	2
1	3	0

■ **Answer:**

STEP 2 Apply Descartes's rule of signs and summarize in a table.

Positive Real Zeros	Negative Real Zeros	Imaginary Zeros
1	2	2
1	0	4

STEP 3 Use the rational zero theorem to
determine the possible rational zeros. $\pm 1, \pm 2$

We know (see table) there is one positive real zero, so let's test the positive possible rational zeros first.

STEP 4 Test possible rational zeros.

1 is a zero: $\qquad P(1) = (1)^5 + 2(1)^4 - (1) - 2 = 0$

Now that we have found *the* positive zero, we can test the other two negative zeros—because either they both are zeros or neither is a zero.

-1 is a zero: $\qquad P(-1) = (-1)^5 + 2(-1)^4 - (-1) - 2 = 0$

At this point, from Descartes's rule of signs we know that -2 must also be a zero, since there are either 2 or 0 negative zeros. Let's confirm this:

-2 is a zero: $\qquad P(-2) = (-2)^5 + 2(-2)^4 - (-2) - 2$
$$= -32 + 32 + 2 - 2 = 0$$

STEP 5 Three of the five zeros have been found: $-1, -2$, and 1.

STEP 6 Write the fifth-degree polynomial as a product of five linear factors:

$$P(x) = x^5 + 2x^4 - x - 2 = (x - 1)(x + 1)(x + 2)(x - c)(x - d)$$

We know that the remaining two zeros are imaginary according to Descartes's rule of signs and complex roots come in conjugate pairs.

STEP 7 Multiply the three known linear factors.

$$(x - 1)(x + 1)(x + 2) = x^3 + 2x^2 - x - 2$$

STEP 8 Rewrite the polynomial.

$$P(x) = x^5 + 2x^4 - x - 2 = (x^3 + 2x^2 - x - 2)(x - c)(x - d)$$

STEP 9 Divide both sides of the equation by $x^3 + 2x^2 - x - 2$.

$$\frac{x^5 + 2x^4 - x - 2}{x^3 + 2x^2 - x - 2} = (x - c)(x - d)$$

STEP 10 Dividing the left side of the equation using long division yields:

$$\frac{x^5 + 2x^4 - x - 2}{x^3 + 2x^2 - x - 2} = x^2 + 1 = (x - i)(x + i)$$

STEP 11 Write the remaining two zeros. $\qquad x = \pm i$

STEP 12 Write $P(x)$ as a product of linear factors.

$$P(x) = (x - 1)(x + 1)(x + 2)(x - i)(x + i)$$

Check: $P(x)$ is a *fifth*-degree polynomial so we expect *five* zeros.

The rational zero theorem gives us possible rational zeros of a polynomial, and Descartes's rule of signs gives us possible combinations of positive and negative real zeros. An additional aid that helps eliminate possible zeros are the *upper and lower bound rules*. These rules can give you an upper and lower bound on the real zeros of a function.

UPPER AND LOWER BOUND RULES

Let $f(x)$ be a polynomial with real coefficients and a positive leading coefficient. Suppose $f(x)$ is divided by $x - c$ using synthetic division.

1. If $c > 0$ and each number in the last row is either positive or zero, c is an **upper bound** for the real zeros of f.

2. If $c < 0$ and the numbers in the last row are alternately positive and negative (zero entries count as either positive or negative), c is a **lower bound** for the real zeros of f.

EXAMPLE 12 Using Upper and Lower Bounds to Eliminate Possible Zeros

Find the real zeros of $f(x) = 4x^3 - x^2 + 36x - 9$.

Solution:

STEP 1 Rational zero theorem gives possible rational zeros.

$$\frac{\text{Factors of } 9}{\text{Factors of } 4} = \frac{\pm 1, \pm 3, \pm 9}{\pm 1, \pm 2, \pm 4}$$

$$= \pm 1, \pm \frac{1}{2}, \pm \frac{1}{4}, \pm \frac{3}{2}, \pm \frac{3}{4}, \pm \frac{9}{2}, \pm \frac{9}{4}, \pm 3, \pm 9$$

STEP 2 Descartes's rule of signs:

$f(x)$ has 3 sign variations.   3 or 1 positive real zeros

$f(-x)$ has no sign variations.   no negative real zeros

STEP 3 Try $x = 1$ by dividing by $x - 1$.

$$\begin{array}{r|rrrr} 1 & 4 & -1 & 36 & -9 \\ & & 4 & 3 & 39 \\ \hline & 4 & 3 & 39 & 30 \end{array}$$

$x = 1$ is not a zero, but because the last row contains all positive entries, $x = 1$ is an *upper* bound.

Since we know there are no negative real zeros, we restrict our search to between 0 and 1.

STEP 4 Try $x = \dfrac{1}{4}$.

$$\begin{array}{r|rrrr} \frac{1}{4} & 4 & -1 & 36 & -9 \\ & & 1 & 0 & 9 \\ \hline & 4 & 0 & 36 & 0 \end{array}$$

$x = \frac{1}{4}$ is a zero, and the quotient is $4x^2 + 36$ which factors to two imaginary zeros. Thus $x = \frac{1}{4}$ is the only real zero.

Note: If $f(x)$ has a common monomial factor, it should be factored out first before applying the bounds rules.

STUDY TIP

If $f(x)$ has a common monomial factor, it should be factored out first before applying the bounds rules.

SECTION 4.4 SUMMARY

A polynomial function, $P(x)$, of degree n with real coefficients has the following properties:

- $P(x)$ has at least one zero (provided degree > 0) and no more than n zeros.
- If $a + bi$ is a zero, then $a - bi$ is also a zero.
- The rational zero theorem gives possible rational zeros as ratios:
  - (Factors of a_0)/(Factors of a_n)
- Descartes's rule of signs leads to the following "tests":
  - The number of sign changes in $P(x)$ is related to the number of positive real zeros.
  - The number of sign changes in $P(-x)$ is related to the number of negative real zeros.
- Upper and lower bounds rules eliminate possible zeros.

These rules and tests can be used to find zeros of polynomials. Once some zeros have been found, synthetic division can be used to divide the polynomial by known linear factors and the remaining quotient can be analyzed similarly.

SECTION 4.4 EXERCISES

■ SKILLS

Find the following values by using synthetic division. Check by substituting the value into the function.

$$f(x) = 3x^4 - 2x^3 + 7x^2 - 8 \qquad g(x) = 2x^3 + x^2 + 1$$

1. $f(1)$ **2.** $f(-1)$ **3.** $g(1)$ **4.** $g(-1)$ **5.** $f(-2)$ **6.** $g(2)$

Determine whether the number given is a zero of the polynomial.

7. -7, $P(x) = x^3 + 2x^2 - 29x + 42$ **8.** 2, $P(x) = x^3 + 2x^2 - 29x + 42$

9. -3, $P(x) = x^3 - x^2 - 8x + 12$ **10.** 1, $P(x) = x^3 - x^2 - 8x + 12$

Given a zero of the polynomial, determine all other zeros and write the polynomial in terms of a product of linear factors.

POLYNOMIAL	ZERO	POLYNOMIAL	ZERO
11. $P(x) = x^3 - 13x + 12$	1	**12.** $P(x) = x^3 + 3x^2 - 10x - 24$	3
13. $P(x) = 2x^3 + x^2 - 13x + 6$	$\dfrac{1}{2}$	**14.** $P(x) = 3x^3 - 14x^2 + 7x + 4$	$-\dfrac{1}{3}$
15. $P(x) = x^4 - 2x^3 - 11x^2 - 8x - 60$	$-2i$	**16.** $P(x) = x^4 - x^3 + 7x^2 - 9x - 18$	$3i$
17. $P(x) = x^4 - 9x^2 + 18x - 14$	$1 + i$	**18.** $P(x) = x^4 - 4x^3 + x^2 + 6x - 40$	$1 - 2i$
19. $P(x) = x^4 + 6x^3 + 13x^2 + 12x + 4$	-2 (multiplicity 2)	**20.** $P(x) = x^4 + 4x^3 - 2x^2 - 12x + 9$	1 (multiplicity 2)

Use Descartes's rule of signs to determine the possible number of positive real zeros, negative real zeros, and imaginary zeros.

21. $P(x) = x^4 - 32$ **22.** $P(x) = x^4 + 32$ **23.** $P(x) = x^5 - 1$ **24.** $P(x) = x^5 + 1$

25. $P(x) = x^5 - 3x^3 - x + 2$ **26.** $P(x) = x^4 + 2x^2 - 9$ **27.** $P(x) = 9x^7 + 2x^5 - x^3 - x$

28. $P(x) = 16x^7 - 3x^4 + 2x - 1$ **29.** $P(x) = x^6 - 16x^4 + 2x^2 + 7$ **30.** $P(x) = -7x^6 - 5x^4 - x^2 + 2x + 1$

31. $P(x) = -3x^4 + 2x^3 - 4x^2 + x - 11$ **32.** $P(x) = 2x^4 - 3x^3 + 7x^2 + 3x + 2$

Use the rational zero theorem to list the *possible* rational zeros.

33. $P(x) = x^4 + 3x^2 - 8x + 4$ **34.** $P(x) = -x^4 + 2x^3 - 5x + 4$ **35.** $P(x) = x^5 - 14x^3 + x^2 - 15x + 12$

36. $P(x) = x^5 - x^3 - x^2 + 4x + 9$ **37.** $P(x) = 2x^6 - 7x^4 + x^3 - 2x + 8$ **38.** $P(x) = 3x^5 + 2x^4 - 5x^3 + x - 10$

39. $P(x) = 5x^5 + 3x^4 + x^3 - x - 20$ **40.** $P(x) = 4x^6 - 7x^4 + 4x^3 + x - 21$

List the possible rational zeros, and test to determine all rational zeros.

41. $P(x) = x^4 + 2x^3 - 9x^2 - 2x + 8$ **42.** $P(x) = x^4 + 2x^3 - 4x^2 - 2x + 3$

43. $P(x) = 2x^3 - 9x^2 + 10x - 3$ **44.** $P(x) = 3x^3 - 5x^2 - 26x - 8$

For each polynomial: (a) use Descartes's rule of signs to determine the possible combinations of positive real zeros, negative real zeros, and imaginary zeros; (b) use the rational zero test to determine possible rational zeros; (c) test for rational zeros; and (d) factor as a product of linear factors.

45. $P(x) = x^3 + 6x^2 + 11x + 6$ **46.** $P(x) = x^3 - 6x^2 + 11x - 6$ **47.** $P(x) = x^3 - 7x^2 - x + 7$

48. $P(x) = x^3 - 5x^2 - 4x + 20$ **49.** $P(x) = x^4 + 6x^3 + 3x^2 - 10x$ **50.** $P(x) = x^4 - x^3 - 14x^2 + 24x$

51. $P(x) = x^4 - 7x^3 + 27x^2 - 47x + 26$ **52.** $P(x) = x^4 - 5x^3 + 5x^2 + 25x - 26$ **53.** $P(x) = 10x^3 - 7x^2 - 4x + 1$

54. $P(x) = 12x^3 - 13x^2 + 2x - 1$ **55.** $P(x) = 6x^3 + 17x^2 + x - 10$ **56.** $P(x) = 6x^3 + x^2 - 5x - 2$

57. $P(x) = x^4 - 2x^3 + 5x^2 - 8x + 4$ **58.** $P(x) = x^4 + 2x^3 + 10x^2 + 18x + 9$

59. $P(x) = x^6 + 12x^4 + 23x^2 - 36$ **60.** $P(x) = x^4 - x^2 - 16x^2 + 16$

61. $P(x) = 4x^4 - 20x^3 + 37x^2 - 24x + 5$ **62.** $P(x) = 4x^4 - 8x^3 + 7x^2 + 30x + 50$

In Exercises 63–66, use the information found in the above Exercises 47, 51, 55, and 61 to assist in generating a plot for the polynomial.

63. Plot the polynomial in Exercise 47.   **64.** Plot the polynomial in Exercise 51.

65. Plot the polynomial in Exercise 55.   **66.** Plot the polynomial in Exercise 61.

 ■ **APPLICATIONS**

67. Profit. A mathematics honorary society wants to sell magazine subscriptions to *Math Weekly*. If there are x hundred subscribers, its monthly revenue and cost are given by:

$$R(x) = 46 - 3x^2 \quad \text{and} \quad C(x) = 20 + 2x$$

a. Determine the profit function. *Hint: P = R - C.*

b. Determine the number of subscribers needed in order to *break even.*

68. Profit. Using the profit equation $P(x) = x^3 - 5x^2 + 3x + 6$, when will the company break even if x represents the units sold?

69. Sports. The University of Central Florida Golden Knights football team develops a play to run when they receive a kickoff with seconds left in the game. Once a receiver catches the kick he runs forward until he is about to get tackled, and then he tosses a lateral (or backwards) pass to a fellow teammate, and they would run until almost tackled and then pass to another, etc. Assume the 50 yard line at midfield is the x-axis. Since the field (from end zone to end zone) is 100 yards long and approximately 53 yards wide, the domain is $[-26.5, 26.5]$ and the range is $[-50, 50]$. Some points that lie along the path of the play are $(-25, -35)$, $(-17, -41)$, $(-6, -13)$, $(0, -25)$, $(5, -5)$, $(10, 0)$, $(25, -10)$, $(25.3, 0)$, $(26.5, 11)$.

What is the lowest degree polynomial that can represent this play?

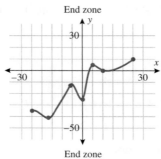

End zone

End zone

70. Sports. A quarterback tosses the ball to his wide receiver. At that point he takes off with the ball to win the game. The path of the wide receiver is given by the following polynomial.

$$y = x^5 - 10x^3 + 9x$$

Graph the polynomial and indicate the zeros if any exist.

■ CATCH THE MISTAKE

In Exercises 71–74, explain the mistake that is made.

71. Given that 1 is a zero of $P(x) = x^3 - 2x^2 - 7x + 6$, find all other zeros.

Step 1: $P(x)$ is a third-degree polynomial, so we expect three zeros.

Step 2: Because 1 is a zero, -1 is a zero, so two linear factors are $(x - 1)$ and $(x + 1)$.

Step 3: Write the polynomial as a product of three linear factors.

$$P(x) = (x - 1)(x + 1)(x - c)$$
$$P(x) = (x^2 - 1)(x - c)$$

Step 4: To find the remaining linear factor, we divide $P(x)$ by $x^2 - 1$.

$$\frac{x^3 - 2x^2 - 7x + 6}{x^2 - 1} = x - 2 - \frac{6x + 8}{x^2 - 1}$$

Which has a nonzero remainder? What went wrong?

72. Factor the polynomial $P(x) = 2x^3 + x^2 + 2x + 1$.

Step 1: Since $P(x)$ is an odd-degree polynomial, we are guaranteed one real zero (since complex zeros come in conjugate pairs).

Step 2: Use the rational zero test to develop a list of potential rational zeros.

Possible zeros: ± 1

Step 3: Test possible zeros.

1 is not a zero: $P(x) = 2(1)^3 + (1)^2 + 2(1) + 1 = 6$

-1 is not a zero: $P(x) = 2(-1)^3 + (-1)^2 + 2(-1) + 1 = -2$

Note: $-\dfrac{1}{2}$ is the real zero. Why did we not find it?

73. Use Descartes's rule of signs to determine the possible combinations of zeros of

$$P(x) = 2x^5 + 7x^4 + 9x^3 + 9x^2 + 7x + 2$$

No sign changes, so no positive real zeros.
$P(x) = 2x^5 + 7x^4 + 9x^3 + 9x^2 + 7x + 2$

Five sign changes, so five negative real zeros.
$P(-x) = -2x^5 + 7x^4 - 9x^3 + 9x^2 - 7x + 2$

74. Determine if $x - 2$ is a factor of $P(x) = x^3 - 2x^2 - 5x + 6$

Solution:

$$
\begin{array}{r|rrrr}
-2 & 1 & -2 & -5 & 6 \\
   &   & -2 & 8 & -6 \\
\hline
   & 1 & -4 & 3 & \boxed{0}
\end{array}
$$

Yes, $x - 2$ is a factor of $P(x)$.

■ CHALLENGE

75. T or F: All zeros of a polynomial correspond to x-intercepts.

76. T or F: A polynomial of degree n, $n > 0$, must have at least one zero.

77. T or F: A polynomial of degree n, $n > 0$, can be written as a product of n linear factors.

78. T or F: The number of sign changes in a polynomial is equal to the number of positive real zeros of that polynomial.

79. Is it possible for an odd-degree polynomial to have all imaginary zeros? Explain.

80. Is it possible for an even-degree polynomial to have all imaginary zeros? Explain.

81. Is it possible for a polynomial to have only one zero that is imaginary? Explain.

82. Given that $x - a$ is a zero of $P(x) = x^3 - (a + b + c)x^2 + (ab + ac + bc)x - abc$, find the other two zeros.

■ TECHNOLOGY

Determine all possible rational zeros of the polynomial. There are many possibilities. Instead of trying them all, use a graphing calculator or software to plot $P(x)$ to help find a zero to test.

83. $P(x) = x^3 - 2x^2 + 16x - 32$

84. $P(x) = x^3 - 3x^2 + 16x - 48$

Determine possible combinations of real and imaginary zeros. Use a graphing calculator or software to plot $P(x)$ and identify any real zeros. Does this agree with your list?

85. $P(x) = x^4 + 13x^2 + 36$

86. $P(x) = x^6 + 2x^4 + 7x^2 - 130x - 288$

SECTION **4.5** Rational Functions

Skills Objectives

■ Find the domain of a rational function.
■ Determine vertical, horizontal, and oblique asymptotes of rational functions.
■ Graph rational functions.

Conceptual Objectives

■ Understand arrow notation.
■ Interpret the behavior of a rational function approaching an asymptote.

In this chapter polynomial functions have been discussed thus far. We now turn our attention to *rational functions*, which are *ratios* of polynomial functions. Short-term memory and drug absorption in the bloodstream are two applications when rational functions are used. Ratios of integers are called *rational numbers*. Similarly, ratios of polynomial functions are called *rational functions*.

> **DEFINITION   RATIONAL FUNCTION**
>
> A function $f(x)$ is a **rational function** if
>
> $$f(x) = \frac{n(x)}{d(x)} \qquad d(x) \neq 0$$
>
> where the numerator, $n(x)$, and the denominator, $d(x)$, are polynomial functions. The domain of $f(x)$ is the set of all real numbers x such that $d(x) \neq 0$.

Domain of Rational Functions

The domain of any polynomial function is the set of all real numbers. When we divide two polynomial functions, the result is a *rational function*, and we must exclude any values of x that make the denominator equal to zero.

EXAMPLE 1 Finding the Domain of a Rational Function with Restrictions

Find the domain of the rational function $f(x) = \dfrac{x+1}{x^2 - x - 6}$. Express the domain in interval notation.

Solution:

Set the denominator equal to zero.	$x^2 - x - 6 = 0$
Factor.	$(x+2)(x-3) = 0$
Solve for x.	$x = -2 \quad$ or $x = 3$
Eliminate these values from the domain.	$x \neq -2 \quad$ or $x \neq 3$
State the domain in interval notation.	$(-\infty, -2) \cup (-2, 3) \cup (3, \infty)$

■ **YOUR TURN** 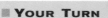 Find the domain of the rational function $f(x) = \dfrac{x-2}{x^2 - 3x - 4}$. Express the domain in interval notation.

It is important to note that there are not always restrictions on the domain. For example, if the denominator is never equal to zero, the domain is the set of all real numbers.

EXAMPLE 2 When the Domain of a Rational Function Is the Set of All Real Numbers

Find the domain of the rational function $g(x) = \dfrac{3x}{x^2 + 9}$. Express the domain in interval notation.

Solution:

Set the denominator equal to zero.	$x^2 + 9 = 0$
Subtract 9 from both sides.	$x^2 = -9$
Solve for x.	$x = -3i$ or $x = 3i$
There are no *real* solutions; therefore the domain has no restrictions.	$\mathbb{R}$, the set of all real numbers
State the domain in interval notation.	$(-\infty, \infty)$

■ **YOUR TURN** Find the domain of the rational function $g(x) = \dfrac{5x}{x^2 + 4}$. Express the domain in interval notation.

It is important to note that $f(x) = \dfrac{x^2 - 4}{x + 2}$ where $x \neq -2$ and $g(x) = x - 2$ are *not* the same function. Although $f(x)$ can be written in the factored form

---

■ **Answer:** The domain is the set of all real numbers such that $x \neq -1$ or $x \neq 4$.
Interval notation: $(-\infty, -1) \cup (-1, 4) \cup (4, \infty)$

■ **Answer:** The domain is the set of all real numbers.   Interval notation: $(-\infty, \infty)$

$$f(x) = \frac{(x-2)(x+2)}{x+2} = x - 2,$$ its domain is different. The domain of $g(x)$ is the set of all real numbers, whereas the domain of $f(x)$ is the set of all real numbers such that $x \neq -2$. If we were to plot $f(x)$ and $g(x)$, they would both look like the line $y = x - 2$. However, $f(x)$ would have a hole, or discontinuity, at the point $x = -2$.

Vertical, Horizontal, and Oblique Asymptotes

If a function is not defined at a point, then it is still useful to know how the function behaves near that point. Let's start with the simple rational function, $f(x) = \dfrac{1}{x}$. This function is defined everywhere except when $x = 0$. We can point plot this function:

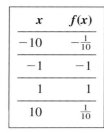

x	$f(x)$
-10	$-\frac{1}{10}$
-1	-1
1	1
10	$\frac{1}{10}$

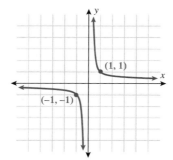

We cannot let $x = 0$ because that point is not in the domain of the function. We should, however, ask the question, "how does $f(x)$ behave as x *approaches* zero?" We use an *arrow* to represent the word *approach* and a *positive* superscript to represent from the *right*, and a *negative* superscript to represent from the *left*. A plot of this function can be generated using point-plotting techniques. The result is the graph above. The following are observations of the graph of $f(x) = \frac{1}{x}$.

WORDS	**MATH**	
As x approaches zero from the *right*, the function $f(x)$ increases without bound.	$x \to 0^+$ $\frac{1}{x} \to \infty$	
As x approaches zero from the *left*, the function $f(x)$ decreases without bound.	$x \to 0^-$ $\frac{1}{x} \to -\infty$	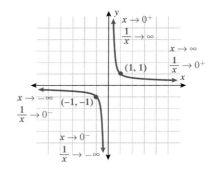
As x approaches infinity (increases without bound), the function $f(x)$ approaches zero from *above*.	$x \to \infty$ $\frac{1}{x} \to 0^+$	
As x approaches negative infinity (decreases without bound), the function $f(x)$ approaches zero from *below*.	$x \to -\infty$ $\frac{1}{x} \to 0^-$	

The symbol ∞ does not represent an actual real number. This symbol represents growing without bound.

1. Notice that the function is not defined at $x = 0$. The y-axis, or the vertical line $x = 0$, represents the *vertical asymptote*.
2. Notice that the value of the function is never equal to zero. The x-axis is never touched by the function. The x-axis, or $y = 0$, is a *horizontal asymptote*.

Asymptotes are lines that a function approaches but does not actually touch. Suppose a football team's defense is on their own 8 yard line and gets an "offsides" penalty that results in loss of "half the distance to the goal." Then the offense would get the ball on the 4 yard line. Suppose the defense gets another penalty on the next play that results in "half the distance to the goal." The offense would then get the ball on the 2 yard line. If the defense received 10 more penalties all resulting in "half the distance to the goal," would the referees *give* the offense a touchdown? No, because although the offense may appear to be snapping the ball from the goal line, technically they have not actually reached the goal line. Asymptotes utilize the same concept: the function approaches an asymptote but never actually touches it.

We will start with *vertical asymptotes.* Although the function $f(x) = \dfrac{1}{x}$ had one vertical asymptote, in general, rational functions can have *none, one, or several* vertical asymptotes. We will first formally define what a vertical asymptote is, and then we will discuss how to find it.

DEFINITION   VERTICAL ASYMPTOTE

The line $x = a$ is a **vertical asymptote** for the graph of a function if $f(x)$ either increases or decreases without bound as x approaches a from either the left or the right.

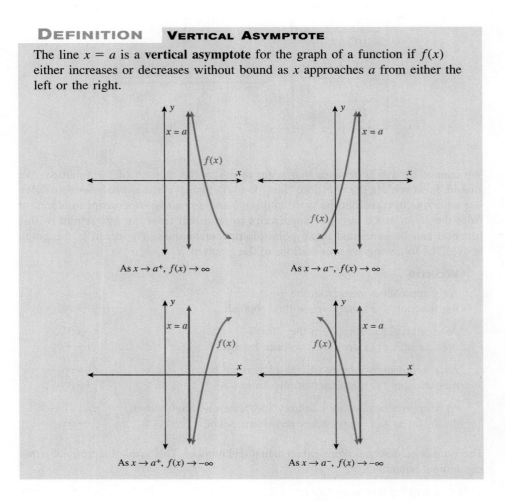

As $x \to a^+$, $f(x) \to \infty$ As $x \to a^-$, $f(x) \to \infty$

As $x \to a^+$, $f(x) \to -\infty$ As $x \to a^-$, $f(x) \to -\infty$

Vertical asymptotes assist us in graphing rational functions since they essentially "steer" the function in a direction. How do we locate the vertical asymptotes of a rational function? Set the denominator equal to zero. If the numerator and denominator have no common factors, then any numbers that are excluded from the domain of a rational function locate vertical asymptotes.

LOCATING VERTICAL ASYMPTOTES

Let f be a rational function defined by

$$f(x) = \frac{n(x)}{d(x)}$$

where $n(x)$ and $d(x)$ are polynomials with no common factors. If a is a real number such that $d(a) = 0$, then $x = a$ is a **vertical asymptote** of f.

EXAMPLE 3 Determining Vertical Asymptotes

Locate any vertical asymptotes of the rational function $f(x) = \dfrac{5x + 2}{6x^2 - x - 2}$.

Solution:

Factor the denominator.
$$f(x) = \frac{5x + 2}{(2x + 1)(3x - 2)}$$

The numerator and denominator have no common factors.

Set the denominator equal to zero. $2x + 1 = 0$ and $3x - 2 = 0$

Solve for x. $x = -\dfrac{1}{2}$ and $x = \dfrac{2}{3}$

The vertical asymptotes are $x = -\dfrac{1}{2}$ and $x = \dfrac{2}{3}$.

■ **YOUR TURN**   Locate any vertical asymptotes of the rational function:

$$f(x) = \frac{3x - 1}{2x^2 - x - 15}$$

EXAMPLE 4 Determining Vertical Asymptotes When the Rational Function Has Common Factors

Locate any vertical asymptotes of the rational function $f(x) = \dfrac{x + 2}{x^3 - 3x^2 - 10x}$.

Solution:

Factor the denominator.
$$x^3 - 3x^2 - 10x = x(x^2 - 3x - 10)$$
$$= x(x - 5)(x + 2)$$

Write the rational function in factored form.
$$f(x) = \frac{(x + 2)}{x(x - 5)(x + 2)}$$

Cancel the common factor $(x + 2)$.
$$f(x) = \frac{1}{x(x - 5)}$$

■ **Answer:** $x = -\dfrac{5}{2}$ and $x = 3$

Find the values when the denominator is equal to zero. $x = 0$ and $x = 5$

The vertical asymptotes are $x = 0$ and $x = 5$.

Note: $x = -2$ is not in the domain of $f(x)$, even though there is no vertical asymptote there.

▪ **YOUR TURN** Locate any vertical asymptotes of the rational function:

$$f(x) = \frac{x^2 - 4x}{x^2 - 7x + 12}$$

We now turn our attention to *horizontal asymptotes*. As we have seen, rational functions can have several vertical asymptotes. However, rational functions can have at most one horizontal asymptote. Horizontal asymptotes imply that a function approaches a constant value as x gets large in either the positive or negative direction. Another difference between vertical and horizontal asymptotes is that the graph of a function never touches a vertical asymptote but, as you will see in the next box, the graph of a function may cross a horizontal asymptote, just not at the ends $(x \to \pm\infty)$.

DEFINITION **HORIZONTAL ASYMPTOTE**

The line $y = b$ is a **horizontal asymptote** of the graph of a function if $f(x)$ approaches b as x increases or decreases without bound. The following are three examples of

$$As\ x \to \infty, f(x) \to b$$

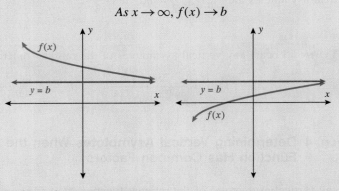

Note: A horizontal asymptote steers a function as x gets large. Therefore, when x is not large, the function may cross the asymptote.

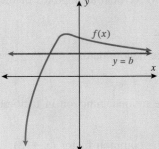

How do we determine if a horizontal asymptote exists? And, if so, how do we locate it? We find horizontal asymptotes by comparing the degree of the numerator to the degree of the denominator. There are three cases to consider:

1. Degree of the numerator is less than the degree of the denominator.
2. Degree of the numerator is equal to the degree of the denominator.
3. Degree of the numerator is greater than the degree of the denominator.

LOCATING HORIZONTAL ASYMPTOTES

Let f be a rational function given by

$$f(x) = \frac{n(x)}{d(x)} = \frac{a_n x^n + a_{n-1} x^{n-1} + \dots + a_1 x + a_0}{b_m x^m + b_{m-1} x^{m-1} + \dots + b_1 x + b_0}$$

where $n(x)$ and $d(x)$ are polynomials with no common factors.

1. When $n < m$, the x-axis $(y = 0)$ is the horizontal asymptote.
2. When $n = m$, the line $y = \frac{a_n}{b_m}$ (ratio of leading coefficients) is the horizontal asymptote.
3. When $n > m$, there is no horizontal asymptote.

In other words,

1. When the degree of the numerator is less than the degree of the denominator, then $y = 0$ is the horizontal asymptote.
2. When the degree of the numerator is the same as the degree of the denominator, then the horizontal asymptote is the ratio of the lead coefficients.
3. If the degree of the numerator is greater than the degree of the denominator, then there is no horizontal asymptote.

EXAMPLE 5 Finding Horizontal Asymptotes

Determine if a horizontal asymptote exists for each of the given rational functions. If so, locate the horizontal asymptote.

1. $f(x) = \dfrac{8x + 3}{4x^2 + 1}$ **2.** $g(x) = \dfrac{8x^2 + 3}{4x^2 + 1}$ **3.** $h(x) = \dfrac{8x^3 + 3}{4x^2 + 1}$

Solution (1):

The degree of the numerator, $8x + 3$, is one.	$n = 1$
The degree of the denominator, $4x^2 + 1$, is two.	$m = 2$
The degree of the numerator is less than the degree of the denominator.	$n < m$
The x-axis is the horizontal asymptote of $f(x)$.	$y = 0$

The line $\boxed{y = 0}$ is the horizontal asymptote of $f(x)$.

Solution (2):

The degree of the numerator, $8x^2 + 3$, is two.	$n = 2$
The degree of the denominator, $4x^2 + 1$, is two.	$m = 2$

TECHNOLOGY TIP

1. Graph of $f(x) = \frac{8x + 3}{4x^2 + 1}$.

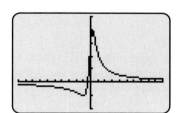

2. Graph of $f(x) = \frac{8x^2 + 3}{4x^2 + 1}$.

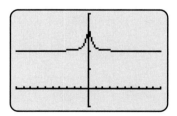

3. Graph of $f(x) = \frac{8x^3 + 3}{4x^2 + 1}$.

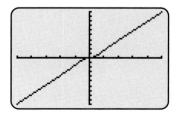

The degree of the numerator is equal to the degree of the denominator.

$n = m$

The ratio of the leading coefficients is the horizontal asymptote of $g(x)$.

$y = \dfrac{8}{4} = 2$

The line $y = 2$ is the horizontal asymptote of $g(x)$.

Solution (3):

The degree of the numerator, $8x^3 + 3$, is three.

$n = 3$

The degree of the denominator, $4x^2 + 1$, is two.

$m = 2$

The degree of the numerator is greater than the degree of the denominator.

$n > m$

The rational function $h(x)$ has   no horizontal asymptote .

■ **YOUR TURN**   Find the horizontal asymptote (if one exists) of the rational function $f(x) = \dfrac{7x^3 + x - 2}{-4x^3 + 1}$.

An asymptote is a line that a function approaches. There are three types of lines: horizontal (slope is zero), vertical (slope is undefined), and slanted (nonzero slope). Similarly, there are three types of asymptotes: horizontal, vertical, and *oblique*.

Recall when dividing polynomials that the degree of the quotient is always the difference between the degree of the numerator and the degree of the denominator. For example, a cubic (third-degree) polynomial divided by a quadratic (second-degree) polynomial results in a linear (first-degree) polynomial. A fifth-degree polynomial divided by a fourth-degree polynomial results in a first-degree (linear) polynomial. When the degree of the numerator is exactly one more than the degree of the denominator, the quotient is linear and represents the *oblique asymptote*.

STUDY TIP

There are three types of asymptotes: horizontal, vertical, and *oblique*.

OBLIQUE ASYMPTOTES

Let f be a rational function given by $f(x) = \dfrac{n(x)}{d(x)}$

where $n(x)$ and $d(x)$ are polynomials and the degree of $n(x)$ is *one more than* the degree of $d(x)$. On dividing $n(x)$ by $d(x)$, the rational function can be expressed as

$$f(x) = mx + b + \frac{r(x)}{d(x)}$$

where the degree of the remainder $r(x)$ is less than the degree of $d(x)$ and the line, $y = mx + b$, is an **oblique asymptote** for the graph of f.

Note that as $x \to -\infty$ or $x \to \infty$, $f(x) \to mx + b$.

■ **Answer:** $y = -\dfrac{7}{4}$ is the horizontal asymptote.

EXAMPLE 6 Finding Oblique Asymptotes

Determine the oblique asymptote of the rational function:

$$f(x) = \frac{4x^3 + x^2 + 3}{x^2 - x + 1}$$

Solution:

Divide the numerator by the denominator using long division.

$$
\begin{array}{r}
4x + 5 \\
x^2 - x + 1 \overline{\smash{\big)}\ 4x^3 + x^2 + 0x + 3} \\
-(4x^3 - 4x^2 + 4x) \\
\hline
5x^2 - 4x + 3 \\
-(5x^2 - 5x + 5) \\
\hline
x - 2
\end{array}
$$

The quotient is the oblique asymptote.

$$y = 4x + 5$$

YOUR TURN Find the oblique asymptote of the rational function:

$$f(x) = \frac{x^2 + 3x + 2}{x - 2}$$

Graphing Rational Functions

We can now graph rational functions using asymptotes as graphing aids. When we have graphed other functions, we have used intercepts, symmetry, and often a few points and then connected those points with a smooth curve. We will use those same techniques in addition to locating asymptotes to assist us in graphing rational functions. The following box summarizes the five-step procedure for graphing rational functions.

GRAPHING RATIONAL FUNCTIONS

Let f be a rational function given by $f(x) = \dfrac{n(x)}{d(x)}$.

Step 1: Find the **intercepts** (if there are any).
- y-intercept: evaluate $f(0)$
- x-intercept: solve the equation $n(x) = 0$

Step 2: Find the **asymptotes** (if there are any).
- vertical asymptotes: solve $d(x) = 0$
- compare the degree of the numerator to the degree of the denominator to determine if either a horizontal or oblique asymptote exists. If one exists, find it.

Step 3: Determine if the graph of f has **symmetry**.
- symmetry about y-axis: $f(-x) = f(x)$
- symmetry about the origin: $f(-x) = -f(x)$

Step 4: Find additional **points** on graph—particularly near asymptotes.

Step 5: **Draw** the intercepts, asymptotes, and additional points and use a smooth curve to complete the graph between and beyond the vertical asymptotes.

Answer: $y = x + 5$

EXAMPLE 7 Graphing a Rational Function with No Horizontal or Oblique Asymptotes

State the asymptotes (if there are any) and graph the rational function

$$f(x) = \frac{x^4 - x^3 - 6x^2}{x^2 - 1}$$

Solution:

STEP 1 Find the **intercepts**.

y-intercept: $\qquad\qquad\qquad f(0) = \dfrac{0}{-1} = 0$

x-intercepts: $\qquad\qquad\quad n(x) = x^4 - x^3 - 6x^2 = 0$

Factor. $\qquad\qquad\qquad x^2(x - 3)(x + 2) = 0$

Solve. $\qquad\qquad\qquad x = 0, \; x = 3, \text{ and } x = -2$

The **intercepts** are the points $(0, 0)$, $(3, 0)$, and $(-2, 0)$.

STEP 2 Find the **asymptotes**.

Vertical asymptote: $\qquad\qquad d(x) = x^2 - 1 = 0$

Factor. $\qquad\qquad\qquad (x + 1)(x - 1) = 0$

Solve. $\qquad\qquad\qquad x = -1 \text{ and } x = 1$

No horizontal asymptote: degree of $n(x) >$ degree of $d(x)$ $\;\;[4 > 2]$

No oblique asymptote: degree of $n(x) -$ degree of $d(x) > 1$ $\;\;[4 - 2 = 2 > 1]$

The **asymptotes** are $x = -1$ and $x = 1$.

STEP 3 No **symmetry** about
y-axis or origin. $\qquad\qquad\quad f(-x) \neq f(x)$
$\qquad\qquad\qquad\qquad\qquad f(-x) \neq -f(x)$

STEP 4 Find **additional points** on graph.

x	-3	-0.5	0.5	2	4
$f(x)$	6.75	1.75	2.08	-5.33	6.4

STEP 5 **Draw** a graph, label the **intercepts** and **asymptotes**, and complete with a smooth curve between and beyond the vertical asymptote.

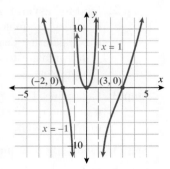

CONCEPT CHECK Does the function $f(x) = \dfrac{x^3 - 2x^2 - 3x}{x + 2}$ have a horizontal asymptote?

■ **YOUR TURN** State the asymptotes (if there are any) and graph the rational function $f(x) = \dfrac{x^3 - 2x^2 - 3x}{x + 2}$.

EXAMPLE 8 Graphing a Rational Function with a Horizontal Asymptote

State the asymptotes (if there are any) and graph the rational function:

$$f(x) = \frac{4x^3 + 10x^2 - 6x}{8 - x^3}$$

Solution:

STEP 1 Find the intercepts.

y-intercept: $\qquad\qquad\qquad\qquad f(0) = \dfrac{0}{8} = 0$

x-intercepts: $\qquad\qquad\qquad\quad n(x) = 4x^3 + 10x^2 - 6x = 0$

    Factor. $\qquad\qquad\qquad\qquad 2x(2x - 1)(x + 3) = 0$

    Solve. $\qquad\qquad\qquad\qquad\quad x = 0,\ x = \dfrac{1}{2},\ \text{and } x = -3$

The **intercepts** are the points $(0, 0)$, $\left(\dfrac{1}{2}, 0\right)$, and $(-3, 0)$.

STEP 2 Find the asymptotes.

    Vertical asymptote: $\qquad\qquad d(x) = 8 - x^3 = 0$

    Solve. $\qquad\qquad\qquad\qquad\quad x = 2$

    Horizontal asymptote: $\qquad$ degree of $n(x) = $ degree of $d(x)$

    Lead terms are $4x^3$ and $-x^3$. $\quad y = \dfrac{4}{-1} = -4$

The **asymptotes** are $x = 2$ and $y = -4$.

STEP 3 No symmetry about y-axis or origin.

$\qquad\qquad\qquad\qquad\qquad\qquad\qquad f(-x) \neq f(x)$
$\qquad\qquad\qquad\qquad\qquad\qquad\qquad f(-x) \neq -f(x)$

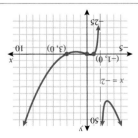

■ **Answer:** Vertical asymptote: $x = -2$. No horizontal or oblique asymptotes.

STEP 4 Find additional points on graph.

x	$f(x)$
-4	-1
-1	1.33
$\frac{1}{4}$	-0.10
1	1.14
3	-9.47

STEP 5 Draw a graph, label the intercepts and asymptotes, and complete with a smooth curve between and beyond the vertical asymptote.

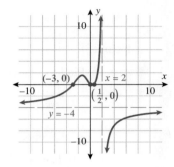

YOUR TURN Graph the rational function $f(x) = \dfrac{2x^2 - 7x + 6}{x^2 - 3x - 4}$. Give equations of the vertical and horizontal asymptotes and state intercepts.

EXAMPLE 9 Graphing a Rational Function with an Oblique Asymptote

Graph the rational function $f(x) = \dfrac{x^2 - 3x - 4}{x + 2}$.

STEP 1 Find the intercepts.

y-intercept: $\qquad\qquad\qquad f(0) = -\dfrac{4}{2} = -2$

x-intercepts: $\qquad\qquad\qquad n(x) = x^2 - 3x - 4 = 0$

Factor. $\qquad\qquad\qquad (x + 1)(x - 4) = 0$

Solve. $\qquad\qquad\qquad x = -1$ and $x = 4$

The **intercepts** are the points $(0, -2)$, $(-1, 0)$, and $(4, 0)$.

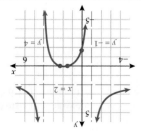

Intercepts: $\left(0, -\dfrac{3}{2}\right)$, $\left(\dfrac{3}{2}, 0\right)$, $(2, 0)$

Horizontal asymptote: $y = 2$

Answer: Vertical asymptotes: $x = 4$, $x = -1$

STEP 2 Find the asymptotes.

Vertical asymptote: $\qquad\qquad d(x) = x + 2 = 0$

Solve. $\qquad\qquad\qquad\qquad\qquad x = -2$

Oblique asymptote: $\qquad\qquad$ degree of $n(x)$ − degree of $d(x) = 1$

Divide $n(x)$ by $d(x)$. $\qquad\qquad f(x) = \dfrac{x^2 - 3x - 4}{x + 2} = x - 5 + \dfrac{6}{x + 2}$

Write the equation of the asymptote. $\quad y = x - 5$

The **asymptotes** are $x = -2$ and $y = x - 5$.

STEP 3 No symmetry about y-axis
or origin. $\qquad\qquad\qquad\qquad f(-x) \neq f(x)$
$\qquad\qquad\qquad\qquad\qquad\qquad f(-x) \neq -f(x)$

STEP 4 Find additional points on graph.

x	$f(x)$
-6	-12.5
-5	-12
-3	-14
5	0.86
6	1.75

STEP 5 Draw a graph, label the intercepts
and asymptotes, and complete with
a smooth curve between and beyond
the vertical asymptote.

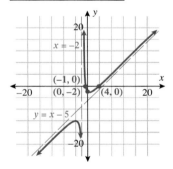

CONCEPT CHECK Does the function $f(x) = \dfrac{x^2 + x - 2}{x - 3}$ have a horizontal or oblique asymptote?

■ **YOUR TURN** For the function $f(x) = \dfrac{x^2 + x - 2}{x - 3}$, state the asymptotes (if any exist) and graph the function.

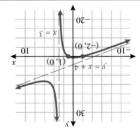

■ **Answer: Horizontal asymptote:** $x = 3$
Oblique asymptote: $y = x + 4$

SECTION 4.5 SUMMARY

In this section, characteristics of rational functions were discussed.

$$f(x) = \frac{n(x)}{d(x)}$$

- Domain: All real numbers except values that make the denominator equal to zero. $\left[d(x) \neq 0\right]$
- Vertical asymptotes: Vertical lines, $x = a$ where $d(a) = 0$, that the graph cannot touch and where the value of the function increases or decreases without bound.
- Horizontal asymptotes: Horizontal lines, $y = b$, that steer the graph as $x \to \pm\infty$.
- Oblique asymptotes: Slanted lines, $y = mx + b$, that steer the graph as $x \to \pm\infty$.

SECTION 4.5 EXERCISES

 ■ SKILLS

In Exercises 1–10, find the domain of each rational function.

1. $f(x) = \dfrac{1}{x + 3}$

2. $f(x) = \dfrac{3}{4 - x}$

3. $f(x) = \dfrac{2x + 1}{(3x + 1)(2x - 1)}$

4. $f(x) = \dfrac{5 - 3x}{(2 - 3x)(x - 7)}$

5. $f(x) = \dfrac{x + 4}{x^2 + x - 12}$

6. $f(x) = \dfrac{x - 1}{x^2 + 2x - 3}$

7. $f(x) = \dfrac{7x}{x^2 + 16}$

8. $f(x) = -\dfrac{2x}{x^2 + 9}$

9. $f(x) = -\dfrac{3(x^2 + x - 2)}{2(x^2 - x - 6)}$

10. $f(x) = \dfrac{5(x^2 - 2x - 3)}{(x^2 - x - 6)}$

In Exercises 11–20, find all vertical asymptotes and horizontal asymptotes (if there are any).

11. $f(x) = \dfrac{1}{x + 2}$

12. $f(x) = \dfrac{1}{5 - x}$

13. $f(x) = \dfrac{7x^3 + 1}{x + 5}$

14. $f(x) = \dfrac{2 - x^3}{2x - 7}$

15. $f(x) = \dfrac{6x^5 - 4x^2 + 5}{6x^2 + 5x - 4}$

16. $f(x) = \dfrac{6x^2 + 3x + 1}{3x^2 - 5x - 2}$

17. $f(x) = \dfrac{\frac{1}{3}x^2 + \frac{1}{3}x - \frac{1}{4}}{x^2 + \frac{1}{9}}$

18. $f(x) = \dfrac{\frac{1}{10}\left(x^2 - 2x + \frac{3}{10}\right)}{2x - 1}$

19. $f(x) = \dfrac{(0.2x - 3.1)(1.2x + 4.5)}{0.7(x - 0.5)(0.2x + 0.3)}$

20. $f(x) = \dfrac{0.8x^4 - 1}{x^2 - 0.25}$

In Exercises 21–26, find the oblique asymptote of each rational function.

21. $f(x) = \dfrac{x^2 + 10x + 25}{x + 4}$

22. $f(x) = \dfrac{x^2 + 9x + 20}{x - 3}$

23. $f(x) = \dfrac{2x^2 + 14x + 7}{x - 5}$

24. $f(x) = \dfrac{3x^3 + 4x^2 - 6x + 1}{x^2 - x - 30}$

25. $f(x) = \dfrac{8x^4 + 7x^3 + 2x - 5}{2x^3 - x^2 + 3x - 1}$

26. $f(x) = \dfrac{2x^6 + 1}{x^5 - 1}$

Match the function to the graph.

27. $f(x) = \dfrac{3}{x - 4}$

28. $f(x) = \dfrac{3x}{x - 4}$

29. $f(x) = \dfrac{3x^2}{x^2 - 4}$

30. $f(x) = -\dfrac{3x^2}{x^2 + 4}$

31. $f(x) = \dfrac{3x^2}{4 - x^2}$

32. $f(x) = \dfrac{3x^2}{x + 4}$

a.
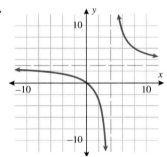

b.
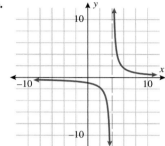

c.
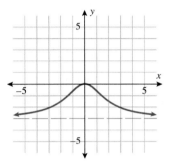

d.
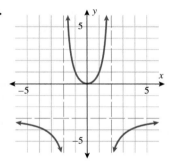

e.

f.
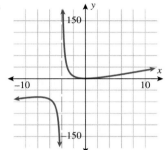

In Exercises 33–52, use the graphing strategy outlined in this section to graph the rational functions.

33. $f(x) = \dfrac{2}{x + 1}$

34. $f(x) = \dfrac{4}{x - 2}$

35. $f(x) = \dfrac{2x}{x - 1}$

36. $f(x) = \dfrac{4x}{x + 2}$

37. $f(x) = \dfrac{x - 1}{x}$

38. $f(x) = \dfrac{2 + x}{x - 1}$

39. $f(x) = \dfrac{2(x^2 - 2x - 3)}{x^2 + 2x}$

40. $f(x) = \dfrac{3(x^2 - 1)}{x^2 - 3x}$

41. $f(x) = \dfrac{x^2}{x + 1}$

42. $f(x) = \dfrac{x^2 - 9}{x + 2}$

43. $f(x) = \dfrac{2x^3 - x^2 - x}{x^2 - 4}$

44. $f(x) = \dfrac{3x^3 + 5x^2 - 2x}{x^2 + 4}$

45. $f(x) = \dfrac{x^2 + 1}{x^2 - 1}$

46. $f(x) = \dfrac{1 - x^2}{x^2 + 1}$

47. $f(x) = \dfrac{7x^2}{(2x + 1)^2}$

48. $f(x) = \dfrac{12x^4}{(3x + 1)^4}$

49. $f(x) = \dfrac{1 - 9x^2}{(1 - 4x^2)^3}$

50. $f(x) = \dfrac{25x^2 - 1}{(16x^2 - 1)^2}$

51. $f(x) = 3x + \dfrac{4}{x}$

52. $f(x) = x - \dfrac{4}{x}$

 ■ APPLICATIONS

53. Medicine. The concentration C of a particular drug in a person's bloodstream t minutes after injection is given by

$$C(t) = \frac{2t}{t^2 + 100}$$

a. What is the concentration in the bloodstream after 1 minute?

b. What is the concentration in the bloodstream after 1 hour?

c. What is the concentration in the bloodstream after 5 hours?

d. Find the horizontal asymptote of $C(t)$. What do you expect the concentration to be after several days?

54. Medicine. The concentration C of aspirin in the bloodstream t hours after consumption is given by $C(t) = \dfrac{t}{t^2 + 40}$.

a. What is the concentration in the bloodstream after a $\frac{1}{2}$ hour?

b. What is the concentration in the bloodstream after 1 hour?

c. What is the concentration in the bloodstream after 4 hours?

d. Find the horizontal asymptote for $C(t)$. What do you expect the concentration to be after several days?

55. Typing. An administrative assistant is hired on graduation from high school and learns to type on the job. The number of words he can type per minute is given by

$$N(t) = \frac{130t + 260}{t + 5} \quad t \geq 0$$

where t is the number of months he has been on the job.

a. How many words per minute can he type the day he starts?

b. How many words per minute can he type after 12 months?

c. How many words per minute can he type after 3 years?

d. How many words per minute would you expect him to type if he worked there until he retired?

56. Memorization. A professor teaching a large lecture course tries to learn students' names. The number of names she can remember, $N(t)$, increases with each week in the semester, t, and is given by the rational function:

$$N(t) = \frac{600t}{t + 20}$$

How many students' names does she know by the third week in the semester? How many students' names should she know by the end of the semester (16 weeks)? According to this function, what are the most names she can remember?

■ CATCH THE MISTAKE

In Exercises 61–64, explain the mistake that is made.

61. Determine the vertical asymptotes of the function $f(x) = \dfrac{x - 1}{x^2 - 1}$.

57. Food. The amount of food that cats typically eat increases as their weight increases. A rational function that describes this is $F(x) = \dfrac{10x^2}{x^2 + 4}$ where the amount of food, $F(x)$, is given in ounces and the weight of the cat, x, is given in pounds. Calculate the horizontal asymptote. How many ounces of food will most adult cats eat?

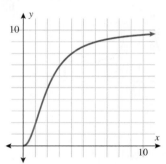

58. Memorization. The *Guinness Book of World Records, 2004* states that Dominic O'Brien (England) memorized on a single sighting a random sequence of 54 separate packs of cards all shuffled together (2808 cards in total) at Simpson's-In-The-Strand, London, England, on May 1, 2002. He memorized the cards in 11 hours 42 minutes, and then recited them in exact sequence in a time of 3 hours 30 minutes. With only a 0.5% margin of error allowed (no more than 14 errors) he broke the record with just 8 errors. If we let x represent the time (hours) it takes to memorize the cards and y represent the number of cards memorized then a rational function that models this event is given by $y = \dfrac{2800x^2 + x}{x^2 + 2}$. According to this model how many cards could be memorized in an hour? What is the greatest number of cards that can be memorized?

59. Gardening. A 500 square foot rectangular garden will be enclosed with fencing. Write a rational function that describes how many linear feet of fence will be needed to enclose the garden as a function of the width of the pen, w.

60. Geometry. A rectangular picture has an area of 414 square inches. A border (matting) is used when framing. If the top and bottom borders are each 4 inches and the side borders are 3.5 inches, write a function that represents the area, $A(l)$, of the entire frame as a function of the length of the picture, l.

Solution:

Set the denominator equal to zero. $x^2 - 1 = 0$

Solve for x. $\qquad\qquad x = \pm 1$

The vertical asymptotes are $x = -1$ and $x = 1$.

The following is a correct graph of the function. Note that only $x = -1$ is an asymptote. What went wrong?

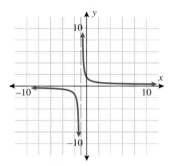

62. Determine the vertical asymptotes of $f(x) = \dfrac{2x}{x^2 + 1}$.

Solution:

Set the denominator equal to zero. $\qquad x^2 + 1 = 0$

Solve for x. $\qquad\qquad\qquad\qquad x = \pm 1$

The vertical asymptotes are $x = -1$ and $x = 1$.

The following is a correct graph of the function. Note that there are no vertical asymptotes. What went wrong?

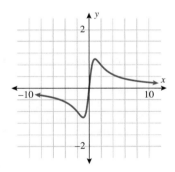

63. Determine if a horizontal or oblique asymptote exists for the function $f(x) = \dfrac{9 - x^2}{x^2 - 1}$. If so, find it.

Solution:

Step 1: The degree of the numerator equals the degree of the denominator so there is a horizontal asymptote.

65. T or F: A rational function can have either a horizontal asymptote or an oblique asymptote, but not both.

Step 2: The horizontal asymptote is the ratio of the lead coefficients: $y = \frac{9}{1} = 9$.

The horizontal asymptote is $y = 9$.

The following is a correct graph of the function. Note that there is not a horizontal asymptote at $y = 9$. What went wrong?

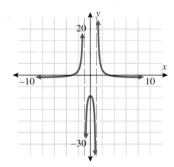

64. Determine if a horizontal or oblique asymptote exists for the function $f(x) = \dfrac{x^2 + 2x - 1}{3x^3 - 2x^2 - 1}$. If so, find it.

Solution:

Step 1: The degree of the denominator is exactly one more than the degree of the numerator, so there is an oblique asymptote.

Step 2: Divide.

$$
\begin{array}{r}
3x - 8 \\
x^2 + 2x - 1\overline{)3x^3 - 2x^2 + 0x - 1} \\
\underline{3x^3 + 6x^2 - 3x} \\
-8x^2 + 3x - 1 \\
\underline{-8x^2 - 16x + 8} \\
19x - 9
\end{array}
$$

The oblique asymptote is $y = 3x - 8$.

The following is the correct graph of the function. Note that $y = 3x - 8$ is not an asymptote. What went wrong?

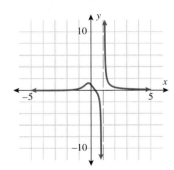

66. T or F: A rational function can have at most one vertical asymptote.

67. T or F: A rational function can cross a vertical asymptote.

68. T or F: A rational function can cross a horizontal or oblique asymptote.

69. Determine the asymptotes of the rational function
$$f(x) = \frac{(x-a)(x+b)}{(x-c)(x+d)}.$$

70. Determine the asymptotes of the rational function
$$f(x) = \frac{3x^2 + b^2}{x^2 + a^2}.$$

71. Write a rational function that has vertical asymptotes at $x = -3$ and $x = 1$ and a horizontal asymptote at $y = 4$.

72. Write a rational function that has no vertical asymptotes, the x-axis is the horizontal asymptote and has an x-intercept $(3, 0)$.

■ TECHNOLOGY

73. Determine the vertical asymptotes of $f(x) = \dfrac{x-4}{x^2 - 2x - 8}$.
Graph this function using a graphing utility. Does the graph confirm the asymptotes?

74. Determine the vertical asymptotes of $f(x) = \dfrac{2x+1}{6x^2 + x - 1}$.
Graph this function using a graphing utility. Does the graph confirm the asymptotes?

75. Find the asymptotes and intercepts of the rational function $f(x) = \dfrac{1}{3x+1} - \dfrac{2}{x}$. Graph this function using a graphing utility. Does the graph confirm what you found?

76. Find the asymptotes and intercepts of the rational function $f(x) = -\dfrac{1}{x^2 + 1} + \dfrac{1}{x}$. Graph this function using a graphing utility. Does the graph confirm what you found?

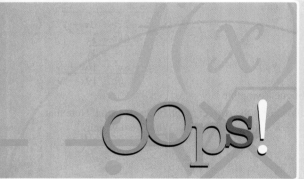

Q: I know that the braking distance (from the time the brakes are applied) for my car is 40 feet when traveling at 30 miles per hour. I thought that if I were traveling twice as fast, my braking distance should also double. This led me to believe that I could stop my car in 80 feet if traveling at 60 miles per hour. However, I experimented on a test track and found that this was not true. Why not?

A: Braking distance is actually a quadratic function of velocity, not a linear function. This means that if I double my speed, I must quadruple my stopping distance.

The function for calculating braking distance D, given the velocity v when the brakes are applied, is of the form

$$D(v) = av^2$$

where a is a constant.

If car A can stop from 30 miles per hour in 37 feet, car B can stop from 40 miles per hour in 71 feet, and car C can stop from 80 miles per hour in 254 feet, determine which car has the shortest braking distance at 50 miles per hour.

TYING IT ALL TOGETHER

Your company produces and sells computer games. The monthly fixed costs are $10,000. In addition, development costs for the game were $40,000 and the company spends $10 to produce each game. If the games are sold for $40 each, then the company sells 5000 per month. For each $1 increase in price, 100 fewer games are sold. At what price should the company sell the game to earn the maximum revenue? What is the maximum monthly revenue? At what price will consumers completely cease buying the game?

SECTION	TOPIC	PAGES	REVIEW EXERCISES	KEY CONCEPTS
4.1	Quadratic functions	264–276	1–28	$f(x) = ax^2 + bx + c$
	Graphs of quadratic functions	264–276	1–16	Parabola $f(x) = a(x - h)^2 + k$ Vertex: (h, k) Open up: $a > 0$ Open down: $a < 0$
4.2	Polynomial functions of higher degree	281–294	29–56	$P(x) = a_n x^n + a_{n-1} x^{n-1} + ... + a_2 x^2 + a_1 x + a_0$ is a polynomial of degree n.
	Graph polynomial functions using transformations	282–287	37–40	$y = x^n$ behaves similar to • $y = x^2$ when n is even • $y = x^3$ when n is odd
	Identify zeros and their multiplicity	287–290	41–50	$P(x) = (x - a)(x - b)^n = 0$ $x = a$ is a zero. $x = b$ is a zero of multiplicity n.
	Graphing polynomials	290–294	51–54	Zeros and end behavior
4.3	Dividing polynomials	299–305	57–70	Use zero placeholders for missing terms.
	Long division	299–303	57–60	Can be used for all polynomial division.
	Synthetic division	303–305	61–64	Can only be used when dividing by $(x \pm a)$.
4.4	Properties and tests of zeros of polynomial functions	308–322	71–100	• $P(x)$ of degree n $(n > 0)$ has at least one zero and at most n zeros. • If $a + bi$ is a zero of $P(x)$, then $a - bi$ is also a zero of $P(x)$.
	Remainder theorem	309–311	71–74	$P(x)$ is divided by $x - a$, then the remainder, r, is $r = P(a)$.
	Factor a polynomial	309–315	75–82	If $r = 0$, then $x - a$ is a factor of $P(x)$.
	Rational zero theorem	316	87–94	Possible zeros $= \dfrac{\text{factors of } a_0}{\text{factors of } a_n}$
	Descartes's rule of signs	317	83–86	Number of positive or negative real zeros is related to the number of sign variations in $P(x)$ or $P(-x)$
4.5	Rational functions	325–338	101–112	$f(x) = \dfrac{n(x)}{d(x)}$
	Vertical asymptotes	327–330	101–106	$x = a$ where $d(a) = 0$
	Horizontal asymptotes	330–332	101–106	degree of $n(x) <$ degree of $d(x)$ $y = 0$ degree of $n(x) >$ degree of $d(x)$ none degree of $n(x) =$ degree of $d(x)$ $\quad y = \dfrac{\text{leading coefficient of } n(x)}{\text{leading coefficient of } d(x)}$
	Oblique asymptotes	332–333	101–106	degree of $n(x) -$ degree of $d(x) = 1$ Divide $n(x)$ by $d(x)$ and the quotient is the oblique asymptote.
	Graphing rational functions	333–337	107–112	Determine intercepts, asymptotes, symmetry, and additional points and complete with smooth curve.

4.1 Quadratic Functions

Match the quadratic function with its graph.

1. $f(x) = -2(x + 6)^2 + 3$ **2.** $f(x) = \dfrac{1}{4}(x - 4)^2 + 2$

3. $f(x) = x^2 + x - 12$ **4.** $f(x) = -3x^2 - 10x + 8$

a.

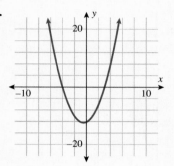

b.

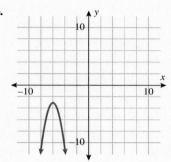

c.

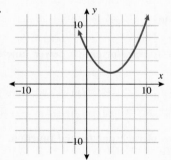

d.

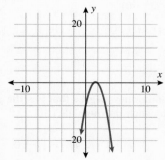

Graph the quadratic function given in standard form.

5. $f(x) = -(x - 7)^2 + 4$

6. $f(x) = (x + 3)^2 - 5$

7. $f(x) = -\dfrac{1}{2}\left(x - \dfrac{1}{3}\right)^2 + \dfrac{2}{5}$

8. $f(x) = 0.6(x - 0.75)^2 + 0.5$

Rewrite the quadratic function in standard form by completing the square.

9. $f(x) = x^2 - 3x - 10$ **10.** $f(x) = x^2 - 2x - 24$

11. $f(x) = 4x^2 + 8x - 7$ **12.** $f(x) = -\dfrac{1}{4}x^2 + 2x - 4$

Graph the quadratic function given in general form.

13. $f(x) = x^2 - 3x + 5$ **14.** $f(x) = -x^2 + 4x + 2$

15. $f(x) = -4x^2 + 2x + 3$ **16.** $f(x) = -0.75x^2 + 2.5$

Find the vertex of the parabola associated with each quadratic function.

17. $f(x) = 13x^2 - 5x + 12$

18. $f(x) = \dfrac{2}{5}x^2 - 4x + 3$

19. $f(x) = -0.45x^2 - 0.12x + 3.6$

20. $f(x) = -\dfrac{3}{4}x^2 + \dfrac{2}{5}x + 4$

Find the quadratic function that has the given vertex and goes through the given point.

21. vertex: $(-2, 3)$ point: $(1, 4)$

22. vertex: $(4, 7)$ point: $(-3, 1)$

23. vertex: $(2.7, 3.4)$ point: $(3.2, 4.8)$

24. vertex: $\left(-\dfrac{5}{2}, \dfrac{7}{4}\right)$ point: $\left(\dfrac{1}{2}, \dfrac{3}{5}\right)$

Applications

25. Profit. The revenue and the cost of a local business are given below as functions of the number of units, x, in thousands produced and sold. Use the cost and the revenue to answer the questions that follow.

$$C(x) = \dfrac{1}{3}x + 2 \text{ and } R(x) = -2x^2 + 12x - 12$$

a. Determine the profit function.

b. State the break-even points.

c. Graph the profit function.

d. What is the range of units to make and sell that will correspond to a profit?

26. Geometry. Given the length of a rectangle is $2x - 4$ and the width is $x + 7$, find the area of the rectangle.

27. Geometry. A triangle has a base of $x + 2$ units and a height of $4 - x$ units. Determine the area of the triangle. What dimensions correspond to the largest area?

28. Geometry. A person standing at a ridge in the Grand Canyon throws a penny upward and toward the pit of the canyon. The height of the penny is given by the function:

$$h(t) = -12t^2 + 80t$$

a. What is the maximum height that the penny will reach?

b. How many seconds will it take the penny to hit the ground below?

4.2 Polynomial Functions of Higher Degree

Determine which functions are polynomials, and for those, state their degree.

29. $f(x) = x^6 - 2x^5 + 3x^2 + 9x - 42$

30. $f(x) = (3x - 4)^3(x + 6)^2$

31. $f(x) = 3x^4 - x^3 + x^2 + \sqrt[4]{x} + 5$

32. $f(x) = 5x^3 - 2x^2 + \dfrac{4x}{7} - 3$

Match the polynomial function with its graph.

33. $f(x) = 2x - 5$

34. $f(x) = -3x^2 + x - 4$

35. $f(x) = x^4 - 2x^3 + x^2 - 6$

36. $f(x) = x^7 - x^5 + 3x^4 + 3x + 7$

a.

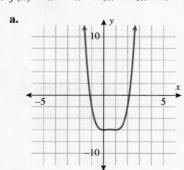

b.

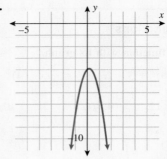

c.

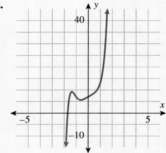

d.

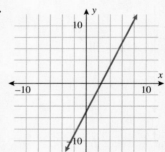

Graph each function by transforming a power function, $y = x^n$.

37. $f(x) = -x^7$

38. $f(x) = (x - 3)^3$

39. $f(x) = x^4 - 2$

40. $f(x) = -6 - (x + 7)^5$

Find all the real zeros of each polynomial function, and state their multiplicity.

41. $f(x) = 3(x + 4)^2(x - 6)^5$

42. $f(x) = 7x(2x - 4)^3(x + 5)$

43. $f(x) = x^5 - 13x^3 + 36x$

44. $f(x) = 4.2x^4 - 2.6x^2$

Find a polynomial of minimum degree that has the given zeros. (There are many such polynomials.)

45. $-3, 0, 4$ **46.** $2, 4, 6, -8$ **47.** $-\dfrac{2}{5}, \dfrac{3}{4}, 0$

48. $2 - \sqrt{5}, 2 + \sqrt{5}$

49. -4.2 (multiplicity of 4), 3.7(multiplicity of 2)

50. $\sqrt{7}$ (multiplicity of 2), -1 (multiplicity of 2), $-\sqrt{7}$(multiplicity of 2)

For each polynomial function given: (a) list each real zero and its multiplicity; (b) determine whether the graph touches or crosses at each x-intercept; (c) find the y-intercept and a few points on the graph; (d) determine the end behavior; and (e) sketch the graph.

51. $f(x) = x^2 - 5x - 14$

52. $f(x) = -(x - 5)^5$

53. $f(x) = 6x^7 + 3x^5 - x^2 + x - 4$

54. $f(x) = -x^4(3x + 6)^3(x - 7)^3$

Applications

55. Salary. Tiffany has started tutoring students x hours per week. The tutoring job corresponds to the following additional income:

$$f(x) = (x - 1)(x - 3)(x - 7)$$

a. Graph the polynomial function.

b. Give any real zeros that occur.

c. How many hours of tutoring are financially beneficial to Tiffany?

56. Profit. The following function is the profit for Walt Disney World, where $P(x)$ represents profit in millions of dollars and x represents the month ($x = 1$ corresponds to January):

$$P(x) = 3(x - 2)^2(x - 5)^2(x - 10)^2 \quad 1 \le x \le 12$$

Graph the polynomial. When are the peak seasons?

4.3 Dividing Polynomials: Long Division and Synthetic Division

Use long division to divide the polynomials. If you choose to use a calculator, do not round off. Keep the exact values instead. Express the answer in the form $Q(x) = \quad r(x) = \quad$.

57. $(x^2 + 2x - 6) \div (x - 2)$

58. $(2x^2 - 5x - 1) \div (2x - 3)$

59. $(4x^4 - 16x^3 + x - 9 + 12x^2) \div (2x - 4)$

60. $(6x^2 + 2x^3 - 4x^4 + 2 - x) \div (2x^2 + x - 4)$

Use synthetic division to divide the polynomial by the linear factor. Indicate the quotient, $Q(x)$, and the remainder, $r(x)$.

61. $(x^4 + 4x^3 + 5x^2 - 2x - 8) \div (x + 2)$

62. $(x^3 - 10x + 3) \div (2 + x)$

63. $(x^6 - 64) \div (x + 8)$

64. $(2x^5 + 4x^4 - 2x^3 + 7x + 5) \div \left(x - \dfrac{3}{4}\right)$

Divide the polynomials using either long division or synthetic division.

65. $(5x^3 + 8x^2 - 22x + 1) \div (5x^2 - 7x + 3)$

66. $(x^4 + 2x^3 - 5x^2 + 4x + 2) \div (x - 3)$

67. $(x^3 - 4x^2 + 2x - 8) \div (x + 1)$

68. $(x^3 - 5x^2 + 4x - 20) \div (x^2 + 4)$

Applications

69. Geometry. The area of a rectangle is given by the polynomial $6x^4 - 8x^3 - 10x^2 + 12x - 16$. If the width is $2x - 4$, what is the length of the rectangle?

70. Volume. A 10 inch by 15 inch rectangular piece of cardboard is used to make a box. Square pieces x inches on a side are cut out from the corners of the cardboard and then the sides are folded up. Find the volume of the box.

4.4 Properties and Tests of Zeros of Polynomial Functions

Find the following values by using synthetic division. Check by substituting the value into the function.

$$f(x) = 6x^5 + x^4 - 7x^2 + x - 1 \quad g(x) = x^3 + 2x^2 - 3$$

71. $f(-2)$ **72.** $f(1)$ **73.** $g(1)$ **74.** $g(-1)$

Determine whether the number given is a zero of the polynomial.

75. $-3, P(x) = x^3 - 5x^2 + 4x + 2$

76. 2 and $-2, P(x) = x^4 - 16$

77. $1, P(x) = 2x^4 - 2x$

78. $4, P(x) = x^4 - 2x^3 - 8x$

Given a zero of the polynomial, determine all other zeros, and write the polynomial in terms of a product of linear factors.

POLYNOMIAL	ZERO
79. $P(x) = x^4 - 6x^3 + 32x$	-2
80. $P(x) = x^3 - 7x^2 + 36$	3
81. $P(x) = x^5 - x^4 - 8x^3 + 12x^2$	0
82. $P(x) = x^4 - 32x^2 - 144$	6

Use Descartes's rule of signs to determine the possible number of positive real zeros, negative real zeros, and imaginary zeros.

83. $P(x) = x^4 + 3x^3 - 16$

84. $P(x) = x^5 + 6x^3 - 4x - 2$

85. $P(x) = x^9 - 2x^7 + x^4 - 3x^3 + 2x - 1$

86. $P(x) = 2x^5 - 4x^3 + 2x^2 - 7$

Use the rational zero theorem to list the possible rational zeros.

87. $P(x) = x^3 - 2x^2 + 4x + 6$

88. $P(x) = x^5 - 4x^3 + 2x^2 - 4x - 8$

89. $P(x) = 2x^4 + 2x^3 - 36x^2 - 32x + 64$

90. $P(x) = -4x^5 - 5x^3 + 4x + 2$

List the possible rational zeros, and test to determine all rational zeros.

91. $P(x) = 2x^3 - 5x^2 + 1$

92. $P(x) = 12x^3 + 8x^2 - 13x + 3$

93. $P(x) = x^4 - 5x^3 + 20x - 16$

94. $P(x) = 24x^4 - 4x^3 - 10x^2 + 3x - 2$

For each polynomial: (a) use Descartes's rule of signs to determine the possible combinations of positive real zeros, negative real zeros, and imaginary zeros; (b) use the rational zero test to determine possible rational zeros; (c) use the upper and lower bounds rules to eliminate possible zeros; (d) test for rational zeros; (e) factor as a product of linear factors if possible; and (f) graph the polynomial function.

95. $P(x) = x^3 + 3x - 5$

96. $P(x) = x^4 - 2x^3 + x + 2$

97. $P(x) = 8x^4 + x^3 - 2x + 1$

98. $P(x) = x^5 - 3x^3 - 6x^2 + 8x$

Applications

99. Farm. The polynomial $y = x^5 + x^3 - 10x + 8$ represents the area of a farm.

   a. Graph the polynomial.

   b. State any turning points of the polynomial.

   c. Identify any zeros of the polynomial.

100. Sports. A kicker for the USC soccer team kicks the ball to his fullback who, at this point, takes off with the ball to score the goal to win the game. The path of the fullback is given by the following polynomial.

$$y = x^5 - 4x^3 + 2x^2 - 7$$

Graph the polynomial and indicate the zeros (midfield) if any exist.

4.5 Rational Functions

Determine the vertical and horizontal (or oblique) asymptotes (if they exist) for the following rational functions.

101. $f(x) = \dfrac{7 - x}{x + 2}$

102. $f(x) = \dfrac{2 - x^2}{(x - 1)^3}$

103. $f(x) = \dfrac{4x^2}{x + 1}$

104. $f(x) = \dfrac{3x^2}{x^2 + 9}$

105. $f(x) = \dfrac{2x^2 - 3x + 1}{x + 4}$

106. $f(x) = \dfrac{4x^3 - 2x^2 + 3x + 5}{2x^2 - 11x + 5}$

Graph the rational functions.

107. $f(x) = -\dfrac{2}{x - 3}$

108. $f(x) = \dfrac{5}{x + 1}$

109. $f(x) = \dfrac{x^2}{x^2 + 4}$

110. $f(x) = \dfrac{x^2 - 36}{x^2 + 25}$

111. $f(x) = \dfrac{x^2 - 49}{x + 7}$

112. $f(x) = \dfrac{2x^2 - 3x - 2}{2x^2 - 5x - 3}$

1. Graph the parabola $y = -(x - 4)^2 + 1$.

2. Write the parabola in standard form $y = -x^2 + 4x - 1$.

3. Find the vertex of the parabola $f(x) = -\dfrac{1}{2}x^2 + 3x - 4$.

4. Find a quadratic function whose vertex is $(-3, -1)$ and whose graph passes through the point $(-4, 1)$.

5. Find a sixth-degree polynomial function with the given zeros:

   2 of multiplicity 3   1 of multiplicity 2   0 of multiplicity 1

6. For the polynomial function $f(x) = x^4 + 6x^3 - 7x$:

   a. List each real zero and its multiplicity.

   b. Determine whether the graph touches or crosses at each x-intercept.

   c. Find the y-intercept and a few points on the graph.

   d. Determine the end behavior.

   e. Sketch the graph.

7. Divide $-4x^4 + 2x^3 - 7x^2 + 5x - 2$ by $2x^2 - 3x + 1$.

8. Divide $17x^5 - 4x^3 + 2x - 10$ by $x + 2$.

9. Is $x - 3$ a factor of $x^4 + x^3 - 13x^2 - x + 12$?

10. Divide $\dfrac{x^{3n} + 3x^n - 4}{x^n - 1}$.

11. Determine if -1 is a zero of $P(x) = x^{21} - 2x^{18} + 5x^{12} + 7x^3 + 3x^2 + 2$.

12. Given that $x - 7$ is a factor of $P(x) = x^3 - 6x^2 - 9x + 14$, factor the polynomial in terms of linear factors.

13. Given that $3i$ is a zero of $P(x) = x^4 - 3x^3 + 19x^2 - 27x + 90$, find all other zeros.

14. Can a polynomial have zeros that are not x-intercepts? Explain.

15. Use the rational zero test to list all possible rational zeros of $P(x) = 3x^4 - 7x^2 + 3x + 12$.

16. Use Descartes's rule of signs to determine the possible combinations of positive real zeros, negative real zeros, and imaginary zeros of $P(x) = 3x^5 + 2x^4 - 3x^3 + 2x^2 - x + 1$.

17. Sports. A football player shows up in August at 300 pounds. After 2 weeks of practice in the hot sun, he is down to 285 pounds. Ten weeks into the season he is up to 315 pounds because of weight training. In the spring he does not work out and is back to 300 pounds by the next August. Plot these points on a graph. What degree polynomial could this be?

18. Profit. The profit of a company is governed by the polynomial $P(x) = x^3 - 13x^2 + 47x - 35$ where x is the number of units sold in thousands. How many units do they have to sell to break even?

19. Interest Rate. The interest rate for a 30-year fixed mortgage fluctuates with the economy. In 1970, the mortgage interest rate was 8%, and in 1988 it peaked at 13%. In 2002 it dipped down to 4%, and in 2005 it was up to 6%. What is the lowest-degree polynomial that can represent this function?

20. Food. On eating a sugary snack the average body almost doubles its glucose level. The percentage increase in glucose level, y, can be approximated by the rational function $y = \dfrac{25x}{x^2 + 50}$, where x represents the number of minutes after eating the snack. Graph the function.

Exponential and Logarithmic Functions

First human fossil found at Olorgesailie (Kenya) field site; Smithsonian anthropologist makes dramatic discovery.

Courtesy Human Origins Program, Smithsonian Institution

Working at the Olorgesailie field site in Kenya during the summer of 2003, Rick Potts, director of the Smithsonian's Human Origins Program at the Museum of Natural History, discovered part of a skull. The fossil is about 900,000 years old. It is the first human fossil ever found at the Olorgesailie site and the first well-dated fossil from a 400,000 year gap between 600,000 and 1 million years old, in the human fossil record of East Africa.

How do archeologists and anthropologists date fossils? One method is carbon testing. The ratio of carbon 12 (the more stable kind of carbon) to carbon 14 at the moment of death is the same as in every other living thing, but the carbon 14 decays and is not replaced. By looking at the ratio of carbon 12 to carbon 14 in the sample and comparing it to the ratio in a living organism, it is possible to determine the age of a formerly living thing fairly precisely.

The amount of carbon in a fossil is a function of how many years it has been dead. The amount is modeled by an *exponential function,* and to determine the age of the fossil the inverse of the exponential function, a *logarithmic function,* is used.

In this chapter, exponential functions and their inverses, logarithmic functions, will be discussed. You will graph these functions and use their properties to solve exponential and logarithmic equations. Many phenomena are modeled by such functions: banking interest, financial investments, populations, pH values in chemistry that rank acids and bases, and the value of a used car.

Exponential and Logarithmic Functions

Functions

- **Exponential Functions**
  - Constant base
  - Base e
- **Logarithmic Functions**
  - Base 10
  - Base e
  - Other bases

Graphs

- **Exponential**
  - Growth
  - Decay
  - Asymptotes/intercepts
- **Logarithmic**
  - Asymptotes/intercepts

Equations

- Solve exponential equations
- Solve logarithmic equations

CHAPTER OBJECTIVES

- Graph exponential and logarithmic functions.
- Understand that exponential functions and logarithmic functions are inverses of each other.
- Evaluate logarithmic functions exactly and by using calculators.
- Model application problems using exponential or logarithmic functions.
- Solve exponential equations and logarithmic equations.

NAVIGATION THROUGH SUPPLEMENTS

DIGITAL VIDEO SERIES #5

STUDENT SOLUTIONS MANUAL CHAPTER 5

BOOK COMPANION SITE
www.wiley.com/college/young

Exponential Functions

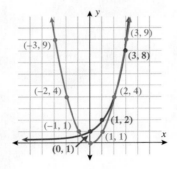

Let's start by noting that the two functions

$$f(x) = 2^x \text{ and } g(x) = x^2$$

are not the same function. Whether the variable, x, appears as an exponent with a constant base or as a base with a constant exponent makes a big difference.

The function $g(x)$ is a quadratic function. The function $f(x)$ is a new type of function called an *exponential* function.

> ### DEFINITION   EXPONENTIAL FUNCTION
>
> An **exponential function** is a function of the form
>
> $$f(x) = b^x \qquad b > 0, b \neq 1$$
>
> where b is called the **base**.

The variable, x, may assume any real value. The domain of f is the set of all real numbers, and the range of f is the set of all positive real numbers. We exclude the base $b = 1$, because this function is the constant function $f(x) = 1^x = 1$. We also exclude negative bases to avoid imaginary numbers such as $f(x) = (-4)^{1/2}$.

How do we evaluate exponential functions? The function $f(x) = 2^x$ can be evaluated for most integer values of x without a calculator using properties of exponents.

$$f(0) = 2^0 = 1 \qquad f(1) = 2^1 = 2 \qquad f(-1) = 2^{-1} = \frac{1}{2}$$

$$f(2) = 2^2 = 4 \qquad f(-2) = 2^{-2} = \frac{1}{2^2} = \frac{1}{4}$$

$$f(3) = 2^3 = 8 \qquad f(-3) = 2^{-3} = \frac{1}{2^3} = \frac{1}{8}$$

What about $f(2.7) = 2^{2.7}$? The button, y^x, on a scientific calculator is used to evaluate exponential functions. Follow the instructions below.

Scientific calculator: [2.] [y^x] [2.7] [=]

Graphing calculator: [2] [^] [2.7] [ENTER]

The answer rounded to three decimal places is 6.498.

EXAMPLE 1 Evaluating Exponential Functions

Evaluate the given exponential functions for $x = 3.2$. Round your answers to two decimal places.

a. $f(x) = 3^x$ **b.** $g(x) = \left(\dfrac{1}{4}\right)^x$ **c.** $h(x) = 10^{x-2.1}$

Solution:

a. $f(3.2) = 3^{3.2} = \boxed{33.63}$ **b.** $g(3.2) = \left(\dfrac{1}{4}\right)^{3.2} = 4^{-3.2} = \boxed{0.01}$

c. $h(3.2) = 10^{3.2-2.1} = 10^{1.1} = \boxed{12.59}$

Graphs of Exponential Functions

Let's graph two exponential functions, $y = 2^x$ and $y = 2^{-x} = \left(\frac{1}{2}\right)^x$, by plotting points.

x	$y = 2^x$	(x, y)
-2	$2^{-2} = \frac{1}{2^2} = \frac{1}{4}$	$\left(-2, \frac{1}{4}\right)$
-1	$2^{-1} = \frac{1}{2^1} = \frac{1}{2}$	$\left(-1, \frac{1}{2}\right)$
0	$2^0 = 1$	$(0, 1)$
1	$2^1 = 2$	$(1, 2)$
2	$2^2 = 4$	$(2, 4)$
3	$2^3 = 8$	$(3, 8)$

x	$y = 2^{-x}$	(x, y)
-3	$2^{-(-3)} = 2^3 = 8$	$(-3, 8)$
-2	$2^{-(-2)} = 2^2 = 4$	$(-2, 4)$
-1	$2^{-(-1)} = 2^1 = 2$	$(-1, 2)$
0	$2^0 = 1$	$(0, 1)$
1	$2^{-1} = \frac{1}{2^1} = \frac{1}{2}$	$\left(1, \frac{1}{2}\right)$
2	$2^{-2} = \frac{1}{2^2} = \frac{1}{4}$	$\left(2, \frac{1}{4}\right)$

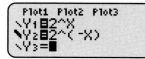

TECHNOLOGY TIP

The graphs of $y_1 = 2^x$ and $y_2 = 2^{-x}$ are shown.

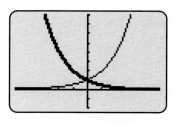

Notice that both graphs' y-intercept is $(0, 1)$ (as shown below) and neither graph has an x-intercept. The x-axis is a horizontal asymptote for both graphs. The following box summarizes general properties of the exponential function.

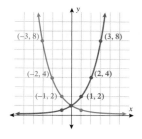

GENERAL PROPERTIES OF THE EXPONENTIAL FUNCTION AND ITS GRAPH

$$f(x) = b^x, \; b > 0, \; b \neq 1$$

Domain: $(-\infty, \infty)$ Range: $(0, \infty)$
x-intercepts: none y-intercept: $(0, 1)$
Horizontal asymptote: x-axis
The graph passes through $(1, b)$.
As x increases, $f(x)$ increases if $b > 1$ and decreases if $0 < b < 1$.
The function f is one-to-one.

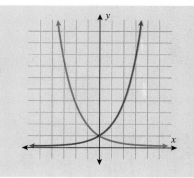

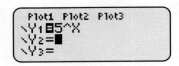

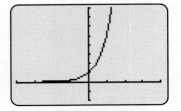

EXAMPLE 2 Graphing Exponential Functions

Graph the function $f(x) = 5^x$.

Solution:

STEP 1 Label the y-intercept $(0, 1)$. $\qquad$ $f(0) = 5^0 = 1$

STEP 2 Label the point $(1, 5)$. $\qquad$ $f(1) = 5^1 = 5$

STEP 3 Draw a smooth curve through those points with the x-axis as a horizontal asymptote.

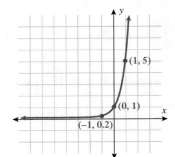

■ **YOUR TURN** Graph the function $f(x) = 5^{-x}$.

Exponential functions, like all functions, can be graphed by point plotting. We can also use transformations (horizontal and vertical shifting and reflection) to graph exponential functions.

EXAMPLE 3 Graphing Exponential Functions Using Transformations

Graph the function $f(x) = 3^{x+1} - 2$.

Solution:

STEP 1 Identify the base function. $\qquad$ $y = 3^x$

STEP 2 Identify base function intercept and asymptotes. $\qquad$ $(0, 1)$ and $y = 0$

STEP 3 State horizontal and vertical shifts of this function to arrive at $f(x) = 3^{x+1} - 2$. $\qquad$ 1 unit to the left; 2 units down

STEP 4 Shift intercept and asymptote 1 unit left and 2 down. $\qquad$ $(0, 1)$ shifts to $(-1, -1)$ $\qquad$ $y = 0$ shift to $y = -2$

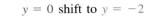

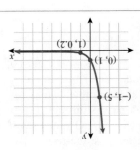

■ **Answer:**

STEP 5 Find additional points on graph.

STEP 6 Graph.

$(0, 1)$ and $(1, 7)$

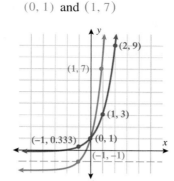

CONCEPT CHECK To graph the function $f(x) = 2^{x+3} - 1$, what is the base, or starting, function and how many units will that function be shifted in each direction?

YOUR TURN Graph $f(x) = 2^{x+3} - 1$.

Applications of Exponential Functions

Exponential functions describe either *growth* or *decay*. Populations and investments are often modeled with exponential growth functions, while the declining value of a used car and the radioactive decay of isotopes are often modeled with exponential decay functions.

In a moderate economy, your investments should double every 10 years. Let's assume that you will retire at the age of 65. There is a saying: *It's not the first time your money doubles, it's the last time that makes such a difference.* As you may already know or as you will soon find, it is important to start investing early.

Suppose Maria invests $5,000 at age 25 and David invests $5,000 at age 35, let's calculate how much will accrue from the initial $5,000 investment by the time they each retire, assuming their money doubles every 10 years.

AGE	MARIA	DAVID
25	$5,000	
35	$10,000	$5,000
45	$20,000	$10,000
55	$40,000	$20,000
65	**$80,000**	**$40,000**

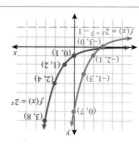

■ **Answer:**

They each made a one-time investment of $5,000. By investing 10 years sooner, Maria made twice what David made.

A measure of growth rate is the *doubling time*, the time it takes for something to double. Often doubling time is used to describe populations.

DOUBLING TIME GROWTH MODEL

The doubling time growth model is given by

$$P = P_0 2^{t/d}$$

where

$$P = \text{Population at time } t$$
$$P_0 = \text{Population at time } t = 0$$
$$d = \text{Doubling time}$$

Note that when $t = d$, $P = 2P_0$ (population is equal to twice the original).

In the investment scenario with Maria and David, $P_0 = 5000$ and $d = 10$, so the model used to predict how much money the original $5,000 investment yielded is $P = 5000(2)^{t/10}$. Maria retired 40 years after the original investment, $t = 40$, and David retired 30 years after the original investment, $t = 30$.

$$\text{Maria: } P = 5000(2)^{40/10} = 5000(2)^4 = 5000(16) = 80,000$$
$$\text{David: } P = 5000(2)^{30/10} = 5000(2)^3 = 5000(8) = 40,000$$

EXAMPLE 4 Doubling Time of Populations

In 2004, the population in Kazakhstan, a country in Eastern Europe, reached 15 million. It is estimated that the population will double in 30 years. If the population continues to grow at the same rate, what will the population be 20 years from now?

Solution:

Write the doubling model. $\qquad P = P_0 2^{t/d}$

Substitute $P_0 = 15$, $d = 30$, and $t = 20$. $\qquad P = 15(2)^{20/30}$

Simplify. $\qquad P = 15(2)^{2/3} \approx 23.81$

In 20 years, there will be approximately 23.81 million people in Kazakhstan.

YOUR TURN What will the approximate population in Kazakhstan be in 40 years?

We now turn our attention from exponential growth to exponential decay, or negative growth. Suppose you buy a brand-new car from a dealership for $24,000. The **half-life** of the car, or the time it takes for the car to depreciate 50%, is 3 years. The exponential decay is described by

$$A = A_0\left(\frac{1}{2}\right)^{t/h}$$

where A_0 is the amount the car is worth when new; that is, when $t = 0$, A is the amount the car is worth after t years, and h is the half-life. In our car scenario, $A_0 = 24{,}000$ and $h = 3$:

$$A = 24{,}000\left(\frac{1}{2}\right)^{t/3}$$

How much is the car worth after three years? Six years? Nine years? Twenty-four years?

$$t = 3: \qquad A = 24{,}000\left(\frac{1}{2}\right)^{3/3} = 24{,}000\left(\frac{1}{2}\right) = 12{,}000$$

$$t = 6: \qquad A = 24{,}000\left(\frac{1}{2}\right)^{6/3} = 24{,}000\left(\frac{1}{2}\right)^{2} = 6000$$

$$t = 9: \qquad A = 24{,}000\left(\frac{1}{2}\right)^{9/3} = 24{,}000\left(\frac{1}{2}\right)^{3} = 3000$$

$$t = 24: \qquad A = 24{,}000\left(\frac{1}{2}\right)^{24/3} = 24{,}000\left(\frac{1}{2}\right)^{8} = 94 \approx 100$$

The car that was worth \$24,000 new is worth \$12,000 in 3 years, \$6,000 in 6 years, \$3,000 in 9 years, and about \$100 in the junkyard in 24 years.

EXAMPLE 5 Radioactive Decay

The radioactive isotope of potassium (^{42}K), used in the diagnosis of brain tumors, has a half-life of 12.36 hours. If 500 milligrams of potassium 42 are taken, how many milligrams will remain after 24 hours?

Solution:

STEP 1 Write the half-life formula. $\qquad\qquad A = A_0\left(\frac{1}{2}\right)^{t/h}$

STEP 2 Substitute $A_0 = 500$, $h = 12.36$, $t = 24$. $\quad A = 500\left(\frac{1}{2}\right)^{24/12.36}$

STEP 3 Simplify. $\qquad\qquad\qquad\qquad\qquad\qquad A \approx 500(0.26) = 130.2$

After 24 hours, there are approximately 130.2 milligrams of potassium 42 left.

✔**CONCEPT CHECK**   Suppose that 500 milligrams of potassium carbonate is ingested. Since this isotope has a half-life of 12.36 hours, would you expect more than or less than 200 milligrams to remain after 1 week?

■ **YOUR TURN**   How much potassium carbonate is expected to be left in the body after 1 week?

■ **Answer:** 0.04 milligram (less than 1 milligram)

In Section 1.2, *simple interest* was discussed where the interest, I, is calculated based on the principal, P, the annual interest rate, r, and the time in years, t: $I = Prt$.

If the interest earned in a period is then reinvested at the same rate, future interest is earned on both the principal and the reinvested interest during the next period. Interest paid on both the principal and interest is called *compound interest*.

COMPOUND INTEREST

If a **principal**, P, is invested at an annual **rate**, r, **compounded** n times a year, then the **amount**, A, in the account at the end of t years is given by

$$A = P\left(1 + \frac{r}{n}\right)^{nt}$$

The annual interest rate r is expressed as a decimal.

The following table lists the typical number of times interest is compounded.

Annually	$n = 1$
Semiannually	$n = 2$
Quarterly	$n = 4$
Monthly	$n = 12$
Weekly	$n = 52$
Daily	$n = 365$

EXAMPLE 6 Compound Interest

If $3,000 is deposited in an account paying 3% compounding quarterly, how much will you have in the account in 7 years?

Solution:

Write the compound interest formula.

$$A = P\left(1 + \frac{r}{n}\right)^{nt}$$

Substitute $P = 3000$, $r = 0.03$, $n = 4$, and $t = 7$.

$$A = 3000\left(1 + \frac{0.03}{4}\right)^{(4)(7)}$$

Simplify.

$$A = 3000(1.0075)^{28} \approx 3698.14$$

You will have $3,698.14 in the account.

■ YOUR TURN If $5,000 is deposited in an account paying 6% compounded annually, how much will you have in the account in 4 years?

SECTION 5.1 SUMMARY

In this section, exponential functions, $f(x) = b^x$, were evaluated and graphed. The common features of these graphs are the y-intercept $(0, 1)$ and the horizontal asymptote, the x-axis $(y = 0)$. Transformations of these graphs follow the same rules use, in Section 3.3. Applications involving doubling time, half-life, and compound interest were solved.

SECTION 5.1 EXERCISES

■ SKILLS

In Exercises 1–12, approximate each number using a calculator. Round your answer to two decimal places.

1. $4^{1.2}$ **2.** $4^{-1.2}$ **3.** $7^{-0.13}$ **4.** $7^{0.13}$ **5.** $5^{\sqrt{2}}$ **6.** $7^{\sqrt{3}}$

7. $5^{2\pi}$ **8.** 5^{π} **9.** $3 \cdot 2^{1.7}$ **10.** $-3 \cdot 2^{-1.7}$ **11.** $1.75^{2/3}$ **12.** $2.3^{3/4}$

In Exercises 13–16, evaluate each exponential function for the given values.

13. $f(x) = 5^{1-x}$ $f(2.7)$ **14.** $f(x) = -5^{x-2}$ $f(-3.1)$ **15.** $f(x) = \left(\dfrac{1}{3}\right)^{2x+1}$ $f\left(\dfrac{3}{4}\right)$ **16.** $f(x) = \left(\dfrac{3}{5}\right)^{1-3x}$ $f\left(-\dfrac{5}{3}\right)$

In Exercises 17–22, match the graph with the function.

17. $y = 5^{x-1}$ **18.** $y = 5^{1-x}$ **19.** $y = -5^x$ **20.** $y = -5^{-x}$ **21.** $y = 1 - 5^{-x}$ **22.** $y = 5^x - 1$

a.

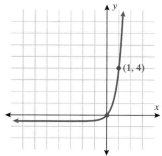

b.

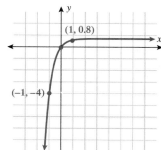

c.

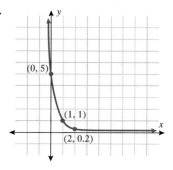

d.

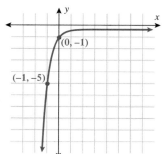

e.

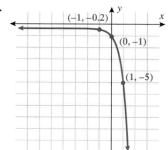

f.

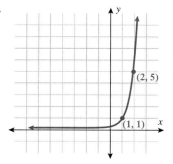

In Exercises 23–32, use transformations to find the y-intercept and horizontal asymptote and to graph the exponential function.

23. $f(x) = 6^x$ **24.** $f(x) = 7^x$ **25.** $f(x) = -4^{-x}$ **26.** $f(x) = 10^{-x}$

27. $f(x) = 2 - 3^{-x}$ **28.** $f(x) = 3^x - 2$ **29.** $f(x) = 1 + \left(\dfrac{1}{2}\right)^{x+2}$ **30.** $f(x) = 2 - \left(\dfrac{1}{3}\right)^{1-x}$

31. $f(x) = 2b^x \quad b > 1$ **32.** $f(x) = 2b^{-x} \quad b > 1$

■ APPLICATIONS

33. Population Growth. In 2002 there were 7.1 million people living in London, England. If the population is expected to double by 2090, what is the expected population in London in 2050?

34. Population Growth. In 2004 the population in Morganton, Georgia, was 43,000. With Publix (a large grocery store chain in the Southeast), Suntrust (a large bank in the Southeast), and the Ritz Carlton all coming to Morganton, its population is expected to double by 2010. If the growth rate remains the same, what is the expected population in Morganton in 2020?

35. Investments. Suppose an investor buys land in a rural area for $1,500 an acre and sells some of it 5 years later at $3,000 an acre and the rest of it 10 years later at $6,000. Write a function that models the value of land in that area assuming the growth rate stays the same. What would the expected cost per acre be 30 years after the initial investment of $1,500?

36. Salaries. Twin brothers, Collin and Cameron, get jobs immediately after graduating from college at the age of 22. Collin opts for the higher starting salary, $55,000, and stays with that company until he retires at 65. His salary doubles every 15 years. Cameron opts for a lower starting salary, $35,000, and changes jobs every 5 years, which results in his doubling of salary every 10 years until he retires at 65. Which brother makes more money when he retires?

37. Radioactive Decay. A radioactive isotope, selenium (^{75}Se), used in the creation of medical images of the pancreas, has a half-life of 119.77 days. If 200 milligrams are given to a patient, how many milligrams are left after 30 days?

38. Radioactive Decay. A radioactive isotope indium 111 (^{111}In), used as a diagnostic tool for locating tumors associated with prostate cancer, has a half-life of 2.807 days. If 300 milligrams are given to a patient how many milligrams will be left after a week?

39. Depreciation of Furniture. A couple buys a new bedroom set for $8,000 and 10 years later sells it for $4,000. If the depreciation continues at the same rate, how much would the bedroom set be worth in 4 more years?

40. Depreciation of a Computer. A student buys a new laptop for $1,500 when she arrives as a freshman. A year later, the computer is worth approximately $750. If the depreciation continues at the same rate, how much will she expect to sell her laptop for when she graduates 4 years after she bought it?

41. Compound Interest. If $3,200 is put in a savings account that earns 2.5% interest per year compounded quarterly, how much is expected to be in that account in 3 years?

42. Compound Interest. If $10,000 is put in a savings account that earns 3.5% interest per year compounded annually, how much is expected to be in that account in 5 years?

43. Compound Interest. How much money should be put in a savings account now that earns 5% a year compounded daily if you want to have $32,000 in 18 years?

44. Compound Interest. How much money should be put in a savings account now that earns 3.0% a year compounded weekly if you want to have $80,000 in 15 years?

■ CATCH THE MISTAKE

In Exercises 45 and 46, explain the mistake that is made.

45. Evaluate the expression: $4^{-1/2}$.

   Solution: $4^{-1/2} = 4^2 = 16$

   The correct value is $\frac{1}{2}$ What mistake was made?

46. Evaluate the function at the given point: $f(x) = 4^x \ f\left(\dfrac{3}{2}\right)$.

   Solution: $f\left(\dfrac{3}{2}\right) = 4^{3/2}$

   $$= \frac{4^3}{4^2} = \frac{64}{16} = 4$$

   The correct value is 8. What mistake was made?

■ CHALLENGE

47. T or F: If $f(x) = b^x$, $b > 1$, then $f(C + D) = f(C) \cdot f(D)$.

48. T or F: If $f(x) = b^x$, $b > 1$, then $f(-x) = \dfrac{1}{f(x)}$.

49. T or F: If $f(x) = b^x$, $b > 1$, then $f(ax) = ab^x$.

50. T or F: If $f(x) = b^x$, $0 < b < 1$, then $f(x)$ increases as $x \to \infty$.

For Exercises 51 and 52, $f(x) = b^{|x|}$, $b > 1$.

51. What are the domain and range of f?

52. Graph f.

53. Graph the function $f(x) = \begin{cases} a^x & x < 0 \\ a^{-x} & x \geq 0 \end{cases}$ where $a > 1$.

54. Graph $f(x) = 3^x$ and its inverse on the same graph.

■ TECHNOLOGY

55. What is the horizontal asymptote of $y = 3.5^{x+1} - 2$? Graph this function using a graphing utility. Does the horizontal asymptote agree with your prediction?

56. What is the horizontal asymptote of $y = 3 + 4.1^{3.2x}$? Graph this function using a graphing utility. Does the horizontal asymptote agree with your prediction?

57. Graph the three functions $y_1 = 0.99^x$, $y_2 = 1^x$, and $y_3 = 1.01^x$. Explain why they are different curves when their bases are so similar.

SECTION 5.2 Exponential Functions with Base *e*

Skills Objectives

■ Define base *e*.
■ Graph exponential functions with base *e*.
■ Revisit growth and decay applications with base *e*.
■ Solve application problems involving interest compounding continuously.

Conceptual Objectives

■ Understand why *e* is the natural base.

The Natural Base e

Most students have heard of the irrational number π, but there is another constant, one just as important and symbolized by the letter *e*, that is approximately equal to 2.72. The number *e* is called the **natural base**. The exponential function with base *e*, $f(x) = e^x$, is called the **natural exponential function**. We use the word *natural* because it arises in many application problems, such as compound interest. What is especially helpful for us is that many operations take on their simplest form when using this base.

Using a calculator, let's find a more accurate approximation using the following steps:

Scientific calculator:

Graphing calculator:

$$e \approx 2.7182818$$

Mathematicians did not pull e out of a hat. The constant e is found by evaluating the following expression (which is closely related to compound interest) for large m.

$$\left(1 + \frac{1}{m}\right)^m$$

m	$\left(1 + \dfrac{1}{m}\right)^m$
1	2
10	2.59374
100	2.70481
1,000	2.71692
10,000	2.71815
100,000	2.71827
1,000,000	**2.71828**
⋮	⋮

Graphing Exponential Functions with Base e

Because $2 < e < 3$, we expect the graph of $f(x) = e^x$ to lie between the graphs of $y = 2^x$ and $y = 3^x$.

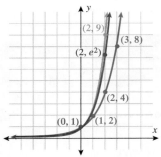

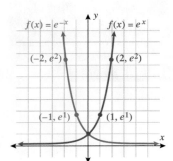

Like all exponential functions of the form $f(x) = b^x$, $f(x) = e^x$ and $f(x) = e^{-x}$ have $(0, 1)$ as their y-intercept and the x-axis as a horizontal asymptote as shown in the figure on the left.

EXAMPLE 1 Graphing Exponential Functions with Base e

Graph the function $f(x) = 3 + e^{2x}$.

Solution:

x	$f(x) = 3 + e^{2x}$	(x, y)
-2	3.02	$(-2, 3.02)$
-1	3.14	$(1, 3.14)$
0	4	$(0, 4)$
1	10.39	$(1, 10.39)$
2	57.60	$(2, 57.60)$

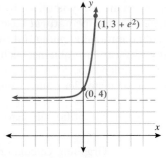

Note: The y-intercept is $(0, 4)$ and the line $y = 3$ is the horizontal asymptote.

TECHNOLOGY TIP
The graph of $y_1 = 3 + e^{2x}$ is shown.

■ **YOUR TURN**   Graph the function $f(x) = e^{x+1} - 2$.

Applications

Exponential functions are rich with applications because growth and decay are common in everyday life, and mathematics can be used to model such functions. Although we will focus on two specific applications, carbon dating and compound interest, you will see in the exercises that exponential functions have a wide variety of applications.

EXAMPLE 2 Carbon Dating

While living, human beings have carbon 14 in their bodies, and at the moment of death, the body no longer takes in carbon. So the amount of carbon 14 begins to decay according to the equation

$$A = A_0 e^{-0.000124t}$$

where A_0 is the amount of carbon at the time of death, $t = 0$, and A is the amount of carbon after t years. If there are 2000 milligrams of carbon 14 in the body while the human being is alive, how much carbon 14 will remain after 10,000 years?

Solution:

Write the formula governing carbon decay. $A = A_0 e^{-0.000124t}$

Substitute $A_0 = 2000$ and $t = 10,000$. $A = 2000 e^{-(0.000124)(10000)}$

Simplify. $A \approx 578.77$

There are approximately 579 milligrams of carbon in the human remains after 10,000 years.

CONCEPT CHECK   How would you confirm that the half-life of carbon 14 is approximately 5700 years?

■ **YOUR TURN**   If 1000 milligrams of carbon 14 are present in a living body, then how many milligrams of carbon 14 are present after 5700 years?

In Section 5.1, the compound interest formula was discussed:

$$A = P\left(1 + \frac{r}{n}\right)^{nt}$$

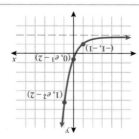

■ **Answer:** 493 milligrams   ■ **Answer:**

where P is the principal, r is the annual interest rate expressed as a decimal, t is time in years, and n is the number of times the interest is compounded a year. As n increases the amount, A, increases. In other words, the more times the interest is compounded per year, the more money you make. Ideally, your bank will compound your interest infinitely many times. This is called *compounding continuously.* We will now show the development of the compounding continuously formula, $A = Pe^{rt}$.

WORDS   **MATH**

Write the compound interest formula. $A = P\left(1 + \dfrac{r}{n}\right)^{nt}$

Note that $\dfrac{r}{n} = \dfrac{1}{\dfrac{n}{r}}$ and $nt = \left(\dfrac{n}{r}\right)rt$. $A = P\left(1 + \dfrac{1}{\left(\dfrac{n}{r}\right)}\right)^{\left(\frac{n}{r}\right)rt}$

Let $m = \dfrac{n}{r}$. $A = P\left(1 + \dfrac{1}{m}\right)^{mrt}$

Use the exponential property: $x^{mrt} = (x^m)^{rt}$. $A = P\left[\left(1 + \dfrac{1}{m}\right)^{m}\right]^{rt}$

In the beginning of this section, we showed that as m increases, $(1 + \frac{1}{m})^m$ approaches e. Therefore, as the number of times the interest is compounded approaches infinity, $n \to \infty$, the amount in an account $A = P(1 + \frac{r}{n})^{nt}$ approaches $A = Pe^{rt}$.

CONTINUOUS COMPOUND INTEREST

If a **principal** P is invested at an annual **rate** r **compounded continuously**, then the **amount** A in the account at the end of t years is given by

$$A = Pe^{rt}$$

The annual interest rate r is expressed as a decimal.

EXAMPLE 3 Continuous Compound Interest

If $3,000 is deposited in a savings account paying 3% a year compounding continuously, how much will you have in the account in 7 years?

Solution:

Write the compound continuous interest formula. $A = Pe^{rt}$

Substitute $P = 3000$, $r = 0.03$, and $t = 7$. $A = 3000e^{(0.03)(7)}$

Simplify. $A \approx 3701.034$

There will be $3,701.03   in the account in 7 years.

Note: In Example 6 in Section 5.1 we worked this same problem compounding *quarterly,* and the result was $3,698.14.

If the number of times per year interest is compounded increases, then the total interest earned that year also increases.

STUDY TIP

If the number of times per year interest is compounded increases, then the total interest earned that year also increases.

CONCEPT CHECK   In the Your Turn following Example 6 in Section 5.1 we found that if $5,000 is deposited in an account paying 6% compounded annually, in 4 years the account will have $6,312.38. If we calculated this based on compounding continuously, would you expect more or less than $6,312.38?

■ **YOUR TURN**    If $5,000 is deposited in an account paying 6% compounded continuously, how much will be in the account in 4 years?

SECTION 5.2   SUMMARY

In this section we discussed the natural base, e, and the natural exponential function, $f(x) = e^x$. The exponential function with base e, like all other exponential functions, has a y-intercept at $(0, 1)$ and the x-axis is a horizontal asymptote.

The exponential function is used to model growth and decay:

- Growth: $y = ce^{kt}$ $c, k > 0$
  - Population of bacteria and animals
  - Growth of money
  - Value of an investment or stock
- Decay: $y = ce^{-kt}$ $c, k > 0$
  - Radioactive decay
  - Depreciation of a car, computer, machine, or other asset

SECTION 5.2   EXERCISES

■ SKILLS

In Exercises 1–6, approximate each number using a calculator. Round your answer to two decimal places.

1. e^2 **2.** $e^{1/2}$ **3.** $e^{1.2}$ **4.** $e^{-1.2}$ **5.** $e^{\sqrt{2}}$ **6.** e^{π}

In Exercises 7–14, state the y-intercept and horizontal asymptote, then graph the exponential function.

7. $y = -e^x$ **8.** $y = -e^{-x}$ **9.** $y = 2 + e^{x-1}$ **10.** $y = e^{x+1} - 4$

11. $y = 2e^{3x}$ **12.** $y = 3e^{x/2}$ **13.** $y = e^{-.05x}$ **14.** $y = e^{.05x}$

■ APPLICATIONS

15. Carbon Dating. Mayans lived in present-day southern Mexico, Guatemala, western Honduras, El Salvador, and northern Belize. Assuming a Mayan had 1000 milligrams of carbon 14 in him when he died, how many milligrams would 1750-year-old remains found in the Mayan ruins contain today? Use the formula $A = A_0 e^{-0.000124t}$ to approximate the amount of carbon 14.

16. Carbon Dating. An archeological site in Chile has been confirmed as dating human settlement in the southernmost part of South America back to at least 12,500 years ago. Assuming these early human beings had 1000 milligrams of carbon 14 in them when they died, how much would still exist in their remains today? Use the formula $A = A_0 e^{-0.000124t}$ to approximate the amount of carbon 14.

17. Population Growth. The population of the Philippines in 2003 was 80 million. Their population increases 2.36% per year. What is the expected population of the Philippines in 2010? Use the formula $N = N_0 e^{rt}$ where N represents the number of people.

18. Population Growth. China's urban population is growing at 2.5 percent a year compounding continuously. If there were 13.7 million people in Shanghai in 1996, approximately how many people will there be in 2016? Use the formula $N = N_0 e^{rt}$ where N represents the number of people.

19. Population. The University of Central Florida is one of the fastest growing universities in the United States. In 1995 there were 26,000 students enrolled at UCF. If the student population was increasing 5% per year, how many students were enrolled at UCF in 2004?

20. Money. Ellen is splurging on weekly manicures at $20 a week and instead decides to get a manicure every other week, saving the $20 for 26 weeks during her freshman year. At the end of her freshman year she puts that saved money in a 3 year CD that earns 3% per year compounded yearly. How much money will have accrued in time for graduation 3 years later?

21. Money. If $3,200 is put in a savings account that pays 2% a year compounded continuously, how much will be in the account in 15 years?

22. Money. If $7,000 is put in a money market account that pays 4.3% a year compounded continuously, how much will be in the account in 10 years?

23. More Money. How much money should be deposited into a money market account that pays 5% a year compounded continuously to have $38,000 in the account in 20 years?

24. More Money. How much money should be deposited into a CD that pays 6% a year compounded continuously to have $80,000 in the account in 18 years?

25. Website Hits. A popular band has a website, and they find that the number of hits they record on their website increases 5% each month. In 1 month they record 7000 hits. How many monthly hits will their website expect to record 18 months later?

26. Online Dating. A college student placed an ad on www.match.com on a Monday, and by Wednesday she

had received 60 e-mail messages from potential suitors. She found that every day the number of e-mails from new potential suitors decreased 10 percent from the day before. How many new e-mails would she expect to receive on the following Sunday?

27. Snake Population. The Golden Lancehead Viper (*Bothrops insularis*) is only found on Ilha Queimada Grande, located 34 km off the southeast coast of Brazil. If the number of snakes that live on the island is given by $S = 3000(1 - e^{-0.02t})$ where S is the number of snakes and t is the number of years beyond the year 2000. How many snakes will be on the island in 2010?

Golden Lancehead Viper, Ilha Queimada Grande

28. Snake Population. In biology, the term *carrying capacity* refers to the maximum number of animals that a system can sustain due to food sources and other environmental factors. In Exercise 27, what is the carrying capacity of the island (the largest number of snakes the island can sustain)?

29. HIV/AIDS. In 2003, there were an estimated 1 million people who had been infected with HIV in the United States. If the infection rate increases at an annual rate of 2.5% a year compounding continuously, how many Americans will be infected with HIV by 2010?

30. HIV/AIDS. In 2003, there were an estimated 25 million people who had been infected with HIV in sub-Saharan Africa. If the infection rate increases at an annual rate of 9% a year compounding continuously, how many Africans will be infected with HIV by 2010?

31. Anesthesia. When a person has a cavity filled, the dentist typically gives a local anesthetic. After leaving the dentist's office, one's mouth often is numb for several more hours. If 100 ml of anesthesia is injected into the bloodstream at the time of the procedure ($t = 0$), and the amount of anesthesia still in the bloodstream t hours after the initial injection is given by $A = 100e^{-0.5t}$, how much is in the bloodstream 4 hours later? 12 hours later?

32. Automotive Sales. A new model BMW convertible coupe is designed and built so that the new models will appear in North America in the fall. BMW Corporation has a limited number of new models available. If the number of purchased new model BMW convertible coupes in North America is given by

$$N = \frac{100{,}000}{1 + 10e^{-2t}},$$

where *t* is the number of weeks after the BMW is released, how many new model BMW convertible coupes will have been purchased 10 weeks after the new model becomes available? How many after 30 weeks? What is the greatest number of new model BMW convertible coupes that will be sold in North America?

33. Lasers. The intensity of a laser beam is given by the ratio of power to area. If a particular laser beam has an intensity function given by $I = e^{-r^2}$ mW/cm^2 where *r* is the radius off the center axis given in centimeters, graph the intensity of the laser beam. Where is the beam brightest (largest intensity)?

34. Grade Distribution. Suppose on your first test in College Algebra your class has a normal, or bell-shaped, grade distribution of test scores, with an average score of 75. An approximate function that models your class's grades on test 1 is $N(x) = 10e^{-(x-75)^2/25}$, where *N* represents the number of students who received the score, *x*. Graph this function. What is the average grade? Approximately how many students scored a 50, and how many students scored 100?

■ **CATCH THE MISTAKE**

In Exercises 35–38, explain the mistake that is made.

35. If \$2,000 is invested in a savings account that earns 2.5% interest compounding continuously, how much will be in the account in 1 year?

Solution:

Write the compound continuous interest formula.	$A = Pe^{rt}$
Substitute $P = 2000$, $r = 2.5$, and $t = 1$.	$A = 2000e^{(2.5)(1)}$
Simplify.	$A = 24{,}364.99$

This is incorrect. What mistake was made?

36. If \$5,000 is invested in a savings account that earns 3% interest compounding continuously, how much will be in the account in 6 months?

Solution:

Write the compound continuous interest formula.	$A = Pe^{rt}$
Substitute $P = 5000$, $r = 0.03$, and $t = 6$.	$A = 5000e^{(0.03)(6)}$
Simplify.	$A = 5986.09$

This is incorrect. What mistake was made?

37. The largest city in Algeria is Algiers whose population has been growing at 3 percent per year compounding continuously. In 2003 there were 1.8 million people in Algiers. Approximately how many people were there in 1983?

Solution:

Write the exponential population model.	$N = N_0 e^{rt}$
Substitute $N_0 = 1.8$, $r = 0.03$, and $t = 20$.	$N = 1.8e^{(0.03)(20)}$
Simplify.	$N \approx 3.3$

There were approximately 3.3 million people in Algiers in 1983.

This is incorrect because the population has been growing. What mistake was made?

38. The third largest city in Algeria is Constantine, whose population has been growing at 2% a year compounding continuously. In 2003, there were 550,000 people in Constantine. Approximately how many people were there in 1983?

Solution:

Write the exponential population model.	$N = N_0 e^{rt}$
Substitute $N_0 = 550{,}000$, $r = 0.03$, and $t = 20$.	$N = 550{,}000e^{(0.03)(20)}$
Simplify.	$N \approx 1{,}002{,}165$

There were approximately 1 million people in Constantine in 1983.

This is incorrect because the population has been growing. What went wrong?

■ CHALLENGE

39. T or F: The function $f(x) = -e^{-x}$ has a y-intercept $(0, 1)$.

40. T or F: The function $f(x) = -e^{-x}$ has a horizontal asymptote along the x-axis.

41. T or F: The functions $y = 3^{-x}$ and $y = \left(\dfrac{1}{3}\right)^x$ have the same graphs.

42. T or F: $e = 2.718$.

43. Find the y-intercept and horizontal asymptote of the function $y = a + be^{x+1}$.

44. Graph $y = e^x$ and its inverse on the same plot.

45. Graph $y = e^{|x|}$.

46. Graph $y = e^{-|x|}$.

■ TECHNOLOGY

47. Plot the function $y = \left(1 + \dfrac{1}{x}\right)^x$. What is the horizontal asymptote as x increases?

48. Plot the functions $y = 2^x$, $y = e^x$, and $y = 3^x$ in the same viewing screen. Explain why $y = e^x$ lies between the other two graphs.

49. Plot $y_1 = e^x$ and $y_2 = 1 + x + \dfrac{x^2}{2} + \dfrac{x^3}{6} + \dfrac{x^4}{24}$ in the same viewing screen. What do you notice?

50. Plot $y_1 = e^{-x}$ and $y_2 = 1 - x + \dfrac{x^2}{2} - \dfrac{x^3}{6} + \dfrac{x^4}{24}$ in the same viewing screen. What do you notice?

SECTION 5.3 Logarithmic Functions and Their Graphs

Skills Objectives

■ Change exponential expressions to logarithmic expressions.
■ Change logarithmic expressions to exponential expressions.
■ Evaluate logarithmic functions.
■ Evaluate common and natural logarithms using a calculator.
■ Graph logarithmic functions.

Conceptual Objectives

■ Interpret logarithmic functions as inverses of exponential functions.
■ Determine domain restrictions on logarithmic functions.

Definition of Logarithmic Functions

Recall that an exponential function, $f(x) = b^x$, has a graph that has its y-intercept at $(0, 1)$ and the x-axis as its horizontal asymptote. This graph passes both the horizontal and vertical line tests, so the exponential function is a one-to-one function. Therefore, its inverse exists. Also recall that if the point (c, d) lies on the graph of the function then the point (d, c) lies on its inverse. In other words, now that we know so much about the exponential function, we also know a lot about its inverse. We now turn our attention to a new class of functions called *logarithmic functions,* which are inverses of exponential functions.

The inverse of the function $y = b^x$ is found by interchanging x and y, and hence, the inverse function is $x = b^y$. This function is so important that it is given a name, the *logarithmic function.*

DEFINITION   **LOGARITHMIC FUNCTION**

A **logarithmic function with base b** is a function of the form

$$y = \log_b x \qquad\qquad x > 0,\ b > 0,\ \text{and}\ b \neq 1$$

where the equations $y = \log_b x$ and $x = b^y$ are equivalent.

In words we can describe this as

$$\text{Exponent} = \log_{\text{base}} x \quad \text{is equivalent to} \quad \text{base}^{\text{exponent}} = x$$

Comparison of Inverse Functions:
$f(x) = \log_b x$ and $f^{-1}(x) = b^x$

EXPONENTIAL FUNCTION	**LOGARITHMIC FUNCTION**
$y = b^x$	$y = \log_b x$
y-intercept $(0, 1)$	x-intercept $(1, 0)$
Domain $(-\infty, \infty)$	Domain $(0, \infty)$
Range $(0, \infty)$	Range $(-\infty, \infty)$
Horizontal asymptote: x-axis	Vertical asymptote: y-axis

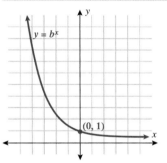
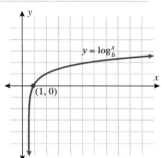

EXAMPLE 1   Changing from Logarithmic to Exponential Form

Write each equation in its equivalent exponential form.

a. $\log_2 8 = 3$ **b.** $\log_9 3 = \dfrac{1}{2}$ **c.** $\log_5\left(\dfrac{1}{25}\right) = -2$

Solution:

a. $\log_2 8 = 3$ is equivalent to $8 = 2^3$.

b. $\log_9 3 = \dfrac{1}{2}$ is equivalent to $3 = 9^{1/2}$.

c. $\log_5\left(\dfrac{1}{25}\right) = -2$ is equivalent to $\dfrac{1}{25} = 5^{-2}$.

■ **YOUR TURN**   Write each equation in its equivalent exponential form.

a. $\log_3 9 = 2$ **b.** $\log_{16} 4 = \dfrac{1}{2}$ **c.** $\log_2\left(\dfrac{1}{8}\right) = -3$

■ **Answer: a.** $9 = 3^2$ **b.** $4 = 16^{1/2}$ **c.** $\dfrac{1}{8} = 2^{-3}$

EXAMPLE 2 Changing from Exponential to Logarithmic Form

Write each equation in its equivalent logarithmic form.

a. $16 = 2^4$ **b.** $9 = \sqrt{81}$ **c.** $\dfrac{1}{9} = 3^{-2}$

Solution:

a. $16 = 2^4$ is equivalent to $\log_2 16 = 4$.

b. $9 = \sqrt{81} = 81^{1/2}$ is equivalent to $\log_{81} 9 = \dfrac{1}{2}$.

c. $\dfrac{1}{9} = 3^{-2}$ is equivalent to $\log_3\left(\dfrac{1}{9}\right) = -2$.

■ **YOUR TURN**   Write each equation in its equivalent logarithmic form.

a. $81 = 9^2$ **b.** $12 = \sqrt{144}$ **c.** $\dfrac{1}{49} = 7^{-2}$

How are logs evaluated? Some can be found exactly, while others must be approximated. The next example illustrates how to find the exact value of a logarithmic function. In the next section, approximating values of logarithmic functions will be discussed.

EXAMPLE 3 Finding the Exact Value of a Logarithmic Function

Find the exact value of

a. $\log_3 81$ **b.** $\log_{169} 13$ **c.** $\log_5\left(\dfrac{1}{5}\right)$

Solution (a):

The logarithm has some value. Let's call it x. $\log_3 81 = x$

Change from logarithmic to exponential form. $3^x = 81$

3 raised to what power is 81? $3^4 = 81$ $x = 4$

Change from exponential to logarithmic form. $\boxed{\log_3 81 = 4}$

Solution (b):

The logarithm has some value. Let's call it x. $\log_{169} 13 = x$

Change from logarithmic to exponential form. $169^x = 13$

169 raised to what power is 13? $169^{1/2} = \sqrt{169} = 13$ $x = \dfrac{1}{2}$

Change from exponential to logarithmic form. $\boxed{\log_{169} 13 = \dfrac{1}{2}}$

■ **Answer: a.** $\log_9 81 = 2$ **b.** $\log_{144} 12 = \dfrac{1}{2}$ **c.** $\log_7\left(\dfrac{1}{49}\right) = -2$

Solution (c):

The logarithm has some value. Let's call it x. $\log_5\left(\dfrac{1}{5}\right) = x$

Change from logarithmic to exponential form. $5^x = \dfrac{1}{5}$

5 raised to what power is $\dfrac{1}{5}$? $5^{-1} = \dfrac{1}{5}$ $x = -1$

Change from exponential to logarithmic form. $\log_5\left(\dfrac{1}{5}\right) = -1$

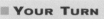 **YOUR TURN** 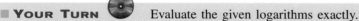 Evaluate the given logarithms exactly.

a. $\log_2 \dfrac{1}{2}$ **b.** $\log_{100} 10$ **c.** $\log_{10} 1000$

Common and Natural Logarithms

Two common bases that arise in logarithms are base 10 and base e. The logarithmic function of base 10 is called the **common logarithmic function**. Since it is common, $f(x) = \log_{10} x$ is often expressed as $f(x) = \log x$. Thus, if no explicit base is indicated, base 10 is implied. The logarithmic function of base e is called the **natural logarithmic function**. The natural logarithmic function $f(x) = \log_e x$ is often expressed as $f(x) = \ln x$. Both the LOG and LN buttons appear on scientific and graphing calculators.

Earlier in this section, we evaluated logarithms exactly by converting to exponential form and identifying the exponent. For example, to evaluate $\log_{10} 100$, we ask the question, 10 raised to what power is 100? The answer is 2. Let us now confirm this with a calculator.

STUDY TIP
$\log_{10}(x) = \log(x)$. No explicit base implies base 10.

Scientific:

Graphing: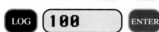

Calculators enable us to approximate logarithms. For example, evaluate $\log_{10} 233$. We are unable to evaluate this exactly by asking the question, 10 raised to what power is 233? Since $10^2 < 10^x < 10^3$ we know the answer, x, must lie between 2 and 3. Instead, we use a calculator to find an approximate value, 2.367.

 **EXAMPLE 4   Using a Calculator to Evaluate Common and Natural Logarithms**

Use a calculator to evaluate the common and natural logarithms. Round your answers to two decimal places.

a. $\log 415$ **b.** $\ln 415$ **c.** $\log 1$
d. $\ln 1$ **e.** $\log(-2)$ **f.** $\ln(-2)$

■ **Answer: a.** $\log_2 \dfrac{1}{2} = -1$ **b.** $\log_{100} 10 = \dfrac{1}{2}$ **c.** $\log_{10} 1000 = 3$

Solution (scientific calculator used):

a. 415		LOG	Answer: 2.62
b. 415		LN	Answer: 6.03
c. 1		LOG	Answer: 0
d. 1		LN	Answer: 0
e. 2	$+/-$	LOG	Answer: Error
f. 2	$+/-$	LN	Answer: Error

Parts c and d in Example 4 illustrate that all logarithmic functions pass through the point $(1, 0)$. Parts e and f in Example 4 illustrate that the domains of logarithmic functions are positive numbers.

Graphs of Logarithmic Functions

The general logarithmic function $y = \log_b x$ is defined as the inverse of the exponential-function $y = b^x$. Therefore, when these two functions are plotted on the same graph, they are symmetric about the line $y = x$.

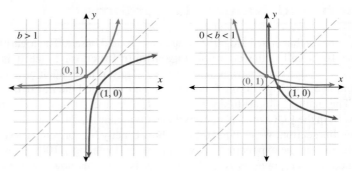

Additionally, the domain of one function is the range of the other and vice versa. When dealing with logarithmic functions, special attention must be paid to the domain of the function. The domain of $y = \log_b x$ is $(0, \infty)$. In other words, you can only take the log of a positive number, $x > 0$.

EXAMPLE 5 Finding the Domain of a Logarithmic Function

Find the domain of the given logarithmic functions.

a. $f(x) = \log_b (x - 4)$ **b.** $g(x) = \log_b (|x + 1|)$

Solution (a):

Determine any domain restrictions.	$x - 4 > 0$
Solve the inequality.	$x > 4$
Write the domain in interval notation.	$(4, \infty)$

Solution (b):

Determine any domain restrictions.	$|x + 1| > 0$
Solve the inequality.	$x \neq -1$
Write the domain in interval notation.	$(-\infty, -1) \cup (-1, \infty)$

■ **YOUR TURN** Find the domain of the given logarithmic functions.
  a. $f(x) = \log_b (x + 2)$ **b.** $g(x) = \log_b (|x - 3|)$

Recall in Section 3.3 a technique for graphing general functions is using transformations of known functions. For example, to graph $f(x) = (x - 3)^2 + 1$, we start with the known parabola $y = x^2$, whose vertex is at $(0, 0)$, and we shift that graph to the right 3 units and up 1 unit. We use the same techniques for graphing logarithmic functions. To graph $y = \log_b (x + 2) - 1$, we start with the graph of $y = \log_b (x)$ and shift the graph to the left 2 units and down 1 unit.

 EXAMPLE 6 **Graphing Logarithmic Functions Using Horizontal and Vertical Shifts**

Graph the functions, and state their domains and ranges.

  a. $y = \log_2 (x - 3)$ **b.** $\log_2 (x) - 3$

Solution:

Identify the base function.	$y = \log_2 x$
Label key features of $y = \log_2 x$.	x-intercept $(1, 0)$
	Vertical asymptote: $x = 0$
	Additional points: $(2, 1)$, $(4, 2)$

a. Shift base function to the *right* 3 units.
  x-intercept $(4, 0)$
  Vertical asymptote: $x = 3$
  Additional points: $(5, 1)$ $(7, 2)$

b. Shift base function *down* 3 units.
  x-intercept: $(1, -3)$
  Vertical asymptote: $x = 0$
  Additional points: $(2, -2)$ $(4, -1)$

a. Domain: $(3, \infty)$ Range: $(-\infty, \infty)$
b. Domain: $(0, \infty)$ Range: $(-\infty, \infty)$

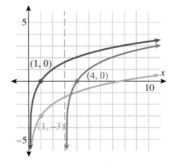

CONCEPT CHECK How is the graph of $f(x + 1)$ related to the graph of $f(x)$? How is the graph of $f(x) + 1$ related to the graph of $f(x)$?

■ **YOUR TURN** Graph the functions and state their domains and ranges.

a. $y = \log_3 x$ **b.** $y = \log_3 (x + 3)$ **c.** $\log_3 (x) + 1$

All of the transformation techniques (shifting, reflection, and compression) discussed in Chapter 3 also apply to logarithmic functions. For example, the graphs of $-\log_2 x$ and $\log_2 (-x)$ are found by reflecting the graph of $y = \log_2 x$ about the x-axis and y-axis, respectively.

EXAMPLE 7 Graphing Logarithmic Functions Using Transformations

Graph the function $f(x) = -\log_2(x - 3)$ and state its domain and range.

Solution:

Graph $y = \log_2 x$.

x-intercept: $(1, 0)$

Vertical asymptote: $x = 0$

Additional points: $(2, 1)$, $(4, 2)$

Graph $y = \log_2 (x - 3)$ by shifting $y = \log_2 x$ to the *right* 3 units.

x-intercept: $(4, 0)$

Vertical asymptote: $x = 3$

Additional points: $(5, 1)$, $(7, 2)$

Graph $y = -\log_2 (x - 3)$ by reflecting $y = \log_2 (x - 3)$ about the x-axis.

x-intercept: $(4, 0)$

Vertical asymptote: $x = 3$

Additional points: $(5, -1)$, $(7, -2)$

Domain: $(3, \infty)$

Range: $(-\infty, \infty)$

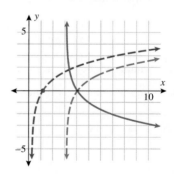

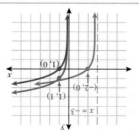

■ **Answer: a. Domain:** $(0, \infty)$ **Range:** $(-\infty, \infty)$

b. Domain: $(-3, \infty)$ **Range:** $(-\infty, \infty)$

c. Domain: $(0, \infty)$ **Range:** $(-\infty, \infty)$

Applications

Logarithms are used to make a large range of numbers manageable. For example, to create a scale to measure a human's ability to hear, we must have a way to measure an explosion, even though the explosion can be more than a trillion times louder than a soft whisper. The values for pH in chemistry, the Richter scale for earthquakes, and decibels that measure sound intensity all are examples of logarithmic functions.

SOUND INTENSITY IN WATTS PER SQUARE METER (W/m²)	SOUND SOURCE
1.0×10^{-12}	Threshold of hearing
1.0×10^{-4}	Vacuum cleaner
1.0×10^{-2}	Walkman
1.0×10^{3}	Jet engine

DEFINITION   **DECIBEL**

The **decibel** is defined as

$$D = 10 \log\left(\frac{I}{I_T}\right)$$

where

D = decibel level of the sound

I = intensity of the sound measured in watts per square meter

I_T = intensity threshold of the least audible sound a human can hear

For average humans, $I_T = 1 \times 10^{-12}$ W/m².

Writing the previous table in terms of decibels (dB), we have

DECIBELS (dB)	SOUND SOURCE
0	Threshold of hearing
80	Vacuum cleaner
100	Walkman
150	Jet engine

EXAMPLE 8 Calculating Decibels of Sounds

Suppose you have front row seats to a concert given by your favorite musical artist. Calculate the approximate decibel levels associated with the sound intensity, given $I = 1 \times 10^{-2}$ W/m².

Solution:

Write the decibel scale formula.

$$D = 10 \log\left(\frac{I}{I_T}\right)$$

Substitute $I = 1 \times 10^{-2}$ W/m² and $I_T = 1 \times 10^{-12}$ W/m².

$$D = 10 \log\left(\frac{1 \times 10^{-2}}{1 \times 10^{-12}}\right)$$

Simplify. $\qquad D = 10 \log \left(10^{10}\right)$

Recall that the implied base for log is 10. $\qquad D = 10 \log_{10} \left(10^{10}\right)$

Evaluate the right side. $\left(\log_{10} \left(10^{10}\right) = 10\right)$ $\qquad D = 10 \cdot 10$

$$D = 100$$

The typical sound level on the front row of a rock concert is $100 \, dB$.

YOUR TURN Calculate the approximate decibels associated with a sound so loud it will cause instant perforation of the eardrums, $I = 1 \times 10^4 \, \text{W/m}^2$.

Another common logarithmic scale, often in the news, is the Richter scale.

DEFINITION **RICHTER SCALE**

The magnitude, M, of an earthquake is measured using the **Richter scale**

$$M = \frac{2}{3} \log \left(\frac{E}{E_0}\right)$$

where

M is the magnitude
E is the seismic energy released by the earthquake (in joules)
E_0 is the energy released by a reference earthquake, $E_0 = 10^{4.4}$ joules

EXAMPLE 9 Calculating the Magnitude of an Earthquake

On October 17, 1989, just moments before game 3 of the World Series between the Oakland A's and the San Francisco Giants was about to start—with 60,000 fans in Candlestick Park—a devastating earthquake erupted. Parts of interstates and bridges collapsed, and President George H. W. Bush declared the area a disaster zone. The earthquake released approximately 1.12×10^{15} joules. Calculate the magnitude of the earthquake using the Richter scale.

Solution:

Write the Richter scale formula. $\qquad M = \frac{2}{3} \log \left(\frac{E}{E_0}\right)$

Substitute $E = 1.12 \times 10^{15}$ and $E_0 = 10^{4.4}$. $\qquad M = \frac{2}{3} \log \left(\frac{1.12 \times 10^{15}}{10^{4.4}}\right)$

Simplify. $\qquad M = \frac{2}{3} \log \left(1.12 \times 10^{10.6}\right)$

Approximate the logarithm using a calculator. $\qquad M \approx \frac{2}{3}(10.65) = 7.1$

The 1989 earthquake in California measured 7.1 on the Richter scale.

▪ **YOUR TURN** On May 3, 1996, Seattle experienced a moderate earthquake. The energy that the earthquake released was approximately 1.12×10^{12} joules. Calculate the magnitude of the 1996 Seattle earthquake using the Richter scale.

SECTION 5.3 SUMMARY

In this section, logarithmic functions, $y = \log_b x$, were defined as inverses of exponential functions, $y = b^x$. The most common bases for logarithmic functions are 10 and e, also called the common and natural logarithms, respectively.

$$\text{Common logarithm:} \quad y = \log x$$

$$\text{Natural logarithm:} \quad y = \ln x$$

Logarithms were evaluated exactly by changing to exponential form first. Common and natural logarithms were evaluated exactly and by using a calculator. The general logarithmic function, $y = \log_b x$, is defined only for positive values of x, has an x-intercept at $(1, 0)$, and the y-axis is a vertical asymptote. Other logarithmic functions can be graphed using transformations.

SECTION 5.3 EXERCISES

▪ SKILLS

In Exercises 1–10, write each logarithmic equation in its equivalent exponential form.

1. $\log_5 125 = 3$ **2.** $\log_3 27 = 3$ **3.** $\log_{81} 3 = \dfrac{1}{4}$ **4.** $\log_{121} 11 = \dfrac{1}{2}$ **5.** $\log_2 \left(\dfrac{1}{32} \right) = -5$

6. $\log_3 \left(\dfrac{1}{81} \right) = -4$ **7.** $\log 0.01 = -2$ **8.** $\log 10{,}000 = 4$ **9.** $\log_{1/4} (64) = -3$ **10.** $\log_{1/6} (36) = -2$

In Exercises 11–20, write each exponential equation in its equivalent logarithmic form.

11. $8^3 = 512$ **12.** $2^6 = 64$ **13.** $0.00001 = 10^{-5}$ **14.** $100{,}000 = 10^5$ **15.** $15 = \sqrt{225}$

16. $7 = \sqrt[3]{343}$ **17.** $\dfrac{8}{125} = \left(\dfrac{2}{5} \right)^3$ **18.** $\dfrac{8}{27} = \left(\dfrac{2}{3} \right)^3$ **19.** $3 = \left(\dfrac{1}{27} \right)^{-1/3}$ **20.** $4 = \left(\dfrac{1}{1024} \right)^{-1/5}$

In Exercises 21–28, evaluate the logarithms exactly.

21. $\log_2 1$ **22.** $\log_5 1$ **23.** $\log_5 3125$ **24.** $\log_3 729$

25. $\log 10^7$ **26.** $\log 10^{-2}$ **27.** $\log_{1/4} 4096$ **28.** $\log_{1/7} 2401$

In Exercises 29–36, approximate (if possible) the common and natural logarithms using a calculator. Round to two decimal places.

29. $\log 29$ **30.** $\ln 29$ **31.** $\ln 380$ **32.** $\log 380$

33. $\log (0)$ **34.** $\ln 0$ **35.** $\ln 0.0003$ **36.** $\log 0.0003$

▪ **Answer:** 5.1

In Exercises 37–42, state the domain of the logarithmic function in interval notation.

37. $f(x) = \log_2(x + 5)$ **38.** $f(x) = \log_3(5 - x)$ **39.** $f(x) = \log(|x|)$

40. $f(x) = \log(|x + 1|)$ **41.** $f(x) = \log(x^2 + 1)$ **42.** $f(x) = \log(1 - x^2)$

In Exercises 43–48, match the graph with the function.

43. $y = \log_5 x$ **44.** $y = \log_5(-x)$ **45.** $y = -\log_5(-x)$

46. $y = \log_5(x + 3) - 1$ **47.** $y = \log_5(1 - x) - 2$ **48.** $y = -\log_5(3 - x) + 2$

a.
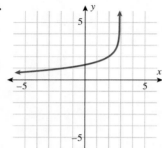

b.
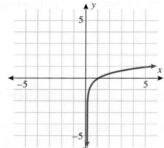

c.
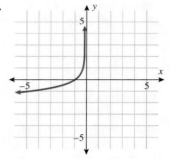

d.
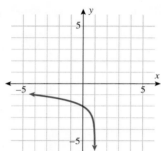

e.
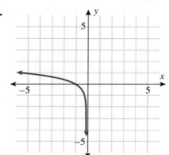

f.
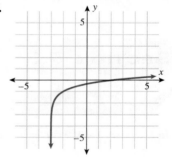

In Exercises 49–56, graph the logarithmic function using transformation techniques.

49. $f(x) = \log_3(x + 2) - 1$ **50.** $f(x) = \log_3(x + 1) - 2$ **51.** $f(x) = -\log(x) + 1$ **52.** $f(x) = \log(-x) + 2$

53. $f(x) = \ln(x + 4)$ **54.** $f(x) = \ln(4 - x)$ **55.** $f(x) = \log(2x)$ **56.** $f(x) = 2\ln(-x)$

■ APPLICATIONS

57. Sound. Calculate the decibels associated with *normal conversation* if the intensity is $I = 1 \times 10^{-6}$ W/m^2.

58. Sound. Calculate the decibels associated with the *onset of pain* if the intensity is $I = 1 \times 10^{1}$ W/m^2.

59. Sound. Calculate the decibels associated with attending *a football game in a loud college stadium* if the intensity is $I = 1 \times 10^{-0.3}$ W/m^2.

60. Sound. Calculate the decibels associated with a *doorbell* if the intensity is $I = 1 \times 10^{-4.5}$ W/m^2.

61. Earthquakes. On Good Friday 1964, one of the most severe North American earthquakes ever recorded struck Alaska. The energy released measured 1.41×10^{17} joules. Calculate the magnitude of the 1964 Alaska earthquake using the Richter scale.

62. Earthquakes. On January 22, 2003, Colima, Mexico experienced a major earthquake. The energy released measured 6.31×10^{15} joules. Calculate the magnitude of the 2003 Mexican earthquake using the Richter scale.

63. Earthquakes. On December 26, 2003 a major earthquake rocked southeastern Iran. In Bam, 30,000 people were killed, and 85% of buildings were damaged or destroyed. The energy released measured 2×10^{14} joules. Calculate the magnitude of the 2003 Iran earthquake using the Richter scale.

64. Earthquakes. On November 1, 1755, Lisbon was destroyed by an earthquake, which killed 90,000 people and destroyed 85% of the city. It was one of the most destructive earthquakes in history. The energy released measured 8×10^{17} joules. Calculate the magnitude of the 1755 Lisbon earthquake using the Richter scale.

In Exercises 65–68, refer to the following:

The pH of a solution is a measure of the molar concentration of hydrogen ions, H^+ in moles per liter, in the solution, which means that it is a measure of the acidity or basicity of the solution. The letters pH stand for "power of hydrogen," and the numerical value is defined as

$$\text{pH} = -\log_{10}[H^+]$$

Very acid corresponds to pH values near 1, neutral corresponds to a pH near 7 (pure water), and very basic corresponds to values near 14. In the next four exercises you will be asked to calculate the pH value of wine, Pepto-Bismol, normal rainwater, tomato juice, acid rain, and bleach. List these six liquids and use your intuition to classify them as neutral, acidic, very acidic, basic, or very basic before you calculate their actual pH values.

65. Chemistry. If wine has an approximate hydrogen ion concentration of 5.01×10^{-4}, calculate its pH value.

66. Chemistry. Pepto-Bismol has a hydrogen ion concentration of about 5.01×10^{-11}. Calculate its pH value.

67. Chemistry. Normal rainwater is slightly acidic and has an approximate hydrogen ion concentration of $10^{-5.6}$. Calculate its pH value. Acid rain and tomato juice have similar approximate hydrogen ion concentrations of 10^{-4}. Calculate the pH value of acid rain and tomato juice.

68. Chemistry. Bleach has an approximate hydrogen ion concentration of 5.0×10^{-13}. Calculate its pH value.

69. Archeology. Carbon dating is a method used to determine the age of a fossil or other organic remains. The age in years, t, is related to the mass (in milligrams) of carbon 14, C, through a logarithmic equation:

$$t = -\frac{\ln\left(\dfrac{C}{500}\right)}{0.0001216}$$

How old is a fossil that contains 100 milligrams of carbon 14?

70. Archeology. Repeat Exercise 63 if the fossil contains 40 milligrams of carbon 14.

71. Broadcasting. Decibels are used to quantify losses associated with atmospheric interference in a communication system. The ratio of the power (watts) received to the power transmitted (watts) is often compared. Often, the watts are transmitted but losses due to the atmosphere typically correspond to milliwatts being received.

$$\text{dB} = 10 \log\left(\frac{\text{power received}}{\text{power transmitted}}\right)$$

If 1 W is transmitted and 3 mW are received, calculate the power loss in dB.

72. Broadcasting. Repeat Exercise 65 assuming 3 W are transmitted and 0.2 mW is received.

■ **CATCH THE MISTAKE**

In Exercises 73–76, explain the mistake that is made.

73. Evaluate the logarithm $\log_2 4$.

Solution:

Set the logarithm equal to x.	$\log_2 4 = x$
Write the logarithm in exponential form.	$x = 2^4$
Simplify.	$x = 16$
Answer:	$\log_2 4 = 16$

This is incorrect. The correct answer is $\log_2 4 = 2$. What went wrong?

74. Evaluate the logarithm $\log_{100} 10$.

Solution:

Set the logarithm equal to x.	$\log_{100} 10 = x$
Express the equation in exponential form.	$10^x = 100$
Solve for ?.	$x = 2$
Answer:	$\log_{100} 10 = 2$

This is incorrect. The correct answer is $\log_{100} 10 = \dfrac{1}{2}$. What went wrong?

75. State the domain of the logarithmic function $f(x) = \log_2(x + 5)$ in interval notation.

Solution:

The domain of all logarithmic functions is $x > 0$.

Interval notation: $(0, \infty)$

This is incorrect. What went wrong?

76. State the domain of the logarithmic function $f(x) = \ln(|x|)$ in interval notation.

Solution:

Since the absolute value eliminates all negative numbers, the domain is the set of all real numbers.

Interval notation: $(-\infty, \infty)$

This is incorrect. What went wrong?

 ■ **CHALLENGE**

77. T or F: The domain of the standard logarithmic function, $y = \ln x$, is the set of nonnegative real numbers.

78. T or F: The horizontal axis is the horizontal asymptote of the graph of $y = \ln x$.

79. T or F: The graphs of $y = \log x$ and $y = \ln x$ have the same x-intercept $(1, 0)$.

80. T or F: The graphs of $y = \log x$ and $y = \ln x$ have the same vertical asymptote, $x = 0$.

81. State the domain, range, and x-intercept of the function $f(x) = -\ln(x - a) + b$ for a and b real positive numbers.

82. State the domain, range, and x-intercept of the function $f(x) = \log(a - x) - b$ for a and b real positive numbers.

83. Graph the function $f(x) = \begin{cases} \ln(-x) & x < 0 \\ \ln(x) & x > 0 \end{cases}$.

84. Graph the function $f(x) = \begin{cases} -\ln(-x) & x < 0 \\ -\ln(x) & x > 0 \end{cases}$.

 ■ **TECHNOLOGY**

85. Using a graphing utility, graph $y = e^x$ and $y = \ln x$ in the same viewing screen. What line are these two graphs symmetric about?

86. Using a graphing utility, graph $y = 10^x$ and $y = \log x$ in the same viewing screen. What line are these two graphs symmetric about?

87. Using a graphing utility, graph $y = \log x$ and $y = \ln x$ in same viewing screen. What are the two common characteristics?

88. Using a graphing utility, graph $y = \log(|x|)$. Is the function defined everywhere?

SECTION 5.4 Properties of Logarithms

Skills Objectives

■ Write a single logarithm as a sum or difference of logarithms.
■ Write a logarithmic expression as a single logarithm.
■ Evaluate logarithms of a general base (other than base 10 or e).

Conceptual Objectives

■ Derive the seven basic logarithmic properties.
■ Derive the change-of-base formula.

Since exponential functions and logarithmic functions are inverses of one another, properties of exponents are related to properties of logarithms. We will start by reviewing properties of exponents, and then proceed to properties of logarithms.

In Chapter 0 properties of exponents were discussed.

PROPERTIES OF EXPONENTS

Let a, b, m, and n be any real numbers and $m > 0$ and $n > 0$. Then the following are true.

$$b^m \cdot b^n = b^{m+n} \qquad b^{-m} = \frac{1}{b^m} = \left(\frac{1}{b}\right)^m \qquad \frac{b^m}{b^n} = b^{m-n}$$

$$(b^m)^n = b^{mn} \qquad (ab)^m = a^m \cdot b^m$$

$$b^0 = 1 \qquad b^1 = b$$

From these properties of exponents we can develop similar properties for logarithms. We list seven basic properties.

Properties of Logarithms

PROPERTIES OF LOGARITHMS

If b, M, and N are positive real numbers, $b \neq 1$, and p and x are real numbers then the following are true:

1. $\log_b 1 = 0$

2. $\log_b b = 1$

3. $\log_b b^x = x$

4. $b^{\log_b x} = x \quad x > 0$

5. $\log_b MN = \log_b M + \log_b N$ Product rule: Log of a product is the sum of the logs.

6. $\log_b \left(\dfrac{M}{N}\right) = \log_b M - \log_b N$ Quotient rule: Log of a quotient is the difference of the logs.

7. $\log_b M^p = p \log_b M$ Power rule: Log of a number raised to an exponent is the exponent times the log of the number.

We will devote this section to proving and illustrating these seven properties.

The first two properties follow directly from the definition of a logarithmic function and properties of exponentials.

Property (1): $\log_b 1 = 0$ since $b^0 = 1$

Property (2): $\log_b b = 1$ since $b^1 = b$

The third and fourth properties follow from the fact that exponential functions and logarithmic functions are inverses of one another. Recall that inverse functions satisfy the relationship that $f^{-1}(f(x)) = x$ for all x in the domain of $f(x)$, and $f(f^{-1}(x)) = x$ for all x in the domain of f^{-1}. Let $f(x) = b^x$ and $f^{-1}(x) = \log_b x$.

Property (3):

Write the inverse identity.	$f^{-1}(f(x)) = x$
Substitute $f^{-1}(x) = \log_b x$.	$\log_b (f(x)) = x$
Substitute $f(x) = b^x$.	$\log_b b^x = x$

Property (4):

Write the inverse identity.	$f(f^{-1}(x)) = x$
Substitute $f(x) = b^x$.	$b^{f^{-1}(x)} = x$
Substitute $f^{-1}(x) = \log_b x,\ x > 0$.	$b^{\log_b x} = x$

The first four properties are summarized for common and natural logarithms.

COMMON AND NATURAL LOGARITHM PROPERTIES

COMMON LOGARITHM (BASE 10)	**NATURAL LOGARITHM** (BASE e)
1. $\log 1 = 0$	**1.** $\ln 1 = 0$
2. $\log 10 = 1$	**2.** $\ln e = 1$
3. $\log 10^x = x$	**3.** $\ln e^x = x$
4. $10^{\log x} = x \qquad x > 0$	**4.** $e^{\ln x} = x \qquad x > 0$

EXAMPLE 1 Using Logarithmic Properties

Use properties 1–4 to simplify the following expressions:

a. $\log_{10} 10$ **b.** $\ln 1$ **c.** $10^{\log (x+8)}$
d. $e^{\ln (2x+5)}$ **e.** $\log 10^{x^2}$ **f.** $\ln e^{x+3}$

Solution:

a. Use property (2). $\log_{10} 10 = 1$

b. Use property (1). $\ln 1 = 0$

c. Use property (4). $10^{\log (x+8)} = x + 8$ $x > -8$

d. Use property (4). $e^{\ln (2x+5)} = 2x + 5$ $x > -\dfrac{5}{2}$

e. Use property (3). $\log 10^{x^2} = x^2$

f. Use property (3). $\ln e^{x+3} = x + 3$

The fifth through seventh properties follow from the properties of exponents and the definition of logarithms. We will prove the product rule and leave the proofs of the quotient and power rules for the exercises.

Property (5): $\log_b MN = \log_b M + \log_b N$

WORDS	**MATH**
Assume two logs that have the same base.	Let $u = \log_b M$ and $v = \log_b N$ $M > 0,\ N > 0$
Change to equivalent exponential forms.	$b^u = M$ and $b^v = N$
Write the log of a product.	$\log_b MN$
Substitute $M = b^u$ and $N = b^v$.	$\log_b (b^u b^v)$

Use properties of exponents. $\log_b (b^{u+v})$

Apply property 3. $u + v$

Substitute $u = \log_b M$, $v = \log_b N$. $\log_b M + \log_b N$

$$\log_b MN = \log_b M + \log_b N$$

In other words, the log of a product is the sum of the logs.

EXAMPLE 2 Writing a Logarithmic Expression as a Sum of Logarithms

Use the logarithmic properties to write the following expression as a sum of simpler logarithms: $\log_b \left(u^2 \sqrt{v} \right)$.

Solution:

Convert the radical to exponential form. $\log_b \left(u^2 \sqrt{v} \right) = \log_b \left(u^2 v^{1/2} \right)$

Use the product property (5). $\log_b \left(u^2 v^{1/2} \right) = \log_b u^2 + \log_b v^{1/2}$

Use the power property (7). $\log_b \left(u^2 v^{1/2} \right) = 2 \log_b u + \dfrac{1}{2} \log_b v$

 CONCEPT CHECK   Express the radical $\sqrt[3]{y}$ in terms of a fractional exponent.

YOUR TURN   Use the logarithmic properties to write the following expression as a sum of simpler logarithms: $\log_b \left(x^4 \sqrt[3]{y} \right)$.

EXAMPLE 3 Writing a Sum of Logarithms as a Single Logarithmic Expression: The Right Way and the Wrong Way

Use properties of logarithms to write the expression $2 \log_b 3 + 4 \log_b u$ as a single logarithmic expression.

COMMON MISTAKE

A common mistake is to write the sum of the logs as a log of the sum.

$$\log_b M + \log_b N \neq \log_b (M + N)$$

 ✓ CORRECT

Use the power property (7).

$2 \log_b 3 + 4 \log_b u = \log_b 3^2 + \log_b u^4$

Simplify.

$$\log_b 9 + \log_b u^4$$

Use the product property (5).

$$= \log_b \left(9u^4 \right)$$

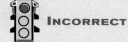 **✗ INCORRECT**

Use the power property (7).

$2 \log_b 3 + 4 \log_b u = \log_b 3^2 + \log_b u^4$

Simplify.

$$\log_b 9 + \log_b u^4$$

Use the product property (5).

$$\neq \log_b (9 + u^4) \quad \textbf{ERROR}$$

 CAUTION

$\log_b M + \log_b N = \log_b (MN)$

$\log_b M + \log_b N \neq \log_b (M + N)$

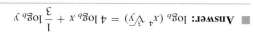 ■ **Answer:** $\log_b \left(x^4 \sqrt[3]{y} \right) = 4 \log_b x + \dfrac{1}{3} \log_b y$

■ **YOUR TURN** Express $2 \ln x + 3 \ln y$ as a single logarithm.

EXAMPLE 4 Writing a Logarithmic Expression as a Difference of Logarithms

Write the expression $\ln\left(\dfrac{x^3}{y^2}\right)$ as a difference of logarithms.

Solution:

Use the quotient property (6). $\qquad$ $\ln\left(\dfrac{x^3}{y^2}\right) = \ln(x^3) - \ln(y^2)$

Use the power property (7). $\qquad\qquad\qquad$ $3 \ln x - 2 \ln y$

■ **YOUR TURN** Write the expression $\log\left(\dfrac{a^4}{b^5}\right)$ as a difference of logarithms.

Another common mistake is misinterpreting the quotient rule.

EXAMPLE 5 Write the Difference of Logarithms as a Logarithm of a Quotient

Write the expression $\dfrac{2}{3} \ln x - \dfrac{1}{2} \ln y$ as a logarithm of a quotient.

CAUTION

$\log_b M - \log_b N =$
$\log_b\left(\dfrac{M}{N}\right)$

$\log_b M - \log_b N \neq$
$\dfrac{\log_b M}{\log_b N}$

COMMON MISTAKE

$$\log_b M - \log_b N \neq \frac{\log_b M}{\log_b N}$$

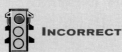

CORRECT

Use the power property (7).

$\dfrac{2}{3} \ln x - \dfrac{1}{2} \ln y = \ln x^{2/3} - \ln y^{1/2}$

Use the quotient property (6).

$\ln\left(\dfrac{x^{2/3}}{y^{1/2}}\right)$

INCORRECT

Use the power property (7).

$\dfrac{2}{3} \ln x - \dfrac{1}{2} \ln y = \ln x^{2/3} - \ln y^{1/2}$

Use the quotient property (6).

$\dfrac{\ln x^{2/3}}{\ln y^{1/2}}$ **ERROR**

■ **YOUR TURN** Write the expression $\frac{1}{2} \log a - 3 \log b$ as a single logarithm.

EXAMPLE 6 Writing Expressions as a Single Logarithm

Write the expression $3 \log_b x + \log_b (2x + 1) - 2 \log_b 4$ as a single logarithm.

Solution:

Use the power property (7) on the first and third terms.	$\log_b x^3 + \log_b (2x + 1) - \log_b 4^2$
Use the product property (5) on the first two terms.	$\log_b \left[x^3(2x + 1) \right] - \log_b 16$
Use the quotient property (6).	$\log_b \left[\dfrac{x^3(2x + 1)}{16} \right]$

■ **YOUR TURN** Write the expression $2 \ln x - \ln (3y) + 3 \ln z$ as a single logarithm.

Change-of-Base Formula

Recall in the last section that we were able to evaluate logarithms two ways: (1) Exactly by writing the logarithm in exponential form and identifying the exponent and (2) Using a calculator if the logarithms were base 10 or e. How do we evaluate a logarithm of general base if we cannot identify the exponent? We use the *change-of-base formula.*

EXAMPLE 7 Using Properties of Logarithms to Change the Base to Evaluate a General Logarithm

Evaluate $\log_3 8$.

Solution:

Let $y = \log_3 8$.	$y = \log_3 8$
Write the logarithm in exponential form.	$3^y = 8$
Take the log of both sides.	$\log 3^y = \log 8$
Use the power property (7).	$y \log 3 = \log 8$
Divide both sides by log 3.	$y = \dfrac{\log 8}{\log 3}$
Use a calculator to approximate.	$y \approx 1.89$
Substitute $y = \log_3 8$.	$\log_3 8 \approx 1.89$

■ **Answer:** $\log \left(\dfrac{a^{1/2}}{b^3} \right)$ ■ **Answer:** $\ln \left(\dfrac{x^2 z^3}{3y} \right)$

Example 7 illustrated our ability to use properties of logarithms to change from base 3 to base 10, which our calculators can handle. This leads to the general change-of-base formula.

CHANGE-OF-BASE FORMULA

For any logarithmic bases a and b, and any positive number M:

$$\log_b M = \frac{\log_a M}{\log_a b}$$

When a is either 10 or e, this relationship becomes

Common logarithms   or   Natural logarithms

$$\log_b M = \frac{\log M}{\log b} \qquad \text{or} \qquad \log_b M = \frac{\ln M}{\ln b}$$

Note: It does not matter which base we select, 10 or e—the ratios will be the same.

EXAMPLE 8 Using the Change-of-Base Formula

Use the change-of-base formula to evaluate $\log_4 17$. Round to two decimal places.

Solution:

We will illustrate this two ways (choosing common and natural logarithms) using a scientific calculator.

Common Logarithms

Use the change-of-base formula with base 10. $\log_4 17 = \dfrac{\log 17}{\log 4}$

Approximate using a calculator. $\boxed{17}\,\boxed{\text{LOG}}\,\boxed{\div}\,\boxed{4}\,\boxed{\text{LOG}}\,\boxed{=}$ ≈ 2.04

Natural Logarithms

Use the change-of-base formula with base e. $\log_4 17 = \dfrac{\ln 17}{\ln 4}$

Approximate using a calculator. $\boxed{17}\,\boxed{\text{LN}}\,\boxed{\div}\,\boxed{4}\,\boxed{\text{LN}}\,\boxed{=}$ ≈ 2.04

■ **YOUR TURN**   Use the change-of-base formula to approximate $\log_7 34$. Round to two decimal places.

■ **Answer:** $\log_7 34 \approx 1.81$

SECTION 5.4 SUMMARY

In this section, we discussed seven properties of logarithms. The first four are results of the definition of a logarithm as the inverse of an exponential. The three additional properties (product rule, quotient rule, and power rule) were discussed, and common mistakes were mentioned. It is important to note that these seven properties only hold when the base is the same. The change of base formula was introduced as a way of evaluating logarithms that are not base 10 or e and that can't be done exactly.

SECTION 5.4 EXERCISES

 ■ SKILLS

In Exercises 1–18, use the properties of logarithms to simplify each expression. Do not use a calculator.

1. $\log_9 1$

2. $\log_{69} 1$

3. $\log_{1/2}\left(\dfrac{1}{2}\right)$

4. $\log_{3.3} 3.3$

5. $\log_{10} 10^8$

6. $\ln e^3$

7. $\log_{10} 0.001$

8. $\log_3 3^7$

9. $\log_2 \sqrt{8}$

10. $\log_5 \sqrt[3]{5}$

11. $8^{\log_8 5}$

12. $2^{\log_2 5}$

13. $e^{\ln(x+5)}$

14. $10^{\log(3x^2+2x+1)}$

15. $5^{3\log_5 2}$

16. $7^{2\log_7 5}$

17. $7^{-2\log 3}$

18. $e^{-2\ln 10}$

In Exercises 19–32, write each expression as a sum or difference of logarithms.

Example: $\log(m^2 n^5) = 2\log m + 5\log n$

19. $\log_b(x^3 y^5)$

20. $\log_b(x^{-3} y^{-5})$

21. $\log_b(x^{1/2} y^{1/3})$

22. $\log_b(\sqrt{r}\sqrt[3]{t})$

23. $\log_b\left(\dfrac{r^{1/3}}{s^{1/2}}\right)$

24. $\log_b\left(\dfrac{r^4}{s^2}\right)$

25. $\log_b\left(\dfrac{x}{yz}\right)$

26. $\log_b\left(\dfrac{xy}{z}\right)$

27. $\log(x^2\sqrt{x+5})$

28. $\log[(x-3)(x+2)]$

29. $\ln\left[\dfrac{x^3(x-2)^2}{\sqrt{x^2+5}}\right]$

30. $\ln\left[\dfrac{\sqrt{x+3}\sqrt[3]{x-4}}{(x+1)^4}\right]$

31. $\log\left[\dfrac{x^2-2x+1}{x^2-9}\right]$

32. $\log\left[\dfrac{x^2-x-2}{x^2+3x-4}\right]$

In Exercises 33–44, write each expression as a single logarithm.

Example: $2\log m + 5\log n = \log(m^2 n^5)$

33. $3\log_b x^3 + 5\log_b y$

34. $2\log_b u + 3\log_b v$

35. $5\log_b u - 2\log_b v$

36. $3\log_b x - \log_b y$

37. $\dfrac{1}{2}\log_b x + \dfrac{2}{3}\log_b y$

38. $\dfrac{1}{2}\log_b x - \dfrac{2}{3}\log_b y$

39. $2\log u - 3\log v - 2\log z$

40. $3\log u - \log 2v - \log z$

41. $\ln(x+1) + \ln(x-1) - 2\ln(x^2+3)$

42. $\ln\sqrt{x-1} + \ln\sqrt{x+1} - 2\ln(x^2-1)$

43. $\dfrac{1}{2}\ln(x+3) - \dfrac{1}{3}\ln(x+2) - \ln(x)$

44. $\dfrac{1}{3}\ln(x^2+4) - \dfrac{1}{2}\ln(x^2-3) - \ln(x-1)$

In Exercises 45–54, evaluate the logarithms using the change-of-base formula.

45. $\log_5 7$

46. $\log_4 19$

47. $\log_{1/2} 5$

48. $\log_5 \dfrac{1}{2}$

49. $\log_{2.7} 5.2$

50. $\log_{7.2} 2.5$

51. $\log_\pi 10$

52. $\log_\pi 2.7$

53. $\log_{\sqrt{3}} 8$

54. $\log_{\sqrt{2}} 9$

 ■ **APPLICATIONS**

55. Sound. Sitting in the front row of a rock concert exposes us to 1×10^{-1} W/m^2 (or 110 dB), and a normal conversation is typically around 1×10^{-6} W/m^2 (or 60 dB). How many decibels are you exposed to if a friend is talking in your ear at a rock concert? (Note: 160 dB causes perforation of the eardrums.)

56. Sound. A whisper corresponds to 1×10^{-10} W/m^2 (or 20 dB) and a normal conversation is typically around 1×10^{-6} W/m^2 (or 60 dB). How many decibels are you exposed to if one friend is whispering in your ear while the other one is talking at a normal level?

In Exercises 57 and 58, refer to the following:

There are two types of waves associated with an earthquake: *compression* and *shear*. The compression, or longitudinal, waves

displace material behind its path. Longitudinal waves travel at great speeds and are often called "primary waves" or simply "P" waves. Shear, or transverse, waves displace material at right angles to its path. Transverse waves do not travel as rapidly through the Earth's crust and mantle as do longitudinal waves, and they are called "secondary" or "S" waves.

57. Earthquakes. If a seismologist records the energy of P waves as 4.5×10^{12} joules and the energy of S waves as 7.8×10^8 joules, what is the total energy? What would the combined effect be on the Richter scale?

58. Earthquakes. Repeat Exercise 57 if the energy associated with the P waves was 5.2×10^{11} joules and the energy associated with the S waves was 4.1×10^9 joules.

 ■ **CATCH THE MISTAKE**

In Exercises 59–62, simplify if possible. Explain the mistake that is made.

59. $3 \log 5 - \log 25$

Solution:

Use the quotient property (6). $\dfrac{3 \log 5}{\log 25}$

Write $25 = 5^2$. $\dfrac{3 \log 5}{\log 5^2}$

Use the power property (7). $\dfrac{3 \log 5}{2 \log 5}$

Simplify. $\dfrac{3}{2}$

This is incorrect. The correct answer is $\log 5$. What mistake was made?

60. $\ln 3 + 2 \ln 4 - 3 \ln 2$

Solution:

Use the power property (7). $\ln 3 + \ln 4^2 - \ln 2^3$

Simplify. $\ln 3 + \ln 16 - \ln 8$

Use property (5). $\ln(3 + 16 - 8)$

Simplify. $\ln 11$

This is incorrect. The correct answer is $\ln 64$. What mistake was made?

61. $\log_2 x + \log_3 y - \log_4 z$

Solution:

Use the product property (5). $\log_6 xy - \log_4 z$

Use the quotient property (6). $\log_{24} xyz$

This is incorrect. What mistake was made?

62. $2(\log 3 - \log 5)$

Solution:

Use the quotient property (6). $2\left(\log \dfrac{3}{5}\right)$

Use the power property (7). $\left(\log \dfrac{3}{5}\right)^2$

Use a calculator to approximate. ≈ 0.0492

This is incorrect. What mistake was made?

■ **CHALLENGE**

63. T or F: $\log e = \dfrac{1}{\ln 10}$

64. T or F: $\ln e = \dfrac{1}{\log 10}$

65. T or F: $\ln (xy)^3 = (\ln x + \ln y)^3$

66. T or F: $\dfrac{\ln a}{\ln b} = \dfrac{\log a}{\log b}$

67. Prove the quotient rule: $\log_b \left(\frac{M}{N}\right) = \log_b M - \log_b N$.

Hint: Let $u = \log_b M$ and $v = \log_b N$, write both in exponential form, and find the quotient, $\log_b \left(\frac{M}{N}\right)$.

68. Prove the power rule: $\log_b M^p = p \log_b M$.

Hint: Let $u = \log_b M$, write this log in exponential form, and find $\log_b M^p$.

69. Write in terms of simpler logarithmic forms:

$$\log_b \left(\sqrt{\dfrac{x^2}{y^3 z^{-5}}} \right)^6$$

70. Show that $\log_b \left(\dfrac{1}{x}\right) = -\log_b x$.

■ **TECHNOLOGY**

71. Use a graphing calculator to plot $y = \ln (2x)$ and $y = \ln 2 + \ln x$. Are they the same graph?

72. Use a graphing calculator to plot $y = \ln (2 + x)$ and $y = \ln 2 + \ln x$. Are they the same graph?

73. Use a graphing calculator to plot $y = \dfrac{\log x}{\log 2}$ and $y = \log x - \log 2$. Are they the same graph?

74. Use a graphing calculator to plot $y = \log \left(\frac{x}{2}\right)$ and $y = \log x - \log 2$. Are they the same graph?

75. Use a graphing calculator to plot $y = \ln (x^2)$ and $y = 2 \ln x$. Are they the same graph?

76. Use a graphing calculator to plot $y = (\ln x)^2$ and $y = 2 \ln x$. Are they the same graph?

SECTION 5.5 Exponential and Logarithmic Equations

Skills Objectives

■ Solve exponential equations.
■ Solve logarithmic equations.
■ Solve application problems using exponential and logarithmic equations.

Conceptual Objectives

■ Understand how exponential and logarithmic equations are solved using properties of one-to-one functions and inverses.

Two Strategies for Solving Exponential and Logarithmic Equations

In this book you have solved algebraic equations such as $x^2 - 9 = 0$, in which the goal is to solve for x by finding the values of x that make the statement true. Exponential and logarithmic equations have the x buried within an exponential or a logarithm, but the goal is the same: Solve for x.

$$\text{Exponential equation:} \quad e^{2x+1} = 5$$

$$\text{Logarithmic equation:} \quad \log (3x - 1) = 7$$

There are two methods for solving exponential and logarithmic equations that are based on the properties of one-to-one functions and inverses. To solve simple exponential and logarithmic equations, we will use one-to-one properties. To solve more complicated exponential and logarithmic equations, we will use properties of inverses. The following box summarizes the one-to-one and inverse properties that hold true when $b > 0$ and $b \neq 1$.

ONE-TO-ONE PROPERTIES

$$b^x = b^y \qquad \text{if and only if} \qquad x = y$$

$$\log_b x = \log_b y \qquad \text{if and only if} \qquad x = y$$

INVERSE PROPERTIES

$$b^{\log_b x} = x \qquad x > 0$$

$$\log_b b^x = x$$

We now outline a strategy for solving exponential and logarithmic equations using the above one-to-one and inverse properties.

STRATEGY FOR SOLVING EXPONENTIAL AND LOGARITHMIC EQUATIONS

SIMPLE EQUATIONS

Rewrite the exponential or logarithmic equation in a form that allows use of the **one-to-one properties to identify** x.

COMPLICATED EQUATIONS

Use one-to-one properties to exponentiate logarithms and take logarithms of exponentials to make use of the **inverse properties to isolate** x.

Solving Exponential Equations

EXAMPLE 1 Solving a Simple Exponential Equation

Solve the exponential equations using the one-to-one property.

a. $3^x = 81$ **b.** $\left(\dfrac{1}{2}\right)^{4y} = 16$

Solution (a):

Substitute $81 = 3^4$. $\qquad\qquad\qquad\qquad\qquad 3^x = 3^4$

Use the one-to-one property to identify x. $\qquad x = 4$

Solution (b):

Substitute $\left(\dfrac{1}{2}\right)^{4y} = \left(\dfrac{1}{2^{4y}}\right) = 2^{-4y}$. $\qquad 2^{-4y} = 16$

Substitute $16 = 2^4$. $\qquad\qquad\qquad\qquad\qquad 2^{-4y} = 2^4$

Use the one-to-one property to identify y. $\qquad y = -1$

■ **YOUR TURN** Solve the equations:

a. $2^{x-1} = 8$ **b.** $\left(\dfrac{1}{3}\right)^y = 27$

In Example 1, we were able to rewrite the equation in a form with the same bases so that we could use the one-to-one property. In Example 2, we will not be able to write them in a form with the same bases. Instead, we will use properties of inverses.

EXAMPLE 2 Solving a More Complicated Exponential Equation with a Base Other Than 10 or e

Solve the exponential equation: $4^{3x+2} = 71$.

Solution:

Take the $\log_4$ of both sides.

$$\log_4\left(4^{3x+2}\right) = \log_4 71$$

Use the property of inverses to simplify the left side.

$$3x + 2 = \log_4 71$$

Subtract 2 from both sides.

$$3x = \log_4 71 - 2$$

Divide both sides by 3.

$$x = \frac{\log_4 71 - 2}{3}$$

Use the change-of-base formula, $\log_4 71 = \dfrac{\ln 71}{\ln 4}$.

$$x = \frac{\dfrac{\ln 71}{\ln 4} - 2}{3}$$

Use a calculator to approximate x to three decimals.

$$x \approx 0.358$$

We could have proceeded in an alternative way by taking either the natural log or common log of both sides and using the power property (instead of using the change-of-base formula) to evaluate the logarithm with base 4.

Take the natural logarithm of both sides.

$$\ln\left(4^{3x+2}\right) = \ln 71$$

Use the power property (7).

$$(3x + 2)\ln 4 = \ln 71$$

Divide by $\ln 4$.

$$3x + 2 = \frac{\ln 71}{\ln 4}$$

Subtract 2 and divide by 3.

$$x = \frac{\dfrac{\ln 71}{\ln 4} - 2}{3}$$

Use a calculator to approximate x.

$$x \approx 0.358$$

■ **YOUR TURN** Solve the equation (round to three decimal places):
$5^{y^2} = 27$.

EXAMPLE 3 Solving a More Complicated Exponential Equation with Base 10 or e

Solve the exponential equation $e^{x^2} = 16$.

Solution:

Take the natural logarithm (ln) of both sides.	$\ln\left(e^{x^2}\right) = \ln 16$
Use the property of inverses to simplify the left side.	$x^2 = \ln 16$
Solve for x using the square root method.	$x = \pm\sqrt{\ln 16}$
Use a calculator to approximate x to three decimal places.	$x \approx \pm 1.665$

■ **YOUR TURN** Solve the equation (round to three decimal places): $10^{2x-3} = 7$.

EXAMPLE 4 Solving an Exponential Equation Quadratic in Form

Solve the equation $e^{2x} - 4e^x + 3 = 0$.

Solution:

Let $u = e^x$. Note: $u^2 = e^x \cdot e^x = e^{2x}$.	$u^2 - 4u + 3 = 0$
Factor.	$(u - 3)(u - 1) = 0$
Solve for u.	$u = 3$ or $u = 1$
Substitute $u = e^x$.	$e^x = 3$ or $e^x = 1$
Take the natural logarithm (ln) of both sides.	$\ln\left(e^x\right) = \ln 3$ or $\ln\left(e^x\right) = \ln 1$
Use properties of logarithms to simplify.	$x = \ln 3$ or $x = \ln 1$
Approximate or evaluate exactly the right sides.	$x \approx 1.10$ or $x = 0$

✓ **CONCEPT CHECK** Does the equation $10^x = -1$ have a solution?

■ **YOUR TURN** Solve the equation $100^x - 10^x - 2 = 0$.

Solving Logarithmic Equations

We will use a similar approach to solving logarithmic equations. We can solve simple logarithmic equations using the property of one-to-one functions. For more complicated logarithmic equations we can use properties of logarithms and properties of inverses. Solutions must be checked to eliminate extraneous solutions.

EXAMPLE 5 Solving a Simple Logarithmic Equation

Solve the equation $\log_4 (2x - 3) = \log_4 (x) + \log_4 (x - 2)$.

Solution:

Use the product property (5) on the right side.

$$\log_4 (2x - 3) = \log_4 \left[x(x - 2)\right]$$

Use the property of one-to-one functions.

$$2x - 3 = x(x - 2)$$

Distribute and simplify.

$$x^2 - 4x + 3 = 0$$

Factor.

$$(x - 3)(x - 1) = 0$$

Solve for x.

$$x = 3 \text{ or } x = 1$$

The possible solution $x = 1$ must be eliminated because it is not in the domain of two of the logarithmic functions.

Answer: $x = 3$

STUDY TIP

Solutions should be checked to eliminate extraneous solutions.

■ YOUR TURN   Solve the equation $\ln (x + 8) = \ln (x) + \ln (x + 3)$.

EXAMPLE 6 Solving a More Complicated Logarithmic Equation

Solve the equation $\log_3 (9x) - \log_3 (x - 8) = 4$.

Solution:

Use the quotient property (6) on the left side.

$$\log_3 \left(\frac{9x}{x - 8}\right) = 4$$

Write in exponential form.

$$\frac{9x}{x - 8} = 3^4$$

Simplify the right side.

$$\frac{9x}{x - 8} = 81$$

Multiply equation by LCD, $x - 8$.

$$9x = 81(x - 8)$$

Eliminate parentheses.

$$9x = 81x - 648$$

Solve for x.

$$-72x = -648$$

$$x = 9$$

■ YOUR TURN   Solve the equation $\log_2 (4x) - \log_2 (2) = 2$.

■ Answer: $x = 2$ **■   ■ Answer:** $x = 2$

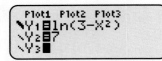
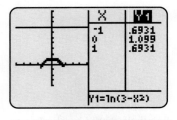
EXAMPLE 7 Solving a Logarithmic Equation with No Solution

Solve the equation $\ln(3 - x^2) = 7$.

Solution:

Exponentiate both sides (base e).

$$e^{\ln(3 - x^2)} = e^7$$

Use the property of inverse functions to simplify the left side.

$$3 - x^2 = e^7$$

Simplify.

$$x^2 = 3 - e^7$$

Note: $3 - e^7$ is negative, so there is   no solution .

Applications

In the chapter opener, we saw that archeologists determine the age of a fossil by how much carbon 14 is present at time of discovery. The number of grams of carbon 14 based on the radioactive decay of the isotope is given by

$$A = A_0 e^{-0.000124t}$$

where A is the number of grams of carbon 14 at the present time, A_0 is the number of grams of carbon 14 while alive, and t is the number of years since death. Using the inverse properties, we can isolate t.

WORDS	MATH
Divide by A_0.	$\dfrac{A}{A_0} = e^{-0.000124t}$
Take the natural logarithm of both sides.	$\ln\left(\dfrac{A}{A_0}\right) = \ln\left(e^{-0.000124t}\right)$
Simplify the right side using properties of inverses.	$\ln\left(\dfrac{A}{A_0}\right) = -0.000124t$
Solve for t.	$t = -\dfrac{1}{0.000124}\ln\left(\dfrac{A}{A_0}\right)$

Let's assume that animals have approximately 1000 mg of carbon 14 in their bodies when they are alive. If a fossil has 200 mg of carbon 14, approximately how old is the fossil? Substituting $A = 200$ and $A_0 = 1000$ into our equation for t we find

$$t = -\frac{1}{0.000124}\ln\left(\frac{1}{5}\right) \approx 12{,}979$$

The fossil is approximately 13,000 years old.

EXAMPLE 8 Calculating How Many Years It Will Take for Money to Double

You save $1,000 from a summer job and put it in a CD earning 5% compounding continuously. How many years will it take for your money to double?

Solution:

Recall the compound continuous interest formula. $\qquad A = Pe^{rt}$

Substitute $P = 1000$, $A = 2000$, and $r = 0.05$. $\qquad 2000 = 1000e^{0.05t}$

Divide by 1000. $\qquad\qquad\qquad\qquad\qquad\qquad\qquad 2 = e^{0.05t}$

Take the natural logarithm of both sides. $\qquad\quad \ln 2 = \ln(e^{0.05t})$

Simplify using the property $\ln e^x = x$. $\qquad\quad\; \ln 2 = 0.05t$

Solve for t. $\qquad\qquad\qquad\qquad\qquad\qquad\qquad t = \dfrac{\ln 2}{0.05} \approx 13.86$

It will take almost $\boxed{14\ \text{years}}$ for your money to double.

■ **YOUR TURN**  How long will it take \$1,000 to triple (become \$3,000) in a savings account earning 10% a year compounding continuously?

SECTION 5.5 SUMMARY

In this section, we solved exponential and logarithmic equations using the properties of one-to-one functions, the properties of inverses, and the properties of exponentials and logarithms. Care must be taken to check extraneous solutions with logarithmic equations due to domain restrictions. Application problems involving exponential and logarithmic equations are solved.

SECTION 5.5 EXERCISES

■ SKILLS

In Exercises 1–10, solve the exponential equations exactly for x.

1. $3^x = 81$ **2.** $5^x = 125$ **3.** $7^x = \dfrac{1}{49}$ **4.** $4^x = \dfrac{1}{16}$ **5.** $2^{x^2} = 16$

6. $169^x = 13$ **7.** $\left(\dfrac{2}{3}\right)^{x+1} = \dfrac{27}{8}$ **8.** $\left(\dfrac{3}{5}\right)^{x+1} = \dfrac{25}{9}$ **9.** $e^{2x+3} = 1$ **10.** $10^{x^2-1} = 1$

In Exercises 11–20, solve the exponential equations. **Round answers to three decimal places.**

11. $10^{2x-3} = 81$ **12.** $2^{3x+1} = 21$ **13.** $e^{3x+4} = 22$ **14.** $e^{x^2} = 73$

15. $3e^{2x} = 18$ **16.** $4(10^x) = 20$ **17.** $e^{2x} + 7e^x - 3 = 0$ **18.** $e^{2x} - 4e^x - 5 = 0$

19. $(3^x - 3^{-x})^2 = 0$ **20.** $(3^x - 3^{-x})(3^x + 3^{-x}) = 0$

■ **Answer:** Approximately 11 years

In Exercises 21–30, solve the logarithmic equations exactly.

21. $\log (2x) = 2$

22. $\log (5x) = 3$

23. $\log_3 (2x + 1) = 4$

24. $\log_2 (3x - 1) = 3$

25. $\ln x^2 - \ln 9 = 0$

26. $\log x^2 + \log x = 3$

27. $\log_5 (x - 4) + \log_5 x = 1$

28. $\log_2 (x + 1) + \log_2 (4 - x) = \log_2 (6x)$

29. $\log_4 (4x) - \log_4 \left(\dfrac{x}{4} \right) = 3$

30. $\log_3 (10 - x) - \log_3 (x + 2) = 1$

In Exercises 31–40, solve the logarithmic equations. Round answers to three decimal places.

31. $\ln x^2 = 5$

32. $\log 3x = 2$

33. $\log (2x + 5) = 2$

34. $\ln (4x - 7) = 3$

35. $\log (2 - 3x) + \log (3 - 2x) = 1.5$

36. $\ln (4x) + \ln (2 + x) = 23$

37. $\log_7 (1 - x) - \log_7 (x + 2) = \log_7 x^2$

38. $\log (\sqrt{1 - x}) - \log (\sqrt{x + 2}) = \log x$

39. $\dfrac{1}{3} \log_b (x^3) + \dfrac{1}{2} \log_b (x^2 - 2x + 1) = 2$

40. $2 \log_b (x) + 2 \log_b (1 - x) = 4$

Solve for x in terms of b.

 ■ APPLICATIONS

41. Money. If money is invested in a savings account earning 3.5% interest compounded yearly, how many years until the money triples?

42. Money. If money is invested in a savings account earning 3.5% interest compounded monthly, how many years until the money triples?

43. Money. If $7,500 is invested in a savings account earning 5% interest compounded quarterly, how many years until there is $20,000?

44. Money. If $9,000 is invested in a savings account earning 6% interest compounded continuously, how many years until there is $15,000?

45. Earthquake. On September 25, 2003 an earthquake that measured 7.4 on the Richter scale shook Hokkaido, Japan. How much energy (joules) did the earthquake emit?

46. Earthquake. Again, on that same day (September 25, 2003), a second earthquake that measured 8.3 on the Richter scale shook Hokkaido, Japan. How much energy (joules) did the earthquake emit?

47. Sound. Matt likes to drive around campus in his classic Mustang with the stereo blaring. If his boom stereo has a sound intensity of 120 dB, how many watts per square meter does the stereo emit?

48. Sound. The New York Philharmonic has a sound intensity of 100 dB. How many watts per square meter does the orchestra emit?

49. HIV/AIDS. In 2003, an estimated 1 million people had been infected with HIV in the United States. If the infection rate increases at an annual rate of 2.5% a year

compounding continuously, how many years until there are 2 million Americans infected with HIV?

50. HIV/AIDS. In 2003, an estimated 25 million people had been infected with HIV in sub-Saharan Africa. If the infection rate increases at an annual rate of 9% a year compounding continuously, how long until there are half a billion sub-Saharan Africans infected with HIV?

51. Anesthesia. When a person has a cavity filled, the dentist typically administers a local anesthetic. After leaving the dentist's office, one's mouth often remains numb for several more hours. If a shot of anesthesia is injected into the bloodstream at the time of the procedure ($t = 0$), and the amount of anesthesia still in the bloodstream t hours after the initial injection is given by $A = A_0 e^{-0.5t}$, how long until there is only 10% of the original anesthetic still in the bloodstream?

52. Introducing a New Car Model. If the number of new model Honda Accord Hybrids purchased in North America is given by $N = \dfrac{100,000}{1 + 10e^{-2t}}$, where t is the number of weeks after Honda releases the new model, how many weeks will it take after the release until there are 50,000 Honda Hybrids from that batch on the road?

53. Earthquakes. A P wave measures 6.2 on the Richter scale, and an S wave measures 3.3 on the Richter scale. What is their combined measure on the Richter scale?

54. Sound. You and a friend get front row seats to a rock concert. The music level is 100 dB, and your normal conversation is 60 dB. If your friend is telling you something during the concert, how many decibels are you subjecting yourself to?

■ CATCH THE MISTAKE

In Exercises 55–58, explain the mistake that is made.

55. Solve the equation: $4e^x = 9$.

Solution:

Take the natural log of
both sides. $\ln(4e^x) = \ln 9$

Use the property of
inverses. $4x = \ln 9$

Solve for x. $x = \dfrac{\ln 9}{4} \approx 0.55$

This is incorrect. What mistake was made?

56. Solve the equation: $\log(x) + \log(3) = 1$.

Solution:

Use the product property (5). $\log(3x) = 1$

Exponentiate (base 10). $10^{\log(3x)} = 1$

Use properties of inverses. $3x = 1$

Solve for x. $x = \dfrac{1}{3}$

This is incorrect. What mistake was made?

57. Solve the equation: $\log(x) + \log(x + 3) = 1$ for x.

Solution:

Use the product property (5). $\log(x^2 + 3x) = 1$

Exponentiate both side (base 10). $10^{\log(x^2+3x)} = 10^1$

Use property of inverses. $x^2 + 3x = 10$

Factor. $(x + 5)(x - 2) = 0$

Solve for x. $x = -5$ and $x = 2$

This is incorrect. What mistake was made?

58. Solve the equation: $\log x + \log 2 = \log 5$.

Solution:

Combine the logarithms on the left. $\log(x + 2) = \log 5$

Use property of one-to-one functions. $x + 2 = 5$

Solve for x. $x = 3$

This is incorrect. What mistake was made?

■ CHALLENGE

59. T or F: The sum of logarithms with the same base is equal to the logarithm of the product.

60. T or F: A logarithm squared is equal to two times the logarithm.

61. T or F: $e^{\log x} = x$.

62. T or F: $e^x = -2$ has no solution.

63. Solve $y = \dfrac{3000}{1 + 2e^{-0.2t}}$ for t in terms of y.

64. State the range of values of x that the following identity holds $e^{\ln(x^2 - a)} = x^2 - a$.

65. A function called the hyperbolic cosine is defined as the average of exponential growth and exponential decay by $y = \dfrac{e^x + e^{-x}}{2}$. Find its inverse.

66. A function called the hyperbolic sine is defined by $y = \dfrac{e^x - e^{-x}}{2}$. Find its inverse.

■ TECHNOLOGY

67. Solve the equation $\ln 3x = \ln(x^2 + 1)$. Using a graphing calculator, plot the graphs $y = \ln(3x)$ and $y = \ln(x^2 + 1)$ in the same viewing rectangle. Zoom in on the point where the graphs intersect. Does this agree with your solution?

68. Solve the equation $10^{x^2} = 0.001^x$. Using a graphing calculator, plot the graphs $y = 10^{x^2}$ and $y = 0.001^x$ in the same viewing rectangle. Does this confirm your solution?

69. Use a graphing utility to help solve $3^x = 5x + 2$.

70. Use a graphing utility to help solve $\log x^2 = \ln(x - 3) + 2$.

Q: An earthquake registering 9.0 on the Richter scale centered off the coast of Indonesia in the Indian Ocean struck on December 26, 2004. The resulting tsunami devastated multiple nations and killed over 200,000 people. In trying to understand the immensity of this earthquake, I compared it to the earthquake of magnitude 7.1 that shook the San Francisco area on October 17, 1989. I concluded that the Indian Ocean earthquake must have had twice the magnitude as the San Francisco earthquake, but my friend who is a physics major told me that I am wrong. Why?

A: Recall from Section 5.3 that the magnitude, M, of an earthquake is $M = \frac{2}{3} \log \left(\frac{E}{E_0} \right)$ on the Richter scale. This means that the difference in energy between the earthquakes is actually exponential.

Using the formula above, calculate the approximate amount of energy (in joules) released by the Indian Ocean earthquake (remember that $E_0 = 10^{4.4}$ joules). If the San Francisco earthquake released approximately 1.12×10^{15} joules of energy, how many times stronger was the magnitude of the Indian Ocean earthquake (round your answer to two significant digits)?

On August 6, 1945 the United States dropped an atomic bomb on the Japanese city of Hiroshima. If this bomb released about 5.5×10^{13} joules of energy, how many times more powerful was the Indian Ocean earthquake (in other words, the energy released by the Indian Ocean earthquake was equivalent to approximately how many of the Hiroshima atomic bombs)? Express your answer using two significant digits.

TYING IT ALL TOGETHER

Often, a population will grow exponentially in its initial stages, but then conditions such as limited food or other resources cause the population to plateau at a value called the carrying capacity for the population. This is called a logistic growth model and can be expressed as $P(t) = \dfrac{c}{1 + ae^{-kt}}$ where $P(t)$ is the population at time t, k is a growth rate, c is the carrying capacity, and $a = \dfrac{c - P_0}{P_0}$ where P_0 is the population at time $t = 0$.

The director of a new nature preserve decides to introduce a population of 50 rabbits. He has already introduced 8 deer into the 25 square mile preserve. Based on the size of the preserve, a study estimates that the carrying capacity would be 500 rabbits and the relative growth rate would be about 0.3. After 2 years the director estimates that there are about 84 rabbits. Write a function $P(t)$ for the population of rabbits at time t. After how many years since the opening of the preserve could the director expect the population to reach 300 rabbits (round to the nearest whole number)?

SECTION	TOPIC	PAGES	REVIEW EXERCISES	KEY CONCEPTS
5.1	Exponential functions and their graphs	352–353	1–20	$f(x) = b^x \quad b > 0, b \neq 1$
	Graphing exponential functions	353–355	9–16	y-intercept $(0, 1)$ Horizontal asymptote: $y = 0$
	Doubling time	355–358	17–20	$P = P_0 2^{t/d}$
	Compound interest			$A = P\left(1 + \dfrac{r}{n}\right)^{nt}$
5.2	Exponential functions with base e	361–362	21–32	$f(x) = e^x$
	Exponential growth Exponential decay	362–364	21–28	$f(x) = ce^{kt} \quad k > 0$ $f(x) = ce^{-kt} \quad k > 0$
	Interest compounded continuously	364–365	29–32	$A = Pe^{rt}$
5.3	Logarithmic functions and their graphs	368–377	33–64	$y = \log_b x \quad x > 0$ $b > 0, b \neq 1$
	Evaluating logarithms	368–371	41–48	$y = \log_b x$ and $x = b^y$
	Common logarithms (base 10) Natural logarithms (base e)	371–372	41–52	$y = \log x$ $y = \ln x$
	Graphing logarithmic functions	372–374	53–60	x-intercept $(1, 0)$ Vertical asymptote: $x = 0$
	Decibel scale	375–377	61–64	$D = 10 \log\left(\dfrac{I}{I_T}\right) \quad I_T = 1 \times 10^{-12} \text{ W/m}^2$
	Richter scale			$M = \dfrac{2}{3} \log\left(\dfrac{E}{E_0}\right) \quad E_0 = 10^{4.4} \text{ joules}$
5.4	Properties of logarithms	380–385	65–78	**1.** $\log_b 1 = 0$ **2.** $\log_b b = 1$ **3.** $\log_b b^x = x$ **4.** $b^{\log_b x} = x \qquad x > 0$ Product property **5.** $\log_b MN = \log_b M + \log_b N$ Quotient property **6.** $\log_b\left(\dfrac{M}{N}\right) = \log_b M - \log_b N$ Power property **7.** $\log_b M^p = p \log_b M$
	Change-of-base formulas	385–386	75–78	$\log_b M = \dfrac{\log M}{\log b}$ or $\log_b M = \dfrac{\ln M}{\ln b}$
5.5	Exponential and logarithmic equations	389–395	79–102	
	One-to-one properties	390–395		$b^x = b^y$ if and only if $x = y$ $\log_b x = \log_b y$ if and only if $x = y$
	Inverse properties	390–395		$b^{\log_b x} = x \quad x > 0$ $\log_b b^x = x$

CHAPTER 5 REVIEW EXERCISES

5.1 Exponential Functions and Their Graphs

Approximate each number using a calculator. Round your answer to two decimal places.

1. $8^{4.7}$ **2.** $\pi^{2/5}$ **3.** $4 \cdot 5^{-2}$ **4.** $1.2^{1.2}$

Evaluate each exponential function for the given values.

5. $f(x) = 2^{4-x}$ $\qquad f(-2.2)$

6. $f(x) = -2^{x+4}$ $\qquad f(1.3)$

7. $f(x) = \left(\dfrac{2}{5}\right)^{1-6x}$ $\qquad f\left(\dfrac{1}{2}\right)$

8. $f(x) = \left(\dfrac{4}{7}\right)^{5x+1}$ $\qquad f\left(\dfrac{1}{5}\right)$

Match the graph with the function.

9. $y = 2^{x-2}$ $\qquad\qquad$ **10.** $y = -2^{2-x}$

11. $y = 2 + 3^{x+2}$ $\qquad$ **12.** $y = -2 - 3^{2-x}$

a.

b.

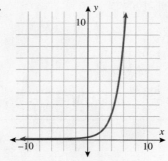

c.

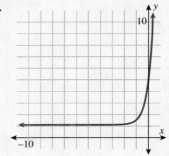

d.

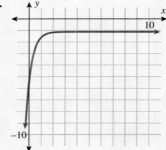

Use transformations to find the *y*-intercept and the horizontal asymptote and to graph the exponential function.

13. $y = -6^{-x}$ $\qquad$ **14.** $y = 4 - 3^x$

15. $y = 1 + 10^{-2x}$ $\qquad$ **16.** $y = 4^x - 4$

Applications

17. Compounded Interest. If \$4,500 is deposited into an account paying 4.5% compounding semiannually, how much will you have in the account in 7 years?

18. Depreciation of Vehicles. A recent college graduate buys a new car for \$22,000 and 4 years later sells it for \$11,000. If the depreciation continues at the same rate, how much would the car be worth in 2 more years?

19. Compound Interest. How much money should be put in a savings account now that earns 4.0% a year compounded quarterly if you want \$25,000 in 8 years?

20. Radioactive Decay. Cobalt is a brittle, hard metal resembling iron or nickel that is used in industry because it doesn't rust. The half-life of the radioactive isotope cobalt 60 is 5.3 years. If 200 grams of cobalt 60 were present in a piece of machinery today, how much would still be present in 10 years?

5.2 Exponential Functions with Base e

Approximate each number using a calculator. Round your answer to two decimal places.

21. $e^{3.2}$ **22.** e^{π} **23.** $e^{\sqrt{\pi}}$ **24.** $e^{-2.5\sqrt{3}}$

State the y-intercept and horizontal asymptote, and graph the exponential function.

25. $y = e^{-2x}$ **26.** $y = e^{x-1}$

27. $y = 3.2e^{x/3}$ **28.** $y = 2 - e^{1-x}$

Applications

29. Money. If \$13,450 is put in a money market account that pays 3.6% a year compounded continuously, how much will be in the account in 15 years?

30. Money. How much money should be invested today in a money market account that pays 2.5% a year compounded continuously if you desire \$15,000 in 10 years?

31. Carrying Capacity. The carrying capacity of a species of beach mice in St. Croix is given by $M = 1000(1 - e^{-0.035t})$ where M is the number of mice and t is time in years ($t = 0$ corresponds to 1998). How many mice will there be in 2010?

32. Population. The city of Brandon, Florida, had 50,000 residents in 1970, and since the crosstown expressway was built, its population has increased 2.3% per year. If the growth continues at the same rate, how many residents will Brandon have in 2030?

5.3 Logarithmic Functions and Their Graphs

Write each logarithmic equation in its equivalent exponential form.

33. $\log_4 64 = 3$ **34.** $\log_4 2 = \dfrac{1}{2}$

35. $\log\left(\dfrac{1}{100}\right) = -2$ **36.** $\log_{16} 4 = \dfrac{1}{2}$

Write each exponential equation in its equivalent logarithmic form.

37. $6^3 = 216$ **38.** $10^{-4} = 0.0001$

39. $\dfrac{4}{169} = \left(\dfrac{2}{13}\right)^2$ **40.** $\sqrt[3]{512} = 8$

Evaluate the logarithms exactly.

41. $\log_7 1$ **42.** $\log_4 256$ **43.** $\log_{1/6} 1296$ **44.** $\log 10^{12}$

Approximate the common and natural logarithms using a calculator. Round to two decimal places.

45. $\log 32$ **46.** $\ln 32$ **47.** $\ln 0.125$ **48.** $\log 0.125$

State the domain of the logarithmic function in interval notation.

49. $f(x) = \log_3 (x + 2)$ **50.** $f(x) = \log_2 (2 - x)$

51. $f(x) = \log (x^2 + 3)$ **52.** $f(x) = \log (3 - x^2)$

Match the graph with the function.

53. $y = \log_7 x$ **54.** $y = -\log_7 (-x)$

55. $y = \log_7 (x + 1) - 3$ **56.** $y = -\log_7 (1 - x) + 3$

a.

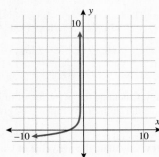

b.

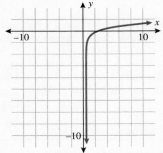

c.

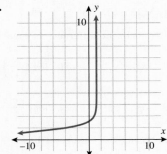

d.

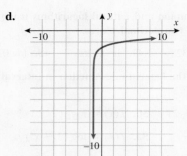

Graph the logarithmic function using transformation techniques.

57. $f(x) = \log_4 (x - 4) + 2$ **58.** $f(x) = \log_4 (x + 4) - 3$

59. $f(x) = -\log_4 (x) - 6$ **60.** $f(x) = -2 \log_4 (-x) + 4$

Applications

61. Everyday Chemistry. Calculate the pH value of milk assuming it has a concentration of hydrogen ions given by $H^+ = 3.16 \times 10^{-7}$.

62. Everyday Chemistry. Calculate the pH value of Coke assuming it has a concentration of hydrogen ions given by $H^+ = 2.0 \times 10^{-3}$.

63. Sound. Calculate the decibels associated with a teacher speaking to a medium-sized class if the sound intensity is 1×10^{-7} W/m².

64. Sound. Calculate the decibels associated with an alarm clock if the sound intensity is 1×10^{-4} W/m².

5.4 Properties of Logarithms

Use the properties of logarithms to simplify each expression.

65. $\log_{2.5} 2.5$ **66.** $\log_2 \sqrt{16}$ **67.** $2.5^{\log_{2.5} 6}$ **68.** $e^{-3\ln 6}$

Write each expression as a sum or difference of logarithms.

69. $\log_c x^a y^b$ **70.** $\log_3 x^2 y^{-3}$ **71.** $\log_j \left(\dfrac{rs}{t^3} \right)$

72. $\log x^c \sqrt{x + 5}$ **73.** $\log \left[\dfrac{a^{1/2}}{b^{3/2} c^{2/5}} \right]$ **74.** $\log_7 \left[\dfrac{c^3 d^{1/3}}{e^6} \right]^{1/3}$

Evaluate the logarithms using the change-of-base formula.

75. $\log_8 3$ **76.** $\log_5 \dfrac{1}{2}$ **77.** $\log_\pi 1.4$ **78.** $\log_{\sqrt{3}} 2.5$

5.5 Exponential and Logarithmic Equations

Solve the exponential equations exactly for x.

79. $4^x = \dfrac{1}{256}$ **80.** $3^{x^2} = 81$ **81.** $e^{3x-4} = 1$

82. $e^{\sqrt{x}} = e^{4.8}$ **83.** $\left(\dfrac{1}{3} \right)^{x+2} = 81$ **84.** $100^{x^2-3} = 10$

Solve the exponential equation. Round your answer to three decimal places.

85. $e^{2x+3} = 13$ **86.** $2^{2x-1} = 14$

87. $e^{2x} + 6e^x + 5 = 0$ **88.** $4e^x = 64$

89. $(2^x - 2^{-x})(2^x + 2^{-x}) = 0$ **90.** $5(2^x) = 25$

Solve the logarithmic equations exactly.

91. $\log (3x) = 2$

92. $\log_3 (x + 2) = 4$

93. $\log_4 x + \log_4 2x = 8$

94. $\log_6 x + \log_6 (2x - 1) = \log_6 3$

Solve the logarithmic equations. Round your answers to three decimal places.

95. $\ln x^2 = 2.2$

96. $\ln (3x - 4) = 7$

97. $\log_3 (2 - x) - \log_3 (x + 3) = \log_3 x$

98. $4 \log (x + 1) - 3 \log (x + 1) = 10$

Applications

99. Managing Finances. If Tania will need $30,000 in a year for a down payment on a new house, how much of that should she put in a 1-year CD earning 5% a year compounding continuously so that she will have exactly $30,000 a year from now?

100. Managing Finances. Jeremy is tracking the stock value of Best Buy (BBY on the NYSE). In 2003, he purchased 100 shares at $28 a share. The stock did not pay dividends because the company reinvested all earnings. In 2005, Jeremy cashed out and sold the stock for $4,000. What was the annual rate of return on BBY?

101. Interest. Money is invested in a savings account earning 4.2% interest compounded quarterly. How many years until the money doubles?

102. Interest. If $9,000 is invested in an investment earning 8% interest compounded continuously, how many years until there is $22,500?

1. Simplify: $\log 10^{x^3}$.

2. Use a calculator to evaluate $\log_5 326$ (round to two decimal places).

3. Find the exact value of $\log_3 81$.

4. Rewrite the expression $\ln \left[\dfrac{e^{5x}}{x(x^4 + 1)} \right]$ in a form with no logarithms of products, quotients, or powers.

In Exercises 5–12, solve for x.

5. $e^{x^2 - 1} = 42$

6. $e^{2x} - 5e^x + 6 = 0$

7. $27e^{0.2x + 1} = 300$

8. $3^x = 15$

9. $3 \ln x = 6$

10. $\log (6x + 5) - \log 3 = \log 2 - \log x$

11. $\ln (\ln x) = 1$

12. $\log_2 (3x - 1) - \log_2 (x - 1) = \log_2 (x + 1)$

13. State the domain of the function $f(x) = \log \left[\dfrac{x}{x^2 - 1} \right]$.

14. State the range of x values for which the following is true: $10^{\log (4x - a)} = 4x - a$.

15. **Interest.** If \$5,000 is invested at a rate of 6% a year, compounded quarterly, what is the amount in the account after 8 years?

16. **Interest.** If \$10,000 is invested at a rate of 5%, compounded continuously, what is the amount in the account after 10 years?

17. **Sound.** A lawn mower's sound intensity is approximately 1×10^{-3} W/m². Assuming your threshold of hearing is 1×10^{-12} W/m², calculate the decibels associated with the lawn mower.

18. **Population.** The population in Seattle, Washington, has been increasing at a rate of 5% a year. If the population continues to grow at that rate, and in 2004 there are 800,000 residents, how many residents will there be in 2010? *Hint: $N = N_0 e^{rt}$.*

19. **Earthquake.** An earthquake is considered moderate if it is between a 5 and 6 on the Richter scale. What is the energy range in joules for a moderate earthquake?

20. **Radioactive Decay.** The mass $m(t)$ remaining after t hours from a 50 gram sample of a radioactive substance is given by the equation $m(t) = 50e^{-0.0578t}$. After how long will only 30 grams of the substance remain? Round your answer to the nearest hour.

Systems of Equations and Inequalities

Hurricane Charley
August 13, 2004

Tim Boyles/Getty Images News and Sport Services

Hurricane Frances
September 4, 2004

©AP/Wide World Photos

Hurricane Ivan
September 16, 2004

Scott Olson/Getty Images News and Sport Services

Hurricane Jeanne
September 26, 2004

Joe Raedle/Getty Images News and Sport Services

On Friday the thirteenth (August 2004), Hurricane Charley whipped through Florida and 3 weeks later Hurricane Frances pummeled Florida. Twelve days after that, Hurricane Ivan decided to make his appearance. There had not been a state hit by four hurricanes in one season since 1886 in Texas. On September 26, 2004 that record was matched as Hurricane Jeanne, the strongest and deadliest of all, made landfall in Florida.

The Federal Emergency Management Agency (FEMA) coordinated the relief effort. FEMA coordinated trucks of relief that hauled generators, water, and tarps. FEMA was faced with an *optimization* problem in mathematics that involved maximizing the number of people aided by the relief subject to the constraints of the weight of the relief items and how much space they would take on the trucks. For example, how many generators, cases of water, and tarps could each truck haul, and how many of each would result in the greatest number of people being aided?

In this chapter, we will solve systems of linear equations, systems of nonlinear equations, and systems of linear inequalities. By the end of the chapter, we will combine these three in linear programming, which will enable you to address the FEMA relief problem as a result of Florida's back-to-back-to-back-to-back hurricanes.

Systems of Equations and Inequalities

Equations

- **Methods of Solving**
  - Elimination
  - Substitution
  - Graphing
- **Types of Solution**
  - One (or more), none, or infinitely many solutions
- **Applications**
  - Partial fractions

Inequalities

- **Graphs of Inequalities**
  - Shaded region
- **Systems of Inequalities**
  - Overlap of shaded regions

Linear Programming

- Minimizing or maximizing functions subject to constraints (inequalities)

CHAPTER OBJECTIVES

- Solve systems of linear equations.
- Understand that systems of linear equations have one, none, or infinitely many solutions.
- Solve systems of nonlinear equations.
- Graph inequalities using shaded regions.

NAVIGATION THROUGH SUPPLEMENTS

DIGITAL VIDEO SERIES #6

STUDENT SOLUTIONS MANUAL CHAPTER 6

BOOK COMPANION SITE
www.wiley.com/college/young

Introduction

Suppose you and a friend decide to go on a road trip to Laughlin, Nevada, for a weekend of jet-skiing and gambling (for blackjack tips read the opener of Chapter 9). You may prefer interstates so that you arrive in the fastest amount of time, or you may prefer the scenic routes that take you through small towns. There are many ways to go, all of which lead you from your university or college to Laughlin. The same analogy is true for solving application problems. Many of the applications we solved in Sections 1.2 and 1.3 could just have easily been solved using a system of linear equations in two variables. In fact, when you solved those problems, you used the method of substitution—you just didn't know it. For example, a movie theater sells 270 tickets to a show. The adult price is $8 and the student price is $6, and the box office took in $2,080. How many of each type of ticket was sold? You already have the tools (Section 1.2) to solve this problem. We will now solve it another way, using two variables.

WORDS	MATH
Select a variable for the number of student tickets sold.	x = number of student tickets
Select a variable for the number of adult tickets sold.	y = number of adult tickets
Cost of a student ticket is $6.	Revenue from students = $6x$
Cost of an adult ticket is $8.	Revenue from adults = $8y$
270 total tickets were sold.	$x + y = 270$
The box office took in $2,080.	$6x + 8y = 2080$

We now have a system of two linear equations in two variables (x and y):

$$x + y = 270$$

$$6x + 8y = 2080$$

The **solution** of this equation is $x = 40$, $y = 230$, or the ordered pair, $(40, 230)$, which is the only point that satisfies *both* equations. In other words, there were 40 student tickets sold and 230 adult tickets sold. In this section, we will solve this problem to illustrate three procedures for solving systems of linear equations: *substitution, elimination,* and *graphing.* Although this particular system of linear equations has **one solution**, some systems of linear equations have **no solution** while others have **infinitely many solutions**, as you will soon find.

Before explaining the three procedures, it is important to note that any of the three procedures can be used to solve systems of linear equations in two variables. However, there are times when one method, is preferred over another method. Typically, substitution and elimination are the preferred methods, and graphing is used to interpret solutions (one solution, no solution, or infinitely many solutions).

Substitution

Solving a system of two linear equations in two variables using the *substitution method* is outlined here using the above-mentioned movie ticket sales.

SUBSTITUTION METHOD

$$\text{Equation (1):} \quad x + y = 270$$
$$\text{Equation (2):} \quad 6x + 8y = 2080$$

Step 1: Solve Equation (1) for y in terms of x.

$$y = 270 - x$$

Step 2: **Substitute** $y = 270 - x$ into Equation (2).

$$6x + 8(270 - x) = 2080$$

Step 3: Solve for x.

$$6x + 2160 - 8x = 2080$$
$$-2x = -80$$
$$x = 40$$

Step 4: **Substitute** $x = 40$ into Equation (1).

$$40 + y = 270$$
$$y = 230$$

Step 5: Express the solution as an ordered pair. $\quad (40, 230)$

Step 6: Check that the solution satisfies *both* equations.

$$\text{Equation (1):} \quad x + y = 270 \qquad\qquad 40 + 230 = 270$$
$$\text{Equation (2):} \quad 6x + 8y = 2080 \qquad 6(40) + 8(230) = 2080$$

It is important to note that in Step 1 we could have also solved for x in terms of y. The general rule of thumb is to solve for the variable that is the easier to isolate. Also, in Step 4, either equation could have been used to find y.

Had this problem been solved in Section 1.2, we would have let x represent the number of student tickets and $270 - x$ represent the number of adult tickets sold, which is Step 1 in the substitution method.

EXAMPLE 1 Using Substitution to Solve a System That Has One Solution

Use the substitution method to solve the system of linear equations.

$$\text{Equation (1):} \quad x + y = 8$$
$$\text{Equation (2):} \quad 3x - y = 4$$

Solution:

STEP 1 Solve Equation (2) for y in terms of x. $\qquad$ $y = 3x - 4$

STEP 2 Substitute $y = 3x - 4$ into Equation (1). $\quad x + (3x - 4) = 8$

STEP 3 Solve for x. $\qquad\qquad\qquad\qquad\qquad x + 3x - 4 = 8$
$$4x = 12$$
$$x = 3$$

STEP 4 Substitute $x = 3$ into Equation (1). $\qquad\qquad 3 + y = 8$
$$y = 5$$

STEP 5 Express the solution as an ordered pair. $\qquad (3, 5)$

STEP 6 Check that the solution satisfies *both* equations.

Equation (1): $\quad x + y = 8$ $\qquad\qquad\qquad 3 + 5 = 8$

Equation (2): $\quad 3x - y = 4$ $\qquad\qquad\qquad 3(3) - 5 = 4$

EXAMPLE 2 Using Substitution to Solve a System That Has No Solution

Use the substitution method to solve the system of linear equations.

$$\text{Equation (1):} \qquad x - y = 2$$
$$\text{Equation (2):} \quad 2x - 2y = 10$$

Solution:

STEP 1 Solve Equation (1) for y in terms of x. $\qquad y = x - 2$

STEP 2 Substitute $y = x - 2$ into Equation (2). $\qquad 2x - 2(x - 2) = 10$

STEP 3 Solve for x. $\qquad\qquad\qquad\qquad\qquad\qquad 2x - 2x + 4 = 10$
$$4 = 10$$

$4 = 10$ is never true and is called an **inconsistent** statement.

There is no solution to this system of linear equations.

EXAMPLE 3 Using Substitution to Solve a System That Has Infinitely Many Solutions

Use the substitution method to solve the system of linear equations.

$$\text{Equation (1):} \qquad x - y = 2$$
$$\text{Equation (2):} \quad -x + y = -2$$

Solution:

STEP 1 Solve Equation (1) for y in terms of x. $\qquad y = x - 2$

STEP 2 Substitute $y = x - 2$ into Equation (2). $\qquad -x + (x - 2) = -2$

STEP 3 Solve for x. $\qquad\qquad\qquad\qquad\qquad\qquad -x + x - 2 = -2$
$$-2 = -2$$

$-2 = -2$ is always true and is called a **consistent** statement. Notice that many points $(2, 0)$, $(4, 2)$, and $(7, 5)$ satisfy both equations. In fact, there are

infinitely many solutions to this system of linear equations. All solutions are in the form (x, y) where $y = x - 2$.

YOUR TURN Use the substitution method to solve the systems of linear equations.

a. $2x + y = 3$ **b.** $x - y = 2$ **c.** $x + 2y = 1$
$4x + 2y = 4$ $4x - 3y = 10$ $2x + 4y = 2$

Elimination

We now turn our attention to another method, *elimination,* which is often preferred over substitution and will later be used in higher order systems. In a system of two linear equations in two variables, the equations can be combined, resulting in a third equation in one variable, thus *eliminating* one of the variables. The following is an example of when elimination would be preferred because the y terms cancel when the two equations are added together.

$$2x - y = 5$$
$$\underline{-x + y = -2}$$
$$x = 3$$

Two rules for generating equivalent equations will guide this method:

- Any equation can be multiplied by a nonzero constant, resulting in an equivalent relation.

$$x = 4 \text{ multiplied by 3 results in } 3x = 12$$

- Two equations can be added together to generate a third equivalent equation.

$$x = 2$$
$$\underline{y = 3}$$
$$x + y = 5$$

Solving a system of two linear equations in two variables using the *elimination method* is outlined here using the earlier mentioned movie ticket sales example.

ELIMINATION METHOD

Equation (1): $x + y = 270$

Equation (2): $6x + 8y = 2080$

In elimination, either variable (x or y) can be eliminated. In this example x is chosen for elimination.

Step 1: Multiply Equation (1) by -6. $\qquad -6x - 6y = -1620$

Step 2: **Eliminate** x by adding $(-6) \cdot$ Equation (1)
to Equation (2).

$$-6x - 6y = -1620$$
$$\underline{6x - 8y = 2080}$$
$$2y = 460$$

Step 3: Solve for y. $\qquad\qquad\qquad\qquad\qquad y = 230$

Step 4: Substitute $y = 230$ into Equation (1). $\qquad x + 230 = 270$

$$x = 40$$

Step 5: Express the solution as an ordered pair. $\qquad (40, 230)$

EXAMPLE 4 Using Elimination to Solve a System That Has One Solution

Use the elimination method to solve the system of linear equations.

$$\text{Equation (1):} \quad 2x - y = -5$$
$$\text{Equation (2):} \quad 4x + y = 11$$

Solution:

Add Equation (1) to Equation (2).

$$2x - y = -5$$
$$\underline{4x + y = 11}$$
$$6x = 6$$

Solve for x. $\qquad\qquad\qquad\qquad\qquad\qquad x = 1$

Substitute $x = 1$ into Equation (2). $\qquad\qquad 4(1) + y = 11$

Solve for y. $\qquad\qquad\qquad\qquad\qquad\qquad y = 7$

Express the answer as an ordered pair. $\qquad (1, 7)$

EXAMPLE 5 Using Elimination to Solve a System That Has No Solution

Use the elimination method to solve the system of linear equations.

$$\text{Equation (1):} \quad -x + y = 7$$
$$\text{Equation (2):} \quad 2x - 2y = 4$$

Solution:

Multiply Equation (1) by 2. $\qquad\qquad\qquad -2x + 2y = 14$

Add Equation (2). $\qquad\qquad\qquad\qquad\quad \underline{2x - 2y = 4}$

$$0 = 18$$

$0 = 18$ is never true so there is no solution to this system of linear equations.

EXAMPLE 6 Using Elimination to Solve a System That Has Infinitely Many Solutions

Use the elimination method to solve the system of linear equations.

$$\text{Equation (1):} \qquad 7x + y = 2$$

$$\text{Equation (2):} \quad -14x - 2y = -4$$

Solution:

Multiply Equation (1) by 2. $\qquad\qquad\qquad 14x + 2y = 4$

Add $(2) \cdot$ Equation (1) and Equation (2).
$$\begin{array}{r} 14x + 2y = 4 \\ -14x - 2y = -4 \\ \hline 0 = 0 \end{array}$$

$0 = 0$ is always true and is a consistent statement. There are infinitely many solutions to this system of linear equations. The solutions are all pairs (x, y) where $y = -7x + 2$.

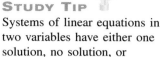

STUDY TIP

Systems of linear equations in two variables have either one solution, no solution, or infinitely many solutions.

YOUR TURN   Use the elimination method to solve the systems of linear equations.

a. $2x + 3y = 1$ $\qquad$ **b.** $\qquad x - 5y = 2$ $\qquad$ **c.** $\quad x - y = 14$
$\quad\;\; 4x - 3y = -7$ $\qquad\qquad -10x + 50y = -20$ $\qquad -x + y = 9$

Graphing

A third way to solve a system of two linear equations in two variables is to graph the two lines. The point where they intersect is the solution. Graphing is typically the most labor-intensive method for solving systems of linear equations in two variables: If aided by a graphing calculator, the graphing method becomes comparable to the other methods. Solving a system of two linear equations in two variables by *graphing* is outlined here using the movie ticket sales example once again.

GRAPHING METHOD

$$\text{Equation (1):} \qquad x + y = 270$$

$$\text{Equation (2):} \qquad 6x + 8y = 2080$$

Step 1: Write each equation in slope–intercept form.

$$\text{Equation (1):} \quad y = -x + 270$$

$$\text{Equation (2):} \quad y = -\frac{3}{4}x + 260$$

Step 2: Plot both lines on the same graph.

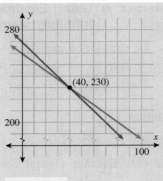

Step 3: Identify the point of intersection.   (40, 230)

EXAMPLE 7 Using Graphing to Solve a System That Has One Solution

Use graphing to solve the system of linear equations.

$$\text{Equation (1):} \qquad x + y = 2$$

$$\text{Equation (2):} \qquad 3x - y = 2$$

Solution:

STEP 1 Write each equation in slope–intercept form.

$$\text{Equation (1):} \qquad y = -x + 2$$

$$\text{Equation (2):} \qquad y = 3x - 2$$

STEP 2 Plot both lines on the same graph.

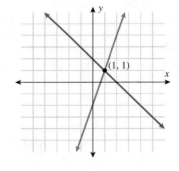

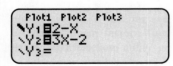
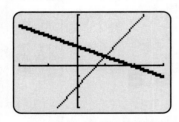
STEP 3 Identify the point of intersection.   (1, 1)

Note: There is one solution, because the two lines intersect at one point.

EXAMPLE 8 Using Graphing to Solve a System That Has No Solution

Use graphing to solve the system of linear equations.

$$\text{Equation (1):} \qquad 2x - 3y = 9$$

$$\text{Equation (2):} \qquad -4x + 6y = 12$$

Solution:

STEP 1 Write each equation in slope–intercept form.

Equation (1): $y = \dfrac{2}{3}x - 3$

Equation (2): $y = \dfrac{2}{3}x + 2$

STEP 2 Plot both lines on the same graph.

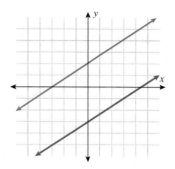

STEP 3 Identify the point of intersection.   None

Note: There is   no solution , because two parallel lines never intersect. Note that both lines have the same slope.

EXAMPLE 9 **Using Graphing to Solve a System That Has Infinitely Many Solutions**

Use graphing to solve the system of linear equations.

Equation (1): $3x + 4y = 12$

Equation (2): $\dfrac{3}{4}x + y = 3$

Solution:

STEP 1 Write each equation in slope–intercept form.

Equation (1): $y = -\dfrac{3}{4}x + 3$

Equation (2): $y = -\dfrac{3}{4}x + 3$

STEP 2 Plot both lines on the same graph.

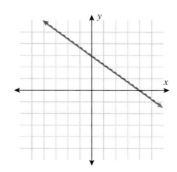

STEP 3 Identify the point of intersection.   Infinitely many points

Note: There are   infinitely many solutions , since the two lines are identical and coincide.

☑ **CONCEPT CHECK** If two equations of lines written in slope–intercept form have the same slope, is it necessary to graph them to determine what type of solution they have?

▮ **YOUR TURN** Use graphing to solve the system of linear equations.

a. $x - 2y = 1$ **b.** $x - 2y = 1$ **c.** $2x + y = 3$
   $2x - 4y = 2$ $2x + y = 7$ $2x + y = 7$

Three Methods and Three Types of Solutions

Given any system of two linear equations in two variables, any of the three methods (substitution, elimination, or graphing) can be used. If the system can easily eliminate a variable by adding multiples of the two equations, then elimination is the preferred choice. If an obvious elimination is not apparent, then substitution is used to solve the system. Graphing is typically not used to solve systems of equations but instead is used to interpret or confirm the solution(s) found using the other two methods. In Example 10, a reason is given to justify choosing a particular method. Realize that any of the three methods can be used on each of the systems.

EXAMPLE 10 Identifying Which Method to Use

State which of the three methods (elimination, substitution, or graphing) would be the preferred method to solve the system of linear equations.

a. $x - 2y = 1$ **b.** $7x - 20y = 1$ **c.** $y = -0.5x + 2.5$

   $-x + y = 2$ $5x + 3y = 18$ $y = -\dfrac{1}{3}x + 2$

Solution:

a. *Elimination,* because the x variable is eliminated when the two equations are added together.

b. *Substitution,* because there is no easy way to eliminate a variable. Instead, isolate either x or y and substitute into the other equation.

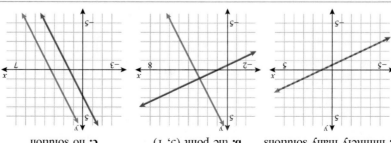

▮ **Answer: a.** infinitely many solutions   **b.** the point (3, 1)   **c.** no solution

c. It's a toss-up between the three methods. Elimination of y is achieved by subtracting one equation from the other. Substitution is achieved by substituting either expression for y into the other equation and solving for x. The equations are given in slope–intercept form so graphing is straightforward as well.

Regardless of which method is used to solve systems of linear equations in two variables, in general, we can summarize the three types of solutions both algebraically and graphically.

THREE TYPES OF SOLUTIONS TO SYSTEMS OF LINEAR EQUATIONS

NUMBER OF SOLUTIONS	GRAPHICAL INTERPRETATION
One solution	The two lines intersect at one point.
No solution	The two lines are parallel.
Infinitely many solutions	The two lines coincide.

Applications

In addition to the types of problems (interest, geometry, and mixtures) we solved in Section 1.2, many other applications involve solving systems of linear equations. Suppose you have two job offers that require sales. One pays a higher base, while the other pays a higher commission. Which job do you take?

EXAMPLE 11 Deciding Which Job to Take

Suppose that upon graduation you are offered a job selling biomolecular devices to laboratories studying DNA. The Beckman-Coulter Company offers you a job selling their DNA sequencer with an annual base salary of $20,000 plus 5% commission on total sales. The MJ Research Corporation offers you a job selling their PCR Machine that makes copies of DNA with an annual base salary of $30,000 plus 3% commission on sales. Determine the required total sales to make the Beckman-Coulter job the better offer.

Solution:

STEP 1 **Identify the question.**

When would these two jobs have equal salaries?

STEP 2 **Make notes.**

Beckman-Coulter   20,000 + 5%
MJ Research   30,000 + 3%

STEP 3 **Set up the equations.**

Let $x =$ total sales and $y =$ salary

Equation (1) Beckman-Coulter: $y = 20{,}000 + 0.05x$

Equation (2) MJ Research: $y = 30{,}000 + 0.03x$

STEP 4 **Solve the system of equations.**
Substitution method*

Substitute Equation (1)
into Equation (2). $20,000 + 0.05x = 30,000 + 0.03x$

Solve for x. $0.02x = 10,000$
$x = 500,000$

If you make \$500,000 worth of sales per year, the jobs will yield equal salaries. If you sell less than \$500,000, the MJ Research job is the better offer, and more than \$500,000, the Beckman-Coulter job is the better offer.

*Elimination method could also have been used.

STEP 5 **Check the solution.**
Equation (1)   Beckman-Coulter: $y = 20,000 + 0.05(500,000) = \$45,000$

Equation (2)   MJ Research: $y = 30,000 + 0.03(500,000) = \$45,000$

EXAMPLE 12 Deciding How Many Pounds of Each Meat at the Deli

The Chi Omega sorority is hosting a party, and the membership committee would like to make sandwiches for the new members. They already have bread and condiments but ask Tara to run to the deli to buy sliced turkey. The membership chair has instructed Tara to buy 5 pounds of the best turkey she can find, and the treasurer has given her \$37 and told her to spend it all. When she arrives at Publix supermarket, she has a choice of two types of turkey: Boar's Head (\$8 per pound) and the generic brand (\$7 per pound). How much of each kind should Tara buy to spend the entire \$37 on the best quality of turkey?

Solution:

Use the five-step procedure for solving word problems from Section 1.2 and apply it to two variables.

STEP 1 **Identify the question.**
How much of each type of sliced turkey should Tara buy?

STEP 2 **Make notes.**
Generic turkey costs \$7/pound.
Boar's Head turkey costs \$8/pound.
Tara has \$37 to spend on 5 pounds of turkey.

STEP 3 **Set up the equations.**
Let x = number of pounds of Boar's Head turkey and y = number of pounds of generic turkey.

Equation (1): $x + y = 5$

Equation (2): $8x + 7y = 37$

STEP 4 **Solve the system of equations.**
Elimination method*

Multiply Equation (1) by -7. $-7x - 7y = -35$

Add $(-7) \cdot$ Equation (1) to Equation (2).

$$-7x - 7y = -35$$
$$8x + 7y = 37$$
$$\overline{ x = 2}$$

Substitute $x = 2$ into the original Equation (1).

$$2 + y = 5$$
$$y = 3$$

Tara should buy 2 pounds of Boar's Head and 3 pounds of generic .

*The Substitution Method could also have been used to solve this system of linear equations.

STEP 5 **Check the solution.**

Equation (1): $2 + 3 = 5$

Equation (2): $8(2) + 7(3) = 37$

SECTION 6.1 SUMMARY

In this section, systems of two linear equations in two variables were discussed. The three methods of solution are substitution, elimination, and graphing. Any of the three methods can be used. Typically, elimination is used if it is convenient to eliminate a variable by adding multiples of the two equations. Substitution is used if it is convenient to isolate a variable. Graphing is used to interpret solutions. There are only three possibilities or three types of solutions: one solution (two lines intersecting), no solution (parallel lines), or infinitely many solutions (same line). Applications previously studied in Section 1.2 can be solved using systems of linear equations.

SECTION 6.1 EXERCISES

 ■ SKILLS

In Exercises 1–10, solve the system of linear equations by substitution.

1. $x - y = 1$
  $x + y = 1$

2. $x - y = 2$
  $x + y = -2$

3. $2x - y = 3$
  $x - 3y = 4$

4. $4x + 3y = 3$
  $2x + y = 1$

5. $2u + 5v = 7$
  $3u - v = 5$

6. $m - 2n = 4$
  $3m + 2n = 1$

7. $2x + y = 7$
  $-2x - y = 5$

8. $3x - y = 2$
  $3x - y = 4$

9. $4r - s = 1$
  $8r - 2s = 2$

10. $-3p + q = -4$
  $6p - 2q = 8$

In Exercises 11–22, solve the system of linear equations by elimination.

11. $x - y = 2$
  $x + y = 4$

12. $x + y = 2$
  $x - y = -2$

13. $5x + 3y = -3$
  $3x - 3y = -21$

14. $-2x + 3y = 1$
  $2x - y = 7$

15. $2x + 5y = 7$
  $3x - 10y = 5$

16. $6x - 2y = 3$
  $-3x + 2y = -2$

17. $2x + 5y = 5$
  $-4x - 10y = -10$

18. $11x + 3y = 3$
  $22x + 6y = 6$

19. $0.02x + 0.05y = 1.25$
$-0.06x - 0.15y = -3.75$

20. $-0.5x + 0.3y = 0.8$
$-1.5x + 0.9y = 2.4$

21. $\frac{1}{3}x + \frac{1}{2}y = 1$
$\frac{1}{5}x + \frac{7}{2}y = 2$

22. $\frac{1}{2}x - \frac{1}{3}y = 0$
$\frac{3}{2}x + \frac{1}{2}y = \frac{3}{4}$

In Exercises 23–26, match the systems of equations with the graphs.

23. $3x - y = 1$
$2x + y = 2$

24. $-x + 2y = -1$
$2x + y = 7$

25. $2x + y = 3$
$2x + y = 7$

26. $x - 2y = 1$
$2x - 4y = 2$

a.
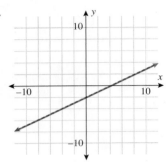

b.
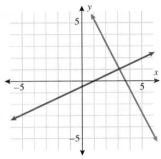

c.
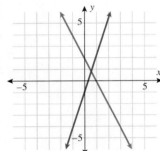

d.
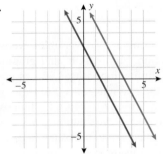

In Exercises 27–34, solve the system of linear equations by graphing.

27. $y = -x$
$y = x$

28. $x - 3y = 0$
$x + 3y = 0$

29. $2x + y = -3$
$x + y = -2$

30. $x - 2y = -1$
$-x - y = -5$

31. $\frac{1}{2}x - \frac{2}{3}y = 4$
$\frac{1}{4}x - y = 6$

32. $\frac{1}{5}x - \frac{5}{2}y = 10$
$\frac{1}{15}x - \frac{5}{6}y = \frac{10}{3}$

33. $1.6x - y = 4.8$
$-0.8x + 0.5y = 1.5$

34. $1.1x - 2.2y = 3.3$
$-3.3x + 6.6y = -6.6$

■ **APPLICATIONS**

35. Mixture. You are in charge of event planning for a major corporation that is having a martini cocktail hour for its guests. Your boss has given you a budget of $1,000 and would like you to order Grey Goose vodka (for the VIPs) and Skyy vodka for the other guests. A bottle of Grey Goose vodka costs $36 and a bottle of Skyy vodka costs $20. You must have 37 bottles total. How many of each type should you buy?

36. Mixture. In chemistry lab, Stephanie has to make a 37 milliliter solution that is 12% HCL. All that is in the lab is 8% and 15% HCL. How many milliliters of each solution should she use to obtain the desired mix?

37. Salary Comparison. Upon graduation with an MIS degree, you decide to work for a company that buys data from states' departments of motor vehicles and sells to banks and car dealerships customized reports detailing how many cars at each dealership are financed through particular banks. Autocount Corporation offers you a $15,000 base salary and 10% commission on your total annual sales. Polk Corporation offers you a base salary of

$30,000 plus a 5% commission on your total annual sales. How many total sales would you have to make per year to make more money at Autocount?

38. Salary Comparison. There are two types of residential realtors: Realtors who sell existing houses (resale) and realtors who sell new homes for developers. Resale of existing homes typically earns 6% commission on every sale, and representing developers in selling new homes typically earns a base salary of $15,000 per year plus an additional 1.5% commission, because they are required to work 5 days a week on site in a new development. How many total dollars would a realtor have to sell per year to make more money in resale than in new homes?

39. Gas Mileage. A Honda Accord gets approximately 26 mpg on the highway and 19 mpg in the city. You drove 349.5 miles on a full tank (16 gallons) of gasoline. Approximately how many miles were driven in the city and on the highway?

40. Wireless Plans. AT&T is offering a 600-minute peak plan with free mobile-to-mobile and weekend minutes at $59 per month plus $0.13 per minute for every minute over 600. The next plan up is the 800-minute plan that costs $79 per month. You think you may go over 600 minutes but are not sure you need 800 minutes. How many minutes would you have to talk for the 800-minute plan to be the better deal?

41. Distance/Rate/Time. A direct flight on Delta Air Lines from Atlanta to Paris is 4000 miles and takes approximately 8 hours going east (Atlanta to Paris) and 10 hours going west (Paris to Atlanta). Although the plane averages the same airspeed, there is a headwind while traveling west and a tailwind while traveling east, resulting in different ground speeds. What is the average ground speed of the plane, and what is the headwind/tailwind?

42. Distance/Rate/Time. A private pilot flies a Cessna 172 on a trip that is 500 miles each way. It takes her approximately 3 hours to get there and 4 hours to return. What is the approximate ground speed of the Cessna, and what are the approximate headwind and tailwind?

43. Investment Portfolio. Leticia has been tracking two volatile stocks. Stock A over the last year has increased 10%, and stock B has increased 14% (using a simple interest model). She has $10,000 to invest and would like to split it between these two stocks. If the stocks continue to perform at the same rate, how much should she invest in each to result in a balance of $11,260?

44. Investment Portfolio. Toby split his savings into two different investments, one earning 5% and the other earning 7%. He put twice as much in the investment earning the higher rate. In 1 year, he earned $665 in interest. How much money did he invest in each account?

■ CATCH THE MISTAKE

In Exercises 45–48, explain the mistake that is made.

45. Solve the system of equations by elimination.

$$2x + y = -3$$
$$3x + y = 8$$

Solution:

Multiply Equation (1) by -1. $\qquad 2x - y = -3$

Add $-1 \cdot$ Equation (1) and Equation (2). $\quad 2x - y = -3$
$$\underline{3x + y = 8}$$
$$5x = 5$$

Solve for x. $\qquad\qquad\qquad\qquad x = 1$

Substitute $x = 1$ into Equation (2). $\quad 3(1) + y = 8$
$$y = 5$$

The answer $(1, 5)$ is incorrect. What mistake was made?

46. Solve the system of equations by elimination.

$$4x - y = 12$$
$$4x - y = 24$$

Solution:

Multiply Equation (1) by -1. $\qquad -4x + y = -12$

Add $-1 \cdot$ Equation (1) and Equation (2). $-4x + y = -12$
$$\underline{4x - y = 24}$$
$$0 = 12$$

Answer: Infinitely many solutions.

This is incorrect. What mistake was made?

47. Solve the system of equations by substitution.

$$2x - y = 6$$
$$3x + y = -4$$

Solution:

Solve Equation (2) for y. $\qquad y = -3x - 4$

Substitute $y = -3x - 4$ into Equation (1). $\qquad 2x - (-3x - 4) = 6$

Solve for x.

$$2x + 3x - 4 = 6$$
$$5x = 10$$
$$x = 2$$

Substitute $x = 2$ into Equation (1).

$$2(2) - y = 6$$
$$y = -2$$

The answer $(2, -2)$ is incorrect. What mistake was made?

48. Solve the system of equations by graphing.

$$2x + 3y = 5$$
$$4x + 6y = 10$$

Solution:

Write both equations in slope–intercept form.

$$y = -\frac{2}{3}x + \frac{5}{3}$$
$$y = -\frac{2}{3}x + \frac{5}{3}$$

Since these lines have the same slope, they are parallel lines.

Parallel lines do not intersect, so there is no solution.

■ CHALLENGE

49. T or F: A system of equations represented by a graph of two lines with the same slope always has no solution.

50. T or F: A system of equations represented by a graph of two lines with slopes that are negative reciprocals always has one solution.

51. T or F: If two lines do not have exactly one point of intersection, then they must be parallel.

52. T or F: The system of equations $Ax - By = 1$ and $-Ax + By = -1$ has no solution.

53. The point $(2, -3)$ is a solution to the system of equations:

$$Ax + By = -29$$
$$Ax - By = 13$$

Find A and B.

54. If you graph the lines

$$x - 50y = 100$$
$$x - 48y = -98$$

they appear to be parallel lines. However, there is a unique solution. Explain how this might be possible.

■ TECHNOLOGY

55. Using a graphing utility, graph the two equations $y = -1.25x + 17.5$ and $y = 2.3x - 14.1$. Approximate the solution to this system of linear equations.

56. Using a graphing utility, graph the two equations $y = 14.76x + 19.43$ and $y = 2.76x + 5.22$. Approximate the solution to this system of linear equations.

57. Using a graphing utility, graph the two equations and determine the solution set: $23x + 15y = 7$ and $46x + 30y = 14$.

58. Using a graphing utility, graph the two equations and determine the solution set: $-3x + 7y = 2$ and $6x - 14y = 3$.

SECTION 6.2 Systems of Multivariable Linear Equations

Skills Objectives

■ Solve systems of three (or more) linear equations.
■ Solve application problems using systems of multivariable linear equations.

Conceptual Objectives

■ Identify three types of solutions: one solution, no solution, or infinitely many solutions.

In the previous section we solved systems of two linear equations with two variables. Two methods we used were elimination and substitution. Solutions were interpreted graphically as two lines intersecting at a point (one solution), parallel lines (no solution), or coinciding lines (infinitely many solutions). In this section we combine the elimination method and substitution method to solve linear systems with more than two equations and variables. We will not interpret the solutions graphically because it would require discussion of three-dimensional graphs. Thus, we will limit our discussion to types of solutions (one, none, or infinitely many).

Solving Multivariable Linear Systems

There are many ways to solve systems of linear equations in more than two variables. One method is to combine the elimination and substitution methods, which will be discussed in this section. Other methods involve matrices, which will be discussed in Chapter 7. We now outline a procedure for solving systems of linear equations in three variables, which can be extended to solve systems of more than three variables. Solutions are usually given as ordered triples using alphabetical order.

SOLVING SYSTEMS OF LINEAR EQUATIONS WITH THREE VARIABLES USING ELIMINATION AND SUBSTITUTION

Step 1: Reduce the system to two equations in two variables using elimination.

Step 2: Solve the resulting system of two equations using elimination or substitution.

Step 3: Substitute the solutions in Step 2 into any of the original equations and solve for the third variable.

Step 4: Check the solution in each of the three original equations.

EXAMPLE 1 Solving a System in Three Variables

Solve the system: Equation (1): $2x + y + 8z = -1$
Equation (2): $x - y + z = -2$
Equation (3): $3x - 2y - 2z = 2$

Solution:

STEP 1 Eliminate y using elimination on Equation (1) and Equation (2) and on Equation (2) and Equation (3).

Equation (1): $2x + y + 8z = -1$

Equation (2): $\underline{x - y + z = -2}$

Add. $3x + 9z = -3$

Multiply Equation (2) by -2. $-2x + 2y - 2z = 4$

Equation (3): $\underline{3x - 2y - 2z = 2}$

Add. $x - 4z = 6$

STEP 2 Solve the systems of two linear equations in two variables.

$$3x + 9z = -3$$
$$x - 4z = 6$$

Substitution method: $x = 4z + 6$

$$3(4z + 6) + 9z = -3$$

Distribute.

$$12z + 18 + 9z = -3$$

Combine like terms.

$$21z = -21$$

Solve for z.

$$z = -1$$

Substitute $z = -1$ into $x = 4z + 6$.

$$x = 4(-1) + 6 = 2$$

$x = 2$ and $z = -1$ are the solutions to the system of two equations.

STEP 3 Substitute $x = 2$ and $z = -1$ into any of the three original equations and solve for y.

Substitute $x = 2$ and $z = -1$ into Equation (2).

$$2 - y - 1 = -2$$

Solve for y.

$$y = 3$$

STEP 4 Check that $x = 2$, $y = 3$, and $z = -1$ satisfy all three equations.

Equation (1): $\quad 2(2) + 3 + 8(-1) = 4 + 3 - 8 = -1$

Equation (2): $\quad 2 - 3 - 1 = -2$

Equation (3): $\quad 3(2) - 2(3) - 2(-1) = 6 - 6 + 2 = 2$

The solution is $\boxed{x = 2, \ y = 3, \ z = -1}$.

 YOUR TURN   Solve the system: $\quad 2x - y + 3z = -1$
$$x + y - z = 0$$
$$3x + 3y - 2z = 1$$

In Example 1 and the Your Turn the variable, y, was eliminated by adding the first and second equations. In practice, any of the three variables can be eliminated, but typically the most convenient variable to eliminate is selected. What if a variable is missing from one of the equations? That variable is the one that must be eliminated when solving the system of the remaining two equations.

EXAMPLE 2   Solving a System in Three Variables When One Variable Is Missing

Solve the system:

Equation (1): $\quad\quad\quad x + z = 1$

Equation (2): $\quad 2x + y - z = -3$

Equation (3): $\quad x + 2y - z = -1$

Since y is missing from Equation (1), y is the variable to be eliminated with Equation (2) and Equation (3).

STEP 1 Eliminate y.

Multiply Equation (2) by -2.

$$-4x - 2y + 2z = 6$$

Equation (3):

$$\underline{x + 2y - z = -1}$$

Add.

$$-3x + z = 5$$

STEP 2 Solve the system of two equations: Equation (1) and the resulting equation in Step 1.

$$x + z = 1$$
$$-3x + z = 5$$

Multiply the second equation by (-1) and add to first equation.

$$4x = -4$$

Solve for x.

$$x = -1$$

Substitute $x = -1$ into Equation (1).

$$-1 + z = 1$$

Solve for z.

$$z = 2$$

STEP 3 Substitute $x = -1$ and $z = 2$ into one of the original equations (Equation 2 or Equation 3) and solve for y.

Substitute $x = -1$ and $z = 2$ into $x + 2y - z = -1$.

$$(-1) + 2y - 2 = -1$$

Gather like terms.

$$2y = 2$$

Solve for y.

$$y = 1$$

STEP 4 Check that $x = -1$, $y = 1$, and $z = 2$ satisfy all three equations.

Equation (1): $(-1) + 2 = 1$

Equation (2): $2(-1) + (1) - (2) = -3$

Equation (3): $(-1) + 2(1) - (2) = -1$

The solution is $\boxed{x = -1, y = 1, z = 2}$.

CONCEPT CHECK   For the system,

$$x + y + z = 0$$
$$2x + z = -1$$
$$x - y - z = 2$$

which variable should be eliminated from the first and third equations?

YOUR TURN   Solve the system:

$$x + y + z = 0$$
$$2x + z = -1$$
$$x - y - z = 2$$

Types of Solutions

When solving systems of linear equations, three types of solutions can be obtained: one solution, no solution, and infinitely many solutions. Examples 1 and 2 each had one solution. Examples 3 and 4 illustrate systems with no solution or infinitely many solutions.

EXAMPLE 3 Solving a Multivariable System with Infinitely Many Solutions

Solve the system:

Equation (1): $2x + y - z = 4$

Equation (2): $x + y = 2$

Equation (3): $3x + 2y - z = 6$

■ **Answer:** $x = 1$, $y = 2$, and $z = -3$

Solution:

Since z is missing from Equation (2), z is the variable to be eliminated from Equation (1) and Equation (3).

STEP 1 Eliminate z.

Multiply Equation (1) by (-1). $\qquad -2x - y + z = -4$

Equation (3): $\qquad\qquad\qquad\qquad\quad \dfrac{3x + 2y - z = 6}{x + y = 2}$

Add.

STEP 2 Solve the system of two equations: Equation (2) and the resulting equation in Step 1.

$$x + y = 2$$
$$x + y = 2$$

Multiply the first equation by (-1) and add to second equation. $\qquad 0 = 0$

This statement is always true; therefore, there are infinitely many solutions.

STEP 3 Let $y = a$, then $x = 2 - a$. Substitute $y = a$ and $x = 2 - a$ into one of the original equations (Equation 1 or Equation 3) and solve for z.

Substitute $y = a$ and $x = 2 - a$ into $2x + y - z = 4$. $\qquad 2(2 - a) + a - z = 4$

Eliminate parentheses. $\qquad\qquad\qquad 4 - 2a + a - z = 4$

Solve for z. $\qquad\qquad\qquad\qquad\qquad\qquad z = -a$

The infinitely many solutions are written as $x = 2 - a$, $y = a$, and $z = -a$ for any real number, a.

STEP 4 Check that $x = 2 - a$, $y = a$, and $z = -a$ satisfy all three equations.

Equation (1): $\quad 2(2 - a) + a - (-a) = 4 - 2a + a + a = 4$

Equation (2): $\quad 2 - a + a = 2$

Equation (3): $\quad 3(2 - a) + 2(a) - (-a) = 6 - 3a + 2a + a = 6$

The solution is $x = 2 - a, y = a, z = -a$.

■ **YOUR TURN** Solve the system: $\quad x + y - 2z = 0$
$$x - z = -1$$
$$x - 2y + z = -3$$

EXAMPLE 4 Solving a Multivariable System with No Solution

Solve the system:

Equation (1): $\qquad x + 2y - z = 3$

Equation (2): $\qquad 2x + y + 2z = -1$

Equation (3): $\quad -2x - 4y + 2z = 5$

STEP 1 Eliminate x.

Multiply Equation (1) by -2.	$-2x - 4y + 2z = -6$
Equation (2):	$\underline{2x + y + 2z = -1}$
Add.	$-3y + 4z = -7$

Equation (2):	$2x + y + 2z = -1$
Equation (3):	$\underline{-2x - 4y + 2z = 5}$
Add.	$-3y + 4z = 4$

STEP 2 Solve the system of two equations:

$$-3y + 4z = -7$$
$$-3y + 4z = 4$$

Multiply the top equation by (-1)
and add to the second equation. $0 = 11$

This is a contradiction, or inconsistent statement, and therefore there
is   no solution .

Applications

EXAMPLE 5 Stock Value

The Oracle Corporation's stock (ORCL) over 3 days (Wednesday, October 13, to
Friday, October 15, 2004) can be approximately modeled by a quadratic function:
$f(t) = at^2 + bt + c$. If Wednesday corresponds to $t = 1$, where t is in days, then
the following data points approximately correspond to the stock value.

t	$f(t)$	Days
1	$12.20	Wednesday
2	$12.00	Thursday
3	$12.20	Friday

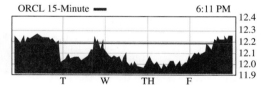

Determine the function that models this behavior.

Solution:

Substitute the points $(1, 12.20)$, $(2, 12.00)$, and $(3, 12.20)$ into $f(t) = at^2 + bt + c$.

$$a(1)^2 + b(1) + c = 12.20$$
$$a(2)^2 + b(2) + c = 12.00$$
$$a(3)^2 + b(3) + c = 12.20$$

Simplify to a system of three equations in three variables (a, b, and c).

Equation (1): $a + b + c = 12.20$
Equation (2): $4a + 2b + c = 12.00$
Equation (3): $9a + 3b + c = 12.20$

Solve for a, b, and c using the technique of this section.

STEP 1 Eliminate c.

Multiply Equation (1) by (-1).

Equation (2):

Add.

$$-a - b - c = -12.20$$
$$\underline{4a + 2b + c = 12.00}$$
$$3a + b = -0.20$$

Multiply Equation (1) by -1.

Equation (3):

Add.

$$-a - b - c = -12.20$$
$$\underline{9a + 3b + c = 12.20}$$
$$8a + 2b = 0$$

STEP 2 Solve the system of two equations.

$$3a + b = -0.20$$
$$8a + 2b = 0$$

Multiply the first equation by -2 and add to the second equation.

Solve for a.

$$2a = 0.4$$
$$a = 0.2$$

Substitute $a = 0.2$ into $8a + 2b = 0$.

$$8(0.2) + 2b = 0$$

Simplify.

Solve for b.

$$2b = -1.6$$
$$b = -0.8$$

STEP 3 Substitute $a = 0.2$ and $b = -0.8$ into one of the original three equations.

Substitute $a = 0.2$ and $b = -0.8$ into $a + b + c = 12.20$.

$$0.2 - 0.8 + c = 12.20$$

Gather like terms.

$$-0.6 + c = 12.20$$

Solve for c.

$$c = 12.80$$

STEP 4 Check that $a = 0.2$, $b = -0.8$, and $c = 12.80$ satisfy all three equations.

Equation (1): $a + b + c = 0.2 - 0.8 + 12.8 = 12.2$

Equation (2): $4a + 2b + c = 4(0.2) + 2(-0.8) + 12.80 = 0.8 - 1.6 + 12.8 = 12.00$

Equation (3): $9a + 3b + c = 9(0.2) + 3(-0.8) + 12.80 = 1.8 - 2.4 + 12.8 = 12.20$

The model is given by $f(t) = 0.2t^2 - 0.8t + 12.80$.

SECTION 6.2 SUMMARY

In this section, systems of linear equations with more than two variables were discussed. There are many techniques for solving multivariable linear systems. A technique was outlined that combined elimination and substitution. Elimination is used to eliminate one variable, resulting in a system of two linear equations. Once that system is solved, substitution is used to solve for the variable that was eliminated. There are three types of solutions: one solution, no solution, and infinitely many solutions. A graphical interpretation was not discussed because it would require graphing in three dimensions. It is important to note that this technique can be extended to solve systems of four or more linear equations. Alternative methods for solving multivariable linear systems will be discussed in Chapter 7 involving matrices.

SECTION 6.2 EXERCISES

■ SKILLS

In Exercises 1–20, solve each system of linear equations.

1. $2x - 3y + 4z = -3$
  $-x + y + 2z = 1$
  $5x - 2y - 3z = 7$

2. $x - 2y + z = 0$
  $-2x + y - z = -5$
  $13x + 7y + 5z = 6$

3. $3y - 4x + 5z = 2$
  $2x - 3y - 2z = -3$
  $3z + 4y - 2x = 1$

4. $2y + z - x = 5$
  $2x + 3z - 2y = 0$
  $-2z + y - 4x = 3$

5. $x - y + z = -1$
  $y - z = -1$
  $-x + y + z = 1$

6. $-y + z = 1$
  $x - y + z = -1$
  $x - y - z = -1$

7. $3x - 2y - 3z = -1$
  $x - y + z = -4$
  $2x + 3y + 5z = 14$

8. $3x - y + z = 2$
  $x - 2y + 3z = 1$
  $2x + y - 3z = -1$

9. $-x + 2y + z = -2$
  $3x - 2y + z = 4$
  $2x - 4y - 2z = 4$

10. $2x - y = 1$
  $-x + z = -2$
  $-2x + y = -1$

11. $x - z - y = 10$
  $2x - 3y + z = -11$
  $y - x + z = -10$

12. $2x + z + y = -3$
  $2y - z + x = 0$
  $x + y + 2z = 5$

13. $3x_1 + x_2 - x_3 = 1$
  $x_1 - x_2 + x_3 = -3$
  $2x_1 + x_2 + x_3 = 0$

14. $2x_1 + x_2 + x_3 = -1$
  $x_1 + x_2 - x_3 = 5$
  $3x_1 - x_2 - x_3 = 1$

15. $2x + 5y = 9$
  $x + 2y - z = 3$
  $-3x - 4y + 7z = 1$

16. $x - 2y + 3z = 1$
  $-2x + 7y - 9z = 4$
  $x + z = 9$

17. $2x_1 - x_2 + x_3 = 3$
  $x_1 - x_2 + x_3 = 2$
  $-2x_1 + 2x_2 - 2x_3 = -4$

18. $x_1 - x_2 - 2x_3 = 0$
  $-2x_1 + 5x_2 + 10x_3 = -3$
  $3x_1 + x_2 = 0$

19. $2x + y - z = 2$
  $x - y - z = 6$

20. $3x + y - z = 0$
  $x + y + 7z = 4$

■ APPLICATIONS

21. Football. On November 29, 2003, the Florida State Seminoles defeated the University of Florida Gators in football by a score of 38–34. The points came from a total of four types of plays: touchdowns (six points), extra points (one point), two-point conversions (two points), and field goals (three points). There were a total of 21 scoring plays. There were three more touchdowns than field goals, and there were six more extra points than two-point conversions. How many touchdowns, extra points, two-point conversions, and field goals were scored in this football game?

22. Basketball. On April 4, 2004, the University of Connecticut Huskies finished the season the same way they started it—as the number one men's basketball team in the NCAA. They defeated the Georgia Tech Yellow Jackets 82–73 in the Final Four championship game. The points came from three types of scoring plays: two-point shots, three-point shots, and 1-point free throws. There were seven more two-point shots made than there were one-point

free throws completed. The number of successful two-point shots was four more than four times the number of successful three point shots. How many two-point, three-point, and one-point free throw shots were made in the finals of the 2004 Final Four NCAA tournament?

Exercises 23 and 24 rely on a selection of Subway sandwiches whose nutrition information is given in the table on the following page.

Suppose you are going to eat only Subway sandwiches for a week (7 days) for lunch and dinner (total of 14 meals).

23. Diet. Your goal is a total of 4940 calories, 525 grams of carbohydrates, and 187 grams of fat. How many of each sandwich would you eat that week to obtain this goal?

24. Diet. Your goal is a total of 5240 calories, 235 grams of fat, and 361 grams of protein. How many of each sandwich would you eat that week to obtain this goal?

Sandwich	Calories	Fat (grams)	Carbohydrates (grams)	Protein (grams)
Mediterranean Chicken	350	18	17	36
Six Inch Tuna	430	19	46	20
Six Inch Roast Beef	290	5	45	19
Turkey-Bacon Wrap	430	27	20	34

www.subway.com

Exercises 25 and 26 involve vertical motion and the effect of gravity on an object.

Because of gravity, an object that is projected upward will eventually reach a maximum height and then fall to the ground. The equation that determines the height, h, of a projectile t seconds after it is shot upward is given by

$$h = \frac{1}{2}at^2 + v_0 t + h_0$$

where a is the acceleration due to gravity, h_0 is the initial height of the object at time $t = 0$, and v_0 is the initial velocity of the object at time $t = 0$. Note that a projectile follows the path of a parabola opening down, so $a < 0$.

25. **Vertical Motion.** An object is thrown upward, and the following table depicts the height of the ball t seconds after the projectile is released.

t Seconds	Height (feet)
1	36
2	40
3	12

Find the initial height, initial velocity, and acceleration due to gravity.

26. **Vertical Motion.** An object is thrown upward, and the following table depicts the height of the ball t seconds after the projectile is released.

t Seconds	Height (feet)
1	84
2	136
3	156

Find the initial height, initial velocity, and acceleration due to gravity.

27. **Data Curve-Fitting.** The average number of minutes that a person spends driving a car can be modeled by a quadratic function, $y = ax^2 + bx + c$ where $a < 0$ and $18 \le x \le 65$. The following table gives the average number of minutes per day that a person spends driving a car.

Age	Average Daily Minutes Driving
20	30
40	60
60	40

Determine the quadratic function that models this quantity.

28. **Data Curve-Fitting.** The average age when a woman gets married began increasing during the last century. In 1930 the average age was 18.6, in 1950 the average age was 20.2, and in 2002 the average age was 25.3. Find a quadratic function, $y = ax^2 + bx + c$, where $a > 0$ and $18 < y < 35$ that models the average age, y, when a woman gets married as a function of the year, x, ($x = 0$ corresponds to 1930). What will the average age be in 2010?

29. **Money.** Tara and Lamar decide to place $20,000 of their savings into investments. They put some in a money market account earning 3% interest, some in a mutual fund that has been averaging 7% a year, and some in a stock that rose 10% last year. If they put $6,000 more in the money market than in the mutual fund, and the mutual fund and stocks have the same growth in the next year as they did in the previous year, they will earn $1,180 in a year. How much money should they put in each of the three investments?

30. Money. Tara talks Lamar into putting less money in the money market and more money in the stock. They place $20,000 of their savings into investments. They put some in a money market account earning 3% interest, some in a mutual fund that has been averaging 7% a year, and some in a stock that rose 10% last year. If they put $6,000 more in the stock than in the mutual fund, and the mutual fund and stock have the same growth in the next year as they did in the previous year, they will earn $1,680 in a year. How much money should they put in each of the three investments?

31. Geometry. The circle given by the equation $x^2 + y^2 + ax + by + c = 0$ passes through the points $(-2, 4)$, $(1, 1)$, and $(-2, -2)$. Find a, b, and c.

32. Geometry. The circle given by the equation $x^2 + y^2 + ax + by + c = 0$ passes through the points $(0, 7)$, $(6, 1)$, and $(5, 4)$. Find a, b, and c.

■ **CATCH THE MISTAKE**

In Exercises 33 and 34, explain the mistake that is made.

33. Solve the system of equations.

Equation (1):	$2x - y + z = 2$
Equation (2):	$x - y = 1$
Equation (3):	$x + z = 1$

Solution:

Equation (2):	$x - y = 1$
Equation (3):	$x + z = 1$
Add Eq. (2) and Eq. (3).	$-y + z = 2$
Multiply Eq. (1) by (-1).	$-2x + y - z = -2$
Add.	$-2x = 0$
Solve for x.	$x = 0$
Substitute $x = 0$ into Equation (2).	$0 - y = 1$
Solve for y.	$y = -1$
Substitute $x = 0$ into Equation (3).	$0 + z = 1$
Solve for z.	$z = 1$

The answer is $x = 0$, $y = -1$, and $z = 1$.

This is incorrect. Although $x = 0$, $y = -1$, and $z = 1$ does satisfy the three original equations, it is only one of infinitely many solutions. What mistake was made?

34. Solve the system of equations:

Equation (1):	$x + 3y + 2z = 4$
Equation (2):	$3x + 10y + 9z = 17$
Equation (3):	$2x + 7y + 7z = 17$

Solution:

Multiply Equation (1) by -3.	$-3x - 9y - 6z = -12$
Equation (2):	$3x + 10y + 9z = 17$
Add.	$y + 3z = 5$
Multiply Equation (1) by -2.	$-2x - 6y - 4z = -8$
Equation (3):	$2x + 7y + 7z = 17$
Add.	$y + 3z = 9$
Solve the system of two equations.	$y + 3z = 5$
	$y + 3z = 9$

Infinitely many solutions.

Let $z = a$, then $y = 5 - 3a$.

Substitute $z = a$ and $y = 5 - 3a$ into Equation (1).	$x + 3y + 2z = 4$
	$x + 3(5 - 3a) + 2a = 4$
Eliminate parentheses.	$x + 15 - 9a + 2a = 4$
Solve for x.	$x = 7a - 11$

The answer is $x = 7a - 11$, $y = 5 - 3a$, and $z = a$.

This is incorrect. There is no solution. What mistake was made?

■ CHALLENGE

35. T or F: A system of linear equations that has more variables than equations cannot have a unique solution.

36. T or F: A system of linear equations that has the same number of equations as variables always has a unique solution.

37. A fourth-degree polynomial, $f(x) = ax^4 + bx^3 + cx^2 + dx + e$, with $a < 0$, can be used to represent the following data on the number of deaths per year due to lightning strikes. Assume 1999 corresponds to $x = -2$ and 2003 corresponds to $x = 2$. Use the data to determine a, b, c, d, and e.

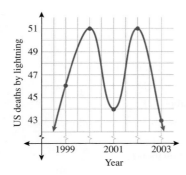

38. A copy machine accepts nickels, dimes, and quarters. After 1 hour, there are 30 coins total, and their value is $4.60. If there are four more quarters than nickels, how many nickels, quarters, and dimes are in the machine?

Solve the system of linear equations.

39.
$$2y + z = 3$$
$$4x - z = -3$$
$$7x - 3y - 3z = 2$$
$$x - y - z = -2$$

40.
$$-2x - y + 2z = 3$$
$$3x - 4z = 2$$
$$2x + y = -1$$
$$-x + y - z = -8$$

41.
$$3x_1 - 2x_2 + x_3 + 2x_4 = -2$$
$$-x_1 + 3x_2 + 4x_3 + 3x_4 = 4$$
$$x_1 + x_2 + x_3 + x_4 = 0$$
$$5x_1 + 3x_2 + x_3 + 2x_4 = -1$$

42.
$$5x_1 + 3x_2 + 8x_3 + x_4 = 1$$
$$x_1 + 2x_2 + 5x_3 + 2x_4 = 3$$
$$4x_1 + x_3 - 2x_4 = -3$$
$$x_2 + x_3 + x_4 = 0$$

■ TECHNOLOGY

For Exercises 43 and 44, use a graphing calculator to solve the system of linear equations (most graphing calculators have the capability of solving linear systems with the user entering the coefficients). The results should agree with answers to Exercises 11 and 12.

43.
$$x - z - y = 10$$
$$2x - 3y + z = -11$$
$$y - x + z = -10$$

44.
$$2x + z + y = -3$$
$$2y - z + x = 0$$
$$x + y + 2z = 5$$

45. Graphing calculators and graphing utilities have the ability to graph in three dimensions (3D) as opposed to the traditional two dimensions (2D). The line must be given in the form $z = ax + by + c$. Rewrite the system of equations in Exercise 43 in this form and graph the three lines in 3D. What is the point of intersection? Compare that with your answer in Exercise 43.

46. Graphing calculators and graphing utilities have the ability to graph in three dimensions (3D) as opposed to the traditional two dimensions (2D). The line must be given in the form $z = ax + by + c$. Rewrite the system of equations in Exercise 44 in this form and graph the three lines in 3D. What is the point of intersection? Compare that with your answer in Exercise 44.

- Decompose rational expressions into sums of partial fractions when the denominator contains
  - distinct linear factors
  - repeated linear factors
  - distinct irreducible quadratic factors
  - repeated irreducible quadratic factors

- Understand the importance of partial fraction decomposition.
- Understand the connection between partial fraction decomposition and systems of linear equations.

Introduction

In Chapter 4 we studied polynomial functions, and in Section 4.5 we discussed ratios of polynomial functions, called rational functions. Rational expressions are of the form:

$$\frac{n(x)}{d(x)} \qquad d(x) \neq 0$$

where the numerator, $n(x)$, and the denominator, $d(x)$, are polynomials. Examples of rational expressions are

$$\frac{4x - 1}{2x + 3} \qquad \frac{2x + 5}{x^2 - 1} \qquad \frac{3x^4 - 2x + 5}{x^2 + 2x + 4}$$

Suppose we are asked to add two rational expressions:

$$\frac{2}{x + 1} + \frac{5}{x - 3}$$

We already possess the skills to accomplish this. We first identify the least common denominator, $(x + 1)(x - 3)$, and combine the fractions into a single expression:

$$\frac{2(x - 3) + 5(x + 1)}{(x + 1)(x - 3)} = \frac{2x - 6 + 5x + 5}{(x + 1)(x - 3)} = \frac{7x - 1}{x^2 - 2x - 3}$$

Therefore we say that

$$\frac{2}{x + 1} + \frac{5}{x - 3} = \frac{7x - 1}{x^2 - 2x - 3}$$

How do we do this in reverse? For example, how do we start with $\dfrac{7x - 1}{x^2 - 2x - 3}$, and write this expression as a sum of two simpler expressions:

Partial Fraction Decomposition

$$\frac{7x - 1}{x^2 - 2x - 3} = \underbrace{\frac{2}{x + 1}}_{\text{Partial Fraction}} + \underbrace{\frac{5}{x - 3}}_{\text{Partial Fraction}}$$

Each of the two expressions on the right is called a **partial fraction**. The sum of these fractions is called the **partial fraction decomposition** of $\dfrac{7x - 1}{x^2 - 2x - 3}$.

You may be asking yourself, "Why did we not cover *partial fraction decomposition* when we were studying rational functions?" Partial fraction decomposition requires the ability to solve systems of linear equations, which is why we have waited to cover this topic until now.

The Basics of Partial Fraction Decomposition

As mentioned earlier, a rational expression is the ratio of two polynomial expressions, $\frac{n(x)}{d(x)}$, and we assume that $n(x)$ and $d(x)$ are polynomials with no common factors. If the degree of $n(x)$ is less than the degree of $d(x)$, then the rational expression $\frac{n(x)}{d(x)}$ is said to be **proper**. If the degree of $n(x)$ is greater than or equal to the degree of $d(x)$, the rational expression is said to be **improper**. If the rational expression is improper, it should first be divided using long division.

$$\frac{n(x)}{d(x)} = Q(x) + \frac{r(x)}{d(x)}$$

The result is the sum of a quotient, $Q(x)$, and a rational expression, which is the ratio of the remainder, $r(x)$, and the divisor, $d(x)$. The rational expression $\frac{r(x)}{d(x)}$ is proper and the techniques outlined in this section can be applied to its partial fraction decomposition.

Partial fraction decomposition of proper rational expressions always begins with factoring the denominator, $d(x)$. The goal is to write $d(x)$ as a product of distinct linear factors, but that may not always be possible. Sometimes $d(x)$ can be factored into a product of linear factors, where one or more are repeated. And, sometimes the factored form of $d(x)$ contains irreducible quadratic factors, such as $x^2 + 1$. There are times when the irreducible quadratic factors are repeated, such as $(x^2 + 1)^2$. A procedure is now outlined for partial fraction decomposition.

PARTIAL FRACTION DECOMPOSITION

To write a rational expression, $\frac{n(x)}{d(x)}$ as a sum of partial fractions:

Step 1: Determine if the rational expression is proper or improper.

- Proper: degree of $n(x) <$ degree of $d(x)$
- Improper: degree of $n(x) \geq$ degree of $d(x)$

Step 2: If proper, proceed to Step 3.

If improper, divide $\frac{n(x)}{d(x)}$ using polynomial (long) division and write the result as $\frac{n(x)}{d(x)} = Q(x) + \frac{r(x)}{d(x)}$ and proceed to Step 3 with $\frac{r(x)}{d(x)}$.

Step 3: Factor $d(x)$. One of four possible cases will arise:

Case 1: Distinct (nonrepeated) *linear* factors: $(ax + b)$

$$d(x) = (3x - 1)(x + 2)$$

Case 2: One or more repeated linear factors: $(ax + b)^m \quad m \geq 2$

$$d(x) = (x + 5)^2(x - 3)$$

Case 3: One or more distinct irreducible ($ax^2 + bx + c = 0$ has no real roots) quadratic factors: ($ax^2 + bx + c$)

$$d(x) = (x^2 + 4)(x + 1)(x - 2)$$

Case 4: One or more repeated irreducible quadratic factors: $(ax^2 + bx + c)^m$

$$d(x) = (x^2 + x + 1)^2(x + 1)(x - 2)$$

Step 4: Decompose the rational expression into a sum of partial fractions according to the procedure outlined in each case in this section.

Step 4 depends on which cases, or types of factors, arise. It is important to note that these four cases are not exclusive and combinations of different types of factors will appear.

Distinct Linear Factors

CASE 1: $d(x)$ HAS ONLY DISCRETE (NONREPEATED) LINEAR FACTORS

If $d(x)$ is a polynomial of degree p, and it can be factored into p linear factors:

$$d(x) = \underbrace{(ax + b)(cx + d) \dots}_{p \text{ linear factors}}$$

where no two factors are the same, then the partial fraction decomposition of $\dfrac{n(x)}{d(x)}$ can be written as

$$\frac{n(x)}{d(x)} = \frac{A}{(ax + b)} + \frac{B}{(cx + d)} + \dots$$

where the numerators, A, B, and so on are constants to be determined.

EXAMPLE 1 Partial Fraction Decomposition with Distinct Linear Factors

Find the partial fraction decomposition of $\dfrac{5x + 13}{x^2 + 4x - 5}$.

Solution:

Factor the denominator.
$$\frac{5x + 13}{(x - 1)(x + 5)}$$

Express as a sum of two partial fractions.
$$\frac{5x + 13}{(x - 1)(x + 5)} = \frac{A}{(x - 1)} + \frac{B}{(x + 5)}$$

Multiply equation by LCD, $(x - 1)(x + 5)$.
$$5x + 13 = A(x + 5) + B(x - 1)$$

Eliminate parentheses.
$$5x + 13 = Ax + 5A + Bx - B$$

Group x's and constants on right.
$$5x + 13 = (A + B)x + (5A - B)$$

Identify like terms.
$$5x + 13 = (A + B)x + (5A - B)$$

Equate coefficients of x.
$$5 = A + B$$

Equate **constant** terms. $13 = 5A - B$

Solve the system of two linear
equations using any method to
solve for A and B. $A = 3, B = 2$

Substitute $A = 3, B = 2$ into
partial fraction decomposition.

$$\frac{5x + 13}{(x - 1)(x + 5)} = \frac{3}{(x - 1)} + \frac{2}{(x + 5)}$$

Check by adding the partial fractions.

$$\frac{3}{(x - 1)} + \frac{2}{(x + 5)} = \frac{3(x + 5) + 2(x - 1)}{(x - 1)(x + 5)} = \frac{5x + 13}{x^2 + 4x - 5}$$

■ **YOUR TURN** 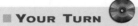 Find the partial fraction decomposition of:

$$\frac{4x - 13}{x^2 - 3x - 10}$$

Repeated Linear Factors

CASE 2: $d(x)$ **HAS AT LEAST ONE**
REPEATED LINEAR FACTOR

If $d(x)$ can be factored into a product of linear factors, then the partial
fraction decomposition will proceed as in Case 1, with the exception of a
repeated factor: $(ax + b)^m$, $m \geq 2$. Any linear factor repeated m times will
result in the sum of m partial fractions:

$$\frac{A}{(ax + b)} + \frac{B}{(ax + b)^2} + \frac{C}{(ax + b)^3} + \cdots + \frac{M}{(ax + b)^m}$$

where the numerators, A, B, C, ..., M are constants to be determined.

Note that if $d(x)$ is degree p, the general form of the decomposition will have p partial
fractions. If some numerator constants turn out to be zero, then the final decomposition
may have less than p partial fractions.

EXAMPLE 2 **Partial Fraction Decomposition with a Repeated**
Linear Factor

Find the partial fraction decomposition of $\dfrac{-3x^2 + 13x - 12}{x^3 - 4x^2 + 4x}$.

Factor the denominator. $$\frac{-3x^2 + 13x - 12}{x(x - 2)^2}$$

Express as a sum of
three partial fractions.

$$\frac{-3x^2 + 13x - 12}{x(x - 2)^2} = \frac{A}{x} + \frac{B}{(x - 2)} + \frac{C}{(x - 2)^2}$$

■ **Answer:** $\dfrac{4x - 13}{x^2 - 3x - 10} = \dfrac{3}{x + 2} + \dfrac{1}{x - 5}$

Multiply equation by LCD, $x(x - 2)^2$.

$$-3x^2 + 13x - 12 = A(x - 2)^2 + Bx(x - 2) + Cx$$

Eliminate parentheses.

$$-3x^2 + 13x - 12 = Ax^2 - 4Ax + 4A + Bx^2 - 2Bx + Cx$$

Group like terms on right.

$$-3x^2 + 13x - 12 = (A + B)x^2 + (-4A - 2B + C)x + 4A$$

Identify like terms on both sides.

$$-3x^2 + 13x - 12 = (A + B)x^2 + (-4A - 2B + C)x + 4A$$

Equate coefficients of x^2. $\qquad -3 = A + B \qquad\qquad$ (1)

Equate coefficients of x. $\qquad 13 = -4A - 2B + C \qquad$ (2)

Equate **constant** terms. $\qquad -12 = 4A \qquad\qquad$ (3)

Solve the system of three equations for A, B, and C.

$\qquad$ Solve (3) for A. $\qquad\qquad\qquad\qquad A = -3$

$\qquad$ Substitute $A = -3$ into (1). $\qquad\qquad B = 0$

$\qquad$ Substitute $A = -3$ and $B = 0$ into (2). $\qquad C = 1$

Substitute $A = -3, B = 0, C = 1$ into partial fraction decomposition.

$$\frac{-3x^2 + 13x - 12}{x(x - 2)^2} = \frac{-3}{x} + \frac{0}{(x - 2)} + \frac{1}{(x - 2)^2}$$

$$\frac{-3x^2 + 13x - 12}{x^3 - 4x^2 + 4x} = \frac{-3}{x} + \frac{1}{(x - 2)^2}$$

Check by adding the partial fractions.

$$\frac{-3}{x} + \frac{1}{(x - 2)^2} = \frac{-3(x - 2)^2 + 1(x)}{x(x - 2)^2} = \frac{-3x^2 + 13x - 12}{x^3 - 4x^2 + 4x}$$

■ **YOUR TURN** Find the partial fraction decomposition of $\dfrac{2x^2 - x - 1}{x^3 + x^2}$.

EXAMPLE 3 Partial Fraction Decomposition with Multiple Repeated Linear Factors

Find the partial fraction decomposition of $\dfrac{2x^3 + 6x^2 + 6x + 9}{x^4 + 6x^3 + 9x^2}$.

Solution:

Factor the denominator. $\dfrac{2x^3 + 6x^2 + 6x + 9}{x^2(x + 3)^2}$

Express as a sum of four partial fractions.

$$\frac{2x^3 + 6x^2 + 6x + 9}{x^2(x + 3)^2} = \frac{A}{x} + \frac{B}{x^2} + \frac{C}{(x + 3)} + \frac{D}{(x + 3)^2}$$

■ **Answer:** $\dfrac{2x^2 - x - 1}{x^3 + x^2} = \dfrac{-1}{x} + \dfrac{2}{x^2} + \dfrac{2}{x + 1}$

Multiply equation by LCD, $x^2(x + 3)^2$.

$$2x^3 + 6x^2 + 6x + 9 = Ax(x + 3)^2 + B(x + 3)^2 + Cx^2(x + 3) + Dx^2$$

Eliminate parentheses.

$$2x^3 + 6x^2 + 6x + 9 = Ax^3 + 6Ax^2 + 9Ax + Bx^2 + 6Bx + 9B + Cx^3 + 3Cx^2 + Dx^2$$

Group like terms on right.

$$2x^3 + 6x^2 + 6x + 9 = (A + C)x^3 + (6A + B + 3C + D)x^2 + (9A + 6B)x + 9B$$

Identify like terms on both sides.

$$2x^3 + 6x^2 + 6x + 9 = (A + C)x^3 + (6A + B + 3C + D)x^2 + (9A + 6B)x + 9B$$

Equate coefficients of x^3.	$2 = A + C$	(1)
Equate coefficients of x^2.	$6 = 6A + B + 3C + D$	(2)
Equate coefficients of x.	$6 = 9A + 6B$	(3)
Equate **constant** terms.	$9 = 9B$	(4)

Solve the system of four equations for A, B, C, and D.

Solve (4) for B.	$B = 1$
Substitute $B = 1$ into (3) and solve for A.	$A = 0$
Substitute $A = 0$ into (1) and solve for C.	$C = 2$
Substitute $A = 0$, $B = 1$, and $C = 2$ into (2) and solve for D.	$D = -1$

Substitute $A = 0, B = 1, C = 2, D = -1$ into partial fraction decomposition.

$$\frac{2x^3 + 6x^2 + 6x + 9}{x^2(x + 3)^2} = \frac{0}{x} + \frac{1}{x^2} + \frac{2}{(x + 3)} + \frac{-1}{(x + 3)^2}$$

$$\frac{2x^3 + 6x^2 + 6x + 9}{x^2(x + 3)^2} = \frac{1}{x^2} + \frac{2}{(x + 3)} - \frac{1}{(x + 3)^2}$$

Check by adding the partial fractions.

$$\frac{1}{x^2} + \frac{2}{(x + 3)} - \frac{1}{(x + 3)^2} = \frac{(x + 3)^2 + 2x^2(x + 3) - 1(x^2)}{x^2(x + 3)^2}$$

$$= \frac{2x^3 + 6x^2 + 6x + 9}{x^4 + 6x^3 + 9x^2}$$

■ **YOUR TURN** Find the partial fraction decomposition of $\dfrac{2x^3 + 2x + 1}{x^4 + 2x^3 + x^2}$.

■ **Answer:** $\dfrac{2x^3 + 2x + 1}{x^4 + 2x^3 + x^2} = \dfrac{1}{x^2} + \dfrac{2}{(x + 1)} - \dfrac{3}{(x + 1)^2}$

Distinct Irreducible Quadratic Factors

There will be times when a polynomial cannot be factored into a product of linear factors with real coefficients. For example, $x^2 + 4$, $x^2 + x + 1$, and $9x^2 + 3x + 2$ are all examples of *irreducible quadratic* expressions. The general form of an **irreducible quadratic factor** is given by:

$$ax^2 + bx + c \qquad \text{where } ax^2 + bx + c = 0 \text{ has no real roots}$$

CASE 3: $d(x)$ HAS A DISTINCT IRREDUCIBLE QUADRATIC FACTOR

If the factored form of $d(x)$ contains an irreducible quadratic factor, $ax^2 + bx + c$, then the partial fraction decomposition will contain a term of the form:

$$\frac{Ax + B}{ax^2 + bx + c}$$

where A and B are constants to be determined.

For example,

$$\frac{7x^2 + 2}{(2x + 1)(x^2 + 1)} = \underbrace{\frac{A}{(2x + 1)}}_{\substack{\text{constant numerator} \\ \text{linear factor}}} + \underbrace{\frac{Bx + C}{(x^2 + 1)}}_{\substack{\text{linear numerator} \\ \text{quadratic factor}}}$$

Constants are used in the numerator when linear factors are in the denominators and linear expressions are used in the numerator when quadratic factors are in the denominators.

EXAMPLE 4 Partial Fraction Decomposition with an Irreducible Quadratic Factor

Find the partial fraction decomposition of $\dfrac{7x^2 + 2}{(2x + 1)(x^2 + 1)}$.

Solution:

The denominator is already in factored form.

$$\frac{7x^2 + 2}{(2x + 1)(x^2 + 1)}$$

Express as a sum of two partial fractions.

$$\frac{7x^2 + 2}{(2x + 1)(x^2 + 1)} = \frac{A}{(2x + 1)} + \frac{Bx + C}{(x^2 + 1)}$$

Multiply equation by LCD, $(2x + 1)(x^2 + 1)$.

$$7x^2 + 2 = A(x^2 + 1) + (Bx + C)(2x + 1)$$

Eliminate parentheses.	$7x^2 + 2 = Ax^2 + A + 2Bx^2 + Bx + 2Cx + C$
Group like terms on right.	$7x^2 + 2 = (A + 2B)x^2 + (B + 2C)x + (A + C)$
Identify like terms on both sides.	$7x^2 + 0x + 2 = (A + 2B)x^2 + (B + 2C)x + (A + C)$
Equate coefficients of x^2.	$7 = A + 2B$
Equate coefficients of x.	$0 = B + 2C$
Equate constant terms.	$2 = A + C$
Solve the system of three equations for A, B, and C.	$A = 3, B = 2, C = -1$

Substitute $A = 3, B = 2, C = -1$ into partial fraction decomposition.

$$\frac{7x^2 + 2}{(2x + 1)(x^2 + 1)} = \frac{3}{(2x + 1)} + \frac{2x - 1}{(x^2 + 1)}$$

Check by adding the partial fractions.

$$\frac{3}{(2x + 1)} + \frac{2x - 1}{(x^2 + 1)} = \frac{3(x^2 + 1) + (2x - 1)(2x + 1)}{(2x + 1)(x^2 + 1)} = \frac{7x^2 + 2}{(2x + 1)(x^2 + 1)}$$

■ **YOUR TURN** Find the partial fraction decomposition of $\dfrac{-2x^2 + x + 6}{(x - 1)(x^2 + 4)}$.

Repeated Irreducible Quadratic Factors

CASE 4: $d(x)$ HAS A REPEATED IRREDUCIBLE QUADRATIC FACTOR

If the factored form of $d(x)$ contains an irreducible quadratic factor, $(ax^2 + bx + c)^m$ where $b^2 - 4ac < 0$, then the partial fraction decomposition will contain a term of the form:

$$\frac{A_1x + B_1}{ax^2 + bx + c} + \frac{A_2x + B_2}{(ax^2 + bx + c)^2} + \frac{A_3x + B_3}{(ax^2 + bx + c)^3} + \cdots + \frac{A_mx + B_m}{(ax^2 + bx + c)^m}$$

where A_i and B_i $i = 1, 2, ..., m$ are constants to be determined.

■ **Answer:** $\dfrac{-2x^2 + x + 6}{(x - 1)(x^2 + 4)} = \dfrac{1}{x - 1} - \dfrac{3x + 2}{x^2 + 4}$

EXAMPLE 5 Partial Fraction Decomposition with a Repeated Irreducible Quadratic Factor

Find the partial fraction decomposition of $\dfrac{x^3 - x^2 + 3x + 2}{(x^2 + 1)^2}$.

Solution:

The denominator is already in factored form.

$$\dfrac{x^3 - x^2 + 3x + 2}{(x^2 + 1)^2}$$

Express as a sum of two partial fractions.

$$\dfrac{x^3 - x^2 + 3x + 2}{(x^2 + 1)^2} = \dfrac{Ax + B}{x^2 + 1} + \dfrac{Cx + D}{(x^2 + 1)^2}$$

Multiply equation by LCD, $(x^2 + 1)^2$.

$$x^3 - x^2 + 3x + 2 = (Ax + B)(x^2 + 1) + Cx + D$$

Eliminate parentheses.

$$x^3 - x^2 + 3x + 2 = Ax^3 + Bx^2 + Ax + B + Cx + D$$

Group like terms on right.

$$x^3 - x^2 + 3x + 2 = Ax^3 + Bx^2 + (A + C)x + (B + D)$$

Identify like terms on both sides.

$$x^3 - x^2 + 3x + 2 = Ax^3 + Bx^2 + (A + C)x + (B + D)$$

Equate coefficients of x^3.

$$1 = A \qquad (1)$$

Equate coefficients of x^2.

$$-1 = B \qquad (2)$$

Equate coefficients of x.

$$3 = A + C \qquad (3)$$

Equate **constant** terms.

$$2 = B + D \qquad (4)$$

Substitute $A = 1$ into (3) and solve for C.

$$C = 2$$

Substitute $B = -1$ into (4) and solve for D.

$$D = 3$$

Substitute $A = 1, B = -1, C = 2, D = 3$ into partial fraction decomposition.

$$\dfrac{x^3 - x^2 + 3x + 2}{(x^2 + 1)^2} = \dfrac{x - 1}{x^2 + 1} + \dfrac{2x + 3}{(x^2 + 1)^2}$$

Check by adding the partial fractions.

$$\dfrac{x - 1}{x^2 + 1} + \dfrac{2x + 3}{(x^2 + 1)^2} = \dfrac{(x - 1)(x^2 + 1) + (2x + 3)}{(x^2 + 1)^2} = \dfrac{x^3 - x^2 + 3x + 2}{(x^2 + 1)^2}$$

■ **YOUR TURN** Find the partial fraction decomposition of $\dfrac{3x^3 + x^2 + 4x - 1}{(x^2 + 4)^2}$.

■ **Answer:** $\dfrac{3x^3 + x^2 + 4x - 1}{(x^2 + 4)^2} = \dfrac{3x + 1}{x^2 + 4} - \dfrac{8x + 5}{(x^2 + 4)^2}$

Combinations of All Four Cases

As you probably can imagine, there are rational expressions that have combinations of all four cases, which can lead to a system of several equations when solving for the numerator constants.

EXAMPLE 6 Partial Fraction Decomposition

Find the partial fraction decomposition of $\dfrac{x^5 + x^4 + 4x^3 - 3x^2 + 4x - 8}{x^2(x^2 + 2)^2}$.

Solution:

The denominator is already in factored form.

$$\dfrac{x^5 + x^4 + 4x^3 - 3x^2 + 4x - 8}{x^2(x^2 + 2)^2}$$

Express as a sum of partial fractions.

There are repeated linear and irreducible quadratic factors.

$$\dfrac{x^5 + x^4 + 4x^3 - 3x^2 + 4x - 8}{x^2(x^2 + 2)^2} = \dfrac{A}{x} + \dfrac{B}{x^2} + \dfrac{Cx + D}{(x^2 + 2)} + \dfrac{Ex + F}{(x^2 + 2)^2}$$

Multiply equation by LCD, $x^2(x^2 + 2)^2$.

$$x^5 + x^4 + 4x^3 - 3x^2 + 4x - 8$$
$$= Ax(x^2 + 2)^2 + B(x^2 + 2)^2 + (Cx + D)x^2(x^2 + 2) + (Ex + F)x^2$$

Eliminate parentheses.

$$x^5 + x^4 + 4x^3 - 3x^2 + 4x - 8$$
$$= Ax^5 + 4Ax^3 + 4Ax + Bx^4 + 4Bx^2 + 4B + Cx^5 + 2Cx^3 + Dx^4 + 2Dx^2 + Ex^3 + Fx^2$$

Group like terms on right.

$$x^5 + x^4 + 4x^3 - 3x^2 + 4x - 8$$
$$= (A + C)x^5 + (B + D)x^4 + (4A + 2C + E)x^3 + (4B + 2D + F)x^2 + 4Ax + 4B$$

Equating coefficients of like terms leads to six equations.

$$A + C = 1$$
$$B + D = 1$$
$$4A + 2C + E = 4$$
$$4B + 2D + F = -3$$
$$4A = 4$$
$$4B = -8$$

Solve this system of equations.

$$A = 1, B = -2, C = 0, D = 3, E = 0, F = -1$$

Substitute $A = 1, B = -2, C = 0, D = 3, E = 0, F = -1$ into partial fraction decomposition.

$$\dfrac{x^5 + x^4 + 4x^3 - 3x^2 + 4x - 8}{x^2(x^2 + 2)^2} = \dfrac{1}{x} + \dfrac{-2}{x^2} + \dfrac{0x + 3}{(x^2 + 2)} + \dfrac{0x + -1}{(x^2 + 2)^2}$$

$$\frac{x^5 + x^4 + 4x^3 - 3x^2 + 4x - 8}{x^2(x^2 + 2)^2} = \frac{1}{x} + \frac{-2}{x^2} + \frac{3}{(x^2 + 2)} - \frac{1}{(x^2 + 2)^2}$$

☐ Check by adding the partial fractions.

SECTION 6.3 SUMMARY

In this section we discussed the partial fraction decomposition of rational expressions, which requires the ability to solve systems of linear equations. Four types of factors arise: Nonrepeated linear factors, repeated linear factors, nonrepeated irreducible quadratic factors, and repeated irreducible quadratic factors.

SECTION 6.3 EXERCISES

■ **SKILLS**

In Exercises 1–6, match the rational expression with the form of the partial fraction decomposition.

1. $\dfrac{3x + 2}{x(x^2 - 25)}$

2. $\dfrac{3x + 2}{x(x^2 + 25)}$

3. $\dfrac{3x + 2}{x^2(x^2 + 25)}$

4. $\dfrac{3x + 2}{x^2(x^2 - 25)}$

5. $\dfrac{3x + 2}{x(x^2 + 25)^2}$

6. $\dfrac{3x + 2}{x^2(x^2 + 25)^2}$

a. $\dfrac{A}{x} + \dfrac{B}{x^2} + \dfrac{Cx + D}{x^2 + 25}$

b. $\dfrac{A}{x} + \dfrac{Bx + C}{x^2 + 25} + \dfrac{Dx + E}{(x^2 + 25)^2}$

c. $\dfrac{A}{x} + \dfrac{Bx + C}{x^2 + 25}$

d. $\dfrac{A}{x} + \dfrac{B}{x + 5} + \dfrac{C}{x - 5}$

e. $\dfrac{A}{x} + \dfrac{B}{x^2} + \dfrac{Cx + D}{x^2 + 25} + \dfrac{Ex + F}{(x^2 + 25)^2}$

f. $\dfrac{A}{x} + \dfrac{B}{x^2} + \dfrac{C}{x + 5} + \dfrac{D}{x - 5}$

In Exercises 7–14, write the form of the partial fraction decomposition. Do not solve for the constants.

7. $\dfrac{9}{x^2 - x - 20}$

8. $\dfrac{8}{x^2 - 3x - 10}$

9. $\dfrac{2x + 5}{x^3 - 4x^2}$

10. $\dfrac{x^2 + 2x - 1}{x^4 - 9x^2}$

11. $\dfrac{2x^3 - 4x^2 + 7x + 3}{(x^2 + x + 5)}$

12. $\dfrac{2x^3 + 5x^2 + 6}{(x^2 - 3x + 7)}$

13. $\dfrac{3x^3 - x + 9}{(x^2 + 10)^2}$

14. $\dfrac{5x^3 + 2x^2 + 4}{(x^2 + 13)^2}$

In Exercises 15–40, find the partial fraction decomposition for the rational function.

15. $\dfrac{1}{x(x + 1)}$

16. $\dfrac{1}{x(x - 1)}$

17. $\dfrac{x}{x(x - 1)}$

18. $\dfrac{x}{x(x + 1)}$

19. $\dfrac{9x - 11}{(x - 3)(x + 5)}$

20. $\dfrac{8x - 13}{(x - 2)(x + 1)}$

21. $\dfrac{4x - 3}{x^2 + 6x + 9}$

22. $\dfrac{3x + 1}{x^2 + 4x + 4}$

23. $\dfrac{4x^2 - 32x + 72}{(x + 1)(x - 5)^2}$

24. $\dfrac{4x^2 - 7x - 3}{(x + 2)(x - 1)^2}$

25. $\dfrac{2x^3 + x^2 - x - 1}{x^4 + x^3}$

26. $\dfrac{-x^3 + 2x - 2}{x^5 - x^4}$

27. $\dfrac{5x^2 + 28x - 6}{(x + 4)(x^2 + 3)}$

28. $\dfrac{x^2 + 5x + 4}{(x - 2)(x^2 + 2)}$

29. $\dfrac{-2x^2 - 17x + 11}{(x - 7)(3x^2 - 7x + 5)}$

30. $\dfrac{14x^2 + 8x + 40}{(x + 5)(2x^2 - 3x + 5)}$

31. $\dfrac{x^3}{(x^2 + 9)^2}$

32. $\dfrac{x^2}{(x^2 + 9)^2}$

33. $\dfrac{2x^3 - 3x^2 + 7x - 2}{(x^2 + 1)^2}$

34. $\dfrac{-x^3 + 2x^2 - 3x + 15}{(x^2 + 8)^2}$

35. $\dfrac{3x + 1}{x^4 - 1}$

36. $\dfrac{2 - x}{x^4 - 81}$

37. $\dfrac{x^5 + 2}{(x^2 + 1)^3}$

38. $\dfrac{x^2 - 4}{(x^2 + 1)^3}$

39. $\dfrac{3x}{x^3 - 1}$

40. $\dfrac{5x + 2}{x^3 - 8}$

 ■ **APPLICATIONS**

41. Optics. The relationship between the distance of an object to a lens, d_0, the distance to the image, d_i, and the focal length, f, of the lens is given by

$$\frac{f(d_i - d_0)}{d_i d_0} = 1$$

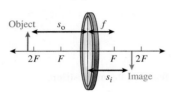

Use partial fraction decomposition to write the lens law in terms of sums of fractions. What does each term represent?

42. Optics. Find the partial fraction decomposition of $\dfrac{1}{n(n + 1)}$, and use it to find the sum of $\dfrac{1}{1 \cdot 2} + \dfrac{1}{2 \cdot 3} + \dfrac{1}{3 \cdot 4} + \cdots$

$+ \dfrac{1}{999 \cdot 1000}$

■ **CATCH THE MISTAKE**

In Exercises 43 and 44, explain the mistake that is made.

43. Find the partial fraction decomposition of: $\dfrac{3x^2 + 3x + 1}{x(x^2 + 1)}$.

Solution:

Write the partial fraction decomposition form.

$$\frac{3x^2 + 3x + 1}{x(x^2 + 1)} = \frac{A}{x} + \frac{B}{x^2 + 1}$$

Multiply by the LCD, $x(x^2 + 1)$.

$$3x^2 + 3x + 1 = A(x^2 + 1) + Bx$$

Eliminate parentheses. $3x^2 + 3x + 1 = Ax^2 + Bx + A$

Matching like terms leads to the equations: $A = 3, B = 3,$ and $A = 1$

This is incorrect. What is the mistake?

44. Find the partial fraction decomposition of: $\dfrac{3x^4 - x - 1}{x(x - 1)}$.

Solution:

Write the partial fraction decomposition form.

$$\frac{3x^4 - x - 1}{x(x - 1)} = \frac{A}{x} + \frac{B}{x - 1}$$

Multiply by the LCD, $x(x - 1)$.

$$3x^4 - x - 1 = A(x - 1) + Bx$$

Eliminate parentheses and group like terms.

$$3x^4 - x - 1 = (A + B)x - A$$

Compare like coefficients: $A = 1, B = -2$

This is incorrect. What is the mistake?

■ **CHALLENGE**

45. T or F: Partial fraction decomposition can only be used when the degree of the numerator is greater than the degree of the denominator.

46. T or F: The degree of the denominator of a reducible rational expression is equal to the number of partial fractions in its decomposition.

47. Find the partial fraction decomposition of:

$$\frac{x^2 + 4x - 8}{x^3 - x^2 - 4x + 4}.$$

48. Find the partial fraction decomposition of:

$$\frac{ax + b}{x^2 - c^2} \quad a, b, c \text{ are real numbers.}$$

■ TECHNOLOGY

49. Using a graphing utility, graph $y_1 = \dfrac{5x + 4}{x^2 + x - 2}$ and

$y_2 = \dfrac{3}{x - 1} + \dfrac{2}{x + 2}$ in the same viewing rectangle. Is y_2 the partial fraction decomposition of y_1?

50. Using a graphing utility, graph $y_1 = \dfrac{2x^2 + 2x - 5}{x^3 + 5x}$ and

$y_2 = \dfrac{3x + 2}{x^2 + 5} - \dfrac{1}{x}$ in the same viewing rectangle. Is y_2 the partial fraction decomposition of y_1?

51. Using a graphing utility, graph $y_1 = \dfrac{x^9 + 8x - 1}{x^5(x^2 + 1)^3}$ and

$y_2 = \dfrac{4}{x} - \dfrac{1}{x^5} + \dfrac{2}{x^2 + 1} - \dfrac{3x + 2}{(x^2 + 1)^2}$ in the same viewing rectangle. Is y_2 the partial fraction decomposition of y_1?

52. Using a graphing utility, graph $y_1 = \dfrac{x^3 + 2x + 6}{(x + 3)(x^2 - 4)^3}$ and

$y_2 = \dfrac{2}{x + 3} + \dfrac{x + 3}{(x^2 - 4)^3}$ in the same viewing rectangle. Is y_2 the partial fraction decomposition of y_1?

SECTION 6.4 Systems of Nonlinear Equations

Skills Objectives

- Solve a system of nonlinear equations using elimination.
- Solve a system of nonlinear equations using substitution.
- Eliminate extraneous solutions.

Conceptual Objectives

- Interpret the algebraic solution graphically.
- Understand the types of solutions: distinct number of solutions, no solution, and infinitely many solutions.

Introduction

Thus far we have only discussed systems of *linear* equations. Let's now consider systems of *nonlinear* equations. We will restrict our discussion to systems of two equations and two variables.

Recall that in Section 6.1 we used elimination, substitution, and graphing to solve systems of two linear equations. The same three methods can *sometimes* be used to solve systems of two nonlinear equations. There is no general procedure that works for solving *all* systems of nonlinear equations.

Another difference between systems of linear equations and systems of nonlinear equations is the number of solutions. Systems of linear equations have one, none, or infinitely many solutions. Systems of nonlinear equations can have more than one solution, such as two or three solutions.

How many points of intersection do two lines have? One, none (parallel lines), or infinitely many (coinciding lines). How many points of intersection do a line and a parabola have? One, two, or no points of intersection correspond to one solution, two solutions, or no solution.

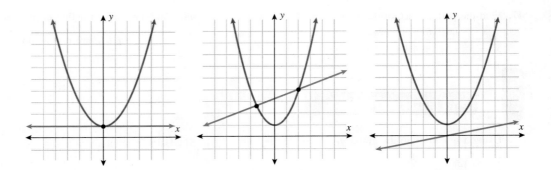

How many points of intersection do a parabola and circle have? One, two, three, four, or no points of intersection correspond to one solution, two solutions, three solutions, four solutions, or no solution, respectively.

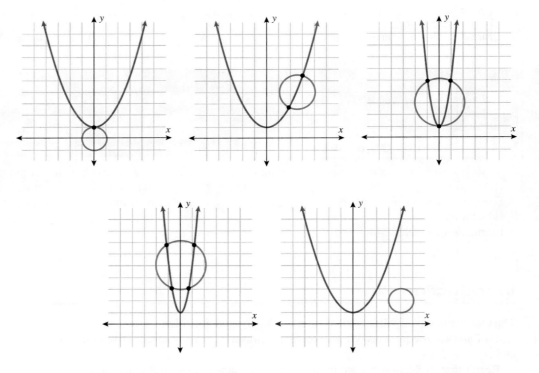

The only system that has infinitely many solutions is a system that has equivalent equations. For example, two identical parabolas or two identical circles would constitute a system of two nonlinear equations with infinitely many solutions.

Using Elimination to Solve Systems of Nonlinear Equations

The first three examples in this section use elimination to solve systems of two nonlinear equations. In linear systems, we can eliminate either variable. In nonlinear systems, the variable to eliminate is the one that is raised to the same power in both equations.

EXAMPLE 1 **Solving a System of Two Nonlinear Equations Using Elimination: One Solution**

Solve the system of equations and graph the corresponding line and parabola to verify the answer.

$$\text{Equation (1): } 2x - y = 3 \qquad \text{Equation (2): } x^2 - y = 2$$

Solution:

Equation (1):	$2x - y = 3$
Multiply Equation (2) by -1.	$-x^2 + y = -2$
Add.	$2x - x^2 = 1$
Gather all terms to one side.	$x^2 - 2x + 1 = 0$
Factor.	$(x - 1)^2 = 0$
Solve for x.	$x = 1$
Substitute $x = 1$ into original Equation (1).	$2(1) - y = 3$
Solve for y.	$y = -1$

The solution is $x = 1$, $y = -1$, or $(1, -1)$.

Graph the line, $y = 2x - 3$, and the parabola, $y = x^2 - 2$, and confirm that the point of intersection is $(1, -1)$.

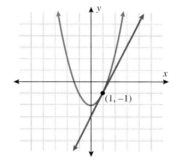

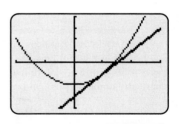

TECHNOLOGY TIP
The graphs of $y_1 = 2x - 3$ and $y_2 = x^2 - 2$ are shown.

EXAMPLE 2 **Solving a System of Two Nonlinear Equations Using Elimination: More Than One Solution**

Solve the system of equations and graph the corresponding parabola and circle to verify the answer.

$$\text{Equation (1): } -x^2 + y = -7 \qquad \text{Equation (2): } x^2 + y^2 = 9$$

Solution:

Equation (1):	$-x^2 + y = -7$
Equation (2):	$x^2 + y^2 = 9$
Add.	$y^2 + y = 2$
Subtract 2.	$y^2 + y - 2 = 0$
Factor.	$(y + 2)(y - 1) = 0$
Solve for y.	$y = -2$ or $y = 1$
Substitute $y = -2$ into Equation (2).	$x^2 + (-2)^2 = 9$
Solve for x.	$x = \pm\sqrt{5}$

TECHNOLOGY TIP
The graphs of $y_1 = x^2 - 7$,
$y_2 = \sqrt{9 - x^2}$, and
$y_3 = -\sqrt{9 - x^2}$ are shown.

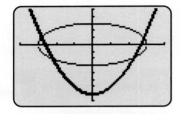

Substitute $y = 1$ into Equation (2). $\qquad x^2 + (1)^2 = 9$

Solve for x. $\qquad\qquad\qquad\qquad x = \pm\sqrt{8} = \pm 2\sqrt{2}$

There are four solutions: $(-\sqrt{5}, -2), (\sqrt{5}, -2)(-2\sqrt{2}, 1)$, and $(2\sqrt{2}, 1)$.

Graph the parabola, $y = x^2 - 7$, and the circle, $x^2 + y^2 = 9$, and confirm the four points of intersection.

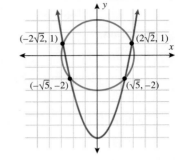

EXAMPLE 3 Solving a System of Two Nonlinear Equations Using Elimination: No Solution

Solve the system of equations and graph the corresponding parabola and circle to verify the answer.

$$\text{Equation (1): } x^2 + y^2 = 1 \qquad \text{Equation (2): } -x^2 + y = 5$$

Solution:

Equation (1):	$x^2 + y^2 = 1$
Equation (2):	$-x^2 + y = 5$
Add.	$y^2 + y = 6$
Subtract 6.	$y^2 + y - 6 = 0$
Factor.	$(y + 3)(y - 2) = 0$
Solve for y.	$y = -3 \text{ or } y = 2$
Substitute $y = -3$ into Equation (1).	$x^2 + (-3)^2 = 1$
Simplify.	$x^2 = -8$

$x^2 = -8$ has no real solution.

Substitute $y = 2$ into Equation (1).	$x^2 + (2)^2 = 1$
Simplify.	$x^2 = -3$

$x^2 = -3$ has no real solution.

There is no solution to this system of nonlinear equations.

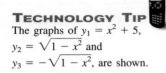

TECHNOLOGY TIP
The graphs of $y_1 = x^2 + 5$,
$y_2 = \sqrt{1 - x^2}$ and
$y_3 = -\sqrt{1 - x^2}$, are shown.

The graphs do not intersect each other. There is no solution to the system.

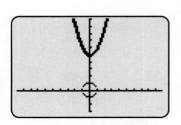

Graph the circle, $x^2 + y^2 = 1$, and the parabola, $y = x^2 + 5$, and confirm there are no points of intersection.

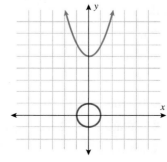

■ **YOUR TURN**   Solve the following systems of nonlinear equations.

a. $-x + y = 3$ **b.** $x^2 + y = 2$
    $x^2 - y = -1$ $-x + y = 5$

Using Substitution to Solve Systems of Nonlinear Equations

Elimination is based on the idea of eliminating one of the variables and solving the remaining equation in one variable. This is not always possible with nonlinear systems. For example, a system consisting of a circle and a line

$$x^2 + y^2 = 5$$
$$-x + y = 1$$

cannot be solved with elimination because both variables are raised to different powers in each equation. We now turn to the substitution method.

EXAMPLE 4 Solving a System of Nonlinear Equations Using Substitution

Solve the system of equations and graph the corresponding parabola and circle to verify the answer.

$$\text{Equation (1): } x^2 + y^2 = 5$$

$$\text{Equation (2): } -x + y = 1$$

Solution:

Rewrite Equation (2) with y isolated.

Equation (1): $x^2 + y^2 = 5$

Equation (2): $y = x + 1$

Substitute Equation (2), $y = x + 1$, into
Equation (1). $x^2 + (x + 1)^2 = 5$

Eliminate parentheses. $x^2 + x^2 + 2x + 1 = 5$

Gather like terms. $2x^2 + 2x - 4 = 0$

Divide by 2. $x^2 + x - 2 = 0$

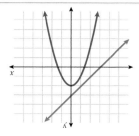

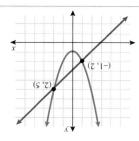

b. no solution

■ **Answer:**   **a.** $(-1, 2)$ and $(2, 5)$

Factor.	$(x + 2)(x - 1) = 0$
Solve for x.	$x = -2$ or $x = 1$
Substitute $x = -2$ into Equation (1).	$(-2)^2 + y^2 = 5$
Solve for y.	$y = -1$ or $y = 1$
Substitute $x = 1$ into Equation (1).	$(1)^2 + y^2 = 5$
Solve for y.	$y = -2$ or $y = 2$

There appears to be four solutions: $(-2, -1)$, $(-2, 1)$, $(1, -2)$, and $(1, 2)$. But a line can intersect a circle in no more than two points. Therefore, at least two solutions are *extraneous*. All four points satisfy Equation (1) but only $(-2, -1)$ and $(1, 2)$ also satisfy Equation (2).

The answer is $(-2, -1)$ and $(1, 2)$.

Graph the circle, $x^2 + y^2 = 5$, and the line, $y = x + 1$, and confirm the two points of intersection.

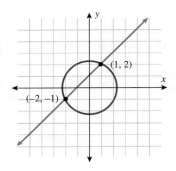

CONCEPT CHECK How many possible points of intersection do a circle and line have?

■ **YOUR TURN** Solve the system of equations: $x^2 + y^2 = 13$ and $x + y = 5$.

EXAMPLE 5 Solving a System of Nonlinear Equations Using Substitution

Solve the system of equations:

$$\text{Equation (1): } x^2 + y^2 = 5$$

$$\text{Equation (2): } \quad xy = 2$$

Solution:

Since Equation (2) tells us that $xy = 2$, we know neither x nor y can be zero.

Solve Equation (2) for y.	$y = \dfrac{2}{x}$
Substitute $y = \dfrac{2}{x}$ into Equation (1).	$x^2 + \left(\dfrac{2}{x}\right)^2 = 5$

Eliminate parentheses.	$x^2 + \dfrac{4}{x^2} = 5$
Multiply by x^2.	$x^4 + 4 = 5x^2$
Collect terms to one side.	$x^4 - 5x^2 + 4 = 0$
Factor.	$(x^2 - 4)(x^2 - 1) = 0$
Solve for x.	$x = \pm 2$ or $x = \pm 1$
Substitute $x = -2$ into Equation (2), $xy = 2$, and solve for y.	$y = -1$
Substitute $x = 2$ into Equation (2), $xy = 2$, and solve for y.	$y = 1$
Substitute $x = -1$ into Equation (2), $xy = 2$, and solve for y.	$y = -2$
Substitute $x = 1$ into Equation (2), $xy = 2$, and solve for y.	$y = 2$

There are four solutions: $(-2, -1), (-1, -2), (2, 1), \text{ and } (1, 2)$.

■ **YOUR TURN**   Solve the system of equations: $x^2 + y^2 = 2$, $xy = 1$

Applications

EXAMPLE 6 Calculating How Much Fence to Buy

A couple buys a rectangular piece of property advertised as 10 acres (approximately 400,000 square feet). They want two fences to divide the land into an internal grazing area and a surrounding riding path. If they want the riding path to be 20 feet wide, one fence will enclose the property and one internal fence will sit 20 feet inside the outer fence. If the internal grazing field is 237,600 square feet, how many linear feet of fencing should they buy?

Solution:

Use the five-step procedure for solving word problems from Section 1.2 and use two variables.

STEP 1   Identify the question.
How many linear feet of fence should they buy? Or, what is the sum of the perimeters of the two fences?

STEP 2   Make notes.

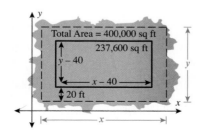

STEP 3 **Set up the equations.**

$x =$ length of property $\qquad x - 40 =$ length of internal field

$y =$ width of property $\qquad y - 40 =$ width of internal field

Equation (1): $\qquad\qquad xy = 400{,}000$

Equation (2): $(x - 40)(y - 40) = 237{,}600$

STEP 4 **Solve the system of equations.**

Substitution Method

Since Equation (1) tells us that $xy = 400{,}000$, we know that neither x nor y can be zero.

Solve Equation (1) for y. $\qquad\qquad\qquad\qquad\qquad y = \dfrac{400{,}000}{x}$

Substitute $y = \dfrac{400{,}000}{x}$ into Equation (2).

$$(x - 40)\left(\frac{400{,}000}{x} - 40\right) = 237{,}600$$

Eliminate parentheses.

$$400{,}000 - 40x - \frac{16{,}000{,}000}{x} + 1600 = 237{,}600$$

Multiply by LCD, x.

$$400{,}000x - 40x^2 - 16{,}000{,}000 + 1600x = 237{,}600x$$

Collect like terms on one side. $\qquad 40x^2 - 164{,}000x + 16{,}000{,}000 = 0$

Divide by 40. $\qquad\qquad\qquad\qquad x^2 - 4100x + 400{,}000 = 0$

Factor. $\qquad\qquad\qquad\qquad\qquad (x - 4000)(x - 100) = 0$

Solve for x. $\qquad\qquad\qquad\qquad x = 4000$ or $x = 100$

Substitute $x = 4000$ into the original Equation (1). $\qquad\qquad\qquad\qquad\qquad 4000y = 400{,}000$

Solve for y. $\qquad\qquad\qquad\qquad\qquad\qquad\qquad y = 100$

Substitute $x = 100$ into the original Equation (1). $\qquad\qquad\qquad\qquad\qquad 100y = 400{,}000$

Solve for y. $\qquad\qquad\qquad\qquad\qquad\qquad\qquad y = 4000$

The two solutions yield the same dimensions: 4000×100. The inner field has the dimensions 3960×60. Therefore, the perimeter of both fences is:

$$2(4000) + 2(100) + 2(3960) + 2(60) = 8000 + 200 + 7920 + 120 = 16{,}240$$

The couple should buy 16,240 linear feet of fencing .

STEP 5 **Check the solution.**

The point (4000, 100) satisfies both Equation (1) and Equation (2).

SECTION 6.4 SUMMARY

In this section, systems of two equations were discussed when at least one of the equations is nonlinear. The substitution method and elimination method from Section 6.1 can *sometimes* be applied to nonlinear systems. When graphing the two equations, the points of intersection are the solutions of the system: Systems of nonlinear equations can have more than one solution. Also, extraneous solutions can appear, so it is important to always check solutions.

SECTION 6.4 EXERCISES

■ **SKILLS**

In Exercises 1–12, solve the system of equations using the elimination method.

1. $x^2 - y = -2$
$-x + y = 4$

2. $x^2 + y = 2$
$2x + y = -1$

3. $x^2 + y = 1$
$2x + y = 2$

4. $x^2 - y = 2$
$-2x + y = -3$

5. $x^2 + y = -5$
$-x + y = 3$

6. $x^2 - y = -7$
$x + y = -2$

7. $x^2 + y^2 = 1$
$x^2 - y = -1$

8. $x^2 + y^2 = 1$
$x^2 + y = -1$

9. $x^2 + y^2 = 3$
$4x^2 + y = 0$

10. $x^2 + y^2 = 6$
$-7x^2 + y = 0$

11. $x^2 + y^2 = -6$
$-2x^2 + y = 7$

12. $x^2 + y^2 = 5$
$3x^2 + y = 9$

In Exercises 13–24, solve the system of equations using the substitution method.

13. $x + y = 2$
$x^2 + y^2 = 2$

14. $x - y = -2$
$x^2 + y^2 = 2$

15. $xy = 4$
$x^2 + y^2 = 10$

16. $xy = -3$
$x^2 + y^2 = 12$

17. $y = x^2 - 3$
$y = -4x + 9$

18. $y = -x^2 + 5$
$y = 3x - 4$

19. $x^2 + xy - y^2 = 5$
$x - y = -1$

20. $x^2 + xy + y^2 = 13$
$x + y = -1$

21. $2x - y = 3$
$x^2 + y^2 - 2x + 6y = -9$

22. $x^2 + y^2 - 2x - 4y = 0$
$-2x + y = -3$

23. $4x^2 + 12xy + 9y^2 = 25$
$-2x + y = 1$

24. $-4xy + 4y^2 = 8$
$3x + y = 2$

In Exercises 25–34, solve the system of equations using any method.

25. $x^3 - y^3 = 63$
$x - y = 3$

26. $x^3 + y^3 = -26$
$x + y = -2$

27. $4x^2 - 3xy = -5$
$-x^2 + 3xy = 8$

28. $2x^2 + 5xy = 2$
$x^2 - xy = 1$

29. $\log_x(2y) = 3$
$\log_x(y) = 2$

30. $\log_x(y) = 1$
$\log_x(2y) = \dfrac{1}{2}$

31. $\dfrac{1}{x^3} + \dfrac{1}{y^2} = 17$
$\dfrac{1}{x^3} - \dfrac{1}{y^2} = -1$

32. $\dfrac{2}{x^2} + \dfrac{3}{y^2} = \dfrac{5}{6}$
$\dfrac{4}{x^2} - \dfrac{9}{y^2} = 0$

33. $2x^2 + 4y^4 = -2$
$6x^2 + 3y^4 = -1$

34. $x^2 + y^2 = -2$
$x^2 + y^2 = -1$

In Exercises 35 and 36, graph each equation and find the point(s) of intersection.

35. The parabola $y = x^2 - 6x + 11$ and the line $y = -x + 7$.

36. The circle $x^2 + y^2 - 4x - 2y + 5 = 0$ and the line $-x + 3y = 6$.

■ APPLICATIONS

37. Numbers. The sum of two numbers is 10 and the difference of their squares is 40. Find the numbers.

38. Numbers. The difference of two numbers is 3 and the difference of their squares is 51. Find the numbers.

39. Numbers. The product of two numbers is equal to the reciprocal of the difference of their reciprocals. The product of the two numbers is 72. Find the numbers.

40. Numbers. The ratio of the sum of two numbers to the difference of two numbers is 9. The product of the two numbers is 80. Find the numbers.

41. Geometry. A rectangle has a perimeter of 36 centimeters and an area of 80 square centimeters. Find the dimensions of the rectangle.

42. Geometry. Two concentric circles have perimeters that add up to 16π and areas that add up to 34π. Find the radii of the two circles.

43. Horse Paddock. An equestrian buys a 5-acre rectangular parcel (approximately 200,000 square feet) and is going to fence in the entire property and then divide the parcel into two halves with a fence. If 2200 linear feet of fencing is required, what are the dimensions of the parcel?

44. Dog Run. A family moves into a new home and decides to fence in the yard to give their dog room to roam. If the area that will be fenced in is rectangular and has an area of 11,250 square feet, and the length is twice as much as the width, how many linear feet of fence should they buy?

45. Footrace. Your college algebra professor and Jeremy Wariner (2004 Olympic Gold Medalist in the men's 400 meter) decided to race. The race was 400 meters and Jeremy gave your professor a 1-minute head start, and still crossed the finish line 1 minute 40 seconds before your professor. If Jeremy ran five times faster than your professor, what was each person's average speed?

46. Footrace. You decided to race Jeremy Wariner for 800 meters. At that distance, Jeremy runs approximately twice as fast as you. He gave you a 1-minute head start and crossed the finish line 20 seconds before you. What were each of your average speeds?

■ CATCH THE MISTAKE

In Exercises 47 and 48, explain the mistake that is made.

47. Solve the system of equations: $\begin{aligned} x^2 + y^2 &= 4 \\ x + y &= 2 \end{aligned}$

Solution:

Multiply the second equation by (-1) and add to the first equation. $\qquad x^2 - x = 2$

Subtract 2. $\qquad x^2 - x - 2 = 0$

Factor. $\qquad (x + 1)(x - 2) = 0$

Solve for x. $\qquad x = -1$ and $x = 2$

Substitute $x = -1$ and $x = 2$ into $x + y = 2$.

$-1 + y = 2$ and $2 + y = 2$

Solve for y. $\qquad y = 3$ and $y = 0$

The answer is $(-1, 3)$ and $(2, 0)$.

This is not correct. What mistake was made?

48. Solve the system of equations: $x^2 + y^2 = 5$
$2x - y = 0$.

Solution:

Solve the second equation for y. $y = 2x$

Substitute $y = 2x$ into the
first equation. $x^2 + (2x)^2 = 5$

Eliminate parentheses. $x^2 + 4x^2 = 5$

Gather like terms. $5x^2 = 5$

Solve for x. $x = -1$ and $x = 1$

Substitute $x = -1$ into the
first equation. $(-1)^2 + y^2 = 5$

Solve for y. $y = -2$ and $y = 2$

Substitute $x = 1$ into the
first equation. $(1)^2 + y^2 = 5$

Solve for y. $y = -2$ and $y = 2$

The answers are $(-1, -2), (-1, 2), (1, -2)$, and $(1, 2)$.

This is not correct. What mistake was made?

■ CHALLENGE

49. T or F: A system of equations representing a line and a parabola can intersect at most at three points.

50. T or F: A system of equations representing a line and a cubic function can intersect in at most three places.

51. T or F: The elimination method can always be used to solve systems of two nonlinear equations.

52. T or F: The substitution method always works for solving systems of nonlinear equations.

53. A circle and a line have at most two points of intersection. A circle and a parabola have at most four points of intersection. What are the most number of points of intersection that a circle and an nth-degree polynomial can have?

54. A line and a parabola have at most two points of intersection. A line and a cubic function have at most three points of intersection. What are the most number of points of intersection that a line and an nth-degree polynomial can have?

55. Find a system of equations representing a line and a parabola that has only one real solution.

56. Find a system of equations representing a circle and a parabola that has only one real solution.

■ TECHNOLOGY

Use a graphing utility to solve the following systems of equations.

57. $y = e^x$
$y = \ln x$

58. $y = 10^x$
$y = \log x$

59. $2x^3 + 4y^2 = 3$
$xy^3 = 7$

60. $3x^4 - 2xy + 5y^2 = 19$
$x^4 y = 5$

SECTION 6.5 Systems of Inequalities

Skills Objectives

- Graph a linear inequality in two variables.
- Graph a nonlinear inequality in two variables.
- Graph a system of inequalities.

Conceptual Objectives

- Interpret the difference between solid lines and dashed lines.
- Interpret an overlapping shaded region as a solution.

Linear Inequalities in Two Variables

To graph linear inequalities in two variables, we will bridge together two concepts that we have already learned: *linear inequalities* (Section 1.5) and *lines* (Section 2.3). Recall that in Section 1.5 we discussed linear inequalities in one variable. For example, $3x - 1 < 8$ has a solution, $x < 3$, which can be represented graphically on a number line as

where the red shaded area to the left of 3 represents the solution.

Recall in Section 2.3 that $y = 2x + 1$ is an *equation in two variables* whose graph is a line in the xy-plane.

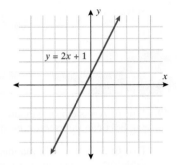

We now turn our attention to **linear inequalities in two variables**. For example, if we change the $=$ in $y = 2x + 1$ to $<$ we get $y < 2x + 1$. The solution to this inequality in two variables is the set of all points (x, y) that make this inequality true. Some solutions to this inequality are $(-2, -5)$, $(0, 0)$, $(3, 4)$, $(5, -1)$, ...

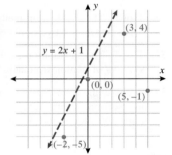

In fact, the entire region *below* the line, $y = 2x + 1$, satisfies the inequality, $y < 2x + 1$.

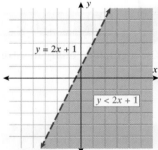

If we reverse the sign of the inequality, $y > 2x + 1$, then the entire region *above* the line $y = 2x + 1$ would represent the solution to the inequality.

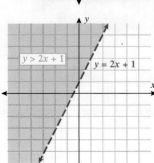

Any line divides the *xy*-plane into two **half-planes**. For example, the line $y = 2x + 1$ divides the *xy*-plane into two half-planes represented as $y > 2x + 1$ and $y < 2x + 1$. Recall that with inequalities in one variable we used the notation of brackets and parentheses to denote the type of inequality (strict or nonstrict). We use a similar notation with linear inequalities in two variables. If the inequality is a strict inequality, $<$ or $>$, then the line is *dashed,* and, if the inequality includes the equal sign, $\le$ or $\ge$, then a *solid* line is used. The following box summarizes the procedure for graphing a linear inequality in two variables.

GRAPHING A LINEAR INEQUALITY IN TWO VARIABLES

Step 1: Change the Sign
   Change the inequality sign, $<$, $\le$, $\ge$, or $>$, to an equal sign, $=$.

Step 2: Draw the Line
   • If the inequality is strict, $<$ or $>$, use a **dashed** line.
   • If the inequality is not strict, $\le$ or $\ge$, use a **solid** line.

Step 3: Test a Point
   • Select a point in one half-plane and test to see if it satisfies the inequality. If it does, then so do all the points in that region (half-plane). If not, then none of the points in that half-plane satisfy the inequality.
   • Repeat this step for the other half-plane.

Step 4: Shade
   Shade the half-plane that satisfies the inequality.

EXAMPLE 1   Graphing a Strict Linear Inequality in Two Variables

Graph the inequality $3x + y < 2$.

Solution:

STEP 1   Change the inequality sign to an
equal sign. $3x + y = 2$

STEP 2   Draw the line.

Convert from standard form to
slope–intercept form. $y = -3x + 2$

Since the inequality, $<$, is a strict
inequality, use a **dashed** line.

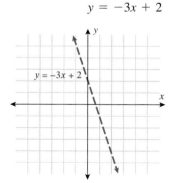

STEP 3   Test points in each half-plane.
   Substitute $(3, 0)$ into $3x + y < 2$. $9 < 2$
   The point $(3, 0)$ does not satisfy the inequality.
   Substitute $(-2, 0)$ into $3x + y < 2$. $-6 < 2$
   The point $(-2, 0)$ does satisfy the inequality.

STEP 4 Shade the region containing the point $(-2, 0)$.

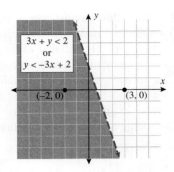

Note: An alternative method is to: (1) write the inequality in slope–intercept form; (2) change inequality to equal; (3) graph the line; (4) shade "above" if $y >$ and "below" if $y <$.

■ YOUR TURN   Graph the inequality $-x + y > -1$.

EXAMPLE 2 Graphing a Nonstrict Linear Inequality in Two Variables

Graph the inequality $2x - 3y \geq 6$.

Solution:

STEP 1 Change the inequality sign to an equal sign.

$$2x - 3y = 6$$

STEP 2 Draw the line.

Convert from standard form to slope–intercept form.

$$y = \frac{2}{3}x - 2$$

Since the inequality, $\geq$, is not a strict inequality, use a **solid** line.

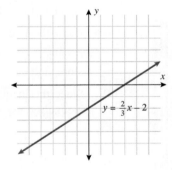

TECHNOLOGY TIP

The graphing calculator can be used to help in shading the linear inequality $3x + y < 2$. However, it will not show whether the line is solid or dashed. First solve for y, $y < -3x + 2$. Then, enter $y_1 = -3x + 2$. Since $y_1 < -3x + 2$, the region below the dashed line is shaded. Use the left arrow key to move the cursor to the left of y_1, press $\boxed{\text{ENTER}}$ until ◥ appears. Set the window at $[-8,8]$ by $[-8,10]$ and press $\boxed{\text{GRAPH}}$.

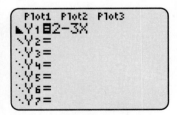

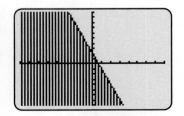

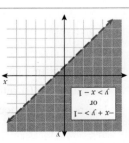

■ Answer:

STEP 3 Test points in each half-plane.

Substitute $(5, 0)$ into $2x - 3y \geq 6$. $10 \geq 6$

The point $(5, 0)$ satisfies the inequality.

Substitute $(0, 0)$ into $2x - 3y \geq 6$. $0 \geq 6$

The point $(0, 0)$ does not satisfy the inequality.

STEP 4 Shade the region containing the point $(5, 0)$.

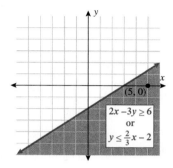

■ **YOUR TURN** Graph the inequality $x - 2y \leq 6$.

Nonlinear Inequalities in Two Variables

Linear inequalities are in the form $Ax + By \leq C$, where any of the four inequalities can be used. Examples of **nonlinear inequalities in two variables** are

$$x^2 + y^2 > 1, \; y \leq -x^2 + 3, \text{ and } y \geq x^3$$

We follow the same procedure as we did with linear inequalities. We change the inequality sign to an equal sign, graph the nonlinear equation, test points from the two regions, and shade the region that makes the inequality true.

EXAMPLE 3 Graphing a Strict Nonlinear Inequality in Two Variables

Graph the inequality $x^2 + y^2 > 1$.

Solution:

STEP 1 Change the inequality sign to an equal sign. $x^2 + y^2 = 1$

STEP 2 Draw the circle.

The center is $(0, 0)$ and the radius is 1.

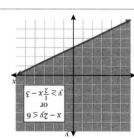

■ **Answer:**

Since the inequality, $>$, is a strict inequality, use a **dashed** curve.

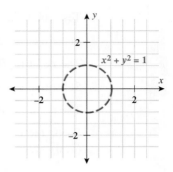

STEP 3 Test points in each region (outside the circle and inside the circle).

Substitute (2, 0) into $x^2 + y^2 > 1$. $\qquad$ $4 \geq 1$

The point (2, 0) satisfies the inequality.

Substitute (0, 0) into $x^2 + y^2 > 1$. $\qquad$ $0 \geq 1$

The point (0, 0) does not satisfy the inequality.

STEP 4 Shade the region containing the point (2, 0).

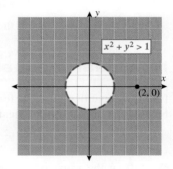

EXAMPLE 4 Graphing a Nonstrict Nonlinear Inequality in Two Variables

Graph the inequality $y \leq -x^2 + 3$.

Solution:

STEP 1 Change the inequality sign to an equal sign. $\quad y = -x^2 + 3$

STEP 2 Graph the parabola.

Reflect the base function, $f(x) = x^2$, about the x-axis and shift up 3 units. Since the inequality, $\leq$, is a nonstrict inequality use a **solid** curve.

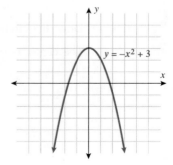

STEP 3 Test points in each region (inside the parabola and outside the parabola).

Substitute (3, 0) into $y \leq -x^2 + 3$. $\qquad$ $0 \leq -6$

The point (3, 0) does not satisfy the inequality.

Substitute $(0, 0)$ into $y \leq -x^2 + 3$. $0 \leq 3$

The point $(0, 0)$ does satisfy the inequality.

STEP 4 Shade the region containing the
point $(0, 0)$.

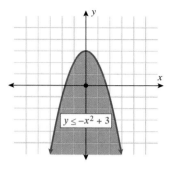

$y \leq -x^2 + 3$

CONCEPT CHECK   What type of curve (solid or dashed) is used for

a. strict inequalities ($<$ or $>$)?

b. nonstrict inequalities ($\leq$ or $\geq$)?

YOUR TURN   Graph the following inequalities:

a. $x^2 + y^2 \leq 9$ **b.** $y > x^3$

Systems of Inequalities in Two Variables

Systems of inequalities are similar to *systems of equations.* In systems of equations we sought the points that satisfied *all* of the equations. The **solution set of a system of inequalities** contains the points that satisfy *all* of the inequalities. The graph of a system of inequalities can be obtained by simultaneously graphing each individual inequality and finding where the shaded regions intersect (or overlap), if at all.

EXAMPLE 5 Graphing a System of Two Linear Inequalities

Graph the system of inequalities.

$$x + y \geq -2$$
$$x + y \leq 2$$

STEP 1 Change the inequality signs to equal signs. $x + y = -2$
                                                           $x + y = 2$

STEP 2 Draw the two lines.

Write the lines in slope–intercept form. $y = -x - 2$ and
                                                $y = -x + 2$

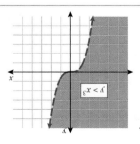

$y < x^3$

b.

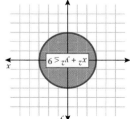

$x^2 + y^2 \leq 9$

Answer: a.

Because the inequality signs are not strict, use solid lines.

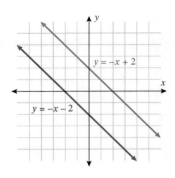

STEP 3 Test points for each inequality.

$x + y \geq -2$

Substitute $(-4, 0)$ into $x + y \geq -2$. $\qquad -4 \geq -2$

The point $(-4, 0)$ does not satisfy the inequality.

Substitute $(0, 0)$ into $x + y \geq -2$. $\qquad 0 \geq -2$

The point $(0, 0)$ does satisfy the inequality.

$x + y \leq 2$

Substitute $(0, 0)$ into $x + y \leq 2$. $\qquad 0 \leq 2$

The point $(0, 0)$ does satisfy the inequality.

Substitute $(4, 0)$ into $x + y \leq 2$. $\qquad 4 \leq 2$

The point $(4, 0)$ does not satisfy the inequality.

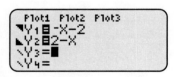

TECHNOLOGY TIP

Enter $y_1 \geq -x - 2$ and $y_2 \leq -x + 2$.

The overlapping region is the solution.

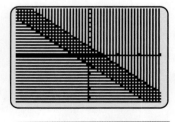

STEP 4 For $x + y \geq -2$, shade the region "above" that includes $(0, 0)$.

For $x + y \leq 2$, shade the region "below" that includes $(0, 0)$.

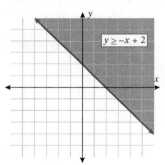

STEP 5 The overlapping region is the solution.

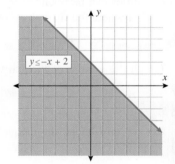

If the inequality signs are reversed in Example 5, there is actually no solution, as you will see in Example 6.

EXAMPLE 6 **Graphing a System of Two Linear Inequalities with No Solution**

Graph the system of inequalities.

$$x + y \le -2$$
$$x + y \ge 2$$

STEP 1 Change the inequality signs to equal signs.

$$x + y = -2$$
$$x + y = 2$$

STEP 2 Draw the two lines.

Write the lines in slope–intercept form.

$$y = -x - 2 \text{ and}$$
$$y = -x + 2$$

Because the inequality signs are not strict, use solid lines.

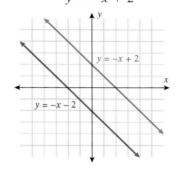

STEP 3 Test points for each inequality.

$x + y \le -2$

Substitute $(-4, 0)$ into $x + y \le -2$. $-4 \le -2$

The point $(-4, 0)$ does satisfy the inequality.

Substitute $(0, 0)$ into $x + y \le -2$. $0 \le -2$

The point $(0, 0)$ does not satisfy the inequality.

$x + y \ge 2$

Substitute $(0, 0)$ into $x + y \ge 2$. $0 \ge 2$

The point $(0, 0)$ does not satisfy the inequality.

Substitute $(4, 0)$ into $x + y \ge 2$. $4 \ge 2$

The point $(4, 0)$ does satisfy the inequality.

STEP 4 For $x + y \ge -2$, shade the region "below" that includes $(-4, 0)$.
For $x + y \le 2$ shade the region "above" that includes $(4, 0)$.

STEP 5 There is no overlapping region. Therefore, no points satisfy both inequalities. We say there is
no solution .

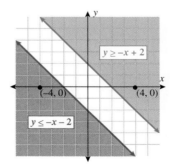

TECHNOLOGY TIP

Solve for y in each inequality first. Enter $y_1 \le -x - 2$ and $y_2 \ge -x + 2$.

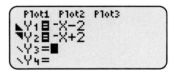

There is no overlapping region. Therefore, there is no solution to the system of inequalities.

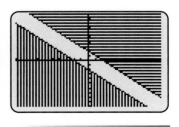

✓**CONCEPT CHECK** What is the solution of a system of inequalities if there is no overlapping region?

■ **YOUR TURN** Graph the solution to the system of inequalities.

a. $y > x + 1$
$\quad y < x - 1$

b. $y < x + 1$
$\quad y > x - 1$

Thus far we have only addressed systems of two linear inequalities. Systems with two nonlinear inequalities or systems with more than two inequalities are treated in a similar manner. The solution is the set of all points that satisfy *all* of the inequalities. When there are nonlinear inequalities or more than two linear inequalities, the solution may be a **bounded** region. If it is not possible graphically, we can algebraically determine where the curves intersect. Although in Examples 7 and 8 we can determine the points of intersection graphically, we will algebraically determine those points as well.

EXAMPLE 7 Graphing a System of Multiple Linear Inequalities

Solve the system of inequalities.
$\quad\quad y \le x$
$\quad\quad y \ge -x$
$\quad\quad y < 3$

STEP 1 Change the inequalities to equal signs.
$\quad\quad y = x$
$\quad\quad y = -x$
$\quad\quad y = 3$

STEP 2 Draw the three lines ($\le$ and $\ge$ correspond to solid lines, $<$ corresponds to dashed line).

To determine the points of intersection, set the y values equal.

Point where $y = x$ and $y = -x$ intersect:
$\quad\quad x = -x$
$\quad\quad x = 0$

Substitute $x = 0$ into $y = x$.
$\quad\quad (0, 0)$

Point where $y = -x$ and $y = 3$ intersect:
$\quad\quad -x = 3$
$\quad\quad x = -3$
$\quad\quad (-3, 3)$

Point where $y = 3$ and $y = x$ intersect:
$\quad\quad x = 3$
$\quad\quad (3, 3)$

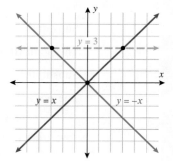

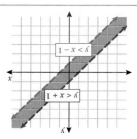

■ **Answer: a.** no solution **b.**

STEP 3 Test points to determine the shaded half-planes corresponding to $y \leq x$, $y \geq -x$, and $y < 3$.

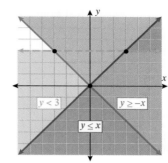

STEP 4 Shade the overlapping region.

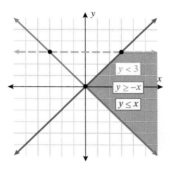

TECHNOLOGY TIP

Solve for y in each inequality first. Enter $y_1 \leq x$, $y_2 \geq -x$, and $y_3 < 3$. Set the window at $[-5, 5]$ by $[-10, 10]$ and press [GRAPH].

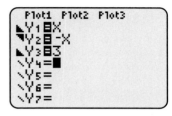

The overlapping region is the solution to the system of inequalities.

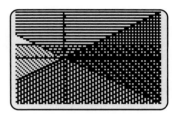

EXAMPLE 8 Graphing a System of Inequalities

Graph the system of inequalities.

$$y \geq x^2 - 1$$
$$y < x + 1$$

STEP 1 Change the inequalities to equal signs.

$$y = x^2 - 1$$
$$y = x + 1$$

STEP 2 Draw the parabola (solid) and the line (dashed).

To determine the points of intersection, set the y values equal.

$$x^2 - 1 = x + 1$$
$$x^2 - x - 2 = 0$$

Factor.

$$(x - 2)(x + 1) = 0$$

Solve for x.

$$x = 2 \text{ or } x = -1$$

Substitute $x = 2$ into $y = x + 1$. $(2, 3)$

Substitute $x = -1$ into $y = x + 1$. $(-1, 0)$

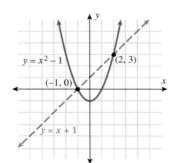

STEP 3 Test points and shade regions.

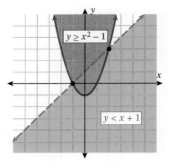

STEP 4 Shade common region.

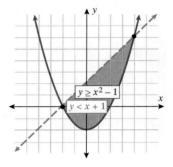

■ **YOUR TURN** Graph the system of inequalities: $\begin{array}{l} x^2 + y^2 < 9 \\ y > 0 \end{array}$.

Applications

Systems of inequalities arise in many applications, for example, the target zone for heart rate during exercise, normal weight ranges for humans, capacity of a room for an event, return on investments, and many others.

EXAMPLE 9 Cost of a Wedding Reception

A couple has invited 300 guests to their wedding. The fixed costs (such as formal wear, entertainment, flowers, and invitations) are $7,000, and the variable costs (party favors, chair covers, food, and drinks) range between $25 and $50 per person depending on the menu. Assuming at least 200 people attend and at most 300 people attend, graph the cost of their wedding as a system of inequalities.

Solution:

Let x represent the number of people attending the wedding and y represent the cost of the wedding. There are four linear inequalities:

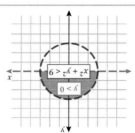

■ **Answer:**

At least 200 guests attend: $x \geq 200$

No more than 300 guests attend: $x \leq 300$

Minimum cost of the wedding: $y \geq 7000 + 25x$

Maximum cost of the wedding: $y \leq 7000 + 50x$

Graph the system of inequalities.

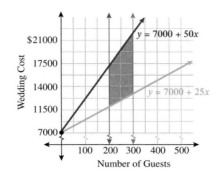

SECTION 6.5 SUMMARY

In this section, we discussed linear and nonlinear inequalities. The difference between a graph of an equation and a graph of an inequality in two variables is the shading of the region that satisfies the inequality. Dashed curves are used for strict inequalities, and solid curves are used when the inequalities are not strict and the curve satisfies the inequality. Systems of inequalities are similar to systems of equations in that we seek the points that satisfy all of the inequalities in the system. The graph of systems of inequalities is obtained by plotting points and shading the individual inequalities where the overlapping shaded regions constitute the solution.

SECTION 6.5 EXERCISES

SKILLS

In Exercises 1–4, match the linear inequality with the correct graph.

1. $y > x$ **2.** $y \geq x$ **3.** $y < x$ **4.** $y \leq x$

a.

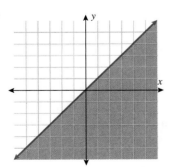

b.

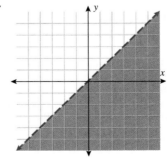

c.

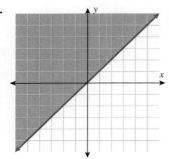

d.

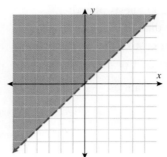

In Exercises 5–12, graph the linear inequality.

5. $y > x - 1$ **6.** $y \geq -x + 1$ **7.** $y \leq -3x + 2$ **8.** $y < 2x + 3$

9. $3x + 4y < 2$ **10.** $2x + 3y > -6$ **11.** $4x - 2y \geq 6$ **12.** $6x - 3y \geq 9$

In Exercises 13–16, match the nonlinear inequality with the correct graph.

13. $x^2 + y^2 < 25$ **14.** $x^2 + y^2 \leq 25$ **15.** $x^2 + y^2 \leq 0$ **16.** $x^2 + y^2 > 0$

a.

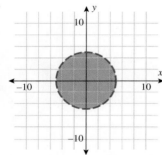

b.

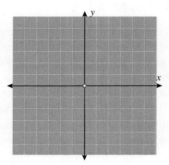

c.

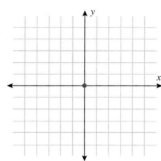

d.

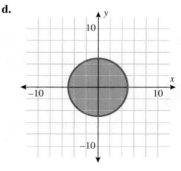

In Exercises 17–26, graph the nonlinear inequality.

17. $y \leq x^2 - 2$ **18.** $y \geq -x^2 + 3$ **19.** $x^2 + y^2 > 4$ **20.** $x^2 + y^2 < 16$

21. $x^2 + y^2 - 2x + 4y + 4 \geq 0$ **22.** $x^2 + y^2 + 2x - 2y - 2 \leq 0$

23. $y \geq e^x$ **24.** $y \leq \ln x$ **25.** $y < -x^3$ **26.** $y > -x^4$

In Exercises 27–48, graph each system of inequalities or indicate that the system has no solution.

27. $y \geq x - 1$
    $y \leq x + 1$

28. $y > x + 1$
    $y < x - 1$

29. $y > 2x + 1$
    $y < 2x - 1$

30. $y \leq 2x - 1$
    $y \geq 2x + 1$

31. $y \geq 2x$
    $y \leq 2x$

32. $y > 2x$
$y < 2x$

33. $x > -2$
$x < 4$

34. $y < 3$
$y > 0$

35. $x \geq 2$
$y \leq x$

36. $y \leq 3$
$y \geq x$

37. $y > x$
$x < 0$
$y < +4$

38. $x + y > 2$
$y < 1$
$x > 0$

39. $-x + y > 1$
$y < 3$
$x > 0$

40. $y < x + 2$
$y > x - 2$
$y < -x + 2$
$y > -x - 2$

41. $x + y > -4$
$-x + y < 2$
$y \geq -1$
$y \leq 1$

42. $-x^2 + y > -1$
$x^2 + y < 1$

43. $x < -y^2 + 1$
$x > y^2 - 1$

44. $y \geq x^2$
$x \geq y^2$

45. $x^2 + y^2 < 36$
$2x + y > 3$

46. $x^2 + y^2 < 36$
$y > 6$

47. $y < e^x$
$y > \ln x$
$x > 0$

48. $y < 10^x$
$y > \log x$
$x > 0$

■ **APPLICATIONS**

49. Area. Find the area enclosed by the system of inequalities:
$$y > |x|$$
$$y < 2$$

50. Area. Find the area enclosed by the system of inequalities:
$$x^2 + y^2 \leq 5$$
$$x \leq 0$$
$$y \geq 0$$

51. Baby Seat. Gloria decides to make baby grocery cart seat covers that have toys attached to protect a baby from germs and entertain them while their parents shop. The materials cost her $17 per cover to make and she is selling them on eBay for $25 per cover (including shipping). Her fixed costs are $120 per month. Write a system of linear inequalities that represents the profit of her small business.

52. Baby Seat. Repeat Exercise 51, but write a system of linear inequalities that represents the loss of her small business. How many covers would she have to sell to break even?

53. Hurricanes. After back-to-back-to-back-to-back hurricanes (Charley, Frances, Ivan, and Jeanne) in Florida in the summer of 2004, FEMA sent disaster relief trucks to Florida. Floridians mainly needed drinking water and generators. Each truck could carry no more than 6000 pounds of cargo or 2400 cubic feet of cargo. Each case of bottled water takes up 1 cubic foot of space and weighs 25 pounds. Each generator takes up 20 cubic feet and weighs 150 pounds. Let x represent the number of cases of water and y represent the number of generators, and write a system of linear inequalities that describes the number of generators and cases of water each truck can haul to Florida.

54. Hurricanes. Repeat Exercise 53 with a smaller truck and different supplies. Suppose the smaller trucks that can haul 2000 pounds and 1500 cubic feet of cargo are used to haul plywood and tarps. A case of plywood is 60 cubic feet and weighs 500 pounds. A case of tarps is 10 cubic feet and weighs 50 pounds. Let x represent the number of cases of plywood and y represent the number of cases of tarps and write a system of linear inequalities that describes the number of cases of tarps and plywood each truck can haul to Florida. Graph the system of linear inequalities.

Andy Washnik

Baby grocery cart seat cover

■ **CATCH THE MISTAKE**

In Exercises 55–58, explain the mistake that is made.

55. Graph the inequality $y \geq 2x + 1$.

Solution:

Graph the line $y = 2x + 1$ with a solid line.

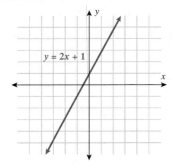

Since the inequality is $\geq$, shade to the *right*.

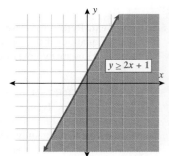

This is incorrect. What mistake was made?

56. Graph the inequality $y < 2x + 1$.

Solution:

Graph the line $y = 2x + 1$ with a solid line.

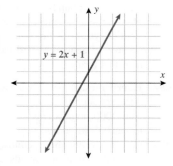

Since the inequality is $<$, shade *below*.

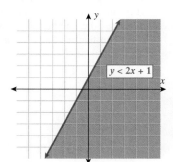

This is incorrect. What mistake was made?

57. Graph the system of inequalities:

$$y > -x + 2$$
$$y > -x - 2$$

Solution:

Graph the lines $y = -x + 2$ and $y = -x - 2$ in dashed lines.

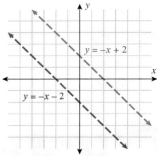

Shade *above* both graphs.

The solution is the common region.

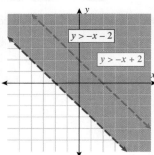

This is incorrect. What mistake was made?

58. Graph the system of inequalities:

$$x^2 + y^2 < 1$$
$$x^2 + y^2 > 4$$

Solution:

Draw the circles $x^2 + y^2 = 1$ and $x^2 + y^2 = 4$.

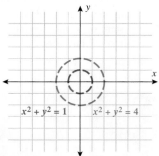

Shade outside $x^2 + y^2 = 1$ and inside $x^2 + y^2 = 4$.

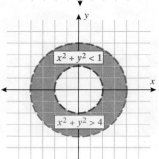

This is incorrect. What mistake was made?

CHALLENGE

59. T or F: A linear inequality always has a solution that is a half-plane.

60. T or F: A nonlinear inequality always has a solution that is a half-plane.

61. T or F: A solid curve is used for strict inequalities.

62. T or F: A system of inequalities always has a solution.

63. For the system of linear inequalities:

$$x \geq a$$
$$x < b$$
$$y > c$$
$$y \leq d$$

describe the solution when $a < b$ and $c < d$. What will the solution be if $a > b$ and $c > d$? Assume a, b, c, and d are real numbers.

64. Given the system of linear inequalities:

$$y \leq ax + b$$
$$y \geq -ax + b$$

what restrictions must be placed on a and b to yield a solution? Graph that solution. For what values of a and b will this system of inequalities have no solution? Assume a and b are real numbers.

65. For the system of nonlinear inequalities:

$$x^2 + y^2 \geq a^2$$
$$x^2 + y^2 \leq b^2$$

what restriction must be placed on the values of a and b for this system to have a solution? Assume a and b are real numbers.

66. Can $x^2 + y^2 < -1$ ever have a real solution? What types of numbers would x and/or y have to be to satisfy this inequality?

TECHNOLOGY

Use a graphing utility to graph the following inequalities.

67. $4x - 2y \geq 6$ (Check with your answer to Exercise 11.)

68. $6x - 3y \geq 9$ (Check with your answer to Exercise 12.)

69. $x^2 + y^2 - 2x + 4y + 4 \geq 0$ (Check with your answer to Exercise 21.)

70. $x^2 + y^2 + 2x - 2y - 2 \leq 0$ (Check with your answer to Exercise 22.)

71. $y \geq e^x$ (Check with your answer to Exercise 23.)

72. $y \leq \ln x$ (Check with your answer to Exercise 24.)

73. $y < e^x$ (Check with your
$y > \ln x$ $x > 0$ answer to Exercise 47.)

74. $y < 10^x$ (Check with your
$y > \log x$ $x > 0$ answer to Exercise 48.)

SECTION **6.6** Linear Programming

Skills Objectives

- Write an objective function that represents a quantity to be minimized or maximized.
- Use inequalities to describe constraints.
- Solve the optimization problem, which combines minimizing or maximizing a function subject to constraints, using linear programming.

Conceptual Objectives

- Understand that linear programming is a graphical method that solves optimization problems.
- Understand why vertices represent maxima or minima.

Optimization: Linear Programming

Often we seek to maximize or minimize a function subject to constraints. This process is called **optimization**. For example, in the chapter opener about Florida's hurricanes in the summer of 2004, FEMA had to determine how many generators, cases of water, and tarps should be in each truck to maximize the number of Floridians given help, yet at the same time factor in the weight and space constraints on the trucks.

When the function we seek to minimize or maximize is linear and the constraints are given in terms of linear inequalities, a graphing approach to such problems is called **linear programming**. In linear programming, we start with a linear equation, called the **objective function**, that represents the quantity that is to be maximized or minimized, for example, the number of Floridians aided by FEMA. The number of people aided, however, is subject to constraints represented as linear inequalities such as how much weight each truck can haul and how much space each truck has for cargo.

In two variables, the constraints, which are represented as a system of linear inequalities, are graphed, and the common shaded region represents **feasible solutions**. If the shaded region is bounded, the vertices represent where the coordinates correspond to the maximum and minimum values of the objective function. If the region is not bounded, then if an optimal solution exists, it will occur at a vertex. A procedure for solving linear programming problems is outlined in the following box.

SOLVING AN OPTIMIZATION PROBLEM USING LINEAR PROGRAMMING

Step 1: Write the objective function. This expression represents the quantity that is to be minimized or maximized.

Step 2: Write the constraints. This is a system of linear inequalities.

Step 3: Graph the feasible solutions. Graph the system of linear inequalities and shade the common region, which contains the feasible solutions.

Step 4: Identify the vertices. The corner points of the shaded region represent maximum or minimum values of the objective function.

Step 5: List the values of the objective function. For each corner point, substitute the coordinates into the objective function and list the value of the objective function.

Step 6: Identify the optimal solution. The largest (maximum) or smallest (minimum) value of the objective function in Step 5 is the optimal solution.

EXAMPLE 1 Maximizing an Objective Function

Find the maximum value of $z = 2x + y$ subject to the constraints:

$$x \geq 1 \qquad x \leq 4 \qquad x + y \leq 5 \qquad y \geq 0$$

Solution:

STEP 1 Write the objective function. $\qquad z = 2x + y$

STEP 2 Write the constraints.

$x \geq 1$
$x \leq 4$
$y \leq -x + 5$
$y \geq 0$

STEP 3 Graph the system of linear inequalities.

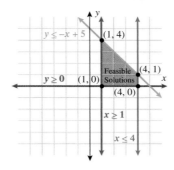

STEP 4 Identify the vertices.　　　$(1, 4), (4, 1), (1, 0), (4, 0)$

STEP 5 List the values of the objective function.

Vertex	x	y	Objective Function: $z = 2x + y$
$(1, 4)$	1	4	$2(1) + 4 = \mathbf{6}$
$(4, 1)$	4	1	$2(4) + 1 = \mathbf{9}$
$(1, 0)$	1	0	$2(1) + 0 = \mathbf{2}$
$(4, 0)$	4	0	$2(4) + 0 = \mathbf{8}$

STEP 6 The maximum value of z is **9**, which occurs when $x = 4$ and $y = 1$.

YOUR TURN Find the maximum value of $z = 2x + y$ subject to the constraints:

$$x \geq 1 \qquad x \leq 4 \qquad y \leq x \qquad y \geq 0$$

EXAMPLE 2　Minimizing an Objective Function

Find the minimum value of $z = 4x + 5y$ subject to the constraints:

$$x \geq 0 \qquad 2x + y \leq 6 \qquad x + y \leq 5 \qquad y \geq 0$$

Solution:

STEP 1 Write the objective function.　　　$z = 4x + 5y$

STEP 2 Write the constraints.

$x \geq 0$
$y \leq -2x + 6$
$y \leq -x + 5$
$y \geq 0$

Answer: The maximum value of z is **12**, which occurs when $x = 4$ and $y = 4$.

STEP 3 Graph the system of linear inequalities.

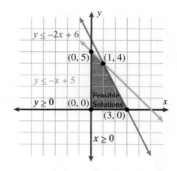

STEP 4 Identify the vertices. $(0, 0), (0, 5), (1, 4), (3, 0)$

STEP 5 List the values of the objective function.

STUDY TIP

Maxima or minima of objective functions occur at vertices of shaded region corresponding to constraints.

Vertex	x	y	Objective Function: $z = 4x + 5y$
$(0, 0)$	0	0	$4(0) + 5(0) = \mathbf{0}$
$(0, 5)$	0	5	$4(0) + 5(5) = \mathbf{25}$
$(1, 4)$	1	4	$4(1) + 5(4) = \mathbf{24}$
$(3, 0)$	3	0	$4(3) + 5(0) = \mathbf{12}$

STEP 6 The minimum value of z is **0**, which occurs when $x = 0$ and $y = 0$.

■ YOUR TURN Find the minimum value of $z = 2x + 3y$ subject to the constraints:

$$x \geq 1 \qquad 2x + y \leq 8 \qquad x + y \geq 4$$

EXAMPLE 3 Solving an Optimization Problem Using Linear Programming: Unbounded Region

Find the maximum value and minimum value of $z = 7x + 3y$ subject to the constraints:

$$y \geq 0 \qquad -2x + y \leq 0 \qquad -x + y \geq -4$$

Solution:

STEP 1 Write the objective function. $\qquad z = 7x + 3y$

STEP 2 Write the constraints. $\qquad y \geq 0$
$\qquad\qquad\qquad\qquad\qquad\qquad\qquad\quad y \leq 2x$
$\qquad\qquad\qquad\qquad\qquad\qquad\qquad\quad y \geq x - 4$

■ Answer: The minimum value of z is 8, which occurs when $x = 4$ and $y = 0$.

STEP 3 Graph the system of linear inequalities.

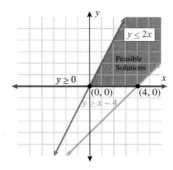

STEP 4 Identify the vertices. $(0, 0), (4, 0)$

STEP 5 List the values of the objective function.

Vertex	x	y	Objective Function: $z = 7x + 3y$
$(0, 0)$	0	0	$7(0) + 3(0) = \mathbf{0}$
$(4, 0)$	4	0	$7(4) + 3(0) = \mathbf{28}$
Additional Point			
$(3, 3)$	3	3	$7(3) + 3(3) = \mathbf{30}$

STEP 6 The minimum value of z is **0**, which occurs when $x = 0$ and $y = 0$.

There is no maximum value.

When the feasible solutions are contained in a bounded region, then a maximum and minimum exist and are located at one of the vertices. If the feasible solutions are contained in an unbounded region, then if a maximum or minimum exists, it is located at one of the vertices.

EXAMPLE 4 Maximizing the Number of People Aided by a Hurricane Relief Effort

After the hurricanes in Florida in the summer of 2004, FEMA sent disaster relief trucks to the state. Each truck could carry no more than 6000 pounds of cargo. Each case of bottled water weighs 25 pounds and each generator weighs 150 pounds. If each generator helps one household and five cases of water help one household, determine the maximum number of Florida households aided by each truck and how many generators and cases of water should be sent in each truck. Due to the number of trucks and the supply of water and generators nationwide, each truck must contain at least 10 times as many cases of water as generators.

Solution:

Let x represent the number of cases of water.

Let y represent the number of generators.

Let z represent the number of households aided per truck.

Step 1 Write the objective function.

$$z = \frac{1}{5}x + y$$

Step 2 Write the constraints.

Number of cases of water is nonnegative. $\quad x \geq 0$

Number of generators is nonnegative. $\quad y \geq 0$

At least 10 times as many cases of
water as generators. $\qquad\qquad\qquad x \geq 10y$

Weight capacity of truck: $\qquad\qquad 25x + 150y \leq 6000$

Step 3 Graph the inequalities and determine
feasible solutions.

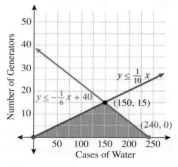

Step 4 Identify the vertices. $\qquad\qquad (0, 0), (150, 15), (240, 0)$

Step 5 List the values of the objective function.

Vertex	x	y	Objective Function: $z = \dfrac{1}{5}x + y$
$(0, 0)$	0	0	**0**
$(150, 15)$	150	15	**45**
$(240, 0)$	240	0	**48**

Step 6 A maximum of **48 households** is aided by each truck. This maximum
occurs when there are **240 cases of water** and **0 generators** on each truck.

Section 6.6 SUMMARY

In this section we discussed optimization problems in which we are asked to maximize or minimize a function (objective function) subject to constraints (system of
linear inequalities). To solve such problems, we used a graphical method called linear programming. In linear programming the constraints are graphed, and the overlapping common shaded region contains all of the feasible solutions. If the region
is bounded, both the maximum and minimum values of the function occur at one
of the vertices of the regions. If the region is unbounded, and if a maximum or minimum exists, it does so at one of the vertices.

SECTION 6.6 EXERCISES

 ■ SKILLS

In Exercises 1–4, find the value of the objective function at each of the vertices. What is the maximum value of the objective function? What is the minimum value of the objective function?

1. Objective function: $z = 2x + 3y$

2. Objective function: $z = 3x + 2y$

3. Objective function: $z = 1.5x + 4.5y$

4. Objective function: $z = \dfrac{2}{3}x + \dfrac{3}{5}y$

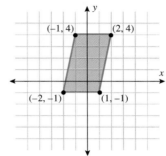

In Exercises 5–12, minimize or maximize the objective function subject to the constraints.

5. Minimize $z = 7x + 4y$ subject to:

$x \geq 0 \quad y \geq 0 \quad -x + y \leq 4$

6. Maximize $z = 7x + 4y$ subject to:

$x \geq 0 \quad y \geq 0 \quad -x + y \leq 4$

7. Maximize $z = 4x + 3y$ subject to:

$x \geq 0 \quad y \leq -x + 4 \quad y \geq -x$

8. Minimize $z = 4x + 3y$ subject to:

$x \geq 0 \quad y \geq 0$
$x + y \leq 10 \quad x + y \geq 0$

9. Minimize $z = 2.5x + 3.1y$ subject to:

$x \geq 0 \quad y \geq 0 \quad x \leq 4$
$-x + y \leq 2 \quad x + y \leq 6$

10. Maximize $z = 2.5x + 3.1y$ subject to:

$x \geq 0 \quad y \geq 0 \quad x \leq 4$
$-x + y \leq 2 \quad x + y \leq 6$

11. Maximize $z = \dfrac{1}{4}x + \dfrac{2}{5}y$ subject to:

$x + y \geq 5 \quad x + y \leq 7$
$-x + y \leq 5 \quad -x + y \geq 3$

12. Minimize $z = \dfrac{1}{4}x + \dfrac{2}{5}y$ subject to:

$x + y \geq 5 \quad x + y \leq 7$
$-x + y \leq 5 \quad -x + y \geq 3$

■ APPLICATIONS

13. Hurricanes. After the 2004 hurricanes in Florida, a student at Valencia Community College decides to create two T-shirts to sell. One T-shirt says, "I survived Charley on Friday the Thirteenth," and the second says, "I survived Charley, Francis, Ivan, and Jeanne." The Charley T-shirt costs him $7 to make and he sells it for $13. The other T-shirt costs him $5 to make and he sells it for $10. He does not want to invest more than $1000. He estimates that the total demand will not exceed 180 T-shirts. Find the number of each type of T-shirt he should make to yield maximum profit.

14. Hurricanes. After Hurricane Charley devastated central Florida unexpectedly, Orlando residents prepared for

Hurricane Frances by boarding up windows and filling up their cars with gas. It took 5 hours of standing in line to get plywood, and lines for gas were just as time-consuming. A student at Seminole Community College decides to do a spoof of the "Got Milk" ads and creates two T-shirts: "Got Plywood," with a line of people in a home improvement store, and "Got Gas," with a street lined with cars waiting to pump gasoline. The "Got Plywood" shirts cost $8 to make, and she sells them for $13. The "Got Gas" shirts cost $6 to make, and she sells them for $10. She decides to limit her costs to $1,400. She estimates that demand for these T-shirts will not exceed 200 T-shirts. Find the number of each type of T-shirt she should make to yield maximum profit.

15. Computer Business. A computer science major and a business major decide to start a small business that builds and sells a desktop computer and a laptop computer. They buy the parts, assemble them, load the operating system, and sell the computers to other students. The costs for parts, time to assemble the computer, and profit are summarized in the following table:

	Desktop	Laptop
Cost of Parts	$700	$400
Time to Assemble	5	3
Profit	$500	$300

They were able to get a small business loan in the amount of $10,000 to cover costs. They plan on making these computers over the summer and selling them the first day of class. They can only dedicate at most 90 hours to assembling these computers. They estimate that the demand for laptops will be at least three times as great as the demand for desktops. How many of each type of computer should they make to maximize profit?

16. Computer Business. Repeat Exercise 15 if they are able to get a loan for $30,000 to cover costs and they can dedicate at most 120 hours to assembling the computers.

17. Passenger Ratio. The Eurostar is a high-speed train that travels between London, Brussels, and Paris. There are 30 cars on each departure. Each train car is designated first class or second class. Based on demand for each type of fare, there should always be at least two but no more than four first class train cars. The management wants to claim that the ratio of first class to second class cars never exceeds 1:8. If the profit on each first class train car is twice as much as the profit of each second class train car, find the number of each class of train car that will generate a maximum profit.

18. Passenger Ratio. Repeat Exercise 17. This time, assume that there has to be at least one first class train car and that the profit from each first class train car is 1.2 times as much as the profit from each second class train car. The ratio of first class to second class cannot exceed 1:10.

■ CATCH THE MISTAKE

In Exercises 19 and 20, explain the mistake that is made.

19. Maximize the objective function: $z = 2x + y$ subject to the following constraints:

$$x \geq 0 \quad y \geq 0$$
$$-x + y \geq 0 \quad x + y \leq 2$$

Solution:

Graph the constraints.

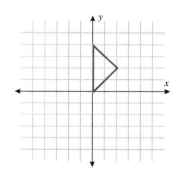

Identify the vertices. $(0, 2), (0, 0), (1, 1)$

Comparing the y-coordinates of the vertices, the largest y-value is 2.

The maximum value occurs at $(0, 2)$.

The objective function at that point is equal to **2**.

This is incorrect. The maximum value of the objective function is not 2. What mistake was made?

20. Maximize the objective function: $z = 2x + y$ subject to the following constraints:

$$x \geq 0 \quad y \geq 0$$
$$-x + y \leq 0 \quad x + y \leq 2$$

Graph the constraints.

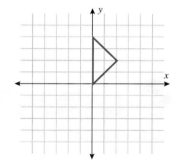

Identify the vertices. $(0, 2), (0, 0), (1, 1)$

Vertex	x	y	Objective Function: $z = 2x + y$
$(0, 2)$	0	2	$2(0) + 2 = 2$
$(0, 0)$	0	0	$2(0) + 0 = 0$
$(1, 1)$	1	1	$2(1) + 1 = 3$

The maximum, **3**, is located at the point $(1, 1)$.

This is incorrect. What mistake was made?

■ **CHALLENGE**

21. T or F: An objective function always has a maximum or minimum.

22. T or F: An objective function subject to constraints that correspond to a bounded region always has a maximum and minimum.

23. Maximize the objective function: $z = 2x + y$ subject to the conditions

$$\begin{aligned} ax + y &\geq -a \\ -ax + y &\leq a \\ ax + y &\leq a \\ -ax + y &\geq -a \end{aligned} \quad \text{where } a > 2$$

24. Maximize the objective function: $z = x + 2y$ subject to the conditions

$$\begin{aligned} x + y &\geq a \\ -x + y &\leq a \\ x + y &\leq a + b \\ -x + y &\geq a - b \end{aligned} \quad \text{where } a > b > 0$$

■ **TECHNOLOGY**

Use a graphing utility to determine the region of feasible solutions and use the zoom feature to find the approximate coordinates of the vertices to solve the linear programming problems in Exercises 25–28.

25. Minimize $z = 2.5x + 3.1y$ subject to:

$$\begin{array}{cc} x \geq 0 & y \geq 0 \quad x \leq 4 \\ -x + y \leq 2 & x + y \leq 6 \end{array}$$

Compare with your answer to Exercise 9.

26. Maximize $z = 2.5x + 3.1y$ subject to:

$$\begin{array}{cc} x \geq 0 & y \geq 0 \quad x \leq 4 \\ -x + y \leq 2 & x + y \leq 6 \end{array}$$

Compare with your answer to Exercise 10.

27. Maximize $z = 17x + 14y$ subject to:

$$y \geq 4.5 \quad -x + y \leq 3.7 \quad x + y \leq 11.2$$

28. Minimize $z = 1.2x + 1.5y$ subject to:

$$\begin{array}{cc} 2.3x + y \leq 14.7 & -2.3x + y \leq 14.7 \\ -5.2x + y \leq 3.7 & -2.3x + y \geq 1.5 \end{array}$$

Q: Jessie and Claudia went to a golf tournament and watched one of the golfers hit the ball on the 7th hole. The fairway slopes downward from the tee to the hole. They know that the tee is at an elevation of 20 feet and the horizontal distance from the tee to where the ball landed is 250 yards. They want to determine the ball's maximum height. Assuming the trajectory of the ball is a parabola, they can find the maximum height if they can find the equation of the trajectory. Jessie finds that the trajectory has the equation $y = x^2 - 750.03x + 20$, while Claudia finds that the equation is $y = 2x^2 - 1500.03x + 20$. Both equations are satisfied by the coordinate pairs $(0, 20)$ and $(750, 0)$, but the equations are not equivalent. Why?

A: When finding the equation of the parabola $y = ax^2 + bx + c$, there are three unknowns (a, b, and c), but only two equations after substituting the coordinate pairs $(0, 20)$ and $(750, 0)$ into $y = ax^2 + bx + c$. In order to find the true equation of the parabola, they would need another coordinate pair from the trajectory.

If a parabola contains the points $(0, -90)$, $(2, 12)$, and $(-10, 0)$, find the equation of the parabola.

TYING IT ALL TOGETHER

Alex owns a business that produces exercise DVDs. He currently sells each DVD for $12, but is considering raising the price to $15 and wants to determine how many DVDs he must sell each month at $15 in order to break even (the point where his costs and revenue are equal). Last month his business sold 5123 DVDs, and the month previous to that the business sold 3815 DVDs. The company's fixed cost each month is $20,000, and the variable cost is $3 per DVD produced. There are currently 60 stores that sell the DVDs. Write an equation for the company's monthly costs and an equation for the monthly revenue if the company raises the retail price of the DVD to $15. How many DVDs would the company have to sell each month to break even? If x represents the number of DVDs sold, for what values of x would the company show a profit?

SECTION	TOPIC	PAGES	REVIEW EXERCISES	KEY CONCEPTS
6.1	Systems of linear equations in two variables	406–417	1–18	$A_1x + B_1y = C_1$ $A_2x + B_2y = C_2$
	Elimination	409–411	1–4	Eliminate a variable by adding multiples of the equations.
	Substitution	407–409	5–8	Solve for one variable in terms of the other and substitute that expression into the other equation.
	Graphing	411–414	9–16	Graph the two lines. The solution is the point of intersection. Parallel lines have no solution and identical lines have infinitely many solutions.
	Applications	415–417	17 and 18	Data curve-fitting.
6.2	Systems of multivariable linear equations	420–426	19–24	Systems consisting of any number of equations and any number of variables.
	Applications	425–426	23 and 24	Data fitting/investments
6.3	Partial fractions	431–441	25–32	$\dfrac{n(x)}{d(x)}$ Factor $d(x)$
	Linear nonrepeated factors	433–434		$\dfrac{n(x)}{d(x)} = \dfrac{A}{(ax+b)} + \dfrac{B}{(cx+d)} + \ldots$
	Linear repeated factors	434–436		$\dfrac{A}{(ax+b)} + \dfrac{B}{(ax+b)^2} + \ldots + \dfrac{M}{(ax+b)^m}$
	Quadratic nonrepeated irreducible factors	436–438		$\dfrac{Ax + B}{ax^2 + bx + c}$
	Quadratic repeated irreducible factors	438–439		$\dfrac{A_1x + B_1}{ax^2 + bx + c} + \dfrac{A_2x + B_2}{(ax^2 + bx + c)^2} + \dfrac{A_3x + B_3}{(ax^2 + bx + c)^3} + \ldots + \dfrac{A_mx + B_m}{(ax^2 + bx + c)^m}$
6.4	Systems of nonlinear equations	443–451	33–44	There is no procedure guaranteed to solve nonlinear equations.
	Elimination	444–447	33–36	Eliminate a variable by either adding or subtracting one equation from the other.
	Substitution	447–449	37–40	Solve for one variable in terms of the other and substitute into second equation.
6.5	Systems of inequalities	453–465	45–58	Solutions are determined graphically by finding the common shaded regions. • $\leq$ or $\geq$ use solid curves • $<$ or $>$ use dashed curves
6.6	Linear programming	469–474	59–64	Finding optimal solutions. Minimizing or maximizing a function subject to constraints (linear inequalities).

6.1 Systems of Linear Equations in Two Variables

Solve the system of linear equations by elimination.

1. $r - s = 3$
$r + s = 3$

2. $3x + 4y = 2$
$x - y = 6$

3. $-4x + 2y = 3$
$4x - y = 5$

4. $0.25x - 0.5y = 0.6$
$0.5x + 0.25y = .8$

Solve the system of linear equations by substitution.

5. $x + y = 3$
$x - y = 1$

6. $3x + y = 4$
$2x + y = 1$

7. $4c - 4d = 3$
$c + d = 4$

8. $5r + 2s = 1$
$r - s = -3$

Match the system of equations with its graph.

9. $2x - 3y = 4$
$x + 4y = 3$

10. $5x - y = 2$
$5x - y = -2$

11. $x + 2y = -6$
$2x + 4y = -12$

12. $5x + 2y = 3$
$4x - 2y = 6$

a.

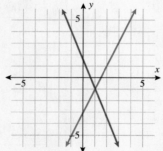

b.

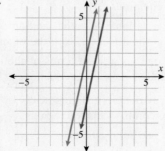

c.

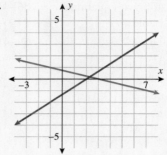

d.

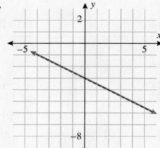

Solve the system of linear equations by graphing.

13. $y = \dfrac{-1}{2}x$

$y = \dfrac{1}{2}x + 2$

14. $2x + 4y = -2$
$4x - 2y = 3$

15. $1.3x - 2.4y = 1.6$
$0.7x - 1.2y = 1.4$

16. $\dfrac{1}{4}x - \dfrac{3}{4}y = 12$

$\dfrac{1}{2}y + \dfrac{1}{4}x = \dfrac{1}{2}$

Applications

17. Chemistry. In chemistry lab Alexandra needs to make a 42 milliliter solution that is 15% NaCl. All that is in the lab is 6% and 18% NaCl. How many milliliters of each solution should she use to obtain the desired mix?

18. Gas Mileage. A Nissan Sentra gets approximately 32 mpg on the highway and 18 mpg in the city. Suppose 265 miles were driven on a full tank (12 gallons) of gasoline. Approximately how many miles were driven in the city and on the highway?

6.2 Systems of Multivariable Linear Equations

Solve the system of linear equations.

19. $x + y + z = 1$
$x - y - z = -3$
$-x + y + z = 3$

20. $x - 2y + z = 3$
$2x - y + z = -4$
$3x - 3y - 5z = 2$

21. $3x + y + z = -4$
$x - 2y + z = -6$

22. $x + z = 3$
$-x + y - z = -1$
$x + y + z = 5$

Applications

23. Fitting a Curve to Data. The average number of flights on a commercial plane that a person takes per year can be modeled by a quadratic function, $y = ax^2 + bx + c$ where $a < 0$ and x represents age: $16 \leq x \leq 65$. The following table gives the average number of flights per year that a person takes on a commercial airline.

Age	Number of Flights per Year
16	2
40	6
65	4

Determine the quadratic function that models this quantity.

24. Investment Portfolio. Danny and Paula decide to invest $20,000 of their savings. They put some in an IRA account earning 4.5% interest, some in a mutual fund that has been averaging 8% a year, and some in a stock that earned 12% last year. If they put $4,000 more in the IRA than in the mutual fund and the mutual fund and stock have the same growth in the next year as they did in the previous year, they will earn $1,525 in a year. How much money should they put in each of the three investments?

6.3 Partial Fractions

Write the form of the partial fraction decomposition. Do not solve for the constants.

25. $\dfrac{3}{x^2 + x - 12}$

26. $\dfrac{x^2 + 3x - 2}{x^3 + 6x^2}$

27. $\dfrac{3x^3 - 5x^2 + 6x - 4}{(x^2 + x - 7)}$

28. $\dfrac{x^3 + 7x^2 + 10}{(x^2 + 13)^2}$

Find the partial fraction decomposition for the rational function.

29. $\dfrac{2}{x^2 + x}$

30. $\dfrac{x}{x(x + 3)}$

31. $\dfrac{5x - 17}{x^2 + 4x + 4}$

32. $\dfrac{x^3}{(x^2 + 64)^2}$

6.4 Systems of Nonlinear Equations

Solve the system of equations using the elimination method.

33. $x^2 + y = -3$
$x - y = 5$

34. $x^2 + y^2 = 4$
$x + y = 2$

35. $x^2 + y^2 = 5$
$2x^2 - y = 0$

36. $x^2 + y^2 = 16$
$6x^2 + y^2 = 16$

Solve the system of equations using the substitution method.

37. $x + y = 3$
$x^2 + y^2 = 4$

38. $xy = 4$
$x^2 + y^2 = 16$

39. $x^2 + xy + y^2 = -12$
$x - y = 2$

40. $3x + y = 3$
$x - y^2 = -9$

Solve the system of equations using any method.

41. $x^3 - y^3 = -19$
$x - y = -1$

42. $2x^2 + 4xy = 9$
$x^2 - 2xy = 0$

43. $\dfrac{2}{x^2} + \dfrac{1}{y^2} = 15$
$\dfrac{1}{x^2} - \dfrac{1}{y^2} = -3$

44. $x^2 + y^2 = 2$
$x^2 + y^2 = 4$

6.5 Systems of Inequalities

Graph the linear inequality.

45. $y \geq -2x + 3$

46. $y < x - 4$

47. $2x + 4y > 5$

48. $5x + 2y \leq 4$

Graph the nonlinear inequality.

49. $y \geq x^2 + 3$

50. $x^2 + y^2 > 16$

51. $y \leq e^x$

52. $y < -x^3 + 2$

Graph each system of inequalities or indicate that the system has no solution.

53. $y \geq x + 2$
$y \leq x - 2$

54. $y \geq 3x$
$y \leq 3x$

55. $x \leq -2$
$y > x$

56. $y \geq x^2 - 2$
$y \leq -x^2 + 2$

57. $x^2 + y^2 \leq 4$
$y \leq x$

58. $x + y > -4$
$x - y < 3$
$y \geq -2$
$x \leq 8$

6.6 Linear Programming

Minimize or maximize the objective function subject to the constraints.

59. Minimize $z = 2x + y$ subject to:

$$x \geq 0 \quad y \geq 0 \quad x + y \leq 3$$

60. Maximize $z = 2x + 3y$ subject to:

$$x \geq 0 \quad y \geq 0$$
$$-x + y \leq 0 \quad x \leq 3$$

61. Maximize $z = 2.5x + 3.2y$ subject to

$$x \geq 0 \quad y \geq 0$$
$$x + y \leq 8 \quad -x + y \geq 0$$

62. Minimize $z = 5x + 11y$ subject to

$$-x + y \leq 10 \quad x + y \geq 2 \quad x \leq 2$$

Applications

Art Business. An art student decides to hand-paint coasters and sell sets at a flea market. She decides to make two types of coaster sets: an ocean watercolor and black-and-white geometric shapes. The cost, profit, and time it takes her to paint each set are summarized in the table below.

	Ocean Watercolor	Geometric Shapes
Cost	$4	$2
Profit	$15	$8
Hours	3	2

63. **Profit.** If her costs cannot exceed $100 and she can only spend 90 hours total painting the coasters, determine the number of each type she should make to maximize her profit.

64. **Profit.** If her costs cannot exceed $300 and she can only spend 90 hours painting, determine the number of each type she should make to maximize her profit.

CHAPTER 6 PRACTICE TEST

In Exercises 1–8, solve the system of linear equations.

1. $x - 2y = 1$
$-x + 3y = 2$

2. $3x + 5y = -2$
$7x + 11y = -6$

3. $x - y = 2$
$-2x + 2y = -4$

4. $3x - 2y = 5$
$6x - 4y = 0$

5. $x + y + z = -1$
$2x + y + z = 0$
$-x + y + 2z = 0$

6. $6x + 9y + z = 5$
$2x - 3y + z = 3$
$10x + 12y + 2z = 9$

7. $3x + 2y - 10z = 2$
$x + y - z = 5$

8. $x + z = 1$
$x + y = -1$
$y + z = 0$

In Exercises 9–12, write the rational expression as a sum of partial fractions.

9. $\dfrac{2x + 5}{x^2 + x}$

10. $\dfrac{1}{2x^2 + 5x - 3}$

11. $\dfrac{5x - 3}{x(x^2 - 9)}$

12. $\dfrac{5x - 3}{x(x^2 + 9)}$

In Exercises 13 and 14, graph the inequalities.

13. $-2x + y < 6$

14. $x^2 + y^2 \geq 4$

In Exercises 15 and 16, graph the system of inequalities.

15. $x + y \leq 4$
$-x + y \geq -2$

16. $y^2 - 1 < x$
$-y^2 + 1 > x$

17. Minimize the function $z = 5x + 7y$ subject to the constraints:

$$x \geq 0 \quad y \geq 0 \quad x + y \leq 3 \quad -x + y \geq 1$$

18. Find the area of a triangle with vertices $(7, 2)$, $(0, 9)$, $(-2, -5)$.

19. Stock Investment. On starting a new job, Cameron gets a $30,000 signing bonus and decides to invest the money. She invests some in a money market account earning 3% and some in two different stocks. The aggressive stock rose 12% last year and the conservative stock rose 6% last year. If she invests $1,000 more in the aggressive stock than in the conservative stock and if the stocks continue to rise at the same rate, then in 1 year she will have earned $1,890 on the investment. How much should she put in the money market, and how much was invested in each stock?

20. Ranch. A rancher buys a rectangular parcel of property that is 245,000 square feet. She fences the entire border and then adds three internal fences so there are four equal rectangular pastures. If the fence required 3150 linear feet, what are the dimensions of the entire property?

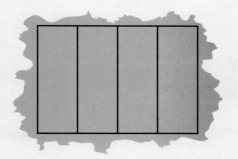

CHAPTER 7

Matrices

Image Source/Getty Images

Ever noticed how some job applicants have a fantastic resume but during the interview don't seem to fit, while others might not look as good on paper but seem to fit in perfectly during the interview? Large companies often have a rubric for scoring job applicants that combines their education, experience, and the interview.

A company may have a few rubrics to assess applicants. Matrices can be used to describe the rubric and the applicant's scores in each qualification area. The product of these two matrices gives the overall scores for each applicant based on each rubric.

In this chapter we will use matrices as a shorthand way of expressing systems of linear equations. Three ways to solve systems of linear equations will be discussed: Gauss–Jordan elimination using augmented matrices, determinants, and matrix equations.

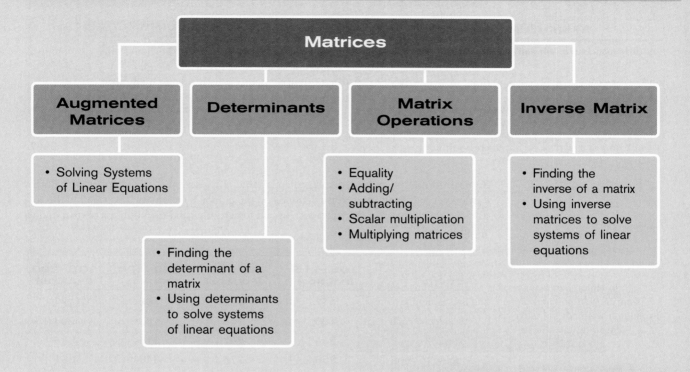

Matrices

Augmented Matrices
- Solving Systems of Linear Equations

Determinants
- Finding the determinant of a matrix
- Using determinants to solve systems of linear equations

Matrix Operations
- Equality
- Adding/ subtracting
- Scalar multiplication
- Multiplying matrices

Inverse Matrix
- Finding the inverse of a matrix
- Using inverse matrices to solve systems of linear equations

 CHAPTER OBJECTIVES

- Solve systems of linear equations using augmented matrices.
- Use determinants to solve systems of linear equations.
- Perform matrix operations: addition, subtraction, and multiplication.
- Use inverse matrices to solve systems of linear equations.

NAVIGATION THROUGH SUPPLEMENTS

DIGITAL VIDEO SERIES #7

STUDENT SOLUTIONS MANUAL CHAPTER 7

BOOK COMPANION SITE
www.wiley.com/college/young

Skills Objectives

- Write a system of linear equations as an augmented matrix.
- Write a system of linear equations as a matrix equation.

Conceptual Objectives

- Understand that matrices are a shorthand way of expressing systems of linear equations.
- Relate matrix methods to previous methods: elimination and substitution.

Some information is best displayed in a table. For example, the number of calories burned per half hour of exercise depends on the person's weight as illustrated in the following table. Note that the rows correspond to activities and the columns correspond to weight.

ACTIVITY	105–115 (POUNDS)	127–137 (POUNDS)	160–170 (POUNDS)	180–200 (POUNDS)
Sex, active	117	129	168	177
Walking, 4 mph	135	156	183	204
Volleyball	234	267	315	348
Jogging, 5 mph	258	276	345	381

Another example is the driving distance in miles from cities in Arizona (columns) to cities outside the state (rows).

CITY	FLAGSTAFF	PHOENIX	TUCSON	YUMA
Albuquerque, NM	325	465	440	650
Las Vegas, NV	250	300	415	295
Los Angeles, CA	470	375	490	285

If we selected only the numbers in the above tables and placed brackets around them, the result would be a *matrix*.

$$\text{Calories:} \begin{bmatrix} 117 & 129 & 168 & 177 \\ 135 & 156 & 183 & 204 \\ 234 & 267 & 315 & 348 \\ 258 & 276 & 345 & 381 \end{bmatrix}$$

$$\text{Miles:} \begin{bmatrix} 325 & 465 & 440 & 650 \\ 250 & 300 & 415 & 295 \\ 470 & 375 & 490 & 285 \end{bmatrix}$$

DEFINITION   MATRIX

A **matrix** is a rectangular array of numbers written within brackets.

$$
\begin{array}{ccccccc}
 & \text{Column 1} & \text{Column 2} & \ldots & \text{Column } j & \ldots & \text{Column } n \\
\text{Row 1} & a_{11} & a_{12} & \ldots & a_{1j} & \ldots & a_{1n} \\
\text{Row 2} & a_{21} & a_{22} & \ldots & a_{2j} & \ldots & a_{2n} \\
\vdots & \vdots & \vdots & & \vdots & & \vdots \\
\text{Row } i & a_{i1} & a_{i2} & \ldots & a_{ij} & \ldots & a_{in} \\
\vdots & \vdots & \vdots & & \vdots & & \vdots \\
\text{Row } m & a_{m1} & a_{m2} & \ldots & a_{mj} & \ldots & a_{mn}
\end{array}
$$

Each number a_{ij} in the matrix is called an **element** of the matrix. The first subscript, i, is the **row index**, and the second subscript, j, is the **column index**. The above matrix contains m rows and n columns.

Matrices come in all sizes. The following table illustrates examples of matrices.

MATRIX	NUMBER OF ROWS	NUMBER OF COLUMNS
$\begin{bmatrix} 2 & 1 \\ 3 & 0 \end{bmatrix}$	2	2
$\begin{bmatrix} 1 & -2 & 5 \\ -1 & 3 & 4 \end{bmatrix}$	2	3
$\begin{bmatrix} 4 & 9 & -\frac{1}{2} & 3 \end{bmatrix}$	1	4
$\begin{bmatrix} 3 & -2 \\ 5 & 1 \\ 0 & -\frac{2}{3} \\ 7 & 6 \end{bmatrix}$	4	2
$\begin{bmatrix} -2 & 5 & 4 \\ 1 & -\frac{1}{3} & 0 \\ 3 & 8 & 1 \end{bmatrix}$	3	3

When the number of rows equals the number of columns, the matrix is a **square matrix**. Matrices can be used as a shorthand way of writing systems of *linear* equations. There are two ways we can represent systems of linear equations with matrices: as *augmented matrices* or with *matrix equations*. It is important to note that matrices cannot be used to solve systems of nonlinear equations.

Augmented Matrices

A particular type of matrix that is used in representing a system of linear equations is an **augmented matrix**. It resembles a matrix with an additional vertical line and column of numbers, hence the name *augmented*. The following table illustrates examples of augmented matrices that represent systems of linear equations.

SYSTEM OF LINEAR EQUATIONS	AUGMENTED MATRIX
$3x + 4y = 1$ $x - 2y = 7$	$\begin{bmatrix} 3 & 4 & 1 \\ 1 & -2 & 7 \end{bmatrix}$
$x - y + z = 2$ $2x + 2y - 3z = -3$ $x + y + z = 6$	$\begin{bmatrix} 1 & -1 & 1 & 2 \\ 2 & 2 & -3 & -3 \\ 1 & 1 & 1 & 6 \end{bmatrix}$
$x + y + z = 0$ $3x + 2y - z = 2$	$\begin{bmatrix} 1 & 1 & 1 & 0 \\ 3 & 2 & -1 & 2 \end{bmatrix}$

- Each row represents an equation.
- The vertical line represents the equal sign.
- The first column represents the coefficients of the variable x.
- The second column represents the coefficients of the variable y.
- The third column (in the second and third systems) represents the coefficients of the variable z.
- The variables are on the left of the equal sign (vertical line), and the constants are on the right.
- Any variable that does not appear in an equation has an implied coefficient of 0.

EXAMPLE 1 Writing a System of Linear Equations as an Augmented Matrix

Write each system of linear equations as an augmented matrix.

a. $\quad 2x - y = 5$
$\quad\quad -x + 2y = 3$

b. $3x - 2y + 4z = 5$
$\quad\quad y - 3z = -2$
$\quad\quad 7x - z = 1$

c. $x_1 - x_2 + 2x_3 - 3 = 0$
$\quad\; x_1 + x_2 - 3x_3 + 5 = 0$
$\quad\; x_1 - x_2 + x_3 - 2 = 0$

Solution:

a. $\begin{bmatrix} 2 & -1 & 5 \\ -1 & 2 & 3 \end{bmatrix}$

b. Note that all missing terms have a 0 coefficient.

$$
\begin{array}{l}
3x - 2y + 4z = 5 \\
0x + y - 3z = -2 \\
7x + 0y - z = 1
\end{array}
\qquad
\begin{bmatrix} 3 & -2 & 4 & 5 \\ 0 & 1 & -3 & -2 \\ 7 & 0 & -1 & 1 \end{bmatrix}
$$

c. Write the constants on the right side of the vertical line in the matrix.

$$
\begin{array}{l}
x_1 - x_2 + 2x_3 = 3 \\
x_1 + x_2 - 3x_3 = -5 \\
x_1 - x_2 + x_3 = 2
\end{array}
\qquad
\begin{bmatrix} 1 & -1 & 2 & 3 \\ 1 & 1 & -3 & -5 \\ 1 & -1 & 1 & 2 \end{bmatrix}
$$

■ **YOUR TURN** 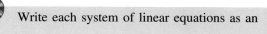 Write each system of linear equations as an augmented matrix.

a. $2x + y - 3 = 0$
$\quad\quad x - y = 5$

b. $y - x + z = 7$
$\quad\quad x - y - z = 2$
$\quad\quad\quad z - y = -1$

In Section 7.2 we will discuss a procedure called Gauss–Jordan elimination that is used with augmented matrices to *solve* systems of linear equations. In this section our goal is only to write the systems of linear equations as matrices.

Matrix Equations

Another way of writing systems of linear equations is using matrix equations. Let A be a matrix with m rows and n columns, which represents the coefficients in the system, X represent the variables in the system, and B represent the constants in the system. Then, a system of linear equations can be written as $AX = B$.

SYSTEM OF LINEAR EQUATIONS	A	X	B	MATRIX EQUATION: AX = B
$3x + 4y = 1$ $x - 2y = 7$	$\begin{bmatrix} 3 & 4 \\ 1 & -2 \end{bmatrix}$	$\begin{bmatrix} x \\ y \end{bmatrix}$	$\begin{bmatrix} 1 \\ 7 \end{bmatrix}$	$\begin{bmatrix} 3 & 4 \\ 1 & -2 \end{bmatrix}\begin{bmatrix} x \\ y \end{bmatrix} = \begin{bmatrix} 1 \\ 7 \end{bmatrix}$
$x - y + z = 2$ $2x + 2y - 3z = -3$ $x + y + z = 6$	$\begin{bmatrix} 1 & -1 & 1 \\ 2 & 2 & -3 \\ 1 & 1 & 1 \end{bmatrix}$	$\begin{bmatrix} x \\ y \\ z \end{bmatrix}$	$\begin{bmatrix} 2 \\ -3 \\ 6 \end{bmatrix}$	$\begin{bmatrix} 1 & -1 & 1 \\ 2 & 2 & -3 \\ 1 & 1 & 1 \end{bmatrix}\begin{bmatrix} x \\ y \\ z \end{bmatrix} = \begin{bmatrix} 2 \\ -3 \\ 6 \end{bmatrix}$
$x + y + z = 0$ $3x + 2y - z = 2$	$\begin{bmatrix} 1 & 1 & 1 \\ 3 & 2 & -1 \end{bmatrix}$	$\begin{bmatrix} x \\ y \\ z \end{bmatrix}$	$\begin{bmatrix} 0 \\ 2 \end{bmatrix}$	$\begin{bmatrix} 1 & 1 & 1 \\ 3 & 2 & -1 \end{bmatrix}\begin{bmatrix} x \\ y \\ z \end{bmatrix} = \begin{bmatrix} 0 \\ 2 \end{bmatrix}$

 EXAMPLE 2 **Writing a System of Linear Equations as a Matrix Equation**

Write each system of linear equations as a matrix equation.

a. $\quad 2x - y = 5$
$\quad\quad -x + 2y = 3$

b. $3x - 2y + 4z = 5$
$\quad\quad\quad y - 3z = -2$
$\quad\quad 7x - z = 1$

c. $x_1 - x_2 + 2x_3 - 3 = 0$
$\quad x_1 + x_2 - 3x_3 + 5 = 0$
$\quad x_1 - x_2 + x_3 - 2 = 0$

Solution:

a. $\begin{bmatrix} 2 & -1 \\ -1 & 2 \end{bmatrix}\begin{bmatrix} x \\ y \end{bmatrix} = \begin{bmatrix} 5 \\ 3 \end{bmatrix}$

■ **Answer: a.** $\begin{bmatrix} 2 & 1 \\ 1 & -1 \end{bmatrix}\begin{bmatrix} x \\ y \end{bmatrix} = \begin{bmatrix} 3 \\ 5 \end{bmatrix}$ **b.** $\begin{bmatrix} 0 & -1 & 1 \\ 1 & -1 & -1 \\ -1 & 1 & 1 \end{bmatrix} = \begin{bmatrix} -1 \\ 2 \\ 7 \end{bmatrix}$

b. Note that all missing terms have a 0 coefficient.

$$\begin{array}{l} 3x - 2y + 4z = 5 \\ 0x + y - 3z = -2 \\ 7x + 0y - z = 1 \end{array} \qquad \begin{bmatrix} 3 & -2 & 4 \\ 0 & 1 & -3 \\ 7 & 0 & -1 \end{bmatrix} \begin{bmatrix} x \\ y \\ z \end{bmatrix} = \begin{bmatrix} 5 \\ -2 \\ 1 \end{bmatrix}$$

c. Write the constants on the right side of the equal sign.

$$\begin{array}{l} x_1 - x_2 + 2x_3 = 3 \\ x_1 + x_2 - 3x_3 = -5 \\ x_1 - x_2 + x_3 = 2 \end{array} \qquad \begin{bmatrix} 1 & -1 & 2 \\ 1 & 1 & -3 \\ 1 & -1 & 1 \end{bmatrix} \begin{bmatrix} x_1 \\ x_2 \\ x_3 \end{bmatrix} = \begin{bmatrix} 3 \\ -5 \\ 2 \end{bmatrix}$$

■ **YOUR TURN** Write each system of linear equations as a matrix equation.

a. $2x + y - 3 = 0$
$x - y = 5$

b. $y - x + z = 7$
$x - y - z = 2$
$z - y = -1$

In Sections 7.4 and 7.5 matrix operations and inverses of matrices will be discussed in order to solve systems of linear equations.

SECTION 7.2 Systems of Linear Equations: Augmented Matrices

Skills Objectives

- Write a system of linear equations as an augmented matrix.
- Identify a reduced matrix.
- Perform row operations on a matrix.
- Solve systems of linear equations using Gauss–Jordan elimination.
- Solve systems of linear equations using nonsquare matrices.

Conceptual Objectives

- Understand that solving systems using augmented matrices is equivalent to the method of elimination.
- Visualize an augmented matrix as a system of linear equations.
- Connect nonsquare matrices with types of solutions.

Introduction

In Section 6.1 we used the *elimination method* to solve systems of two linear equations such as

$$3x + 4y = 1$$
$$x - 2y = 7$$

which has the solution $x = 3$, $y = -2$ or $(3, -2)$.

■ **Answer: a.** $\begin{bmatrix} 2 & 1 \\ 1 & -1 \end{bmatrix} \begin{bmatrix} x \\ y \end{bmatrix} = \begin{bmatrix} 3 \\ 5 \end{bmatrix}$ **b.** $\begin{bmatrix} -1 & 1 & 1 \\ 1 & -1 & -1 \\ 0 & -1 & 1 \end{bmatrix} \begin{bmatrix} x \\ y \\ z \end{bmatrix} = \begin{bmatrix} 7 \\ 2 \\ -1 \end{bmatrix}$

A more effective way of organizing our work to do elimination is with *augmented matrices*. A system of linear equations is written as an *augmented matrix. Row operations* are performed to reduce the matrix. The solution can then be identified.

SYSTEM OF EQUATIONS	AUGMENTED MATRIX	REDUCED MATRIX	SOLUTION
$3x + 4y = 1$ $x - 2y = 7$	$\begin{bmatrix} 3 & 4 & \vert & 1 \\ 1 & -2 & \vert & 7 \end{bmatrix}$	$\begin{bmatrix} 1 & 0 & \vert & 3 \\ 0 & 1 & \vert & -2 \end{bmatrix}$	$x = 3$ $y = -2$

Relying on matrices for performing elimination is especially useful when the system has more than two equations and two variables, such as

$$x - y + z = 2$$

$$2x + 2y - 3z = -3$$

$$x + y + z = 6$$

The solution $x = 1$, $y = 2$, $z = 3$ is the only set of values that satisfies all three equations.

SYSTEM OF EQUATIONS	AUGMENTED MATRIX	REDUCED MATRIX	BACK SUBSTITUTION
$x - y + z = 2$ $2x + 2y - 3z = -3$ $x + y + z = 6$	$\begin{bmatrix} 1 & -1 & 1 & \vert & 2 \\ 2 & 2 & -3 & \vert & -3 \\ 1 & 1 & 1 & \vert & 6 \end{bmatrix}$	$\begin{bmatrix} 1 & 0 & 0 & \vert & 1 \\ 0 & 1 & 0 & \vert & 2 \\ 0 & 0 & 1 & \vert & 3 \end{bmatrix}$	$x = 1$ $y = 2$ $z = 3$

Augmented Matrices

In solving systems of linear equations, the procedure is simplified if the variables and equal signs can be eliminated and written in the form of an augmented matrix:

System of linear equations: $\quad \begin{array}{l} 3x + 4y = 1 \\ x - 2y = 7 \end{array}$

Augmented matrix representation of the system: $\quad \begin{bmatrix} 3 & 4 & \vert & 1 \\ 1 & -2 & \vert & 7 \end{bmatrix}$

- The first column represents the coefficients of the variable x.
- The second column represents the coefficients of the variable y.
- The vertical line represents the equal sign.
- The third column represents the constants on the right side of the equation.
- Each row represents an equation.

Row Operations on a Matrix

Now that you are comfortable using augmented matrix notation, let's return to the opening example. *Why* did we transform the augmented matrix: $\begin{bmatrix} 3 & 4 & \vert & 1 \\ 1 & -2 & \vert & 7 \end{bmatrix}$ to an equivalent *reduced matrix:* $\begin{bmatrix} 1 & 0 & \vert & 3 \\ 0 & 1 & \vert & -2 \end{bmatrix}$ and *how* did we do it? Let us begin by defining a **reduced matrix**.

DEFINITION   REDUCED MATRIX

A matrix is in **reduced form** if all of the following conditions are true:

1. The leftmost nonzero element in each row is 1.
2. The column containing the leftmost 1 of a given row has 0's above and below the 1.
3. The leftmost 1 in any row is to the right of the leftmost 1 in the row above it.
4. Any row consisting entirely of 0's is below any row having a nonzero element.

The goal in solving systems of linear equations with augmented matrices is to transform the augmented matrix into a reduced matrix so that the solution can be easily obtained.

Translating the reduced matrix $\begin{bmatrix} 1 & 0 & | & 3 \\ 0 & 1 & | & -2 \end{bmatrix}$ back to a system of equations, we find $x = 3$, $y = -2$. Now that we know *why* this reduced form is desired, we will address *how* it is done. Such a transformation can be achieved using matrix *row operations*. The selection of row operations is guided by a method called *Gauss–Jordan elimination*, which is the matrix equivalent to the method of elimination in Section 6.1.

EXAMPLE 1 Determining If a Matrix Is Reduced

The following two matrices are not in reduced form. State which condition is not satisfied.

a. $\begin{bmatrix} 1 & 1 & 0 & | & -3 \\ 0 & 1 & 0 & | & -1 \\ 0 & 0 & 1 & | & 2 \end{bmatrix}$ **b.** $\begin{bmatrix} 1 & 0 & 1 & 0 & | & 2 \\ 0 & 0 & 0 & 1 & | & 4 \\ 0 & 1 & 0 & 0 & | & 5 \end{bmatrix}$

Solution:

a. Condition 2 is not satisfied because the 1 in row 1/column 2 should be a zero.

b. Condition 3 is not satisfied. Note that if we interchanged rows 2 and 3 the matrix would be in reduced form.

STUDY TIP

A reduced matrix can have more than one number in a row and still be reduced.

$\begin{bmatrix} 1 & 0 & 1 & 0 & | & 2 \\ 0 & 1 & 0 & 0 & | & 5 \\ 0 & 0 & 0 & 1 & | & 4 \end{bmatrix}$

■ **YOUR TURN**   For each matrix below determine if the matrix is in reduced form. If it is not reduced, state the condition that is not satisfied.

a. $\begin{bmatrix} 1 & 0 & 1 & 1 & | & 2 \\ 0 & 1 & 0 & 2 & | & 3 \\ 0 & 0 & 0 & 1 & | & 5 \end{bmatrix}$ **b.** $\begin{bmatrix} 1 & 2 & 0 & 2 & | & 3 \\ 0 & 0 & 1 & 3 & | & 7 \end{bmatrix}$

There are three *row operations* that can be used to transform a matrix to a reduced matrix. Because each row in a matrix represents an equation, the operations that produce

---

equivalent systems of equations that are used in the elimination method will also produce equivalent augmented matrices.

ROW OPERATIONS

1. Interchange any two rows.
2. Multiply a row by a nonzero constant.
3. Replace a row by the sum of that row and a constant multiple of another row.

The following symbols describe these row operations:

1. $R_i \leftrightarrow R_j$ Interchange row i with row j.
2. $cR_i \rightarrow R_i$ Multiply row i by the constant c.
3. $R_j + cR_i \rightarrow R_j$ Multiply row i by the constant c and add to row j, writing the results in row j.

WORDS

MATH

Let's now use row operations to transform:

$$\begin{bmatrix} 3 & 4 & | & 1 \\ 1 & -2 & | & 7 \end{bmatrix} \text{ to } \begin{bmatrix} 1 & 0 & | & 3 \\ 0 & 1 & | & -2 \end{bmatrix}$$

We start with:

$$\begin{bmatrix} 3 & 4 & | & 1 \\ 1 & -2 & | & 7 \end{bmatrix}$$

Step 1: Interchange rows 1 and 2.

$R_1 \leftrightarrow R_2$

$$\begin{bmatrix} 1 & -2 & | & 7 \\ 3 & 4 & | & 1 \end{bmatrix}$$

Step 2: Multiply row 1 by -3 and add to row 2, writing the results in row 2. Note: Row 1 remains unchanged.

$R_2 - 3R_1 \rightarrow R_2$

$$\begin{bmatrix} 1 & -2 & | & 7 \\ 3-3(1) & 4-3(-2) & | & 1-3(7) \end{bmatrix}$$

$$\begin{bmatrix} 1 & -2 & | & 7 \\ 0 & 10 & | & -20 \end{bmatrix}$$

Step 3: Multiply row 2 by $\frac{1}{10}$.

$\frac{1}{10}R_2 \rightarrow R_2$

$$\begin{bmatrix} 1 & -2 & & 7 \\ \frac{1}{10}(0) & \frac{1}{10}(10) & & \frac{1}{10}(-20) \end{bmatrix}$$

$$\begin{bmatrix} 1 & -2 & | & 7 \\ 0 & 1 & | & -2 \end{bmatrix}$$

Step 4: Multiply row 2 by 2 and add to row 1, writing the results in row 1. Note: Row 2 remains unchanged.

$R_1 + 2R_2 \rightarrow R_1$

$$\begin{bmatrix} 1+2(0) & -2+2(1) & | & 7+2(-2) \\ 0 & 1 & | & -2 \end{bmatrix}$$

$$\begin{bmatrix} 1 & 0 & | & 3 \\ 0 & 1 & | & -2 \end{bmatrix}$$

Notice that once we have the augmented matrix in the form, $\begin{bmatrix} 1 & 0 & | & 3 \\ 0 & 1 & | & -2 \end{bmatrix}$, we are able to identify the solution $x = 3$, $y = -2$. How did we know what row operations would transform the original augmented matrix into reduced form? There is a procedure called *Gauss–Jordan elimination* that will be outlined soon. Before we start solving systems of equations with augmented matrices using Gauss–Jordan elimination, let's first practice row operations.

EXAMPLE 2 Applying a Row Operation to an Augmented Matrix

For each matrix, perform the given operation.

a. $\begin{bmatrix} -1 & 0 & 1 & | & -2 \\ 3 & -1 & 2 & | & 3 \\ 0 & 1 & 3 & | & 1 \end{bmatrix}$ $R_2 + 3R_1 \rightarrow R_2$

b. $\begin{bmatrix} 1 & 2 & 0 & 2 & | & 2 \\ 0 & 1 & 2 & 3 & | & 5 \end{bmatrix}$ $R_1 - 2R_2 \rightarrow R_1$

Solution:

a. Multiply row 1 by 3 and add to row 2, writing the results in row 2.

$R_2 + 3R_1 \rightarrow R_2$ $\begin{bmatrix} -1 & 0 & 1 & | & -2 \\ 3+3(-1) & -1+3(0) & 2+3(1) & | & 3+3(-2) \\ 0 & 1 & 3 & | & 1 \end{bmatrix}$

$\begin{bmatrix} -1 & 0 & 1 & | & -2 \\ 0 & -1 & 5 & | & -3 \\ 0 & 1 & 3 & | & 1 \end{bmatrix}$

b. Row 1 minus 2 times row 2, and write the answer in row 1.

$R_1 - 2R_2 \rightarrow R_1$ $\begin{bmatrix} 1-2(0) & 2-2(1) & 0-2(2) & 2-2(3) & | & 2-2(5) \\ 0 & 1 & 2 & 3 & | & 5 \end{bmatrix}$

$\begin{bmatrix} 1 & 0 & -4 & -4 & | & -8 \\ 0 & 1 & 2 & 3 & | & 5 \end{bmatrix}$

■ **YOUR TURN** Perform the operation $R_1 + 2R_3 \rightarrow R_1$ on the matrix:

$\begin{bmatrix} 1 & 0 & -2 & | & -3 \\ 0 & 1 & 2 & | & 3 \\ 0 & 0 & 1 & | & 2 \end{bmatrix}$

Solving Systems Using Augmented Matrices: Gauss–Jordan Elimination

Let's put it all together to solve systems of linear equations. We know how to write a system of linear equations as an augmented matrix. We are armed with row operations to transform a matrix, and when we finally have the matrix in reduced form, we will know

■ **Answer:** $\begin{bmatrix} 1 & 0 & 0 & | & 1 \\ 0 & 1 & 2 & | & 3 \\ 0 & 0 & 1 & | & 2 \end{bmatrix}$

it when we get there. Once the matrix is in reduced form, we can identify the solution. One piece of the puzzle that is still missing is how to select the appropriate combinations of row operations that will result in a reduced matrix. The procedure that will guide us is called **Gauss–Jordan elimination**.

When the matrix is **square**, which means the number of equations equals the number of variables, and the corresponding solution is unique, the reduced matrix will have 1's along the diagonal (elements in the ith row and ith column) and zeros above and below those 1's. As we saw earlier, not all matrices that are in fact reduced have this feature, but we can think of it as the goal. When there are infinitely many solutions, there will be at least one all-zero row in the bottom of the matrix.

GAUSS–JORDAN ELIMINATION

Use row operations to obtain the following:

Step 1: Get a 1 in the top left element of the matrix (row 1/column 1).

$$\left[\begin{array}{ccc|c} 1 & * & * & * \\ * & * & * & * \\ * & * & * & * \end{array}\right]$$

Step 2: Get 0's below that 1.

$$\left[\begin{array}{ccc|c} 1 & * & * & * \\ 0 & * & * & * \\ 0 & * & * & * \end{array}\right]$$

Step 3: Repeat Steps 1 and 2 until the diagonals are all 1's.

$$\left[\begin{array}{ccc|c} 1 & * & * & * \\ 0 & 1 & * & * \\ 0 & * & * & * \end{array}\right] \quad \left[\begin{array}{ccc|c} 1 & * & * & * \\ 0 & 1 & * & * \\ 0 & 0 & * & * \end{array}\right] \quad \left[\begin{array}{ccc|c} 1 & * & * & * \\ 0 & 1 & * & * \\ 0 & 0 & 1 & * \end{array}\right]$$

Step 4: Start with the leftmost 1 in the bottom (nonzero) row and get 0's above that 1 all the way up that column. Repeat this step with the leftmost 1 in the next row up until the top row is reached.

$$\left[\begin{array}{ccc|c} 1 & * & * & * \\ 0 & 1 & 0 & * \\ 0 & 0 & 1 & * \end{array}\right] \quad \left[\begin{array}{ccc|c} 1 & * & 0 & * \\ 0 & 1 & 0 & * \\ 0 & 0 & 1 & * \end{array}\right] \quad \left[\begin{array}{ccc|c} 1 & 0 & 0 & * \\ 0 & 1 & 0 & * \\ 0 & 0 & 1 & * \end{array}\right]$$

If at any time a row appears containing all 0's, move that row to the bottom.

EXAMPLE 3 Using Gauss–Jordan Elimination to Transform a Matrix into Reduced Form

Use Gauss–Jordan elimination to transform the matrix to reduced form.

$$\left[\begin{array}{ccc|c} 2 & 1 & 8 & -1 \\ 1 & -1 & 1 & -2 \\ 3 & -2 & -2 & 2 \end{array}\right]$$

Solution:

STEP 1 Get a 1 in the top left.

$$R_1 \leftrightarrow R_2 \qquad \begin{bmatrix} 1 & -1 & 1 & | & -2 \\ 2 & 1 & 8 & | & -1 \\ 3 & -2 & -2 & | & 2 \end{bmatrix}$$

STEP 2 Get 0's below by:

$$R_2 - 2R_1 \rightarrow R_2 \qquad \begin{bmatrix} 1 & -1 & 1 & | & -2 \\ 0 & 3 & 6 & | & 3 \\ 3 & -2 & -2 & | & 2 \end{bmatrix}$$

$$R_3 - 3R_1 \rightarrow R_3 \qquad \begin{bmatrix} 1 & -1 & 1 & | & -2 \\ 0 & 3 & 6 & | & 3 \\ 0 & 1 & -5 & | & 8 \end{bmatrix}$$

STEP 3 Repeat Steps 1 and 2 along the diagonal.

Make the 3 a 1.

$$\tfrac{1}{3}R_2 \rightarrow R_2 \qquad \begin{bmatrix} 1 & -1 & 1 & | & -2 \\ 0 & 1 & 2 & | & 1 \\ 0 & 1 & -5 & | & 8 \end{bmatrix}$$

Make the 1 in
row 3/column 2 a 0.

$$R_3 - R_2 \rightarrow R_3 \qquad \begin{bmatrix} 1 & -1 & 1 & | & -2 \\ 0 & 1 & 2 & | & 1 \\ 0 & 0 & -7 & | & 7 \end{bmatrix}$$

Make the -7 a 1.

$$-\tfrac{1}{7}R_3 \rightarrow R_3 \qquad \begin{bmatrix} 1 & -1 & 1 & | & -2 \\ 0 & 1 & 2 & | & 1 \\ 0 & 0 & 1 & | & -1 \end{bmatrix}$$

Note: There are 1's along the diagonal and 0's below them. Proceed to Step 4.

STEP 4 Get 0's above the 1 in row 3/column 3.

$$R_2 - 2R_3 \rightarrow R_2 \qquad \begin{bmatrix} 1 & -1 & 1 & | & -2 \\ 0 & 1 & 0 & | & 3 \\ 0 & 0 & 1 & | & -1 \end{bmatrix}$$

$$R_1 - R_3 \rightarrow R_1 \qquad \begin{bmatrix} 1 & -1 & 0 & | & -1 \\ 0 & 1 & 0 & | & 3 \\ 0 & 0 & 1 & | & -1 \end{bmatrix}$$

Get a 0 above the
1 in row 2/column 2.

$$R_1 + R_2 \rightarrow R_1 \qquad \begin{bmatrix} 1 & 0 & 0 & | & 2 \\ 0 & 1 & 0 & | & 3 \\ 0 & 0 & 1 & | & -1 \end{bmatrix}$$

This matrix is in reduced form.

STUDY TIP

On the way *down* the stairs use rows *above*. On the way *up* the stairs use rows *below*.

To avoid changing particular 1's and 0's that you already have in place, it is helpful to remember the following. Think of Steps 1–3 as climbing *down* a set of stairs and Step 4 as climbing back *up* the stairs. On the way *down* the stairs always use operations with rows *above* where you currently are, and on the way back *up* the stairs always use rows *below* where you currently are.

YOUR TURN   Use Gauss–Jordan elimination to transform the matrix to reduced form:

$$\begin{bmatrix} 2 & -1 & 3 & | & -1 \\ 1 & 1 & -1 & | & 0 \\ 3 & 3 & -2 & | & 1 \end{bmatrix}$$

Once a matrix is in reduced form, the solution can be identified. As we found in Section 6.1, systems of linear equations can have *one solution, no solution,* or *infinitely many solutions.* The procedure for solving systems of linear equations using augmented matrices is outlined below.

PROCEDURE FOR SOLVING SYSTEMS OF LINEAR EQUATIONS USING AUGMENTED MATRICES

Step 1: Write the system of equations as an augmented matrix.

Step 2: Use Gauss–Jordan elimination to transform the matrix into reduced form.

Step 3: Identify the solution.

    a. Unique solution: The reduced matrix has 1's along the diagonal (elements in the ith row and ith column) and zeros above and below.

$$\begin{bmatrix} 1 & 0 & 0 & | & 2 \\ 0 & 1 & 0 & | & 3 \\ 0 & 0 & 1 & | & -1 \end{bmatrix} \qquad \begin{matrix} x = 2 \\ y = 3 \\ z = -1 \end{matrix}$$

    b. No solution: The reduced matrix has a contradiction.

The third row implies $0 = -1$.
$$\begin{bmatrix} 1 & 0 & 0 & | & 2 \\ 0 & 1 & 0 & | & 3 \\ 0 & 0 & 0 & | & -1 \end{bmatrix}$$

    c. Infinitely many solutions.

Reduced matrix: $\begin{bmatrix} 1 & 0 & -2 & | & 2 \\ 0 & 1 & 1 & | & 3 \\ 0 & 0 & 0 & | & 0 \end{bmatrix}$ $\qquad \begin{matrix} x - 2z = 2 \\ y + z = 3 \\ 0 = 0 \end{matrix}$

We let $z = t$ and write the infinitely many solutions as $\quad \begin{matrix} x = 2t + 2 \\ y = -t + 3 \\ z = t \end{matrix}$

EXAMPLE 4 Solving a System of Two Linear Equations Using Augmented Matrices

Solve the system of equations using an augmented matrix.

$$2x + 5y = -4$$
$$4x - 3y = 18$$

Solution:

STEP 1 Write the system as an augmented matrix.
$\begin{bmatrix} 2 & 5 & | & -4 \\ 4 & -3 & | & 18 \end{bmatrix}$

STEP 2 Use Gauss–Jordan elimination to reduce the matrix.

$\frac{1}{2}R_1 \rightarrow R_1$ $\begin{bmatrix} 1 & \frac{5}{2} & | & -2 \\ 4 & -3 & | & 18 \end{bmatrix}$

$R_2 - 4R_1 \rightarrow R_2$ $\begin{bmatrix} 1 & \frac{5}{2} & | & -2 \\ 0 & -13 & | & 26 \end{bmatrix}$

$-\frac{1}{13}R_2 \rightarrow R_2$ $\begin{bmatrix} 1 & \frac{5}{2} & | & -2 \\ 0 & 1 & | & -2 \end{bmatrix}$

$R_1 - \frac{5}{2}R_2 \rightarrow R_1$ $\begin{bmatrix} 1 & 0 & | & 3 \\ 0 & 1 & | & -2 \end{bmatrix}$

STEP 3 Identify the solution.
$$x = 3$$
$$y = -2$$

The solution to this system of equations is $\boxed{x = 3, y = -2}$.

CONCEPT CHECK Use the method of elimination or substitution from Section 6.1 to solve the system of equations $\begin{array}{l} 2x - y = 15 \\ x - 3y = 10 \end{array}$.

YOUR TURN Solve the system of equations using an augmented matrix:

$$2x - y = 15$$
$$x - 3y = 10$$

EXAMPLE 5 Solving a System of Three Linear Equations Using Augmented Matrices

Solve the system of equations using an augmented matrix.

$$x - y + 2z = -1$$
$$3x + 2y - 6z = 1$$
$$2x + 3y + 4z = 8$$

Solution:

STEP 1 Write the system as an augmented matrix.
$\begin{bmatrix} 1 & -1 & 2 & | & -1 \\ 3 & 2 & -6 & | & 1 \\ 2 & 3 & 4 & | & 8 \end{bmatrix}$

■ **Answer:** $x = 7, y = -1$

STEP 2 Use Gauss–Jordan elimination to reduce the matrix.

Get 0's below the 1 in column 1.

$R_2 - 3R_1 \rightarrow R_2$
$R_3 - 2R_1 \rightarrow R_3$

$$\left[\begin{array}{ccc|c} 1 & -1 & 2 & -1 \\ 0 & 5 & -12 & 4 \\ 0 & 5 & 0 & 10 \end{array}\right]$$

Get a 1 in row 2/ column 2.

$R_2 \leftrightarrow R_3$

$$\left[\begin{array}{ccc|c} 1 & -1 & 2 & -1 \\ 0 & 5 & 0 & 10 \\ 0 & 5 & -12 & 4 \end{array}\right]$$

$\frac{1}{5}R_2 \rightarrow R_2$

$$\left[\begin{array}{ccc|c} 1 & -1 & 2 & -1 \\ 0 & 1 & 0 & 2 \\ 0 & 5 & -12 & 4 \end{array}\right]$$

Get a 0 in row 3/ column 2.

$R_3 - 5R_2 \rightarrow R_3$

$$\left[\begin{array}{ccc|c} 1 & -1 & 2 & -1 \\ 0 & 1 & 0 & 2 \\ 0 & 0 & -12 & -6 \end{array}\right]$$

Get a 1 in row 3/ column 3.

$-\frac{1}{12}R_3 \rightarrow R_3$

$$\left[\begin{array}{ccc|c} 1 & -1 & 2 & -1 \\ 0 & 1 & 0 & 2 \\ 0 & 0 & 1 & \frac{1}{2} \end{array}\right]$$

Get 0's above the 1 in row 3/column 3.

$R_1 - 2R_3 \rightarrow R_1$

$$\left[\begin{array}{ccc|c} 1 & -1 & 0 & -2 \\ 0 & 1 & 0 & 2 \\ 0 & 0 & 1 & \frac{1}{2} \end{array}\right]$$

Get a 0 in row 1/ column 2.

$R_1 + R_2 \rightarrow R_1$

$$\left[\begin{array}{ccc|c} 1 & 0 & 0 & 0 \\ 0 & 1 & 0 & 2 \\ 0 & 0 & 1 & \frac{1}{2} \end{array}\right]$$

STEP 3 Identify the solution.

$x = 0$
$y = 2$
$z = \frac{1}{2}$

The solution to this system of linear equations is $\boxed{x = 0, y = 2, z = \frac{1}{2}}$.

■ **YOUR TURN**   Use an augmented matrix and Gauss–Jordan elimination to solve the system of equations:

$$-3x + y - z = 2$$
$$x + 2y - 3z = -6$$
$$2x - y + z = -1$$

EXAMPLE 6 Solving a System of Four Linear Equations Using Augmented Matrices

Solve the system of equations using an augmented matrix.

$$x_1 + x_2 - x_3 + 3x_4 = 3$$
$$3x_2 - 2x_4 = 4$$
$$2x_1 - 3x_3 = -1$$
$$4x_4 + 2x_1 = -6$$

■ **Answer:** $x = -1, y = 2, z = 3$

Solution:

STEP 1 Write the system as an augmented matrix.

$$\begin{bmatrix} 1 & 1 & -1 & 3 & | & 3 \\ 0 & 3 & 0 & -2 & | & 4 \\ 2 & 0 & -3 & 0 & | & -1 \\ 2 & 0 & 0 & 4 & | & -6 \end{bmatrix}$$

STEP 2 Use Gauss–Jordan elimination to reduce the matrix.

Get 0's below the 1 in row 1/column 1.

$R_3 - 2R_1 \rightarrow R_3$
$R_4 - 2R_1 \rightarrow R_4$

$$\begin{bmatrix} 1 & 1 & -1 & 3 & | & 3 \\ 0 & 3 & 0 & -2 & | & 4 \\ 0 & -2 & -1 & -6 & | & -7 \\ 0 & -2 & 2 & -2 & | & -12 \end{bmatrix}$$

Get a 1 in row 2/column 2.

$R_2 \leftrightarrow R_4$

$$\begin{bmatrix} 1 & 1 & -1 & 3 & | & 3 \\ 0 & -2 & 2 & -2 & | & -12 \\ 0 & -2 & -1 & -6 & | & -7 \\ 0 & 3 & 0 & -2 & | & 4 \end{bmatrix}$$

$-\frac{1}{2}R_2 \leftrightarrow R_2$

$$\begin{bmatrix} 1 & 1 & -1 & 3 & | & 3 \\ 0 & 1 & -1 & 1 & | & 6 \\ 0 & -2 & -1 & -6 & | & -7 \\ 0 & 3 & 0 & -2 & | & 4 \end{bmatrix}$$

Get 0's below the 1 in row 2/column 2.

$R_3 + 2R_2 \rightarrow R_3$
$R_4 - 3R_2 \rightarrow R_4$

$$\begin{bmatrix} 1 & 1 & -1 & 3 & | & 3 \\ 0 & 1 & -1 & 1 & | & 6 \\ 0 & 0 & -3 & -4 & | & 5 \\ 0 & 0 & 3 & -5 & | & -14 \end{bmatrix}$$

Get a 1 in row 3/column 3.

$-\frac{1}{3}R_3 \rightarrow R_3$

$$\begin{bmatrix} 1 & 1 & -1 & 3 & | & 3 \\ 0 & 1 & -1 & 1 & | & 6 \\ 0 & 0 & 1 & \frac{4}{3} & | & -\frac{5}{3} \\ 0 & 0 & 3 & -5 & | & -14 \end{bmatrix}$$

Get a 0 in row 4/column 3.

$R_4 - 3R_3 \rightarrow R_4$

$$\begin{bmatrix} 1 & 1 & -1 & 3 & | & 3 \\ 0 & 1 & -1 & 1 & | & 6 \\ 0 & 0 & 1 & \frac{4}{3} & | & -\frac{5}{3} \\ 0 & 0 & 0 & -9 & | & -9 \end{bmatrix}$$

Get a 1 in row 4/column 4.

$-\frac{1}{9}R_4 \rightarrow R_4$

$$\begin{bmatrix} 1 & 1 & -1 & 3 & | & 3 \\ 0 & 1 & -1 & 1 & | & 6 \\ 0 & 0 & 1 & \frac{4}{3} & | & -\frac{5}{3} \\ 0 & 0 & 0 & 1 & | & 1 \end{bmatrix}$$

Get 0's above the 1 in row 4/column 4.

$R_3 - \frac{4}{3}R_4 \rightarrow R_3$
$R_2 - R_4 \rightarrow R_2$
$R_1 - 3R_4 \rightarrow R_1$

$$\begin{bmatrix} 1 & 1 & -1 & 0 & | & 0 \\ 0 & 1 & -1 & 0 & | & 5 \\ 0 & 0 & 1 & 0 & | & -3 \\ 0 & 0 & 0 & 1 & | & 1 \end{bmatrix}$$

Get 0's above the 1 in row 3/column 3.

$R_2 + R_3 \rightarrow R_2$
$R_1 + R_3 \rightarrow R_1$

$$\begin{bmatrix} 1 & 1 & 0 & 0 & | & -3 \\ 0 & 1 & 0 & 0 & | & 2 \\ 0 & 0 & 1 & 0 & | & -3 \\ 0 & 0 & 0 & 1 & | & 1 \end{bmatrix}$$

Get a 0 in row 1/
column 2.

$$R_1 - R_2 \rightarrow R_1 \quad \begin{bmatrix} 1 & 0 & 0 & 0 & | & -5 \\ 0 & 1 & 0 & 0 & | & 2 \\ 0 & 0 & 1 & 0 & | & -3 \\ 0 & 0 & 0 & 1 & | & 1 \end{bmatrix}$$

STEP 3 Identify the solution.

$$x_1 = -5$$
$$x_2 = 2$$
$$x_3 = -3$$
$$x_4 = 1$$

The solution to this system of equations is

$$x_1 = -5, x_2 = 2, x_3 = -3, x_4 = 1 \; .$$

All of the previous examples have had a unique solution. Systems of equations may not have a solution, or they may have infinitely many solutions. Example 7 illustrates a contradiction in a matrix that indicates that the system has no solution. Example 8 shows how to identify when there are infinitely many solutions and how to express them.

EXAMPLE 7 Solving a System of Linear Equations Using Augmented Matrices: No Solution

Solve the system of equations using Gauss–Jordan elimination on an augmented matrix:

$$x + 2y - z = 3$$
$$2x + y + 2z = -1$$
$$-2x - 4y + 2z = 5$$

Solution:

STEP 1 Write the system of equations
as an augmented matrix.

$$\begin{bmatrix} 1 & 2 & -1 & | & 3 \\ 2 & 1 & 2 & | & -1 \\ -2 & -4 & 2 & | & 5 \end{bmatrix}$$

STEP 2 Reduce the matrix using Gauss–Jordan elimination.

Get 0's below the 1
in column 1.

$$R_2 - 2R_1 \rightarrow R_2$$
$$R_3 + 2R_1 \rightarrow R_3$$

$$\begin{bmatrix} 1 & 2 & -1 & | & 3 \\ 0 & -3 & 4 & | & -7 \\ 0 & 0 & 0 & | & 11 \end{bmatrix}$$

There is no need to continue because row 3 is a contradiction.

$$0x + 0y + 0z = 11 \quad \text{or} \quad 0 = 11$$

Since this is inconsistent, there is no solution to this system of equations.

EXAMPLE 8 Solving a System of Linear Equations Using Augmented Matrices: Infinitely Many Solutions

Solve the system of equations using Gauss–Jordan elimination on an augmented matrix:

$$x + z = 3$$
$$2x + y + 4z = 8$$
$$3x + y + 5z = 11$$

Solution:

STEP 1 Write the system of equations as an augmented matrix.

$$\begin{bmatrix} 1 & 0 & 1 & | & 3 \\ 2 & 1 & 4 & | & 8 \\ 3 & 1 & 5 & | & 11 \end{bmatrix}$$

STEP 2 Reduce the matrix using Gauss–Jordan elimination.

Get 0's below the 1 in column 1.

$$\begin{aligned} R_2 - 2R_1 &\rightarrow R_2 \\ R_3 - 3R_1 &\rightarrow R_3 \end{aligned}$$

$$\begin{bmatrix} 1 & 0 & 1 & | & 3 \\ 0 & 1 & 2 & | & 2 \\ 0 & 1 & 2 & | & 2 \end{bmatrix}$$

Get a 0 in row 3/ column 2.

$$R_3 - R_2 \rightarrow R_3$$

$$\begin{bmatrix} 1 & 0 & 1 & | & 3 \\ 0 & 1 & 2 & | & 2 \\ 0 & 0 & 0 & | & 0 \end{bmatrix}$$

This matrix is in reduced form. Note: The last row is consistent, $0 = 0$.

STEP 3 Identify the solution.

$$x + z = 3$$
$$y + 2z = 2$$

Let $z = t$ and substitute this into the two equations. We find that $x = 3 - t$ and $y = 2 - 2t$. The answer is written as

$$x = 3 - t, y = 2 - 2t, z = t \quad \text{for } t \text{ any real number.}$$

STUDY TIP

Unique solution

$$\begin{matrix} 2z = 0 \\ z = 0 \end{matrix} \quad \begin{bmatrix} * & * & * & | & * \\ * & * & * & | & * \\ * & * & 2 & | & 0 \end{bmatrix}$$

No solution

$$0 = 2 \quad \begin{bmatrix} * & * & * & | & * \\ * & * & * & | & * \\ 0 & 0 & 0 & | & 2 \end{bmatrix}$$

Infinitely many solutions

$$\begin{bmatrix} * & * & * & | & * \\ * & * & * & | & * \\ 0 & 0 & 0 & | & 0 \end{bmatrix}$$

COMMON MISTAKE

A common mistake that is made is to identify a unique solution when one of the variables is equal to zero as no solution.

What is the difference between $\begin{bmatrix} 1 & 0 & 2 & | & 1 \\ 0 & 1 & 3 & | & 2 \\ 0 & 0 & 3 & | & 0 \end{bmatrix}$ and $\begin{bmatrix} 1 & 0 & 2 & | & 1 \\ 0 & 1 & 3 & | & 2 \\ 0 & 0 & 0 & | & 3 \end{bmatrix}$? The

first matrix has a unique solution, whereas the second matrix has no solution. The third row of the first matrix corresponds to the equation $3z = 0$, which implies that $z = 0$. The third row of the second matrix corresponds to the equation $0x + 0y + 0z = 3$ or $0 = 3$, which is inconsistent and therefore the system has no solution.

In the previous examples we have only solved square matrices. In a square matrix the number of equations is equal to the number of variables. When the number of equations differs from the number of variables in a system of linear equations, the corresponding matrix is called a **nonsquare matrix**. A nonsquare matrix cannot have a unique solution unless there are *at least* as many equations as there are variables.

EXAMPLE 9 **Solving a Nonsquare System of Linear Equations Using Augmented Matrices**

Solve the system of linear equations using an augmented matrix.

$$2x + y + z = 8$$
$$x + y - z = -3$$

Solution:

STEP 1 Write the system of equations as an augmented matrix.

$$\begin{bmatrix} 2 & 1 & 1 & | & 8 \\ 1 & 1 & -1 & | & -3 \end{bmatrix}$$

STEP 2 Reduce the matrix using Gauss–Jordan elimination.

Get a 1 in row 1/ column 1. $R_1 \leftrightarrow R_2$

$$\begin{bmatrix} 1 & 1 & -1 & | & -3 \\ 2 & 1 & 1 & | & 8 \end{bmatrix}$$

Get a 0 in row 2/ column 1. $R_2 - 2R_1 \rightarrow R_2$

$$\begin{bmatrix} 1 & 1 & -1 & | & -3 \\ 0 & -1 & 3 & | & 14 \end{bmatrix}$$

Get a 1 in row 2/ column 2. $-R_2 \rightarrow R_2$

$$\begin{bmatrix} 1 & 1 & -1 & | & -3 \\ 0 & 1 & -3 & | & -14 \end{bmatrix}$$

Get a 0 in row 1/ column 2. $R_1 - R_2 \rightarrow R_1$

$$\begin{bmatrix} 1 & 0 & 2 & | & 11 \\ 0 & 1 & -3 & | & -14 \end{bmatrix}$$

This matrix is in reduced form.

STEP 3 Identify the answer.

$$x + 2z = 11$$
$$y - 3z = -14$$

Let $z = t$, where t is any real number. Substituting $z = t$ into these two equations gives the infinitely many solutions,

$x = 11 - 2t, y = 3t - 14, z = t$, for t any real number.

CONCEPT CHECK How many equations and how many variables does the following system of equations have?

$$x + y + z = 0$$
$$3x + 2y - z = 2$$

What type of solution do you expect to find?

■ **YOUR TURN** Solve the system of equations using an augmented matrix.

$$x + y + z = 0$$
$$3x + 2y - z = 2$$

Applications

Many times in the real world we see a relationship that looks like a particular function such as a quadratic function and we know particular data points, but we do not know the function. We start with the general function, fit the curve to particular data points, and solve a system of linear equations to determine the specific function parameters.

■ **Answer:** $x = 3t + 2, y = -4t - 2, z = t$

EXAMPLE 10 Fitting a Curve to Data

The amount of money awarded in medical malpractice suits is rising.

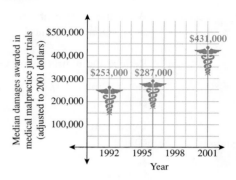

This can be modeled with a quadratic function $y = at^2 + bt + c$, where $t > 0$ and $a > 0$. Graphically, this function corresponds to the right half of a parabola opening upward. Determine a quadratic function that passes through these three points.

Solution:

Let 1992 correspond to $t = 0$ and y represent the number of dollars awarded for malpractice suits. The following data are reflected in the illustration above:

Year	t	y (thousands of dollars)	(t, y)
1992	0	253	(0, 253)
1996	4	287	(4, 287)
2001	9	431	(9, 431)

Substitute the three points, $(0, 253)$, $(4, 287)$, and $(9, 431)$ into the general quadratic equation: $y = at^2 + bt + c$,

Point	$y = at^2 + bt + c$	System of Equations
(0, 253)	$253 = a(0)^2 + b(0) + c$	$0a + 0b + c = 253$
(4, 287)	$287 = a(4)^2 + b(4) + c$	$16a + 4b + c = 287$
(9, 431)	$431 = a(9)^2 + b(9) + c$	$81a + 9b + c = 431$

Note: The first equation implies that $c = 253$, so substituting this value for c into the second and third equations yields the following system of two equations and two variables:

$$16a + 4b = 34$$
$$81a + 9b = 178$$

STEP 1 Write this system of linear equations as an augmented matrix.

$$\begin{bmatrix} 16 & 4 & | & 34 \\ 81 & 9 & | & 178 \end{bmatrix}$$

STEP 2 Reduce the matrix using Gauss–Jordan elimination.

$$\frac{1}{16}R_1 \rightarrow R_1 \qquad \begin{bmatrix} 1 & \frac{1}{4} & \Big| & \frac{17}{8} \\ 81 & 9 & \Big| & 178 \end{bmatrix}$$

$$R_2 - 81R_1 \rightarrow R_2 \qquad \begin{bmatrix} 1 & \frac{1}{4} & \Big| & \frac{17}{8} \\ 0 & -\frac{45}{4} & \Big| & \frac{47}{8} \end{bmatrix}$$

$$-\frac{4}{45}R_2 \rightarrow R_2 \qquad \begin{bmatrix} 1 & \frac{1}{4} & \Big| & \frac{17}{8} \\ 0 & 1 & \Big| & -\frac{47}{90} \end{bmatrix}$$

$$R_1 - \frac{1}{4}R_2 \rightarrow R_1 \qquad \begin{bmatrix} 1 & 0 & \Big| & \frac{203}{90} \\ 0 & 1 & \Big| & -\frac{47}{90} \end{bmatrix}$$

STEP 3 Identify the answer.
$$a = \frac{203}{90}$$
$$b = -\frac{47}{90}$$

Substituting $a = \frac{203}{90}$, $b = -\frac{47}{90}$, $c = 253$ into $y = at^2 + bt + c$, we find that the thousands of dollars spent on malpractice suits as a function of year is given by

$$y = \frac{203}{90}t^2 - \frac{47}{90}t + 253 \qquad \text{1992 is } t = 0$$

Notice that all three points lie on this curve.

SECTION 7.2 SUMMARY

In this section, we used augmented matrices to represent a system of linear equations. Using row operations, we readily transformed a matrix to reduced form using the Gauss–Jordan elimination technique. This technique can be used for any size matrix. Square matrices, when the number of equations equals the number of variables, can have one of three solutions: unique, none, or infinitely many. When the number of variables exceeds the number of equations, there is no unique solution. Curve-fitting is one of many applications involving systems of linear equations.

SECTION 7.2 EXERCISES

■ **SKILLS**

In Exercises 1–8, write the augmented matrix for each system of linear equations.

1. $3x - 2y = 7$
$-4x + 6y = -3$

2. $-x + y = 2$
$x - y = -4$

3. $2x - 3y + 4z = -3$
$-x + y + 2z = 1$
$5x - 2y - 3z = 7$

4. $x - 2y + z = 0$
$-2x + y - z = -5$
$13x + 7y + 5z = 6$

5. $x + y = 3$
$x - z = 2$
$y + z = 5$

6. $x - y = -4$
$y + z = 3$

7. $3y - 4x + 5z - 2 = 0$
$2x - 3y - 2z = -3$
$3z + 4y - 2x - 1 = 0$

8. $2y + z - x - 3 = 2$
$2x + 3z - 2y = 0$
$-2z + y - 4x - 3 = 0$

In Exercises 9–18, indicate whether each matrix is in reduced form. If not, state the condition that is violated.

9. $\begin{bmatrix} 1 & 0 & | & 3 \\ 1 & 1 & | & 2 \end{bmatrix}$

10. $\begin{bmatrix} 0 & 1 & | & 3 \\ 1 & 0 & | & 2 \end{bmatrix}$

11. $\begin{bmatrix} 1 & 0 & -1 & | & -3 \\ 0 & 1 & 3 & | & 14 \end{bmatrix}$

12. $\begin{bmatrix} 1 & 0 & 0 & | & -3 \\ 0 & 1 & 3 & | & 14 \end{bmatrix}$

13. $\begin{bmatrix} 1 & 0 & 1 & | & 3 \\ 0 & 0 & 0 & | & 0 \\ 0 & 1 & 2 & | & 2 \end{bmatrix}$

14. $\begin{bmatrix} 1 & 0 & 1 & | & 3 \\ 0 & 1 & 2 & | & 2 \\ 0 & 0 & 0 & | & 0 \end{bmatrix}$

15. $\begin{bmatrix} 1 & 0 & 0 & | & 3 \\ 0 & 1 & 0 & | & 2 \\ 0 & 0 & 1 & | & 5 \end{bmatrix}$

16. $\begin{bmatrix} -1 & 0 & 0 & | & 3 \\ 0 & -1 & 0 & | & 2 \\ 0 & 0 & -1 & | & 5 \end{bmatrix}$

17. $\begin{bmatrix} 1 & 0 & 0 & 1 & | & 3 \\ 0 & 1 & 0 & 3 & | & 2 \\ 0 & 0 & 1 & 0 & | & 5 \\ 0 & 0 & 0 & 1 & | & 0 \end{bmatrix}$

18. $\begin{bmatrix} 1 & 0 & 0 & 1 & | & 3 \\ 0 & 1 & 0 & 3 & | & 2 \\ 0 & 0 & 1 & 0 & | & 5 \\ 0 & 0 & 0 & 0 & | & 0 \end{bmatrix}$

In Exercises 19–28, perform the indicated row operations on each matrix.

19. $\begin{bmatrix} 1 & -2 & | & -3 \\ 2 & 3 & | & -1 \end{bmatrix}$ $\quad R_2 - 2R_1 \rightarrow R_2$

20. $\begin{bmatrix} 2 & -3 & | & -4 \\ 1 & 2 & | & 5 \end{bmatrix}$ $\quad R_1 \leftrightarrow R_2$

21. $\begin{bmatrix} 1 & -2 & -1 & | & 3 \\ 2 & 1 & -3 & | & 6 \\ 3 & -2 & 5 & | & -8 \end{bmatrix}$ $\quad R_2 - 2R_1 \rightarrow R_2$

22. $\begin{bmatrix} 1 & -2 & 1 & | & 3 \\ 0 & 1 & -2 & | & 6 \\ -3 & 0 & -1 & | & -5 \end{bmatrix}$ $\quad R_3 + 3R_1 \rightarrow R_3$

23. $\begin{bmatrix} 1 & -2 & 5 & -1 & | & 2 \\ 0 & 3 & 0 & -1 & | & -2 \\ 0 & -2 & 1 & -2 & | & 5 \\ 0 & 0 & 1 & -1 & | & -6 \end{bmatrix}$ $\quad R_3 + R_2 \rightarrow R_2$

24. $\begin{bmatrix} 1 & 0 & 5 & -10 & | & 15 \\ 0 & 1 & 2 & -3 & | & 4 \\ 0 & 2 & -3 & 0 & | & -1 \\ 0 & 0 & 1 & -1 & | & -3 \end{bmatrix}$ $\quad R_2 - \dfrac{1}{2}R_3 \rightarrow R_3$

25. $\begin{bmatrix} 1 & 0 & 5 & -10 & | & -5 \\ 0 & 1 & 2 & -3 & | & -2 \\ 0 & 2 & -3 & 0 & | & -1 \\ 0 & -3 & 2 & -1 & | & -3 \end{bmatrix}$ $\quad \begin{array}{l} R_3 - 2R_2 \rightarrow R_3 \\ R_4 + 3R_2 \rightarrow R_4 \end{array}$

26. $\begin{bmatrix} 1 & 0 & 4 & 0 & | & 1 \\ 0 & 1 & 2 & 0 & | & -2 \\ 0 & 0 & 1 & 0 & | & 0 \\ 0 & 0 & 0 & 1 & | & -3 \end{bmatrix}$ $\quad \begin{array}{l} R_2 - 2R_3 \rightarrow R_2 \\ R_1 - 4R_3 \rightarrow R_1 \end{array}$

27. $\begin{bmatrix} 1 & 0 & 4 & 8 & | & 3 \\ 0 & 1 & 2 & -3 & | & -2 \\ 0 & 0 & 1 & 6 & | & 3 \\ 0 & 0 & 0 & 1 & | & -3 \end{bmatrix}$ $\quad \begin{array}{l} R_3 - 6R_4 \rightarrow R_3 \\ R_2 + 3R_4 \rightarrow R_2 \\ R_1 - 8R_4 \rightarrow R_1 \end{array}$

28. $\begin{bmatrix} 1 & 0 & -1 & 5 & | & 2 \\ 0 & 1 & 2 & 3 & | & -5 \\ 0 & 0 & 1 & -2 & | & 2 \\ 0 & 0 & 0 & 1 & | & 1 \end{bmatrix}$ $\quad \begin{array}{l} R_3 + 2R_4 \rightarrow R_3 \\ R_2 - 3R_4 \rightarrow R_2 \\ R_1 - 5R_4 \rightarrow R_1 \end{array}$

In Exercises 29–38, use row operations to transform each matrix to reduced form.

29. $\begin{bmatrix} 1 & 2 & | & 4 \\ 2 & 3 & | & 2 \end{bmatrix}$

30. $\begin{bmatrix} 1 & -1 & | & 3 \\ -3 & 2 & | & 2 \end{bmatrix}$

31. $\begin{bmatrix} 1 & -1 & 1 & | & -1 \\ 0 & 1 & -1 & | & -1 \\ -1 & 1 & 1 & | & 1 \end{bmatrix}$

32. $\begin{bmatrix} 0 & -1 & 1 & | & 1 \\ 1 & -1 & 1 & | & -1 \\ 1 & -1 & -1 & | & -1 \end{bmatrix}$

33. $\begin{bmatrix} 3 & -2 & -3 & | & -1 \\ 1 & -1 & 1 & | & -4 \\ 2 & 3 & 5 & | & 14 \end{bmatrix}$

34. $\begin{bmatrix} 3 & -1 & 1 & | & 2 \\ 1 & -2 & 3 & | & 1 \\ 2 & 1 & -3 & | & -1 \end{bmatrix}$

35. $\begin{bmatrix} 2 & 1 & -6 & | & 4 \\ 1 & -2 & 2 & | & -3 \end{bmatrix}$

36. $\begin{bmatrix} -3 & -1 & 2 & | & -1 \\ -1 & -2 & 1 & | & -3 \end{bmatrix}$

37. $\begin{bmatrix} -1 & 2 & 1 & | & -2 \\ 3 & -2 & 1 & | & 4 \\ 2 & -4 & -2 & | & 4 \end{bmatrix}$

38. $\begin{bmatrix} 2 & -1 & 0 & | & 1 \\ -1 & 0 & 1 & | & -2 \\ -2 & 1 & 0 & | & -1 \end{bmatrix}$

In Exercises 39–58, solve the system of linear equations using augmented matrices.

39. $2x + 3y = 1$
$\quad x + y = -2$

40. $3x + 2y = 11$
$\quad x - y = 12$

41. $-x + 2y = 3$
$\quad 2x - 4y = -6$

42. $3x - y = -1$
$\quad 2y + 6x = 2$

43. $\dfrac{2}{3}x + \dfrac{1}{3}y = \dfrac{8}{9}$

$\dfrac{1}{2}x + \dfrac{1}{4}y = \dfrac{3}{4}$

44. $\quad 0.4x - 0.5y = 2.08$

$-0.3x + 0.7y = 1.88$

45. $\quad x - z - y = 10$

$2x - 3y + z = -11$

$y - x + z = -10$

46. $2x + z + y = -3$

$2y - z + x = 0$

$x + y + 2z = 5$

47. $3x_1 + x_2 - x_3 = 1$

$x_1 - x_2 + x_3 = -3$

$2x_1 + x_2 + x_3 = 0$

48. $2x_1 + x_2 + x_3 = -1$

$x_1 + x_2 - x_3 = 5$

$3x_1 - x_2 - x_3 = 1$

49. $\quad\quad 2x + 5y = 9$

$x + 2y - z = 3$

$-3x - 4y + 7z = 1$

50. $\quad x - 2y + 3z = 1$

$-2x + 7y - 9z = 4$

$x + z = 9$

51. $\quad 2x_1 - x_2 + x_3 = 3$

$x_1 - x_2 + x_3 = 2$

$-2x_1 + 2x_2 - 2x_3 = -4$

52. $\quad x_1 - x_2 - 2x_3 = 0$

$-2x_1 + 5x_2 + 10x_3 = -3$

$3x_1 + x_2 = 0$

53. $2x + y - z = 2$

$x - y - z = 6$

54. $3x + y - z = 0$

$x + y + 7z = 4$

55. $\quad\quad 2y + z = 3$

$4x - z = -3$

$7x - 3y - 3z = 2$

$x - y - z = -2$

56. $-2x - y + 2z = 3$

$3x - 4z = 2$

$2x + y = -1$

$-x + y - z = -8$

57. $3x_1 - 2x_2 + x_3 + 2x_4 = -2$

$-x_1 + 3x_2 + 4x_3 + 3x_4 = 4$

$x_1 + x_2 + x_3 + x_4 = 0$

$5x_1 + 3x_2 + x_3 + 2x_4 = -1$

58. $5x_1 + 3x_2 + 8x_3 + x_4 = 1$

$x_1 + 2x_2 + 5x_3 + 2x_4 = 3$

$4x_1 + x_3 - 2x_4 = -3$

$x_2 + x_3 + x_4 = 0$

 ■ **APPLICATIONS**

59. **Football.** In Superbowl XXXVIII, the New England Patriots defeated the Carolina Panthers 32–29. The points came from a total of four types of plays: touchdowns (6 points), extra points (1 point), two-point conversions (2 points), and field goals (3 points). There were a total of 16 scoring plays. There were four times as many touchdowns as field goals, and there were five times as many extra points as two-point conversions. How many touchdowns, extra points, two-point conversions, and field goals were scored in Superbowl XXXVIII?

60. **Basketball.** In the 2004 Summer Olympics in Athens, Greece, the USA men's basketball team consisting of NBA superstars was defeated by the Puerto Rican team 92–73. The points came from three types of scoring plays: two-point shots, three-point shots, and one-point free throws. There were six more two-point shots made than there were one-point free throws. The number of successful two-point shots was three less than four times the number of successful three-point shots. How many two-point, three-point, and one-point free throw shots were made in that Olympic competition?

Exercises 61 and 62 rely on a selection of Subway sandwiches whose nutrition information is given in the table.

Sandwich	Calories	Fat (grams)	Carbohydrates (grams)	Protein (grams)
Mediterranean Chicken	350	18	17	36
6 Inch Tuna	430	19	46	20
6 Inch Roast Beef	290	5	45	19
Turkey-Bacon Wrap	430	27	20	34

www.subway.com

Suppose you are going to eat only Subway sandwiches for a week (7 days) for lunch and dinner (total of 14 meals).

61. Diet. Your goal is a low fat diet consisting of 526 grams of carbohydrates, 168 grams of fat, and 332 grams of protein. How many of each sandwich would you eat that week to obtain this goal?

62. Diet. Your goal is a low carb diet consisting of 5180 calories, 335 grams of carbohydrates, and 263 grams of fat. How many of each sandwich would you eat that week to obtain this goal?

Exercises 63 and 64 involve vertical motion and the effect of gravity on an object.

Because of gravity, an object that is projected upward will eventually reach a maximum height and then fall to the ground. The equation that relates the height, h, of a projectile t seconds after it is shot upward is given by

$$h = \frac{1}{2}at^2 + v_0 t + h_0$$

where a is the acceleration due to gravity, h_0 is the initial height of the object at time $t = 0$, and v_0 is the initial velocity of the object at time $t = 0$. Note that a projectile follows the path of a parabola opening down, so $a < 0$.

63. Vertical Motion. An object is thrown upward, and the following table depicts the height of the ball t seconds after the projectile is released.

t Seconds	Height (feet)
1	34
2	36
3	6

Find the initial height, initial velocity, and acceleration due to gravity.

64. Vertical Motion. An object is thrown upward, and the following table depicts the height of the ball t seconds after the projectile is released.

t Seconds	Height (feet)
1	54
2	66
3	46

Find the initial height, initial velocity, and acceleration due to gravity.

65. Data Curve-Fitting. The average number of minutes that a person spends driving a car can be modeled by a

quadratic function, $y = ax^2 + bx + c$ where $a < 0$ and $16 \le x \le 65$. The following table gives the average number of minutes a day that a person spends driving a car.

Age	Average Daily Minutes Driving
16	25
40	64
65	40

Determine the quadratic function that models this quantity.

66. Data Curve-Fitting. The average age when a woman gets married has been increasing during the last century. In 1920 the average age was 18.4, in 1960 the average age was 20.3, and in 2002 the average age was 25.30. Find a quadratic function, $y = ax^2 + bx + c$, where $a > 0$ and $0 \le x \le 100$ that models the average age, y, when a woman gets married as a function of the year, x, ($x = 0$ corresponds to 1920). What will the average age be in 2010?

67. Money. Gary and Ginger decide to place $10,000 of their savings into investments. They put some in a money market account earning 3% interest, some in a mutual fund that has been averaging 7% a year, and some in a stock that rose to 10% last year. If they put $3,000 more in the money market than in the mutual fund, and the mutual fund and stocks have the same growth in the next year as they did in the previous year, they will earn $540 in a year. How much money should they put in each of the three investments?

68. Money. Ginger talks Gary into putting less money in the money market and more money in the stock. They place $10,000 of their savings into investments. They put some in a money market account earning 3% interest, some in a mutual fund that has been averaging 7% a year, and some in a stock that rose 10% last year. If they put $3,000 more in the stock than in the mutual fund, and the mutual fund and stock have the same growth in the next year as they did in the previous year, they will earn $840 in a year. How much money should they put in each of the three investments?

69. Geometry. The circle given by the equation $x^2 + y^2 + ax + by + c = 0$ passes through the points $(4, 4)$, $(-3, -1)$, and $(1, -3)$. Find a, b, and c.

70. Geometry. The circle given by the equation $x^2 + y^2 + ax + by + c = 0$ passes through the points $(0, 7)$, $(6, 1)$, and $(5, 4)$. Find a, b, and c.

■ CATCH THE MISTAKE

In Exercises 71–74, explain the mistake that is made.

71. Solve the system of equations using augmented matrices:

$$y - x + z = 2$$
$$x - 2z + y = -3$$
$$x + y + z = 6$$

Solution:

Step 1: Write as an augmented matrix.

$$\begin{bmatrix} 1 & -1 & 1 & | & 2 \\ 1 & -2 & 1 & | & -3 \\ 1 & 1 & 1 & | & 6 \end{bmatrix}$$

Step 2: Reduce the matrix using Gauss–Jordan elimination.

$$\begin{bmatrix} 1 & -1 & 1 & | & 2 \\ 0 & 1 & 0 & | & 5 \\ 0 & 0 & 0 & | & -6 \end{bmatrix}$$

Step 3: Identify the solution. Row 3 is inconsistent, so there is no solution.

This is incorrect. The correct answer is $x = 1$, $y = 2$, $z = 3$. What mistake was made?

72. Perform the indicated row operations on the matrix:

$$\begin{bmatrix} 1 & -1 & 1 & | & 2 \\ 2 & -3 & 1 & | & 4 \\ 3 & 1 & 2 & | & -6 \end{bmatrix}$$

a. $R_2 - 2R_1 \rightarrow R_2$ **b.** $R_3 - 3R_1 \rightarrow R_3$

Solution:

a. $\begin{bmatrix} 1 & -1 & 1 & | & 2 \\ 0 & -3 & 1 & | & 4 \\ 3 & 1 & 2 & | & -6 \end{bmatrix}$ **b.** $\begin{bmatrix} 1 & -1 & 1 & | & 2 \\ 2 & -3 & 1 & | & 4 \\ 0 & 1 & 2 & | & -6 \end{bmatrix}$

These are not correct. What mistakes were made?

73. Solve the system of equations using augmented matrices.

$$3x - 2y + z = -1$$
$$x + y - z = 3$$
$$2x - y + 3z = 0$$

Solution:

Step 1: Write the system as an augmented matrix.

$$\begin{bmatrix} 3 & -2 & 1 & | & -1 \\ 1 & 1 & -1 & | & 3 \\ 2 & -1 & 3 & | & 0 \end{bmatrix}$$

Step 2: Reduce the matrix using Gauss–Jordan elimination.

$$\begin{bmatrix} 1 & 0 & 0 & | & 1 \\ 0 & 1 & 0 & | & 2 \\ 0 & 0 & 1 & | & 0 \end{bmatrix}$$

Step 3: Identify the answer. Row 3 is inconsistent $1 = 0$; therefore there is no solution.

This is incorrect. What mistake was made?

74. Solve the system of equations using augmented matrices.

$$x + 3y + 2z = 4$$
$$3x + 10y + 9z = 17$$
$$2x + 7y + 7z = 17$$

Solution:

Step 1: Write the system as an augmented matrix.

$$\begin{bmatrix} 1 & 3 & 2 & | & 4 \\ 3 & 10 & 9 & | & 17 \\ 2 & 7 & 7 & | & 17 \end{bmatrix}$$

Step 2: Reduce the matrix using Gauss–Jordan elimination.

$$\begin{bmatrix} 1 & 0 & -7 & | & -11 \\ 0 & 1 & 3 & | & 5 \\ 0 & 0 & 0 & | & 4 \end{bmatrix}$$

Step 3: Identify the answer. Infinitely many solutions.

$$x = 7t - 11, y = -3t + 5, z = t$$

This is incorrect. What mistake was made?

■ CHALLENGE

75. T or F: A nonsquare matrix cannot have a unique solution.

76. T or F: The procedure for Gauss–Jordan elimination can only be used for square matrices.

77. T or F: A square matrix that has a unique solution has a reduced matrix with 1's along the diagonal and 0's above and below the 1's.

78. T or F: A square matrix with an all-zero row has infinitely many solutions.

79.

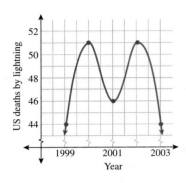

A fourth-degree polynomial,

$$f(x) = ax^4 + bx^3 + cx^2 + dx + e,$$

with $a < 0$, can be used to represent the graph on the left on the number of deaths per year due to lightning strikes (assume 1999 corresponds to $x = 0$).

Use the data to determine a, b, c, d, and e.

80. A copy machine accepts nickels, dimes, and quarters. After 1 hour, there are 30 coins total, and their value is $4.60. How many nickels, quarters, and dimes are in the machine?

■ TECHNOLOGY

Exercises 81–84 rely on the system of equations in Exercises 45 and 46.

$$
\begin{array}{ll}
45. & x - z - y = 10 \\
& 2x - 3y + z = -11 \\
& y - x + z = -10
\end{array}
\qquad
\begin{array}{ll}
46. & 2x + z + y = -3 \\
& 2y - z + x = 0 \\
& x + y + 2z = 5
\end{array}
$$

81. In Exercise 45, you were asked to solve the system of equations using an augmented matrix. A graphing calculator or graphing utility can solve systems of linear equations by entering the coefficients of the matrix. Solve this system and confirm your answer with the calculator's answer.

82. In Exercise 46, you were asked to solve the system of equations using an augmented matrix. A graphing calculator or graphing utility can solve systems of linear equations by entering the coefficients of the matrix. Solve this system and confirm your answer with the calculator's answer.

83. Graphing calculators and graphing utilities have the ability to graph in three dimensions (3D) as opposed to

the traditional two dimensions (2D). The line must be given in the form $z = ax + by + c$. Rewrite the system of equations in Exercise 45 in this form and graph the three lines in 3D. What is the point of intersection? Compare that with your answer in Exercise 45.

84. Graphing calculators and graphing utilities have the ability to graph in 3D as opposed to the traditional 2D. The line must be given in the form $z = ax + by + c$. Rewrite the system of equations in Exercise 46 in this form and graph the three lines in 3D. What is the point of intersection? Compare that with your answer in Exercise 46.

SECTION 7.3 Systems of Linear Equations: Determinants

Skills Objectives

- Evaluate 2×2 determinants.
- Solve a system of two linear equations, with two variables, using Cramer's rule.
- Evaluate 3×3 determinants.
- Solve a system of three linear equations, with three variables, using Cramer's rule.
- Use determinants to identify inconsistent systems that have no solution and dependent systems that have infinitely many solutions.

Conceptual Objectives

- Derive Cramer's rule.
- Understand the purpose of a determinant and how to apply determinants to a system of equations.

In Section 7.2, we used Gauss–Jordan elimination on augmented matrices to solve systems of linear equations. Augmented matrices can be used to solve systems of equations when the number of equations and the number of variables are not the same. In this section we will describe another method for solving linear systems called *Cramer's rule*, which uses *determinants*. However, this method can only be used on square matrices, when the number of equations equals the number of variables and is typically only used for systems with either two or three equations.

2 × 2 Determinants

Every square matrix has a number associated with it called its *determinant*. A system of two equations and two variables has a corresponding 2 × 2 matrix.

DEFINITION **DETERMINANT OF A 2 × 2 MATRIX**

The **determinant** of the 2 × 2 matrix $\begin{bmatrix} a & b \\ c & d \end{bmatrix}$

is given by

$$D = \begin{vmatrix} a & b \\ c & d \end{vmatrix} = ad - bc$$

Notice that brackets, [ ], are used to denote a matrix, whereas vertical lines similar to absolute value, | |, are used to denote the determinant of a matrix.

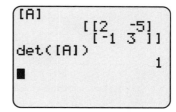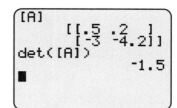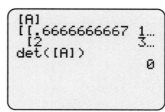

The following illustrations may help you remember the operations involved in calculating the determinant of a 2 × 2 matrix. The result is the difference of the product of the diagonal elements.

$$D = \begin{vmatrix} a & b \\ c & d \end{vmatrix} = ad - bc \qquad D = \begin{vmatrix} a & b \\ c & d \end{vmatrix} = ad - bc$$

$$\text{minus} \quad ad$$

EXAMPLE 1 Evaluating the Determinant of a 2 × 2 Matrix

Find the determinant of each matrix.

a. $\begin{bmatrix} 2 & -5 \\ -1 & 3 \end{bmatrix}$ **b.** $\begin{bmatrix} 0.5 & 0.2 \\ -3 & -4.2 \end{bmatrix}$ **c.** $\begin{bmatrix} \frac{2}{3} & 1 \\ 2 & 3 \end{bmatrix}$

Solution:

a. $\begin{vmatrix} 2 & -5 \\ -1 & 3 \end{vmatrix} = (2)(3) - (-1)(-5) = 6 - 5 = \boxed{1}$

b. $\begin{vmatrix} 0.5 & 0.2 \\ -3 & -4.2 \end{vmatrix} = (0.5)(-4.2) - (-3)(0.2) = -2.1 + 0.6 = \boxed{-1.5}$

c. $\begin{vmatrix} \frac{2}{3} & 1 \\ 2 & 3 \end{vmatrix} = \left(\frac{2}{3}\right)(3) - (2)(1) = 2 - 2 = \boxed{0}$

> **YOUR TURN** Evaluate the determinant: $\begin{vmatrix} -2 & 1 \\ -3 & 2 \end{vmatrix}$.

In Example 1, we see that determinants are real numbers that can be positive, negative, or zero. Although evaluating determinants of 2×2 matrices is a simple process, one **common mistake** is reversing the difference: $\begin{vmatrix} a & b \\ c & d \end{vmatrix} \neq bc - ad$.

Cramer's Rule for Systems of Two Equations

Let's now use determinants of 2×2 matrices to solve systems of two linear equations and two variables. We begin by solving the general system of two linear equations in two variables:

$$\text{Equation (1):} \quad a_1 x + b_1 y = c_1$$

$$\text{Equation (2):} \quad a_2 x + b_2 y = c_2$$

WORDS	MATH

Step 1: Solve for x using elimination (eliminate y).

Multiply Equation (1) by b_2. $b_2 a_1 x + b_2 b_1 y = b_2 c_1$

Multiply Equation (2) by $-b_1$. $-b_1 a_2 x - b_1 b_2 y = -b_1 c_2$

Add two new equations to eliminate y. $(b_2 a_1 - b_1 a_2)x = (b_2 c_1 - b_1 c_2)$

Divide both sides by $(a_1 b_2 - a_2 b_1)$. $x = \dfrac{(b_2 c_1 - b_1 c_2)}{(b_2 a_1 - b_1 a_2)}$

Write both the numerator and the denominator as determinants. $x = \dfrac{\begin{vmatrix} c_1 & b_1 \\ c_2 & b_2 \end{vmatrix}}{\begin{vmatrix} a_1 & b_1 \\ a_2 & b_2 \end{vmatrix}} = \dfrac{D_x}{D}$

Step 2: Solve for y using elimination (eliminate x).

Multiply Equation (1) by $-a_2$. $-a_2 a_1 x - a_2 b_1 y = -a_2 c_1$

Multiply Equation (2) by a_1. $a_1 a_2 x + a_1 b_2 y = a_1 c_2$

Add two new equations to eliminate x. $(a_1 b_2 - a_2 b_1)y = (a_1 c_2 - a_2 c_1)$

Divide both sides by $(a_1 b_2 - a_2 b_1)$. $y = \dfrac{(a_1 c_2 - a_2 c_1)}{(a_1 b_2 - a_2 b_1)}$

Write both the numerator and the denominator as determinants. $y = \dfrac{\begin{vmatrix} a_1 & c_1 \\ a_2 & c_2 \end{vmatrix}}{\begin{vmatrix} a_1 & b_1 \\ a_2 & b_2 \end{vmatrix}} = \dfrac{D_y}{D}$

Note: The denominator cannot equal zero, $D \neq 0$. The answer can be written in terms of determinants and is called Cramer's rule.

CRAMER'S RULE: SOLUTION FOR SYSTEMS OF TWO EQUATIONS AND TWO VARIABLES

The system of linear equations
$$\begin{aligned} a_1x + b_1y &= c_1 \\ a_2x + b_2y &= c_2 \end{aligned}$$

has the solution
$$x = \frac{D_x}{D}, \; y = \frac{D_y}{D}$$

where the determinants are given by

$$D = \begin{vmatrix} a_1 & b_1 \\ a_2 & b_2 \end{vmatrix} \neq 0 \qquad D_x = \begin{vmatrix} c_1 & b_1 \\ c_2 & b_2 \end{vmatrix} \qquad D_y = \begin{vmatrix} a_1 & c_1 \\ a_2 & c_2 \end{vmatrix}$$

Notice that the determinants D_x and D_y are similar to the determinant D. A three-step procedure is outlined for setting up the three determinants for a system of two linear equations in two variables.

$$a_1x + b_1y = c_1$$

$$a_2x + b_2y = c_2$$

Step 1: Set up D.

Use the coefficients of x and y.
$$D = \begin{vmatrix} a_1 & b_1 \\ a_2 & b_2 \end{vmatrix}$$

Step 2: Set up D_x.

Start with D and replace the coefficients of x (column 1) with the constants on the right side of the equal sign.
$$D_x = \begin{vmatrix} c_1 & b_1 \\ c_2 & b_2 \end{vmatrix}$$

Step 3: Set up D_y.

Start with D and replace the coefficients of y (column 2) with the constants on the right side of the equal signs.
$$D_y = \begin{vmatrix} a_1 & c_1 \\ a_2 & c_2 \end{vmatrix}$$

EXAMPLE 2 Using Cramer's Rule to Solve a System of Two Equations

Use Cramer's rule to solve the system: $\begin{array}{l} x + 3y = 1 \\ 2x + y = -3 \end{array}$.

Solution:

STEP 1 Set up the three determinants.
$$D = \begin{vmatrix} 1 & 3 \\ 2 & 1 \end{vmatrix}$$

$$D_x = \begin{vmatrix} 1 & 3 \\ -3 & 1 \end{vmatrix} \qquad D_y = \begin{vmatrix} 1 & 1 \\ 2 & -3 \end{vmatrix}$$

STEP 2 Evaluate the determinants.
$$D = 1 - 6 = -5$$

$$D_x = 1 - (-9) = 10, \; D_y = -3 - 2 = -5$$

STEP 3 Solve for x and y.
$$x = \frac{D_x}{D} = \frac{10}{-5} = -2, \; y = \frac{D_y}{D} = \frac{-5}{-5} = 1$$

$$x = -2, \; y = 1$$

TECHNOLOGY TIP

A graphing calculator can be used to solve the system using Cramer's rule. Enter the matrix A for the determinant D_x, B for D_y, C for D.

```
[A]
        [[1   3]
         [-3  1]]
[B]
        [[1   1]
         [2  -3]]
```

```
[C]
        [[1  3]
         [2  1]]
```

To solve for x and y, enter D_x/D as A/C for x and D_y/D as B/C for y.

```
det([A])/det([C]
)
                -2
det([B])/det([C]
)
                 1
```

■ **YOUR TURN** Use Cramer's rule to solve the system: $\begin{array}{l} 2x + y = 5 \\ -3x - 2y = -3 \end{array}$.

Recall in Section 6.1 that systems of two equations and two variables have one of three possibilities: a unique solution (two lines intersecting at a single point), no solution (parallel lines), and infinitely many solutions (identical lines). When $D = 0$, Cramer's rule does not apply and the system is either inconsistent (no solution) or contains dependent equations (infinitely many solutions).

No solution: $D = 0$ and $D_x \neq 0$ and $D_y \neq 0$

Infinitely many solutions: $D = 0$ and either $D_x = 0$ or $D_y = 0$

EXAMPLE 3 Using Cramer's Rule to Solve a System of Two Equations: Infinitely Many Solutions

Use Cramer's rule to solve the system:

$$2x - y = 3$$

$$-x + \frac{1}{2}y = -\frac{3}{2}$$

Solution:

STEP 1 Set up the three determinants.

$$D = \begin{vmatrix} 2 & -1 \\ -1 & \frac{1}{2} \end{vmatrix}$$

$$D_x = \begin{vmatrix} 3 & -1 \\ -\frac{3}{2} & \frac{1}{2} \end{vmatrix} \qquad D_y = \begin{vmatrix} 2 & 3 \\ -1 & -\frac{3}{2} \end{vmatrix}$$

STEP 2 Evaluate the determinants.

$$D = 1 - 1 = 0$$

$$D_x = \tfrac{3}{2} - \tfrac{3}{2} = 0, \; D_y = -3 + 3 = 0$$

STEP 3 Solve for x and y. Infinitely many solutions

EXAMPLE 4 Using Cramer's Rule to Solve a System of Two Equations: No Solution

Use Cramer's rule to solve the system:

$$0.5x + y = 0.3$$

$$2x + 4y = 1$$

Solution:

STEP 1 Set up the three determinants.

$$D = \begin{vmatrix} 0.5 & 1 \\ 2 & 4 \end{vmatrix}$$

$$D_x = \begin{vmatrix} 0.3 & 1 \\ 1 & 4 \end{vmatrix} \qquad D_y = \begin{vmatrix} 0.5 & 0.3 \\ 2 & 1 \end{vmatrix}$$

■ **Answer:** $x = 7, y = -9$

STEP 2 Evaluate the
determinants.

$$D = 2 - 2 = 0$$

$$D_x = 1.2 - 1 = 0.2, \; D_y = 0.5 - 0.6 = -0.1$$

STEP 3 Solve for x and y.   No solution

■ **YOUR TURN** Use Cramer's rule to solve the systems:

a. $x + y = 1$ b. $-4x + 2y = -6$
   $x + y = 3$ $2x - y = 3$

3 × 3 Determinants

As mentioned earlier, every square matrix has a number associated with it called a *determinant*. A system of three equations and three variables has a corresponding 3×3 matrix.

DEFINITION   **DETERMINANT OF A 3 × 3 MATRIX**

The **determinant** of the 3×3 matrix $\begin{bmatrix} a_1 & b_1 & c_1 \\ a_2 & b_2 & c_2 \\ a_3 & b_3 & c_3 \end{bmatrix}$

is given by $\begin{vmatrix} a_1 & b_1 & c_1 \\ a_2 & b_2 & c_2 \\ a_3 & b_3 & c_3 \end{vmatrix} = a_1 b_2 c_3 - a_1 b_3 c_2 - b_1 a_2 c_3 + b_1 a_3 c_2 + c_1 a_2 b_3 - c_1 a_3 b_2$

It turns out that this 3×3 determinant can be written six different ways (expanding down any row or column) in terms of combinations of 2×2 determinants. Although any of the three columns or three rows can be expanded to yield the determinant, usually either the first row or first column is used.

3×3 Determinant using row 1 expansion:

$$\begin{vmatrix} a_1 & b_1 & c_1 \\ a_2 & b_2 & c_2 \\ a_3 & b_3 & c_3 \end{vmatrix} = a_1 \begin{vmatrix} b_2 & c_2 \\ b_3 & c_3 \end{vmatrix} - b_1 \begin{vmatrix} a_2 & c_2 \\ a_3 & c_3 \end{vmatrix} + c_1 \begin{vmatrix} a_2 & b_2 \\ a_3 & b_3 \end{vmatrix}$$

$$= a_1(b_2 c_3 - b_3 c_2) - b_1(a_2 c_3 - a_3 c_2) + c_1(a_2 b_3 - a_3 b_2)$$

$$= a_1 b_2 c_3 - a_1 b_3 c_2 - b_1 a_2 c_3 + b_1 a_3 c_2 + c_1 a_2 b_3 - c_1 a_3 b_2$$

3×3 Determinant using column 1 expansion:

$$\begin{vmatrix} a_1 & b_1 & c_1 \\ a_2 & b_2 & c_2 \\ a_3 & b_3 & c_3 \end{vmatrix} = a_1 \begin{vmatrix} b_2 & c_2 \\ b_3 & c_3 \end{vmatrix} - a_2 \begin{vmatrix} b_1 & c_1 \\ b_3 & c_3 \end{vmatrix} + a_3 \begin{vmatrix} b_1 & c_1 \\ b_2 & c_2 \end{vmatrix}$$

$$= a_1(b_2 c_3 - b_3 c_2) - a_2(b_1 c_3 - b_3 c_1) + a_3(b_1 c_2 - b_2 c_1)$$

$$= a_1 b_2 c_3 - a_1 b_3 c_2 - a_2 b_1 c_3 + a_2 b_3 c_1 + a_3 b_1 c_2 - a_3 b_2 c_1$$

Whichever row or column is expanded, an alternating sign scheme is used.

Notice that in either of the expansions above the 2×2 determinant obtained is found by crossing out the row and column containing the element that is multiplying the determinant.

Row 1 expansion:

$$\begin{vmatrix} \cancel{a_1} & \cancel{b_1} & \cancel{c_1} \\ \cancel{a_2} & b_2 & c_2 \\ \cancel{a_3} & b_3 & c_3 \end{vmatrix} \quad \begin{vmatrix} \cancel{a_1} & \cancel{b_1} & \cancel{c_1} \\ a_2 & \cancel{b_2} & c_2 \\ a_3 & \cancel{b_3} & c_3 \end{vmatrix} \quad \begin{vmatrix} \cancel{a_1} & \cancel{b_1} & \cancel{c_1} \\ a_2 & b_2 & \cancel{c_2} \\ a_3 & b_3 & \cancel{c_3} \end{vmatrix}$$

$$\begin{vmatrix} a_1 & b_1 & c_1 \\ a_2 & b_2 & c_2 \\ a_3 & b_3 & c_3 \end{vmatrix} = a_1 \begin{vmatrix} b_2 & c_2 \\ b_3 & c_3 \end{vmatrix} - b_1 \begin{vmatrix} a_2 & c_2 \\ a_3 & c_3 \end{vmatrix} + c_1 \begin{vmatrix} a_2 & b_2 \\ a_3 & b_3 \end{vmatrix}$$

Column 1 expansion:

$$\begin{vmatrix} \cancel{a_1} & \cancel{b_1} & \cancel{c_1} \\ \cancel{a_2} & b_2 & c_2 \\ \cancel{a_3} & b_3 & c_3 \end{vmatrix} \quad \begin{vmatrix} \cancel{a_1} & b_1 & c_1 \\ \cancel{a_2} & \cancel{b_2} & \cancel{c_2} \\ \cancel{a_3} & b_3 & c_3 \end{vmatrix} \quad \begin{vmatrix} \cancel{a_1} & b_1 & c_1 \\ \cancel{a_2} & b_2 & c_2 \\ \cancel{a_3} & \cancel{b_3} & \cancel{c_3} \end{vmatrix}$$

$$\begin{vmatrix} a_1 & b_1 & c_1 \\ a_2 & b_2 & c_2 \\ a_3 & b_3 & c_3 \end{vmatrix} = a_1 \begin{vmatrix} b_2 & c_2 \\ b_3 & c_3 \end{vmatrix} - a_2 \begin{vmatrix} b_1 & c_1 \\ b_3 & c_3 \end{vmatrix} + a_3 \begin{vmatrix} b_1 & c_1 \\ b_2 & c_2 \end{vmatrix}$$

EXAMPLE 5 Evaluating the Determinant of a 3 × 3 Matrix

Find the determinant of the matrix: $\begin{bmatrix} 2 & 0 & 3 \\ -1 & 5 & -2 \\ 1 & 7 & 4 \end{bmatrix}$.

Solution:

Set up the 3×3 determinant.
$$\begin{vmatrix} 2 & 0 & 3 \\ -1 & 5 & -2 \\ 1 & 7 & 4 \end{vmatrix}$$

Write the 3×3 determinant in terms of combinations of 2×2 determinants by expanding across row 1.

$$\begin{vmatrix} 2 & 0 & 3 \\ -1 & 5 & -2 \\ 1 & 7 & 4 \end{vmatrix} = 2\begin{vmatrix} 5 & -2 \\ 7 & 4 \end{vmatrix} - 0\begin{vmatrix} -1 & -2 \\ 1 & 4 \end{vmatrix} + 3\begin{vmatrix} -1 & 5 \\ 1 & 7 \end{vmatrix}$$

Evaluate 2×2 determinants.

$$= 2[(5)(4) - (7)(-2)] - 0[(-1)(4) - (1)(-2)] + 3[(-1)(7) - (1)(5)]$$
$$= 2(34) - 0 + 3(-12)$$
$$= 32$$

$$\begin{vmatrix} 2 & 0 & 3 \\ -1 & 5 & -2 \\ 1 & 7 & 4 \end{vmatrix} = 32$$

TECHNOLOGY TIP
Enter the matrix as A and find the determinant.

```
[A]
     [[2   0  3 ]
      [-1  5  -2]
      [1   7  4 ]]
det([A])
                  32
```

■ **YOUR TURN** Evaluate the determinant: $\begin{vmatrix} 1 & -2 & 0 \\ -1 & 0 & 3 \\ -4 & 1 & 2 \end{vmatrix}$.

■ **Answer:** 17

Cramer's Rule for Systems of Three Equations

Cramer's rule can also be used to solve systems of three equations and three variables.

CRAMER'S RULE: SOLUTION FOR SYSTEMS OF THREE EQUATIONS AND THREE VARIABLES

The system of linear equations
$$a_1x + b_1y + c_1z = d_1$$
$$a_2x + b_2y + c_2z = d_2$$
$$a_3x + b_3y + c_3z = d_3$$

has the solution
$$x = \frac{D_x}{D}, \; y = \frac{D_y}{D}, \; z = \frac{D_z}{D}$$

where the determinants are given by

$$D = \begin{vmatrix} a_1 & b_1 & c_1 \\ a_2 & b_2 & c_2 \\ a_3 & b_3 & c_3 \end{vmatrix} \neq 0$$
Coefficients of x, y, and z.

$$D_x = \begin{vmatrix} d_1 & b_1 & c_1 \\ d_2 & b_2 & c_2 \\ d_3 & b_3 & c_3 \end{vmatrix}$$
Replace the coefficients of x (column 1) in D with the constants on the right side of the equal sign.

$$D_y = \begin{vmatrix} a_1 & d_1 & c_1 \\ a_2 & d_2 & c_2 \\ a_3 & d_3 & c_3 \end{vmatrix}$$
Replace the coefficients of y (column 2) in D with the constants on the right side of the equal sign.

$$D_z = \begin{vmatrix} a_1 & b_1 & d_1 \\ a_2 & b_2 & d_2 \\ a_3 & b_3 & d_3 \end{vmatrix}$$
Replace the coefficients of z (column 3) in D with the constants on the right side of the equal sign.

EXAMPLE 6 Using Cramer's Rule to Solve a System of Three Equations

Use Cramer's rule to solve the system.
$$3x - 2y + 3z = -3$$
$$5x + 3y + 8z = -2$$
$$x + y + 3z = 1$$

Solution:

STEP 1 Set up the four determinants.

Coefficients of x, y, and z: $D = \begin{vmatrix} 3 & -2 & 3 \\ 5 & 3 & 8 \\ 1 & 1 & 3 \end{vmatrix}$

Replace a column with constants on right side of equation.

$$D_x = \begin{vmatrix} -3 & -2 & 3 \\ -2 & 3 & 8 \\ 1 & 1 & 3 \end{vmatrix} \quad D_y = \begin{vmatrix} 3 & -3 & 3 \\ 5 & -2 & 8 \\ 1 & 1 & 3 \end{vmatrix} \quad D_z = \begin{vmatrix} 3 & -2 & -3 \\ 5 & 3 & -2 \\ 1 & 1 & 1 \end{vmatrix}$$

STEP 2 Evaluate the determinants.

$$D = 3(9 - 8) - (-2)(15 - 8) + 3(5 - 3) = 23$$

$$D_x = -3(9 - 8) - (-2)(-6 - 8) + 3(-2 - 3) = -46$$

$$D_y = 3(-6 - 8) - (-3)(15 - 8) + 3(5 + 2) = 0$$

$$D_z = 3(3 + 2) - (-2)(5 + 2) - 3(5 - 3) = 23$$

STEP 3 Solve for x, y, and z.

$$x = \frac{D_x}{D} = \frac{-46}{23} = -2, \quad y = \frac{D_y}{D} = \frac{0}{23} = 0, \quad z = \frac{D_z}{D} = \frac{23}{23} = 1$$

$$x = -2, y = 0, z = 1$$

■ **YOUR TURN** Use Cramer's rule to solve the system:

$$2x + 3y + z = -1$$
$$x - y - z = 0$$
$$-3x - 2y + 3z = 10$$

As it was the case in two equations, when $D = 0$ Cramer's rule does not apply, and the system of three equations is either inconsistent (no solution) or contains dependent equations (infinitely many solutions).

No solution: $D = 0$ and $D_x \neq 0$ and $D_y \neq 0$ and $D_z \neq 0$

Infinitely many solutions: $D = 0$ and $D_x = 0$ or $D_y = 0$ or $D_z = 0$

SECTION 7.3 ## SUMMARY

In this section determinants were discussed for square matrices. Only when the number of equations equals the number of variables does the determinant exist. Cramer's rule, which involves determinants, can be used to solve systems of equations. Cramer's rule was developed for 2×2 and 3×3 matrices but it can be extended to general $n \times n$ matrices. When the coefficient determinant is equal to zero, $D = 0$, then the system is either inconsistent (and has no solution) or it represents dependent equations (and has infinitely many solutions).

SECTION 7.3 EXERCISES

■ SKILLS

In Exercises 1–10, evaluate each 2×2 determinant.

1. $\begin{vmatrix} 1 & 2 \\ 3 & 4 \end{vmatrix}$

2. $\begin{vmatrix} 1 & -2 \\ -3 & -4 \end{vmatrix}$

3. $\begin{vmatrix} 7 & 9 \\ -5 & -2 \end{vmatrix}$

4. $\begin{vmatrix} -3 & -11 \\ 7 & 15 \end{vmatrix}$

5. $\begin{vmatrix} 0 & 7 \\ 4 & -1 \end{vmatrix}$

6. $\begin{vmatrix} 0 & 0 \\ 1 & 0 \end{vmatrix}$

7. $\begin{vmatrix} -1.2 & 2.4 \\ -0.5 & 1.5 \end{vmatrix}$

8. $\begin{vmatrix} -1.0 & 1.4 \\ 1.5 & -2.8 \end{vmatrix}$

9. $\begin{vmatrix} \frac{3}{4} & \frac{1}{3} \\ 2 & \frac{8}{9} \end{vmatrix}$

10. $\begin{vmatrix} -\frac{1}{2} & \frac{1}{4} \\ \frac{2}{3} & -\frac{8}{9} \end{vmatrix}$

■ **Answer:** $x = 1, y = -2, z = 3$

In Exercises 11–26, use Cramer's rule to solve each system of equations or to determine if the system is inconsistent (no solution) or dependent (infinitely many solutions).

11. $x + y = -1$
$x - y = 11$

12. $x + y = -1$
$x - y = -9$

13. $3x + 2y = -4$
$-2x + y = 5$

14. $5x + 3y = 1$
$4x - 7y = -18$

15. $3x - 5y = 7$
$-6x + 10y = -21$

16. $3x - 5y = 7$
$6x - 10y = 14$

17. $2x - 3y = 4$
$-10x + 15y = -20$

18. $2x - 3y = 2$
$10x - 15y = 20$

19. $3x + \dfrac{1}{2}y = 1$
$4x + \dfrac{1}{3}y = \dfrac{5}{3}$

20. $\dfrac{3}{2}x + \dfrac{9}{4}y = \dfrac{9}{8}$
$\dfrac{1}{3}x + \dfrac{1}{4}y = \dfrac{1}{12}$

21. $0.3x - 0.5y = -0.6$
$0.2x + y = 2.4$

22. $0.5x - 0.4y = -3.6$
$10x + 3.6y = -14$

23. $y = 17x + 7$
$y = -15x + 7$

24. $9x = -45 - 2y$
$4x = -3y - 20$

25. $\dfrac{2}{x} - \dfrac{3}{y} = 2$
$\dfrac{5}{x} - \dfrac{6}{y} = 7$

26. $\dfrac{2}{x} - \dfrac{3}{y} = -12$
$\dfrac{3}{x} + \dfrac{1}{2y} = 7$

In Exercises 27–36, evaluate each 3×3 determinant.

27. $\begin{vmatrix} 3 & 1 & 0 \\ 2 & 0 & -1 \\ -4 & 1 & 0 \end{vmatrix}$

28. $\begin{vmatrix} 1 & 1 & 0 \\ 0 & 2 & -1 \\ 0 & -3 & 5 \end{vmatrix}$

29. $\begin{vmatrix} 2 & 1 & -5 \\ 3 & 0 & -1 \\ 4 & 0 & 7 \end{vmatrix}$

30. $\begin{vmatrix} 2 & 1 & -5 \\ 3 & -7 & 0 \\ 4 & -6 & 0 \end{vmatrix}$

31. $\begin{vmatrix} 1 & 1 & -5 \\ 3 & -7 & -4 \\ 4 & -6 & 9 \end{vmatrix}$

32. $\begin{vmatrix} -3 & 2 & -5 \\ 1 & 8 & 2 \\ 4 & -6 & 9 \end{vmatrix}$

33. $\begin{vmatrix} 1 & 3 & 4 \\ 2 & -1 & 1 \\ 3 & -2 & 1 \end{vmatrix}$

34. $\begin{vmatrix} -7 & 2 & 5 \\ \frac{7}{8} & 3 & 4 \\ -1 & 4 & 6 \end{vmatrix}$

35. $\begin{vmatrix} a & 0 & 0 \\ 0 & b & 0 \\ 0 & 0 & c \end{vmatrix}$

36. $\begin{vmatrix} a_1 & b_1 & c_1 \\ 0 & b_2 & c_2 \\ 0 & 0 & c_3 \end{vmatrix}$

In Exercises 37–46, use Cramer's rule to solve each system of equations or to determine if the system is inconsistent (no solution) or dependent (infinitely many solutions).

37. $x + y - z = 0$
$x - y + z = 4$
$x + y + z = 10$

38. $-x + y + z = -4$
$x + y - z = 0$
$x + y + z = 2$

39. $3x + 8y + 2z = 28$
$-2x + 5y + 3z = 34$
$4x + 9y + 2z = 29$

40. $7x + 2y - z = -1$
$6x + 5y + z = 16$
$-5x - 4y + 3z = -5$

41. $3x + 5z = 11$
$4y + 3z = -9$
$2x - y = 7$

42. $3x - 2z = 7$
$4x + z = 24$
$6x - 2y = 10$

43. $x + y - z = 5$
$x - y + z = -1$
$-2x - 2y + 2z = -10$

44. $x + y - z = 3$
$x - y + z = -2$
$-2x - 2y + 2z = -6$

45. $x + y + z = 9$
$x - y + z = 3$
$-x + y - z = 5$

46. $x + y + z = 6$
$x - y - z = 0$
$-x + y + z = 7$

 ■ **APPLICATIONS**

For Exercises 47–50, the area of a triangle with vertices (x_1, y_1), (x_2, y_2), and (x_3, y_3) is given by

$$\text{Area} = \pm\frac{1}{2}\begin{vmatrix} x_1 & y_1 & 1 \\ x_2 & y_2 & 1 \\ x_3 & y_3 & 1 \end{vmatrix}$$

where the sign is chosen so that the area is positive.

47. Geometry. Use determinants to find the area of a triangle with vertices $(3, 2)$, $(5, 2)$, and $(3, -4)$. Check your answer by plotting these vertices in a Cartesian plane and using the area of a right triangle.

48. Geometry. Use determinants to find the area of a triangle with vertices $(2, 3)$, $(7, 3)$, and $(7, 7)$. Check your answer by plotting these vertices in a Cartesian plane and using the area of a right triangle.

49. Geometry. Use determinants to find the area of a triangle with vertices $(1, 2)$, $(3, 4)$, and $(-2, 5)$.

50. Geometry. Use determinants to find the area of a triangle with vertices $(-1, -2)$, $(3, 4)$, and $(2, 1)$.

51. Geometry. An equation of a line that passes through two points (x_1, y_1) and (x_2, y_2) can be expressed as a determinant:

$$\begin{vmatrix} x & y & 1 \\ x_1 & y_1 & 1 \\ x_2 & y_2 & 1 \end{vmatrix} = 0$$

Use the determinant to write an equation of the line passing through the points $(1, 2)$ and $(2, 4)$. Expand the determinant and express the equation of the line in slope–intercept form.

52. Geometry. If three points (x_1, y_1), (x_2, y_2), and (x_3, y_3) are collinear (or lie on the same line) then the following determinant must be satisfied:

$$\begin{vmatrix} x_1 & y_1 & 1 \\ x_2 & y_2 & 1 \\ x_3 & y_3 & 1 \end{vmatrix} = 0$$

Determine if $(0, 5)$, $(2, 0)$, and $(1, 2)$ are collinear.

■ CATCH THE MISTAKE

In Exercises 53–56, explain the mistake that is made.

53. Evaluate the determinant: $\begin{vmatrix} 2 & 1 & 3 \\ -3 & 0 & 2 \\ 1 & 4 & -1 \end{vmatrix}$.

Solution:

Expand the 3 × 3 determinant in terms of 2 × 2 determinants.

$$\begin{vmatrix} 2 & 1 & 3 \\ -3 & 0 & 2 \\ 1 & 4 & -1 \end{vmatrix} = 2\begin{vmatrix} 0 & 2 \\ 4 & -1 \end{vmatrix} + 1\begin{vmatrix} -3 & 2 \\ 1 & -1 \end{vmatrix} + 3\begin{vmatrix} -3 & 0 \\ 1 & 4 \end{vmatrix}$$

Expand the 2 × 2 determinants. $= 2(0 - 8) + 1(3 - 2) + 3(-12 - 0)$

Simplify. $= -16 + 1 - 36 = -51$

This is incorrect. What is the mistake?

54. Evaluate the determinant: $\begin{vmatrix} 2 & 1 & 3 \\ -3 & 0 & 2 \\ 1 & 4 & -1 \end{vmatrix}$.

Solution:

Expand the 3 × 3 determinant in terms of 2 × 2 determinants.

$$\begin{vmatrix} 2 & 1 & 3 \\ -3 & 0 & 2 \\ 1 & 4 & -1 \end{vmatrix} = 2\begin{vmatrix} 0 & 2 \\ 4 & -1 \end{vmatrix} - 1\begin{vmatrix} -3 & 2 \\ 1 & -1 \end{vmatrix} + 3\begin{vmatrix} -3 & 2 \\ 1 & -1 \end{vmatrix}$$

Expand the 2 × 2 determinants. $= 2(0 - 8) - 1(3 - 2) + 3(3 - 2)$

Simplify. $= -16 - 1 + 3 = -14$

This is incorrect. What is the mistake?

55. Solve the system of linear equations: $\begin{aligned} 2x + 3y &= 6 \\ -x - y &= -3 \end{aligned}$.

Solution:

Set up the determinants.

$$D = \begin{vmatrix} 2 & 3 \\ -1 & -1 \end{vmatrix}, \quad D_x = \begin{vmatrix} 2 & 6 \\ -1 & -3 \end{vmatrix}, \text{ and } D_y = \begin{vmatrix} 6 & 3 \\ -3 & -1 \end{vmatrix}$$

Evaluate the determinants.

$$D = 1, \quad D_x = 0, \text{ and } D_y = 3$$

Solve for x and y.

$$x = \frac{D_x}{D} = \frac{0}{1} = 0 \text{ and } y = \frac{D_y}{D} = \frac{3}{1} = 3$$

$x = 0, y = 3$ is incorrect. What is the mistake?

56. Solve the system of linear equations: $\begin{aligned} 4x - 6y &= 0 \\ 4x + 6y &= 4 \end{aligned}$.

Solution:

Set up the determinants.

$$D = \begin{vmatrix} 4 & -6 \\ 4 & 6 \end{vmatrix}, \quad D_x = \begin{vmatrix} 0 & -6 \\ 4 & 6 \end{vmatrix}, \text{ and } D_y = \begin{vmatrix} 4 & 0 \\ 4 & 4 \end{vmatrix}$$

Evaluate the determinants.

$$D = 48, \quad D_x = 24, \text{ and } D_y = 16$$

Solve for x and y.

$$x = \frac{D}{D_x} = \frac{48}{24} = 2 \text{ and } y = \frac{D_y}{D} = \frac{48}{16} = 3$$

$x = 2, y = 3$ is incorrect. What is the mistake?

■ CHALLENGE

57. T or F: The value of a determinant changes sign if any two rows are interchanged. (Assume value of determinant $\neq 0$.)

58. T or F: If all the entries in any column are equal to zero, the value of the determinant is 0.

59. T or F: $\begin{vmatrix} 2 & 6 & 4 \\ 0 & 2 & 8 \\ 4 & 0 & 10 \end{vmatrix} = 2 \begin{vmatrix} 1 & 3 & 2 \\ 0 & 1 & 4 \\ 2 & 0 & 5 \end{vmatrix}$.

60. T or F: $\begin{vmatrix} 3 & 1 & 2 \\ 0 & 2 & 8 \\ 3 & 1 & 2 \end{vmatrix} = 0$.

61. Evaluate the determinant: $\begin{vmatrix} 1 & -2 & -1 & 3 \\ 4 & 0 & 1 & 2 \\ 0 & 3 & 2 & 4 \\ 1 & -3 & 5 & -4 \end{vmatrix}$.

62. For the system of equations: $\begin{array}{l} 3x + 2y = 5 \\ ax - 4y = 1 \end{array}$, find a that guarantees no unique solution.

63. Show that $\begin{vmatrix} a_1 & b_1 & c_1 \\ a_2 & b_2 & c_2 \\ a_3 & b_3 & c_3 \end{vmatrix} = \begin{array}{l} a_1 b_2 c_3 + b_1 c_2 a_3 + c_1 a_2 b_3 \\ -a_3 b_2 c_1 - b_3 c_2 a_1 - c_3 b_1 a_2 \end{array}$ by expanding down the second column.

64. Show that $\begin{vmatrix} a_1 & b_1 & c_1 \\ a_2 & b_2 & c_2 \\ a_3 & b_3 & c_3 \end{vmatrix} = \begin{array}{l} a_1 b_2 c_3 + b_1 c_2 a_3 + c_1 a_2 b_3 \\ -a_3 b_2 c_1 - b_3 c_2 a_1 - c_3 b_1 a_2 \end{array}$ by expanding across the third row.

■ TECHNOLOGY

Use a graphing utility to evaluate the following determinants.

65. $\begin{vmatrix} 1 & 1 & -5 \\ 3 & -7 & -4 \\ 4 & -6 & 9 \end{vmatrix}$ (Compare with your answer to Exercise 31.)

66. $\begin{vmatrix} -3 & 2 & -5 \\ 1 & 8 & 2 \\ 4 & -6 & 9 \end{vmatrix}$ (Compare with your answer to Exercise 32.)

67. $\begin{vmatrix} -3 & 2 & -1 & 3 \\ 4 & 1 & 5 & 2 \\ 17 & 2 & 2 & 8 \\ 13 & -4 & 10 & -11 \end{vmatrix}$

68. $\begin{vmatrix} -3 & 21 & 19 & 3 \\ 4 & 1 & 16 & 2 \\ 17 & 31 & 2 & 5 \\ 13 & -4 & 10 & 2 \end{vmatrix}$

SECTION **7.4** Matrix Algebra

Skills Objectives

- Add and subtract matrices.
- Perform scalar multiplication.
- Multiply two matrices.
- Find the inverse of a matrix.
- Solve systems of linear equations using inverse matrices.

Conceptual Objectives

- Understand what is meant by equal matrices.
- Understand why multiplication of some matrices is undefined.
- Realize that not every matrix has an inverse.

Introduction

Matrices are a convenient way to represent data.

Average Marriage Age on the Rise

A survey conducted by the Census Bureau (Shannon Reilly and Kevin Kepple: www.usatoday.com) found that the average marriage age in the United States is rising. This data can be represented by a 2×2 matrix $\begin{bmatrix} 22.8 & 26.9 \\ 20.3 & 25.3 \end{bmatrix}$, with two rows (gender) and two columns (year).

	1960	2002
Men	22.8	26.9
Women	20.3	25.3

Esbin-Anderson Photography/ Getty/Brand X Pictures

The U.S. War with Iraq

In a survey (The Harris Poll, number 60, August 20, 2004) adult Americans were asked the question, "Do you think the invasion of Iraq strengthened or weakened the war on terrorism?"

The results were along party lines:

	REPUBLICAN	**DEMOCRAT**
Strengthened the war on terrorism	75%	34%
Weakened the war on terrorism	19%	53%
Not sure	6%	13%

Main battle tank from Bravo Company of the 185th Armor conducting an area reconnaissance around Balad, Iraq, Sept. 4, 2004.

Photo of Staff Sgt. Shane A. Cuomo, US Air Force/Department of Defense/Still Media Record Center

This data can be represented with a 3×2 matrix $\begin{bmatrix} 75 & 34 \\ 19 & 53 \\ 6 & 13 \end{bmatrix}$, with three rows (opinions of war) and two columns (registered political party).

In general, a matrix with m rows and n columns is called an $m \times n$ (**m by n**) **matrix**. The number of elements in a matrix is equal to $m \cdot n$. When $m = n$ we say the matrix is **square**. The following table summarizes examples of matrices. Typically matrices are represented by capital letters such as A, B, C, and D.

MATRIX	NUMBER OF ROWS (m)	NUMBER OF COLUMNS (n)	NUMBER OF ELEMENTS
$A = \begin{bmatrix} -3 & 5 & 7 \\ -2 & 0 & 1 \end{bmatrix}$	2	3	6
$B = \begin{bmatrix} -4 & 3 \\ 6 & 11 \\ -17 & 5 \end{bmatrix}$	3	2	6
$C = \begin{bmatrix} 2 \\ 3 \\ -5 \\ 7 \end{bmatrix}$	4	1	4
$D = \begin{bmatrix} 1 & 0 & 3 & -7 \end{bmatrix}$	1	4	4

There is an entire field of study called **linear algebra** (matrix algebra) that treats matrices similar to functions and variables in traditional algebra. This section serves as an introduction to matrix algebra. We will discuss how to add, subtract, and multiply matrices and how to find the inverse of a matrix. We will use these operations as yet another way to solve systems of linear equations.

It is important to pay special attention to the **order** (number of rows and columns) of a matrix, because it determines if certain operations are defined. The following table summarizes the order requirement for operations to be defined for matrices A and B.

SYMBOL	OPERATION	ORDER REQUIREMENTS
$A = B$	Equality	The orders must be the same: $A(m \times n)$ and $B(m \times n)$.
$A \pm B$	Addition/subtraction	The orders must be the same: $A(m \times n)$ and $B(m \times n)$.
AB	Matrix multiplication	The orders must satisfy the relationship: $A(m \times n)$ and $B(n \times p)$, resulting in $AB(m \times p)$.
A^{-1}	Inverse matrix	Only square matrices have inverses: $A(n \times n)$ and $A^{-1}(n \times n)$

Equality of Matrices

Two matrices are equal if and only if they have the same order, $m \times n$, and all of their corresponding elements are equal.

DEFINITION   **EQUALITY OF MATRICES**

Two matrices, A and B, are **equal**, written as

$$A = B$$

if and only if *both* of the following are true:

- A and B have the same order $m \times n$, and
- every corresponding element is equal: $a_{ij} = b_{ij}$ for all $i = 1, 2, ..., m$ and $j = 1, 2, ..., n$.

EXAMPLE 1 Determining If Two Matrices Are Equal

State whether the given matrices are equal to $A = \begin{bmatrix} 3 & \frac{1}{10} \\ -3 & 0.5 \end{bmatrix}$

a. $B = \begin{bmatrix} 3 & 0.1 \\ \sqrt[3]{-27} & \frac{1}{2} \end{bmatrix}$ **b.** $C = \begin{bmatrix} 3 & 1 \\ -3 & 2 \end{bmatrix}$ **c.** $D = \begin{bmatrix} 3 & \frac{1}{10} \\ -3 & 1 \end{bmatrix}$

Solution:

a. $A = B$, because $\sqrt[3]{-27} = -3, 0.1 = \frac{1}{10}$, and $\frac{1}{2} = 0.5$.

b. $A \neq C$, because the elements in column 2 are not equal.

c. $A \neq D$, because $a_{22} \neq d_{22}$ $(0.5 \neq 1)$.

If A and B are both $m \times n$ matrices, then both A and B each have $m \cdot n$ elements, and the statement $A = B$ represents a system of $m \cdot n$ equations.

EXAMPLE 2 Using Equality of Matrices to Solve a System of Equations

Given that $\begin{bmatrix} 2x & y - 1 \\ 1 & 2 \end{bmatrix} = \begin{bmatrix} 6 & 7 \\ 1 & 2 \end{bmatrix}$, solve for x and y.

Solution:

Equate corresponding elements. $\begin{bmatrix} 2x & y - 1 \\ 1 & 2 \end{bmatrix} = \begin{bmatrix} 6 & 7 \\ 1 & 2 \end{bmatrix}$

$2x = 6, y - 1 = 7, 1 = 1,$ and $2 = 2$

Solve for x and y. $x = 3$ and $y = 8$

Matrix Addition and Subtraction

Two matrices, A and B, can be added or subtracted only if they have the same order. Suppose A and B are both $m \times n$; then the **sum**, $A + B$, is found by adding corresponding elements, $a_{ij} + b_{ij}$. The **difference**, $A - B$, is found by subtracting the elements in B from the corresponding elements in A, $a_{ij} - b_{ij}$.

EXAMPLE 3 Adding and Subtracting Matrices

Given that $A = \begin{bmatrix} -1 & 3 & 4 \\ -5 & 2 & 0 \end{bmatrix}$ and $B = \begin{bmatrix} 2 & 1 & -3 \\ 0 & -5 & 4 \end{bmatrix}$ find the following:

a. $A + B$ **b.** $A - B$

Solution:

a. Write the sum. $A + B = \begin{bmatrix} -1 & 3 & 4 \\ -5 & 2 & 0 \end{bmatrix} + \begin{bmatrix} 2 & 1 & -3 \\ 0 & -5 & 4 \end{bmatrix}$

Add corresponding elements. $\begin{bmatrix} -1 + 2 & 3 + 1 & 4 + (-3) \\ -5 + 0 & 2 + (-5) & 0 + 4 \end{bmatrix}$

Simplify.
$$\begin{bmatrix} 1 & 4 & 1 \\ -5 & -3 & 4 \end{bmatrix}$$

$$\begin{bmatrix} 1 & 4 & 1 \\ -5 & -3 & 4 \end{bmatrix}$$

b. Write the difference.
$$A - B = \begin{bmatrix} -1 & 3 & 4 \\ -5 & 2 & 0 \end{bmatrix} - \begin{bmatrix} 2 & 1 & -3 \\ 0 & -5 & 4 \end{bmatrix}$$

Subtract corresponding elements.
$$\begin{bmatrix} -1-2 & 3-1 & 4-(-3) \\ -5-0 & 2-(-5) & 0-4 \end{bmatrix}$$

Simplify.
$$\begin{bmatrix} -3 & 2 & 7 \\ -5 & 7 & -4 \end{bmatrix}$$

$$\begin{bmatrix} -3 & 2 & 7 \\ -5 & 7 & -4 \end{bmatrix}$$

TECHNOLOGY TIP
Enter the matrices as A and B.

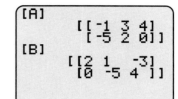

Now enter $[A] + [B]$ and $[A] - [B]$.

It is important to note that only matrices of the same order can be added or subtracted. For example, if $A = \begin{bmatrix} -1 & 3 & 4 \\ -5 & 2 & 0 \end{bmatrix}$ and $B = \begin{bmatrix} 5 & -3 \\ 12 & 1 \end{bmatrix}$, the sum and difference of these matrices is undefined because A is a 2×3 matrix and B is a 2×2 matrix.

■ **YOUR TURN** Perform the indicated matrix operations, if possible.

a. $\begin{bmatrix} -4 & 0 \\ 1 & 2 \end{bmatrix} - \begin{bmatrix} 2 & 3 \\ -4 & 0 \end{bmatrix}$ **b.** $\begin{bmatrix} 2 & 9 & 5 & -1 \end{bmatrix} + \begin{bmatrix} 0 \\ -3 \\ 4 \\ 2 \end{bmatrix}$

A matrix whose elements are all equal to 0 is called a **zero matrix**. The following are all examples of zero matrices:

2×2 square zero matrix: $\begin{bmatrix} 0 & 0 \\ 0 & 0 \end{bmatrix}$

3×2 zero matrix: $\begin{bmatrix} 0 & 0 \\ 0 & 0 \\ 0 & 0 \end{bmatrix}$

1×4 zero matrix: $\begin{bmatrix} 0 & 0 & 0 & 0 \end{bmatrix}$

If A, an $m \times n$ matrix, is added to the $m \times n$ zero matrix, the result is A.

$$A + 0 = A$$

For example, $\begin{bmatrix} 1 & -3 \\ 2 & 5 \end{bmatrix} + \begin{bmatrix} 0 & 0 \\ 0 & 0 \end{bmatrix} = \begin{bmatrix} 1 & -3 \\ 2 & 5 \end{bmatrix}$

■ **Answer: a.** $\begin{bmatrix} -6 & -3 \\ 5 & 2 \end{bmatrix}$ **b.** not defined

Because of this result, an $m \times n$ zero matrix is called the **additive identity** for $m \times n$ matrices. Similarly, for any matrix A, there exists an **additive inverse**, $-A$, such that every element of $-A$ is the negative of every element of A. For example, $A = \begin{bmatrix} 1 & -3 \\ 2 & 5 \end{bmatrix}$ and $-A = \begin{bmatrix} -1 & 3 \\ -2 & -5 \end{bmatrix}$, and adding these two matrices results in a zero matrix: $A + (-A) = 0$.

The same properties that hold for adding real numbers also hold for adding matrices, provided that addition of matrices is defined.

PROPERTIES OF MATRIX ADDITION

If A, B, and C are all $m \times n$ matrices and 0 is the $m \times n$ zero matrix then the following are true:

1. Commutative property of addition		$A + B = B + A$
2. Associative property of addition		$(A + B) + C = A + (B + C)$
3. Additive identity property		$A + 0 = A$
4. Additive inverse property		$A + (-A) = 0$

Scalar and Matrix Multiplication

TECHNOLOGY TIP

Enter matrices as A and B.

```
[A]
        [[-1 2]
         [-3 4]]
[B]
        [[0  1]
         [-2 3]]
```

Now enter $2A$, $-3B$, $2A - 3B$.

```
2[A]
        [[-2 4]
         [-6 8]]
-3[B]
        [[0 -3]
         [6 -9]]
■
```

```
2[A]-3[B]
        [[-2 1 ]
         [0  -1]]
■
```

There are two types of multiplication involving matrices: *scalar multiplication* and *matrix multiplication*. A **scalar** is any real number. *Scalar multiplication* is when a matrix is multiplied by a scalar, or real number, and is defined for all matrices. *Matrix multiplication* is when two matrices are multiplied and is only defined for certain matrix products, depending on the order of each matrix.

To multiply a matrix A by a scalar k, multiply every element in A by k.

$$3\begin{bmatrix} -1 & 0 & 4 \\ 7 & 5 & -2 \end{bmatrix} = \begin{bmatrix} 3(-1) & 3(0) & 3(4) \\ 3(7) & 3(5) & 3(-2) \end{bmatrix} = \begin{bmatrix} -3 & 0 & 12 \\ 21 & 15 & -6 \end{bmatrix}$$

EXAMPLE 4 Multiplying a Matrix by a Scalar

Given that $A = \begin{bmatrix} -1 & 2 \\ -3 & 4 \end{bmatrix}$ and $B = \begin{bmatrix} 0 & 1 \\ -2 & 3 \end{bmatrix}$ perform the operations:

a. $2A$ **b.** $-3B$ **c.** $2A - 3B$

Solution:

a. Write the scalar multiplication. $2A = 2\begin{bmatrix} -1 & 2 \\ -3 & 4 \end{bmatrix}$

Multiply all elements of A by 2. $\begin{bmatrix} 2(-1) & 2(2) \\ 2(-3) & 2(4) \end{bmatrix}$

Simplify. $\begin{bmatrix} -2 & 4 \\ -6 & 8 \end{bmatrix}$

b. Write the scalar multiplication.

$$-3B = -3 \begin{bmatrix} 0 & 1 \\ -2 & 3 \end{bmatrix}$$

Multiply all elements of B by -3.

$$\begin{bmatrix} -3(0) & -3(1) \\ -3(-2) & -3(3) \end{bmatrix}$$

Simplify.

$$\begin{bmatrix} 0 & -3 \\ 6 & -9 \end{bmatrix}$$

c. Add

$$2A - 3B = 2A + -3B$$

$$\begin{bmatrix} -2 & 4 \\ -6 & 8 \end{bmatrix} + \begin{bmatrix} 0 & -3 \\ 6 & -9 \end{bmatrix}$$

Add corresponding elements.

$$\begin{bmatrix} -2 + 0 & 4 + (-3) \\ -6 + 6 & 8 + (-9) \end{bmatrix}$$

Simplify.

$$\begin{bmatrix} -2 & 1 \\ 0 & -1 \end{bmatrix}$$

■ **YOUR TURN** For the matrices A and B given in Example 4, find $-5A + 2B$.

Scalar multiplication is straightforward because it is defined for all matrices and is performed by multiplying every element in the matrix by the scalar. Addition of matrices is also an element-by-element operation. Matrix multiplication, on the other hand, is not as straightforward because we *do not multiply the corresponding elements* and it is not defined for all matrices. Matrices are multiplied using a row-by-column method.

We will start by multiplying two 2×2 matrices and then later define matrix multiplication in general.

$$
AB = \begin{bmatrix} a_{11} & a_{12} \\ a_{21} & a_{22} \end{bmatrix}\begin{bmatrix} b_{11} & b_{12} \\ b_{21} & b_{22} \end{bmatrix} = \begin{bmatrix} \overbrace{a_{11}b_{11} + a_{12}b_{21}}^{\substack{\text{Row 1 of } A \text{ Times} \\ \text{Column 1 of } B}} & \overbrace{a_{11}b_{12} + a_{12}b_{22}}^{\substack{\text{Row 1 of } A \text{ Times} \\ \text{Column 2 of } B}} \\ \underbrace{a_{21}b_{11} + a_{22}b_{21}}_{\substack{\text{Row 2 of } A \text{ Times} \\ \text{Column 1 of } B}} & \underbrace{a_{21}b_{12} + a_{22}b_{22}}_{\substack{\text{Row 2 of } A \text{ Times} \\ \text{Column 2 of } B}} \end{bmatrix}
$$

Let us focus on only the entry in row 1/column 1 of the resulting matrix: $a_{11}b_{11} + a_{12}b_{21}$

Multiply the **first** element in the **row** by the **first** element in the **column**.

$$\begin{bmatrix} a_{11} & a_{12} \end{bmatrix}\begin{bmatrix} b_{11} \\ b_{21} \end{bmatrix} = \begin{bmatrix} a_{11}b_{11} & + & a_{12}b_{21} \\ \text{Add } \textbf{firsts} \text{ and } \textbf{seconds} \end{bmatrix}$$

Multiply the **second** element in the **row** by the **second** element in the **column**.

Continue the procedure for the remaining three combinations.

■ **Answer:** $-5A + 2B = \begin{bmatrix} 5 & -8 \\ 11 & -14 \end{bmatrix}$

EXAMPLE 5 Multiplication of Two 2 × 2 Matrices

Given $A = \begin{bmatrix} 1 & 2 \\ 3 & 4 \end{bmatrix}$ and $B = \begin{bmatrix} 5 & 6 \\ 7 & 8 \end{bmatrix}$, find AB.

COMMON MISTAKE

Do not multiply element by element.

CORRECT

Write the product of the two matrices, A and B.

$$AB = \begin{bmatrix} 1 & 2 \\ 3 & 4 \end{bmatrix}\begin{bmatrix} 5 & 6 \\ 7 & 8 \end{bmatrix}$$

Perform the row-by-column multiplication.

$$AB = \begin{bmatrix} (1)(5) + (2)(7) & (1)(6) + (2)(8) \\ (3)(5) + (4)(7) & (3)(6) + (4)(8) \end{bmatrix}$$

Simplify.

$$AB = \begin{bmatrix} 19 & 22 \\ 43 & 50 \end{bmatrix}$$

INCORRECT

Write the product of the two matrices, A and B.

$$AB = \begin{bmatrix} 1 & 2 \\ 3 & 4 \end{bmatrix}\begin{bmatrix} 5 & 6 \\ 7 & 8 \end{bmatrix}$$

Multiply corresponding elements.

ERROR $AB \neq \begin{bmatrix} (1)(5) & (2)(6) \\ (3)(7) & (4)(8) \end{bmatrix}$

Simplify.

INCORRECT $AB \neq \begin{bmatrix} 5 & 12 \\ 21 & 32 \end{bmatrix}$

CAUTION

Do not multiply element by element. Matrices are multiplied using a row-by-column method.

■ **YOUR TURN** For matrices A and B given in Example 5, find BA.

Compare the products obtained in Example 5 and the above Your Turn. Note that $AB \neq BA$. Therefore, there is no commutative property for matrix multiplication.

Let's define the general matrix multiplication procedure for matrices. For the product of two matrices to be defined, **the number of columns in the first matrix must equal the number of rows in the second matrix**.

Matrix: A B AB

Order: $m \times n$ $n \times p$ $m \times p$

DEFINITION   MATRIX MULTIPLICATION

If A is an $m \times n$ matrix and B is an $n \times p$ matrix, then their product AB is an $m \times p$ matrix whose elements are found according to:

$$(ab)_{ij} = a_{i1}b_{1j} + a_{i2}b_{2j} + \ldots + a_{in}b_{nj}$$

In other words, the element in row i/column j of AB is the sum of the products of the corresponding elements in the ith row of A and the jth column of B.

■ **Answer:** $BA = \begin{bmatrix} 5 & 6 \\ 7 & 8 \end{bmatrix}\begin{bmatrix} 1 & 2 \\ 3 & 4 \end{bmatrix} = \begin{bmatrix} 23 & 34 \\ 31 & 46 \end{bmatrix}$

The reason that the number of rows (n) in A must be equal to the number of columns (n) in B is that, for elements to correspond, there must be an equal number.

EXAMPLE 6 Determining If the Product of Two Matrices Is Defined

Given the matrices:

$$A = \begin{bmatrix} 1 & -2 & 0 \\ 5 & -1 & 3 \end{bmatrix} \quad B = \begin{bmatrix} 2 & 3 \\ 0 & 7 \\ 4 & 9 \end{bmatrix} \quad C = \begin{bmatrix} 6 & -1 \\ 5 & 2 \end{bmatrix} \quad D = \begin{bmatrix} -3 & -2 \end{bmatrix}$$

State whether the following products exist. If the product exists, state the order of the product matrix.

a. AB **b.** AC **c.** BC **d.** CD **e.** DC

Solution:

Label the order of each matrix: $A_{2 \times 3}$, $B_{3 \times 2}$, $C_{2 \times 2}$, and $D_{1 \times 2}$.

a. AB is defined, because A has 3 columns and B has 3 rows. AB is order 2×2.

b. AC is not defined, because A has 3 columns and C has 2 rows.

c. BC is defined, because B has 2 columns and C has 2 rows. BC is order 3×2.

d. CD is not defined, because C has 2 columns and D has 1 row.

e. DC is defined, because D has 2 columns and C has 2 rows. DC is order 1×2.

■ **YOUR TURN** For the matrices given in Example 6, state whether the following products exist. If the product exists, state the order of the product matrix.

a. DA **b.** CB **c.** BA

EXAMPLE 7 Multiplying Matrices Resulting in a Square Matrix

For $A = \begin{bmatrix} -1 & 2 & -3 \\ -2 & 0 & 4 \end{bmatrix}$ and $B = \begin{bmatrix} 2 & 0 \\ 1 & 3 \\ -1 & -2 \end{bmatrix}$ find AB.

Solution:

Since A is order 2×3 and B is order 3×2, the product AB is defined and has order 2×2.

Write the product of the two matrices. $\quad AB = \begin{bmatrix} -1 & 2 & -3 \\ -2 & 0 & 4 \end{bmatrix} \begin{bmatrix} 2 & 0 \\ 1 & 3 \\ -1 & -2 \end{bmatrix}$

TECHNOLOGY TIP

Enter the matrices as A and B and calculate AB.

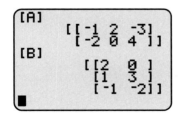

```
[A]
      [[-1  2  -3]
       [-2  0   4]]
[B]
      [[2   0 ]
       [1   3 ]
       [-1  -2]]
■
```

```
[A][B]
      [[3   12]
       [-8  -8]]
```

Perform the row-by-column multiplication.

$$AB = \begin{bmatrix} (-1)(2) + (2)(1) + (-3)(-1) & (-1)(0) + (2)(3) + (-3)(-2) \\ (-2)(2) + (0)(1) + (4)(-1) & (-2)(0) + (0)(3) + (4)(-2) \end{bmatrix}$$

Simplify.
$$AB = \begin{bmatrix} 3 & 12 \\ -8 & -8 \end{bmatrix}$$

■ **YOUR TURN** For $A = \begin{bmatrix} 1 & 0 & 2 \\ -3 & -1 & 4 \end{bmatrix}$ and $B = \begin{bmatrix} 0 & -1 \\ 1 & 2 \\ 0 & -2 \end{bmatrix}$, find AB.

EXAMPLE 8 Multiplying Matrices Resulting in a Nonsquare Matrix

For $A = \begin{bmatrix} 1 & 0 & 3 \\ -2 & 5 & -1 \end{bmatrix}$ and $B = \begin{bmatrix} -2 & 0 & 1 \\ -3 & -1 & 4 \\ 0 & 2 & 5 \end{bmatrix}$, find AB.

Solution:

Since A is order 2×3 and B is order 3×3, the product AB is defined and has order 2×3.

Write the product of the two matrices.
$$AB = \begin{bmatrix} 1 & 0 & 3 \\ -2 & 5 & -1 \end{bmatrix} \begin{bmatrix} -2 & 0 & 1 \\ -3 & -1 & 4 \\ 0 & 2 & 5 \end{bmatrix}$$

Perform the row-by-column multiplication.

$$AB = \begin{bmatrix} A\ \text{row1}/B\ \text{column1} & A\ \text{row1}/B\ \text{column2} & A\ \text{row1}/B\ \text{column3} \\ A\ \text{row2}/B\ \text{column1} & A\ \text{row2}/B\ \text{column2} & A\ \text{row2}/B\ \text{column3} \end{bmatrix}$$

$$AB = \begin{bmatrix} (1)(-2) + (0)(-3) + (3)(0) & (1)(0) + (0)(-1) + (3)(2) & (1)(1) + (0)(4) + (3)(5) \\ (-2)(-2) + (5)(-3) + (-1)(0) & (-2)(0) + (5)(-1) + (-1)(2) & (-2)(1) + (5)(4) + (-1)(5) \end{bmatrix}$$

Simplify.
$$AB = \begin{bmatrix} -2 & 6 & 16 \\ -11 & -7 & 13 \end{bmatrix}$$

✓**CONCEPT CHECK** Given $A = \begin{bmatrix} 1 \\ 2 \\ 3 \end{bmatrix}$ and $B = \begin{bmatrix} 4 & 5 \end{bmatrix}$ explain why BA does not exist. What is the order of AB?

■ **YOUR TURN** Using A and B from the Concept Check, find AB.

■ **Answer:** $AB = \begin{bmatrix} 0 & -5 \\ -1 & -7 \end{bmatrix}$ ■ **Answer:** $AB = \begin{bmatrix} 4 & 5 \\ 8 & 10 \\ 12 & 15 \end{bmatrix}$

Although we have shown repeatedly that there is no commutative property for matrices, matrices do have an associative property of multiplication and a distributive property of multiplication similar to that for real numbers.

Finding the Inverse of a Matrix

Recall from Chapter 3 (Functions and Their Graphs) that not all functions have inverses. Only one-to-one functions have inverses. Similarly, not all matrices have *inverses*. Only square matrices can have inverses, and as you will soon see, even some square matrices do not have inverses. And recall that the composition of a function and its inverse yield the identity (*x*) and vice versa (Sections 3.4 and 3.5). Matrices behave similarly. Let's start by defining the *identity matrix*. A square matrix with 1's along the **diagonal** (a_{ii}) and zeros for all other elements is called the **multiplicative identity matrix, I_n.**

$$I_2 = \begin{bmatrix} 1 & 0 \\ 0 & 1 \end{bmatrix} \qquad I_3 = \begin{bmatrix} 1 & 0 & 0 \\ 0 & 1 & 0 \\ 0 & 0 & 1 \end{bmatrix} \qquad I_4 = \begin{bmatrix} 1 & 0 & 0 & 0 \\ 0 & 1 & 0 & 0 \\ 0 & 0 & 1 & 0 \\ 0 & 0 & 0 & 1 \end{bmatrix}$$

Recall that a real number multiplied by 1 is itself: $a \cdot 1 = a$. Similarly, a matrix multiplied by the appropriate identity matrix should result in itself.

IDENTITY PROPERTY

Let A be an $m \times n$ matrix, then

$$AI_n = A \text{ and } I_m A = A$$

If A is a square matrix, then $AI_n = I_n A = A$.

EXAMPLE 9 Multiplying a Matrix by the Multiplicative Identity Matrix, I_n

For $A = \begin{bmatrix} -2 & 4 & 1 \\ 3 & 7 & -1 \end{bmatrix}$, find $I_2 A$.

Solution:

Write the two matrices. $A = \begin{bmatrix} -2 & 4 & 1 \\ 3 & 7 & -1 \end{bmatrix} \qquad I_2 = \begin{bmatrix} 1 & 0 \\ 0 & 1 \end{bmatrix}$

Find the product: $I_2 A$. $I_2 A = \begin{bmatrix} 1 & 0 \\ 0 & 1 \end{bmatrix} \begin{bmatrix} -2 & 4 & 1 \\ 3 & 7 & -1 \end{bmatrix}$

$$= \begin{bmatrix} (1)(-2) + (0)(3) & (1)(4) + (0)(7) & (1)(1) + (0)(-1) \\ (0)(-2) + (1)(3) & (0)(4) + (1)(7) & (0)(1) + (1)(-1) \end{bmatrix}$$

$$= \begin{bmatrix} -2 & 4 & 1 \\ 3 & 7 & -1 \end{bmatrix} = A$$

■ **YOUR TURN:** For A in Example 9, find AI_3.

The identity matrix, I_n, will assist us in developing the concept of an inverse of a matrix. Recall the property of real numbers that $a \cdot \frac{1}{a} = 1$. The term $\frac{1}{a}$ is the multiplicative inverse of a, and their product is 1. Note that we could also write this property as $a \cdot a^{-1} = 1$. We will define the *multiplicative inverse of a matrix* the same way.

DEFINITION INVERSE OF A MATRIX

Let A be a square $n \times n$ matrix. If there exists another square $n \times n$ matrix, A^{-1}, such that the following are true:

$$AA^{-1} = I_n \text{ and } A^{-1}A = I_n$$

then A^{-1}, stated as "A inverse," is the **multiplicative inverse** of A.

Only square matrices can have inverses, but as you will soon see, not all square matrices have inverses.

EXAMPLE 10 Multiplying a Matrix by Its Inverse

Verify that the inverse of $A = \begin{bmatrix} 1 & 3 \\ 2 & 5 \end{bmatrix}$ is $A^{-1} = \begin{bmatrix} -5 & 3 \\ 2 & -1 \end{bmatrix}$.

Solution:

Show that both $AA^{-1} = I_2$ and $A^{-1}A = I_2$.

Find the product: AA^{-1}.
$$\begin{bmatrix} 1 & 3 \\ 2 & 5 \end{bmatrix}\begin{bmatrix} -5 & 3 \\ 2 & -1 \end{bmatrix}$$

$$\begin{bmatrix} (1)(-5) + (3)(2) & (1)(3) + (3)(-1) \\ (2)(-5) + (5)(2) & (2)(3) + (5)(-1) \end{bmatrix}$$

$$\begin{bmatrix} 1 & 0 \\ 0 & 1 \end{bmatrix} = I_2$$

Find the product: $A^{-1}A$.
$$\begin{bmatrix} -5 & 3 \\ 2 & -1 \end{bmatrix}\begin{bmatrix} 1 & 3 \\ 2 & 5 \end{bmatrix}$$

$$\begin{bmatrix} (-5)(1) + (3)(2) & (-5)(3) + (3)(5) \\ (2)(1) + (-1)(2) & (2)(3) + (-1)(5) \end{bmatrix}$$

$$\begin{bmatrix} 1 & 0 \\ 0 & 1 \end{bmatrix} = I_2$$

■ **YOUR TURN** Verify that the inverse of $A = \begin{bmatrix} 1 & 4 \\ 2 & 9 \end{bmatrix}$ is $A^{-1} = \begin{bmatrix} 9 & -4 \\ -2 & 1 \end{bmatrix}$.

■ **Answer:** $AI_3 = \begin{bmatrix} -2 & 4 & 1 \\ 3 & 7 & -1 \end{bmatrix} = A$ ■ **Answer:** $AA^{-1} = A^{-1}A = I_2$

Now that we can show that two matrices are inverses of one another, let us now describe the process for how to find an inverse, if it exists. If an inverse exists, A^{-1}, then the matrix, A, is said to be **nonsingular**. If the inverse does not exist, then the matrix, A, is said to be **singular**.

Let $A = \begin{bmatrix} 1 & -1 \\ 2 & -3 \end{bmatrix}$ and the inverse be $A^{-1} = \begin{bmatrix} w & x \\ y & z \end{bmatrix}$, where $w, x, y,$ and z are values to be determined. A matrix and its inverse must satisfy the identity $AA^{-1} = I_2$.

WORDS	**MATH**
The product of a matrix and its inverse is the identity matrix.	$\begin{bmatrix} 1 & -1 \\ 2 & -3 \end{bmatrix}\begin{bmatrix} w & x \\ y & z \end{bmatrix} = \begin{bmatrix} 1 & 0 \\ 0 & 1 \end{bmatrix}$
Multiply the two matrices on the left.	$\begin{bmatrix} w - y & x - z \\ 2w - 3y & 2x - 3z \end{bmatrix} = \begin{bmatrix} 1 & 0 \\ 0 & 1 \end{bmatrix}$
Equate corresponding matrix elements.	$\begin{bmatrix} w - y & x - z \\ 2w - 3y & 2x - 3z \end{bmatrix} = \begin{bmatrix} 1 & 0 \\ 0 & 1 \end{bmatrix}$
	$\begin{array}{c} w - y = 1 \\ 2w - 3y = 0 \end{array}$ and $\begin{array}{c} x - z = 0 \\ 2x - 3z = 1 \end{array}$

Notice there are two systems of equations, which can be solved by several methods (elimination, substitution, graphing, determinants, or augmented matrices). We will find that $w = 3, x = -1, y = 2, z = -1$. Therefore we know the inverse is $A^{-1} = \begin{bmatrix} 3 & -1 \\ 2 & -1 \end{bmatrix}$. But let us use augmented matrices instead, in order to develop the general procedure. Write the two systems of equations as two augmented matrices.

$$\begin{array}{cc} w & y \end{array}$$
$$\left[\begin{array}{cc|c} 1 & -1 & 1 \\ 2 & -3 & 0 \end{array}\right] \qquad \begin{array}{cc} x & z \end{array}$$
$$\left[\begin{array}{cc|c} 1 & -1 & 0 \\ 2 & -3 & 1 \end{array}\right]$$

Since the left side of each of the augmented matrices is the same, we can combine these two matrices into one matrix, thereby simultaneously solving both systems of equations.

$$\left[\begin{array}{cc|cc} 1 & -1 & 1 & 0 \\ 2 & -3 & 0 & 1 \end{array}\right]$$

Notice that the right side of the vertical line is the identity matrix, I_2. Using Gauss–Jordan elimination, reduce the matrix on the left.

$$R_2 - 2R_1 \rightarrow R_2 \qquad \left[\begin{array}{cc|cc} 1 & -1 & 1 & 0 \\ 0 & -1 & -2 & 1 \end{array}\right]$$

$$-R_2 \rightarrow R_2 \qquad \left[\begin{array}{cc|cc} 1 & -1 & 1 & 0 \\ 0 & 1 & 2 & -1 \end{array}\right]$$

$$R_1 + R_2 \rightarrow R_1 \qquad \left[\begin{array}{cc|cc} 1 & 0 & \mathbf{3} & \mathbf{-1} \\ 0 & 1 & \mathbf{2} & \mathbf{-1} \end{array}\right]$$

The matrix on the right of the vertical line is the inverse: $A^{-1} = \begin{bmatrix} 3 & -1 \\ 2 & -1 \end{bmatrix}$

PROCEDURE FOR FINDING THE INVERSE OF A NONSINGULAR MATRIX

To find the inverse of an $n \times n$ nonsingular matrix A:

Step 1: Form the matrix $[A|I_n]$.

Step 2: Use row operations to transform this matrix to $[I_n|A^{-1}]$. This is done by using Gauss–Jordan elimination to reduce A to the identity matrix, I_n. If this is not possible, then A is a singular matrix and no inverse exists.

Step 3: Verify the result by showing that $AA^{-1} = I_n$ and $A^{-1}A = I_n$.

EXAMPLE 11 Finding the Inverse of a Matrix

Find the inverse of $A = \begin{bmatrix} 1 & 2 & -1 \\ 0 & 1 & -1 \\ -1 & 0 & -2 \end{bmatrix}$.

Solution:

STEP 1 Form the matrix $[A|I_3]$.

$$\left[\begin{array}{ccc|ccc} 1 & 2 & -1 & 1 & 0 & 0 \\ 0 & 1 & -1 & 0 & 1 & 0 \\ -1 & 0 & -2 & 0 & 0 & 1 \end{array}\right]$$

STEP 2 Use Gauss–Jordan elimination to reduce A.

$R_3 + R_1 \to R_3$
$$\left[\begin{array}{ccc|ccc} 1 & 2 & -1 & 1 & 0 & 0 \\ 0 & 1 & -1 & 0 & 1 & 0 \\ 0 & 2 & -3 & 1 & 0 & 1 \end{array}\right]$$

$R_3 - 2R_2 \to R_3$
$$\left[\begin{array}{ccc|ccc} 1 & 2 & -1 & 1 & 0 & 0 \\ 0 & 1 & -1 & 0 & 1 & 0 \\ 0 & 0 & -1 & 1 & -2 & 1 \end{array}\right]$$

$-R_3 \to R_3$
$$\left[\begin{array}{ccc|ccc} 1 & 2 & -1 & 1 & 0 & 0 \\ 0 & 1 & -1 & 0 & 1 & 0 \\ 0 & 0 & 1 & -1 & 2 & -1 \end{array}\right]$$

$R_2 + R_3 \to R_2$
$R_1 + R_3 \to R_1$
$$\left[\begin{array}{ccc|ccc} 1 & 2 & 0 & 0 & 2 & -1 \\ 0 & 1 & 0 & -1 & 3 & -1 \\ 0 & 0 & 1 & -1 & 2 & -1 \end{array}\right]$$

$R_1 - 2R_2 \to R_1$
$$\left[\begin{array}{ccc|ccc} 1 & 0 & 0 & 2 & -4 & 1 \\ 0 & 1 & 0 & -1 & 3 & -1 \\ 0 & 0 & 1 & -1 & 2 & -1 \end{array}\right]$$

Identify the inverse.
$$A^{-1} = \begin{bmatrix} 2 & -4 & 1 \\ -1 & 3 & -1 \\ -1 & 2 & -1 \end{bmatrix}$$

STEP 3 Check.

$$AA^{-1} = \begin{bmatrix} 1 & 2 & -1 \\ 0 & 1 & -1 \\ -1 & 0 & -2 \end{bmatrix}\begin{bmatrix} 2 & -4 & 1 \\ -1 & 3 & -1 \\ -1 & 2 & -1 \end{bmatrix} = \begin{bmatrix} 1 & 0 & 0 \\ 0 & 1 & 0 \\ 0 & 0 & 1 \end{bmatrix} = I_3$$

$$A^{-1}A = \begin{bmatrix} 2 & -4 & 1 \\ -1 & 3 & -1 \\ -1 & 2 & -1 \end{bmatrix}\begin{bmatrix} 1 & 2 & -1 \\ 0 & 1 & -1 \\ -1 & 0 & -2 \end{bmatrix} = \begin{bmatrix} 1 & 0 & 0 \\ 0 & 1 & 0 \\ 0 & 0 & 1 \end{bmatrix} = I_3$$

TECHNOLOGY TIP

A graphing calculator can be used to find the inverse of A.

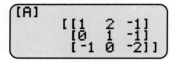

```
[A]
    [[1   2  -1]
     [0   1  -1]
     [-1  0  -2]]
```

```
[A]⁻¹
    [[2   -4  1 ]
     [-1  3   -1]
     [-1  2   -1]]
```

■ **YOUR TURN**   Find the inverse of $A = \begin{bmatrix} 1 & 1 & 0 \\ -1 & 0 & 1 \\ 2 & 0 & -1 \end{bmatrix}$.

This Gauss–Jordan elimination procedure works for all square nonsingular matrices. However, in practice it is only used for $n \geq 3$. For the special case of a 2×2 matrix, there is a formula (that will be derived in the exercises) for finding the inverse.

Let $A = \begin{bmatrix} a & b \\ c & d \end{bmatrix}$ represent any 2×2 matrix. If the *determinant* of A is nonzero, $ad - bc \neq 0$, then the inverse is given by $A^{-1} = \dfrac{1}{ad - bc} \begin{bmatrix} d & -b \\ -c & a \end{bmatrix}$.

EXAMPLE 12 Finding the Inverse of a 2 × 2 Matrix

Find the inverse of $A = \begin{bmatrix} 3 & 1 \\ 5 & 2 \end{bmatrix}$.

Solution:

Write the formula for A^{-1}.
$$A^{-1} = \frac{1}{ad - bc} \begin{bmatrix} d & -b \\ -c & a \end{bmatrix}$$

Substitute $a = 3, b = 1$, $c = 5, d = 2$ into formula.
$$\frac{1}{(3)(2) - (1)(5)} \begin{bmatrix} 2 & -(1) \\ -(5) & 3 \end{bmatrix}$$

Simplify.
$$\begin{bmatrix} 2 & -1 \\ -5 & 3 \end{bmatrix}$$

Check.
$$AA^{-1} = \begin{bmatrix} 3 & 1 \\ 5 & 2 \end{bmatrix} \begin{bmatrix} 2 & -1 \\ -5 & 3 \end{bmatrix} = \begin{bmatrix} 1 & 0 \\ 0 & 1 \end{bmatrix}$$
$$A^{-1}A = \begin{bmatrix} 2 & -1 \\ -5 & 3 \end{bmatrix} \begin{bmatrix} 3 & 1 \\ 5 & 2 \end{bmatrix} = \begin{bmatrix} 1 & 0 \\ 0 & 1 \end{bmatrix}$$

■ **YOUR TURN**   Find the inverse of $A = \begin{bmatrix} 2 & 3 \\ 5 & 8 \end{bmatrix}$.

Solving Systems of Linear Equations Using Matrix Algebra

We can solve systems of linear equations using matrix algebra. We will use a system of three equations and three variables to demonstrate the procedure. However, it can be extended to any square system.

LINEAR SYSTEM OF EQUATIONS

$$a_1 x + b_1 y + c_1 z = d_1$$
$$a_2 x + b_2 y + c_2 z = d_2$$
$$a_3 x + b_3 y + c_3 z = d_3$$

MATRIX FORM OF THE SYSTEM

$$\underbrace{\begin{bmatrix} a_1 & b_1 & c_1 \\ a_2 & b_2 & c_2 \\ a_3 & b_3 & c_3 \end{bmatrix}}_{A} \underbrace{\begin{bmatrix} x \\ y \\ z \end{bmatrix}}_{X} = \underbrace{\begin{bmatrix} d_1 \\ d_2 \\ d_3 \end{bmatrix}}_{B}$$

■ **Answer:** $A^{-1} = \begin{bmatrix} 8 & -3 \\ -5 & 2 \end{bmatrix}$

■ **Answer:** $A^{-1} = \begin{bmatrix} 0 & 1 & 1 \\ 1 & -1 & -1 \\ 0 & 2 & 1 \end{bmatrix}$

The matrix, A, contains the coefficients of the system, the matrix X contains the variables of the system, and the matrix B contains the constants of the system. Recall that a system of linear equations either has a unique solution, no solution, or infinitely many solutions. If a system has a unique solution, it can be found using the following procedure.

SOLVING A SYSTEM OF LINEAR EQUATIONS USING MATRIX ALGEBRA

If a system of linear equations is represented by $AX = B$:

Step 1: Multiply both sides of the equation by A^{-1}. $A^{-1}AX = A^{-1}B$

Step 2: Note that $A^{-1}A = I$. $IX = A^{-1}B$

Step 3: Note that $IX = X$. $X = A^{-1}B$

EXAMPLE 13 Solving a System of Linear Equations Using Inverses

Solve the system of equations using inverses:

$$x + y + z = 2$$
$$x + z = 1$$
$$x - y - z = -4$$

Solution:

Write the system in matrix form. $AX = B$

$$A = \begin{bmatrix} 1 & 1 & 1 \\ 1 & 0 & 1 \\ 1 & -1 & -1 \end{bmatrix} \quad X = \begin{bmatrix} x \\ y \\ z \end{bmatrix} \quad B = \begin{bmatrix} 2 \\ 1 \\ -4 \end{bmatrix}$$

Find the inverse of A.

Form the matrix, $[A | I_3]$.

$$\begin{bmatrix} 1 & 1 & 1 & | & 1 & 0 & 0 \\ 1 & 0 & 1 & | & 0 & 1 & 0 \\ 1 & -1 & -1 & | & 0 & 0 & 1 \end{bmatrix}$$

$$\begin{matrix} R_2 - R_1 \rightarrow R_2 \\ R_3 - R_1 \rightarrow R_3 \end{matrix} \quad \begin{bmatrix} 1 & 1 & 1 & | & 1 & 0 & 0 \\ 0 & -1 & 0 & | & -1 & 1 & 0 \\ 0 & -2 & -2 & | & -1 & 0 & 1 \end{bmatrix}$$

$$-R_2 \rightarrow R_2 \quad \begin{bmatrix} 1 & 1 & 1 & | & 1 & 0 & 0 \\ 0 & 1 & 0 & | & 1 & -1 & 0 \\ 0 & -2 & -2 & | & -1 & 0 & 1 \end{bmatrix}$$

$$R_3 + 2R_2 \rightarrow R_3 \quad \begin{bmatrix} 1 & 1 & 1 & | & 1 & 0 & 0 \\ 0 & 1 & 0 & | & 1 & -1 & 0 \\ 0 & 0 & -2 & | & 1 & -2 & 1 \end{bmatrix}$$

$$-\tfrac{1}{2}R_3 \rightarrow R_3 \quad \begin{bmatrix} 1 & 1 & 1 & | & 1 & 0 & 0 \\ 0 & 1 & 0 & | & 1 & -1 & 0 \\ 0 & 0 & 1 & | & -\frac{1}{2} & 1 & -\frac{1}{2} \end{bmatrix}$$

$$R_1 - R_3 \rightarrow R_1 \quad \begin{bmatrix} 1 & 1 & 0 & \bigm| & \frac{3}{2} & -1 & \frac{1}{2} \\ 0 & 1 & 0 & \bigm| & 1 & -1 & 0 \\ 0 & 0 & 1 & \bigm| & -\frac{1}{2} & 1 & -\frac{1}{2} \end{bmatrix}$$

$$R_1 - R_2 \rightarrow R_1 \quad \begin{bmatrix} 1 & 0 & 0 & \bigm| & \frac{1}{2} & 0 & \frac{1}{2} \\ 0 & 1 & 0 & \bigm| & 1 & -1 & 0 \\ 0 & 0 & 1 & \bigm| & -\frac{1}{2} & 1 & -\frac{1}{2} \end{bmatrix}$$

Identify the inverse.

$$A^{-1} = \begin{bmatrix} \frac{1}{2} & 0 & \frac{1}{2} \\ 1 & -1 & 0 \\ -\frac{1}{2} & 1 & -\frac{1}{2} \end{bmatrix}$$

The solution to the system is $X = A^{-1}B$.

$$X = A^{-1}B = \begin{bmatrix} \frac{1}{2} & 0 & \frac{1}{2} \\ 1 & -1 & 0 \\ -\frac{1}{2} & 1 & -\frac{1}{2} \end{bmatrix} \begin{bmatrix} 2 \\ 1 \\ -4 \end{bmatrix}$$

Simplify.

$$X = \begin{bmatrix} x \\ y \\ z \end{bmatrix} = \begin{bmatrix} -1 \\ 1 \\ 2 \end{bmatrix}$$

$$x = -1, y = 1, z = 2$$

✓**CONCEPT CHECK** If $A = \begin{bmatrix} 1 & 1 & -1 \\ 0 & 1 & 1 \\ 2 & 3 & 1 \end{bmatrix}$ find A^{-1}.

■ **YOUR TURN** Solve the system of equations using inverses.

$$x + y - z = 3$$
$$y + z = 1$$
$$2x + 3y + z = 5$$

SECTION 7.4 SUMMARY

Matrices are a convenient way of expressing data. They also lend another technique for solving systems of linear equations, particularly systems with many equations and variables. Equality, addition, and subtraction of matrices are defined for matrices of the same order, *rows × columns*. Scalar multiplication is defined for all matrices, while matrix multiplication is only defined if the number of columns in the first matrix is equal to the number of rows in the second matrix. The same properties that hold for multiplication of real numbers hold for multiplication of matrices with one exception: there is no commutative property, that is $AB \neq BA$, so particular attention must be paid to order of operations. Only square matrices have inverses. A procedure was outlined for finding the inverse of a square matrix that is valid for all orders, and a particular formula for calculating inverses of 2×2 matrices was given. Inverses can be used to solve systems of equations that have an equal number of equations and variables.

■ **Answer:** $x = 0, y = 2, z = -1$

SECTION 7.4 EXERCISES

■ SKILLS

In Exercises 1–20, calculate the given expression if possible.

$$A = \begin{bmatrix} -1 & 3 & 0 \\ 2 & 4 & 1 \end{bmatrix} \quad B = \begin{bmatrix} 0 & 2 & 1 \\ 3 & -2 & 4 \end{bmatrix} \quad C = \begin{bmatrix} 0 & 1 \\ 2 & -1 \\ 3 & 1 \end{bmatrix} \quad D = \begin{bmatrix} 3 & -1 \\ 2 & 5 \end{bmatrix}$$

1. $A + B$

2. $A - B$

3. $B + C$

4. $C + D$

5. $2A$

6. $-5D$

7. $2A + 3B$

8. $2B - 3A$

9. CD

10. BC

11. AB

12. CA

13. $(A + B)C$

14. $D(A + B)$

15. $BC + CD$

16. $DA + BC$

17. $C(I_2 + 0)$ where 0 is the 2×2 zero matrix.

18. $(A - B)CI_2$

19. BB

20. DD

In Exercises 21–30, determine if B is the multiplicative inverse of A using $AA^{-1} = I$.

21. $A = \begin{bmatrix} 8 & -11 \\ -5 & 7 \end{bmatrix}$ $\quad B = \begin{bmatrix} 7 & 11 \\ 5 & 8 \end{bmatrix}$

22. $A = \begin{bmatrix} 7 & -9 \\ -3 & 4 \end{bmatrix}$ $\quad B = \begin{bmatrix} 4 & 9 \\ 3 & 7 \end{bmatrix}$

23. $A = \begin{bmatrix} 3 & 1 \\ 1 & -2 \end{bmatrix}$ $\quad B = \begin{bmatrix} \frac{2}{7} & \frac{1}{7} \\ \frac{1}{7} & -\frac{3}{7} \end{bmatrix}$

24. $A = \begin{bmatrix} 2 & 3 \\ 1 & -1 \end{bmatrix}$ $\quad B = \begin{bmatrix} \frac{1}{5} & \frac{3}{5} \\ \frac{1}{5} & -\frac{5}{5} \end{bmatrix}$

25. $A = \begin{bmatrix} 1 & 2 \\ 3 & 4 \end{bmatrix}$ $\quad B = \begin{bmatrix} 4 & -2 \\ -3 & 1 \end{bmatrix}$

26. $A = \begin{bmatrix} 1 & 2 \\ 3 & 4 \end{bmatrix}$ $\quad B = \begin{bmatrix} 1 & \frac{1}{2} \\ \frac{1}{3} & \frac{1}{4} \end{bmatrix}$

27. $A = \begin{bmatrix} 1 & -1 & 1 \\ 1 & 0 & -1 \\ 0 & 1 & -1 \end{bmatrix}$ $\quad B = \begin{bmatrix} 1 & 0 & 1 \\ 1 & -1 & 2 \\ 1 & -1 & 1 \end{bmatrix}$

28. $A = \begin{bmatrix} -1 & 0 & -1 \\ -1 & 1 & -2 \\ -1 & 1 & -1 \end{bmatrix}$ $\quad B = \begin{bmatrix} -1 & 1 & -1 \\ -1 & 0 & 1 \\ 0 & -1 & 1 \end{bmatrix}$

29. $A = \begin{bmatrix} 2 & 0 & 1 \\ 0 & 3 & 1 \\ 0 & 2 & -1 \end{bmatrix}$ $\quad B = \begin{bmatrix} 0 & 2 & 1 \\ 0 & 3 & 0 \\ 2 & 0 & 2 \end{bmatrix}$

30. $A = \begin{bmatrix} 1 & 0 & 0 \\ 0 & 2 & 0 \\ 0 & 0 & 3 \end{bmatrix}$ $\quad B = \begin{bmatrix} 1 & 0 & 0 \\ 0 & \frac{1}{2} & 0 \\ 0 & 0 & \frac{1}{3} \end{bmatrix}$

In Exercises 31–36, find A^{-1}. Use the fact that for a 2×2 matrix, $A = \begin{bmatrix} a & b \\ c & d \end{bmatrix}$, if $ad - bc \neq 0$, the inverse is given by $A^{-1} = \frac{1}{ad - bc} \begin{bmatrix} d & -b \\ -c & a \end{bmatrix}$.

31. $A = \begin{bmatrix} 2 & 1 \\ -1 & 0 \end{bmatrix}$

32. $A = \begin{bmatrix} 3 & 1 \\ 2 & 1 \end{bmatrix}$

33. $A = \begin{bmatrix} \frac{1}{3} & 2 \\ 5 & \frac{3}{4} \end{bmatrix}$

34. $A = \begin{bmatrix} \frac{1}{4} & 2 \\ \frac{1}{3} & \frac{2}{3} \end{bmatrix}$

35. $A = \begin{bmatrix} 1.3 & 2.4 \\ 5.3 & 1.7 \end{bmatrix}$

36. $A = \begin{bmatrix} -2.3 & 1.1 \\ 4.6 & -3.2 \end{bmatrix}$

In Exercises 37–44, find A^{-1} given A, by forming the matrix $[A|I_n]$ and using Gauss–Jordan elimination to reduce A.

37. $A = \begin{bmatrix} 1 & 1 & 1 \\ 1 & -1 & -1 \\ -1 & 1 & -1 \end{bmatrix}$

38. $A = \begin{bmatrix} 1 & -1 & 1 \\ 1 & 1 & 1 \\ -1 & 2 & -3 \end{bmatrix}$

39. $A = \begin{bmatrix} 1 & 0 & 1 \\ 0 & 1 & 1 \\ 1 & -1 & 0 \end{bmatrix}$

40. $A = \begin{bmatrix} 1 & 2 & -3 \\ 1 & -1 & -1 \\ 1 & 0 & -4 \end{bmatrix}$

41. $A = \begin{bmatrix} 2 & 4 & 1 \\ 1 & 1 & -1 \\ 1 & 1 & 0 \end{bmatrix}$
42. $A = \begin{bmatrix} 1 & 0 & 1 \\ 1 & 1 & -1 \\ 2 & 1 & -1 \end{bmatrix}$
43. $A = \begin{bmatrix} 1 & 1 & -1 \\ 1 & -1 & 1 \\ 2 & -1 & -1 \end{bmatrix}$
44. $A = \begin{bmatrix} 1 & -1 & -1 \\ 1 & 1 & -3 \\ 3 & -5 & 1 \end{bmatrix}$

In Exercises 45–52, the coefficient matrices are those in Exercises 37–44. Use the inverses found in Exercises 37–44 to solve the system of linear equations using matrix algebra.

45.
$x + y + z = 1$
$x - y - z = -1$
$-x + y - z = -1$

46.
$x - y + z = 0$
$x + y + z = 2$
$-x + 2y - 3z = 1$

47.
$x + z = 3$
$y + z = 1$
$x - y = 2$

48.
$x + 2y - 3z = 1$
$x - y - z = 3$
$x - 4z = 0$

49.
$2x + 4y + z = -5$
$x + y - z = 7$
$x + y = 0$

50.
$x + z = 3$
$x + y - z = -3$
$2x + y - z = -5$

51.
$x + y - z = 4$
$x - y + z = 2$
$2x - y - z = -3$

52.
$x - y - z = 0$
$x + y - 3z = 2$
$3x - 5y + z = 4$

 ■ **APPLICATIONS**

53. Smoking. On January 6 and 10, 2000, the Harris Poll conducted a survey of adult smokers in the United States. When asked, "Have you ever tried to quit smoking?," 70% said yes and 30% said no. Write a 2 × 1 matrix—call it *A*—that represents those smokers. When asked what consequences smoking would have on their lives, 89% believed it would increase their chance of getting lung cancer and 84% believed smoking would shorten their lives. Write a 2 × 1 matrix—call it *B*—that represents those smokers. If there are 46 million adult smokers in the United States:

a. What does 46*A* tell us? **b.** What does 46*B* tell us?

54. Women in Science. According to the study of science and engineering indicators by the National Science Foundation (www.nsf.gov), the number of female graduate students in science and engineering disciplines has increased over the last 30 years. In 1981, 24% of mathematics graduate students were female and 23% of graduate students in computer science were female. In 1991 32% of mathematics graduate students and 21% of computer science graduate students were female. In 2001 38% of mathematics graduate students and 30% of computer science graduate students were female. Write three 2 × 1 matrices representing the percentage of female graduate students.

$$A = \begin{bmatrix} \%\text{female–math–1981} \\ \%\text{female–C.S.–1981} \end{bmatrix}$$

$$B = \begin{bmatrix} \%\text{female–math–1991} \\ \%\text{female–C.S.–1991} \end{bmatrix}$$

$$C = \begin{bmatrix} \%\text{female–math–2001} \\ \%\text{female–C.S.–2001} \end{bmatrix}$$

What does $C - B$ tell us? What does $B - A$ tell us? What can you conclude about the number of women pursuing mathematics and computer science graduate degrees?

55. Registered Voters. According to the U.S. Census Bureau (www.census.gov) in the 2000 national election, 58.9% of the men over the age of 18 were registered voters but only 41.4% voted, and 62.8% of women over 18 were registered voters but only 43% actually voted. Write a 2 × 2 matrix with the following data:

$$A = \begin{bmatrix} \text{Percentage of} & \text{Percentage of} \\ \text{Registered Male} & \text{Registered Female} \\ \text{Voters} & \text{Voters} \\ \text{Percent of Males} & \text{Percent of Females} \\ \text{That Voted} & \text{That Voted} \end{bmatrix}$$

If we let *B* be a 2 × 1 matrix representing the total population of males and females over the age of 18 in the United States: $B = \begin{bmatrix} 100 \text{ M} \\ 110 \text{ M} \end{bmatrix}$ what does *AB* tell us?

56. Job Applications. A company has two rubrics for scoring job applicants based on weighting education, experience, and the interview differently.

		Rubric 1	Rubric 2
Matrix *A*	Education	0.5	0.6
	Experience	0.3	0.1
	Interview	0.2	0.3

Applicants receive a score from 1 to 10 in each category (education, experience, and interview). Two applicants are shown in the matrix *B*:

	Education	Experience	Interview
Matrix *B* Applicant 1	8	7	5
Applicant 2	6	8	8

What is the order of *BA*? What does each entry in *BA* tell us?

■ CATCH THE MISTAKE

In Exercises 57–60, explain the mistake that is made.

57. Multiply: $\begin{bmatrix} 3 & 2 \\ 1 & 4 \end{bmatrix} \begin{bmatrix} -1 & 3 \\ -2 & 5 \end{bmatrix}$.

Solution:

Multiply corresponding elements.

$$\begin{bmatrix} 3 & 2 \\ 1 & 4 \end{bmatrix} \begin{bmatrix} -1 & 3 \\ -2 & 5 \end{bmatrix} = \begin{bmatrix} (3)(-1) & (2)(3) \\ (1)(-2) & (4)(5) \end{bmatrix}$$

Simplify.

$$\begin{bmatrix} 3 & 2 \\ 1 & 4 \end{bmatrix} \begin{bmatrix} -1 & 3 \\ -2 & 5 \end{bmatrix} = \begin{bmatrix} -3 & 6 \\ -2 & 20 \end{bmatrix}$$

This is incorrect. What is the mistake?

58. Multiply: $\begin{bmatrix} 3 & 2 \\ 1 & 4 \end{bmatrix} \begin{bmatrix} -1 & 3 \\ -2 & 5 \end{bmatrix}$.

Solution:

Multiply using column-by-row method.

$$\begin{bmatrix} 3 & 2 \\ 1 & 4 \end{bmatrix} \begin{bmatrix} -1 & 3 \\ -2 & 5 \end{bmatrix} = \begin{bmatrix} (3)(-1) + (1)(3) & (2)(-1) + (4)(3) \\ (3)(-2) + (1)(5) & (2)(-2) + (4)(5) \end{bmatrix}$$

Simplify.

$$\begin{bmatrix} 3 & 2 \\ 1 & 4 \end{bmatrix} \begin{bmatrix} -1 & 3 \\ -2 & 5 \end{bmatrix} = \begin{bmatrix} 0 & 10 \\ -1 & 16 \end{bmatrix}$$

This is incorrect. What is the mistake?

59. Find the inverse of $A = \begin{bmatrix} 1 & 0 & 1 \\ -1 & 0 & -1 \\ 1 & 2 & 0 \end{bmatrix}$.

Solution:

Write the matrix $[A|I_3]$.

$$\left[\begin{array}{ccc|ccc} 1 & 0 & 1 & 1 & 0 & 0 \\ -1 & 0 & -1 & 0 & 1 & 0 \\ 1 & 2 & 0 & 0 & 0 & 1 \end{array}\right]$$

Use Gauss–Jordan elimination to reduce A.

$$\begin{array}{c} R_2 + R_1 \rightarrow R_2 \\ R_3 - R_1 \rightarrow R_3 \end{array} \left[\begin{array}{ccc|ccc} 1 & 0 & 1 & 1 & 0 & 0 \\ 0 & 0 & 0 & 1 & 1 & 0 \\ 0 & 2 & -1 & -1 & 0 & 1 \end{array}\right]$$

$$R_2 \leftrightarrow R_3 \quad \left[\begin{array}{ccc|ccc} 1 & 0 & 1 & 1 & 0 & 0 \\ 0 & 2 & -1 & -1 & 0 & 1 \\ 0 & 0 & 0 & 1 & 1 & 0 \end{array}\right]$$

$$\tfrac{1}{2} R_2 \rightarrow R_2 \quad \left[\begin{array}{ccc|ccc} 1 & 0 & 1 & 1 & 0 & 0 \\ 0 & 1 & -\tfrac{1}{2} & -\tfrac{1}{2} & 0 & \tfrac{1}{2} \\ 0 & 0 & 0 & 1 & 1 & 0 \end{array}\right]$$

$$A^{-1} = \begin{bmatrix} 1 & 0 & 0 \\ -\tfrac{1}{2} & 0 & \tfrac{1}{2} \\ 1 & 1 & 0 \end{bmatrix} \text{ is incorrect because } AA^{-1} \neq I_3.$$

What is the mistake?

60. Find the inverse of A given that $A = \begin{bmatrix} 2 & 5 \\ 3 & 10 \end{bmatrix}$.

$$A^{-1} = \frac{1}{A}$$

$$A^{-1} = \frac{1}{\begin{bmatrix} 2 & 5 \\ 3 & 10 \end{bmatrix}}$$

Simplify.

$$A^{-1} = \begin{bmatrix} \tfrac{1}{2} & \tfrac{1}{5} \\ \tfrac{1}{3} & \tfrac{1}{10} \end{bmatrix}$$

This is incorrect. What mistake was made?

■ CHALLENGE

61. T or F: If $A = \begin{bmatrix} a_{11} & a_{12} \\ a_{21} & a_{22} \end{bmatrix}$ and $B = \begin{bmatrix} b_{11} & b_{12} \\ b_{21} & b_{22} \end{bmatrix}$ then

$AB = \begin{bmatrix} a_{11}b_{11} & a_{12}b_{12} \\ a_{21}b_{21} & a_{22}b_{22} \end{bmatrix}$.

62. T or F: If $A = \begin{bmatrix} a_{11} & a_{12} \\ a_{21} & a_{22} \end{bmatrix}$ then $A^{-1} = \begin{bmatrix} \tfrac{1}{a_{11}} & \tfrac{1}{a_{12}} \\ \tfrac{1}{a_{21}} & \tfrac{1}{a_{22}} \end{bmatrix}$.

63. T or F: AB is only defined if the number of columns in A equals the number of rows in B.

64. T or F: All square matrices have inverses.

65. Verify that $A^{-1} = \dfrac{1}{ad - bc} \begin{bmatrix} d & -b \\ -c & a \end{bmatrix}$ is the inverse of $A = \begin{bmatrix} a & b \\ c & d \end{bmatrix}$ if the determinant of A is nonzero.

66. Let $A = \begin{bmatrix} a & b \\ c & d \end{bmatrix}$ and form the matrix $[A|I_2]$. Use Gauss-Jordan elimination to transform $[I_2|A^{-1}]$ where $A^{-1} = \dfrac{1}{ad - bc} \begin{bmatrix} d & -b \\ -c & a \end{bmatrix}$ such that $ad - bc \neq 0$ and $a \neq 0$.

67. Let $A = \begin{bmatrix} a & 0 & 0 \\ 0 & b & 0 \\ 0 & 0 & c \end{bmatrix}$, find A^{-1}.

68. For what values of x does the inverse of A not exist, given $A = \begin{bmatrix} x & 6 \\ 3 & 2 \end{bmatrix}$?

69. Why does the square matrix $A = \begin{bmatrix} 2 & 3 \\ 4 & 6 \end{bmatrix}$ not have an inverse?

70. Why does the square matrix $A = \begin{bmatrix} 1 & 2 & -1 \\ 2 & 4 & -2 \\ 0 & 1 & 3 \end{bmatrix}$ not have an inverse?

 ■ **TECHNOLOGY**

In Exercises 71–74, use a graphing utility to perform the indicated matrix operations.

$$A = \begin{bmatrix} 1 & 7 & 9 & 2 \\ -3 & -6 & 15 & 11 \\ 0 & 3 & 2 & 5 \\ 9 & 8 & -4 & 1 \end{bmatrix} \qquad B = \begin{bmatrix} 7 & 9 \\ 8 & 6 \\ -4 & -2 \\ 3 & 1 \end{bmatrix}$$

71. AB **72.** AA **73.** Find A^{-1}. **74.** AA^{-1}

Q: Mike is trying to solve the system of equations $\begin{cases} x^2 + y^2 = 6 \\ y = x^2 - 4 \end{cases}$.
He rewrites the second equation as $x^2 - y = 4$ and sets up the matrix $\begin{bmatrix} 1 & 1 & 6 \\ 1 & -1 & 4 \end{bmatrix}$. He then writes the matrix in reduced form and arrives at the solution $x = 5, y = 1$. However, he decides to double-check his work by graphing the equations and notices that the parabola and circle intersect at four points. He looks back at his Gauss–Jordan elimination, but cannot find his error. Why is his answer incorrect?

A: A matrix can only be used to solve a system of linear equations. However, the system that Mike was solving was nonlinear. The method of substitution will work well on this problem.

Solve the system of equations that Mike was trying to solve. What is the actual system of equations that the matrix represents?

TYING IT ALL TOGETHER

Sam has saved $5,000 and wants to invest some of it, and put the rest in his savings account. Last year he opened his savings account and deposited $2,000, but only earned $36 in interest. He has been saving at least 10% of his income, but was also able to save 30% of his annual bonus of $1,800.

After doing some investigating, Sam finds a stock that has consistently risen at a rate of about 14% per year for the last 2 years. He also finds a mutual fund that pays an average of about 6% per year. If Sam invests $2,500 into the mutual fund and his savings account combined, then how much should he invest in savings, the stock, and the mutual fund in order to earn $450 over the next year from his investments? Assume that the stock and mutual fund continue to grow at the same rates. Round all answers to two decimal places.

SECTION	TOPIC	PAGES	REVIEW EXERCISES	KEY CONCEPTS
7.1	Matrices and systems of linear equations	486–490		
7.2	Systems of linear equations: Augmented matrices	490–505	1–22	Systems consisting of any number of equations and any number of variables.
	Augmented matrices	491–494	1–4	$\begin{bmatrix} A_1 & B_1 & \mid & C_1 \\ A_2 & B_2 & \mid & C_2 \end{bmatrix}$ or $\begin{bmatrix} A_1 & B_1 & C_1 & \mid & D_1 \\ A_2 & B_2 & C_2 & \mid & D_2 \\ A_3 & B_3 & C_3 & \mid & D_3 \end{bmatrix}$
	Reduced matrices	492–494	5–8	**5.** The leftmost nonzero element in each row is 1. **6.** The column containing the leftmost 1 of a given row has 0's above and below the 1. **7.** The leftmost 1 in any row is the right of the leftmost 1 in the row above it. **8.** Any row consisting entirely of 0's is below any row having a nonzero element.
	Row operations	493–494	9–16	**1.** $R_i \leftrightarrow R_j$ Interchange row i with row j. **2.** $cR_i \rightarrow R_i$ Multiply row i by the constant c. **3.** $R_j + cR_i \rightarrow R_j$ Multiply row i by the constant c and add to row j, writing the results in row j.
	Gauss–Jordan elimination	494–503	17–20	**Step 1:** Write the system of equations as an augmented matrix. **Step 2:** Use Gauss–Jordan elimination to transform the matrix into reduced form. **Step 3:** Identify the solution.
	Applications	503–505	21 and 22	
7.3	Systems of linear equations: Determinants	510–518	23–40	Determinants can only be used to solve square systems (number of equations and variables is equal).
	Evaluate a 2×2 determinant	511–512	23–26	For $\begin{bmatrix} a & b \\ c & d \end{bmatrix}$, $D = \begin{vmatrix} a & b \\ c & d \end{vmatrix} = ad - bc$
	Cramer's rule: 2×2	512–515	27–30	The system $\begin{aligned} a_1 x + b_1 y &= c_1 \\ a_2 x + b_2 y &= c_2 \end{aligned}$ has the solution $x = \dfrac{D_x}{D}, \quad y = \dfrac{D_y}{D}$ where the determinants are given by $D = \begin{vmatrix} a_1 & b_1 \\ a_2 & b_2 \end{vmatrix} \neq 0$ $D_x = \begin{vmatrix} c_1 & b_1 \\ c_2 & b_2 \end{vmatrix} \quad D_y = \begin{vmatrix} a_1 & c_1 \\ a_2 & c_2 \end{vmatrix}$

SECTION	TOPIC	PAGES	REVIEW EXERCISES	KEY CONCEPTS
	Evaluate a 3×3 determinant	515–516	31–34	For $\begin{bmatrix} a_1 & b_1 & c_1 \\ a_2 & b_2 & c_2 \\ a_3 & b_3 & c_3 \end{bmatrix}$ the determinant is given by $$\begin{vmatrix} a_1 & b_1 & c_1 \\ a_2 & b_2 & c_2 \\ a_3 & b_3 & c_3 \end{vmatrix} = \begin{matrix} a_1(b_2c_3 - b_3c_2) - b_1(a_2c_3 - a_3c_2) \\ + c_1(a_2b_3 - a_3b_2) \end{matrix}$$
	Cramer's rule	517–518	35–38	The system $\begin{matrix} a_1x + b_1y + c_1z = d_1 \\ a_2x + b_2y + c_2z = d_2 \\ a_3x + b_3y + c_3z = d_3 \end{matrix}$ has the solution $x = \dfrac{D_x}{D}, y = \dfrac{D_y}{D}, z = \dfrac{D_z}{D}$.
7.4	Matrix algebra	521–537	41–58	$\begin{bmatrix} & \text{Column 1} & \text{Column 2} & \dots & \text{Column } j & \dots & \text{Column } n \\ \text{Row 1} & a_{11} & a_{12} & \dots & a_{1j} & \dots & a_{1n} \\ \text{Row 2} & a_{21} & a_{22} & \dots & a_{2j} & \dots & a_{2n} \\ \vdots & \vdots & \vdots & & \vdots & & \vdots \\ \text{Row } i & a_{i1} & a_{i2} & \dots & a_{ij} & \dots & a_{in} \\ \vdots & \vdots & \vdots & & \vdots & & \vdots \\ \text{Row } m & a_{m1} & a_{m2} & \dots & a_{mj} & \dots & a_{mn} \end{bmatrix}$
	Equality and adding and subtracting matrices	523–526	41–46	The orders must be the same: $A(m \times n)$ and $B(m \times n)$. Perform operation element by element.
	Multiplying matrices	526–531	41–46	The orders must satisfy the relationship: $A(m \times n)$ and $B(n \times p)$, resulting in $AB(m \times p)$. Perform multiplication row-by-column.
	Finding the inverse of a matrix	531–535	47–54	Only square matrices $n \times n$ have inverses. $A^{-1}A = I_n$ Step 1: Form the matrix $[A\,\vert\,I_n]$. Step 2: Use row operations to transform this matrix to $[I_n\,\vert\,A^{-1}]$.
	Solving square systems of linear equations using matrix algebra	535–537	55–58	Linear system: $AX = B$. Multiply both sides of the equation by A^{-1}, which results in $X = A^{-1}B$.

7.2 Systems of Linear Equations: Augmented Matrices

Write the augmented matrix for each system of linear equations.

1. $5x + 7y = 2$
$3x - 4y = -2$

2. $2.3x - 4.5y = 6.8$
$-0.4x + 2.1y = -9.1$

3. $2x - z = 3$
$y - 3z = -2$
$x + 4z = -3$

4. $2y - x + 3z = 1$
$4z - 2y + 3x = -2$
$x - y - 4z = 0$

Indicate whether each matrix is in reduced form. If not, state the condition that is violated.

5. $\begin{bmatrix} 1 & 1 & | & 0 \\ 0 & 1 & | & 2 \end{bmatrix}$

6. $\begin{bmatrix} 1 & 2 & | & 0 \\ 0 & 0 & | & 1 \end{bmatrix}$

7. $\begin{bmatrix} 2 & 0 & 0 & | & 1 \\ 0 & -2 & 0 & | & 2 \\ 0 & 0 & 2 & | & 3 \end{bmatrix}$

8. $\begin{bmatrix} 1 & 0 & 1 & 0 & | & 2 \\ 0 & 0 & 1 & 1 & | & -3 \\ 0 & 1 & 0 & 0 & | & 2 \\ 0 & 0 & 0 & 1 & | & 1 \end{bmatrix}$

Perform the indicated row operations on each matrix.

9. $\begin{bmatrix} 1 & -2 & | & 1 \\ 0 & -2 & | & 2 \end{bmatrix}$ $-\dfrac{1}{2}R_2 \rightarrow R_2$

10. $\begin{bmatrix} 1 & 4 & | & 1 \\ 2 & -2 & | & 3 \end{bmatrix}$ $R_2 - 2R_1 \rightarrow R_2$

11. $\begin{bmatrix} 1 & -2 & 0 & | & 1 \\ 0 & -2 & 3 & | & -2 \\ 0 & 1 & -4 & | & 8 \end{bmatrix}$ $R_2 + R_1 \rightarrow R_1$

12. $\begin{bmatrix} 1 & 1 & 1 & 6 & | & 0 \\ 0 & 2 & -2 & 3 & | & -2 \\ 0 & 0 & 1 & -2 & | & 4 \\ 0 & -1 & 3 & 3 & | & 3 \end{bmatrix}$ $\begin{matrix} -2R_1 + R_2 \rightarrow R_1 \\ R_4 + R_3 \rightarrow R_4 \end{matrix}$

Use row operations to transform each matrix to reduced form.

13. $\begin{bmatrix} 1 & 3 & | & 0 \\ 3 & 4 & | & 1 \end{bmatrix}$

14. $\begin{bmatrix} 1 & 2 & -1 & | & 0 \\ 0 & 1 & 1 & | & -1 \\ -2 & 0 & 1 & | & -2 \end{bmatrix}$

15. $\begin{bmatrix} 4 & 1 & -2 & | & 0 \\ 1 & 0 & -1 & | & 0 \\ -2 & 1 & 1 & | & 12 \end{bmatrix}$

16. $\begin{bmatrix} 2 & 3 & 2 & | & 1 \\ 0 & -1 & 1 & | & -2 \\ 1 & 1 & -1 & | & 6 \end{bmatrix}$

Solve the system of linear equations using augmented matrices.

17. $3x - 2y = 2$
$-2x + 4y = 1$

18. $0.5x - 0.4y = -2.4$
$-0.25x + 0.2y = 3.6$

19. $x - 2y + z = 3$
$2x - y + z = -4$
$3x - 3y - 5z = 2$

20. $3x + y + z = -4$
$x - 2y + z = -6$

Applications

21. Fitting a Curve to Data. The average number of flights on a commercial plane that a person takes a year can be modeled by a quadratic function, $y = ax^2 + bx + c$ where $a < 0$ and x represents age: $16 < x < 65$. The following table gives the average number of flights per year that a person takes on a commercial airline.

Age	Number of Flights per Year
16	2
40	6
65	4

Determine the quadratic function that models this quantity.

22. Investment Portfolio. Danny and Paula decide to invest $20,000 of their savings. They put some in an IRA account earning 4.5% interest, some in a mutual fund that has been averaging 8% a year, and some in a stock that earned 12% last year. If they put $3,000 more in the mutual fund than in the IRA, and the mutual fund and stock have the same growth in the next year as they did in the previous year, they will earn $1,877.50 in a year. How much money should they put in each of the three investments?

7.3 Systems of Linear Equations: Determinants

Evaluate each 2 × 2 determinant.

23. $\begin{vmatrix} 2 & 4 \\ 3 & 2 \end{vmatrix}$

24. $\begin{vmatrix} -2 & -4 \\ -3 & 2 \end{vmatrix}$

25. $\begin{vmatrix} 2.4 & -2.3 \\ 3.6 & -1.2 \end{vmatrix}$

26. $\begin{vmatrix} -\frac{1}{4} & 4 \\ \frac{3}{4} & -4 \end{vmatrix}$

Use Cramer's rule to solve each system of equations or to determine if the system is inconsistent (no solution) or dependent (infinitely many solutions).

27. $x - y = 2$
$x + y = 4$

28. $x + 2y = 6$
$x - 2y = 6$

29. $-x + 3y = 4$
$2x - 6y = -5$

30. $-3x = 40 - 2y$
$2x = 25 + y$

Evaluate each 3 × 3 determinant.

31. $\begin{vmatrix} 1 & 2 & 2 \\ 0 & 1 & 3 \\ 2 & -1 & 0 \end{vmatrix}$

32. $\begin{vmatrix} 0 & -2 & 1 \\ 0 & -3 & 7 \\ 1 & -1 & -3 \end{vmatrix}$

33. $\begin{vmatrix} a & 0 & -b \\ -a & b & c \\ 0 & 0 & -d \end{vmatrix}$ **34.** $\begin{vmatrix} -2 & -4 & 6 \\ 2 & 0 & 3 \\ -1 & 2 & \frac{3}{4} \end{vmatrix}$

Use Cramer's rule to solve each system of equations or to determine if the system is inconsistent (no solution) or dependent (infinitely many solutions).

35. $x + y - 2z = -2$
$2x - y + z = 3$
$x + y + z = 4$

36. $-x - y + z = 3$
$x + 2y - 2z = 8$
$2x + y + 4z = -4$

37. $3x + 4z = -1$
$x + y + 2z = -3$
$y - 4z = -9$

38. $x + y + z = 0$
$-x - 3y + 5z = -2$
$2x + y - 3z = -4$

Applications

39. Geometry. Use determinants to find the area of a triangle with vertices $(2, 4)$, $(4, 4)$, and $(-4, 3)$.

40. Geometry. If three points (x_1, y_1), (x_2, y_2), and (x_3, y_3) are collinear (lie on the same line), then the following determinant must be satisfied:

$$\begin{vmatrix} x_1 & y_1 & 1 \\ x_2 & y_2 & 1 \\ x_3 & y_3 & 1 \end{vmatrix} = 0$$

Determine if $(0, -3)$, $(3, 0)$, and $(1, 6)$ are collinear.

7.4 Matrix Algebra

Calculate the given expression if possible.

$A = \begin{bmatrix} 1 & 0 & 3 \\ 0 & -4 & 3 \\ -2 & 6 & -5 \end{bmatrix}$ $B = \begin{bmatrix} -1 & -4 & -2 \\ 6 & -2 & -3 \end{bmatrix}$

$C = \begin{bmatrix} 1 & -4 \\ -3 & 5 \end{bmatrix}$ $D = \begin{bmatrix} -3 & 5 \\ 2 & 4 \\ 1 & -3 \end{bmatrix}$

41. $A + C$ **42.** $B + A$ **43.** $2D - 3A$

44. BA **45.** AB **46.** $A(BD)$

Determine if B is the multiplicative inverse of A using $AA^{-1} = I$.

47. $A = \begin{bmatrix} 6 & 4 \\ 4 & 2 \end{bmatrix}$ $B = \begin{bmatrix} -0.5 & 1 \\ 1 & -1.5 \end{bmatrix}$

48. $A = \begin{bmatrix} 1 & -2 \\ 2 & -4 \end{bmatrix}$ $B = \begin{bmatrix} 1 & 2 \\ 2 & -2 \end{bmatrix}$

49. $A = \begin{bmatrix} 1 & -2 & 6 \\ 2 & 3 & -2 \\ 0 & -1 & 1 \end{bmatrix}$ $B = \begin{bmatrix} -\frac{1}{7} & \frac{4}{7} & 2 \\ \frac{2}{7} & -\frac{1}{7} & -2 \\ \frac{2}{7} & -\frac{1}{7} & -1 \end{bmatrix}$

50. $A = \begin{bmatrix} 0 & 7 & 6 \\ 1 & 0 & -4 \\ -2 & 1 & 0 \end{bmatrix}$ $B = \begin{bmatrix} 1 & 1 & 1 \\ -2 & -2 & -2 \\ 2 & 0 & 6 \end{bmatrix}$

Find A^{-1}. Use the fact that for a 2×2 matrix, $A = \begin{bmatrix} a & b \\ c & d \end{bmatrix}$, if $ad - bc \neq 0$. The inverse is given by

$$A^{-1} = \frac{1}{ad - bc}\begin{bmatrix} d & -b \\ -c & a \end{bmatrix}.$$

51. $A = \begin{bmatrix} 1 & 2 \\ -3 & 4 \end{bmatrix}$ **52.** $A = \begin{bmatrix} -2 & 7 \\ -4 & 6 \end{bmatrix}$

53. $A = \begin{bmatrix} 0 & 1 \\ -2 & 0 \end{bmatrix}$ **54.** $A = \begin{bmatrix} 3 & -1 \\ -2 & 2 \end{bmatrix}$

Find A^{-1} given A, by forming the matrix $[A \,|\, I_n]$ and using Gauss-Jordan elimination to reduce A.

55. $A = \begin{bmatrix} 1 & 3 & -2 \\ 2 & 1 & -1 \\ 0 & 1 & -3 \end{bmatrix}$ **56.** $A = \begin{bmatrix} 0 & 1 & 0 \\ 4 & 1 & 2 \\ -3 & -2 & 1 \end{bmatrix}$

57. $A = \begin{bmatrix} -1 & 1 & 0 \\ -2 & 1 & 2 \\ 1 & 2 & 4 \end{bmatrix}$ **58.** $A = \begin{bmatrix} -4 & 4 & 3 \\ 1 & 2 & 2 \\ 3 & -1 & 6 \end{bmatrix}$

Write each of the following systems of linear equations as an augmented matrix and as a matrix equation (if possible).

1. $x - 2y = 1$
$-x + 3y = 2$

2. $3x + 5y = -2$
$7x + 11y = -6$

3. $6x + 9y + z = 5$
$2x - 3y + z = 3$
$10x + 12y + 2z = 9$

4. $3x + 2y - 10z = 2$
$x + y - z = 5$

5. Perform the following row operations:

$$\begin{bmatrix} 1 & 3 & 5 \\ 2 & 7 & -1 \\ -3 & -2 & 0 \end{bmatrix} \quad \begin{matrix} R_2 - 2R_1 \rightarrow R_2 \\ R_3 + 3R_1 \rightarrow R_3 \end{matrix}$$

6. Reduce the following matrix: $\begin{bmatrix} 2 & -1 & 1 & | & 3 \\ 1 & 1 & -1 & | & 0 \\ 3 & 2 & -2 & | & 1 \end{bmatrix}$.

In Exercises 7 and 8, solve the systems of linear equations using Gauss–Jordan elimination.

7. $6x + 9y + z = 5$
$2x - 3y + z = 3$
$10x + 12y + 2z = 9$

8. $3x + 2y - 10z = 2$
$x + y - z = 5$

In Exercises 9 and 10, calculate the determinant.

9. $\begin{vmatrix} 7 & -5 \\ 2 & -1 \end{vmatrix}$

10. $\begin{vmatrix} 1 & -2 & -1 \\ 3 & -5 & 2 \\ 4 & -1 & 0 \end{vmatrix}$

In Exercises 11 and 12, solve the system of linear equations using determinants.

11. $x - 2y = 1$
$-x + 3y = 2$

12. $3x + 5y = -2$
$7x + 11y = -6$

13. Multiply the matrices $\begin{bmatrix} 1 & -2 & 5 \\ 0 & -1 & 3 \end{bmatrix} \begin{bmatrix} 0 & 4 \\ 3 & -5 \\ -1 & 1 \end{bmatrix}$ if possible.

14. Add the matrices $\begin{bmatrix} 1 & -2 & 5 \\ 0 & -1 & 3 \end{bmatrix} + \begin{bmatrix} 0 & 4 \\ 3 & -5 \\ -1 & 1 \end{bmatrix}$ if possible.

15. Find the inverse of $\begin{bmatrix} 4 & 3 \\ 5 & -1 \end{bmatrix}$ if it exists.

16. Find the inverse of $\begin{bmatrix} 1 & -3 & 2 \\ 4 & 2 & 0 \\ -1 & 2 & 5 \end{bmatrix}$ if it exists.

17. Find the inverse of $\begin{bmatrix} 3 & 1 & 0 \\ 5 & 2 & -1 \end{bmatrix}$ if it exists.

18. Solve the system of linear equations using matrix algebra (inverses).

$$x + 3y = -1$$
$$-2x - 5y = 4$$

19. Job Applicants. A company has two rubrics for scoring job applicants based on weighting education, experience, and the interview differently.

	Rubric 1	Rubric 2
Education	0.4	0.6
Experience	0.5	0.1
Interview	0.1	0.3

Matrix A corresponds to the table above (Education, Experience, Interview rows).

Applicants receive a score from 1 to 10 in each category (education, experience, and interview). Two applicants are shown in the matrix B:

	Education	Experience	Interview
Applicant 1	4	7	3
Applicant 2	6	5	4

What is the order of BA? What does each entry in BA tell us?

20. Stock Investments. A college student inherits $15,000 from his favorite aunt. He decides to invest it with a diversified scheme. He divides the money into a money market account paying 2% annual simple interest, a conservative stock that rose 4% last year, and an aggressive stock that rose 22% last year. He places $1,000 more in the conservative stock than in the money market and twice as much in the aggressive stock than in the money market. If the stocks perform the same way this year he will make $1,790 in 1 year assuming a simple interest model. How much did he put in each of the three investments?

CHAPTER 8

Conics

The Ejection Seat (Parabola)

Courtesy Technical Park, ride manufacturer, Italy

The Zipper (Ellipse)

Courtesy Chance Morgan, Inc.

The Hurricane (Hyperbola)

©David Burton

What is your favorite ride at an amusement park? It probably has the shape of conic sections. There are three main conic sections that we will discuss in this chapter: Parabolas, ellipses, and hyperbolas. We have already covered the circle (Section 2.4), which is a special kind of ellipse. The Ejection Seat is in the shape of a *parabola*, the Zipper is in the shape of an *ellipse*, and the Hurricane is in the shape of a *hyperbola*.

In this chapter, we will discuss conic sections such as parabolas, ellipses, and hyperbolas both geometrically and algebraically. We will define these conic sections using geometric definitions and graph them. We will then develop the algebraic equations associated with parabolas, ellipses, and hyperbolas. The goal of this chapter is to graph conic sections given equations and also to develop an equation of the conic based on the graph, which then will give us two strategies to solve application problems using conics.

Conics

Parabola

- **Geometric Definition**
  - Vertex
  - Focus
- **Graph of a parabola**
- **Equation of a parabola**
- **Applications**
  - Satellite dish
  - Mirrors

Ellipse

- **Geometric Definition**
  - Center
  - Foci
  - Major axis
  - Minor axis
- **Graph of an ellipse**
- **Equation of an ellipse**
- **Applications**
  - Orbits
  - Racetracks

Hyperbola

- **Geometric Definition**
  - Center
  - Foci
  - Vertices
  - Transverse axis
- **Graph of a hyperbola**
- **Equation of a hyperbola**
- **Applications**
  - Navigation
  - Calibration of instruments

 CHAPTER OBJECTIVES

- Understand geometric definition of a parabola, ellipse, and hyperbola.
- Graph parabolas, ellipses, and hyperbolas given characteristics such as vertices and foci.
- Determine equations of conics given the characteristics of a graph.
- Solve applications involving conics.

NAVIGATION THROUGH SUPPLEMENTS

DIGITAL VIDEO SERIES #8

STUDENT SOLUTIONS MANUAL CHAPTER 8

BOOK COMPANION SITE
www.wiley.com/college/young

- Learn the names of each conic section.
- Define conics geometrically.
- Recognize the algebraic equation associated with each conic.

- Relate each conic to a cross section of a plane and a cone.
- Understand how the three equations of the conic sections are related to the general form of a second-degree equation in two variables.

Names of Conics

The word *conic* is derived from the word *cone*. Let's start with a (right circular) **double cone** (see figure on the left).

   Conic sections are curves that result from the intersection of a plane and a double cone. The four conic sections are a **circle**, **ellipse**, **parabola**, and **hyperbola**. **"Conics"** is an abbreviation for conic sections.

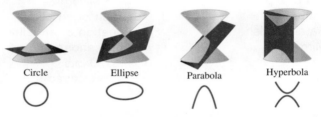

Circle Ellipse Parabola Hyperbola

   In Section 2.4, circles were discussed, and it will be shown that a circle is a particular type of an ellipse. Now we will discuss parabolas, ellipses, and hyperbolas. There are two ways in which we usually describe conics: graphically and algebraically. An entire section will be devoted to each of the three conics, but in this section we will summarize the geometric and algebraic definitions of a parabola, ellipse, and hyperbola whose vertex and centers are at the origin.

STUDY TIP

A circle is a special type of ellipse. All circles are ellipses, but not all ellipses are circles.

Geometric Definition

You already know that a circle consists of all points equidistant (radius) from a point (center). Ellipses, parabolas, and hyperbolas have similar type definitions in that they all have a constant distance (or a sum of distances) to some reference point.

   A **parabola** is the set of all points that are **equidistant to both a line and a point.** An **ellipse** is the set of all points whose **sum of the distances to two fixed points is constant**. A **hyperbola** is the set of all points whose **difference of the distances to two fixed points is a constant**.

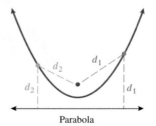

Parabola

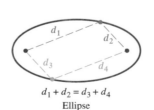

$d_1 + d_2 = d_3 + d_4$

Ellipse

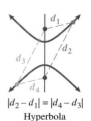

$|d_2 - d_1| = |d_4 - d_3|$

Hyperbola

Algebraic Definition

The general form of a second-degree equation in two variables, x and y, is given by

$$Ax^2 + Bxy + Cy^2 + Dx + Ey + F = 0$$

If we let $A = 1, B = 0, C = 1, D = 0, E = 0$, and $F = -r^2$, this general equation reduces to the equation of a circle centered at the origin: $x^2 + y^2 = r^2$. In fact, all three conics (parabolas, ellipses, and hyperbolas) are special cases of the general second-degree equation.

Recall in Section 1.4 (Quadratic Equations) that the discriminant, $b^2 - 4ac$, determined what type of solutions resulted from solving a second-degree equation in one variable. If the discriminant was positive the solutions were two distinct real roots, if the discriminant was zero the solution was a real repeated root, and if the discriminant was negative the solutions were two complex conjugate roots.

Similarly, the value of $B^2 - 4AC$ determines which conic the equation represents.

CONIC	$B^2 - 4AC$
Ellipse	$B^2 - 4AC < 0$
Parabola	$B^2 - 4AC = 0$
Hyperbola	$B^2 - 4AC > 0$

We now identify conics from the general form of a second-degree equation in two variables.

EXAMPLE 1 Determining the Type of Conic

Determine what type of conic corresponds to the following equations.

a. $\dfrac{x^2}{a^2} + \dfrac{y^2}{b^2} = 1$ **b.** $y = x^2$ **c.** $\dfrac{x^2}{a^2} - \dfrac{y^2}{b^2} = 1$

Solution:

Write the general form of the second-degree equation:

$$Ax^2 + Bxy + Cy^2 + Dx + Ey + F = 0$$

a. Identify A, B, C, D, E, and F.

$$A = \frac{1}{a^2}, B = 0, C = \frac{1}{b^2},$$
$$D = 0, E = 0, F = -1$$

Calculate the discriminant.

$$B^2 - 4AC = -\frac{4}{a^2 b^2} < 0$$

Since the discriminant is negative, the equation, $\dfrac{x^2}{a^2} + \dfrac{y^2}{b^2} = 1$, is an **ellipse**.

Notice that if $a = b = r$, then this equation of an ellipse reduces to the general equation of a circle, $x^2 + y^2 = r^2$, centered at the origin, with radius r.

b. Identify A, B, C, D, E, and F.

$$A = 1, B = 0, C = 0,$$
$$D = 0, E = -1, F = 0$$

Calculate the discriminant.

$$B^2 - 4AC = 0$$

Since the discriminant is zero, the equation, $y = x^2$, is a **parabola**.

c. Identify A, B, C, D, E, and F.

$$A = \frac{1}{a^2}, B = 0, C = -\frac{1}{b^2},$$
$$D = 0, E = 0, F = -1$$

Calculate the discriminant.

$$B^2 - 4AC = \frac{4}{a^2 b^2} > 0$$

Since the discriminant is positive, the equation, $\dfrac{x^2}{a^2} - \dfrac{y^2}{b^2} = 1$, is a **hyperbola**.

Notice the subtle (sign) difference between the equation of an ellipse and an equation of a hyperbola. The ellipse has a plus sign, and the hyperbola has a minus sign. In the next three sections, we will discuss the standard form of equations of parabolas, ellipses, and hyperbolas and their graphs.

■ **YOUR TURN** Determine what type of conic corresponds to the following equations.

a. $2x^2 + y^2 = 4$ **b.** $2x^2 = y^2 + 4$ **c.** $2y^2 = x$

SECTION 8.1 **SUMMARY**

In this section we defined a conic section both geometrically and algebraically with respect to the general form of a second-degree equation in two variables:

$$Ax^2 + Bxy + Cy^2 + Dx + Ey + F = 0$$

The following table summarizes the three conics: ellipse, parabola, and hyperbola.

CONIC	GEOMETRIC DEFINITION: THE SET OF ALL POINTS...	$B^2 - 4AC$
Ellipse	Sum of the distances to two fixed points is constant.	Negative
Parabola	Equidistant to both a line and a point.	Zero
Hyperbola	Difference of the distances to two fixed points is a constant.	Positive

It is important to note that a circle is a special type of ellipse and that the equations of ellipses and hyperbolas only differ by a minus sign.

■ **Answer: a. Ellipse** **b. Hyperbola** **c. Parabola**

SECTION 8.1 EXERCISES

■ **SKILLS**

In Exercises 1–8, identify the conic section as a parabola, ellipse, circle, or hyperbola.

1. $x^2 + xy - y^2 + 2x = -3$ **2.** $x^2 + xy + y^2 + 2x = -3$ **3.** $2x^2 + 2y^2 = 10$ **4.** $x^2 - 4x + y^2 + 2y = 4$

5. The set of all points exactly 2 units from the point $(0, 0)$.

6. The set of all points exactly 2 units from the point $(0, 1)$ and the line $y = -1$.

7. The set of all points whose distance to point A plus the distance to point B is 5 units.

8. The set of all points whose distance to point A minus the distance to point B is 4 units.

SECTION 8.2 The Parabola

Skills Objectives	Conceptual Objectives
■ Graph a parabola given the focus, directrix, and vertex.	■ Derive the general equation of a parabola.
■ Find the equation of a parabola whose vertex is at the origin.	■ Identify, draw, and use the focus, directrix, and axis of symmetry.
■ Find the equation of a parabola whose vertex is at the point (h, k).	
■ Solve applied problems that involve parabolas.	

Definition of a Parabola

Recall in Section 4.1 that the graphs of quadratic functions such as

$$f(x) = a(x - h)^2 + k \text{ or } y = ax^2 + bx + c$$

were *parabolas* that either opened upward or downward. We now give *parabolas* a geometric definition that allows for parabolas to open to the right or left.

We did not discuss these types of parabolas before because they are not functions. They fail the vertical line test. Here is a formal geometric definition for a parabola.

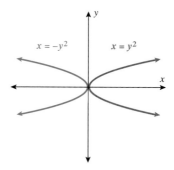

DEFINITION   PARABOLA

A **parabola** is the set of all points in a plane that are equidistant from a fixed line, the **directrix**, and a fixed point not on the line, the **focus**. The line through the focus and perpendicular to the directrix is the **axis of symmetry**.

The **vertex** of the parabola is located at the midpoint between the directrix and the focus along the axis of symmetry.

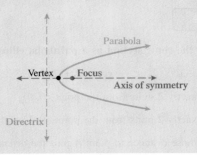

The simplest parabola to sketch is one whose vertex is located at the origin. Recall the general second-degree equation in two variables: $Ax^2 + Bxy + Cy^2 + Dx + Ey + F = 0$. If $A = -1$, $E = 4p$, and $B = C = D = F = 0$, the resulting equation is the standard equation of a vertical parabola $x^2 = 4py$, with vertex at the origin. If $C = -1$, $D = 4p$, and $A = B = E = F = 0$, the resulting equation is the standard equation of a horizontal parabola, $y^2 = 4px$, with vertex at the origin.

EQUATION OF A PARABOLA WITH VERTEX AT THE ORIGIN

The **standard form of the equation of a parabola** with vertex at the origin is given by:

Equation	$y^2 = 4px$	$x^2 = 4py$
Vertex	$(0, 0)$	$(0, 0)$
Focus	$(p, 0)$	$(0, p)$
Directrix	$x = -p$	$y = -p$
Axis of symmetry	x-axis	y-axis
$p > 0$	Opens to the right	Opens upward
$p < 0$	Opens to the left	Opens downward
Graph ($p > 0$)		

EXAMPLE 1 Finding the Focus and Directrix of a Parabola Whose Vertex Is Located at the Origin

Find the focus and directrix of a parabola whose equation is $y^2 = 8x$.

Solution:

Compare this parabola with the general equation of a parabola.

$$y^2 = 4px \quad y^2 = 8x$$

Solve for p.

$$4p = 8$$
$$p = 2$$

The focus of a parabola of the form $y^2 = 4px$ is $(p, 0)$.

The directrix of a parabola of the form $y^2 = 4px$ is $x = -p$.

The focus is $(2, 0)$ and the directrix is $x = -2$.

CONCEPT CHECK A parabola is the set of all points equidistant from the ___ and ___?

■ **YOUR TURN** Find the focus and directrix of a parabola whose equation is $y^2 = 16x$.

Graphing a Parabola Whose Vertex Is at the Origin

When a seamstress starts with a pattern for a custom-made suit, the pattern is used as a guide. The pattern is not sewn into the suit, but rather removed once it is used to determine the exact shape and size of the fabric to be sewn together. The focus and directrix of a parabola are similar to the pattern. Although the focus and directrix define a parabola, they do not appear on the graph of a parabola. An approximate sketch of a parabola whose vertex is at the origin can be drawn with three pieces of information. We know that the vertex is located at $(0, 0)$. Additional information that we seek is the direction in which the parabola opens and approximately how wide or narrow to draw the parabolic curve. The direction to open the parabola is found from the equation. An equation of the form $y^2 = 4px$ opens either left or right. It opens right if $p > 0$ and opens left if $p < 0$. An equation of the form $x^2 = 4py$ opens either up or down. It opens up if $p > 0$ and opens down if $p < 0$. How narrow or wide should the parabolic curve be drawn? If we select a few points that satisfy the equation, we can use those as graphing aids.

In Example 1, we found that the focus of that parabola is located at $(2, 0)$. If we select the x-coordinate of the focus, $x = 2$, and substitute that value into the equation of the parabola, $y^2 = 8x$, we find the corresponding y values to be $y = -4$ and $y = 4$. If we plot the three points, $(0, 0)$, $(2, -4)$, and $(2, 4)$, and then connect the points with a parabolic curve, we get the graph on the right.

STUDY TIP

The focus and directrix define a parabola but do not appear on its graph.

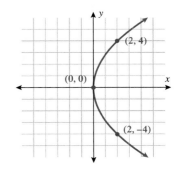

The line segment that passes through the focus, $(2, 0)$, is parallel to the directrix, $x = -2$, and whose endpoints are on the parabola is called the **latus rectum**. The latus rectum in this case has length 8.

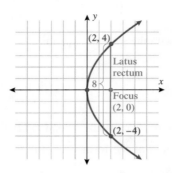

The latus rectum is a graphing aid that assists us in finding the width of a parabola.

In general, the points on a parabola, $y^2 = 4px$, that lie above and below the focus, $(p, 0)$, satisfy the equation $y^2 = 4p^2$, and are located at $(p, -2p)$ and $(p, 2p)$. The latus rectum will have length $4|p|$. Similarly, a parabola of the form $x^2 = 4py$ will have a horizontal latus rectum of length $4|p|$. We will use the latus rectum as a graphing aid to determine the parabola's width.

TECHNOLOGY TIP

To graph $x^2 = -12y$ using a graphing calculator, solve for y first. That is, $y = -\frac{1}{12}x^2$

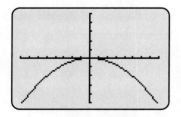

EXAMPLE 2 Graphing a Parabola Whose Vertex Is at the Origin Using the Focus, Directrix, and Latus Rectum as Graphing Aids

Determine the focus, directrix, and length of the latus rectum of the parabola $x^2 = -12y$. Use these to assist in graphing the parabola.

Solution:

Compare this parabola with the general equation of a parabola.
$$x^2 = 4py \quad x^2 = -12y$$

Solve for p.
$$4p = -12$$
$$p = -3$$

A parabola of the form $x^2 = 4py$ has focus $(0, p)$, directrix $y = -p$, and a latus rectum of length $4|p|$. For this parabola, $p = -3$; therefore, the focus is $(0, -3)$,

directrix is $y = 3$, and the length of the

latus rectum is 12. The graph is given on the right.

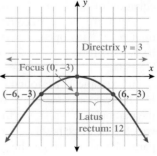

CONCEPT CHECK In which direction does a parabola given by $y^2 = -8x$ open?

YOUR TURN Find the focus, directrix, and length of the latus rectum, and use these to graph the parabola $y^2 = -8x$.

Finding the Equation of a Parabola Whose Vertex Is at the Origin

Thus far we have started with the equation of a parabola and then determined its focus and directrix. Let's now go the other way. For example, if we know the focus and directrix of a parabola, find the equation of the parabola. If we are given the focus and directrix, then we can find the vertex, which is the midpoint between the focus and the directrix. If the vertex is at the origin, then we know the general equation of the parabola that corresponds with the focus.

EXAMPLE 3 Finding the Equation of a Parabola Whose Vertex Is at the Origin

Find the equation of a parabola whose focus is at the point $\left(0, \dfrac{1}{2}\right)$ and whose directrix is $y = -\dfrac{1}{2}$. Graph the equation.

Solution:

The midpoint of the segment joining the focus and the directrix along the axis of symmetry is the vertex.

Calculate the midpoint between $\left(0, \dfrac{1}{2}\right)$ and $\left(0, -\dfrac{1}{2}\right)$.

$$\text{Vertex} = \left(\frac{0 + 0}{2}, \frac{\frac{1}{2} - \frac{1}{2}}{0}\right) = (0, 0).$$

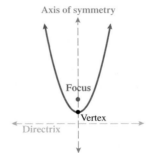

A parabola with vertex at $(0, 0)$, focus at $(0, p)$, and directrix $y = -p$ corresponds to the equation $x^2 = 4py$.

Identify p given that the focus is $(0, p) = \left(0, \dfrac{1}{2}\right)$. $\qquad p = \dfrac{1}{2}$

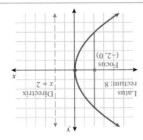

Answer: The focus is $(-2, 0)$.
The directrix is $x = 2$.
The length of latus rectum is 8.

Substitute $p = \frac{1}{2}$ into the standard equation of a parabola with vertex at the origin $x^2 = 4py$. $\qquad x^2 = 2y$

Now that the equation is known, a few points can be selected, and the parabola can be point plotted. Alternatively, the length of the latus rectum can be calculated to sketch the approximate width of the parabola.

To graph $x^2 = 2y$, first calculate the latus rectum. $\qquad 4|p| = 4\left(\dfrac{1}{2}\right) = 2$

Label the focus, directrix, and latus rectum and draw a parabolic curve whose vertex is at the origin that intersects with the latus rectum's endpoints.

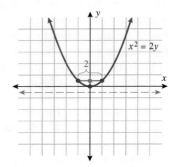

■ **YOUR TURN** Find the equation of a parabola whose focus is at the point $(-5, 0)$ and whose directrix is $x = 5$.

Before we proceed to parabolas with general vertices, let's first make a few observations: The larger the latus rectum the wider the parabola. An alternative graphing approach to finding the latus rectum is to plot a few points that satisfy the equation, which is the approach in most textbooks.

Parabola with a Vertex at (h, k)

Recall in Section 4.1 that the standard form for a quadratic function, or a parabola with vertex (h, k), is

$$f(x) = a(x - h)^2 + k$$

One way to graph this function is to start with the parabola $y = x^2$, and shift to the right h units, up k units, and the parabola opens up or down depending on the sign of a. Note that the parabola can be compressed or stretched depending on if $|a| > 1$ or $0 < |a| < 1$.

These same translations can be used for all four types of parabolas. The characteristics of the parabola, such as the focus, directrix, axis of symmetry, and latus rectum, also shift with the parabola. The following table summarizes parabolas with a vertex at (h, k). It is assumed that $p > 0$. The length of the latus rectum is $4|p|$ for all of the parabolas in the table.

PARABOLA WITH VERTEX (h, k)

		GRAPHS
VERTEX	(h, k)	
FOCUS	$(h, k + p)$	
DIRECTRIX	$y = -p + k$	
AXIS OF SYMMETRY	$x = h$	
DIRECTION	Up	
EQUATION	$(x - h)^2 = 4p(y - k)$	

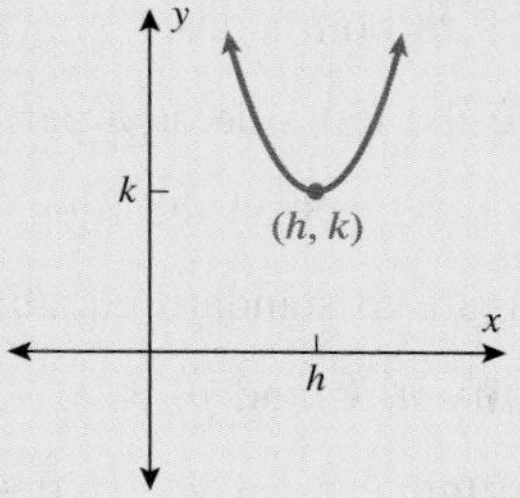

VERTEX	(h, k)	
FOCUS	$(h, k - p)$	
DIRECTRIX	$y = p + k$	
AXIS OF SYMMETRY	$x = h$	
DIRECTION	Down	
EQUATION	$(x - h)^2 = -4p(y - k)$	

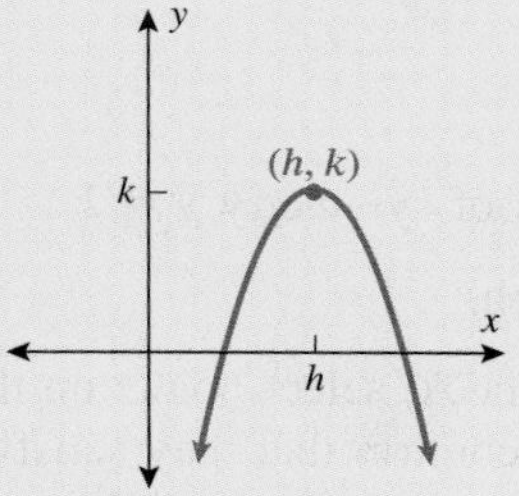

VERTEX	(h, k)	
FOCUS	$(h + p, k)$	
DIRECTRIX	$x = -p + h$	
AXIS OF SYMMETRY	$y = k$	
DIRECTION	Right	
EQUATION	$(y - k)^2 = 4p(x - h)$	

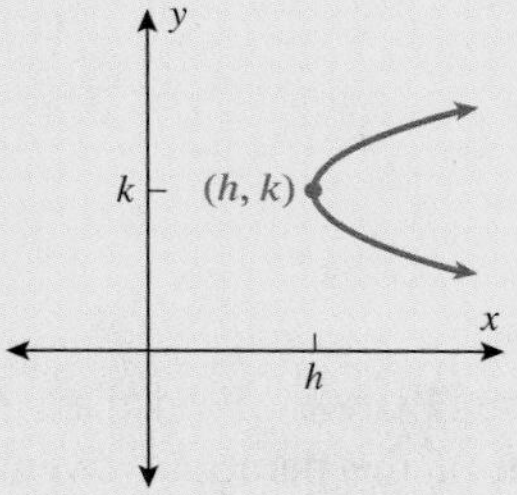

VERTEX	(h, k)	
FOCUS	$(h - p, k)$	
DIRECTRIX	$x = p + h$	
AXIS OF SYMMETRY	$y = k$	
DIRECTION	Left	
EQUATION	$(y - k)^2 = -4p(x - h)$	

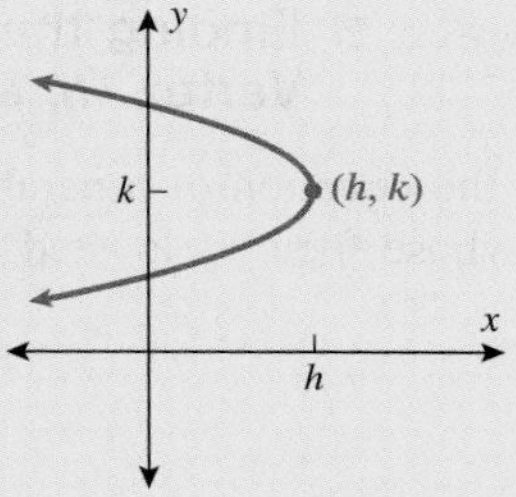

EXAMPLE 4 Graphing a Parabola with Vertex (h, k)

Graph the parabola given by the equation $y^2 - 6y + 8x + 17 = 0$.

Solution:

Since the equation $y^2 - 6y + 8x + 17 = 0$ is quadratic in y, we seek to transform this equation to the form $(y - k)^2 = 4p(x - h)$ or $(y - k)^2 = -4p(x - h)$.

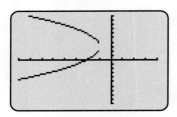

Complete the square on $y^2 - 6y + 8x + 17 = 0$ for y.

Isolate the y terms.	$y^2 - 6y = -8x - 17$
Add 9 to both sides.	$y^2 - 6y + 9 = -8x - 17 + 9$
Write the left side as a perfect square.	$(y - 3)^2 = -8x - 8$
Factor out -8 on the right.	$(y - 3)^2 = -8(x + 1)$
Compare to standard equation.	$(y - k)^2 = -4p(x - h)$
Identify h, k, and p.	$h = -1, k = 3, p = 2$

Substitute $h = -1, k = 3, p = 2$ into the characteristics of $(y - k)^2 = -4p(x - h)$.

Vertex $(h, k) = (-1, 3)$

Focus $(h - p, k) = (-3, 3)$

Directrix $x = p + h = 1$

Axis of symmetry $y = k = 3$ and latus rectum length $4|p| = 8$.

Graph.

To check, select points on the graph and confirm that they satisfy the equation of the parabola.

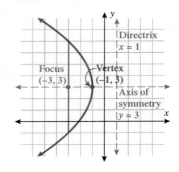

YOUR TURN Find the vertex, focus, directrix, and length of the latus rectum of the parabola given by $x^2 + 2x - 12y + 25 = 0$ and graph.

EXAMPLE 5 Finding the Equation of a Parabola with Vertex (h, k)

Find the equation of a parabola whose vertex is located at the point $(2, -3)$ and whose focus is located at the point $(5, -3)$.

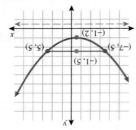

■ Answer: Vertex: $(-1, 2)$
Focus: $(-1, 5)$
Directrix: $x = -1$
Latus rectum: 12

Solution:

Draw a Cartesian plane and label the vertex and focus. The vertex and focus share the same axis of symmetry, $y = -3$ and indicate a parabola opening to the right.

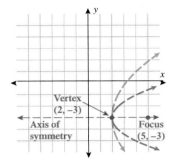

Write the standard equation of a parabola opening to the right.

$$(y - k)^2 = 4p(x - h)$$

Substitute the vertex, $(h, k) = (2, -3)$, into the standard equation.

$$(y - (-3))^2 = 4p(x - 2)$$

Find p.

The general form of the vertex is (h, k) and the focus is $(h + p, k)$.

For this parabola the vertex is $(2, -3)$ and the focus is $(5, -3)$.

Find p by taking the difference of the x coordinates.

$$p = 3$$

Substitute $p = 3$ into $(y - (-3))^2 = 4p(x - 2)$. $(y + 3)^2 = 4(3)(x - 2)$

Eliminate parentheses. $y^2 + 6y + 9 = 12x - 24$

Simplify. $y^2 + 6y - 12x + 33 = 0$

■ **YOUR TURN**   Find the equation of the parabola whose vertex is located at $(2, -3)$ and whose focus is located at $(0, -3)$.

Applications

If we start with a parabola in the xy-plane and rotate it around its axis of symmetry, the result will be a three-dimensional paraboloid.

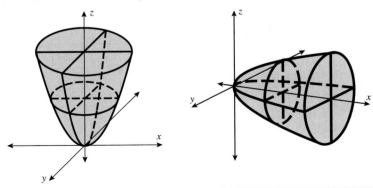

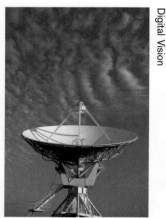

A satellite dish is in the shape of a paraboloid. Functioning as an antenna, the parabolic dish collects all of the signals and reflects them to a single point, the focal point, which is where the receiver is located.

EXAMPLE 6 Finding the Location of the Receiver in a Satellite Dish

A satellite dish is 24 feet in diameter at its opening and 4 feet deep in its center. Where should the receiver be placed?

Solution:

Draw a parabola with a vertex at the origin representing the center cross section of the satellite dish.

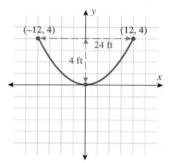

Satellite dish

Write the standard equation of a parabola opening up with vertex at the $(0, 0)$.

$$x^2 = 4py$$

The point $(12, 4)$ lies on the parabola, so substitute into $x^2 = 4py$.

$$(12)^2 = 4p(4)$$

Simplify.

$$144 = 16p$$

Solve for p.

$$p = 9$$

Substitute $p = 9$ into the focus, $(0, p)$.

$$\text{focus} = (0, 9)$$

The receiver should be placed 9 feet from the vertex of the dish.

SECTION 8.2 **SUMMARY**

In this section we discussed the geometric definition of a parabola, the equation of a parabola, and how to graph a parabola. The quantities that characterize a parabola are the vertex, focus, directrix, latus rectum, and axis of symmetry. Four types of parabolas were discussed: opening up, down, to the right, and to the left.

SECTION 8.2 EXERCISES

■ **SKILLS**

In Exercises 1–4, match the parabola to the equation.

1. $y^2 = 4x$ **2.** $y^2 = -4x$ **3.** $x^2 = -4y$ **4.** $x^2 = 4y$

a.

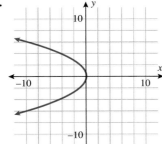

b.

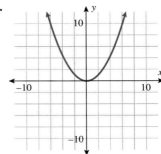

c.

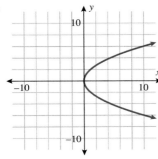

d.

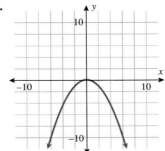

In Exercises 5–8, match the parabola to the equation.

5. $(y - 1)^2 = 4(x - 1)$ **6.** $(y + 1)^2 = -4(x - 1)$ **7.** $(x + 1)^2 = -4(y + 1)$ **8.** $(x - 1)^2 = 4(y - 1)$

a.

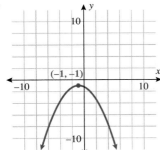

b.

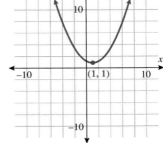

c.

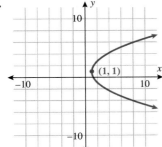

d.

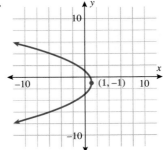

In Exercises 9–20, find an equation for the parabola described.

9. vertex at $(0, 0)$; focus at $(0, 3)$ **10.** vertex at $(0, 0)$; focus at $(2, 0)$ **11.** vertex at $(0, 0)$; focus at $(-5, 0)$

12. vertex at $(0, 0)$; focus at $(0, -4)$ **13.** vertex at $(3, 5)$; focus at $(3, 7)$ **14.** vertex at $(3, 5)$; focus at $(7, 5)$

15. vertex at $(2, 4)$; focus at $(0, 4)$ **16.** vertex at $(2, 4)$; focus at $(2, -1)$ **17.** focus at $(2, 4)$; directrix at $y = -2$

18. focus at $(2, -2)$; directrix at $y = 4$ **19.** focus at $(3, -1)$; directrix at $x = 1$ **20.** focus at $(-1, 5)$; directrix at $x = 5$

In Exercises 21–24, write an equation for each parabola.

21.

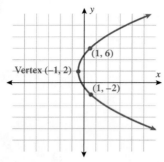

22.

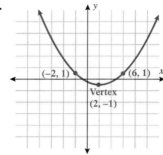

23.

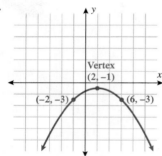

24.

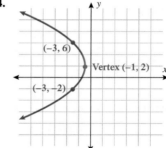

In Exercises 25–40, find the focus, vertex, directrix, and length of latus rectum and graph the parabola.

25. $x^2 = 8y$ **26.** $x^2 = -12y$ **27.** $y^2 = -2x$ **28.** $y^2 = 6x$

29. $(y - 2)^2 = 4(x + 3)$ **30.** $(y + 2)^2 = -4(x - 1)$ **31.** $(x - 3)^2 = -8(y + 1)$ **32.** $(x + 3)^2 = -8(y - 2)$

33. $(x + 5)^2 = -2y$ **34.** $y^2 = -16(x + 1)$ **35.** $y^2 - 4y - 2x + 4 = 0$ **36.** $x^2 - 6x + 2y + 9 = 0$

37. $y^2 + 2y - 8x - 23 = 0$ **38.** $x^2 - 6x - 4y + 10 = 0$ **39.** $x^2 - x + y - 1 = 0$ **40.** $y^2 + y - x + 1 = 0$

■**APPLICATIONS**

41. Satellite Dish. A satellite dish measures 8 feet across its opening and 2 feet deep at its center. The receiver should be placed at the focus of the parabolic dish. Where should the receiver be placed?

42. Satellite Dish. A satellite dish measures 30 feet across its opening and 5 feet deep at its center. The receiver should be placed at the focus of the parabolic dish. Where should the receiver be placed?

43. Eyeglass Lens. Eyeglass lenses can be thought of as very wide parabolic curves. If the focus occurs 2 centimeters

from the center of the lens, and the lens at its opening is 5 centimeters, find an equation that governs the shape of the center cross section of the lens.

44. Optical Lens. A parabolic lens focuses light onto a focal point 3 centimeters from the vertex of the lens. How wide is the lens 0.5 centimeter from the vertex?

Parabolic shapes are often used to generate intense heat by collecting sun rays and focusing all of them at a focal point. Exercises 45 and 46 are examples of solar cookers.

45. Solar Cooker. The parabolic cooker MS-ST10 is delivered as a kit, handily packed in a single carton, with complete assembly instructions and even the necessary tools.

Solar cooker, Ubuntu Village, Johannesburg, South Africa

Thanks to the reflector diameter of 1 meter, it develops an immense power of close to 500 watts under 1 kW/m² of incoming radiation. Under a clear sky, 1 liter of water boils in significantly less than half an hour, $2\frac{1}{2}$ liters in around 40 minutes. If the rays are focused 40 centimeters from the vertex, find the equation for the parabolic cooker.

46. Le Four Solaire at Font-Romeur "Mirrors of the Solar Furnace" There is a reflector in the Pyrenees Mountains that is eight stories high. It cost two million dollars and it took 10 years to build it. It is made of 9000 mirrors arranged in a parabolic mirror. It can reach 6000 degrees Fahrenheit just from the Sun!

Solar furnace, Odellio, France

If the diameter of the parabolic mirror is 100 meters and the sunlight is focused 25 meters from the vertex, find the equation for the parabolic dish.

47. Sailing under a Bridge. A bridge with a parabolic shape has an opening 80 feet wide at the base of the bridge (where the bridge meets the water), and the height in the center of the bridge is 20 feet. A sailboat whose mast reaches 17 feet above the water is traveling under the bridge 10 feet from the center of the bridge. Will it clear the bridge without scraping its mast? Justify your answer.

Bridge, Baden-Württemberg, Heidelberg, Germany

48. Driving under a Bridge. A bridge with a parabolic shape reaches a height of 25 feet in the center of the road, and the width of the bridge opening at ground level is 20 feet combined (both lanes). If an RV is 10 feet tall and is 8 feet wide, it won't make it under the bridge if it hugs the center line. What if it straddles the center line? Will it make it? Justify your answer.

49. Braking Distance. The *Virginia Department of Motor Vehicles Manual* shows braking distances. Braking distances are quadratic functions of speed. Determine the equation that relates braking distances to speed.

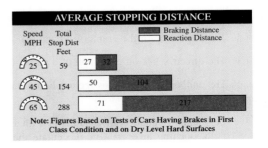

AVERAGE STOPPING DISTANCE		
Speed MPH	Total Stop Dist Feet	▪ Braking Distance ☐ Reaction Distance
25	59	27 32
45	154	50 104
65	288	71 217

Note: Figures Based on Tests of Cars Having Brakes in First Class Condition and on Dry Level Hard Surfaces

50. Suspension Bridge. If one parabolic segment of a suspension bridge is 300 feet and if the cables at the vertex are suspended 10 feet above the bridge, whereas the height of the cables 150 feet from the vertex reach 60 feet high, find the equation of the parabolic path of the suspension cables.

Pont de Terenez in Finistere, Brittany, France

E. Streichan/Masterfile

■ CATCH THE MISTAKE

In Exercises 51 and 52, explain the mistake that is made.

51. Find an equation for a parabola whose vertex is at the origin and whose focus is at the point $(3, 0)$.

Solution:

Write the general equation for a parabola whose vertex is at the origin. $x^2 = 4py$

The focus of this parabola is $(p, 0) = (3, 0)$. $p = 3$

Substitute $p = 3$ into $x^2 = 4py$. $x^2 = 12y$

This is incorrect. What mistake was made?

52. Find an equation for a parabola whose vertex is at the point $(3, 2)$ and whose focus is located at $(5, 2)$.

Solution:

Write the equation associated with a parabola whose vertex is $(3, 2)$. $(x - h)^2 = 4p(y - k)$

Substitute $(3, 2)$ into $(x - h)^2 = 4p(y - k)$. $(x - 3)^2 = 4p(y - 2)$

The focus is located at $(5, 2)$; therefore $p = 5$.

Substitute $p = 5$ into $(x - 3)^2 = 4p(y - 2)$. $(x - 3)^2 = 20(y - 2)$

This is not correct. What mistake (s) were made?

■ CHALLENGE

53. T or F: The vertex lies on the graph of a parabola.

54. T or F: The focus lies on the graph of a parabola.

55. T or F: The directrix lies on the graph of a parabola.

56. T or F: The endpoints of the latus rectum lie on the graph of a parabola.

57. Derive the standard equation of a parabola with vertex at the origin, opening up $x^2 = 4py$. [Calculate the distance, d_1, from any point on the parabola, (x, y), to the focus, $(0, p)$. Calculate the distance, d_2, from any point on the parabola, (x, y), to the directrix $(-p, y)$. Set $d_1 = d_2$.]

58. Derive the standard equation of a parabola opening right $y^2 = 4px$. [Calculate the distance, d_1, from any point on the parabola, (x, y), to the focus, $(p, 0)$. Calculate the distance, d_2, from any point on the parabola, (x, y), to the directrix $(x, -p)$. Set $d_1 = d_2$.]

59. Using a graphing utility, plot the parabola $x^2 - x + y - 1 = 0$. Compare with the sketch you drew for Exercise 39.

60. Using a graphing utility, plot the parabola $y^2 + y - x + 1 = 0$. Compare with the sketch you drew for Exercise 40.

61. In your mind, picture the parabola given by $(y + 3.5)^2 = 10(x - 2.5)$. Where is the vertex? Which way does this parabola open? Now plot the parabola using a graphing utility.

62. In your mind, picture the parabola given by $(x + 1.4)^2 = -5(y + 1.7)$. Where is the vertex? Which way does this parabola open? Now plot the parabola using a graphing utility.

SECTION 8.3 The Ellipse

Skills Objectives	Conceptual Objectives
■ Graph an ellipse given the center, major axis, and minor axis.	■ Derive the general equation of an ellipse.
■ Find the equation of an ellipse centered at the origin.	■ Understand the meaning of major and minor axes and foci.
■ Find the equation of an ellipse centered at the point (h, k).	■ Understand the difference between a circle and an ellipse.
■ Solve applied problems that involve ellipses.	

Definition of an Ellipse

If we were to take a piece of string, tie loops at both ends, and tack the ends down so that the string had lots of slack, we would have the picture on the right.

If we then took a pencil and pulled the string taut and traced our way around for one full rotation the result would be an ellipse. See second figure on the right.

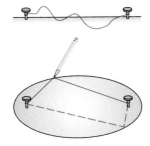

DEFINITION ELLIPSE

An **ellipse** is the set of all points whose sum of their distances from two fixed points is constant. These two fixed points are called **foci** (plural of focus). The line through the foci, called the **major axis**, intersects the ellipse at the **vertices**. The midpoint of the line segment joining the vertices is called the **center**. The line segment that joins two points on the ellipse and is perpendicular to the major axis is called the **minor axis**.

Let's start with an ellipse whose center is located at the origin. Using graph-shifting techniques, we can later extend the characteristics of an ellipse centered at a point other than the origin. Ellipses can take the shape of circles or be stretchy to resemble a racetrack. It can be shown that the standard equation of an ellipse with its center at the origin is given by one of two forms, depending on if the orientation of the major axis of the ellipse is horizontal or vertical. If it is horizontal, then the equation is given by $\frac{x^2}{a^2} + \frac{y^2}{b^2} = 1$. If the ellipse is vertical, then the equation is given by $\frac{x^2}{b^2} + \frac{y^2}{a^2} = 1$, for $a > b > 0$. The horizontal and vertical ellipses are summarized in the table below.

EQUATION OF AN ELLIPSE WITH CENTER AT THE ORIGIN

The **standard form of the equation of an ellipse** with its center at the origin is given by:

ORIENTATION OF MAJOR AXIS	EQUATION	GRAPH	FOCI	VERTICES
Horizontal Along the x-axis	$\frac{x^2}{a^2} + \frac{y^2}{b^2} = 1$		$(-c, 0)$ $(c, 0)$	$(-a, 0)$ $(a, 0)$
Vertical Along the y-axis	$\frac{x^2}{b^2} + \frac{y^2}{a^2} = 1$		$(0, -c)$ $(0, c)$	$(0, -a)$ $(0, a)$

Where $a > b > 0$, $c^2 = a^2 - b^2$, the length of the major axis is $2a$ and the length of the minor axis is $2b$.

Graphing an Ellipse with Center at the Origin

The equation of an ellipse in standard form can be used to graph an ellipse. Although an ellipse is defined in terms of the foci, the foci are not part of the graph.

EXAMPLE 1 Graphing an Ellipse with a Horizontal Major Axis

Graph the ellipse given by $\frac{x^2}{25} + \frac{y^2}{9} = 1$.

Solution:

Since $25 > 9$ this ellipse is horizontal. $a^2 = 25$ and $b^2 = 9$

Solve for a and b. $a = 5$ and $b = 3$

Identify the vertices: $(-a, 0)$ and $(a, 0)$.

$(-5, 0)$ and $(5, 0)$

Identify the endpoints of the minor axis: $(0, -b)$ and $(0, b)$.

$(0, -3)$ and $(0, 3)$

Graph by labeling the points, $(-5, 0)$, $(5, 0)$, $(0, -3)$, and $(0, 3)$, and connect with a smooth elliptic curve.

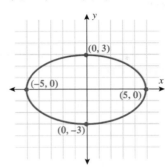

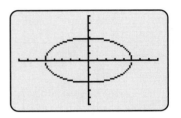

TECHNOLOGY TIP
Use a graphing calculator to check the graph of $\frac{x^2}{25} + \frac{y^2}{9} = 1$. Solve for y first. That is,
$y_1 = 3\sqrt{1 - \frac{x^2}{25}}$ or
$y_2 = -3\sqrt{1 - \frac{x^2}{25}}$.

If the denominator of x^2 is larger than the denominator of y^2, then the major axis is horizontal along the x-axis, as in Example 1. If the denominator of y^2 is larger than the denominator of x^2, then the major axis is vertical along the y-axis, as you will see in Example 2.

EXAMPLE 2 Graphing an Ellipse with a Vertical Major Axis

Graph the ellipse given by $16x^2 + y^2 = 16$.

Solution:

Write the equation in standard form by dividing by 16.

$$\frac{x^2}{1} + \frac{y^2}{16} = 1$$

Since $16 > 1$, this ellipse is vertical.

$a^2 = 16$ and $b^2 = 1$

Solve for a and b.

$a = 4$ and $b = 1$

Identify the vertices: $(0, -a)$ and $(0, a)$.

$(0, -4)$ and $(0, 4)$

Identify the endpoints of the minor axis: $(-b, 0)$ and $(b, 0)$.

$(-1, 0)$ and $(1, 0)$

Graph by labeling the points, $(0, -4)$, $(0, 4)$, $(-1, 0)$, and $(1, 0)$, and connect with a smooth elliptic curve.

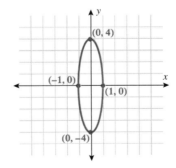

CONCEPT CHECK What direction (horizontal or vertical) is the major axis of each of the following ellipses:

$$\frac{x^2}{8} + \frac{y^2}{12} = 1 \text{ and } \frac{x^2}{8} + \frac{y^2}{4} = 1$$

YOUR TURN Graph the ellipses:

a. $\dfrac{x^2}{9} + \dfrac{y^2}{4} = 1$ **b.** $\dfrac{x^2}{9} + \dfrac{y^2}{36} = 1$

Finding the Equation of an Ellipse Centered at the Origin

What if we know the vertices and the foci of an ellipse and want to find the equation that it corresponds to? Whatever axis the foci and vertices are located on is the major axis. Therefore, we will have the standard equation of an ellipse, and a will be known (from the vertices). Since c is known from the foci, we can use the relation $c^2 = a^2 - b^2$ to determine the unknown b.

EXAMPLE 3 Finding the Equation of an Ellipse Centered at the Origin

Find the standard form of the equation of an ellipse with foci at $(-3, 0)$ and $(3, 0)$ and vertices $(-4, 0)$ and $(4, 0)$.

Solution:

The major axis is the x-axis, since it contains the foci and vertices.

Write the corresponding general equation of an ellipse. $\dfrac{x^2}{a^2} + \dfrac{y^2}{b^2} = 1$

Identify a from the vertices.

Match vertices, $(-4, 0) = (-a, 0)$ and $(4, 0) = (a, 0)$ $a = 4$

Identify c from the foci.

Match foci, $(-3, 0) = (-c, 0)$ and $(3, 0) = (c, 0)$. $c = 3$

Substitute $a = 4$ and $c = 3$ into $b^2 = a^2 - c^2$. $b^2 = 4^2 - 3^2$

Simplify. $b^2 = 7$

Substitute $a^2 = 16$ and $b^2 = 7$ into $\dfrac{x^2}{a^2} + \dfrac{y^2}{b^2} = 1$. $\dfrac{x^2}{16} + \dfrac{y^2}{7} = 1$

The equation of the ellipse is $\dfrac{x^2}{16} + \dfrac{y^2}{7} = 1$.

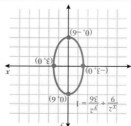

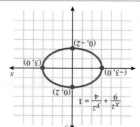

Answer: a.

b.

YOUR TURN   Find the standard form of the equation of an ellipse with vertices at $(0, -6)$ and $(0, 6)$ and foci $(0, -5)$ and $(0, 5)$.

An Ellipse Centered at the Point (*h*, *k*)

We can use graph-shifting techniques to graph ellipses that are centered at a point other than the origin. For example, to graph $\dfrac{(x-h)^2}{a^2} + \dfrac{(y-k)^2}{b^2} = 1$, start with the graph of $\dfrac{x^2}{a^2} + \dfrac{y^2}{b^2} = 1$ and shift to the right h units and up k units. The center, the vertices, the foci, and the major and minor axes all shift. In other words, the two ellipses are identical in shape and size, except that the ellipse, $\dfrac{(x-h)^2}{a^2} + \dfrac{(y-k)^2}{b^2} = 1$ is centered at the point (h, k).

The following table summarizes the characteristics of ellipses centered at a point other than the origin.

EQUATION OF AN ELLIPSE WITH CENTER AT THE POINT (*h*, *k*)

The **standard form of the equation of an ellipse** with its center at the point (h, k) is given by:

ORIENTATION OF MAJOR AXIS	EQUATION	GRAPH	FOCI
Horizontal (Parallel to the *x*-axis)	$\dfrac{(x-h)^2}{a^2} + \dfrac{(y-k)^2}{b^2} = 1$		$(h - c, k)$ $(h + c, k)$
Vertical (Parallel to the *y*-axis)	$\dfrac{(x-h)^2}{b^2} + \dfrac{(y-k)^2}{a^2} = 1$		$(h, k - c)$ $(h, k + c)$

Where $a > b > 0$, $c^2 = a^2 - b^2$, the length of the major axis is $2a$ and the length of the minor axis is $2b$.

■ **Answer:** $\dfrac{x^2}{11} + \dfrac{y^2}{36} = 1$

EXAMPLE 4 Graphing an Ellipse with Center (h, k), Given the Equation in Standard Form

Graph the ellipse given by $\dfrac{(x-2)^2}{9} + \dfrac{(y+1)^2}{16} = 1$.

Solution:

Write the equation in the form

$$\dfrac{(x-h)^2}{b^2} + \dfrac{(y-k)^2}{a^2} = 1. \qquad\qquad \dfrac{(x-2)^2}{3^2} + \dfrac{(y-(-1))^2}{4^2} = 1$$

Identify a, b, and the center (h, k). $a = 4$, $b = 3$, and center $= (2, -1)$

Draw a graph and label the center: $(2, -1)$

Since $a = 4$, the vertices are up 4 and down 4 units from the center: $(2, -5)$ and $(2, 3)$

Since $b = 3$, the endpoints of the minor axis are to the left and right 3 units: $(-1, -1)$ and $(5, -1)$

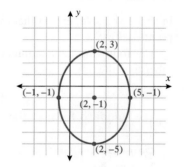

■ YOUR TURN Graph the ellipse given by $\dfrac{(x+1)^2}{9} + \dfrac{(y-3)^2}{1} = 1$.

All active members of the Lambda Chi fraternity are college students, but not all college students are members of the Lambda Chi fraternity. Similarly, all circles are ellipses but not all ellipses are circles. When $a = b$, the standard equation of an ellipse simplifies to a standard equation of a circle. Recall that when we are given the equation of a circle in general form, we first complete the square in order to

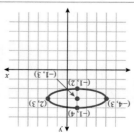

■ Answer:

express the equation in standard form, which allows the center and radius to be iden-tified. That same approach is used when the equation of an ellipse is given in a gen-eral form.

EXAMPLE 5 Graphing an Ellipse with Center (h, k), Given an Equation in General Form

Graph the ellipse given by $4x^2 + 24x + 25y^2 - 50y - 39 = 0$.

Solution:

Transform general equation into standard form.

Group x terms together and y terms together and add 39 to both sides.

$$(4x^2 + 24x) + (25y^2 - 50y) = 39$$

Factor out the 4 common to x terms and the 25 common to y terms.

$$4(x^2 + 6x) + 25(y^2 - 2y) = 39$$

Complete the square on x and y.

$$4(x^2 + 6x + 9) + 25(y^2 - 2y + 1) = 39 + 4(9) + 25(1)$$

Simplify.

$$4(x + 3)^2 + 25(y - 1)^2 = 100$$

Divide by 100.

$$\frac{(x + 3)^2}{25} + \frac{(y - 1)^2}{4} = 1.$$

Since $25 > 4$, this is a horizontal ellipse.

Now that the equation of the ellipse is in standard form, compare to $\frac{(x - h)^2}{a^2} + \frac{(y - k)^2}{b^2} = 1$ and identify a, b, h, k.

$a = 5$, $b = 2$

Center at $(-3, 1)$.

Since $a = 5$, the vertices are 5 units to the left and right of the center.

$(-8, 1)$ and $(2, 1)$

Since $b = 2$, the endpoints of the minor axis are up and down 2 units from the center.

$(-3, -1)$ and $(-3, 3)$

Graph.

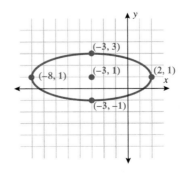

■ **YOUR TURN** Write the equation $4x^2 + 32x + y^2 - 2y + 61 = 0$ in standard form. Identify the center, vertices, and endpoints of the minor axis and graph.

Applications

There are many examples of ellipses all around us. On Earth we have racetracks, and in our solar system, the planets have elliptical orbits with the Sun as a focus. Satellites have elliptical orbits around the Earth. Most communications satellites are in a *geosynchronous* (GEO) orbit—orbit circles the Earth once each day.

To stay over the same spot on Earth, a *geostationary* satellite also has to be directly above the equator.

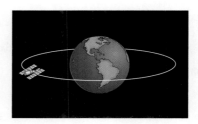

Otherwise, from the Earth the satellite would appear to move in a north–south line every day. The footprints of GEO satellites do not cover the polar regions of Earth. So communications satellites in north–south elliptical orbits cover the areas in the high northern and southern hemispheres that are not covered by GEO satellites, as shown in the figure on the left.

If we start with an ellipse in the *xy*-plane and rotate it around its major axis, the result would be a three-dimensional ellipsoid.

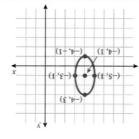

■ **Answer:**

$$\frac{(x + 4)^2}{1} + \frac{(y - 1)^2}{4} = 1$$

Center: $(-4, 1)$
Vertices: $(-4, -1)$ and $(-4, 3)$
Endpoints of minor axis: $(-5, 1)$ and $(-3, 1)$

A football and a blimp are two examples of ellipsoids. The ellipsoid shape allows for a more aerodynamic path.

PhotoDisc, Inc.

Peter Phipp/Age Fotostock America, Inc.

EXAMPLE 6 An Official NFL Football

A cross section (that includes the two vertices and the center) of an official Wilson NFL football is an ellipse. The cross section is approximately 11 inches long and 7 inches wide. Write an equation governing the elliptical cross section.

PhotoDisc, Inc.

Solution:

Locate the center of the ellipse at the origin and orient the football horizontally.

Write the general equation of a circle centered at the origin.
$$\frac{x^2}{a^2} + \frac{y^2}{b^2} = 1$$

The length of the major axis is 11 inches. $2a = 11$

Solve for a. $a = 5.5$

The length of the minor axis is 7 inches. $2b = 7$

Solve for b. $b = 3.5$

Substitute $a = 5.5$ and $b = 3.5$ into $\frac{x^2}{a^2} + \frac{y^2}{b^2} = 1$.
$$\frac{x^2}{5.5^2} + \frac{y^2}{3.5^2} = 1$$

SECTION 8.3 SUMMARY

In this section an ellipse was defined geometrically as a set of points whose sum of the distances to two fixed points, foci, is constant. There are two types of ellipses centered at the origin: horizontal and vertical. Equations for both horizontal and vertical ellipses centered at the origin were given. Although an ellipse is defined in terms of its foci, the foci are not part of the graph. The graph can be obtained directly from the equation of an ellipse. Horizontal and vertical graph-shifting techniques can be used to graph an ellipse centered at the point (h, k).

SECTION 8.3 EXERCISES

■ SKILLS

In Exercises 1–4, match the equation to the ellipse.

1. $\dfrac{x^2}{36} + \dfrac{y^2}{16} = 1$
2. $\dfrac{x^2}{16} + \dfrac{y^2}{36} = 1$
3. $\dfrac{x^2}{8} + \dfrac{y^2}{72} = 1$
4. $4x^2 + y^2 = 1$

a.

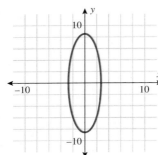

b.

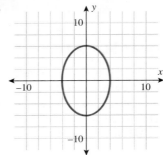

c.

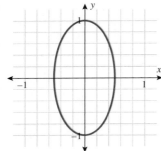

d.

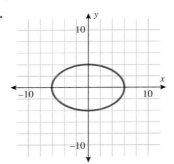

In Exercises 5–16, graph each ellipse.

5. $\dfrac{x^2}{25} + \dfrac{y^2}{16} = 1$
6. $\dfrac{x^2}{49} + \dfrac{y^2}{9} = 1$
7. $\dfrac{x^2}{16} + \dfrac{y^2}{64} = 1$
8. $\dfrac{x^2}{25} + \dfrac{y^2}{144} = 1$

9. $\dfrac{x^2}{100} + y^2 = 1$
10. $9x^2 + 4y^2 = 36$
11. $\dfrac{x^2}{\frac{9}{4}} + \dfrac{y^2}{\frac{1}{81}} = 1$
12. $\dfrac{4}{25}x^2 + \dfrac{100}{9}y^2 = 1$

13. $4x^2 + y^2 = 16$
14. $x^2 + y^2 = 81$
15. $8x^2 + 16y^2 = 32$
16. $10x^2 + 25y^2 = 50$

In Exercises 17–24, find the standard form of an equation of the ellipse with the given characteristics.

17. foci: $(-4, 0)$ and $(4, 0)$ vertices: $(-6, 0)$ and $(6, 0)$
18. foci: $(-1, 0)$ and $(1, 0)$ vertices: $(-3, 0)$ and $(3, 0)$

19. foci: $(0, -3)$ and $(0, 3)$ vertices: $(0, -4)$ and $(0, 4)$
20. foci: $(0, -1)$ and $(0, 1)$ vertices: $(0, -2)$ and $(0, 2)$

21. Major axis vertical with length of 8, minor axis length of 4 and centered at $(0, 0)$.

22. Major axis horizontal with length of 10, minor axis length of 2 and centered at $(0, 0)$.

23. Vertices $(0, -7)$ and $(0, 7)$ and endpoints of minor axis $(-3, 0)$ and $(3, 0)$.

24. Vertices $(-9, 0)$ and $(9, 0)$ and endpoints of minor axis $(0, -4)$ and $(0, 4)$.

In Exercises 25–28, match each equation with the ellipse.

25. $\dfrac{(x - 3)^2}{4} + \dfrac{(y + 2)^2}{25} = 1$ **26.** $\dfrac{(x + 3)^2}{4} + \dfrac{(y - 2)^2}{25} = 1$ **27.** $\dfrac{(x - 3)^2}{25} + \dfrac{(y + 2)^2}{4} = 1$ **28.** $\dfrac{(x + 3)^2}{25} + \dfrac{(y - 2)^2}{4} = 1$

a.

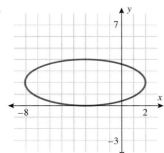

b.

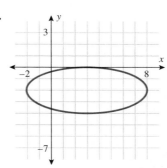

c.

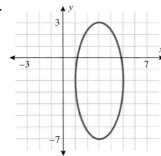

d.

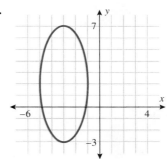

In Exercises 29–38, graph each ellipse.

29. $\dfrac{(x - 1)^2}{16} + \dfrac{(y - 2)^2}{4} = 1$

30. $\dfrac{(x + 1)^2}{36} + \dfrac{(y + 2)^2}{9} = 1$

31. $10(x + 3)^2 + (y - 4)^2 = 80$

32. $3(x + 3)^2 + 12(y - 4)^2 = 36$

33. $x^2 + 4y^2 - 24y + 32 = 0$

34. $25x^2 + 2y^2 - 4y - 48 = 0$

35. $x^2 - 2x + 2y^2 - 4y - 5 = 0$

36. $9x^2 - 18x + 4y^2 - 27 = 0$

37. $5x^2 + 20x + y^2 + 6y - 21 = 0$

38. $9x^2 + 36x + y^2 + 2y + 36 = 0$

In Exercises 39–46, find the standard form of an equation of the ellipse with the given characteristics.

39. foci: $(-2, 5)$ and $(6, 5)$ vertices: $(-3, 5)$ and $(7, 5)$ **40.** foci: $(2, -2)$ and $(4, -2)$ vertices: $(0, -2)$ and $(6, -2)$

41. foci: $(4, -7)$ and $(4, -1)$ vertices: $(4, -8)$ and $(4, 0)$ **42.** foci: $(2, -6)$ and $(2, -4)$ vertices: $(2, -7)$ and $(2, -3)$

43. Major axis vertical with length of 8, minor axis length of 4 and centered at $(3, 2)$.

44. Major axis horizontal with length of 10, minor axis length of 2 and centered at $(-4, 3)$.

45. Vertices $(-1, -9)$ and $(-1, 1)$ and endpoints of minor axis $(-4, -4)$ and $(2, -4)$.

46. Vertices $(-2, 3)$ and $(6, 3)$ and endpoints of minor axis $(2, 1)$ and $(2, 5)$.

■ APPLICATIONS

47. Carnival Ride. The Zipper, a favorite carnival ride, has a major axis of 150 feet and a minor axis of 30 feet. Assuming it is centered at the origin, find an equation for the ellipse.

Courtesy Chance Morgan, Inc.

Zipper

48. Carnival Ride. A Ferris wheel has a major and minor axis of 180 feet. Assuming it is centered at the origin, find an equation for the ellipse (circle).

Tina Buckman/Index Stock

Ferris wheel, Barcelona, Spain

For Exercises 49 and 50, refer to the following information.

A high school wants to build a football field surrounded by an elliptical track. A regulation football field must be 120 yards long and 30 yards wide.

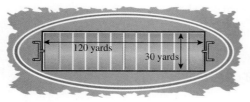

120 yards
30 yards

49. Sports Field. Suppose the elliptical track is centered at the origin and has a horizontal major axis of length 150 yards and a minor axis length of 40 yards.

   a. Write an equation for the ellipse.

   b. Find the width of the track at the end of the field. Will the track completely enclose the football field?

50. Sports Field. Suppose the elliptical track is centered at the origin and has a horizontal major axis of length 150 yards, how long should the minor axis be in order to enclose the field?

For Exercises 51 and 52, refer to orbits in our solar system.

The planets have elliptical orbits with the Sun as one of the foci.

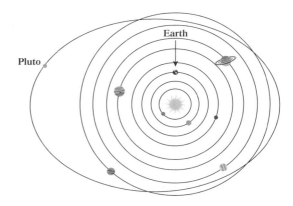

Pluto (orange), the planet furthest from the Sun, has a very elliptical orbit, whereas Earth (royal blue) has almost a circular orbit. Because of Pluto's elliptical path it is not always the planet furthest from the Sun.

51. Planetary Orbits. Pluto's orbit is summarized in the picture below. Use the fact that the Sun is a focus to determine an equation for Pluto's elliptical orbit around the Sun. Round to the nearest million kilometers.

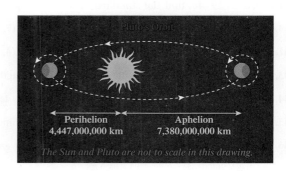

Pluto's Orbit

Perihelion
4,447,000,000 km

Aphelion
7,380,000,000 km

The Sun and Pluto are not to scale in this drawing.

52. Planetary Orbits. Earth's orbit is summarized in the picture below. Use the fact that the Sun is a focus to determine an equation for Earth's elliptical orbit around the Sun. Round to the nearest thousand kilometers.

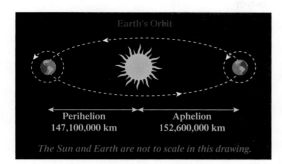

Earth's Orbit

Perihelion 147,100,000 km   Aphelion 152,600,000 km

The Sun and Earth are not to scale in this drawing.

Asteroids, meteors, and comets all orbit the Sun in elliptical patterns and often cross paths with the Earth's orbit, making life a little tense now and again. Asteroids are rocks (or mini planets), meteors are fine sand particles, and comets have debris. A few asteroids have orbits which cross the Earth's orbit—called "Apollos" or "Earth-crossing asteroids." In recent years asteroids have passed within 100,000 kilometers of the Earth!

53. Asteroids. The asteroid 433 Eros is the second largest near-Earth asteroid. The semimajor axis is 150 million kilometers and the eccentricity is 0.223, where eccentricity is defined as $e = \sqrt{1 - \dfrac{b^2}{a^2}}$ where a is the semimajor axis, or $2a$ is the major axis, and b is the semiminor axis, or $2b$ is the minor axis. Find the equation of 433 Eros' orbit. Round to the nearest millionth kilometer.

54. Asteroids. The asteroid Toutatis is the largest near-Earth asteroid. The semimajor axis is 350 million kilometers and the eccentricity is 0.634, where eccentricity is defined as $e = \sqrt{1 - \dfrac{b^2}{a^2}}$ where a is the semimajor axis, or $2a$ is the major axis, and b is the semimajor axis, or $2b$ is the minor axis. On September 29, 2004 it missed Earth by 961,000 miles. Find the equation of Toutatis' orbit.

■ CATCH THE MISTAKE

In Exercises 55 and 56, explain the mistake that is made.

55. Graph the ellipse given by $\dfrac{x^2}{6} + \dfrac{y^2}{4} = 1$.

Solution:

Write the standard form of the equation of an ellipse.
$$\frac{x^2}{a^2} + \frac{y^2}{b^2} = 1$$

Identify a and b.
$$a = 6, b = 4$$

Label the vertices and the endpoints of the minor axis, $(-6, 0), (6, 0), (0, -4)$ $(0, 4)$, and connect with an elliptical curve.

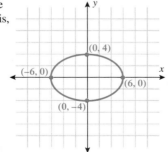

This is incorrect. What mistake was made?

56. Determine the foci of the ellipse $\dfrac{x^2}{16} + \dfrac{y^2}{9} = 1$.

Solution:

Write the general equation of a horizontal ellipse.
$$\frac{x^2}{a^2} + \frac{y^2}{b^2} = 1$$

Identify a and b.
$$a = 4, b = 3$$

Substitute $a = 4, b = 3$ into $c^2 = a^2 + b^2$.
$$c^2 = 4^2 + 3^2$$

Solve for c.
$$c = 5$$

Foci are located at $(-5, 0)$ and $(5, 0)$.

The points $(-5, 0)$ and $(5, 0)$ are located outside of the ellipse. What mistake was made?

■ CHALLENGE

57. T or F: If you know the vertices of an ellipse, you can determine the equation for the ellipse.

58. T or F: If you know the foci and the endpoints of the minor axis, you can determine the equation for the ellipse.

59. T or F: Ellipses centered at the origin have symmetry with respect to the x-axis, y-axis, and the origin.

60. T or F: All ellipses are circles, but not all circles are ellipses.

61. The eccentricity of an ellipse is defined as $e = \frac{c}{a}$. Compare the eccentricity of Pluto to Earth (refer to Exercises 51 and 52).

62. The eccentricity of an ellipse is defined as $e = \frac{c}{a}$. Since $a > c > 0$, then $0 < e < 1$. Describe the shape of an ellipse when

a. e is close to zero

b. e is close to one

c. $e = 0.5$

■ TECHNOLOGY

63. Graph the following three ellipses: $x^2 + y^2 = 1$, $x^2 + 5y^2 = 1$, and $x^2 + 10y^2 = 1$. What can be said to happen to the ellipse $x^2 + cy^2 = 1$ as c increases?

64. Graph the following three ellipses: $x^2 + y^2 = 1$, $5x^2 + y^2 = 1$, and $10x^2 + y^2 = 1$. What can be said to happen to the ellipse $cx^2 + y^2 = 1$ as c increases?

65. Graph the following three ellipses: $x^2 + y^2 = 1$, $5x^2 + 5y^2 = 1$, and $10x^2 + 10y^2 = 1$. What can be said to happen to the ellipse $cx^2 + cy^2 = 1$ as c increases?

66. Graph the equation $\frac{x^2}{9} - \frac{y^2}{16} = 1$. Notice what a difference the sign makes. Is this an ellipse?

SECTION 8.4 The Hyperbola

Skills Objectives

■ Find a hyperbola's foci and vertices.
■ Find the equation of a hyperbola centered at the origin.
■ Graph a hyperbola using asymptotes as graphing aids.
■ Find the equation of a hyperbola centered at the point (h, k).
■ Solve applied problems that involve hyperbolas.

Conceptual Objectives

■ Derive the general equation of a hyperbola.
■ Identify, use, and graph the transverse axis, vertices, and foci.

Definition of a Hyperbola

The definition of a hyperbola is similar to the definition of an ellipse. An ellipse is the set of all points whose *sum* of the distances between the foci and a point on the ellipse is constant. A *hyperbola* is the set of all points whose *difference* of the distances between the foci and a point on the hyperbola is constant. What distinguishes their equations is a minus sign.

$$\text{Ellipse centered at the origin:} \quad \frac{x^2}{a^2} + \frac{y^2}{b^2} = 1$$

$$\text{Hyperbola centered at the origin:} \quad \frac{x^2}{a^2} - \frac{y^2}{b^2} = 1$$

DEFINITION   HYPERBOLA

A **hyperbola** is the set of all points whose difference of their distances from two fixed points is a positive constant. These two fixed points are called **foci** (plural of focus). The line containing the foci is called the **transverse axis**. The hyperbola has two separate curves called **branches**. The **transverse axis** intersects the hyperbola at two points called the **vertices**. The midpoint of the line segment joining the vertices is called the **center**. The line through the center and perpendicular to the transverse axis is the **conjugate axis**.

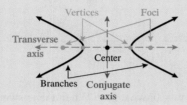

Hyperbolas Centered at the Origin

Let's start with the special case when the center is located at the origin. It can be shown that the standard equation of a hyperbola with its center at the origin is given by one of two forms, depending on if the orientation of the transverse axis of the hyperbola is horizontal or vertical. If it is horizontal, then the equation is given by $\dfrac{x^2}{a^2} - \dfrac{y^2}{b^2} = 1$. If the transverse axis is vertical, then the equation is given by $\dfrac{y^2}{a^2} - \dfrac{x^2}{b^2} = 1$. In addition to vertices, foci, and a center, a hyperbola also has two slant asymptotes that intersect at the center to assist in guiding the graph. The two types of hyperbolas are summarized in the following table.

EQUATION OF A HYPERBOLA WITH CENTER AT THE ORIGIN

The **standard form of the equation of a hyperbola** with its center at the origin is given by:

ORIENTATION OF TRANSVERSE AXIS	EQUATION	GRAPH	FOCI	ASYMPTOTES
Horizontal Along the x-axis	$\dfrac{x^2}{a^2} - \dfrac{y^2}{b^2} = 1$		$(-c, 0)$ $(c, 0)$	$y = \dfrac{b}{a}x$ and $y = -\dfrac{b}{a}x$

Where $c^2 = a^2 + b^2$.

ORIENTATION OF TRANSVERSE AXIS	EQUATION	GRAPH	FOCI	ASYMPTOTES
Vertical Along the y-axis	$\dfrac{y^2}{a^2} - \dfrac{x^2}{b^2} = 1$	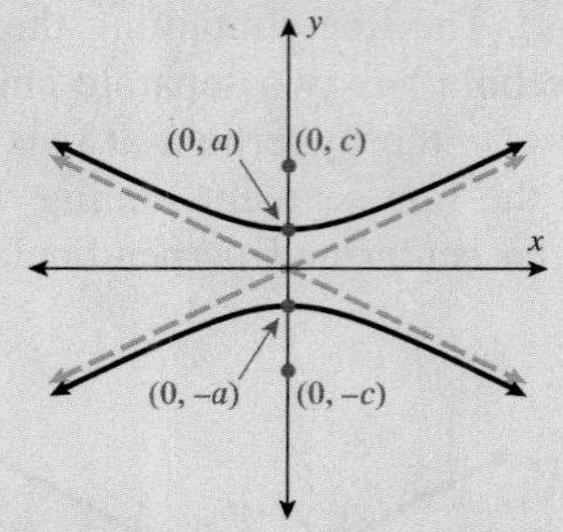	$(0, -c)$ $(0, c)$	$y = \dfrac{a}{b}x$ and $y = -\dfrac{a}{b}x$

Where $c^2 = a^2 + b^2$.

When $x = 0$, what is y? If $y = 0$, what is x? When the answer is a real number, you get the vertices. For example, in $\dfrac{x^2}{a^2} - \dfrac{y^2}{b^2} = 1$, if $x = 0$, $\dfrac{-y^2}{b^2} = 1$, which yields an imaginary number for y. But when $y = 0$, $\dfrac{x^2}{a^2} = 1$, therefore $x = \pm a$. The vertices for this hyperbola are $(-a, 0)$ and $(a, 0)$.

EXAMPLE 1 Finding the Foci and Vertices of a Hyperbola Given the Equation

Find the foci and vertices of the hyperbola given by $\dfrac{x^2}{9} - \dfrac{y^2}{4} = 1$.

Solution:

Compare to the standard equation
of a hyperbola, $\dfrac{x^2}{a^2} - \dfrac{y^2}{b^2} = 1$.　　　　　　　　$a^2 = 9, b^2 = 4$

Solve for a and b.　　　　　　　　　　　　$a = 3, \ b = 2$

Substitute $a = 3$ into the vertices,
$(-a, 0)$ and $(a, 0)$.　　　　　　　　　　$(-3, 0)$ and $(3, 0)$

Substitute $a = 3, b = 2$ into $c^2 = a^2 + b^2$.　　$c^2 = 3^2 + 2^2$

Solve for c.　　　　　　　　　　　　　　$c^2 = 13$
　　　　　　　　　　　　　　　　　　　　$c = \sqrt{13}$

Substitute $c = \sqrt{13}$ into the foci, $(-c, 0)$
and $(c, 0)$.　　　　　　　　　　　　$(-\sqrt{13}, 0)$ and $(\sqrt{13}, 0)$

The vertices are $(-3, 0)$ and $(3, 0)$, and the foci are $(-\sqrt{13}, 0)$ and $(\sqrt{13}, 0)$.

CONCEPT CHECK　Is the transverse axis of the hyperbola given by $\dfrac{y^2}{16} - \dfrac{x^2}{20} = 1$ horizontal or vertical?

■ **YOUR TURN**   Find the vertices and foci of the hyperbola $\dfrac{y^2}{16} - \dfrac{x^2}{20} = 1$.

EXAMPLE 2   Finding the Equation of a Hyperbola Given Foci and Vertices

Find the equation of a hyperbola whose vertices are located at $(0, -4)$ and $(0, 4)$ and whose foci are located at $(0, -5)$ and $(0, 5)$.

Solution:

The center is located at the midpoint of the segment joining the vertices.

$$\left(\frac{0 + 0}{2}, \frac{-4 + 4}{2}\right) = (0, 0)$$

Since the foci and vertices are located on the y-axis, the standard equation is given by

$$\frac{y^2}{a^2} - \frac{x^2}{b^2} = 1$$

The vertices $(0, \pm a)$ and the foci $(0, \pm c)$ can be used to identify a and c.

$$a = 4, c = 5$$

Substitute $a = 4, c = 5$ into $b^2 = c^2 - a^2$.

$$b^2 = 5^2 - 4^2$$

Solve for b.

$$b^2 = 25 - 16 = 9$$

$$b = 3$$

Substitute $a = 4$ and $b = 3$ into $\dfrac{y^2}{a^2} - \dfrac{x^2}{b^2} = 1$.

$$\frac{y^2}{16} - \frac{x^2}{9} = 1$$

■ **YOUR TURN**   Find the equations of a hyperbola whose vertices are located at $(-2, 0)$ and $(2, 0)$ and whose foci are located at $(-4, 0)$ and $(4, 0)$.

To graph a hyperbola, we use the vertices and asymptotes. The asymptotes are found by the equations $y = \pm\dfrac{b}{a}$ or $y = \pm\dfrac{a}{b}$ depending on if the transverse axis is horizontal or vertical. An easy way to draw these graphing aids is to first draw the rectangular box that passes through the vertices and the points, $(0, \pm b)$ or $(\pm b, 0)$, located on the conjugate axis. The asymptotes pass through the center of the hyperbola and the corners of the rectangular box.

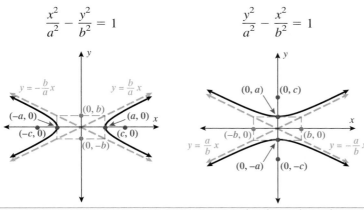

EXAMPLE 3 Graphing a Hyperbola Centered at the Origin with a Horizontal Transverse Axis

Graph the hyperbola given by $\dfrac{x^2}{4} - \dfrac{y^2}{9} = 1$.

Solution:

Compare $\dfrac{x^2}{2^2} - \dfrac{y^2}{3^2} = 1$ to the general equation $\dfrac{x^2}{a^2} - \dfrac{y^2}{b^2} = 1$.

Identify a and b. $a = 2$ and $b = 3$

The transverse axis of this hyperbola is the x-axis.

Label the vertices, $(-a, 0) = (-2, 0)$ and $(a, 0) = (2, 0)$, and the points $(0, -b) = (0, -3)$ and $(0, b) = (0, 3)$. Draw the rectangular box that passes through those points. Draw the asymptotes that pass through the center and the corners of the rectangle.

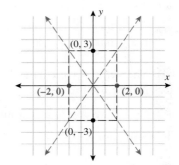

Draw the two branches of the hyperbola, each passing through a vertex and guided by the asymptotes.

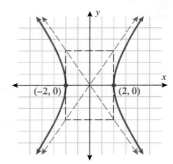

In Example 3, if we let $y = 0$, then $\dfrac{x^2}{4} = 1$ or $x = \pm 2$. So the vertices are $(-2, 0)$ and $(2, 0)$, and the x-axis is the transverse axis. Note that if $x = 0$, $y = \pm 3i$.

EXAMPLE 4 Graphing a Hyperbola Centered at the Origin with a Vertical Transverse Axis

Graph the hyperbola given by $\dfrac{y^2}{16} - \dfrac{x^2}{4} = 1$.

Solution:

Compare $\dfrac{y^2}{4^2} - \dfrac{x^2}{2^2} = 1$ to the general equation $\dfrac{y^2}{a^2} - \dfrac{x^2}{b^2} = 1$.

Identify a and b. $a = 4$ and $b = 2$

The transverse axis of this hyperbola is the y-axis.

Label the vertices, $(0, -a) = (0, -4)$ and $(0, a) = (0, 4)$, and the points $(-b, 0) = (-2, 0)$ and $(b, 0) = (2, 0)$. Draw the rectangular box that passes through those points. Draw the asymptotes that pass through the center and the corners of the rectangle.

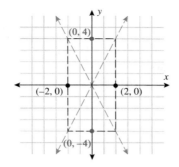

Draw the two branches of the hyperbola, each passing through a vertex and guided by the asymptotes.

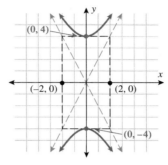

CONCEPT CHECK What is the transverse axis of each hyperbola?

a. $\dfrac{y^2}{1} - \dfrac{x^2}{4} = 1$ **b.** $\dfrac{x^2}{4} - \dfrac{y^2}{1} = 1$

YOUR TURN Graph the hyperbolas:

a. $\dfrac{y^2}{1} - \dfrac{x^2}{4} = 1$ **b.** $\dfrac{x^2}{4} - \dfrac{y^2}{1} = 1$

Hyperbolas Centered at the Point (*h*, *k*)

We can use graph-shifting techniques to graph hyperbolas that are centered at a point other than the origin, say (h, k). For example, to graph $\dfrac{(x - h)^2}{a^2} - \dfrac{(y - k)^2}{b^2} = 1$, start with the graph of $\dfrac{x^2}{a^2} - \dfrac{y^2}{b^2} = 1$ and shift to the right h units and up k units. The center, the vertices, the foci, the transverse and conjugate axes, and the asymptotes all shift. The following table summarizes the characteristics of hyperbolas centered at a point other than the origin.

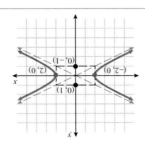

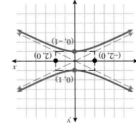

Answer: a.

EQUATION OF A HYPERBOLA WITH CENTER AT THE POINT (h, k)

The **standard form of the equation of a hyperbola** with its center at the point (h, k) is given by:

ORIENTATION OF TRANSVERSE AXIS	EQUATION	GRAPH	VERTICES	FOCI
Horizontal Parallel to the x-axis	$\dfrac{(x-h)^2}{a^2} - \dfrac{(y-k)^2}{b^2} = 1$		$(h-a, k)$ $(h+a, k)$	$(h-c, k)$ $(h+c, k)$
Vertical Parallel to the y-axis	$\dfrac{(y-k)^2}{a^2} - \dfrac{(x-h)^2}{b^2} = 1$		$(h, k-a)$ $(h, k+a)$	$(h, k-c)$ $(h, k+c)$

Where $c^2 = a^2 + b^2$.

EXAMPLE 5 Graphing a Hyperbola with Center Not at the Origin

Graph the hyperbola $\dfrac{(y-2)^2}{16} - \dfrac{(x-1)^2}{9} = 1$.

Solution:

Compare $\dfrac{(y-2)^2}{4^2} - \dfrac{(x-1)^2}{3^2} = 1$ to the general equation $\dfrac{(y-k)^2}{a^2} - \dfrac{(x-h)^2}{b^2} = 1$.

Identify a, b, and (h, k). $\qquad\qquad\qquad a = 4,\ b = 3,$ and $(h, k) = (1, 2)$

The transverse axis of this hyperbola is $x = 2$, which is parallel to the y-axis.

Label the vertices, $(h, k - a) = (1, -2)$ and $(h, k + a) = (1, 6)$, and the points $(h - b, k) = (-2, 2)$ and $(h + b, k) = (4, 2)$. Draw the rectangular box that passes through those points. Draw the asymptotes that pass through the center, $(h, k) = (1, 2)$, and the corners of the rectangle. Draw the two branches of the hyperbola, each passing through a vertex and guided by the asymptotes.

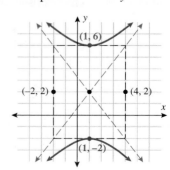

Applications

Nautical navigation is assisted by hyperbolas. For example, suppose two radio stations on a coast are emitting simultaneous signals. If a boat is at sea, it will be slightly closer to one station than the other station, which results in a small time difference between the received signals from each station. If the boat follows the path associated with a constant time difference, that path will be hyperbolic.

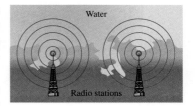

The synchronized signals would intersect one another in associated hyperbolas. Each time difference corresponds to a different path. The radio stations are the foci of the hyperbolas. This principle forms the basis of a hyperbolic radio navigation system known as *loran* (**L**ong-**R**ange **N**avigation).

There are navigational charts that correspond to different time differences. A ship selects the hyperbolic path that will take it to the desired port, and the loran chart lists the corresponding time difference.

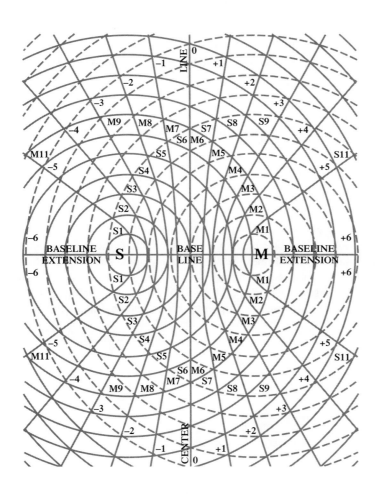

EXAMPLE 6 Nautical Navigation Using Loran

Two loran stations are located 200 miles apart along a coast. If a ship records a time difference of 0.00043 second and continues on the hyperbolic path corresponding to that difference, where would it reach shore?

Solution:

Draw the xy-plane and the two stations corresponding to the foci at $(-100, 0)$ and $(100, 0)$. Draw the ship somewhere in quadrant I.

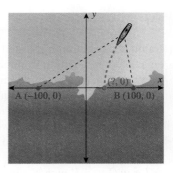

The hyperbola corresponds to a path where the difference of the distance between the ship and each station remains constant. The constant is $2a$, where $(a, 0)$ is a vertex. Find that difference by using $d = rt$. Assume the speed of the radio signal is 186,000 miles per second.

Substitute $r = 186,000$ miles/second and $t = 0.00043$ second into $d = rt$.

$$d = (186,000 \text{ miles/second})(0.00043 \text{ second}) \approx 80 \text{ miles}$$

Set the constant equal to $2a$. $2a = 80$

Find a vertex, $(a, 0)$. $(40, 0)$

The ship reaches shore between the two stations, 60 miles from station B.

SECTION 8.4 SUMMARY

Although equations of hyperbolas and ellipses only differ by a minus sign, the graphs are very different. Hyperbolas have two branches. The vertices and foci are located along horizontal or vertical lines. Vertices and asymptotes are the graphing aids. In particular, the asymptotes pass through the center and the corners of the rectangular box connecting the vertices and two points on the conjugate axes. Graph-shifting techniques can be used to graph hyperbolas centered at a point other than the origin.

SECTION 8.4 EXERCISES

■ **SKILLS**

In Exercises 1–4, match each equation with the corresponding hyperbola.

1. $\dfrac{x^2}{36} - \dfrac{y^2}{16} = 1$ **2.** $\dfrac{y^2}{36} - \dfrac{x^2}{16} = 1$ **3.** $\dfrac{x^2}{8} - \dfrac{y^2}{72} = 1$ **4.** $4y^2 - x^2 = 1$

a.

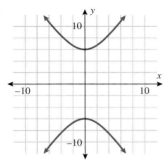

b.

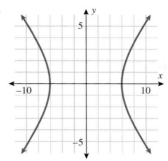

c.

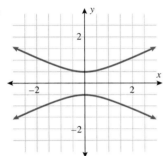

d.

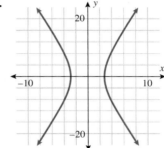

In Exercises 5–16, graph each hyperbola.

5. $\dfrac{x^2}{25} - \dfrac{y^2}{16} = 1$

6. $\dfrac{x^2}{49} - \dfrac{y^2}{9} = 1$

7. $\dfrac{y^2}{16} - \dfrac{x^2}{64} = 1$

8. $\dfrac{y^2}{144} - \dfrac{x^2}{25} = 1$

9. $\dfrac{x^2}{100} - y^2 = 1$

10. $9y^2 - 4x^2 = 36$

11. $\dfrac{4y^2}{9} - 81x^2 = 1$

12. $\dfrac{4}{25}x^2 - \dfrac{100}{9}y^2 = 1$

13. $4x^2 - y^2 = 16$

14. $y^2 - x^2 = 81$

15. $8y^2 - 16x^2 = 32$

16. $10x^2 - 25y^2 = 50$

In Exercises 17–24, find the standard form of an equation of the hyperbola with the given characteristics.

17. vertices: $(-4, 0)$ and $(4, 0)$ foci: $(-6, 0)$ and $(6, 0)$

18. vertices: $(-1, 0)$ and $(1, 0)$ foci: $(-3, 0)$ and $(3, 0)$

19. vertices: $(0, -3)$ and $(0, 3)$ foci: $(0, -4)$ and $(0, 4)$

20. vertices: $(0, -1)$ and $(0, 1)$ foci: $(0, -2)$ and $(0, 2)$

21. center: $(0, 0)$; transverse: x-axis; asymptotes: $y = x$ and $y = -x$

22. center: $(0, 0)$; transverse: y-axis; asymptotes: $y = x$ and $y = -x$

23. center: $(0, 0)$; transverse axis: y-axis; asymptotes: $y = 2x$ and $y = -2x$

24. center: $(0, 0)$; transverse axis: x-axis; asymptotes: $y = 2x$ and $y = -2x$

In Exercises 25–28, match each equation with the hyperbola.

25. $\dfrac{(x-3)^2}{4} - \dfrac{(y+2)^2}{25} = 1$

26. $\dfrac{(x+3)^2}{4} - \dfrac{(y-2)^2}{25} = 1$

27. $\dfrac{(y-3)^2}{25} - \dfrac{(x+2)^2}{4} = 1$

28. $\dfrac{(y+3)^2}{25} - \dfrac{(x-2)^2}{4} = 1$

a.

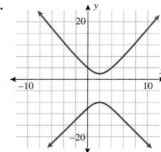

b.

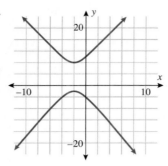

c.

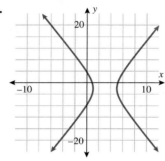

d.

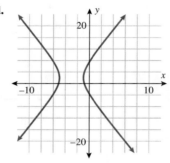

In Exercises 29–38, graph each hyperbola.

29. $\dfrac{(x-1)^2}{16} - \dfrac{(y-2)^2}{4} = 1$

30. $\dfrac{(y+1)^2}{36} - \dfrac{(x+2)^2}{9} = 1$

31. $10(y+3)^2 - (x-4)^2 = 80$

32. $3(x+3)^2 - 12(y-4)^2 = 36$

33. $x^2 - 4x - 4y^2 = 0$

34. $-9x^2 + y^2 + 2y - 8 = 0$

35. $-9x^2 - 18x + 4y^2 - 8y - 41 = 0$

36. $25x^2 - 50x - 4y^2 - 8y - 79 = 0$

37. $x^2 - 6x - 4y^2 - 16y - 8 = 0$

38. $-4x^2 - 16x + y^2 - 2y - 19 = 0$

In Exercises 39–42, find the standard form of an equation of the hyperbola with the given characteristics.

39. vertices: $(-2, 5)$ and $(6, 5)$ foci: $(-3, 5)$ and $(7, 5)$

40. vertices: $(1, -2)$ and $(3, -2)$ foci: $(0, -2)$ and $(4, -2)$

41. vertices: $(4, -7)$ and $(4, -1)$ foci: $(4, -8)$ and $(4, 0)$

42. vertices: $(2, -6)$ and $(2, -4)$ foci: $(2, -7)$ and $(2, -3)$

 ■ **APPLICATIONS**

43. Ship Navigation. Two loran stations are located 150 miles apart along a coast. If a ship records a time difference of 0.0005 second and continues on the hyperbolic path corresponding to that difference, where would it reach shore?

44. Ship Navigation. Two loran stations are located 300 miles apart along a coast. If a ship records a time difference of 0.0007 second and continues on the hyperbolic path corresponding to that difference, where would it reach shore? Round to the nearest mile.

45. Ship Navigation. If the ship in Exercise 43 wants to reach shore between the stations and 30 miles from one of them, what time difference should they look for?

46. Ship Navigation. If the ship in Exercise 44 wants to reach shore between the stations and 50 miles from one of them, what time difference should they look for?

47. Light. If the light from a lamp casts a hyperbolic pattern on the wall, calculate the equation of the hyperbola if the distance between the vertices is 2 feet and the foci are half a foot from the vertices.

Andy Washnik

48. Special Ops. A military special ops team is calibrating their recording devices used for passive ascertaining of enemy location. They place two recording stations, alpha and bravo, 3000 feet apart (alpha is due east of bravo). They detonate small explosives 300 feet west of alpha and record the time it takes each station to register an explosion. They set up a second set of explosives directly north of the alpha station. How many feet north of alpha should they set off the explosives if they want to record the same times as on the first explosion?

 ■ **CATCH THE MISTAKE**

In Exercises 49 and 50, explain the mistake that is made.

49. Graph the hyperbola $\dfrac{y^2}{4} - \dfrac{x^2}{9} = 1$.

Solution:

Compare the equation to the standard form and solve for a and b. $a = 2, b = 3$

Label the vertices $(-a, 0)$ and $(a, 0)$. $(-2, 0)$ and $(2, 0)$.

Label the points $(0, -b)$ and $(0, b)$. $(0, -3)$ and $(0, 3)$.

Draw the rectangle connecting these four points and align the asymptotes so that they pass through the center and the corner of the boxes. Then draw the hyperbola using the vertices and asymptotes.

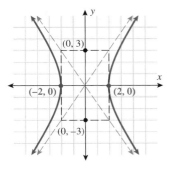

This is not correct. What mistake was made?

50. Graph the hyperbola $\dfrac{x^2}{1} - \dfrac{y^2}{4} = 1$.

Solution:

Compare with the equation to the general form and solve for a and b. $a = 2, b = 1$

Label the vertices $(-a, 0)$ and $(a, 0)$. $(-2, 0)$ and $(2, 0)$.

Label the points $(0, -b)$ and $(0, b)$. $(0, -1)$ and $(0, 1)$.

Draw the rectangle connecting these four points and align the asymptotes so that they pass through the center and the corner of the boxes. Then draw the hyperbola using the vertices and asymptotes.

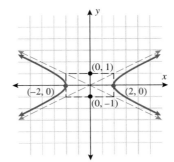

This is not correct. What mistake was made?

■ CHALLENGE

51. T or F: If you know the vertices of a hyperbola, you can determine the equation for the hyperbola.

52. T or F: If you know the foci and vertices, you can determine the equation for the hyperbola.

53. T or F: Hyperbolas centered at the origin have symmetry with respect to the x-axis, y-axis, and the origin.

54. T or F: The center and foci are part of the graph of a hyperbola.

55. Find the general equation of a hyperbola whose asymptotes are perpendicular.

56. Find the general equation of a hyperbola whose vertices are $(3, -2)$ and $(-1, -2)$ and asymptotes are the lines $y = 2x - 4$ and $y = -2x$.

■ TECHNOLOGY

57. Graph the following three hyperbolas: $x^2 - y^2 = 1$, $x^2 - 5y^2 = 1$, and $x^2 - 10y^2 = 1$. What can be said to happen to the hyperbola $x^2 - cy^2 = 1$ as c increases?

58. Graph the following three hyperbolas: $x^2 - y^2 = 1$, $5x^2 - y^2 = 1$, and $10x^2 - y^2 = 1$. What can be said to happen to the hyperbola $cx^2 - y^2 = 1$ as c increases?

Q: Kyra works for a trucking company that is starting a new route. Along this route is a tunnel that has a height of 20 feet. Kyra is asked to determine whether the company's truck, which is 12 feet tall and 8 feet wide, will be able to successfully travel through the tunnel without hitting the interior of the tunnel. Kyra believes that the tunnel is a semicircle, so she calculates that if the truck travels 1 foot to the right of the center line, then it will still have about 17.9 feet of clearance. A week later, the truck driver arrives at the tunnel and immediately notices that his truck will hit the tunnel even if he tries to drive right alongside the centerline. Why was Kyra's calculation incorrect?

A: Kyra assumed that the tunnel was semicircular, when in fact it was parabolic.

If the width of the tunnel along the ground is 24 feet, what is the height of the tunnel along the side edges of the truck if the truck straddles the centerline of the tunnel? Will it be able to drive through the tunnel?

TYING IT ALL TOGETHER

An interesting architectural application of the ellipse is the whisper chamber, which is a room found in many science museums. The ends of an ellipsoid inscribed into the parameters of the room are placed at opposite ends of the room, and boxes are placed at each of the foci. Sound waves leaving one of the foci are reflected directly to the opposite focus. The result is that even words that are whispered at one of the foci can be heard at the other focus but cannot be heard by people in the middle of the room.

The local science museum is building a whisper chamber into an existing room that is 20 feet long and 16 feet wide. The boxes centered at each of the foci will measure 2 feet by 2 feet and will clearly mark the location of the focus The door will be located halfway along the length of the room (10 feet from each end). Find the equation of the elliptical cross section of the largest ellipsoid that will fit in the room. Take the cross section parallel to the floor halfway between the floor and ceiling. Assume that the center of the room is the origin of the coordinate system. How far from the center of the room are the foci located?

For the parabola (8.2):

$$x^2 = 4py \qquad \text{Focus } (0, p), \quad \text{Directrix } y = -p$$
$$\text{Up: } p > 0 \qquad \text{Down: } p < 0$$

$$y^2 = 4px \qquad \text{Focus } (p, 0), \quad \text{Directrix } x = -p$$
$$\text{Right: } p > 0 \qquad \text{Left: } p < 0$$

For the ellipse (8.3):

$$\frac{x^2}{a^2} + \frac{y^2}{b^2} = 1 \qquad c^2 = a^2 - b^2$$

$$\frac{x^2}{b^2} + \frac{y^2}{a^2} = 1 \qquad c^2 = a^2 - b^2$$

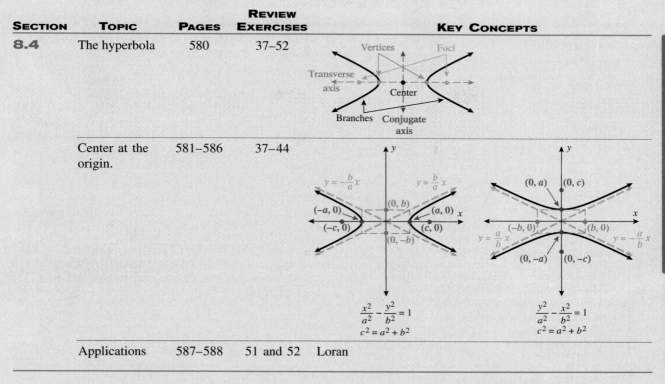

$$\frac{x^2}{a^2} - \frac{y^2}{b^2} = 1$$
$$c^2 = a^2 + b^2$$

$$\frac{y^2}{a^2} - \frac{x^2}{b^2} = 1$$
$$c^2 = a^2 + b^2$$

8.2 The Parabola

Find an equation for the parabola described.

1. vertex at $(0, 0)$; focus at $(3, 0)$

2. vertex at $(0, 0)$; focus at $(0, 2)$

3. vertex at $(0, 0)$; directrix at $x = 5$

4. vertex at $(0, 0)$; directrix at $y = 4$

5. vertex at $(2, 3)$; focus at $(2, 5)$

6. vertex at $(-1, -2)$; focus at $(1, -2)$

7. focus at $(1, 5)$; directrix at $y = 7$

8. focus at $(2, 2)$; directrix at $x = 0$

Find the focus, vertex, directrix, and length of the latus rectum and graph the parabola.

9. $x^2 = -12y$

10. $x^2 = 8y$

11. $y^2 = x$

12. $y^2 = -6x$

13. $(y + 2)^2 = 4(x - 2)$

14. $(y - 2)^2 = -4(x + 1)$

15. $(x + 3)^2 = -8(y - 1)$

16. $(x - 3)^2 = -8(y + 2)$

17. $x^2 + 5x + 2y + 25 = 0$

18. $y^2 + 2y - 16x + 1 = 0$

Applications

19. **Satellite Dish.** A satellite dish measures 10 feet across its opening and 2 feet deep at its center. The receiver should be placed at the focus of the parabolic dish. Where should the receiver be placed?

20. **Clearance under a Bridge.** A bridge with a parabolic shape reaches a height of 40 feet in the center of the road, and the width of the bridge opening at ground level is 30 feet combined (both lanes). If an RV is 14 feet tall and is 8 feet wide, will it make it through the tunnel?

8.3 The Ellipse

Graph each ellipse.

21. $\dfrac{x^2}{9} + \dfrac{y^2}{64} = 1$

22. $\dfrac{x^2}{81} + \dfrac{y^2}{49} = 1$

23. $25x^2 + y^2 = 25$

24. $4x^2 + 8y^2 = 64$

Find the standard form of an equation of the ellipse with the given characteristics.

25. foci: $(-3, 0)$ and $(3, 0)$ vertices: $(-5, 0)$ and $(5, 0)$

26. foci: $(0, -2)$ and $(0, 2)$ vertices: $(0, -3)$ and $(0, 3)$

27. Major axis vertical with length of 16, minor axis length of 6 and centered at $(0, 0)$.

28. Major axis horizontal with length of 30, minor axis length of 20 and centered at $(0, 0)$.

Graph each ellipse.

29. $\dfrac{(x - 7)^2}{100} + \dfrac{(y + 5)^2}{36} = 1$

30. $20(x + 3)^2 + (y - 4)^2 = 120$

31. $4x^2 - 16x + 12y^2 + 72y + 123 = 0$

32. $4x^2 - 8x + 9y^2 - 72y + 147 = 0$

Find the standard form of an equation of the ellipse with the given characteristics.

33. foci: $(-1, 3)$ and $(7, 3)$ vertices: $(-2, 3)$ and $(8, 3)$

34. foci: $(1, -3)$ and $(1, -1)$ vertices: $(1, -4)$ and $(1, 0)$

Applications

35. **Planetary Orbits.** Jupiter's orbit is summarized in the picture. Use the fact that the Sun is a focus to determine an equation for Jupiter's elliptical orbit around the Sun. Round to the nearest hundred thousand kilometers.

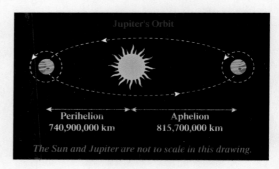

Jupiter's Orbit

Perihelion
740,900,000 km

Aphelion
815,700,000 km

The Sun and Jupiter are not to scale in this drawing.

36. **Planetary Orbits.** Mars' orbit is summarized in the picture below. Use the fact that the Sun is a focus to determine an equation for Mars' elliptical orbit around the Sun. Round to the nearest million kilometers.

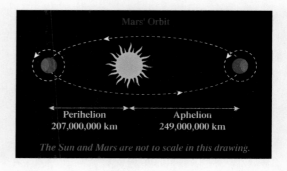

Mars' Orbit

Perihelion
207,000,000 km

Aphelion
249,000,000 km

The Sun and Mars are not to scale in this drawing.

8.4 The Hyperbola

Graph each hyperbola.

37. $\dfrac{x^2}{9} - \dfrac{y^2}{64} = 1$ **38.** $\dfrac{x^2}{81} - \dfrac{y^2}{49} = 1$

39. $x^2 - 25y^2 = 25$ **40.** $8y^2 - 4x^2 = 64$

Find the standard form of an equation of the hyperbola with the given characteristics.

41. vertices: $(-3, 0)$ and $(3, 0)$ foci: $(-5, 0)$ and $(5, 0)$

42. vertices: $(0, -1)$ and $(0, 1)$ foci: $(0, -3)$ and $(0, 3)$

43. center: $(0, 0)$; transverse: y-axis; asymptotes: $y = 3x$ and $y = -3x$

44. center: $(0, 0)$; transverse axis: y-axis; asymptotes: $y = \frac{1}{2}x$ and $y = -\frac{1}{2}x$

Graph each hyperbola.

45. $\dfrac{(y-1)^2}{36} - \dfrac{(x-2)^2}{9} = 1$

46. $3(x + 3)^2 - 12(y - 4)^2 = 72$

47. $8x^2 - 32x - 10y^2 - 60y - 138 = 0$

48. $2x^2 + 12x - 8y^2 + 16y + 6 = 0$

Find the standard form of an equation of the hyperbola with the given characteristics.

49. vertices: $(0, 3)$ and $(8, 3)$ foci: $(-1, 3)$ and $(9, 3)$

50. vertices: $(4, -2)$ and $(4, 0)$ foci: $(4, -3)$ and $(4, 1)$

Applications

51. Ship Navigation. Two loran stations are located 220 miles apart along a coast. If a ship records a time difference of 0.00048 second and continues on the hyperbolic path corresponding to that difference, where would it reach shore? Assume sound travels at $r = 186{,}000$ miles per second.

52. Ship Navigation. Two loran stations are located 400 miles apart along a coast. If a ship records a time difference of 0.0008 second and continues on the hyperbolic path corresponding to that difference, where would it reach shore?

Match the equation to the graph.

1. $x = 16y^2$

2. $y = 16x^2$

3. $x^2 + 16y^2 = 1$

4. $x^2 - 16y^2 = 1$

5. $16x^2 + y^2 = 1$

6. $16y^2 - x^2 = 1$

a.

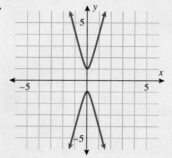

b.

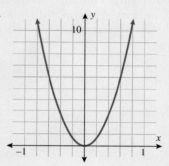

c.

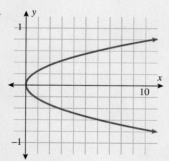

d.

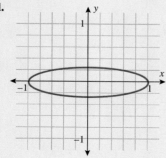

e.

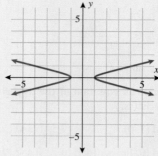

f.

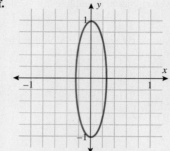

Find the equation of the conic with given characteristics.

7. parabola   vertex: $(0, 0)$ focus: $(-4, 0)$

8. parabola   vertex: $(0, 0)$ directrix: $y = 2$

9. parabola   vertex: $(-1, 5)$ focus: $(-1, 2)$

10. parabola   vertex: $(2, -3)$ directrix: $x = 0$

11. ellipse   center: $(0, 0)$ vertices: $(0, -4), (0, 4)$
foci: $(0, -3), (0, 3)$

12. ellipse   center: $(0, 0)$ vertices: $(-3, 0), (3, 0)$
foci: $(-1, 0), (1, 0)$

13. ellipse   vertices: $(2, -6), (2, 6)$
foci: $(2, -4), (2, 4)$

14. ellipse   vertices: $(-7, -3), (-4, -3)$
foci: $(-6, -3), (-5, -3)$

15. hyperbola   vertices: $(-1, 0)$ and $(1, 0)$
asymptotes: $y = -2x$ and $y = 2x$

16. hyperbola   vertices: $(0, -1)$ and $(0, 1)$
asymptotes: $y = -\frac{1}{3}x$ and $y = \frac{1}{3}x$

17. hyperbola   foci: $(2, -6), (2, 6)$
vertices: $(2, -4), (2, 4)$

18. hyperbola   foci: $(-7, -3), (-4, -3)$
vertices: $(-6, -3), (-5, -3)$

Graph the following equations.

19. $9x^2 + 18x - 4y^2 + 16y - 43 = 0$

20. $4x^2 - 8x + y^2 + 10y + 28 = 0$

21. $y^2 + 4y - 16x + 20 = 0$

22. **Planetary Orbits.** The planet Uranus' orbit is described in the following picture with the Sun as a focus of the elliptical orbit. Write an equation for the orbit.

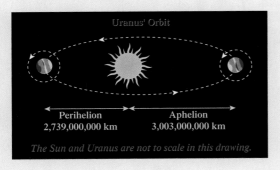

Uranus' Orbit

Perihelion
2,739,000,000 km

Aphelion
3,003,000,000 km

The Sun and Uranus are not to scale in this drawing.

23. **Eyeglass Lens.** Eyeglass lenses can be thought of as very wide parabolic curves. If the focus occurs 1.5 centimeters from the center of the lens, and the lens at its opening is 4 centimeters across, find an equation that governs the shape of the center cross section of the lens.

CHAPTER 9

Sequences, Series, and Probability

A BASIC STRATEGY FOR BLACKJACK

◄──── Dealer's Up Card ────►

Your Hand		2	3	4	5	6	7	8	9	10	A
I	17+	S	S	S	S	S	S	S	S	S	S
	16	S	S	S	S	S	H	H	H	H	H
	15	S	S	S	S	S	H	H	H	H	H
	14	S	S	S	S	S	H	H	H	H	H
	13	S	S	S	S	S	H	H	H	H	H
	12	H	H	S	S	S	H	H	H	H	H
II	11	D	D	D	D	D	D	D	D	D	H
	10	D	D	D	D	D	D	D	D	H	H
	9	H	D	D	D	D	H	H	H	H	H
	5 - 8	H	H	H	H	H	H	H	H	H	H
III	A, 8 - 10	S	S	S	S	S	S	S	S	S	S
	A, 7	S	D	D	D	D	S	S	H	H	H
	A, 6	H	D	D	D	D	H	H	H	H	H
	A, 5	H	H	D	D	D	H	H	H	H	H
	A, 4	H	H	D	D	D	H	H	H	H	H
	A, 3	H	H	H	D	D	H	H	H	H	H
	A, 2	H	H	H	D	D	H	H	H	H	H
IV	A, A; 8, 8	SP	SP	SP	SP	SP	SP	SP	SP	SP	SP
	10, 10	S	S	S	S	S	S	S	S	S	S
	9, 9	SP	SP	SP	SP	SP	S	SP	SP	S	S
	7, 7	SP	SP	SP	SP	SP	SP	H	H	H	H
	6, 6	H	SP	SP	SP	SP	H	H	H	H	H
	5, 5	D	D	D	D	D	D	D	D	H	H
	4, 4	H	H	H	H	H	H	H	H	H	H
	3, 3	H	H	SP	SP	SP	SP	H	H	H	H
	2, 2	H	H	SP	SP	SP	SP	H	H	H	H

When surrender is allowed, surrender 9, 7 or 10, 6 vs 9, 10, A; 9, 6 or 10, 5 vs 10

When doubling down after splitting is allowed, split: 2's, 3's, 7's vs 2-7; 4's vs 5 or 6; 6's vs 2-6

HIT STAND DOUBLE DOWN SPLIT

Insurance is NOT Recommended

Have you ever been to a casino and played black-jack? It is the only game in the casino that you can win based on the law of large numbers. In the early 1990s a group of math and science majors from the Massachusetts Institute of Technology (MIT) devised a fool-proof scheme to win at blackjack. A professor at MIT developed a basic strategy outlined in the figure on the left that is based on the probability of combinations of particular cards being dealt given certain cards already showing. To play blackjack (also called 21), each person is dealt two cards with the option of taking additional cards. The goal is to get a combination of cards that is worth 21 points (or less) without going over (called a bust). You have to avoid going over 21 or staying too far below 21. All face cards (jacks, queens, and kings) are worth 10 points, and an ace in blackjack is worth either 1 or 11 points. The students used the professor's strategy along with a card-counting technique to place higher bets when there were more high-value cards left in the deck.

It is reported that in 1992 the team won $4,000,000 from Las Vegas casinos. The casinos caught on, and the students were all banned within 2 years.

In this chapter, we will discuss counting and probability in addition to three other topics: sequences and series, mathematical induction, and the binomial theorem.

Sequences, Series, and Probability

Sequences and Series

- **Sequence Notation**
- **Factorial Notation**
- **Summation Notation**
  - Arithmetic series
  - Geometric series

Mathematical Induction

- Mathematical Proof

Binomial Theorem

- Raising a Binomial to an Integer Power

Counting and Probability

- Combinations
- Permutations
- Probability of an Event

CHAPTER OBJECTIVES

- Understand the difference between sequences and series.
- Find the general, n^{th}, term of a sequence or series.
- Prove a mathematical statement using induction.
- Use the binomial theorem to expand a binomial raised to an integer power.
- Understand the difference between permutations and combinations.
- Calculate the probability of an event.

NAVIGATION THROUGH SUPPLEMENTS

DIGITAL VIDEO SERIES #9

STUDENT SOLUTIONS MANUAL CHAPTER 9

BOOK COMPANION SITE
www.wiley.com/college/young

Skills Objectives	Conceptual Objectives
▪ Find terms of a sequence given the general term. ▪ Look for a pattern in a sequence and find the general term. ▪ Use factorial notation. ▪ Use recursion formulas. ▪ Use summation (sigma) notation to represent a series. ▪ Find the sum of a series.	▪ Understand the difference between a sequence and a series.

Sequences

The word *sequence* means to follow one thing after another in succession (in a certain order). In mathematics, it means the same thing. For example, if we write $x, 2x^2, 3x^3, 4x^4, 5x^5, ?$, what would the next term in the *sequence* be, the one where the question mark now stands? The answer is $6x^6$.

DEFINITION **A SEQUENCE**

A **sequence** is a function whose domain is a set of positive integers. The function values, or **terms,** of the sequence are written as

$$a_1, a_2, a_3, \ldots, a_n, \ldots$$

Rather than using function notation, sequences are usually written with subscript (or index) notation, $a_{subscript}$.

A **finite sequence** has the domain $\{1, 2, 3, \ldots, n\}$ for some positive integer n.

An **infinite sequence** has the domain of all positive integers $\{1, 2, 3, \ldots\}$.

There are times when it is convenient to start the indexing at 0 instead of 1:

$$a_0, a_1, a_2, a_3, \ldots, a_n, \ldots$$

Sometimes a pattern in the sequence can be obtained and the sequence can be written using a general term. In the previous example, $x, 2x^2, 3x^3, 4x^4, 5x^5, 6x^6, \ldots$, each term has the same exponent and coefficient. We can write this sequence as $a_n = nx^n$, $n = 1, 2, 3, 4, 3, 6, \ldots$, where a_n is called the **general term**.

EXAMPLE 1 Finding the Sequence Given the General Term

Find the first four ($n = 1, 2, 3, 4$) terms of the sequences given the general term.

a. $a_n = 2n - 1$ **b.** $b_n = \dfrac{(-1)^n}{n + 1}$

Solution (a):

Find the first term, $n = 1$. $a_1 = 2(1) - 1 = 1$

Find the second term, $n = 2$. $a_2 = 2(2) - 1 = 3$

Find the third term, $n = 3$. $a_3 = 2(3) - 1 = 5$

Find the fourth term, $n = 4$. $a_4 = 2(4) - 1 = 7$

The first four terms of the sequence are $1, 3, 5, 7$.

Solution (b):

Find the first term, $n = 1$. $b_1 = \dfrac{(-1)^1}{1 + 1} = -\dfrac{1}{2}$

Find the second term, $n = 2$. $b_2 = \dfrac{(-1)^2}{2 + 1} = \dfrac{1}{3}$

Find the third term, $n = 3$. $b_3 = \dfrac{(-1)^3}{3 + 1} = -\dfrac{1}{4}$

Find the fourth term, $n = 4$. $b_4 = \dfrac{(-1)^4}{4 + 1} = \dfrac{1}{5}$

The first four terms of the sequence are $-\dfrac{1}{2}, \dfrac{1}{3}, -\dfrac{1}{4}, \dfrac{1}{5}$.

■ **YOUR TURN**   Find the first four terms of the sequence $a_n = \dfrac{e^n}{n}$.

EXAMPLE 2 Finding the General Term Given Several Terms of the Sequence

Find the general term of the sequence given the first five terms.

a. $1, \dfrac{1}{4}, \dfrac{1}{9}, \dfrac{1}{16}, \dfrac{1}{25}, \ldots$ **b.** $-1, 4, -9, 16, -25, \ldots$

Solution (a):

Write 1 as $\dfrac{1}{1}$. $\dfrac{1}{1}, \dfrac{1}{4}, \dfrac{1}{9}, \dfrac{1}{16}, \dfrac{1}{25}, \ldots$

Notice that each denominator is an integer squared. $\dfrac{1}{1^2}, \dfrac{1}{2^2}, \dfrac{1}{3^2}, \dfrac{1}{4^2}, \dfrac{1}{5^2}, \ldots$

Identify the general term. $a_n = \dfrac{1}{n^2}$ $n = 1, 2, 3, 4, 5, \ldots$

■ **Answer:** $e, \dfrac{e^2}{2}, \dfrac{e^3}{3}, \dfrac{e^4}{4}$

Solution (b):

Notice that each term includes an integer squared.

$$-1^2, 2^2, -3^2, 4^2, -5^2, \ldots$$

Identify the general term.

$$b_n = (-1)^n n^2 \qquad n = 1, 2, 3, 4, 5, \ldots$$

■ **YOUR TURN** Find the general term of the sequence given the first five terms:

$$-\frac{1}{2}, \frac{1}{4}, -\frac{1}{6}, \frac{1}{8}, -\frac{1}{10}$$

Parts b in both Example 1 and Example 2 are called **alternating** sequences, because the terms alternate signs (positive and negative). If the odd terms, $a_1, a_3, a_5, \ldots$ are negative and the even terms, $a_2, a_4, a_6, \ldots$, are positive, we include $(-1)^n$ in the general term. If the opposite is true, and the odd terms are positive and the even terms are negative, we include $(-1)^{n+1}$ in the general term.

Factorial Notation

Many important sequences that arise in mathematics involve terms that are defined with products of consecutive positive integers. The products are expressed in *factorial notation*.

DEFINITION   FACTORIAL

If n is a positive integer, then $n!$ (stated as "n factorial") is the product of all positive integers from n down to 1.

$$n! = n(n-1)(n-2) \cdots 3 \cdot 2 \cdot 1 \qquad n \geq 2$$

By definition, $0! = 1$ and $1! = 1$.

The values of $n!$ for the first six nonnegative integers are

$$0! = 1$$
$$1! = 1$$
$$2! = 2 \cdot 1 = 2$$
$$3! = 3 \cdot 2 \cdot 1 = 6$$
$$4! = 4 \cdot 3 \cdot 2 \cdot 1 = 24$$
$$5! = 5 \cdot 4 \cdot 3 \cdot 2 \cdot 1 = 120$$

Notice that $4! = 4 \cdot \underbrace{3 \cdot 2 \cdot 1}_{3!} = 4 \cdot 3!$. In general we can use the formula: $n! = n[(n-1)!]$.

Often the brackets are not used, and the notation, $n! = n(n-1)!$, implies calculating

■ **Answer:** $a_n = \dfrac{(-1)^n}{2n}$

the factorial, $(n - 1)!$, and then multiplying that quantity by n. For example, to find $6!$, we use the relationship, $n! = n(n - 1)!$, and set $n = 6$:

$$6! = 6 \cdot 5! = 6 \cdot 120 = 720$$

EXAMPLE 3 Finding the Terms of a Sequence Involving Factorials

Find the first four terms of the sequence given the general term.

$$a_n = \frac{x^n}{n!}$$

Solution:

Find the first term, $n = 1$. $\qquad\qquad a_1 = \dfrac{x^1}{1!} = x$

Find the second term, $n = 2$. $\qquad\quad a_2 = \dfrac{x^2}{2!} = \dfrac{x^2}{2 \cdot 1} = \dfrac{x^2}{2}$

Find the third term, $n = 3$. $\qquad\qquad a_3 = \dfrac{x^3}{3!} = \dfrac{x^3}{3 \cdot 2 \cdot 1} = \dfrac{x^3}{6}$

Find the fourth term, $n = 4$. $\qquad\quad a_4 = \dfrac{x^4}{4!} = \dfrac{x^4}{4 \cdot 3 \cdot 2 \cdot 1} = \dfrac{x^4}{24}$

The first four terms of the sequence are $\boxed{x, \dfrac{x^2}{2}, \dfrac{x^3}{6}, \dfrac{x^4}{24}}$.

EXAMPLE 4 Evaluating Expressions with Factorials

Evaluate each factorial expression:

a. $\dfrac{6!}{2! \cdot 3!}$ $\qquad$ **b.** $\dfrac{(n + 1)!}{(n - 1)!}$

Solution (a):

Expand each factorial in the numerator and denominator. $\qquad \dfrac{6!}{2! \cdot 3!} = \dfrac{6 \cdot 5 \cdot 4 \cdot 3 \cdot 2 \cdot 1}{2 \cdot 1 \cdot 3 \cdot 2 \cdot 1}$

Cancel the $3 \cdot 2 \cdot 1$ in both the numerator and denominator. $\qquad \dfrac{6 \cdot 5 \cdot 4}{2 \cdot 1}$

Simplify. $\qquad\qquad\qquad\qquad\qquad \dfrac{6 \cdot 5 \cdot 2}{1} = 60$

$$\boxed{\dfrac{6!}{2! \cdot 3!} = 60}$$

Solution (b):

Expand each factorial in the numerator and denominator.

$$\frac{(n+1)!}{(n-1)!} = \frac{(n+1)(n)(n-1)(n-2)\cdots 3\cdot 2\cdot 1}{(n-1)(n-2)\cdots 3\cdot 2\cdot 1}$$

Cancel the $(n-1)(n-2)\cdots 3\cdot 2\cdot 1$ in the numerator and denominator.

$$\frac{(n+1)!}{(n-1)!} = (n+1)(n)$$

$$\frac{(n+1)!}{(n-1)!} = n^2 + n$$

STUDY TIP
In general, $m!n! \neq (mn)!$

COMMON MISTAKE

In Example 4 we found $\dfrac{6!}{2!\cdot 3!} = 60$. It is important to note that $2!\cdot 3! \neq 6!$

■ **YOUR TURN**   Evaluate the factorial expression: $\dfrac{3!\cdot 4!}{2!\cdot 6!}$.

Recursion Formulas

Another way to define a sequence is **recursively**, or using a **recursion formula**. The first few terms are listed, and the recursion formula determines the remaining terms based on previous terms. For example, the famous Fibonacci sequence is 1, 1, 2, 3, 5, 8, 13, 21, 34, 55, 89, … Each term in the Fibonacci sequence is found by adding the previous two terms.

$1 + 1 = 2$	$1 + 2 = 3$	$2 + 3 = 5$
$3 + 5 = 8$	$5 + 8 = 13$	$8 + 13 = 21$
$13 + 21 = 34$	$21 + 34 = 55$	$34 + 55 = 89$

We can define the Fibonacci sequence using a general term:

$$a_1 = 1, \, a_2 = 1, \text{ and } a_n = a_{n-2} + a_{n-1} \quad n \geq 3$$

The Fibonacci sequence is found in places we least expect them (bananas, pineapples, broccoli, and flowers). If you cut a banana in half (as if to break it in half), the number of flat sides is a Fibonacci number. The number of petals in a flower is a Fibonacci number. For example, a wild rose has 5 petals, lilies and irises have 3 petals, and daisies have 34, 55, or even 89 petals. The number of spirals in an Italian broccoli is a Fibonacci number (13).

■ **Answer:** $\dfrac{1}{10}$

Tamara Lischka/Graphistock/Images.com

Italian broccoli (broccoflower)

EXAMPLE 5 Using a Recursion Formula to Find a Sequence

Find the first four terms of the sequence: $a_1 = 2$ and $a_n = 2a_{n-1} - 1$ $n \geq 2$.

Solution:

Write the first term, $n = 1$. $a_1 = 2$

Find the second term, $n = 2$. $a_2 = 2a_1 - 1 = 2(2) - 1 = 3$

Find the third term, $n = 3$. $a_3 = 2a_2 - 1 = 2(3) - 1 = 5$

Find the fourth term, $n = 4$. $a_4 = 2a_3 - 1 = 2(5) - 1 = 9$

The first four terms of the sequence are $2, 3, 5, 9$.

■ **YOUR TURN**   Find the first four terms of the sequence: $a_1 = 1$ and

$$a_n = \frac{a_{n-1}}{n!}   n \geq 2$$

Sums and Series

It is often important to be able to find the sum of the terms in a sequence, which is called a *series*.

DEFINITION   SERIES

Given the infinite sequence $a_1, a_2, a_3, \ldots, a_n, \ldots$

■ The sum of all of the terms in the infinite sequence is called an **infinite series** and is denoted by

$$a_1 + a_2 + a_3 + \cdots + a_n + \cdots$$

■ The sum of only the first n terms is called a **finite series**, or **nth partial sum**, and is denoted, S_n.

$$a_1 + a_2 + a_3 + \cdots + a_n = S_n$$

■ **Answer:** $1, \dfrac{1}{2}, \dfrac{1}{12}, \dfrac{1}{288}$

The Greek letter, Σ (sigma), represents the same sound as a capital S in our alphabet. Therefore, we use Σ as a shorthand way to represent a sum (series). For example, the sum of the first five terms of the sequence 1, 4, 9, 16, 25, ..., n^2, ..., can be represented using **sigma (or summation) notation**:

$$\sum_{n=1}^{5} n^2 = (1)^2 + (2)^2 + (3)^2 + (4)^2 + (5)^2$$
$$= 1 + 4 + 9 + 16 + 25$$

This is read "the sum as n goes from 1 to 5 of (n^2)." The letter n is called the **index of summation**, and often other letters are used instead of n. It is important to note that the sum can start at other numbers besides 1.

If we wanted the sum of all of the terms in the sequence, we would represent that infinite series using summation notation as

$$\sum_{n=1}^{\infty} n^2 = 1 + 4 + 9 + 16 + 25 + \cdots$$

EXAMPLE 6 Writing a Series Using Sigma Notation

Write the following series using sigma notation.

a. $1 + 1 + \dfrac{1}{2} + \dfrac{1}{6} + \dfrac{1}{24} + \dfrac{1}{120}$ **b.** $8 + 27 + 64 + 125 + \cdots$

Solution (a):

Write 1 as $\dfrac{1}{1}$.

$$\frac{1}{1} + \frac{1}{1} + \frac{1}{2} + \frac{1}{6} + \frac{1}{24} + \frac{1}{120}$$

Notice that we can write the denominators using factorials.

$$\frac{1}{1} + \frac{1}{1} + \frac{1}{2!} + \frac{1}{3!} + \frac{1}{4!} + \frac{1}{5!}$$

Recall that $0! = 1$ and $1! = 1$.

$$\frac{1}{0!} + \frac{1}{1!} + \frac{1}{2!} + \frac{1}{3!} + \frac{1}{4!} + \frac{1}{5!}$$

Identify the general term.

$$a_n = \frac{1}{n!} \quad n = 0, 1, 2, 3, 4, 5$$

Write the finite series using sigma notation.

$$\sum_{n=0}^{5} \frac{1}{n!}$$

Solution (b):

Write the infinite series as a sum of terms cubed.

$$8 + 27 + 64 + 125 + \cdots$$
$$= 2^3 + 3^3 + 4^3 + 5^3 + \cdots$$

Identify the general term of the series.

$$a_n = n^3 \quad n \geq 2$$

Write the infinite series using sigma notation.

$$\sum_{n=2}^{\infty} n^3$$

Now that we are comfortable with sigma (summation) notation, let's turn our attention to calculating the sums of series. You can always find the sum of a finite series. However, you cannot always find the sum of an infinite series.

 EXAMPLE 7 **Finding the Sum of a Finite Series**

Find the sum of the series $\displaystyle\sum_{i=0}^{4}(2i+1)$.

Solution:

Write out the partial sum.

$$\overset{(i=1)\qquad(i=3)}{\underset{(i=0)\quad(i=2)\quad(i=4)}{\sum_{i=0}^{4}(2i+1) = 1 + \underset{\downarrow}{3} + 5 + \underset{\downarrow}{7} + 9}}$$

$$= 25$$

Simplify.

$$\sum_{i=0}^{4}(2i+1) = 25$$

■ **YOUR TURN** Find the sum of the series $\displaystyle\sum_{n=1}^{5}(-1)^{n}\,n$.

STUDY TIP
The sum of a finite series always exists. The sum of an infinite series may or may not exist.

Infinite sums may or may not sum to a single value. For example, if we keep adding $1 + 1 + 1 + 1...$ then there is no number that the series sums to because the sum continues to grow. However, if we add $1 + .1 + 0.01 + 0.001 + 0.0001 + \cdots$, this sum is $1.11111\overline{1}$.

EXAMPLE 8 **Finding the Sum of an Infinite Series (If Possible)**

Find the sum of the following infinite series (if possible):

a. $\displaystyle\sum_{n=1}^{\infty}\frac{3}{10^{n}}$ **b.** $\displaystyle\sum_{n=1}^{\infty}n^{2}$

Solution (a):

Expand the series.

$$\sum_{n=1}^{\infty}\frac{3}{10^{n}} = \frac{3}{10} + \frac{3}{100} + \frac{3}{1000} + \frac{3}{10,000} + \cdots$$

Write in decimal form.

$$\sum_{n=1}^{\infty}\frac{3}{10^{n}} = 0.3 + 0.03 + 0.003 + 0.0003 + \cdots$$

■ **Answer:** -3

Calculate the sum.
$$\sum_{n=1}^{\infty}\frac{3}{10^n} = 0.333333\overline{3} = \frac{1}{3}$$

$$\sum_{n=1}^{\infty}\frac{3}{10^n} = \frac{1}{3}$$

Solution (b):

Expand the series.
$$\sum_{n=1}^{\infty} n^2 = 1 + 4 + 9 + 16 + 25 + 36 + \cdots$$

This sum cannot be calculated since it continues to grow without end.

Applications

The annual sales at Home Depot from 2000 to 2002 can be approximated by the model $a_n = 45.7 + 9.5n - 1.6n^2$ where a_n is the yearly sales in billions of dollars and $n = 0, 1, 2$.

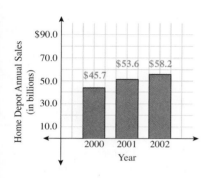

YEAR	n	$a_n = 45.7 + 9.5n - 1.6n^2$	TOTAL SALES IN BILLIONS
2000	0	$a_0 = 45.7 + 9.5(0) - 1.6(0)^2$	$45.7
2001	1	$a_1 = 45.7 + 9.5(1) - 1.6(1)^2$	$53.6
2002	2	$a_2 = 45.7 + 9.5(2) - 1.6(2)^2$	$58.3

What does the finite series $\dfrac{1}{3}\sum_{n=0}^{2} a_n$ tell us? It tells us the average sales over 3 years.

SECTION 9.1 **SUMMARY**

In this section, we defined sequences and used general terms to represent them. Factorial notation and recursion formulas were also discussed. When the terms of a sequence are added together, the resulting sum is called a series. When the sequence is infinite, the series is called an infinite series, and when the sequence has n terms, the finite series is also called the nth partial sum. Summation (sigma) notation was introduced to represent a series. Finite series always sum to some number, whereas infinite series might not.

SECTION 9.1 EXERCISES

■ SKILLS

In Exercises 1–12, write the first four terms of the sequence. Assume n starts at 1.

1. $a_n = n$ **2.** $a_n = n^2$ **3.** $a_n = 2n - 1$ **4.** $a_n = x^n$

5. $a_n = \dfrac{n}{(n+1)}$ **6.** $a_n = \dfrac{(n+1)}{n}$ **7.** $a_n = \dfrac{2^n}{n!}$ **8.** $a_n = \dfrac{n!}{(n+1)!}$

9. $a_n = (-1)^n x^{n+1}$ **10.** $a_n = (-1)^{n+1} n^2$ **11.** $a_n = \dfrac{(-1)^n}{(n+1)(n+2)}$ **12.** $a_n = \dfrac{(n-1)^2}{(n+1)^2}$

In Exercises 13–20, find the indicated term of the sequence.

13. $a_n = \left(\dfrac{1}{2}\right)^n \quad a_9 = ?$ **14.** $a_n = \dfrac{n}{(n+1)^2} \quad a_{15} = ?$ **15.** $a_n = \dfrac{(-1)^n n!}{(n+2)!} \quad a_{19} = ?$

16. $a_n = \dfrac{(-1)^{n+1}(n-1)(n+2)}{n} \quad a_{13} = ?$ **17.** $a_n = \left(1 + \dfrac{1}{n}\right)^2 \quad a_{100} = ?$

18. $a_n = 1 - \dfrac{1}{n^2} \quad a_{10} = ?$ **19.** $a_n = \log 10^n \quad a_{23} = ?$ **20.** $a_n = e^{\ln n} \quad a_{49} = ?$

In Exercises 21–28, write an expression for the nth term of the given sequence.

21. 2, 4, 6, 8, 10, ... **22.** 3, 6, 9, 12, 15, ... **23.** $\dfrac{1}{2 \cdot 1}, \dfrac{1}{3 \cdot 2}, \dfrac{1}{4 \cdot 3}, \dfrac{1}{5 \cdot 4}, \dfrac{1}{6 \cdot 5}, \cdots$ **24.** $\dfrac{1}{2}, \dfrac{1}{4}, \dfrac{1}{8}, \dfrac{1}{16}, \dfrac{1}{32}, \cdots$

25. $-\dfrac{2}{3}, \dfrac{4}{9}, -\dfrac{8}{27}, \dfrac{16}{81}, \cdots$ **26.** $\dfrac{1}{2}, \dfrac{3}{4}, \dfrac{9}{8}, \dfrac{27}{16}, \dfrac{81}{32}, \cdots$ **27.** 1, −1, 1, −1, 1, ... **28.** $\dfrac{1}{3}, -\dfrac{2}{4}, \dfrac{3}{5}, -\dfrac{4}{6}, \dfrac{5}{7}, \cdots$

In Exercises 29–36, simplify the ratio of factorials.

29. $\dfrac{9!}{7!}$ **30.** $\dfrac{4!}{6!}$ **31.** $\dfrac{29!}{27!}$ **32.** $\dfrac{32!}{30!}$ **33.** $\dfrac{(n-1)!}{(n+1)!}$ **34.** $\dfrac{(n+2)!}{n!}$

35. $\dfrac{(2n+3)!}{(2n+1)!}$ **36.** $\dfrac{(2n+2)!}{(2n-1)!}$

In Exercises 37–46, write the first four terms of the sequence defined by the recursion formula. Assume the sequence begins at 1.

37. $a_1 = 7 \quad a_n = a_{n-1} + 3$ **38.** $a_1 = 2 \quad a_n = a_{n-1} + 1$ **39.** $a_1 = 1 \quad a_n = n \cdot a_{n-1}$

40. $a_1 = 2 \quad a_n = (n+1) \cdot a_{n-1}$ **41.** $a_1 = 100 \quad a_n = \dfrac{a_{n-1}}{n!}$ **42.** $a_1 = 20 \quad a_n = \dfrac{a_{n-1}}{n^2}$

43. $a_1 = 1, a_2 = 2 \quad a_n = a_{n-1} \cdot a_{n-2}$ **44.** $a_1 = 1, a_2 = 2 \quad a_n = \dfrac{a_{n-2}}{a_{n-1}}$

45. $a_1 = 1, a_2 = -1 \quad a_n = (-1)^n \left[a_{n-1}^2 + a_{n-2}^2\right]$ **46.** $a_1 = 1, a_2 = -1 \quad a_n = (n-1)a_{n-1} + (n-2)a_{n-2}$

In Exercises 47–60, find the sum of the finite series.

47. $\displaystyle\sum_{n=1}^{5} 2$ **48.** $\displaystyle\sum_{n=1}^{5} 7$ **49.** $\displaystyle\sum_{n=0}^{4} n^2$ **50.** $\displaystyle\sum_{n=1}^{4} \dfrac{1}{n}$ **51.** $\displaystyle\sum_{n=1}^{6} (2n-1)$

52. $\displaystyle\sum_{n=1}^{6} (n+1)$ **53.** $\displaystyle\sum_{n=0}^{4} 1^n$ **54.** $\displaystyle\sum_{n=0}^{4} 2^n$ **55.** $\displaystyle\sum_{n=0}^{3} (-x)^n$ **56.** $\displaystyle\sum_{n=0}^{3} (-x)^{n+1}$

57. $\displaystyle\sum_{k=0}^{5} \dfrac{2^k}{k!}$ **58.** $\displaystyle\sum_{k=0}^{5} \dfrac{(-1)^k}{k!}$ **59.** $\displaystyle\sum_{k=0}^{4} \dfrac{x^k}{k!}$ **60.** $\displaystyle\sum_{k=0}^{4} \dfrac{(-1)^k x^k}{k!}$

In Exercises 61–64, find the sum of the infinite series (if possible).

61. $\displaystyle\sum_{j=0}^{\infty} 2 \cdot (0.1)^j$ **62.** $\displaystyle\sum_{i=0}^{\infty} 5 \cdot \left(\dfrac{1}{10}\right)^i$ **63.** $\displaystyle\sum_{j=0}^{\infty} n^j \quad n \geq 1$ **64.** $\displaystyle\sum_{j=0}^{\infty} 1^j$

In Exercises 65–72, use sigma notation to write the sum.

65. $1 - \dfrac{1}{2} + \dfrac{1}{4} - \dfrac{1}{8} + \cdots + \dfrac{1}{64}$

66. $1 + \dfrac{1}{2} + \dfrac{1}{4} + \dfrac{1}{8} + \cdots + \dfrac{1}{64} + \cdots$

67. $1 - 2 + 3 - 4 + 5 - 6 + \cdots$

68. $1 + 2 + 3 + 4 + 5 + \cdots + 21 + 22 + 23$

69. $\dfrac{2 \cdot 1}{1} + \dfrac{3 \cdot 2 \cdot 1}{1} + \dfrac{4 \cdot 3 \cdot 2 \cdot 1}{2 \cdot 1} + \dfrac{5 \cdot 4 \cdot 3 \cdot 2 \cdot 1}{3 \cdot 2 \cdot 1} + \dfrac{6 \cdot 5 \cdot 4 \cdot 3 \cdot 2 \cdot 1}{4 \cdot 3 \cdot 2 \cdot 1}$

70. $1 + \dfrac{2}{1} + \dfrac{2^2}{2 \cdot 1} + \dfrac{2^3}{3 \cdot 2 \cdot 1} + \dfrac{2^4}{4 \cdot 3 \cdot 2 \cdot 1} + \cdots$

71. $1 - x + \dfrac{x^2}{2} - \dfrac{x^3}{6} + \dfrac{x^4}{24} - \dfrac{x^5}{120} + \cdots$

72. $x + x^2 + \dfrac{x^3}{2} + \dfrac{x^4}{6} + \dfrac{x^5}{24} + \dfrac{x^6}{120}$

 ■ APPLICATIONS

73. Money. On graduation, Jessica receives a commission from the U.S. Navy to become an officer, and she receives a $20,000 signing bonus for selecting aviation. She puts the entire bonus in an account that earns 6% interest compounded monthly. The balance in the account after n months is

$$A_n = 20{,}000\left(1 + \frac{0.06}{12}\right)^n \qquad n = 1, 2, 3, \ldots$$

Her commitment to the Navy is 6 years. Calculate A_{72}. What does A_{72} represent?

74. Money. Dylan sells his car freshman year and puts $7,000 in an account that earns 5% interest compounded quarterly. The balance in the account after n quarters is

$$A_n = 7000\left(1 + \frac{0.05}{4}\right)^n \qquad n = 1, 2, 3, \ldots$$

Calculate A_{12}. What does A_{12} represent?

75. Salary. An attorney is trying to calculate the costs associated with going into private practice. If she hires a paralegal to assist her, she will have to pay the paralegal $20.00 per hour. To be competitive with most firms, she will have to give her paralegal a $2 raise per year. Find a general term of a sequence, a_n, which represents the hourly salary of a paralegal with n years of experience. What will be the paralegal's salary with 20 years of experience?

76. NFL Salaries. A player in the NFL typically has a career that lasts 3 years. The practice squad makes the league

minimum of $275,000 (2004) in their first year, with $75,000 raises per year. Write the general term, a_n, of a sequence that represents the salary of an NFL player making the league minimum during his entire career. Assuming $n = 1$ corresponds to the first year, what does

$$\sum_{n=1}^{3} a_n$$

represent?

77. Salary. On graduation Sheldon decides to go to work for a local police department. His starting salary is $30,000 per year, and he expects to get 3% raises every year. Write the recursion formula for a sequence that represents his salary n years on the job. Assume $n = 0$ represents his first year making $30,000.

78. Escherichia Coli. A single cell of bacteria reproduces through a process called binary fission. *Escherichia coli* cells divide into two every 20 minutes. Suppose the same rate of division is maintained for 12 hours after the original cell enters the body. How many *E. coli* bacteria cells would be in the body 12 hours later? Suppose there is an infinite nutrient source so that the *E. coli* bacteria maintain the same rate of division for 48 hours after the original cell enters the body. How many *E. coli* bacteria cells would be in the body 48 hours later?

79. AIDS/HIV. A typical person has 500 to 1500 T cells per drop of blood in their body. HIV destroys a T cell count at a rate of 50–100 cells per year depending on how aggressive it is in the body. Once the body's T cell count drops below 200 generally is when the onset of AIDS occurs. Write a sequence that represents the total number of T cells in a person infected with HIV.

Assume that before infection the person has a 1000 T cell count, $a_1 = 1000$, and the rate at which the infection spreads corresponds to a loss of 75 T cells per year. How long until this person has full-blown AIDS?

80. Company Sales. Lowe's reported total sales from 2003 through 2004 in the billions. The sequence, $a_n = 3.8 + 1.6n$, represents the total sales in billions of dollars. Assuming $n = 3$ corresponds to 2003, what were the reported sales in 2003 and 2004? What does $\frac{1}{2} \cdot \sum_{n=3}^{4} a_n$ represent?

81. Cost of Eating Out. A college student tries to save money by bringing a bag lunch instead of eating out. He will be able to save $100 per month. He puts the money into his savings account, which draws 1.2% interest and is compounded monthly. The balance in his account after n months of bagging his lunch is

$$A_n = 100{,}000\big[(1.001)^n - 1\big] \qquad n = 1, 2, \ldots$$

Calculate the first four terms of this sequence. Calculate the amount after 3 years (36 months).

82. Cost of Acrylic Nails. A college student tries to save money by growing her own nails out and not spending $50 per month on acrylic fills.

She will be able to save $50 per month. She puts the money into her savings account, which draws 1.2% interest and is compounded monthly.

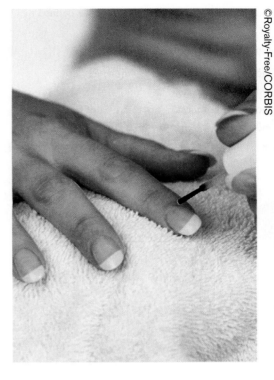

The balance in her account after n months of natural nails is

$$A_n = 50{,}000\big[(1.001)^n - 1\big] \quad n = 1, 2, \ldots$$

Calculate the first four terms of this sequence. Calculate the amount after 4 years (48 months).

■ CATCH THE MISTAKE

In Exercises 83–86, explain the mistake that is made.

83. Simplify the ratio of factorials: $\dfrac{(3!)(5!)}{6!}$.

Solution:

Express 6! in factored form.

$$\frac{(3!)(5!)}{(3!)(2!)}$$

Cancel the 3! in numerator and denominator.

$$\frac{(5!)}{(2!)}$$

Write out the factorials.

$$\frac{5 \cdot 4 \cdot 3 \cdot 2 \cdot 1}{2 \cdot 1}$$

Simplify.

$$5 \cdot 4 \cdot 3 = 60$$

$\dfrac{(3!)(5!)}{(3!)(2!)} \neq 60$. What mistake was made?

84. Simplify the factorial expression: $\dfrac{2n(2n - 2)!}{(2n + 2)!}$.

Solution:

Express factorials in factored form.

$$\frac{2n(2n - 2)(2n - 4)(2n - 6) \cdots}{(2n + 2)(2n)(2n - 2)(2n - 4)(2n - 6) \cdots}$$

Cancel common terms.

$$\frac{1}{2n + 2}$$

This is incorrect. What mistake was made?

85. Find the first four terms of the sequence defined by $a_n = (-1)^{n+1} n^2$.

Solution:

Find the $n = 1$ term.　　　$a_1 = -1$

Find the $n = 2$ term.　　　$a_2 = 4$

Find the $n = 3$ term.　　　$a_3 = -9$

Find the $n = 4$ term.　　　$a_4 = 16$

The sequence $-1, 4, -9, 16, \ldots$, is incorrect. What mistake was made?

86. Find the sum of the series $\sum\limits_{k=0}^{3} (-1)^k k^2$.

Write out the sum.　　$\sum\limits_{k=0}^{3} (-1)^k k^2 = -1 + 4 - 9$

Simplify the sum.　　$\sum\limits_{k=0}^{3} (-1)^k k^2 = -6$

This is incorrect. What mistake was made?

■ **CHALLENGE**

87. T or F: $\sum\limits_{k=0}^{n} cx^k = c \sum\limits_{k=0}^{n} x^k$

88. T or F: $\sum\limits_{i=1}^{n} a_i + b_i = \sum\limits_{i=1}^{n} a_i + \sum\limits_{i=1}^{n} b_i$

89. T or F: $\sum\limits_{k=1}^{n} a_k b_k = \sum\limits_{k=1}^{n} a_k \cdot \sum\limits_{k=1}^{n} b_k$

90. T or F: $(a!)(b!) = (ab)!$

91. Write the first four terms of the sequence defined by the recursion formula:

$$a_1 = C \qquad a_n = a_{n-1} + D \qquad D \neq 0$$

92. Write the first four terms of the sequence defined by the recursion formula:

$$a_1 = C \qquad a_n = D a_{n-1} \qquad D \neq 0$$

■ **TECHNOLOGY**

93. The sequence defined by $a_n = \left(1 + \dfrac{1}{n}\right)^n$ approaches the number e as n gets large. Use a graphing calculator to find a_{100}, a_{1000}, $a_{10,000}$, and keep increasing n until the terms in the sequence approach 2.7183.

94. The Fibonacci sequence is defined by $a_1 = 1$, $a_2 = 1$, and $a_n = a_{n-2} + a_{n-1}$. The two ratios $\dfrac{a_n}{a_{n+1}}$ and $\dfrac{a_{n+1}}{a_n}$ are both called the golden ratio, because they approach a constant, phi, as n gets large. Find the golden ratio using a graphing utility.

SECTION 9.2 Arithmetic Sequences and Series

Skills Objectives

- Recognize an arithmetic sequence.
- Find the general, nth term, of an arithmetic sequence.
- Find the sum of a finite arithmetic series.
- Use arithmetic sequences and series to model real-world problems.

Conceptual Objectives

- Understand the difference between an arithmetic sequence and an arithmetic series.

Arithmetic Sequences

The word *arithmetic* often implies adding or subtracting of numbers. *Arithmetic sequences* are sequences whose terms are found by adding a constant to each previous term. The sequence $1, 3, 5, 7, 9, \ldots$, is arithmetic because each successive term is found by adding 2 to the previous term.

> **DEFINITION**   **ARITHMETIC SEQUENCES**
>
> A sequence is **arithmetic** if each term in the sequence is found by adding a number, d, to the previous term, $a_{n+1} = a_n + d$. Because $a_{n+1} - a_n = d$, the number d is called the **common difference**.

EXAMPLE 1   **Identifying the Common Difference in Arithmetic Sequences**

Find the common difference for each of the arithmetic sequences.

a. $5, 9, 13, 17, \ldots$ **b.** $18, 9, 0, -9, \ldots$ **c.** $\dfrac{1}{2}, \dfrac{5}{4}, 2, \dfrac{11}{4}, \ldots$

Solution (a):

Label the terms. $a_1 = 5, a_2 = 9, a_3 = 13, a_4 = 17, \ldots$

Find the difference, $d = a_{n+1} - a_n$. $d = a_2 - a_1 = 9 - 5 = 4$

$d = a_3 - a_2 = 13 - 9 = 4$

$d = a_4 - a_3 = 17 - 13 = 4$

The common difference is 4 . Therefore, each term in the arithmetic sequence is found by adding 4 to the previous term.

Solution (b):

Label the terms. $a_1 = 18, a_2 = 9, a_3 = 0, a_4 = -9, \ldots$

Find the difference, $d = a_{n+1} - a_n$. $d = a_2 - a_1 = 9 - 18 = -9$

$d = a_3 - a_2 = 0 - 9 = -9$

$d = a_4 - a_3 = -9 - 0 = -9$

The common difference is -9 . Therefore, each term in the arithmetic sequence is found by subtracting 9 from the previous term.

Solution (c):

Label the terms. $a_1 = \dfrac{1}{2}, a_2 = \dfrac{5}{4}, a_3 = 2, a_4 = \dfrac{11}{4}, \ldots$

Find the difference, $d = a_{n+1} - a_n$ $d = a_2 - a_1 = \dfrac{5}{4} - \dfrac{1}{2} = \dfrac{3}{4}$

$d = a_3 - a_2 = 2 - \dfrac{5}{4} = \dfrac{3}{4}$

$d = a_4 - a_3 = \dfrac{11}{4} - 2 = \dfrac{3}{4}$

The common difference is $\frac{3}{4}$. Therefore, each term in the arithmetic sequence is found by adding $\frac{3}{4}$ to the previous term.

■ **YOUR TURN** Find the common difference for each of the arithmetic sequences:

a. $7, 2, -3, -8, \ldots$ **b.** $1, \dfrac{5}{3}, \dfrac{7}{3}, 3, \ldots$

The General (*n*th) Term of an Arithmetic Sequence

To find a formula for the general, or *n*th, term of an arithmetic sequence, write out the first several terms and look for a pattern.

First term, $n = 1$. $\qquad a_1$
Second term, $n = 2$. $\qquad a_2 = a_1 + d$
Third term, $n = 3$. $\qquad a_3 = a_2 + d = (a_1 + d) + d = a_1 + 2d$
Fourth term, $n = 4$. $\qquad a_4 = a_3 + d = (a_1 + 2d) + d = a_1 + 3d$

In general, the *n*th term is given by $a_n = a_1 + (n - 1)d$.

THE *N*TH TERM OF AN ARITHMETIC SEQUENCE

The ***n*th term** of an arithmetic sequence with common difference d is given by

$$a_n = a_1 + (n - 1)d \qquad \text{for } n \geq 1$$

EXAMPLE 2 Finding the *n*th Term of an Arithmetic Sequence

Find the 13th term of the sequence $2, 5, 8, 11, \ldots$

Solution:

Identify the common difference. $\qquad\qquad d = 5 - 2 = 3$

Identify the first ($n = 1$) term. $\qquad\qquad a_1 = 2$

Substitute $a_1 = 2$ and $d = 3$
into $a_n = a_1 + (n - 1)d$. $\qquad\qquad a_n = 2 + 3(n - 1)$

Substitute $n = 13$ into $a_n = 2 + 3(n - 1)$. $\quad a_{13} = 2 + 3(13 - 1) = 38$

The 13th term of the arithmetic sequence is 38 .

■ **YOUR TURN** Find the 10th term of the sequence $3, 10, 17, 24, \ldots$

■ **Answer: a.** -5 **b.** $\dfrac{2}{3}$ ■ **Answer:** 66

EXAMPLE 3 Finding the Arithmetic Sequence

The 4th term of an arithmetic sequence is 16, and the 21st term is 67. Find a_1 and d and construct the sequence.

Solution:

Write the 4th and 21st terms.	$a_4 = 16$ and $a_{21} = 67$
Adding d 17 times to a_4 would result in a_{21}.	$a_{21} = a_4 + 17d$
Substitute $a_4 = 16$ and $a_{21} = 67$.	$67 = 16 + 17d$
Solve for d.	$d = 3$
Substitute $d = 3$ into $a_n = a_1 + (n - 1)d$.	$a_n = a_1 + 3(n - 1)$
Let $a_4 = 16$.	$16 = a_1 + 3(4 - 1)$
Solve for a_1.	$a_1 = 7$

The arithmetic sequence that starts at 7 and has a common difference of 3 is

$7, 10, 13, 16, \ldots$

CONCEPT CHECK The 7th term of an arithmetic sequence is 26, and the 13th term is 50. What is the common difference of this arithmetic sequence?

■ **YOUR TURN** Construct the arithmetic sequence whose 7th term is 26 and 13th term is 50.

The Sum of an Arithmetic Sequence

The sum of the first n terms of an arithmetic sequence is called the ***n*th partial sum**, or **finite arithmetic series**, and is denoted by S_n. An arithmetic sequence can be found by starting at the first term and adding the common difference to each successive term, and so the nth partial sum, or finite series, can be found the same way but terminating the sum at the nth term:

$$S_n = a_1 + \overbrace{(a_1 + d)}^{a_2} + \overbrace{(a_1 + 2d)}^{a_3} + \overbrace{(a_1 + 3d)}^{a_4} + \cdots + (a_n)$$

Similarly, we can start with the nth term and find terms going backwards by subtracting the common difference until we arrive at the first term:

$$S_n = a_n + \overbrace{(a_n - d)}^{a_{n-1}} + \overbrace{(a_n - 2d)}^{a_{n-2}} + \overbrace{(a_n - 3d)}^{a_{n-3}} + \cdots + (a_1)$$

■ **Answer:** $2, 6, 10, 14, \ldots$

Add these two representations of the nth partial sum. Notice that the d terms are eliminated:

$$S_n = a_1 + (a_1 + d) + (a_1 + 2d) + (a_1 + 3d) + \cdots + (a_n)$$

$$S_n = a_n + (a_n - d) + (a_n - 2d) + (a_n - 3d) + \cdots + (a_1)$$

$$2S_n = \underbrace{(a_1 + a_n) + (a_1 + a_n) + (a_1 + a_n) + \cdots + (a_1 + a_n)}_{n(a_1 + a_n)}$$

$$2S_n = n(a_1 + a_n)$$

$$S_n = \frac{n}{2}(a_1 + a_n)$$

SUM OF A FINITE ARITHMETIC SERIES

The sum of the first n terms of an arithmetic sequence, called a **finite arithmetic series**, is given by the formula

$$S_n = \frac{n}{2}(a_1 + a_n)$$

EXAMPLE 4 Finding the Sum of a Finite Arithmetic Series

Find the sum of the finite arithmetic series: $\displaystyle\sum_{k=1}^{100} k$.

Solution:

Expand the arithmetic series. $\qquad \displaystyle\sum_{k=1}^{100} k = 1 + 2 + 3 + \cdots + 99 + 100$

This is the sum of an arithmetic sequence of numbers with a common difference of 1.

Identify the parameters of the arithmetic sequence. $\qquad a_1 = 1,\ a_n = 100,\ \text{and}\ n = 100$

Substitute these values into $S_n = \frac{n}{2}(a_1 + a_n)$. $\quad S_{100} = \frac{100}{2}(1 + 100)$

Simplify. $\qquad\qquad\qquad\qquad\qquad\qquad S_{100} = 5050$

The sum of the first 100 natural numbers is 5050.

YOUR TURN Find the sum of the finite arithmetic series: $\displaystyle\sum_{k=1}^{30} k$.

EXAMPLE 5 Finding the nth Partial Sum of an Arithmetic Sequence

Find the sum of the first 20 terms of the arithmetic sequence 3, 8, 13, 18, 23, ...

Solution:

Note that the first term of the arithmetic sequence is 3. $\qquad a_1 = 3$

This is an arithmetic sequence with a common difference of 5. $\qquad d = 5$

Answer: $1 + 2 + \cdots + 29 + 30 = 465$

Recall that the general, nth, term of an arithmetic sequence is given by:

$$a_n = a_1 + (n - 1)d$$

Substitute $a_1 = 3$ and $d = 5$ into $a_n = a_1 + (n - 1)d$.

$$a_n = 3 + (n - 1)5$$

Substitute $n = 20$ to find the 20th term.

$$a_{20} = 3 + (20 - 1)5$$
$$= 98$$

Recall the partial sum formula.

$$S_n = \frac{n}{2}(a_1 + a_n)$$

Find the 20th partial sum of this arithmetic sequence.

$$S_{20} = \frac{20}{2}(a_1 + a_{20})$$

Substitute $a_1 = 3$ and $a_{20} = 98$ into the partial sum.

$$S_{20} = 10(3 + 98) = 1010$$

The sum of the first 20 terms of this arithmetic sequence is 1010.

CONCEPT CHECK   What is the nth term, a_n, of the arithmetic sequence 2, 6, 10, 14, 18, …?

YOUR TURN   Find the sum of the first 25 terms of the arithmetic sequence 2, 6, 10, 14, 18, …

Applications

EXAMPLE 6 Marching Band Formation

Suppose a band has 18 members in the first row, 22 members in the second row, 26 members in the third row, and continues with that pattern for a total of nine rows. How many marchers are there all together?

David Young-Wolff/PhotoEdit

UC Berkeley marching band

Answer: 1250

Solution:

The number of members in each row forms an arithmetic sequence with a common difference of 4 and the first row has 18 members.

$$a_1 = 18 \quad d = 4$$

Calculate the nth term of the sequence:
$a_n = a_1 + (n-1)d.$

$$a_n = 18 + (n-1)4$$

Find the 9th term, $n = 9$.

$$a_9 = 18 + (9-1)4 = 50$$

Calculate the sum, $S_n = \dfrac{n}{2}(a_1 + a_n)$, of the nine rows.

$$S_9 = \dfrac{9}{2}(a_1 + a_9)$$

$$= 4.5(18 + 50)$$

$$= 306$$

There are 306 members in the marching band .

SECTION 9.2 SUMMARY

In this section, arithmetic sequences were defined as sequences where each successive term was found by adding the same constant to the previous term. Formulas were developed for the general, or nth, term of an arithmetic sequence, and for the nth partial sum of an arithmetic sequence, also called a finite arithmetic series.

SECTION 9.2 EXERCISES

■ **SKILLS**

In Exercises 1–10, determine if the sequence is arithmetic. If it is, find the common difference.

1. $2, 5, 8, 11, 14, \ldots$ **2.** $9, 6, 3, 0, -3, -6, \ldots$ **3.** $1^2 + 2^2 + 3^2 + \cdots$ **4.** $1! + 2! + 3! + \cdots$

5. $3.33, 3.30, 3.27, 3.24, \ldots$ **6.** $0.7, 1.2, 1.7, 2.2, \ldots$ **7.** $4, \dfrac{14}{3}, \dfrac{16}{3}, 6, \ldots$ **8.** $2, \dfrac{7}{3}, \dfrac{8}{3}, 3, \ldots$

9. $10^1, 10^2, 10^3, 10^4, \ldots$ **10.** $120, 60, 30, 15, \ldots$

In Exercises 11–20, find the first four terms of the sequence. Determine if the sequence is arithmetic, and if so find the common difference.

11. $a_n = -2n + 5$ **12.** $a_n = 3n - 10$ **13.** $a_n = n^2$ **14.** $a_n = \dfrac{n^2}{n!}$

15. $a_n = 5n - 3$ **16.** $a_n = -4n + 5$ **17.** $a_n = 10(n-1)$ **18.** $a_n = 8n - 4$

19. $a_n = (-1)^n \, n$ **20.** $a_n = (-1)^{n+1} \, 2n$

In Exercises 21–28, find the general, or nth, term of the arithmetic sequence given the first term and the common difference.

21. $a_1 = 11$ $d = 5$ **22.** $a_1 = 5$ $d = 11$ **23.** $a_1 = -4$ $d = 2$ **24.** $a_1 = 2$ $d = -4$

25. $a_1 = 0$ $d = \dfrac{2}{3}$ **26.** $a_1 = -1$ $d = -\dfrac{3}{4}$ **27.** $a_1 = 0$ $d = e$ **28.** $a_1 = 1.1$ $d = -0.3$

In Exercises 29–32, find the specified term for each arithmetic sequence.

29. The 10th term of the sequence $7, 20, 33, 46, \ldots$

30. The 19th term of the sequence $7, 1, -5, -11, \ldots$

31. The 100th term of the sequence $9, 2, -5, -12, \ldots$

32. The 90th term of the sequence $13, 19, 25, 31, \ldots$

In Exercises 33–38, for each arithmetic sequence described below, find a_1 and d and construct the sequence by stating the general, or nth, term.

33. The 5th term is 44 and the 17th term is 152.

34. The 9th term is -19 and the 21st term is -55.

35. The 7th term is -1 and the 17th term is -41.

36. The 8th term is 47 and the 21st term is 112.

37. The 4th term is 3 and the 22nd term is 15.

38. The 11th term is -3 and the 31st term is -13.

In Exercises 39–50, find the sum.

39. $\displaystyle\sum_{k=1}^{23} 2k$ **40.** $\displaystyle\sum_{k=0}^{20} 5k$ **41.** $\displaystyle\sum_{n=1}^{30} -2n + 5$ **42.** $\displaystyle\sum_{n=0}^{17} 3n - 10$ **43.** $\displaystyle\sum_{j=3}^{14} 0.5j$ **44.** $\displaystyle\sum_{j=1}^{33} \dfrac{j}{4}$

45. $2 + 7 + 12 + 17 + \cdots + 62$

46. $1 - 3 - 7 - \cdots - 75$

47. $4 + 7 + 10 + \cdots + 151$

48. $2 + 0 - 2 - \cdots - 56$

49. $\dfrac{1}{6} - \dfrac{1}{6} - \dfrac{1}{2} - \cdots - \dfrac{13}{2}$

50. $\dfrac{11}{12} + \dfrac{7}{6} + \dfrac{17}{12} + \cdots + \dfrac{14}{3}$

 ■ **APPLICATIONS**

51. Comparing Salaries. Colin and Camden are twin brothers graduating with B.S. degrees in biology. Colin takes a job at the San Diego Zoo making $28,000 for his first year with a $1,500 raise per year every year after that. Camden accepts a job at Florida Fish and Wildlife making $25,000 with a guaranteed $2,000 raise per year. How much will each of the brothers have made in a total of 10 years?

52. Comparing Salaries. On graduating with a Ph.D. in optical sciences, Jasmine and Megan choose different career paths. Jasmine accepts a faculty position at University of Arizona making $80,000 with a guaranteed $2,000 raise every year. Megan takes a job with the Boeing Corporation making $90,000 with a guaranteed $5,000 raise each year. Calculate how many total dollars each woman will have made after 15 years.

53. Theater Seating. You walk into the premiere of Brad Pitt's new movie, and the theater is packed, with almost every seat filled. You want to estimate the number of people in the theater. You quickly count to find that there are 22 seats in the front row, and there are 25 rows in the theater. Each row appears to have 1 more seat than the row in front of it. How many seats are in that theater?

54. Field of Tulips. Every spring the Skagit County Tulip Festival plants more than 100,000 bulbs. In honor of the Tri-Delta sorority that has sent 120 sisters from the University of Washington to volunteer for the festival, Skagit County has planted tulips in the shape of $\Delta\Delta\Delta$. In each of the triangles there are 20 rows of tulips, each row having one less than the row before. How many tulips are planted in each delta if there is 1 tulip in the first row?

World's largest champagne fountain, standing 7.02 meters high, weighing 3,308 tons, and made with 33,081 champagne glasses. Built by Leclerc Briant. Feb. 12, 2004, Roissy, France.

55. World's Largest Champagne Fountain. From December 28 to 30, 1999, Luuk Broos, director of Maison Luuk-Chalet Fontain, constructed a 56-story champagne fountain at the Steigenberger Kurhaus Hotel, Scheveningen, Netherlands. The fountain consisted of 30,856 champagne glasses. Assuming there was one glass at the top, how many were on the bottom row? How many glasses less did each successive row have?

56. Stacking of Logs. If 25 logs are laid side by side on the ground, and 24 logs are placed on top of those, and 23 logs are placed on the 3rd row, and the pattern continues until there is a single log on the 25th row, how many logs are in the stack?

57. Falling Object. When a skydiver jumps out of an airplane, she falls approximately 16 feet in the 1st second, 48 feet during the 2nd second, 80 feet during the 3rd second, 112 feet during the 4th second, and 144 feet during the 5th second, and this pattern continues. If she deploys her parachute after 10 seconds have elapsed, how far will she have fallen during those 10 seconds?

58. Falling Object. If a penny is dropped out of a plane, it falls approximately 4.9 meters during the 1st second, 14.7 meters during the 2nd second, 24.5 meters during the 3rd second, and 34.3 meters during the 4th second. Assuming this pattern continues, how many meters will the penny have fallen after 10 seconds?

■ **CATCH THE MISTAKE**

In Exercises 59–62, explain the mistake that is made.

59. Find the general, or *n*th, term of the arithmetic sequence: 3, 4, 5, 6, 7, …

Solution:

The common difference of this sequence is 1.	$d = 1$
The first term is 3.	$a_1 = 3$
The general term is $a_n = a_1 + nd$.	$a_n = 3 + n$

This is incorrect. What mistake was made?

60. Find the general, or *n*th, term of the arithmetic sequence: 10, 8, 6, …

Solution:

The common difference of this sequence is 2.	$d = 2$
The first term is 10.	$a_1 = 10$
The general term is $a_n = a_1 + (n - 1)d$.	$a_n = 10 + 2(n - 1)$

This is incorrect. What mistake was made?

61. Find the sum $\displaystyle\sum_{k=0}^{10} 2n + 1$.

Solution:

The sum is given by $S_n = \dfrac{n}{2}(a_1 + a_n)$, where $n = 10$.

Identify the 1st and 10th terms. $a_1 = 1 \quad a_{10} = 21$

Substitute $a_1 = 1$, $a_{10} = 21$,

and $n = 10$ into

$S_n = \dfrac{n}{2}(a_1 + a_n)$. $S_{10} = \dfrac{10}{2}(1 + 21) = 110$

This is incorrect. What mistake was made?

62. Find the sum $3 + 9 + 15 + 21 + 27 + 33 + \cdots + 87$

Solution:

This is an arithmetic sequence with common difference of 6. $d = 6$

The general term is given by $a_n = a_1 + (n - 1)d$. $a_n = 3 + (n - 1)6$

87 is the 15th term of the series. $a_{15} = 3 + (15 - 1)6 = 87$

The sum of the series is $S_n = \dfrac{n}{2}(a_n - a_1)$. $S_{15} = \dfrac{15}{2}(87 - 3) = 630$

This is incorrect. What mistake was made?

 ■ **CHALLENGE**

63. T or F: An arithmetic sequence and a finite arithmetic series are the same.

64. T or F: The sum of all infinite and finite arithmetic series can always be found.

65. T or F: An alternating sequence cannot be an arithmetic sequence.

66. T or F: The common difference of an arithmetic sequence is always positive.

67. Find the sum of
$a + (a + b) + (a + 2b) + \cdots + (a + nb)$.

68. Find the sum of $\displaystyle\sum_{k=-29}^{30} \ln e^k$.

 ■ **TECHNOLOGY**

69. Use a graphing calculator "SUM" to sum the natural numbers from 1 to 100.

70. Use a graphing calculator to sum the even natural numbers from 1 to 100.

71. Use a graphing calculator to sum the odd natural numbers from 1 to 100.

72. Use a graphing calculator to sum $\displaystyle\sum_{n=1}^{30} -2n + 5$.
Compare with your answer from Exercise 41.

SECTION 9.3 Geometric Sequences and Series

Skills Objectives

- Recognize a geometric sequence.
- Find the general, nth term, of a geometric sequence.
- Find the sum of a finite geometric series.
- Find the sum of an infinite geometric series, if it exists.
- Use geometric sequences and series to model real-world problems.

Conceptual Objectives

- Understand the difference between a geometric sequence and a geometric series.
- Distinguish between an arithmetic sequence and a geometric sequence.
- Understand why it is not possible to find the sum of all infinite geometric series.

Geometric Sequences

In Section 9.2 we discussed *arithmetic* sequences where successive terms had a *common difference*. In other words, each term was found by adding the same constant to the previous term. In this section, we discuss *geometric* sequences where successive terms have a *common ratio*. In other words, each term is found by multiplying the same constant by the previous term. The sequence 4, 12, 36, 108, ... , is geometric because each successive term is found by multiplying the previous term by 3.

DEFINITION   **GEOMETRIC SEQUENCES**

A sequence is **geometric** if each term in the sequence is found by multiplying the previous term by a number, r, $a_{n+1} = r \cdot a_n$. Because $\dfrac{a_{n+1}}{a_n} = r$, the number r is called the **common ratio**.

EXAMPLE 1   **Identifying the Common Ratio in Geometric Sequences**

Find the common ratio for each of the geometric sequences.

a. 5, 20, 80, 320, ...   **b.** $1, -\dfrac{1}{2}, \dfrac{1}{4}, -\dfrac{1}{8}, \cdots$ **c.** \$5,000, \$5,500, \$6,050, \$6,655, ...

Solution (a):

Label the terms. $a_1 = 5, a_2 = 20, a_3 = 80, a_4 = 320, \ldots$

Find the ratio, $r = \dfrac{a_{n+1}}{a_n}$. $r = \dfrac{a_2}{a_1} = \dfrac{20}{5} = 4$

$$r = \dfrac{a_3}{a_2} = \dfrac{80}{20} = 4$$

$$r = \dfrac{a_4}{a_3} = \dfrac{320}{80} = 4$$

The common ratio is 4 . Therefore, each term in the geometric sequence is found by multiplying the previous term by 4.

Solution (b):

Label the terms. $a_1 = 1, a_2 = -\dfrac{1}{2}, a_3 = \dfrac{1}{4}, a_4 = -\dfrac{1}{8}, \cdots$

Find the ratio, $r = \dfrac{a_{n+1}}{a_n}$. $r = \dfrac{a_2}{a_1} = \dfrac{-\dfrac{1}{2}}{1} = -\dfrac{1}{2}$

$$r = \dfrac{a_3}{a_2} = \dfrac{\dfrac{1}{4}}{-\dfrac{1}{2}} = -\dfrac{1}{2}$$

$$r = \dfrac{a_4}{a_3} = \dfrac{-\dfrac{1}{8}}{\dfrac{1}{4}} = -\dfrac{1}{2}$$

The common ratio is $-\frac{1}{2}$. Therefore, each term in the geometric sequence is found by multiplying the previous term by $-\frac{1}{2}$.

Solution (c):

Label the terms. $\qquad a_1 = \$5,000, a_2 = \$5,500, a_3 = \$6,050, a_4 = \$6,655, \ldots$

Find the ratio, $r = \dfrac{a_{n+1}}{a_n}$. $\qquad r = \dfrac{a_2}{a_1} = \dfrac{\$5,500}{\$5,000} = 1.1$

$$r = \dfrac{a_3}{a_2} = \dfrac{\$6,050}{\$5,500} = 1.1$$

$$r = \dfrac{a_4}{a_3} = \dfrac{\$6,655}{\$6,050} = 1.1$$

The common ratio is 1.1. Therefore, each term in the geometric sequence is found by multiplying the previous term by 1.1. This could correspond to interest of 10%.

■ **YOUR TURN** Find the common ratios of each geometric series.

a. $1, -3, 9, -27, \ldots$ **b.** $320, 80, 20, 5, \ldots$

The General Term, (nth), of a Geometric Sequence

To find a formula for the general, or nth, term of a geometric sequence, write out the first several terms and look for a pattern.

First term, $n = 1$. $\qquad a_1$
Second term, $n = 2$. $\qquad a_2 = a_1 \cdot r$
Third term, $n = 3$. $\qquad a_3 = a_2 \cdot r = (a_1 \cdot r) \cdot r = a_1 \cdot r^2$
Fourth term, $n = 4$. $\qquad a_4 = a_3 \cdot r = (a_1 \cdot r^2) \cdot r = a_1 \cdot r^3$

In general, the nth term is given by $a_n = a_1 \cdot r^{n-1}$.

THE NTH TERM OF A GEOMETRIC SEQUENCE

The **nth term** of a geometric sequence with common ratio r is given by

$$a_n = a_1 \cdot r^{n-1} \text{ for } n \geq 1$$

EXAMPLE 2 Finding the nth Term of a Geometric Sequence

Find the 7th term of the sequence $2, 10, 50, 250, \ldots$

Solution:

Identify the common ratio. $\qquad r = \dfrac{10}{2} = \dfrac{50}{10} = \dfrac{250}{50} = 5$

Identify the first ($n = 1$) term. $\qquad a_1 = 2$

■ **Answer: a.** -3 **b.** $\dfrac{1}{4}$ or 0.25

Substitute $a_1 = 2$ and $r = 5$ into
$a_n = a_1 \cdot r^{n-1}$.

$a_n = 2 \cdot 5^{n-1}$

Substitute $n = 7$ into $a_n = 2 \cdot 5^{n-1}$.

$a_7 = 2 \cdot 5^{7-1} = 2 \cdot 5^6 = 31{,}250$

The 7th term of the geometric sequence is $31{,}250$.

YOUR TURN Find the 8th term of the sequence $3, 12, 48, 192, \ldots$

EXAMPLE 3 Finding the Geometric Sequence

Find the geometric sequence whose 5th term is 0.01 and whose common ratio is 0.1.

Solution:

Label the common ratio and 5th term.

$a_5 = 0.01$ and $r = 0.1$

Substitute $a_5 = 0.01$, $n = 5$, and $r = 0.1$ into $a_n = a_1 \cdot r^{n-1}$.

$0.01 = a_1 \cdot (0.1)^{5-1}$

Solve for a_1.

$a_1 = \dfrac{0.01}{(0.1)^4} = \dfrac{0.01}{0.0001} = 100$

The geometric sequence that starts at 100 and has a common ratio of 0.1 is

$100, 10, 1, 0.1, 0.01, \ldots$

CONCEPT CHECK The 4th term of a geometric sequence is 3, and the 5th term is 1. What is the common ratio of this geometric sequence?

YOUR TURN Construct the geometric sequence whose 4th term is 3 and 5th term is 1.

Geometric Series

The sum of the terms of a geometric sequence is called a **geometric series**.

$$a_1 + a_1 \cdot r + a_1 \cdot r^2 + a_1 \cdot r^3 + \cdots$$

If we only sum the first n terms of a geometric sequence, the result is a **finite geometric series**, given by

$$S_n = a_1 + a_1 \cdot r + a_1 \cdot r^2 + a_1 \cdot r^3 + \cdots + a_1 \cdot r^{n-1}$$

To develop a formula for the nth partial sum, we multiply the above equation by r,

$$r \cdot S_n = a_1 \cdot r + a_1 \cdot r^2 + a_1 \cdot r^3 + \cdots + a_1 \cdot r^{n-1} + a_1 \cdot r^n$$

Subtract the second equation from the first equation and we find that all of the terms on the right side drop out except the *first* term in the first equation and the *last* term in the second equation:

$$S_n = a_1 + a_1 \cdot r + a_1 \cdot r^2 + \cdots + a_1 r^{n-1}$$
$$\underline{-rS_n = \qquad -a_1 \cdot r - a_1 \cdot r^2 - \cdots - a_1 r^{n-1} - a_1 r^n}$$
$$S_n - rS_n = a_1 \qquad\qquad\qquad\qquad\qquad\qquad\qquad -a_1 r^n$$

Factor the S_n out of the left side and the a_1 out of the right side:

$$S_n(1 - r) = a_1(1 - r^n)$$

Divide both sides by $(1 - r)$, assuming $r \neq 1$, and the result is a general formula for the sum of a finite geometric series:

$$S_n = a_1 \frac{(1 - r^n)}{(1 - r)}$$

SUM OF A FINITE GEOMETRIC SERIES

The sum of the first n terms of a geometric sequence, called a **finite geometric series**, is given by the formula

$$S_n = a_1 \frac{(1 - r^n)}{(1 - r)} \qquad r \neq 1$$

It is important to note that a finite geometric series can also be written in sigma (summation) notation:

$$S_n = \sum_{k=1}^{n} a_1 \cdot r^{k-1} = a_1 + a_1 \cdot r + a_1 \cdot r^2 + a_1 \cdot r^3 + \cdots + a_1 \cdot r^{n-1}$$

EXAMPLE 4 Finding the Sum of a Finite Geometric Series

Find the sum of the finite geometric series:

a. $\displaystyle\sum_{k=1}^{13} 3 \cdot (0.4)^{k-1}$

b. The first nine terms of the series $1 + 2 + 4 + 8 + 16 + 32 + 64 + \cdots$

Solution (a):

Identify a_1, n, and r. $\qquad\qquad\qquad\qquad a_1 = 3, n = 13,$ and $r = 0.4$

Substitute $a_1 = 3$, $n = 13$, and $r = 0.4$
into $s_n = a_1 \dfrac{(1 - r^n)}{(1 - r)}$. $\qquad\qquad S_{13} = 3\dfrac{(1 - 0.4^{13})}{(1 - 0.4)}$

Simplify. $\qquad\qquad\qquad\qquad\qquad\qquad S_{13} \approx 4.99997$

Solution (b):

Identify the first term and common ratio. $\qquad a_1 = 1$ and $r = 2$

Substitute $a_1 = 1$ and $r = 2$ into $S_n = a_1 \dfrac{(1 - r^n)}{(1 - r)}$.

$$S_n = \frac{(1 - 2^n)}{(1 - 2)}$$

To sum the first nine terms, let $n = 9$.

$$S_9 = \frac{(1 - 2^9)}{(1 - 2)}$$

Simplify.

$$S_9 = 511$$

The sum of an infinite geometric sequence is called an **infinite geometric series**. Some infinite geometric series have a sum, and some do not. For example,

$$\frac{1}{2} + \frac{1}{4} + \frac{1}{8} + \frac{1}{16} + \frac{1}{32} + \cdots + \frac{1}{2^n} + \cdots \text{ sums to } 1$$

$$2 + 4 + 8 + 16 + 32 + \cdots + 2^n + \cdots \text{ does not sum to a single value}$$

For some geometric series the partial sum, S_n, approaches a single number as n gets large. The formula for the sum of a finite geometric series

$$S_n = a_1 \frac{(1 - r^n)}{(1 - r)}$$

can be extended to an infinite geometric series for certain values of r. If $|r| < 1$, then when r is raised to a power it continues to get smaller, approaching 0. For those values of r, the infinite geometric series has a sum.

Let $n \to \infty$, then $a_1 \dfrac{(1 - r^n)}{(1 - r)} \to a_1 \dfrac{(1 - 0)}{(1 - r)} = a_1 \dfrac{1}{1 - r}$ if $|r| < 1$.

SUM OF AN INFINITE GEOMETRIC SERIES

The **sum of an infinite geometric series** is given by the formula

$$S_\infty = a_1 \frac{1}{(1 - r)} \qquad |r| < 1$$

EXAMPLE 5 **Determining if the Sum of an Infinite Series Exists**

Determine if the sum exists for each of the geometric series:

a. $3 + 15 + 75 + 375 + \cdots$ **b.** $8 + 4 + 2 + 1 + \dfrac{1}{2} + \dfrac{1}{4} + \dfrac{1}{8} + \cdots$

Solution (a):

Identify the common ratio. $\qquad\qquad\qquad\qquad r = 5$

Since 5 is greater than 1, the *sum does not exist.* $\qquad r = 5 > 1$

Solution (b):

Identify the common ratio. $\qquad\qquad\qquad\qquad r = \dfrac{1}{2}$

Since $\frac{1}{2}$ is less than 1, the *sum exists.* $\qquad\qquad r = \dfrac{1}{2} < 1$

■ **YOUR TURN**  Determine if the sum exists for the following geometric series.

a. $81, 9, 1, \dfrac{1}{9}, \ldots$ **b.** $1, 5, 25, 125, \ldots$

Do you expect $\frac{1}{4} + \frac{1}{12} + \frac{1}{36} + \frac{1}{64} + \cdots$ and $\frac{1}{4} - \frac{1}{12} + \frac{1}{36} - \frac{1}{64} + \cdots$ to sum to the same number? The answer is no, because the second series is an alternating series and terms are both added and subtracted. Hence, we would expect the second number to sum to a smaller number than the first series sums to.

EXAMPLE 6 Finding the Sum of an Infinite Geometric Series

Find the sum of each infinite geometric series:

a. $1 + \dfrac{1}{3} + \dfrac{1}{9} + \dfrac{1}{27} + \cdots$ **b.** $1 - \dfrac{1}{3} + \dfrac{1}{9} - \dfrac{1}{27} + \cdots$

Solution (a):

Identify the first term and the common ratio. $a_1 = 1 \quad r = \dfrac{1}{3}$

Since $|r| = \left|\dfrac{1}{3}\right| < 1$ the sum of the series exists.

Substitute $a_1 = 1$ and $r = \dfrac{1}{3}$ into $S_\infty = a_1 \dfrac{1}{(1 - r)}$. $S_\infty = \dfrac{1}{1 - \dfrac{1}{3}}$

Simplify. $S_\infty = \dfrac{1}{\dfrac{2}{3}} = \dfrac{3}{2}$

$$1 + \dfrac{1}{3} + \dfrac{1}{9} + \dfrac{1}{27} + \cdots = \dfrac{3}{2}$$

Solution (b):

Identify the first term and the common ratio. $a_1 = 1 \quad r = -\dfrac{1}{3}$

Since $|r| = \left|-\dfrac{1}{3}\right| < 1$ the sum of the series exists.

Substitute $a_1 = 1$ and $r = -\dfrac{1}{3}$ into $S_\infty = a_1 \dfrac{1}{(1 - r)}$. $S_\infty = \dfrac{1}{1 + \dfrac{1}{3}}$

Simplify.

$$S_\infty = \dfrac{1}{\dfrac{4}{3}} = \dfrac{3}{4}$$

$$1 - \frac{1}{3} + \frac{1}{9} - \frac{1}{27} + \cdots = \frac{3}{4}$$

Notice that the alternating series summed to $\frac{3}{4}$, whereas the positive series summed to $\frac{3}{2}$.

Applications

Suppose you are given a job offer with a guaranteed percentage raise per year. What will your annual salary be 10 years from now? That answer can be obtained using a geometric sequence. Suppose you want to make voluntary contributions to a retirement account directly debited from your paycheck every month. Suppose the account earns a fixed percentage rate, how much will you have in 30 years if you deposit $50 a month? What is the difference in the total you will have in 30 years if you deposit $100 a month instead? These important questions about your personal finances can be answered using geometric sequences and series.

EXAMPLE 7 Future Salary: Geometric Sequence

Suppose you are offered a job as an event planner for the PGA Tour. The starting salary is $45,000, and they give employees a 5% raise per year. What will your annual salary be during the 10th year with the PGA Tour?

Solution:

Every year the salary is 5% more than the previous year.

Label year 1 salary. $\qquad a_1 = 45{,}000$

Calculate year 2 salary. $\qquad a_2 = 1.05 \cdot a_1$

Calculate year 3 salary. $\qquad a_3 = 1.05 \cdot a_2$
$\qquad\qquad\qquad = 1.05(1.05 \cdot a_1) = (1.05)^2\, a_1$

Calculate year 4 salary. $\qquad a_4 = 1.05 \cdot a_3$
$\qquad\qquad\qquad = 1.05(1.05)^2\, a_1 = (1.05)^3\, a_1$

Identify the year n salary. $\qquad a_n = 1.05^{n-1}\, a_1$

Substitute $n = 10$ and $a_1 = 45{,}000$. $\qquad a_{10} = (1.05)^9 \cdot 45{,}000$

Simplify. $\qquad a_{10} \approx 69{,}809.77$

During your 10th year with the company your salary will be $69,809.77 .

■ **YOUR TURN** Suppose you are offered a job with AT&T at $37,000 per year with a guaranteed raise of 4% after every year. What will your annual salary be after 15 years with the company?

61. Find the sum $\displaystyle\sum_{k=0}^{10} 2n + 1$.

Solution:

The sum is given by $S_n = \dfrac{n}{2}(a_1 + a_n)$, where $n = 10$.

Identify the 1st and 10th terms. $a_1 = 1$ $a_{10} = 21$

Substitute $a_1 = 1$, $a_{10} = 21$,

and $n = 10$ into

$S_n = \dfrac{n}{2}(a_1 + a_n)$. $S_{10} = \dfrac{10}{2}(1 + 21) = 110$

This is incorrect. What mistake was made?

62. Find the sum $3 + 9 + 15 + 21 + 27 + 33 + \cdots + 87$

Solution:

This is an arithmetic sequence
with common difference of 6. $d = 6$

The general term is given
by $a_n = a_1 + (n - 1)d$. $a_n = 3 + (n - 1)6$

87 is the 15th term of the
series. $a_{15} = 3 + (15 - 1)6 = 87$

The sum of the series
is $S_n = \dfrac{n}{2}(a_n - a_1)$. $S_{15} = \dfrac{15}{2}(87 - 3) = 630$

This is incorrect. What mistake was made?

■ **CHALLENGE**

63. T or F: An arithmetic sequence and a finite arithmetic series are the same.

64. T or F: The sum of all infinite and finite arithmetic series can always be found.

65. T or F: An alternating sequence cannot be an arithmetic sequence.

66. T or F: The common difference of an arithmetic sequence is always positive.

67. Find the sum of
$a + (a + b) + (a + 2b) + \cdots + (a + nb)$.

68. Find the sum of $\displaystyle\sum_{k=-29}^{30} \ln e^k$.

■ **TECHNOLOGY**

69. Use a graphing calculator "SUM" to sum the natural numbers from 1 to 100.

70. Use a graphing calculator to sum the even natural numbers from 1 to 100.

71. Use a graphing calculator to sum the odd natural numbers from 1 to 100.

72. Use a graphing calculator to sum $\displaystyle\sum_{n=1}^{30} -2n + 5$.
Compare with your answer from Exercise 41.

SECTION 9.3 Geometric Sequences and Series

Skills Objectives

■ Recognize a geometric sequence.
■ Find the general, nth term, of a geometric sequence.
■ Find the sum of a finite geometric series.
■ Find the sum of an infinite geometric series, if it exists.
■ Use geometric sequences and series to model real-world problems.

Conceptual Objectives

■ Understand the difference between a geometric sequence and a geometric series.
■ Distinguish between an arithmetic sequence and a geometric sequence.
■ Understand why it is not possible to find the sum of all infinite geometric series.

Geometric Sequences

In Section 9.2 we discussed *arithmetic* sequences where successive terms had a *common difference*. In other words, each term was found by adding the same constant to the previous term. In this section, we discuss *geometric* sequences where successive terms have a *common ratio*. In other words, each term is found by multiplying the same constant by the previous term. The sequence 4, 12, 36, 108, ... , is geometric because each successive term is found by multiplying the previous term by 3.

> **DEFINITION** **GEOMETRIC SEQUENCES**
>
> A sequence is **geometric** if each term in the sequence is found by multiplying the previous term by a number, r, $a_{n+1} = r \cdot a_n$. Because $\dfrac{a_{n+1}}{a_n} = r$, the number r is called the **common ratio**.

EXAMPLE 1 **Identifying the Common Ratio in Geometric Sequences**

Find the common ratio for each of the geometric sequences.

a. 5, 20, 80, 320, ... **b.** $1, -\dfrac{1}{2}, \dfrac{1}{4}, -\dfrac{1}{8}, \cdots$ **c.** \$5,000, \$5,500, \$6,050, \$6,655, ...

Solution (a):

Label the terms. $\qquad\qquad a_1 = 5, a_2 = 20, a_3 = 80, a_4 = 320, \ldots$

Find the ratio, $r = \dfrac{a_{n+1}}{a_n}$. $\qquad r = \dfrac{a_2}{a_1} = \dfrac{20}{5} = 4$

$$r = \dfrac{a_3}{a_2} = \dfrac{80}{20} = 4$$

$$r = \dfrac{a_4}{a_3} = \dfrac{320}{80} = 4$$

The common ratio is 4 . Therefore, each term in the geometric sequence is found by multiplying the previous term by 4.

Solution (b):

Label the terms. $\qquad\qquad a_1 = 1, a_2 = -\dfrac{1}{2}, a_3 = \dfrac{1}{4}, a_4 = -\dfrac{1}{8}, \cdots$

Find the ratio, $r = \dfrac{a_{n+1}}{a_n}$. $\qquad r = \dfrac{a_2}{a_1} = \dfrac{-\dfrac{1}{2}}{1} = -\dfrac{1}{2}$

$$r = \dfrac{a_3}{a_2} = \dfrac{\dfrac{1}{4}}{-\dfrac{1}{2}} = -\dfrac{1}{2}$$

$$r = \dfrac{a_4}{a_3} = \dfrac{-\dfrac{1}{8}}{\dfrac{1}{4}} = -\dfrac{1}{2}$$

The common ratio is $-\frac{1}{2}$. Therefore, each term in the geometric sequence is found by multiplying the previous term by $-\frac{1}{2}$.

Solution (c):

Label the terms. $a_1 = \$5,000,\ a_2 = \$5,500,\ a_3 = \$6,050,\ a_4 = \$6,655, \ldots$

Find the ratio, $r = \dfrac{a_{n+1}}{a_n}$. $r = \dfrac{a_2}{a_1} = \dfrac{\$5,500}{\$5,000} = 1.1$

$r = \dfrac{a_3}{a_2} = \dfrac{\$6,050}{\$5,500} = 1.1$

$r = \dfrac{a_4}{a_3} = \dfrac{\$6,655}{\$6,050} = 1.1$

The common ratio is 1.1 . Therefore, each term in the geometric sequence is found by multiplying the previous term by 1.1. This could correspond to interest of 10%.

■ **YOUR TURN**   Find the common ratios of each geometric series.

a. $1, -3, 9, -27, \ldots$ **b.** $320, 80, 20, 5, \ldots$

The General Term, (*n*th), of a Geometric Sequence

To find a formula for the general, or *n*th, term of a geometric sequence, write out the first several terms and look for a pattern.

First term, $n = 1$. a_1
Second term, $n = 2$. $a_2 = a_1 \cdot r$
Third term, $n = 3$. $a_3 = a_2 \cdot r = (a_1 \cdot r) \cdot r = a_1 \cdot r^2$
Fourth term, $n = 4$. $a_4 = a_3 \cdot r = (a_1 \cdot r^2) \cdot r = a_1 \cdot r^3$

In general, the *n*th term is given by $a_n = a_1 \cdot r^{n-1}$.

THE *n*TH TERM OF A GEOMETRIC SEQUENCE

The ***n*th term** of a geometric sequence with common ratio r is given by

$$a_n = a_1 \cdot r^{n-1} \text{ for } n \geq 1$$

EXAMPLE 2 Finding the *n*th Term of a Geometric Sequence

Find the 7th term of the sequence $2, 10, 50, 250, \ldots$

Solution:

Identify the common ratio. $r = \dfrac{10}{2} = \dfrac{50}{10} = \dfrac{250}{50} = 5$

Identify the first ($n = 1$) term. $a_1 = 2$

■ **Answer: a.** -3 **b.** $\dfrac{1}{4}$ or 0.25

Substitute $a_1 = 2$ and $r = 5$ into
$a_n = a_1 \cdot r^{n-1}$. $\qquad\qquad a_n = 2 \cdot 5^{n-1}$

Substitute $n = 7$ into $a_n = 2 \cdot 5^{n-1}$. $\qquad a_7 = 2 \cdot 5^{7-1} = 2 \cdot 5^6 = 31{,}250$

The 7th term of the geometric sequence is $31{,}250$.

■ **YOUR TURN** Find the 8th term of the sequence $3, 12, 48, 192, \ldots$

EXAMPLE 3 Finding the Geometric Sequence

Find the geometric sequence whose 5th term is 0.01 and whose common ratio is 0.1.

Solution:

Label the common ratio and 5th term. $\qquad a_5 = 0.01$ and $r = 0.1$

Substitute $a_5 = 0.01$, $n = 5$, and $r = 0.1$
into $a_n = a_1 \cdot r^{n-1}$. $\qquad\qquad 0.01 = a_1 \cdot (0.1)^{5-1}$

Solve for a_1. $\qquad\qquad a_1 = \dfrac{0.01}{(0.1)^4} = \dfrac{0.01}{0.0001} = 100$

The geometric sequence that starts at 100 and has a common ratio of 0.1 is

$100, 10, 1, 0.1, 0.01, \ldots$

✓**CONCEPT CHECK** The 4th term of a geometric sequence is 3, and the 5th term is 1. What is the common ratio of this geometric sequence?

■ **YOUR TURN** Construct the geometric sequence whose 4th term is 3 and 5th term is 1.

Geometric Series

The sum of the terms of a geometric sequence is called a **geometric series**.

$$a_1 + a_1 \cdot r + a_1 \cdot r^2 + a_1 \cdot r^3 + \cdots$$

If we only sum the first n terms of a geometric sequence, the result is a **finite geometric series**, given by

$$S_n = a_1 + a_1 \cdot r + a_1 \cdot r^2 + a_1 \cdot r^3 + \cdots + a_1 \cdot r^{n-1}$$

To develop a formula for the nth partial sum, we multiply the above equation by r,

$$r \cdot S_n = a_1 \cdot r + a_1 \cdot r^2 + a_1 \cdot r^3 + \cdots + a_1 \cdot r^{n-1} + a_1 \cdot r^n$$

■ **Answer:** $81, 27, 9, 3, 1, \ldots$ ■ **Answer:** $49{,}152$

Subtract the second equation from the first equation and we find that all of the terms on the right side drop out except the *first* term in the first equation and the *last* term in the second equation:

$$S_n = a_1 + a_1 \cdot r + a_1 \cdot r^2 + \cdots + a_1 r^{n-1}$$
$$-rS_n = \quad -a_1 \cdot r - a_1 \cdot r^2 - \cdots - a_1 r^{n-1} - a_1 r^n$$
$$\overline{S_n - rS_n = a_1 \qquad\qquad\qquad\qquad\qquad\qquad -a_1 r^n}$$

Factor the S_n out of the left side and the a_1 out of the right side:

$$S_n(1 - r) = a_1(1 - r^n)$$

Divide both sides by $(1 - r)$, assuming $r \neq 1$, and the result is a general formula for the sum of a finite geometric series:

$$S_n = a_1 \frac{(1 - r^n)}{(1 - r)}$$

SUM OF A FINITE GEOMETRIC SERIES

The sum of the first n terms of a geometric sequence, called a **finite geometric series**, is given by the formula

$$S_n = a_1 \frac{(1 - r^n)}{(1 - r)} \qquad r \neq 1$$

It is important to note that a finite geometric series can also be written in sigma (summation) notation:

$$S_n = \sum_{k=1}^{n} a_1 \cdot r^{k-1} = a_1 + a_1 \cdot r + a_1 \cdot r^2 + a_1 \cdot r^3 + \cdots + a_1 \cdot r^{n-1}$$

EXAMPLE 4 Finding the Sum of a Finite Geometric Series

Find the sum of the finite geometric series:

a. $\displaystyle\sum_{k=1}^{13} 3 \cdot (0.4)^{k-1}$

b. The first nine terms of the series $1 + 2 + 4 + 8 + 16 + 32 + 64 + \cdots$

Solution (a):

Identify a_1, n, and r. $\qquad\qquad\qquad\qquad$ $a_1 = 3$, $n = 13$, and $r = 0.4$

Substitute $a_1 = 3$, $n = 13$, and $r = 0.4$
into $s_n = a_1 \dfrac{(1 - r^n)}{(1 - r)}$. $\qquad\qquad\qquad$ $S_{13} = 3\dfrac{(1 - 0.4^{13})}{(1 - 0.4)}$

Simplify. $\qquad\qquad\qquad\qquad\qquad\qquad$ $S_{13} \approx 4.99997$

Solution (b):

Identify the first term and common ratio. $\qquad$ $a_1 = 1$ and $r = 2$

Substitute $a_1 = 1$ and $r = 2$ into $S_n = a_1 \dfrac{(1 - r^n)}{(1 - r)}$. $\qquad S_n = \dfrac{(1 - 2^n)}{(1 - 2)}$

To sum the first nine terms, let $n = 9$. $\qquad S_9 = \dfrac{(1 - 2^9)}{(1 - 2)}$

Simplify. $\qquad S_9 = 511$

The sum of an infinite geometric sequence is called an **infinite geometric series**. Some infinite geometric series have a sum, and some do not. For example,

$$\frac{1}{2} + \frac{1}{4} + \frac{1}{8} + \frac{1}{16} + \frac{1}{32} + \cdots + \frac{1}{2^n} + \cdots \text{ sums to } 1$$

$$2 + 4 + 8 + 16 + 32 + \cdots + 2^n + \cdots \text{ does not sum to a single value}$$

For some geometric series the partial sum, S_n, approaches a single number as n gets large. The formula for the sum of a finite geometric series

$$S_n = a_1 \frac{(1 - r^n)}{(1 - r)}$$

can be extended to an infinite geometric series for certain values of r. If $|r| < 1$, then when r is raised to a power it continues to get smaller, approaching 0. For those values of r, the infinite geometric series has a sum.

$$\text{Let } n \to \infty, \text{ then } a_1 \frac{(1 - r^n)}{(1 - r)} \to a_1 \frac{(1 - 0)}{(1 - r)} = a_1 \frac{1}{1 - r} \text{ if } |r| < 1.$$

SUM OF AN INFINITE GEOMETRIC SERIES

The **sum of an infinite geometric series** is given by the formula

$$S_\infty = a_1 \frac{1}{(1 - r)} \qquad |r| < 1$$

EXAMPLE 5 Determining if the Sum of an Infinite Series Exists

Determine if the sum exists for each of the geometric series:

a. $3 + 15 + 75 + 375 + \cdots$ $\qquad$ **b.** $8 + 4 + 2 + 1 + \dfrac{1}{2} + \dfrac{1}{4} + \dfrac{1}{8} + \cdots$

Solution (a):

Identify the common ratio. $\qquad r = 5$

Since 5 is greater than 1, the *sum does not exist*. $\qquad r = 5 > 1$

Solution (b):

Identify the common ratio. $\qquad r = \dfrac{1}{2}$

Since $\frac{1}{2}$ is less than 1, the *sum exists*. $\qquad r = \dfrac{1}{2} < 1$

■ **YOUR TURN**   Determine if the sum exists for the following geometric series.

a. $81, 9, 1, \dfrac{1}{9}, \ldots$ **b.** $1, 5, 25, 125, \ldots$

Do you expect $\frac{1}{4} + \frac{1}{12} + \frac{1}{36} + \frac{1}{64} + \cdots$ and $\frac{1}{4} - \frac{1}{12} + \frac{1}{36} - \frac{1}{64} + \cdots$ to sum to the same number? The answer is no, because the second series is an alternating series and terms are both added and subtracted. Hence, we would expect the second number to sum to a smaller number than the first series sums to.

EXAMPLE 6 Finding the Sum of an Infinite Geometric Series

Find the sum of each infinite geometric series:

a. $1 + \dfrac{1}{3} + \dfrac{1}{9} + \dfrac{1}{27} + \cdots$ **b.** $1 - \dfrac{1}{3} + \dfrac{1}{9} - \dfrac{1}{27} + \cdots$

Solution (a):

Identify the first term and the common ratio. $a_1 = 1 \quad r = \dfrac{1}{3}$

Since $|r| = \left|\dfrac{1}{3}\right| < 1$ the sum of the series exists.

Substitute $a_1 = 1$ and $r = \dfrac{1}{3}$ into $S_\infty = a_1 \dfrac{1}{(1 - r)}$. $S_\infty = \dfrac{1}{1 - \dfrac{1}{3}}$

Simplify. $S_\infty = \dfrac{1}{\dfrac{2}{3}} = \dfrac{3}{2}$

$$1 + \dfrac{1}{3} + \dfrac{1}{9} + \dfrac{1}{27} + \cdots = \dfrac{3}{2}$$

Solution (b):

Identify the first term and the common ratio. $a_1 = 1 \quad r = -\dfrac{1}{3}$

Since $|r| = \left|-\dfrac{1}{3}\right| < 1$ the sum of the series exists.

Substitute $a_1 = 1$ and $r = -\dfrac{1}{3}$ into $S_\infty = a_1 \dfrac{1}{(1 - r)}$. $S_\infty = \dfrac{1}{1 + \dfrac{1}{3}}$

Simplify.

$$S_\infty = \dfrac{1}{\frac{4}{3}} = \dfrac{3}{4}$$

$$1 - \frac{1}{3} + \frac{1}{9} - \frac{1}{27} + \cdots = \frac{3}{4}$$

Notice that the alternating series summed to $\frac{3}{4}$, whereas the positive series summed to $\frac{3}{2}$.

Applications

Suppose you are given a job offer with a guaranteed percentage raise per year. What will your annual salary be 10 years from now? That answer can be obtained using a geometric sequence. Suppose you want to make voluntary contributions to a retirement account directly debited from your paycheck every month. Suppose the account earns a fixed percentage rate, how much will you have in 30 years if you deposit $50 a month? What is the difference in the total you will have in 30 years if you deposit $100 a month instead? These important questions about your personal finances can be answered using geometric sequences and series.

EXAMPLE 7 Future Salary: Geometric Sequence

Suppose you are offered a job as an event planner for the PGA Tour. The starting salary is $45,000, and they give employees a 5% raise per year. What will your annual salary be during the 10th year with the PGA Tour?

Solution:

Every year the salary is 5% more than the previous year.

Label year 1 salary. $a_1 = 45{,}000$

Calculate year 2 salary. $a_2 = 1.05 \cdot a_1$

Calculate year 3 salary. $a_3 = 1.05 \cdot a_2$
$$= 1.05(1.05 \cdot a_1) = (1.05)^2\, a_1$$

Calculate year 4 salary. $a_4 = 1.05 \cdot a_3$
$$= 1.05(1.05)^2\, a_1 = (1.05)^3\, a_1$$

Identify the year n salary. $a_n = 1.05^{n-1}\, a_1$

Substitute $n = 10$ and $a_1 = 45{,}000$. $a_{10} = (1.05)^9 \cdot 45{,}000$

Simplify. $a_{10} \approx 69{,}809.77$

During your 10th year with the company your salary will be $69,809.77 .

■ **YOUR TURN** Suppose you are offered a job with AT&T at $37,000 per year with a guaranteed raise of 4% after every year. What will your annual salary be after 15 years with the company?

■ **Answer:** $64,072.03

EXAMPLE 8 Savings Growth: Geometric Series

Karen has maintained acrylic nails by paying for them with money earned from a part-time job. After hearing a lecture from her economics professor on the importance of investing early in life, she decides to remove the acrylic nails, which cost $50 per month, and do her own manicures herself. She has that $50 automatically debited from her checking account on the first of every month and put into a money market account that receives 3% interest compounded monthly. What will the balance be in the money market account exactly 2 years from the day of her initial $50 deposit?

Solution:

Recall the compound interest formula.

$$A = P\left(1 + \frac{r}{n}\right)^{nt}$$

Substitute $r = 0.03$ and $n = 12$ into the compound interest formula.

$$A = P\left(1 + \frac{0.03}{12}\right)^{12t}$$

$$= P(1.0025)^{12t}$$

If we let $t = n/12$, where n is the number of months of the investment:

$$A_n = P(1.0025)^n$$

The first deposit of $50 will gain interest for 24 months. $A_{24} = 50(1.0025)^{24}$

The second deposit of $50 will gain interest for 23 months.

$$A_{23} = 50(1.0025)^{23}$$

The third deposit of $50 will gain interest for 22 months. $A_{22} = 50(1.0025)^{22}$

The last deposit of $50 will gain interest for 1 month. $A_1 = 50(1.0025)^1$

Sum the amounts accrued from the 24 deposits.

$$A_1 + A_2 + \cdots + A_{24} = 50(1.0025) + 50(1.0025)^2 + 50(1.0025)^3 + \cdots + 50(1.0025)^{24}$$

Identify the first term and common ratio. $a_1 = 50(1.0025)$ and $r = 1.0025$

Sum the first n terms of a geometric series. $S_n = a_1\dfrac{(1 - r^n)}{(1 - r)}$

Substitute $n = 24$, $a_1 = 50(1.0025)$, and $r = 1.0025$.

$$S_{24} = 50(1.0025)\frac{(1 - 1.0025^{24})}{(1 - 1.0025)}$$

Simplify. $S_{24} \approx 1238.23$

Karen will have $1,238.23 saved in her money market account in 2 years .

SECTION 9.3 SUMMARY

In this section, we discussed geometric sequences, in which each successive term is found by multiplying the previous term by a constant. That constant is called the common ratio. The sum of the terms of a geometric sequence is called a geometric series. Finite geometric series sum to a number. Infinite geometric series sum to a number if the absolute value of the common ratio is less than 1. If the absolute value of the common ratio is greater than or equal to 1, the sum of the infinite geometric series does not exist. Many real-world applications involve geometric sequences and series such as growth of salaries and annuities through percentage increases.

SECTION 9.3 EXERCISES

 ■ **SKILLS**

In Exercises 1–8, determine if the sequence is geometric. If it is, find the common ratio.

1. $1, 3, 9, 27, \ldots$

2. $2, 4, 8, 16, \ldots$

3. $1, 4, 9, 16, 25, \ldots$

4. $1, \dfrac{1}{4}, \dfrac{1}{9}, \dfrac{1}{16}, \ldots$

5. $8, 4, 2, 1, \ldots$

6. $8, -4, 2, -1, \ldots$

7. $800, 1360, 2312, 3930.4, \ldots$

8. $7, 15.4, 33.88, 74.536, \ldots$

In Exercises 9–16, write the first five terms of the geometric series.

9. $a_1 = 6 \quad r = 3$

10. $a_1 = 17 \quad r = 2$

11. $a_1 = 1 \quad r = -4$

12. $a_1 = -3 \quad r = -2$

13. $a_1 = 10{,}000 \quad r = 1.06$

14. $a_1 = 10{,}000 \quad r = 0.8$

15. $a_1 = \dfrac{2}{3} \quad r = \dfrac{1}{2}$

16. $a_1 = \dfrac{1}{10} \quad r = -\dfrac{1}{5}$

In Exercises 17–24, write the formula for the *n*th term of the geometric series.

17. $a_1 = 5 \quad r = 2$

18. $a_1 = 12 \quad r = 3$

19. $a_1 = 1 \quad r = -3$

20. $a_1 = -4 \quad r = -2$

21. $a_1 = 1000 \quad r = 1.07$

22. $a_1 = 1000 \quad r = 0.5$

23. $a_1 = \dfrac{16}{3} \quad r = -\dfrac{1}{4}$

24. $a_1 = \dfrac{1}{200} \quad r = 5$

In Exercises 25–30, find the indicated term of the geometric sequence.

25. 7th term of the sequence $-2, 4, -8, 16, \ldots$

26. 10th term of the sequence $1, -5, 25, -225, \ldots$

27. 13th term of the sequence $\dfrac{1}{3}, \dfrac{2}{3}, \dfrac{4}{3}, \dfrac{8}{3}, \ldots$

28. 9th term of the sequence $100, 20, 4, 0.8, \ldots$

29. 15th term of the sequence $1000, 50, 2.5, 0.125, \ldots$

30. 8th term of the sequence $1000, -800, 640, -512, \ldots$

In Exercises 31–40, find the sum of the finite geometric series.

31. $\dfrac{1}{3} + \dfrac{2}{3} + \dfrac{2^2}{3} + \cdots + \dfrac{2^{12}}{3}$

32. $1 + \dfrac{1}{3} + \dfrac{1}{3^2} + \dfrac{1}{3^3} + \cdots + \dfrac{1}{3^{10}}$

33. $2 + 6 + 18 + 54 + \cdots + 2(3^9)$

34. $1 + 4 + 16 + 64 + \cdots 4^9$

35. $\displaystyle\sum_{n=0}^{10} 2(0.1)^n$

36. $\displaystyle\sum_{n=0}^{11} 3(0.2)^n$

37. $\displaystyle\sum_{n=1}^{8} 2(3)^{n-1}$

38. $\displaystyle\sum_{n=1}^{9} \dfrac{2}{3}(5)^{n-1}$

39. $\displaystyle\sum_{k=0}^{13} 2^k$

40. $\displaystyle\sum_{k=0}^{13} \left(\dfrac{1}{2}\right)^k$

In Exercises 41–50, find the sum of the infinite geometric series, if possible.

41. $\displaystyle\sum_{n=0}^{\infty} \left(\dfrac{1}{2}\right)^n$

42. $\displaystyle\sum_{n=1}^{\infty} \left(\dfrac{1}{3}\right)^n$

43. $\displaystyle\sum_{n=1}^{\infty} \left(-\dfrac{1}{3}\right)^n$

44. $\displaystyle\sum_{n=0}^{\infty} \left(-\dfrac{1}{2}\right)^n$

45. $\displaystyle\sum_{n=0}^{\infty} 1^n$

46. $\displaystyle\sum_{n=0}^{\infty} 1.01^n$

47. $\displaystyle\sum_{n=0}^{\infty} -9\left(\dfrac{1}{3}\right)^n$

48. $\displaystyle\sum_{n=0}^{\infty} -8\left(-\dfrac{1}{2}\right)^n$

49. $\displaystyle\sum_{n=0}^{\infty} 10{,}000(0.05)^n$

50. $\displaystyle\sum_{n=0}^{\infty} 200(0.04)^n$

■ APPLICATIONS

51. Salary. Jeremy is offered a government job with the Department of Commerce. He is hired on the "GS" scale at a base rate of $34,000 with 2.5% increases in his salary per year. Calculate what his salary will be after he has been with the Department of Commerce for 12 years.

52. Salary. Alison is offered a job with a small start-up company that wants to promote loyalty to the company and incentives for employees to stay with the company. The company offers her a starting salary of $22,000 with a guaranteed 15% raise per year. What will her salary be after she has been with the company for 10 years?

53. Depreciation. A graduating senior in high school, Brittany receives a laptop computer as a graduation gift from her Aunt Jeanine so that she can use it when she gets to the University of Alabama. If the laptop costs $2,000 new and depreciates 50% per year, write a formula for the value of the laptop n years after it was purchased. How much will the laptop be worth when Brittany graduates from college (assuming she will graduate in 4 years)? How much will it be worth when she takes it to graduate school? Assume graduate school is another 3 years.

54. Depreciation. Derek is deciding between a new top-of-the-line Honda Accord and the BMW 325 series. The BMW costs $35,000 and the Honda costs $25,000. If the BMW depreciates at 20% per year and the Honda depreciates at 10% per year, find formulas for the value of each car n years after it is purchased. Which car is worth more in 10 years?

55. Bungee Jumping. A bungee jumper rebounds 70% of the height jumped. Assuming the bungee jump is made with a

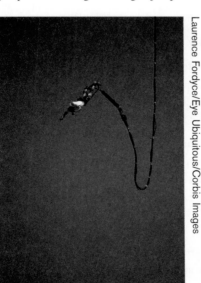

cord that stretches to 100 feet, how far will the bungee jumper travel upward on the fifth rebound? How many rebounds will it take until the bungee jumper is at rest?

56. Bungee Jumping. A bungee jumper rebounds 65% of the height jumped. Assuming the bungee cord stretches 200 feet, how far will the bungee jumper travel upward on the eighth rebound? How many rebounds will it take until the bungee jumper is at rest?

57. Population Growth. One of the fastest-growing universities in the country is the University of Central Florida. The student populations starting in 2000 were 36,000, 37,800, 39,690, 41,675, ... If this rate continues, how many students will be at UCF in 2010?

58. Website Hits. The website for Matchbox 20 (www.matchboxtwenty.com) has noticed that every week the number of hits to its website increases 5%. If there were 20,000 hits this week, how many will there be exactly 52 weeks from now?

59. Rich Man's Promise. A rich man promises that he will give you $1,000 on January 1, and every day after that, he will pay you 90% of what he paid you the day before. How many days will it take before you are making less than $1? How much will the rich man pay out for the entire month of January?

60. Poor Man's Clever Deal. A poor man promises to work for you for $.01 the first day, $.02 on the second day, $.04 on the third day, and his salary will continue to double. If he did this in January, how much would he be paid to work on January 31? How much total will he make during the month?

61. Investing Lunch. A newlywed couple decides to stop going out to lunch every day and instead bring their lunch. They estimate it will save them $100 per month. They invest that $100 on the first of every month into an account that is compounded monthly and pays

5% interest. How much will be in the account at the end of 3 years?

62. Pizza as an Investment. A college freshman decides to stop ordering late night pizzas (both for health and cost reasons). He realizes that he has been spending $50 a week on pizzas. Instead, he deposits $50 into an account that compounds weekly and pays 4% interest. How much money will be in the account in 52 weeks?

63. Tax-Deferred Annuity. Dr. Wesely contributes $500 from her paycheck (weekly) to a tax-deferred investment account. Assuming the investment earns 6% and is compounded weekly, how much will be in the account in 26 weeks? 52 weeks?

64. Saving for a House. If a new graduate decides she wants to save for a house and she is able to put $300 every month into an account that earns 5% compounded monthly, how much will she have in the account in 5 years?

■ CATCH THE MISTAKE

In Exercises 65–68, explain the mistake that is made.

65. Find the nth term of the geometric sequence:

$$-1, \frac{1}{3}, -\frac{1}{9}, \frac{1}{27}, \ldots$$

Solution:

Identify the first term and common ratio. $\qquad a_1 = -1$ and $r = \frac{1}{3}$

Substitute $a_1 = -1$ and $r = \frac{1}{3}$ into $a_n = a_1 \cdot r^{n-1}$. $\qquad a_n = (-1) \cdot \left(\frac{1}{3}\right)^{n-1}$

Simplify. $\qquad a_n = \frac{-1}{3^{n-1}}$

This is incorrect. What mistake was made?

66. Find the sum of the first n terms of the finite geometric series: $2, 4, 8, 16, \ldots$

Write the sum in sigma notation. $\qquad \sum_{k=1}^{n} (2)^k$

Identify the first term and common ratio. $\qquad a_1 = 1$ and $r = 2$

Substitute $a_1 = 1$ and $r = 2$ into $S_n = a_1 \frac{(1 - r^n)}{(1 - r)}$. $\qquad S_n = 1\frac{(1 - 2^n)}{(1 - 2)}$

Simplify. $\qquad S_n = 2^n - 1$

This is incorrect. What mistake was made?

67. Find the sum of the finite geometric series: $\sum_{n=1}^{8} 4\,(-3)^n$.

Solution:

Identify the first term and common ratio. $\qquad a_1 = 4$ and $r = -3$

Substitute $a_1 = 4$ and $r = 3$

into $S_n = a_1 \frac{(1 - r^n)}{(1 - r)}$. $\qquad S_n = 4\frac{(1 - 3^n)}{(1 - 3)}$

Simplify. $\qquad S_n = -2(1 - 3^n)$

Substitute $n = 8$. $\qquad S_8 = -2(1 - 3^8) = 13{,}120$

This is incorrect. What mistake was made?

68. Find the sum of the infinite geometric series: $\sum_{n=1}^{\infty} 2 \cdot 3^{n-1}$.

Solution:

Identify the first term and common ratio. $\qquad a_1 = 2$ and $r = 3$

Substitute $a_1 = 2$ and $r = 3$

into $S_\infty = a_1 \frac{1}{(1 - r)}$. $\qquad S_\infty = 2\frac{1}{(1 - 3)}$

Simplify. $\qquad S_\infty = -1$

This is incorrect. The series does not sum to -1. What mistake was made?

69. T or F: An alternating sequence cannot be a geometric sequence.

70. T or F: The sum of a finite or infinite geometric series can always be found.

71. T or F: The common ratio of a geometric sequence can be positive or negative.

72. T or F: The sum of an infinite geometric series can be found if the common ratio is less than or equal to 1.

73. State the conditions for the sum $a + a \cdot b + a \cdot b^2 + \cdots + a \cdot b^n + \cdots$ to exist. Assuming those conditions are met, find the sum.

74. Find the sum of $\sum_{k=0}^{20} \log 10^{2^k}$.

■ **TECHNOLOGY**

75. Sum the series: $\sum_{k=1}^{50} 1(-2)^{k-1}$. Use a graphing utility to confirm your answer.

76. Does the sum of the infinite series, $\sum_{n=0}^{\infty} \left(\frac{1}{3}\right)^n$ exist? Use a graphing calculator to find it.

77. Use a graphing utility to plot $y_1 = 1 + x + x^2 + x^3 + x^4$ and $y_2 = \dfrac{1}{1-x}$ and let x range from $[-0.5, 0.5]$. Based on what you see, what do you expect the geometric series, $\sum_{n=0}^{\infty} x^n$ to sum to?

SECTION **9.4** Mathematical Induction

Skills Objectives	Conceptual Objectives
■ Know the steps required to prove by mathematical induction. ■ Prove mathematical statements using mathematical induction.	■ Understand that just because there appears to be a pattern, the pattern may not be true for all values. ■ Understand that mathematical ideas are accepted because they can be proved.

Introduction

Is the expression $n^2 - n + 41$ *always* a prime number if n is a natural number? Your instinct may lead you to try a few values for n.

n	n² − n + 41	PRIME?
1	41	Yes
2	43	Yes
3	47	Yes
4	53	Yes
5	61	Yes

It appears that the statement might be true for all natural numbers. However, what about when $n = 41$?

$$n^2 - n + 41 = (41)^2 - 41 + 41 = 41^2$$

We find that when $n = 41$, $n^2 - n + 41$ is not prime. The moral of the story is that just because a pattern seems to exist for *some* values doesn't mean that it is true for *all* values. We must look for a way to show a statement is true for all values. In this section we talk about *mathematical induction*, which is a way to show a statement is true for all values.

Mathematical Induction

Mathematics is based on logic and proof (not assumptions or belief). One of the most famous mathematical statements was Fermat's Last Theorem. You may have even seen it on television. Pierre de Fermat (1601–1665) made the conjecture that there are no integer values for x, y, and z such that $x^n + y^n = z^n$ if $n \geq 3$. Although mathematicians *believed* that this theorem was true, no one was able to *prove* it until 350 years after the assumption was made. Professor Andrew Wiles at Princeton University received a $50,000 prize for successfully proving Fermat's Last Theorem in 1994.

Mathematical induction is a technique used in college algebra and even in very advanced mathematics to prove mathematical statements. In this section, you will use it to prove statements like if $x > 1$ then $x^n > 1$ for all natural numbers, n.

The principle of mathematical induction can be illustrated by dominos. We make two assumptions:

1. The first domino is knocked down.
2. If a domino is knocked down, then the domino immediately following it will also be knocked down.

If both of these assumptions are true, then all of the dominos will fall.

> **PRINCIPLE OF MATHEMATICAL INDUCTION**
>
> Let S_n be a statement involving the positive integer n. To prove that S_n is true for all positive integers, the following steps are required.
>
> **Step 1:** Show that S_1 is true.
>
> **Step 2:** Assume it is true for S_k and show it is true for S_{k+1} (k = integer).
>
> Combining Steps 1 and 2 proves the statement is true for all positive integers.

EXAMPLE 1 Using Mathematical Induction

Use the principle of mathematical induction to prove this statement:

If $x > 1$, then $x^n > 1$ for all natural numbers, n.

Solution:

STEP 1 Show the statement is true for $n = 1$.　　$x^1 > 1$ because $x > 1$

STEP 2 Assume the statement is true for $n = k$.　　$x^k > 1$

Show the statement is true for $k + 1$.

Multiply both sides by x.　　　　$x^k \cdot x > 1 \cdot x$

(Since $x > 1$, this step does not reverse the inequality sign.)

Simplify. $x^{k+1} > x$

Recall that $x > 1$. $x^{k+1} > x > 1$

Therefore, we have shown that $x^{k+1} > 1$.

This completes the induction proof. Thus, the following statement is true.

"If $x > 1$ then $x^n > 1$ for **all** natural numbers, n."

EXAMPLE 2 Using Mathematical Induction

Use mathematical induction to prove that $n^2 + n$ is divisible by 2 for all natural numbers, n.

Solution:

STEP 1 Show the statement we
are testing is true for $n = 1$. $1^2 + 1 = 2$

2 is divisible by 2. $\dfrac{2}{2} = 1$

STEP 2 Assume the statement is true
for $n = k$. $\dfrac{k^2 + k}{2} = \text{integer}$

Show it is true for $k + 1$. $\dfrac{(k+1)^2 + (k+1)}{2} \overset{?}{=} \text{integer}$

                                                $\dfrac{k^2 + 2k + 1 + k + 1}{2} \overset{?}{=} \text{integer}$

Regroup terms. $\dfrac{(k^2 + k) + 2(k+1)}{2} \overset{?}{=} \text{integer}$

                                                $\dfrac{(k^2 + k)}{2} + \dfrac{2(k+1)}{2} \overset{?}{=} \text{integer}$

We assumed $\dfrac{k^2 + k}{2} = \text{integer}$. $\text{integer} + (k+1) \overset{?}{=} \text{integer}$

Since k is a natural number: $\text{integer} + \text{integer} = \text{integer}$

This completes the induction proof. The following statement is true:

"$n^2 + n$ is divisible by 2 for all natural numbers, n."

Mathematical induction is often used to prove formulas for partial sums.

EXAMPLE 3 Using Mathematical Induction to Prove a Partial Sum Formula

Use mathematical induction to prove the following partial sum formula:

$$1 + 2 + 3 + \cdots + n = \frac{n(n+1)}{2} \quad \text{for all positive integers, } n$$

Solution:

STEP 1 Show the formula is true for
$n = 1$. $1 = \dfrac{1(1+1)}{2} = \dfrac{2}{2} = 1$

STEP 2 Assume the formula is
true for $n = k$.

$$1 + 2 + 3 + \cdots + k = \frac{k(k+1)}{2}$$

Show it is
true for
$n = k + 1$.

$$1 + 2 + 3 + \cdots + k + (k+1) \stackrel{?}{=} \frac{(k+1)(k+2)}{2}$$

$$\underbrace{1 + 2 + 3 + \cdots + k}_{\frac{k(k+1)}{2}} + (k+1) \stackrel{?}{=} \frac{(k+1)(k+2)}{2}$$

$$\frac{k(k+1)}{2} + (k+1) \stackrel{?}{=} \frac{(k+1)(k+2)}{2}$$

$$\frac{k(k+1) + 2(k+1)}{2} \stackrel{?}{=} \frac{(k+1)(k+2)}{2}$$

$$\frac{k^2 + 3k + 2}{2} \stackrel{?}{=} \frac{(k+1)(k+2)}{2}$$

$$\frac{(k+1)(k+2)}{2} = \frac{(k+1)(k+2)}{2}$$

This completes the induction proof. The following statement is true:

$$1 + 2 + 3 + \cdots + n = \frac{n(n+1)}{2} \text{ for all positive integers, } n.$$

SECTION 9.4 SUMMARY

Just because we believe something is true does not mean that it is. In mathematics we rely on proof. In this section we discussed *mathematical induction,* a process of proving mathematical statements. The two-step procedure for mathematical induction is to (1) show the statement is true for $n = 1$, then (2) assume the statement is true for $n = k$ and show the statement must be true for $n = k + 1$. The combination of Steps 1 and 2 proves the statement.

SECTION 9.4 EXERCISES

 ■ SKILLS

In Exercises 1–24, prove the statements using mathematical induction for all positive integers, n.

1. $n^2 \leq n^3$ **2.** If $0 < x < 1$, then $0 < x^n < 1$.   **3.** $2n \leq 2^n$ **4.** $5^n < 5^{n+1}$

5. $n! > 2^n$ $n \geq 4$ (Show it is true for $n = 4$, instead of $n = 1$.)   **6.** $(1 + c)^n \geq nc$ $c > 1$

7. $n(n + 1)(n - 1)$ is divisible by 3.

8. $n^3 - n$ is divisible by 3.

9. $n^2 + 3n$ is divisible by 2.

10. $n(n + 1)(n + 2)$ is divisible by 6.

11. $2 + 4 + 6 + 8 + \cdots + 2n = n(n + 1)$

12. $1 + 3 + 5 + 7 + \cdots + (2n - 1) = n^2$

13. $1 + 3 + 3^2 + 3^3 + \cdots + 3^n = \dfrac{3^{n+1} - 1}{2}$

14. $2 + 4 + 8 + \cdots + 2^n = 2^{n+1} - 2$

15. $1^2 + 2^2 + 3^2 + \cdots + n^2 = \dfrac{n(n + 1)(2n + 1)}{6}$

16. $1^3 + 2^3 + 3^3 + \cdots + n^3 = \dfrac{n^2(n + 1)^2}{4}$

17. $\dfrac{1}{1 \cdot 2} + \dfrac{1}{2 \cdot 3} + \dfrac{1}{3 \cdot 4} + \cdots + \dfrac{1}{n(n + 1)} = \dfrac{n}{n + 1}$

18. $\dfrac{1}{2 \cdot 3} + \dfrac{1}{3 \cdot 4} + \cdots + \dfrac{1}{(n + 1)(n + 2)} = \dfrac{n}{2(n + 2)}$

19. $(1 \cdot 2) + (2 \cdot 3) + (3 \cdot 4) + \cdots + n(n + 1) = \dfrac{n(n + 1)(n + 2)}{3}$

20. $(1 \cdot 3) + (2 \cdot 4) + (3 \cdot 5) + \cdots + n(n + 2) = \dfrac{n(n + 1)(2n + 7)}{6}$

21. $1 + x + x^2 + x^3 + \cdots + x^{n-1} = \dfrac{1 - x^n}{1 - x} \qquad x \neq 1$

22. $\dfrac{1}{2} + \dfrac{1}{4} + \dfrac{1}{8} + \cdots + \dfrac{1}{2^n} = 1 - \dfrac{1}{2^n}$

23. The sum of an arithmetic sequence: $a_1 + (a_1 + d) + (a_1 + 2d) + \cdots + [a_1 + (n - 1)d] = \dfrac{n}{2}[2a_1 + (n - 1)d]$

24. The sum of a geometric sequence: $a_1 + a_1 r + a_1 r^2 + \cdots + a_1 r^{n-1} = a_1\left(\dfrac{1 - r^n}{1 - r}\right)$

 ■ **APPLICATIONS**

The Tower of Hanoi. This is a game with three pegs and n disks (largest on the bottom and smallest on the top). The goal is to move this entire tower of discs to another peg (in the same order). The challenge is that you may only move one disk at a time, and at no time can a larger disk be resting on a smaller disk.

25. What is the smallest number of moves needed if there are three disks?

26. What is the smallest number of moves needed if there are four disks?

27. What is the smallest number of moves needed if there are five disks?

28. What is the smallest number of moves needed if there are n disks? Prove it by mathematical induction.

Andy Washnik

Tower of Hanoi

■ **CHALLENGE**

29. T or F: Assume S_k is true. If it can be shown that S_{k+1} is true, then it is true for all S_n where n is any positive integer.

30. T or F: Assume S_1 is true. Then if it can be shown that S_2 and S_3 are true, then it is true for all S_n where n is any positive integer.

31. Use mathematical induction to prove:

$$\sum_{k=1}^{n} k^4 = \frac{n(n+1)(2n+1)(3n^2+3n-1)}{30}$$

32. Use mathematical induction to prove:

$$\sum_{k=1}^{n} k^5 = \frac{n^2(n+1)^2(2n^2+2n-1)}{12}$$

33. Use mathematical induction to prove:

$$\left(1+\frac{1}{1}\right)\left(1+\frac{1}{2}\right)\left(1+\frac{1}{3}\right)\cdots\left(1+\frac{1}{n}\right) = n+1$$

34. Use mathematical induction to prove that $x+y$ is a factor of $x^{2n} - y^{2n}$.

35. Use mathematical induction to prove:

$$\ln(c_1 \cdot c_2 \cdot c_3 \cdots c_n) = \ln c_1 + \ln c_2 + \cdots + \ln c_n$$

SECTION 9.5 The Binomial Theorem

Skills Objectives

■ Evaluate a binomial coefficient using the binomial theorem.
■ Evaluate a binomial coefficient using Pascal's triangle.
■ Expand a binomial raised to a power.
■ Find a particular term of a binomial expansion.

Conceptual Objectives

■ Recognize patterns in binomial expansions.

Introduction

A **binomial** is a polynomial that has two terms. The following are all examples of binomials:

$$x^2 + 2y \qquad a + 3b \qquad 4x^2 + 9$$

In this section, we will develop a formula for raising a binomial to a power n, where n is a positive integer.

$$(x^2 + 2y)^6 \qquad (a + 3b)^4 \qquad (4x^2 + 9)^5$$

To begin, let's start by writing out the expansions of $(a+b)^n$ for several values of n.

$$(a+b)^0 = 1$$
$$(a+b)^1 = a+b$$
$$(a+b)^2 = a^2 + 2ab + b^2$$
$$(a+b)^3 = a^3 + 3a^2b + 3ab^2 + b^3$$

$$(a + b)^4 = a^4 + 4a^3b + 6a^2b^2 + 4ab^3 + b^4$$
$$(a + b)^5 = a^5 + 5a^4b + 10a^3b^2 + 10a^2b^3 + 5ab^4 + b^5$$

There are several *patterns* that all of the **binomial expansions** have.

1. The number of terms in each resulting polynomial is always *one more* than the power of the binomial, *n*. Thus, there are $n + 1$ terms in each expansion.

$$n = 3: \qquad (a + b)^3 = \underbrace{a^3 + 3a^2b + 3ab^2 + b^3}_{\text{four terms}}$$

2. Each expansion has symmetry. For example, a and b can be interchanged, and you will arrive at the same expansion. Furthermore, the powers of a decrease by 1 in each successive term, and the powers of b increase by 1 in each successive term.

$$(a + b)^3 = a^3b^0 + 3a^2b^1 + 3a^1b^2 + a^0b^3$$

3. The sum of the powers of each term in the expansion is *n*.

$$n = 3: \qquad (a + b)^3 = \overset{3+0=3}{a^3b^0} + \overset{2+1=3}{3a^2b^1} + \overset{1+2=3}{3a^1b^2} + \overset{0+3=3}{a^0b^3}$$

4. The coefficients increase and decrease in a symmetric manner.

$$(a + b)^5 = 1a^5 + 5a^4b + 10a^3b^2 + 10a^2b^3 + 5ab^4 + 1b^5$$

Using these patterns, we can develop a generalized formula for $(a + b)^n$.

$$(a + b)^n = \_a^n + \_a^{n-1}b + \_a^{n-2}b^2 + \cdots + \_a^2b^{n-2} + \_ab^{n-1} + \_b^n$$

We know that there are $n + 1$ terms in the expansion. We also know that the sum of the powers of each term must equal *n*. The powers increase and decrease by 1 in each successive term, and if we interchanged a and b, the result would be the same expansion. The question that remains is, what coefficients go in the blanks?

Binomial Coefficients

We know that the coefficients must increase and then decrease in a symmetric order (similar to walking up and then down a hill). It turns out that the *binomial coefficients* are represented by a symbol that we will now define.

DEFINITION   **BINOMIAL COEFFICIENTS**

For nonnegative integers n and k, where $n \geq k$, the symbol $\dbinom{n}{k}$ is called the **binomial coefficient** and is defined by

$$\binom{n}{k} = \frac{n!}{(n - k)!k!} \qquad \binom{n}{k} \text{ is read "n choose k."}$$

TECHNOLOGY TIP

Use calculators to find the binomial coefficients,

$$\binom{6}{4}, \binom{5}{5}, \binom{4}{0}, \binom{10}{9}:$$

Many scientific calculators:

Press	Display
6 [nCr] 4	15
5 [nCr] 5	1
4 [nCr] 0	1
10 [nCr] 9	10

Graphing calculators:

```
6 nCr 4
           15
5 nCr 5
            1
4 nCr 0
            1
```

```
10 nCr 9
           10
■
```

EXAMPLE 1 Evaluating a Binomial Coefficient

Evaluate the following binomial coefficients.

a. $\binom{6}{4}$ **b.** $\binom{5}{5}$ **c.** $\binom{4}{0}$ **d.** $\binom{10}{9}$

Solution:

Select the top number as n and the bottom number as k and substitute into the binomial coefficient formula $\binom{n}{k} = \dfrac{n!}{(n-k)!k!}$.

a. $\binom{6}{4} = \dfrac{6!}{(6-4)!4!} = \dfrac{6!}{2!4!} = \dfrac{6 \cdot 5 \cdot 4 \cdot 3 \cdot 2 \cdot 1}{(2 \cdot 1)(4 \cdot 3 \cdot 2 \cdot 1)} = \dfrac{6 \cdot 5}{2} = \boxed{15}$

b. $\binom{5}{5} = \dfrac{5!}{(5-5)!5!} = \dfrac{5!}{0!5!} = \dfrac{1}{0!} = \dfrac{1}{1} = \boxed{1}$

c. $\binom{4}{0} = \dfrac{4!}{(4-0)!0!} = \dfrac{4!}{4!0!} = \dfrac{1}{0!} = \boxed{1}$

d. $\binom{10}{9} = \dfrac{10!}{(10-9)!9!} = \dfrac{10!}{1!9!} = \dfrac{10 \cdot 9!}{9!} = \boxed{10}$

■ **YOUR TURN**   Evaluate the following binomial coefficients.

a. $\binom{9}{6}$ **b.** $\binom{8}{6}$

Parts b and c of Example 1 lead to the general formulas:

$$\binom{n}{n} = 1 \qquad \text{and} \qquad \binom{n}{0} = 1$$

Binomial Expansion

Let's return to the question of the binomial expansion and how to determine the coefficients:

$$(a+b)^n = \_a^n + \_a^{n-1}b + \_a^{n-2}b^2 + \cdots + \_a^2b^{n-2} + \_ab^{n-1} + \_b^n$$

The symbol $\binom{n}{k}$ is called a binomial coefficient because the coefficients in the blanks in the binomial expansion are equivalent to this symbol.

THE BINOMIAL THEOREM

Let a and b be real numbers; then for any positive integer n,

$$(a+b)^n = \binom{n}{0}a^n + \binom{n}{1}a^{n-1}b + \binom{n}{2}a^{n-2}b^2 + \cdots + \binom{n}{n-2}a^2b^{n-2} + \binom{n}{n-1}ab^{n-1} + \binom{n}{n}b^n$$

or in sigma (summation) notation as

$$(a+b)^n = \sum_{k=0}^{n} \binom{n}{k}a^{n-k}b^k$$

■ **Answer: a.** 84   **b.** 28

Let's use the binomial theorem to write out binomial expansions.

 EXAMPLE 2 Using the Binomial Theorem

Expand $(x + 2)^3$ using the binomial theorem.

Solution:

Substitute $a = x$, $b = 2$, $n = 3$
into binomial theorem.

$$(x + 2)^3 = \sum_{k=0}^{3} \binom{3}{k} x^{3-k} 2^k$$

Expand the summation.

$$\binom{3}{0}x^3 + \binom{3}{1}x^2 \cdot 2 + \binom{3}{2}x \cdot 2^2 + \binom{3}{3}2^3$$

Find the binomial coefficients.

$$x^3 + 3x^2 \cdot 2 + 3x \cdot 2^2 + 2^3$$

Simplify.

$$x^3 + 6x^2 + 12x + 8$$

$$(x + 2)^3 = x^3 + 6x^2 + 12x + 8$$

■ **YOUR TURN** Expand $(x + 1)^4$ using the binomial theorem.

EXAMPLE 3 Using the Binomial Theorem

Expand $(2x - 3)^4$ using the binomial theorem.

Solution:

Substitute $a = 2x$,
$b = -3$, $n = 4$ into
the binomial theorem.

$$(2x - 3)^4 = \sum_{k=0}^{4} \binom{4}{k} (2x)^{4-k} (-3)^k$$

Expand the summation.

$$\binom{4}{0}(2x)^4 + \binom{4}{1}(2x)^3(-3) + \binom{4}{2}(2x)^2(-3)^2 + \binom{4}{3}(2x)(-3)^3 + \binom{4}{4}(-3)^4$$

Find the binomial coefficients.

$$(2x)^4 + 4(2x)^3(-3) + 6(2x)^2(-3)^2 + 4(2x)(-3)^3 + (-3)^4$$

Simplify.

$$16x^4 - 96x^3 + 216x^2 - 216x + 81$$

$$(2x - 3)^4 = 16x^4 - 96x^3 + 216x^2 - 216x + 81$$

■ **YOUR TURN** Expand $(2x + 1)^4$ using the binomial theorem.

■ **Answer:** $16x^4 + 32x^3 + 24x^2 + 8x + 1$ ■ **Answer:** $x^4 + 4x^3 + 6x^2 + 4x + 1$

Pascal's Triangle

Instead of writing out the binomial theorem and calculating the binomial coefficients using factorials every time you want to do a binomial expansion, we now present an alternative, more convenient way of remembering the binomial coefficients.

Let's arrange values of $\binom{n}{k}$ in a triangular pattern.

$$\binom{0}{0}$$

$$\binom{1}{0} \quad \binom{1}{1}$$

$$\binom{2}{0} \quad \binom{2}{1} \quad \binom{2}{2}$$

$$\binom{3}{0} \quad \binom{3}{1} \quad \binom{3}{2} \quad \binom{3}{3}$$

$$\binom{4}{0} \quad \binom{4}{1} \quad \binom{4}{2} \quad \binom{4}{3} \quad \binom{4}{4}$$

$$\binom{5}{0} \quad \binom{5}{1} \quad \binom{5}{2} \quad \binom{5}{3} \quad \binom{5}{4} \quad \binom{5}{5}$$

$$\binom{6}{0} \quad \binom{6}{1} \quad \binom{6}{2} \quad \binom{6}{3} \quad \binom{6}{4} \quad \binom{6}{5} \quad \binom{6}{6}$$

This is called **Pascal's triangle**. If we evaluate the binomial coefficients, Pascal's triangle becomes

$$
\begin{array}{ccccccccccc}
 & & & & & 1 & & & & & \\
 & & & & 1 & & 1 & & & & \\
 & & & 1 & & 2 & & 1 & & & \\
 & & 1 & & 3 & & 3 & & 1 & & \\
 & 1 & & 4 & & 6 & & 4 & & 1 & \\
1 & & 5 & & 10 & & 10 & & 5 & & 1
\end{array}
$$

Notice that the first and last number in every row is 1. Each of the other numbers is found by adding the two numbers directly above it. For example,

$$3 = 2 + 1 \qquad 4 = 1 + 3 \qquad 10 = 6 + 4$$

It turns out that these numbers in Pascal's triangle are exactly the coefficients in a binomial expansion.

$$(a + b)^0 = 1$$
$$(a + b)^1 = 1a + 1b$$
$$(a + b)^2 = 1a^2 + 2ab + 1b^2$$
$$(a + b)^3 = 1a^3 + 3a^2b + 3ab^2 + 1b^3$$
$$(a + b)^4 = 1a^4 + 4a^3b + 6a^2b^2 + 4ab^3 + 1b^4$$
$$(a + b)^5 = 1a^5 + 5a^4b + 10a^3b^2 + 10a^2b^3 + 5ab^4 + 1b^5$$

The top row is called the zero row because it corresponds to the binomial raised to the zero power, $n = 0$. Since each row in Pascal's triangle starts and ends with a 1 and all other values are found by adding the two numbers directly above it, we can now easily calculate the sixth row.

$$(a + b)^5 = 1a^5 + 5a^4b + 10a^3b^2 + 10a^2b^3 + 5ab^4 + 1b^5$$

$$(a + b)^6 = 1a^6 + 6a^5b + 15a^4b^2 + 20a^3b^3 + 15a^2b^4 + 6ab^5 + 1b^6$$

EXAMPLE 4 Using Pascal's Triangle in a Binomial Expansion

Use Pascal's triangle to determine the binomial expansion of $(x + 2)^5$.

Solution:

Write the binomial expansion with blanks for coefficients.

$$(x + 2)^5 = \_x^5 + \_x^4 \cdot 2 + \_x^3 \cdot 2^2 + \_x^2 \cdot 2^3 + \_x \cdot 2^4 + \_2^5$$

Write the binomial coefficients in the *fifth* row of Pascal's triangle.

$$1, 5, 10, 10, 5, 1$$

Substitute these coefficients into the blanks of the binomial expansion.

$$(x + 2)^5 = 1x^5 + 5x^4 \cdot 2 + 10x^3 \cdot 2^2 + 10x^2 \cdot 2^3 + 5x \cdot 2^4 + 1 \cdot 2^5$$

Simplify. $$(x + 2)^5 = x^5 + 10x^4 + 40x^3 + 80x^2 + 80x + 32$$

■ **YOUR TURN** Use Pascal's triangle to determine the binomial expansion of $(x + 3)^4$.

EXAMPLE 5 Using Pascal's Triangle in Binomial Expansion

Use Pascal's triangle to determine the binomial expansion of $(2x + 5)^4$.

Solution:

Write the binomial expansion with blanks for coefficients.

$$(2x + 5)^4 = \_(2x)^4 + \_(2x)^3 \cdot 5 + \_(2x)^2 \cdot 5^2 + \_(2x) \cdot 5^3 + \_5^4$$

Write the binomial coefficients in the *fourth* row of Pascal's triangle.

$$1, 4, 6, 4, 1$$

Substitute these coefficients into the blanks of the binomial expansion.

$$(2x + 5)^4 = 1(2x)^4 + 4(2x)^3 \cdot 5 + 6(2x)^2 \cdot 5^2 + 4(2x) \cdot 5^3 + 1 \cdot 5^4$$

Simplify. $$(2x + 5)^4 = 16x^4 + 160x^3 + 600x^2 + 1000x + 625$$

■ **YOUR TURN** Use Pascal's triangle to determine the binomial expansion of $(3x + 2)^3$.

■ **Answer:** $(3x + 2)^3 = 27x^3 + 54x^2 + 36x + 8$
■ **Answer:** $(x + 3)^4 = x^4 + 12x^3 + 54x^2 + 108x + 81$

Finding a Particular Term of a Binomial Expansion

What if we don't want to find the entire expansion, but instead just a single term? For example, what is the fourth term of $(a + b)^5$?

WORDS	**MATH**
Recall the sigma notation.	$(a + b)^n = \sum_{k=0}^{n} \binom{n}{k} a^{n-k} b^k$
Let $n = 5$.	$(a + b)^5 = \sum_{k=0}^{5} \binom{5}{k} a^{5-k} b^k$
Expand.	$(a + b)^5 = \binom{5}{0} a^5 + \binom{5}{1} a^4 b + \binom{5}{2} a^3 b^2 +$ $\underbrace{\binom{5}{3} a^2 b^3}_{\text{fourth term}} + \binom{5}{4} ab^4 + \binom{5}{5} b^5$
Simplify fourth term.	$10 a^2 b^3$

FINDING A PARTICULAR TERM OF A BINOMIAL EXPANSION

The $(r + 1)$ term of the expansion $(a + b)^n$ is $\binom{n}{r} a^{n-r} b^r$.

EXAMPLE 6 Finding a Particular Term of a Binomial Expansion

Find the fifth term of the binomial expansion of $(2x - 7)^6$.

Solution:

Recall that the $r + 1$ term of $(a + b)^n$ is $\binom{n}{r} a^{n-r} b^r$.

For the fifth term, let $r = 4$. $\binom{n}{4} a^{n-4} b^4$

For this expansion let $a = 2x$, $b = -7$, $n = 6$. $\binom{6}{4} (2x)^{6-4} (-7)^4$

Note that $\binom{6}{4} = 15$. $15 (2x)^2 (-7)^4$

Simplify. $144{,}060 x^2$

■ **YOUR TURN**   What is the third term of the binomial expansion of $(3x + 2)^5$?

SECTION 9.5 SUMMARY

In this section, we developed a formula for raising a binomial expression to a power, $n \geq 0$. The patterns that surfaced were symmetry between the two terms of the binomial, every expansion has $n + 1$ terms, the powers sum to n, and the coefficients, called binomial coefficients, are ratios of factorials. Also, Pascal's triangle, a shortcut method for evaluating the binomial coefficients, was discussed. The patterns in the triangle are that every row begins and ends with 1 and all other numbers are found by adding the two numbers above the entry. Lastly, a formula was given for finding a particular term of a binomial expansion.

SECTION 9.5 EXERCISES

 ■ SKILLS

In Exercises 1–10, evaluate the binomial coefficients.

1. $\dbinom{7}{3}$ 2. $\dbinom{8}{2}$ 3. $\dbinom{10}{8}$ 4. $\dbinom{23}{21}$ 5. $\dbinom{17}{0}$

6. $\dbinom{100}{0}$ 7. $\dbinom{99}{99}$ 8. $\dbinom{52}{52}$ 9. $\dbinom{48}{45}$ 10. $\dbinom{29}{26}$

In Exercises 11–32, expand the expression using the binomial theorem.

11. $(x + 2)^4$ 12. $(x + 3)^5$ 13. $(y - 3)^5$ 14. $(y - 4)^4$ 15. $(x + y)^5$

16. $(x - y)^6$ 17. $(x + 3y)^3$ 18. $(2x - y)^3$ 19. $(5x - 2)^3$ 20. $(a - 7b)^3$

21. $\left(\dfrac{1}{x} + 5y\right)^4$ 22. $\left(2x + \dfrac{3}{y}\right)^4$ 23. $(x^2 + y^2)^4$ 24. $(r^3 - s^3)^3$ 25. $(ax + by)^5$

26. $(ax - by)^5$ 27. $(\sqrt{x} + 2)^6$ 28. $(3 + \sqrt{y})^4$ 29. $(a^{3/4} + b^{1/4})^4$ 30. $(x^{2/3} + y^{1/3})^3$

31. $(x^{1/4} + 2\sqrt{y})^4$ 32. $(\sqrt{x} - 3y^{1/4})^8$

In Exercises 33–36, expand the expression using Pascal's triangle.

33. $(r - s)^4$ 34. $(x^2 + y^2)^7$ 35. $(ax + by)^6$ 36. $(x + 3y)^4$

In Exercises 37–44, find the coefficient, C, of the term in the binomial expansion.

	BINOMIAL	TERM		BINOMIAL	TERM		BINOMIAL	TERM
37.	$(x + 2)^{10}$	Cx^6	38.	$(3 + y)^9$	Cy^5	39.	$(y - 3)^8$	Cy^4
40.	$(x - 1)^{12}$	Cx^5	41.	$(2x + 3y)^7$	Cx^3y^4	42.	$(3x - 5y)^9$	Cx^2y^7
43.	$(x^2 + y)^8$	Cx^8y^4	44.	$(r - s^2)^{10}$	Cr^6s^8			

■ APPLICATIONS

45. Lottery. In a state lottery in which six numbers are drawn from a possible 40 numbers, the number of possible six-number combinations is equal to $\binom{40}{6}$. How many possible combinations are there?

46. Lottery. In a state lottery in which six numbers are drawn from a possible 60 numbers, the number of possible six-number combinations is equal to $\binom{60}{6}$. How many possible combinations are there?

47. Poker. With a deck of 52 cards, 5 cards are dealt in a game of poker. There are a total of $\binom{52}{5}$ different 5-card poker hands that can be dealt. How many possible hands are there?

48. Canasta. In the card game canasta, two decks of cards including the jokers are used and 11 cards are dealt to each person. There are a total of $\binom{108}{11}$ different 11-card canasta hands that can be dealt. How many possible hands are there?

■ CATCH THE MISTAKE

In Exercises 49 and 50, explain the mistake that is made.

49. Evaluate the expression $\binom{7}{5}$.

Solution:

Write out the binomial coefficient in terms of factorials. $\binom{7}{5} = \frac{7!}{5!}$

Write out the factorials. $\binom{7}{5} = \frac{7!}{5!} = \frac{7 \cdot 6 \cdot 5 \cdot 4 \cdot 3 \cdot 2 \cdot 1}{5 \cdot 4 \cdot 3 \cdot 2 \cdot 1}$

Simplify. $\binom{7}{5} = \frac{7!}{5!} = \frac{7 \cdot 6}{1} = 42$

This is incorrect. What mistake was made?

50. Expand $(x + 2y)^4$.

Write out with blanks.

$(x + 2y)^4 = \_x^4 + \_x^3y + \_x^2y^2 + \_xy^3 + \_y^4$

Write out the terms from the fifth row of Pascal's triangle.

1, 4, 6, 4, 1

Substitute these coefficients into the binomial expansion.

$(x + 2y)^4 = x^4 + 4x^3y + 6x^2y^2 + 4xy^3 + y^4$

This is incorrect. What mistake was made?

■ CHALLENGE

51. T or F: The binomial expansion of $(x + y)^{10}$ has 10 terms.

52. T or F: The binomial expansion of $(x^2 + y^2)^{15}$ has 16 terms.

53. T or F: $\binom{n}{n} = 1$

54. T or F: $\binom{n}{-n} = -1$

55. Show that $\binom{n}{k} = \binom{n}{n-k}$ if $0 \le k \le n$.

56. Show that if n is a positive integer, then:

$$\binom{n}{0} + \binom{n}{1} + \binom{n}{2} + \cdots + \binom{n}{n} = 2^n$$

Hint: Let $2^n = (1 + 1)^n$ and use the binomial theorem to expand.

57. Using a graphing utility, plot $y_1 = 1 - 3x + 3x^2 - x^3$, $y_2 = -1 + 3x - 3x^2 + x^3$, and $y_3 = (1 - x)^3$ in the same viewing screen. What is the binomial expansion of $(1 - x)^3$?

58. Using a graphing utility, plot $y_1 = (x + 3)^4$, $y_2 = x^4 + 4x^3 + 6x^2 + 4x + 1$, and $y_3 = x^4 + 12x^3 + 54x^2 + 108x + 81$. What is the binomial expansion of $(x + 3)^4$?

59. Using a graphing utility, plot $y_1 = 1 - 3x$, $y_2 = 1 - 3x + 3x^2$, $y_3 = 1 - 3x + 3x^2 - x^3$, and $y_4 = (1 - x)^3$ for $-1 < x < 1$. What do you notice

happening each time an additional term is added? Now, let $1 < x < 2$. Does the same thing happen?

60. Using a graphing utility, plot $y_1 = 1 - \dfrac{3}{x}$, $y_2 = 1 - \dfrac{3}{x} + \dfrac{3}{x^2}$, $y_3 = 1 - \dfrac{3}{x} + \dfrac{3}{x^2} - \dfrac{1}{x^3}$, and $y_4 = \left(1 - \dfrac{1}{x}\right)^3$ for $1 < x < 2$. What do you notice

happening each time an additional term is added? Now, let $0 < x < 1$. Does the same thing happen?

SECTION 9.6 Counting, Permutations, and Combinations

Skills Objectives

■ Use the fundamental counting principle to solve counting problems.
■ Use permutations to solve counting problems.
■ Use combinations to solve counting problems.

Conceptual Objectives

■ Understand the difference between permutations and combinations.

The Fundamental Counting Principle

You are traveling through Europe for the summer and decide the best packing option is to select separates that can be mixed and matched. You pack one denim miniskirt and one pair of khaki pants. You pack a pair of Teva sport sandals and a pair of sneakers. You have three shirts (red, blue, and white). How many different outfits can be worn using only the clothes mentioned above?

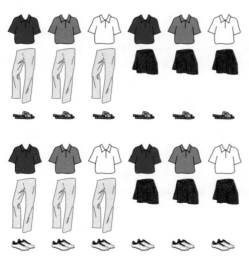

The answer is 12. There are two options for bottoms (pants or skirt), three options for shirts, and two options for shoes. The product of these is

$$2 \cdot 3 \cdot 2 = 12$$

The general formula for counting possibilities is given by the *fundamental counting principle*.

FUNDAMENTAL COUNTING PRINCIPLE

Let E_1 and E_2 be two events. The first event, E_1, can occur in m_1 ways. The second event, E_2, can occur in m_2 ways. The number of ways that the combination of the two events can occur is $m_1 \cdot m_2$.

In other words, *the number of ways in which successive things can occur is found by multiplying the number of ways each thing can occur.*

EXAMPLE 1 Possible Meals Served at a Restaurant

A restaurant is rented for a retirement party. They offer an appetizer, entrée, and dessert for a set price. The following are the choices that people attending the party may choose from. How many possible dinners could be served that night?

Appetizers:	calamari, stuffed mushrooms, or caesar salad
Entrées:	tortellini alfredo, shrimp scampi, eggplant parmesan, or chicken marsala
Desserts:	tiramisu or flan

Solution:

There are three possible appetizers, four possible entrees, and two possible desserts.

Write the product of possible options. $\qquad\qquad$ $3 \cdot 4 \cdot 2 = 24$

There are $\boxed{24 \text{ possible dinners}}$ for the retirement party.

▮ **YOUR TURN** In Example 1, the restaurant will lower the cost per person for the retirement party if the number of appetizers and entrées is reduced. Suppose the appetizers are reduced to either soup or salad and the entrees are reduced to either tortellini or eggplant parmesan, how many possible dinners could be served at the party?

EXAMPLE 2 Telephone Numbers (When to Require 10-Digit Dialing)

In many towns in the United States, residents can call one another using a 7-digit dialing system. In some large cities, 10-digit dialing is required because more than one area code coexists. Determine how many telephone numbers can be allocated in a 7-digit dialing area.

▮ **Answer: Eight**

Solution:

With 7-digit telephone numbers, the first number cannot be a 0 or a 1, and the following six numbers can be 0, 1, 2, 3, 4, 5, 6, 7, 8, or 9.

First number: 2, 3, 4, 5, 6, 7, 8, or 9.	8 possible digits
Second number: 0, 1, 2, 3, 4, 5, 6, 7, 8, or 9.	10 possible digits
Third number:	10 possible digits
Fourth number:	10 possible digits
Fifth number:	10 possible digits
Sixth number:	10 possible digits
Seventh number:	10 possible digits
Counting Principle:	$8 \cdot 10 \cdot 10 \cdot 10 \cdot 10 \cdot 10 \cdot 10$
Possible telephone numbers:	8,000,000

Eight million 7-digit telephone numbers can be allocated within one area code.

YOUR TURN If the first digit of an area code cannot be 0 or 1, but the second and third numbers of an area code can be 0, 1, 2, 3, 4, 5, 6, 7, 8, or 9, how many 10-digit telephone numbers can be allocated in the United States?

In the fundamental counting theorem, once an event is used, it cannot be used again. We now introduce two other concepts, *permutations* and *combinations,* which allow events to be used again. For example, in Example 2, the possible telephone numbers can have a number repeat, such as 555-1212. However, in the lottery, once a number is selected, it cannot be used again.

An important distinction between a *permutation* and a *combination* is that in a *permutation* order matters, but in a combination order does not matter. For example, the Florida winning lotto numbers one week could be 2–3–5–11–19–27. This would be a *combination* because the order in which they are drawn does not matter. However, if you were betting on a trifecta at the Kentucky Derby, to win you must not only select the first, second, and third place horses, you must select them in the order in which they finished. This would be a *permutation.*

Permutations

DEFINITION   **PERMUTATION**

A **permutation** is an *ordered* arrangement of distinct objects without repetition.

EXAMPLE 3 Finding the Number of Permutations of *n* Objects

How many permutations are possible for the letters A, B, C, and D?

Solution:

ABCD	ABDC	ACBD	ACDB	ADCB	ADBC
BACD	BADC	BCAD	BCDA	BDCA	BDAC
CABD	CADB	CBAD	CBDA	CDAB	CDBA
DABC	DACB	DBCA	DBAC	DCAB	DCBA

 There are 24 (or 4!) possible permutations of the letters A, B, C, and D.

Notice in the first row of permutations in Example 3, A was selected for the first space. That left one of the remaining three letters to fill the second space. Once that was selected there remained two letters to choose from for the third space and then the last space was filled with the unselected letter. In general, there are *n*! ways to order *n* objects.

NUMBER OF PERMUTATIONS OF *n* OBJECTS

The number of permutations of *n* objects is

$$n! = n \cdot (n-1) \cdot (n-2) \cdots 2 \cdot 1$$

EXAMPLE 4 Running Order of Dogs

In an AKC sponsored field trial, the dogs compete in random order depending on when their application was submitted and what the Dow Jones closed at the Tuesday before a competition. If there are nine dogs competing in the open (all age stake), how many possible running orders are there?

Courtesy Cynthia Young, University of Central Florida

Solution:

There are nine dogs that will run.   *n* = 9

The number of possible running orders is $n!$. $n! = 9! = 362,880$

There are 362,880 different possible running orders of nine dogs.

YOUR TURN Five contestants in the Miss America pageant reach the live television interview round. How many possible orders can the contestants compete in the interview round?

In Examples 3 and 4, we were interested in all possible permutations. Sometimes, we are interested in only some permutations. For instance, 20 horses usually run in the Kentucky Derby. If you bet on a trifecta, you must pick the top three places in the correct order to win. Therefore, we do not consider all possible permutations of 20 horses finishing (places 1–20). Instead, we only consider the possible permutations of first, second, and third place finishes of the 20 horses. We would call this *a permutation of 20 objects taken 3 at a time*. In general, this ordering is called *a permutation of n objects taken r at a time*.

If 20 horses are entered in the Kentucky Derby, there are 20 possible first place finishers. We consider the permutations with one horse in first place. That leaves 19 possible horses for second place and then 18 possible horses for third place. Therefore, there are $20 \cdot 19 \cdot 18 = 6840$ possible winning scenarios for the trifecta. This can also be represented as $\dfrac{20!}{(20 - 3)!} = \dfrac{20!}{17!}$.

NUMBER OF PERMUTATIONS OF *n* OBJECTS TAKEN *r* AT A TIME

The number of permutations of n objects taken r at a time is

$$_nP_r = \frac{n!}{(n - r)!} = n(n - 1)(n - 2) \cdots (n - r + 1)$$

EXAMPLE 5 Starting Lineup for a Volleyball Team

A six-woman volleyball team has to start in a particular order (1–6). If there are 13 women on the team, how many possible starting lineups are there?

Solution:

Identify the total number of players. $n = 13$

Identify the total number of starters in the lineup. $r = 6$

Substitute $n = 13$ and $r = 6$ into $_nP_r = \dfrac{n!}{(n - r)!}$.

$$_{13}P_6 = \frac{13!}{(13 - 6)!} = \frac{13!}{7!} = 13 \cdot 12 \cdot 11 \cdot 10 \cdot 9 \cdot 8 = 1,235,520$$

There are 1,235,520 possible combinations.

TECHNOLOGY TIP

Use calculators to find the term $_{13}P_6$:

Many scientific calculators:

	Press	*Display*
13	nPr 6 =	1235520

Graphing calculators:

Press

6 MATH PRB 2: nPr

ENTER 6 ENTER

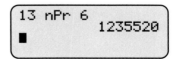

13 nPr 6
                1235520

> ■ **YOUR TURN** A softball team has 12 players, 10 of whom will be in the starting lineup (batters 1–10). How many possible starting lineups are there for this team?

Combinations

The difference between a permutation and a combination is that the permutation has an order associated with it, whereas a *combination* does not have an order associated with it.

> **DEFINITION** **COMBINATION**
>
> A **combination** is an arrangement, without specific order, of distinct objects without repetition.

The six winning Florida lotto numbers and the NCAA men's Final Four basketball tournament are examples of combinations (six numbers and four teams) without regard to order. The number of combinations of n objects taken r at a time is equal to the binomial coefficient.

> **NUMBER OF COMBINATIONS OF n OBJECTS TAKEN r AT A TIME**
>
> The number of combinations of n objects taken r at a time is
>
> $$_nC_r = \frac{n!}{(n-r)!r!}$$

Compare the number of permutations, $_nP_r = \dfrac{n!}{(n-r)!}$, to the number of combinations $_nC_r = \dfrac{n!}{(n-r)!r!}$. It makes sense that the number of combinations is less than the number of permutations because there is no order associated with a combination.

EXAMPLE 6 Possible Combinations to the Lottery

If there are a possible 59 numbers, and the lottery officials draw 6 numbers, how many possible combinations are there?

Solution:

Identify how many numbers are in the drawing. $n = 59$

Identify how many numbers are chosen. $r = 6$

Substitute $n = 59$ and $r = 6$ into $_nC_r = \dfrac{n!}{(n-r)!r!}$.

$$_{59}C_6 = \frac{59!}{(59-6)!6!}$$

Simplify the binomial coefficient.

$$\frac{59 \cdot 58 \cdot 57 \cdot 56 \cdot 55 \cdot 54 \cdot (53)!}{53! \cdot 6!}$$

$$= \frac{59 \cdot 58 \cdot 57 \cdot 56 \cdot 55 \cdot 54}{6 \cdot 5 \cdot 4 \cdot 3 \cdot 2}$$

$$= 45,057,474$$

There are 45,057,474 possible combinations.

TECHNOLOGY TIP
Use calculators to find the term $_{59}C_6$:

Many scientific calculators:

Press	Display
59 [nCr] 6 [=]	45057474

Many graphing calculators:

Press

59 [MATH] [PRB] [3: nCr]

[ENTER] 6 [ENTER]

```
59 nCr 6
         45057474
```

YOUR TURN What are the possible combinations for a lottery with 49 possible numbers and 6 drawn numbers?

Permutations with Repetition

Permutations and combinations are arrangements of distinct (nonrepeated) objects. A permutation in which some of the objects are repeated is called a **permutation with repetition** or a **nondistinguishable permutation**. For example, if a sack has three red marbles, two blue marbles, and one white marble, how many possible permutations would there be when drawing six marbles, one at a time?

This is a different problem from writing the numbers 1 through 6 on pieces of paper, putting them in a hat, and drawing them out. The reason the problems are different is that the two blue balls are indistinguishable and the three red balls are also indistinguishable. The possible combinations for drawing numbers out of the hat is $6!$, whereas the possible combinations for drawing balls out of the sack is given by $\dfrac{6!}{3! \cdot 2! \cdot 1!}$.

NUMBER OF DISTINGUISHABLE PERMUTATIONS

If a set of n objects has n_1 of one kind of object, n_2 of another kind of object, n_3 of a third kind of object, and so on for k different types of objects so that $n = n_1 + n_2 + \cdots + n_k$, then the number of **distinguishable permutations** of the n objects is

$$\frac{n!}{n_1! \cdot n_2! \cdot n_3! \cdots n_k!}$$

In our sack of marbles, there were six marbles, $n = 6$. Specifically, there were three red marbles, $n_1 = 3$, two blue marbles, $n_2 = 2$, and one white marble, $n_3 = 1$. Notice that $n = n_1 + n_2 + n_3$ and that the number of distinguishable permutations is equal to $\dfrac{6!}{3! \cdot 2! \cdot 1!}$.

EXAMPLE 7 Peg Game at Cracker Barrel

The peg game on the tables at Cracker Barrel is a triangle with 15 holes drilled in it, in which pegs are placed. There are 5 red pegs, 5 white pegs, 3 blue pegs, and 2 yellow pegs. If all 15 pegs are in the holes, how many different ways can the pegs be aligned?

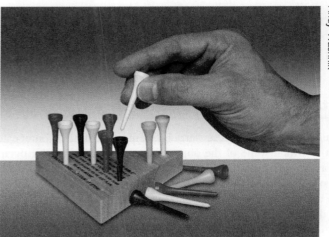

Andy Washnik

Solution:

There are four different-colored pegs (red, white, blue, and yellow).

$$5 \text{ red pegs:} \qquad n_1 = 5$$
$$5 \text{ white pegs:} \qquad n_2 = 5$$
$$3 \text{ blue pegs:} \qquad n_3 = 3$$
$$2 \text{ yellow pegs:} \qquad n_4 = 2$$

There are 15 pegs total: $n = 15$

Substitute $n = 15$, $n_1 = 5$, $n_2 = 5$, $n_3 = 3$, and $n_4 = 2$ into

$$\dfrac{n!}{n_1! \cdot n_2! \cdot n_3! \cdots n_k!} \qquad\qquad \dfrac{15!}{5! \cdot 5! \cdot 3! \cdot 2!}$$

Simplify. $\qquad\qquad\qquad\qquad\qquad 7{,}567{,}560$

There are a possible $7{,}567{,}560$ ways to fill 15 colored pegs at the Cracker Barrel.

■ **YOUR TURN** Suppose a similar game to the peg game at Cracker Barrel is set up with only six holes in a triangle. With two red pegs, two white pegs, and two blue pegs, how many different permutations can fill that board?

SECTION 9.6 SUMMARY

In this section we discussed counting principles. The fundamental counting principle is used in determining possible attire combinations, telephone numbers, and menus, since objects can be repeated or used in more than one scenario. When objects cannot be repeated and order matters, as in contestant running orders, permutations are used. When objects cannot be repeated and order doesn't matter, as in lotto numbers, combinations are used. When there are indistinguishable objects and when order does matter, as in colored marbles, then we use the formula for distinguishable permutations.

SECTION 9.6 EXERCISES

■ **SKILLS**

In Exercises 1–8, use the formula for $_nP_r$ to evaluate each expression.

1. $_6P_4$ **2.** $_7P_3$ **3.** $_9P_5$ **4.** $_9P_4$ **5.** $_8P_8$ **6.** $_6P_6$ **7.** $_{13}P_3$ **8.** $_{20}P_3$

In Exercises 9–18, use the formula for $_nC_r$ to evaluate each expression.

9. $_{10}C_5$ **10.** $_9C_4$ **11.** $_{50}C_6$ **12.** $_{50}C_{10}$ **13.** $_7C_7$

14. $_8C_8$ **15.** $_{30}C_4$ **16.** $_{13}C_5$ **17.** $_{45}C_8$ **18.** $_{30}C_4$

■ **APPLICATIONS**

19. Computers. At the www.dell.com website, a customer can "build" a system. If there are four monitors to choose from, three different computers, and two different keyboards, how many possible system configurations are there?

20. Houses. In a "new home" community a person can select from one of four models, five paint colors, three tile selections, and two landscaping options. How many different houses (interior and exterior) are there to choose from?

21. Wedding Invitations. An engaged couple is ordering wedding invitations. The wedding invitations come in white or ivory. The writing can be printed, embossed, or engraved. The envelopes can come with liners or without. How many possible combinations of invitations do they have to choose from?

22. Dinner. Siblings are planning their father's 65th birthday dinner and have to select one of four main courses (baked chicken, grilled mahi–mahi, beef Wellington, or lasagna), one of two starches (rosemary potatoes or rice), one of three vegetables (green beans, carrots, or zucchini), and one of five appetizers (soup, salad, pot stickers, artichoke dip, or calamari). How many possible dinner combinations are there?

23. **PIN Number.** Most banks require a four-digit ATM PIN code for each customer's bank card. How many possible combinations are there to choose from?

24. **Password.** All e-mail accounts require passwords. If a four-character password is required that can contain letters (but no numbers), how many possible passwords can there be?

25. **Leadership.** There are 15 professors in a department and there are four leadership positions (chair, assistant chair, undergraduate coordinator, and graduate coordinator). How many possible leadership teams are there?

26. **Fraternity Elections.** A fraternity is having elections. There are three men running for president, two men running for vice-president, four men running for secretary, and one man running for treasurer. How many possible outcomes do the elections have?

27. **Multiple-Choice Tests.** There are 20 questions on a multiple-choice exam, and each question has four possible answers (A, B, C, and D). Assuming no answers are left blank, how many different ways can you answer the questions on the exam?

28. **Multiple-Choice Tests.** There are 25 questions on a multiple-choice exam, and each question has five possible answers (A, B, C, D, and E). Assuming no answers are left blank, how many different ways can you answer the questions on the exam?

29. **Zip Codes.** In the United States a five-digit zip code is used to route mail. How many possible five-digit zip codes are possible if all digits can be used? If 0's are eliminated from the first and last digits, how many possible zip codes are there?

30. **License Plates.** In a particular state there are six characters in a license plate: three letters followed by three numbers. If 0's and 1's are eliminated from possible numbers and O's and I's are eliminated from possible letters, how many different license plates can be made?

31. **Class Seating.** If there are 30 students in a class and there are exactly 30 seats, how many possible seating charts can be made assuming all 30 students are present?

32. **Season Tickets.** Four friends buy four season tickets to the Green Bay Packers. To be fair, they change the seating arrangement every game. How many different seating arrangements are there for the four friends?

33. **Combination Lock.** A combination lock on most lockers will open when the correct choice of three numbers (1 to 40) is selected and entered in the correct order. Therefore, a combination lock should really be called a permutation lock. How many possible permutations are there assuming no numbers can be repeated?

34. **Safe.** A safe will open when the correct choice of three numbers (1 to 50) is selected and entered in the correct order. How many possible permutations are there assuming no numbers can be repeated?

35. **Raffle.** A fundraiser raffle is held to benefit cystic fibrosis research, and 1000 raffle tickets are sold. There are three prizes raffled off. First prize is a round-trip ticket on Delta Air Lines, second prize is a round of golf for four people at a local golf course, and the third prize is a $50 gift certificate to Chili's. How many possible winning scenarios are there if all 1000 tickets were sold to different people?

36. **Ironman Triathlon.** If 100 people compete in an Ironman triathlon, how many possible placings are there (first, second, and third place)?

37. **Lotto.** If a state lottery picks from 53 numbers, and 6 numbers are selected, how many possible 6-number combinations are there?

38. **Lotto.** If a state lottery picks from 53 numbers, and 5 numbers are selected, how many possible 5-number combinations are there?

39. **Cards.** In a deck of 52 cards, how many different 5-card hands can be dealt?

40. **Cards.** In a deck of 52 cards, how many different 7-card hands can be dealt?

41. **Blackjack.** In a single-deck blackjack game (52 cards), how many different 2-card combinations are there?

42. **Blackjack.** In a single deck, how many two-card combinations are there that equal 21: ace (worth 11) and a 10 or face card—jack, queen, or king?

43. **March Madness.** Every spring, the NCAA men's basketball tournament starts with 64 teams. After two rounds, it is down to the Sweet Sixteen, and after two more rounds, it is reduced to the Final Four. Once 64 teams are selected (but not yet put in brackets), how many possible scenarios are there for the Sweet Sixteen?

44. **March Madness.** Every spring, the NCAA men's basketball tournament starts with 64 teams. After two rounds, it is down to the Sweet Sixteen, and after two more rounds it is reduced to the Final Four. Once the 64 teams are identified (but not yet put in brackets), how many possible scenarios are there for the Final Four?

45. NFL Playoffs. There are 32 teams in the National Football League (16 AFC and 16 NFC). How many possible combinations are there for the Superbowl?

46. NFL Play-offs. After the regular season in the National Football League, 12 teams make the play-offs (6 from the AFC and 6 from the NFC). How many possible combinations are there for the Superbowl once the 6 teams in each conference are identified?

47. Voting Someone Off the Island. On the television show *Survivor,* one person is voted off the island every week. When it is down to six contestants, how many possible

voting combinations remain, if no one will vote themselves off the island? Assume that the order (who votes for whom) makes a difference. How many total possible voting outcomes are there?

48. American Idol. On the television show *American Idol,* a young rising star is eliminated from the competition every week. The first week, each of the 12 contestants sings one song. How many possible ways could they be ordered 1–12? To alternate female/male, and if there are six men and six women, how many possible ways could they be ordered?

■ CATCH THE MISTAKE

In Exercises 49 and 50, explain the mistake that is made.

49. In a lottery that picks from 30 numbers, how many five-number combinations are there?

   Solution:

   Let $n = 30$ and $r = 5$.

   Calculate $_nP_r = \dfrac{n!}{(n-r)!}$. $_{30}P_5 = \dfrac{30!}{25!}$

   Simplify. $_{30}P_5 = 17,100,720$

   This is incorrect. What mistake was made?

50. A home owners association has 12 members on the board of directors. How many ways can they elect a president, vice-president, secretary, and treasurer?

   Solution:

   Let $n = 12$ and $r = 4$.

   Calculate $_nC_r = \dfrac{n!}{(n-r)! \cdot r!}$. $_{12}C_4 = \dfrac{12!}{8! \cdot 4!}$

   Simplify. $_{12}C_4 = 495$

   This is incorrect. What mistake was made?

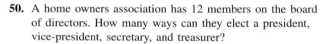

■ CHALLENGE

51. T or F: The number of permutations of n objects is always greater than the number of combinations of n objects if the objects are distinct.

52. T or F: The number of permutations of n objects is always greater than the number of combinations of n objects even when the objects are indistinguishable.

53. T or F: The number of four-letter permutations of the letters A, B, C, and D is equal to the number of four-letter permutations of the word ABBA.

54. T or F: The possible answers to a true/false question is a permutation problem.

55. What is the relationship between $_nC_r$ and $_nC_{r+1}$?

56. What is the relationship between $_nP_r$ and $_nP_{r-1}$?

■ TECHNOLOGY

57. Use a graphing utility with a $_nP_r$ feature and compare with answers from Exercises 1–8.

58. Use a graphing utility with a $_nC_r$ feature and compare with answers from Exercises 9–18.

Skills Objectives

- Work with the rules of probability.
- Find the probability of an event.
- Find the probability that an event will not occur.
- Find the probability of mutually exclusive events.
- Find the probability of independent events.

Conceptual Objectives

- Understand that the mathematics of probability gives us a good sense of how likely it is that a certain event will happen.
- Understand that probability affects our decision making in our daily lives.
- Understand the difference between the probability of:
  - event 1 *and* event 2
  - event 1 *or* event 2

Introduction

You are sitting at a blackjack table at Caesar's Palace, and the dealer is showing a 7. You have a 9 and a 7, should you hit? Will it rain today? What will the lotto numbers be this week? Will the coin toss at the Superbowl result in a head or a tail? Will Alex Rodriguez get a hit at his next trip to the plate? These are all questions where *probability* is used to guide us.

Anything that happens for which the result is uncertain is called an **experiment**. The possible results of the experiment are called **outcomes**. All of the possible outcomes of an experiment constitute the **sample space**. The term **event** is used to describe a set of possible outcomes of the experiment.

For example, a coin toss is an experiment. The possible outcomes are heads or tails. The sample space is {heads, tails}. The event could be heads or tails. Sitting at the blackjack table after you have received your two cards, you are interested in what the next card would be. The "hitting" of the card is the experiment. The possible outcomes are ace (1 or 11), 2, 3, 4, 5, 6, 7, 8, 9, 10, jack (worth 10), queen (worth 10), or king (worth 10). The event you may be interested in is if the next card has a value of 10. Thus the event would be 10, jack, queen, or king.

The result of one experiment has no certain outcome. However, if the experiment was performed many times, the results will produce regular patterns. For example, if you toss a coin, you don't know if it will come up heads or tails. You can toss a coin 10 times and get 10 heads. However, if you tossed 1,000,000 coins you would get about 500,000 heads and 500,000 tails. Therefore, since we assume a head is equally likely as a tail and there are only two possible events, we assign a *probability of a head* equal to $\frac{1}{2}$ and the *probability of a tail* equal to $\frac{1}{2}$.

EXAMPLE 1 Finding the Sample Space

Find the sample space for each of the following:

a. Tossing a coin once   **b.** Tossing a coin twice   **c.** Tossing a coin three times

Solution (a):

Tossing a coin one time will result in either a head (H) or a tail (T).

The sample space, S, is written as $S = \{H, T\}$.

Solution (b):

Tossing a coin twice can result in one of four possible outcomes:

HH = heads on the first coin and heads on the second coin

HT = heads on the first coin and tails on the second coin

TH = tails on the first coin and heads on the second coin

TT = tails on the first coin and tails on the second coin

The sample space consists of all possible outcomes: $S = \{HH, HT, TH, TT\}$.

Note that TH and HT are two different outcomes.

Solution (c):

There are eight possible outcomes when a coin is tossed three times:

$$S = \{HHH, HHT, HTH, HTT, TTT, TTH, THT, THH\}$$

■ **YOUR TURN** Find the sample space associated with having three children (B, boys or G, girls).

Probability of an Event

To calculate the probability of an event, start by counting the number of outcomes in the event and the number of outcomes in the sample space. The ratio is equal to the probability if all outcomes are equally likely.

DEFINITION   PROBABILITY OF AN EVENT

If an event E has $n(E)$ equally likely outcomes and its sample space, S, has $n(S)$ equally likely outcomes, then the **probability of event** E, denoted $P(E)$, is

$$P(E) = \frac{n(E)}{n(S)} = \frac{\text{number of outcomes in event } E}{\text{number of outcomes in sample space } S}$$

Since the number of outcomes in an event must be less than or equal to the number of outcomes in the sample space, the probability of an event must be a number between 0 and 1, $0 \leq P(E) \leq 1$. If $P(E) = 0$, then the event can never happen, and if $P(E) = 1$, the event is certain to happen.

EXAMPLE 2 Finding the Probability of Two Girls

If two children are born, what is the probability that they are both girls?

Solution:

The event is both children being girls. $\qquad E = \{GG\}$

The sample space is all four possible
combinations. $\qquad S = \{BB, BG, GB, GG\}$

■ **Answer:** $S = \{GGG, GGB, GBG, GBB, BBB, BBG, BGB, BGG\}$

The number of outcomes in the event is 1. $\qquad n(E) = 1$

The number of events in the sample space is 4. $\qquad n(S) = 4$

Compute the probability using $P(E) = \dfrac{n(E)}{n(S)}$. $\qquad P(E) = \dfrac{1}{4}$

The probability that both children are girls is $\frac{1}{4}$, or 0.25 .

EXAMPLE 3 Finding the Probability of Drawing a Face Card

Find the probability of drawing a face card (jack, queen, or king) out of a 52-card deck.

Solution:

There are 12 face cards in a deck. $\qquad n(E) = 12$

There are 52 cards in a deck. $\qquad n(S) = 52$

Compute the probability using $P(E) = \dfrac{n(E)}{n(S)}$. $\qquad P(E) = \dfrac{12}{52} = \dfrac{3}{13}$

The probability of drawing a face card out of a 52-card deck is $\frac{3}{13}$ or ≈ 0.23 .

■ YOUR TURN

Find the probability that an ace is drawn from a deck of 52 cards.

EXAMPLE 4 Finding the Probability of Rolling a 7 or 11

If you bet on the "pass line" at a craps table and the person's first roll is a 7 or 11, using a pair of dice, then you win. Find the probability of winning a pass line bet on the first roll.

Solution:

The fundamental counting theorem tells us that there will be $6 \cdot 6 = 36$ possible rolls of the pair of dice. $\qquad n(S) = 36$

Draw a table listing possible combinations of the dice.

Dice Value	1	2	3	4	5	6
1	2	3	4	5	6	7
2	3	4	5	6	7	8
3	4	5	6	7	8	9
4	5	6	7	8	9	10
5	6	7	8	9	10	11
6	7	8	9	10	11	12

■ **Answer:** $\dfrac{4}{52} = \dfrac{1}{13}$

Of the 36 rolls, there are 8 rolls that will
produce a 7 or 11. $\qquad\qquad\qquad\qquad\qquad\qquad n(E) = 8$

Compute the probability using $P(E) = \dfrac{n(E)}{n(S)}$. $\qquad\quad P(E) = \dfrac{8}{36} = \dfrac{2}{9}$

The probability of winning a pass line bet is $\frac{2}{9}$ or ≈ 0.22 (22%) .

CONCEPT CHECK If a 2, 3, or 12 is rolled on the first roll, then the pass
line bet loses. How many possible rolls will produce a 2, 3, or 12?

YOUR TURN Find the probability of losing a pass line bet.

Probability of an Event Not Occurring

The sum of the probabilities of all possible outcomes is 1. For example, when a die is
rolled, if the outcomes are equally likely, then the probabilities are all $\frac{1}{6}$:

$$P(1) = \frac{1}{6}, \; P(2) = \frac{1}{6}, \; P(3) = \frac{1}{6}, \; P(4) = \frac{1}{6}, \; P(5) = \frac{1}{6}, \; P(6) = \frac{1}{6}$$

The sum of these six probabilities is 1.

Since the sum of the probabilities of all possible outcomes sums to 1, we can find the
probability that an event won't occur by subtracting the probability that the event will
occur from 1.

$$P(E) + P(\text{not } E) = 1 \qquad \text{or} \qquad P(\text{not } E) = 1 - P(E)$$

PROBABILITY OF AN EVENT NOT OCCURRING

The **probability** that an event E will **not occur** is equal to 1 minus the
probability that E will occur.

$$P(\text{not } E) = 1 - P(E)$$

EXAMPLE 5 Probability of Not Winning the Lottery

Find the probability of not winning the lottery if six numbers are selected from
1 to 49.

Solution:

Calculate the number of possible
six-number combinations. $\qquad\qquad {}_{49}C_6 = \dfrac{49!}{43! \cdot 6!} = 13{,}983{,}816$

Calculate the probability
of winning. $\qquad\qquad\qquad P(\text{winning}) = \dfrac{1}{13{,}983{,}816}$

Answer: $\frac{1}{9}$ or ≈ 0.11

Calculate the probability of
not winning.

$$P(\text{not winning}) = 1 - P(\text{winning})$$

$$P(\text{not winning}) = 1 - \frac{1}{13{,}983{,}816}$$

$$P(\text{not winning}) = \frac{13{,}983{,}816}{13{,}983{,}816} - \frac{1}{13{,}983{,}816}$$

$$P(\text{not winning}) = \frac{13{,}983{,}815}{13{,}983{,}816} \approx 0.999999928$$

The probability of *not* winning the lottery is very close to 1. This is why the
states want you to play their lotteries—and the reason why you should have
doubts about lotteries.

Mutually Exclusive Events

Recall the definition of union and intersection in Section 1.5. The probability of one
event, E_1, or a second event, E_2, occurring is given by the probability of the union of the
two events.

$$P(E_1 \cup E_2)$$

If there is any overlap between the two events, we must be careful not to count those
twice. For example, what is the probability of drawing *either* a face card *or* a spade out
of a deck of 52 cards? We must be careful not to count any face cards that are spades
twice.

PROBABILITY OF THE UNION OF TWO EVENTS

If E_1 and E_2 are two events in the same sample space, the probability of *either*
E_1 *or* E_2 occurring is given by

$$P(E_1 \cup E_2) = P(E_1) + P(E_2) - P(E_1 \cap E_2)$$

If E_1 and E_2 are **mutually exclusive**, then $E_1 \cap E_2 = 0$. In that case the
probability of *either* E_1 *or* E_2 occurring is given by

$$P(E_1 \cup E_2) = P(E_1) + P(E_2)$$

**EXAMPLE 6 Finding the Probability of Drawing a Face Card
or a Spade**

Find the probability of drawing either a face card or a spade out of a deck of
52 cards.

Solution:

The deck has 12 face cards. $P(\text{face card}) = \dfrac{12}{52}$

The deck has 13 spades. $P(\text{spade}) = \dfrac{13}{52}$

The deck has 3 face
cards that are spades. $P(\text{face card and spade}) = \dfrac{3}{52}$

Apply the probability
formula. $P(E_1 \cup E_2) = P(E_1) + P(E_2) - P(E_1 \cap E_2)$

$P(\text{face card or a spade}) = \dfrac{12}{52} + \dfrac{13}{52} - \dfrac{3}{52}$

Simplify. $P(\text{face card or a spade}) = \dfrac{22}{52} = \dfrac{11}{26}$

The probability of either a face card or a spade being drawn is $\dfrac{11}{26} \approx 0.42$.

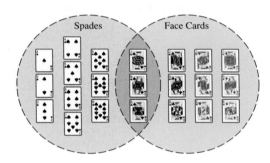

EXAMPLE 7 Finding the Probability of Mutually Exclusive Events

Find the probability of drawing either an ace or a joker in a 54-card deck
(a deck with two jokers).

Solution:

Drawing an ace or a joker are two mutually exclusive events since a card
cannot be both an ace and a joker.

The deck has four aces. $P(\text{ace}) = \dfrac{4}{54}$

The deck has two jokers. $P(\text{joker}) = \dfrac{2}{54}$

Apply the probability formula. $P(E_1 \cup E_2) = P(E_1) + P(E_2)$

$P(\text{ace or a joker}) = \dfrac{4}{54} + \dfrac{2}{54}$

Simplify. $P(\text{ace or a joker}) = \dfrac{6}{54} = \dfrac{1}{9} \approx 0.11\overline{1}$

The probability of drawing either an ace or a joker is $\dfrac{3}{27} \approx 0.11$.

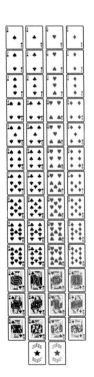

CONCEPT CHECK Is the following question referring to mutually exclusive events?

"If there are 10 women mathematicians, 8 men mathematicians, 6 women engineers, and 12 men engineers, what is the probability that a selected person is *either* a woman *or* an engineer?"

▪ **YOUR TURN** Calculate the probability that a selected person is *either* a woman *or* an engineer.

Independent Events

Suppose you have two children. The sex of the second child is not affected by the sex of the first child. For example, if your first child is a boy, then the second child being a boy is not any less likely. We say that two events are **independent** if the occurrence of either of them has no effect on the probability of the other occurring.

> **PROBABILITIES OF INDEPENDENT EVENTS**
>
> If E_1 and E_2 are independent events, then the probability of both occurring is the product of the individual probabilities:
>
> $$P(E_1 \text{ and } E_2) = P(E_1) \cdot P(E_2)$$

Scientists used to believe that one gene controlled human eye color. Each parent gives one chromosome (either blue, green, or brown), and the result is a child with an eye color gene composed of combinations, with brown being dominant over blue and blue being dominant over green. The genetic basis for eye color is actually far more complex, but we use this simpler model in the next example.

EXAMPLE 8 Probabilities of Blue-Eyed Children of Brown-Eyed Parents

If two brown-eyed parents have a blue-eyed child, then the parents must each have one blue and one brown-eyed gene. In order to have blue eyes, the child must get the blue-eyed genes from both parents. What is the probability that two brown-eyed parents can have three children, all with blue eyes?

Solution:

The sample space for childrens' eye color genes from these parents are

$$S = \{\text{Blue/Blue, Blue/Brown, Brown/Brown, Brown/Blue}\}$$

▪ **Answer:** $\frac{7}{9}$ or ≈ 0.78

The only way for a child to have blue eyes is if that child inherits two blue genes.

$$P(\text{blue-eyed child}) = \frac{1}{4}$$

Each child is independent of the other. The probability of having three blue-eyed children is the product of the three individual probabilities.

$$P(\text{all three children with blue eyes}) = P(\text{blue-eyed child}) \cdot P(\text{blue-eyed child}) \cdot P(\text{blue-eyed child})$$

$$= \left(\frac{1}{4}\right) \cdot \left(\frac{1}{4}\right) \cdot \left(\frac{1}{4}\right)$$

$$= \frac{1}{64}$$

The probability of two brown-eyed parents having three blue-eyed children is

$$\frac{1}{64} \approx .016 \ .$$

■ **YOUR TURN** Find the probability of the brown-eyed parents in Example 8 having three brown-eyed children.

SECTION 9.7 SUMMARY

In this section, we discussed the probability, or likelihood, of an event. It is found by dividing the total number of possible equally likely outcomes in the event, by all of the possible outcomes. Probability is a number between 0 and 1. The probability of an event not occurring is 1 minus the probability of the event occurring. The probability of one event *or* another event occurring is found by adding the individual probabilities of each event and subtracting the probability of both. If two events are mutually exclusive, they have no outcomes in common. The probability of two events occurring is the product of the individual probabilities, provided the two events are independent or do not affect one another.

SECTION 9.7 EXERCISES

■ **SKILLS**

In Exercises 1–6, find the sample space for each experiment.

1. The sum of two dice rolled simultaneously.

2. A coin tossed three times in a row.

■ **Answer:** $\frac{27}{64} \approx 0.422$

3. The sex (boy or girl) of four children born to the same parents.

4. Tossing a coin and rolling a die.

5. Two balls are selected from a container that has three red balls, two blue balls, and one white ball.

6. The grade (freshman, sophomore, or junior) of two high school students who work at a local restaurant.

Heads or Tails. In Exercises 7–10, find the probability for the experiment of tossing a coin three times.

7. Getting all heads.

8. Getting exactly one heads.

9. Getting at least one head.

10. Getting more than one head.

Tossing a Die. In Exercises 11–16, find the probability for the experiment of tossing two dice.

11. The sum is 3.

12. The sum is odd.

13. The sum is even.

14. The sum is prime.

15. The sum is more than 7.

16. The sum is less than 7.

Drawing a Card. In Exercises 17–20, find the probability for the experiment of drawing a single card from a deck of 52 cards.

17. Drawing a non-face card.

18. Drawing a black card.

19. Drawing a 2, 4, 6, or 8.

20. Drawing a 3, 5, 7, 9, or ace.

In Exercises 21–26, let $P(E_1) = \dfrac{1}{4}$ and $P(E_2) = \dfrac{1}{2}$ and find the probability of the event.

21. Probability of E_1 not occurring.

22. Probability of E_2 not occurring.

23. Probability of either E_1 or E_2 occurring if E_1 and E_2 are mutually exclusive.

24. Probability of either E_1 or E_2 occurring if E_1 and E_2 are not mutually exclusive and $P(E_1 \cap E_2) = \dfrac{1}{8}$.

25. Probability of both E_1 and E_2 occurring if E_1 and E_2 are mutually exclusive.

26. Probability of both E_1 and E_2 occurring if E_1 and E_2 are independent.

■ **APPLICATIONS**

27. Cards. A deck of 52 cards is dealt.
   a. How many possible combinations of four-card hands are there?
   b. What is the probability of having all spades?
   c. What is the probability of having four of a kind?

28. Blackjack. A deck of 52 cards is dealt for blackjack.
   a. How many possible combinations of two-card hands are there?
   b. What is the probability of having 21 points (ace with a 10 or face card)?

29. Cards. With a 52-card deck, what is the probability of drawing a 7 or an 8?

30. Cards. With a 52-card deck, what is the probability of drawing a red 7 or a black 8?

31. Cards. By drawing twice, what is the probability of drawing a 7 and then an 8?

32. Cards. By drawing twice, what is the probability of drawing a red 7 and then a black 8?

33. Children. What is the probability of having five daughters in a row and no sons?

34. Children. What is the probability of having four sons in a row and no daughters?

35. Children. What is the probability that of five children at least one is a boy?

36. Children. What is the probability that of six children at least one is a girl?

37. Roulette. In roulette, there are 38 numbered slots (1–36, 0, and 00). Eighteen are red, 18 are black, and the 0 and 00 are green. What is the probability of having 4 reds in a row?

38. Roulette. What is the probability of having two greens in a row on a roulette table?

39. Item Defectiveness. In a batch of 10 DVD players 1 is defective. A particular company has ordered 8 DVD players. The 8 are randomly taken off the truck. What is the probability that none of the 8 DVD players is defective?

40. Item Defectiveness. A shipment of 100 generators was delivered to Home Depot. Of those 20 are defective. If a small company buys 10 generators, what is the probability that none of the generators is defective?

41. Number Generator. A random generator (computer program that selects numbers in no particular order) is used to select two numbers between 1 and 10. What is the probability that both numbers are even?

42. Number Generator. A random generator is used to select two numbers between 1 and 15. What is the probability that both numbers are odd?

43. Blackjack. What is the probability of getting dealt a blackjack (any ace and any face card) with a single deck?

44. Blackjack. What is the probability of getting dealt two blackjacks in a row with a single deck?

45. Sports. With the salary cap in the NFL, it is said that on "any given Sunday" any team could beat any other team. If we assume every week a team has a 50% chance of winning, what is the probability that a team will go 16–0?

46. Sports. With the salary cap in the NFL, it is said that on "any given Sunday" any team could beat any other team. If we assume every week a team has a 50% chance of winning, what is the probability that a team will have at least 1 win?

■ **CATCH THE MISTAKE**

In Exercises 47 and 48, explain the mistake that is made.

47. Calculate the probability of drawing a 2 or a spade from a deck of 52 cards.

Solution:

The probability of drawing a 2 from a deck of 52 cards is $\frac{4}{52}$.

The probability of drawing a spade from a deck of 52 cards is $\frac{13}{52}$.

The probability of drawing a 2 or a spade is $\frac{4}{52} + \frac{13}{52} = \frac{17}{52}$.

This is incorrect. What mistake was made?

48. Calculate the probability of having two boys and one girl.

Solution:

The probability of having a boy is $\frac{1}{2}$.

The probability of having a girl is $\frac{1}{2}$.

These are independent, so the probability of having two boys and a girl is $\left(\frac{1}{2}\right)\left(\frac{1}{2}\right)\left(\frac{1}{2}\right) = \frac{1}{8}$.

This is incorrect. What mistake was made?

■ **CHALLENGE**

49. T or F: If $P(E_1) = 0.5$ and $P(E_2) = 0.4$, then $P(E_3)$ must equal 0.1 if there are three possible events and they are all mutually exclusive.

50. T or F: If two events are mutually exclusive, then they cannot be independent.

51. T or F: If two events are independent, then they are not mutually exclusive.

52. T or F: The probability of having five sons and no daughters is one minus the probability of having five daughters and no sons.

53. If two people are selected at random, what is the probability that they have the same birthday? Assume 365 days per year.

54. If 30 people are selected at random, what is the probability that at least two of them will have the same birthday?

55. If one die is weighted so that 3 and 4 are the only numbers that the die will roll, and the other die is equally loaded, what is the probability of rolling two dice that sum to 2, 5, or 6?

56. If one die is weighted so that 3 and 4 are the only numbers that the dice will roll, and 3 comes up twice as much as 4, what is the probability of rolling a 3?

■ TECHNOLOGY

57. Use a random-number generator on a graphing utility to select two numbers between 1 and 10. Run this generator 50 times. How many times (out of 50 trials) were both of the two numbers even? Compare with your answer from Exercise 41.

58. Use a random-number generator on a graphing utility to select two numbers between 1 and 15. Run this 50 times. How many times (out of 50 trials) were both of the two numbers odd? Compare with your answer from Exercise 42.

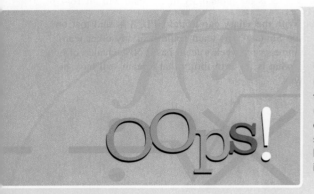

Q: Alex interviewed 100 students walking through the student union at his 4-year university to collect data for a probability and statistics project. His data show that 51 of the students interviewed are under age 20, 22 are smokers, and 52 are women. As he tried to calculate the probability that a student chosen at random from those surveyed is a woman or under age 20 or a smoker, he realized that his value of $\frac{125}{100}$ is incorrect because it exceeds 1. What did he do incorrectly?

A: Alex added together the probability that a randomly chosen participant from his survey is under age 20 ($\frac{51}{100}$), the probability that the participant is a smoker ($\frac{22}{100}$), and the probability that the participant is a woman ($\frac{52}{100}$). However, he forgot to take into account that these events are not mutually exclusive. A randomly chosen participant could have two or all three of these characteristics.

The probability of event A, B, or C occurring is

$$P(A \cup B \cup C) = P(A) + P(B) + P(C) - P(A \cap B) - P(B \cap C) - P(A \cap C) + P(A \cap B \cap C)$$

Of the students that Alex interviewed, 10 of those under age 20 are smokers, 23 of those under age 20 are women, 11 of the smokers are women, and 4 of those surveyed are women under age 20 who smoke. (a) Calculate the probability that a randomly chosen participant in the survey is a woman, a smoker, or under age 20. (b) What is the probability that a randomly chosen participant is not under age 20 or a smoker?

TYING IT ALL TOGETHER

Ann drops a rubber ball from a height of 90 feet within a building that is 120 feet tall. The ball weighs half an ounce and has an initial velocity of 2 feet per second (Ann gently tosses the ball from a window). If the ball rebounds $\frac{2}{3}$ of the height of the previous drop, calculate the total distance the ball has traveled when it hits the ground for the sixth time.

SECTION	TOPIC	PAGES	REVIEW EXERCISES	KEY CONCEPTS
	Pascal's triangle	644–645	89 and 90	Shortcut way of remembering binomial coefficients. Each term is found by adding the two numbers above it.
	Particular term of binomial expansion	646–647	91–94	The $(r + 1)$ term of the expansion $(a + b)^n$ is $\binom{n}{r} a^{n-r} b^r$.
9.6	Counting, permutations, and combinations	649–657	97–114	
	Fundamental counting principle	649–651	105, 106, 108	The number of ways in which successive things can occur is found by multiplying the number of ways each thing can occur.
	Permutations	651–654 655–657	97–100, 107, 109–111	The number of permutations of n objects is $$n! = n \cdot (n - 1) \cdot (n - 2) \cdots 2 \cdot 1$$ The number of permutations of n objects taken r at a time is $$_nP_r = \frac{n!}{(n-r)!} = n(n-1)(n-2)\cdots(n-r+1)$$ The number of distinguishable permutations of n objects is $$\frac{n!}{n_1! \cdot n_2! \cdot n_3! \cdots n_k!}$$
	Combinations	654–655	101–104 112–114	The number of combinations of n objects taken r at a time $$_nC_r = \frac{n!}{(n - r)! \, r!}$$
	Applications	649–657	105–114	
9.7	Probability	660	115–126	
	Probability of an event	661–663	115, 116, 118	The probability of event E, denoted $P(E)$, is $$P(E) = \frac{n(E)}{n(S)} = \frac{\text{number of outcomes in event } E}{\text{number of outcomes in sample space } S}$$
	Probability of an event *not* occurring.	663–664	117 and 119	$P(\text{not } E) = 1 - P(E)$
	Mutually exclusive events	664–666	115, 120, 121, and 123	The probability of *either* E_1 *or* E_2 occurring $$P(E_1 \cup E_2) = P(E_1) + P(E_2) - P(E_1 \cap E_2)$$ If E_1 and E_2 are mutually exclusive $E_1 \cap E_2 = 0$.
	Independent events	666–667	122, 124–126	If E_1 and E_2 are independent events, then the probability of both occurring is the product of the individual probabilities: $$P(E_1 \text{ and } E_2) = P(E_1) \cdot P(E_2)$$

9.1 Sequences and Series

Write the first four terms of the sequence. Assume n starts at 1.

1. $a_n = n^3$

2. $a_n = \dfrac{n!}{n}$

3. $a_n = 3n + 2$

4. $a_n = (-1)^n x^{n+2}$

Find the indicated term of the sequence.

5. $a_n = \left(\dfrac{2}{3}\right)^n \quad a_5 = ?$

6. $a_n = \dfrac{n^2}{3^n} \quad a_8 = ?$

7. $a_n = \dfrac{(-1)^n(n-1)!}{n(n+1)!} \quad a_{15} = ?$

8. $a_n = 1 + \dfrac{1}{n} \quad a_{10} = ?$

Write an expression for the nth term of the given sequence.

9. $3, -6, 9, -12, \ldots$

10. $1, \dfrac{1}{2}, 3, \dfrac{1}{4}, \ldots$

11. $-1, 1, -1, 1, \ldots$

12. $1, 10, 10^2, 10^3, \ldots$

Simplify the ratio of factorials.

13. $\dfrac{8!}{6!}$

14. $\dfrac{20!}{23!}$

15. $\dfrac{n(n-1)!}{(n+1)!}$

16. $\dfrac{(n-2)!}{n!}$

Write the first four terms of the sequence defined by the recursion formula.

17. $a_1 = 5 \quad a_n = a_{n-1} - 2$

18. $a_1 = 1 \quad a_n = n^2 \cdot a_{n-1}$

19. $a_1 = 1, a_2 = 2 \quad a_n = (a_{n-1})^2 \cdot (a_{n-2})$

20. $a_1 = 1, a_2 = 2 \quad a_n = \dfrac{a_{n-2}}{(a_{n-1})^2}$

Find the sum of the finite series.

21. $\displaystyle\sum_{n=1}^{5} 3$

22. $\displaystyle\sum_{n=1}^{4} \dfrac{1}{n^2}$

23. $\displaystyle\sum_{n=1}^{6} (3n + 1)$

24. $\displaystyle\sum_{k=0}^{5} \dfrac{2^{k+1}}{k!}$

Use sigma (summation) notation to write the sum.

25. $-1 + \dfrac{1}{2} - \dfrac{1}{4} + \dfrac{1}{8} + \cdots - \dfrac{1}{64}$

26. $2 + 4 + 6 + 8 + 10 + \cdots + 20$

27. $1 + x + \dfrac{x^2}{2} + \dfrac{x^3}{6} + \dfrac{x^4}{24} + \cdots$

28. $x - x^2 + \dfrac{x^3}{2} - \dfrac{x^4}{6} + \dfrac{x^5}{24} - \dfrac{x^6}{120} + \cdots$

Applications

29. Marines Investment. With the prospect of continued fighting in Iraq, in December 2004, the Marine Corps offered bonuses of as much as $30,000—in some cases, tax-free—to persuade enlisted personnel with combat experience and training to re-enlist. Suppose a Marine put her entire $30,000 re-enlistment bonus in an account that earns 4% interest compounded monthly. The balance in the account after n months is

$$A_n = 30{,}000\left(1 + \dfrac{0.04}{12}\right)^n \qquad n = 1, 2, 3, \ldots$$

Her commitment with the Marines is 5 years. Calculate A_{60}. What does A_{60} represent?

30. Sports. The league minimum salary for a rookie is $180,000 (2004). Suppose a rookie comes into the league making the minimum and gets a $30,000 raise every year he or she plays. Write the general term, a_n, of a sequence that represents the salary of an NFL player making the league minimum during his or her entire career. Assuming $n = 1$ corresponds to the first year, what does $\displaystyle\sum_{n=1}^{4} a_n$ represent?

9.2 Arithmetic Sequences and Series

Determine if the sequence is arithmetic. If it is, find the common difference.

31. $7, 5, 3, 1, -1, \ldots$

32. $1^3 + 2^3 + 3^3 + \cdots$

33. $1, \dfrac{3}{2}, 2, \dfrac{5}{2}, \ldots$

34. $a_n = -n + 3$

35. $a_n = \dfrac{(n+1)!}{n!}$

36. $a_n = 5(n-1)$

Find the general, or nth, term of the arithmetic sequence given the first term and the common difference.

37. $a_1 = -4 \quad d = 5$

38. $a_1 = 5 \quad d = 6$

39. $a_1 = 1 \quad d = -\dfrac{2}{3}$

40. $a_1 = 0.001 \quad d = 0.01$

For each arithmetic sequence described below, find a_1 and d and construct the sequence by stating the general, or nth, term.

41. The 5th term is 13 and the 17th term is 37.

42. The 7th term is -14 and the 10th term is -23.

43. The 8th term is 52 and the 21st term is 130.

44. The 11th term is -30 and the 21st term is -80.

Find the sum.

45. $\sum_{k=1}^{20} 3k$

46. $\sum_{n=1}^{15} n + 5$

47. $2 + 8 + 14 + 20 + \cdots + 68$

48. $\dfrac{1}{4} - \dfrac{1}{4} - \dfrac{3}{4} - \cdots - \dfrac{31}{4}$

Applications

49. **Salary.** On graduating with a M.B.A. Bob and Tania opt for different career paths. Bob accepts a job for the U.S. Department of Transportation making $45,000 with a guaranteed $2,000 raise every year. Tania takes a job with Templeton Corporation making $38,000 with a guaranteed $4,000 raise every year. Calculate how many total dollars both Bob and Tania will have each made after 15 years.

50. **Gravity.** When a skydiver jumps out of an airplane, she falls approximately 16 feet in the 1st second, 48 feet during the 2nd second, 80 feet during the 3rd second, 112 feet during the 4th second, and 144 feet during the 5th second, and this pattern continues. If she deploys her parachute after 5 seconds have elapsed, how far will she have fallen during those 5 seconds?

9.3 Geometric Sequences and Series

Determine if the sequence is geometric. If it is, find the common ratio.

51. $2, -4, 8, -16, \ldots$

52. $1, \dfrac{1}{2^2}, \dfrac{1}{3^2}, \dfrac{1}{4^2}, \ldots$

53. $20, 10, 5, \dfrac{5}{2}, \ldots$

54. $\dfrac{1}{100}, \dfrac{1}{10}, 1, 10, \ldots$

Write the first five terms of the geometric series.

55. $a_1 = 3 \quad r = 2$

56. $a_1 = 10 \quad r = \dfrac{1}{4}$

57. $a_1 = 100 \quad r = -4$

58. $a_1 = -60 \quad r = -\dfrac{1}{2}$

Write the formula for the nth term of the geometric series.

59. $a_1 = 7 \quad r = 2$

60. $a_1 = 12 \quad r = \dfrac{1}{3}$

61. $a_1 = 1 \quad r = -2$

62. $a_1 = \dfrac{32}{5} \quad r = -\dfrac{1}{4}$

Find the indicated term of the geometric sequence.

63. 25th term of the sequence $2, 4, 8, 16, \ldots$

64. 10th term of the sequence $\dfrac{1}{2}, 1, 2, 4, \ldots$

65. 12th term of the sequence $100, -20, 4, -0.8, \ldots$

66. 11th term of the sequence $1000, -500, 250, -125, \ldots$

Find the sum of the geometric series (if possible).

67. $\dfrac{1}{2} + \dfrac{3}{2} + \dfrac{3^2}{2} + \cdots + \dfrac{3^8}{2}$

68. $1 + \dfrac{1}{2} + \dfrac{1}{2^2} + \dfrac{1}{2^3} + \cdots + \dfrac{1}{2^{10}}$

69. $\sum_{n=1}^{8} 5(3)^{n-1}$

70. $\sum_{n=1}^{7} \dfrac{2}{3}(5)^n$

71. $\sum_{n=0}^{\infty} \left(\dfrac{2}{3}\right)^n$

72. $\sum_{n=1}^{\infty} \left(-\dfrac{1}{5}\right)^{n+1}$

Applications

73. **Salary.** Murad is fluent in four languages and is offered a job with the U.S. government as a translator. He is hired on the "GS" scale at a base rate of $48,000 with 2% increases in his salary per year. Calculate what his salary will be *after* he has been with the U.S. government for 12 years.

74. **Boat Depreciation.** On graduating from Auburn University, Philip and Steve get jobs at Disney Ride and Show Engineering and decide to buy a ski boat together. If the boat costs $15,000 new, and depreciates 20% per year, write a formula for the value of the boat n years after it was purchased. How much will the boat be worth when Philip and Steve have been working at Disney for 3 years?

9.4 Mathematical Induction

Prove the statements using mathematical induction for all positive integers, n.

75. $3n \le 3^n$

76. $4^n < 4^{n+1}$

77. $2 + 7 + 12 + 17 + \cdots + (5n - 3) = \dfrac{n}{2}(5n - 1)$

78. $2n^2 > (n + 1)^2 \quad n \ge 3$

674

9.5 The Binomial Theorem

Evaluate the binomial coefficients.

79. $\begin{pmatrix} 11 \\ 8 \end{pmatrix}$ **80.** $\begin{pmatrix} 10 \\ 0 \end{pmatrix}$ **81.** $\begin{pmatrix} 22 \\ 22 \end{pmatrix}$ **82.** $\begin{pmatrix} 47 \\ 45 \end{pmatrix}$

Expand the expression using the binomial theorem.

83. $(x - 5)^4$ **84.** $(x + y)^5$ **85.** $(2x - 5)^3$

86. $(x^2 + y^3)^4$ **87.** $(\sqrt{x} + 1)^5$ **88.** $(x^{2/3} + y^{1/3})^6$

Expand the expression using Pascal's triangle.

89. $(r - s)^5$ **90.** $(ax + by)^4$

Find the coefficient, C, of the term in the binomial expansion.

	BINOMIAL	TERM
91.	$(x - 2)^8$	Cx^6
92.	$(3 + y)^7$	Cy^4
93.	$(2x + 5y)^6$	Cx^2y^4
94.	$(r^2 - s)^8$	Cr^8s^4

Applications

95. Lottery. In a state lottery in which 6 numbers are drawn from a possible 53 numbers, the number of possible 6-number combinations is equal to $\begin{pmatrix} 53 \\ 6 \end{pmatrix}$. How many possible combinations are there?

96. Canasta. In the card game canasta, two decks of cards including the jokers are used, and 13 cards are dealt to each person. A total of $\begin{pmatrix} 108 \\ 13 \end{pmatrix}$ different 13-card canasta hands can be dealt. How many possible hands are there?

9.6 Counting, Permutations, and Combinations

Use the formula for $_nP_r$ to evaluate each expression.

97. $_7P_4$ **98.** $_9P_9$ **99.** $_{12}P_5$ **100.** $_{10}P_1$

Use the formula for $_nC_r$ to evaluate each expression.

101. $_{12}C_7$ **102.** $_{40}C_5$ **103.** $_9C_9$ **104.** $_{53}C_6$

Applications

105. Car Options. A new Honda Accord comes in three models (LX, VX, and EX). Each of those models comes with either a cloth or leather interior, and the exterior comes in either silver, white, black, red, or blue. How many different cars (models, interior seat upholstery, and exterior color) are there to choose from?

106. Email Passwords. All e-mail accounts require passwords. If a six-character password is required that can contain letters (but no numbers), how many possible passwords can there be if letters can be repeated?

107. Team Arrangements. There are 10 candidates for the board of directors, and there are four leadership positions (president, vice-president, secretary, and treasurer). How many possible leadership teams are there?

108. License Plates. In a particular state, there are six characters in a license plate consisting of letters and numbers. If 0's and 1's are eliminated from possible numbers and O's and I's are eliminated from possible letters, how many different license plates can be made?

109. Seating Arrangements. Five friends buy five season tickets to the Philadelphia Eagles. To be fair, they change the seating arrangement every game. How many different seating arrangements are there for the five friends? How many seasons would they have to buy tickets in order to sit in all of the combinations (each season has eight home games)?

110. Safe. A safe will open when the correct choice of three numbers (1 to 60) is selected in a specific order. How many possible permutations are there?

111. Raffle. A fundraiser raffle is held to benefit the Make a Wish Foundation, and 100 raffle tickets are sold. Four prizes are raffled off. First prize is a round-trip ticket on American Airlines, second prize is a round of golf for four people at a Links golf course, the third prize is a $100 gift certificate to the Outback Steakhouse, and the fourth prize is a half hour massage. How many possible winning scenarios are there if all 100 tickets were sold to different people?

112. Sports. There are 117 Division 1-A football teams in the United States. At the end of the regular season is the Bowl Championship Series, and the top two teams play each other in the championship game. Assuming that any two Division 1-A teams can advance to the championship, how many possible matchups are there for the championship game?

113. Cards. In a deck of 52 cards, how many different 6-card hands can be dealt?

114. Blackjack. In a game of two-deck blackjack (104 cards) how many 2-card combinations are there that equal 21, that is, ace and a 10 or face card—jack, queen, or king?

9.7 Probability

115. Coin Tossing. For the experiment of tossing a coin four times, what is the probability of getting all heads?

116. Dice. For an experiment of tossing two dice, what is the probability that the sum of the dice is odd?

117. **Dice.** For an experiment of tossing two dice, what is the probability of not rolling a combined 7?

118. **Cards.** For a deck of 52 cards, what is the probability of drawing a diamond?

For Exercises 199–122, let $P(E_1) = \frac{1}{3}$ and $P(E_2) = \frac{1}{2}$ and find the probability of the event.

119. **Probability.** Find the probability of an event, E_1, not occurring.

120. **Probability.** Find the probability of either E_1 or E_2 occurring if E_1 and E_2 are mutually exclusive.

121. **Probability.** Find the probability of either E_1 or E_2 occurring if E_1 and E_2 are not mutually exclusive and $P(E_1 \cap E_2) = \frac{1}{4}$.

122. **Probability.** Find the probability of both E_1 and E_2 occurring if E_1 and E_2 are independent.

123. **Cards.** With a 52-card deck, what is the probability of drawing an ace or a 2?

124. **Cards.** By drawing twice, what is the probability of drawing an ace and then a 2?

125. **Children.** What is the probability that in a family of five children at least one is a girl?

126. **Sports.** With the salary cap in the NFL, it is said that on "any given Sunday" any team could beat any other team. If we assume every week a team has a 50% chance of winning, what is the probability that a team will go 11–1?

For Exercises 1–5, use the sequence

$$1, x, x^2, x^3, \ldots$$

1. Write the nth term of the sequence.

2. Classify this sequence as arithmetic, geometric, or neither.

3. Find the nth partial sum of the series, S_n.

4. Assuming this sequence is infinite, write the series using sigma notation.

5. Assuming this sequence is infinite, what condition would have to be satisfied in order for the sum to exist?

6. Use mathematical induction to prove that $2 + 4 + 6 + \cdots + 2n = n^2 + n$.

In Exercises 7–10, evaluate expressions.

7. $\dbinom{15}{12}$ 8. $\dbinom{k}{k}$ 9. $_{14}P_3$ 10. $_{200}C_3$

11. Expand the expression: $\left(x^2 + \dfrac{1}{x}\right)^5$.

12. What is the coefficient of the x^5y^2 in the expansion of $(x - y)^7$?

13. Explain why there are always more permutations than combinations.

14. What is the probability of not winning a trifecta (selecting the first, second, and third place finishers) in a horse race with 15 horses?

For Exercises 15–17, refer to a roulette wheel with 18 red, 18 black, and 2 green slots.

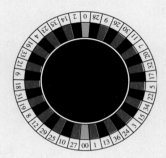

15. **Roulette.** What is the probability of the ball landing in a red slot?

16. **Roulette.** What is the probability of the ball landing in a red slot 5 times in a row?

17. **Roulette.** If the four previous rolls landed on red, what is the probability that the next roll will land on red?

18. **Marbles.** If there are four red marbles, three blue marbles, two green marbles, and one black marble in a sack, find the probability of pulling out the following order: black; blue; red; red; green.

19. **Cards.** What is the probability of drawing an ace or a diamond from a deck of 52 cards?

20. **Children.** Genetically, green eyes are the most recessive (brown dominates both blue and green, and blue dominates green). What is the probability that two brown-eyed parents have two green-eyed children? Assume each parent is equally likely to pass on one of each of the 3 genes.

ANSWERS TO ODD NUMBERED EXERCISES*

Chapter 0

Section 0.1
1. rational
3. irrational
5. rational
7. irrational
9. **a.** 7.347 **b.** 7.347
11. **a.** 2.995 **b.** 2.994
13. 4
15. 26
17. 3
19. $x + y - z$
21. $-3x - y$
23. $\frac{3}{5}$
25. $\frac{19}{12}$
27. $\frac{1}{2}$
29. $\frac{x}{3}$
31. $\frac{4}{3}$
33. $\frac{3}{35}$
35. $\frac{b^2}{a}$
37. \$7,724,007,000,000
39. \$26,121
41. The mistake is rounding the number that is used in the rounding. Look to the right of the number; if it is less than 5, round down.
43. false
45. true
47. irrational

Section 0.2
1. -3
3. -2
5. 1
7. x^5
9. x^6
11. $64a^3$
13. $\frac{y^2}{x^2}$
15. $\frac{16}{b^4}$
17. $\frac{1}{a^6 b^2}$
19. 2.76×10^7
21. 5.67×10^{-8}
23. 47,000,000
25. 0.000041
27. $x^3 y^4$
29. $x^{7/6} y^2$
31. $-4\sqrt{2}$
33. $\sqrt{21}$
35. $\frac{\sqrt{3}}{3}$
37. $-\frac{3}{4} - \frac{3\sqrt{5}}{4}$
39. $-3 - 2\sqrt{2}$
41. $4\sqrt{5} \approx 8.9$ seconds
43. Forgot to square the 4.
45. false
47. false
49. $-\frac{2}{11} + \frac{5\sqrt{5}}{11}$
51. 3.317

Section 0.3
1. $-7x^4 - 2x^3 + 5x^2 + 16$ degree 4
3. 15 degree 0
5. $-x^2 + 5x + 5$
7. $2z^2 + 2z - 3$
9. $x^2 - x + 2$
11. $-4t^2 + 6t - 1$
13. $35x^2 y^3$
15. $2x^5 - 2x^4 + 2x^3$
17. $2y^4 - 9y^3 + 5y^2 + 7y$
19. $6x^2 - 5x - 4$
21. $x^2 - 4$
23. $4x^2 - 9$
25. $t^2 - 4t + 4$
27. $z^2 + 4z + 4$
29. $6y^3 + 5y^2 - 4y$
31. $x^4 - 1$
33. $x - y$
35. $11x - 100$
37. The negative was not distributed through the second polynomial.
39. true
41. false
43. $2401x^4 - 1568x^2 y^4 + 256y^8$

Section 0.4
1. $5(x + 5)$
3. $2(2t^2 - 1)$
5. $3x(x^2 - 3x + 4)$
7. $(x - 5)(x - 1)$
9. $(y - 3)(y + 1)$
11. $(2y + 1)(y - 3)$
13. $(3t + 1)(t + 2)$
15. $(x^2 + 1)^2$
17. $(x + 3)(x - 3)$
19. not possible
21. $(p + q)^2$
23. $(t + 3)(t^2 - 3t + 9)$
25. $(y - 4)(y^2 + 4y + 16)$
27. $x(3x + 1)(x - 2)$
29. $x(x - 3)(x + 3)$
31. $(x^2 + 2)(x - 3)$
33. $(x^2 - 9) \neq (x - 3)^2$. Instead, $x^2 - 9 = (x - 3)(x + 3)$.
35. false
37. $(a^n - b^n)(a^n + b^n)$

Section 0.5
1. $x \neq 0$
3. $x \neq -1$
5. $p \neq \pm 1$
7. all real numbers
9. $\frac{x - 3}{2}$ $x \neq -1$
11. $\frac{y + 1}{5}$ $y \neq \frac{1}{5}$
13. $x + 2$ $x \neq 2$
15. 1 $x \neq -7$
17. $\frac{2}{3}$
19. $\frac{t - 3}{3(t + 2)}$
21. $\frac{x}{4}$
23. $\frac{1}{2(1 - p)}$
25. $\frac{2t(t - 3)}{5}$
27. $\frac{13}{5x}$
29. $\frac{4}{5x - 1}$

31. $\frac{x^2 + 4x - 6}{x^2 - 4}$
33. $\frac{1 - x}{x - 2}$ $x \neq 0, 2$
35. $x \neq -1$
37. false
39. false
41. $\frac{(x + a)(x + d)}{(x + b)(x + c)}$ $x \neq -b, -c, -d$

Section 0.6
1. $x = 5$
3. $z = -5$
5. $x = \frac{1}{4}$
7. $z = -4$
9. $x = 5$
11. $y = -3$
13. $x = 12$
15. $x = \pm 5$
17. no real solution
19. $x = 0$ $x = 2$
21. $x = -3$ $x = 0$ $x = 5$
23. $x = \pm 2$ $x = \pm 4$
25. $x = -1$ $x = \pm\sqrt{2}$
27. Width is 1 inch.
29. 5 feet
31. When $3x$ was subtracted and 10 was added, we forgot to do these on *both* sides.
33. The original assumption was $x = 1$, and we divided by $(x - 1)$.
35. false
37. $x = \frac{c - b}{a}$
39. $x = 2.79$

Section 0.7
1. $-i$
3. 1
5. $4i$
7. $2i\sqrt{5}$
9. $2 - 9i$
11. $2 - 2i$
13. $5 - i$
15. 65
17. $\frac{3}{10} + \frac{1}{10}i$
19. $\frac{18}{53} - \frac{43}{53}i$
21. $5i$
23. $13 + i$
25. $\sqrt{-4} = 2i$. Convert square roots to imaginary numbers before multiplication.
27. Multiplied by the denominator, $4 - i$, instead of the conjugate, $4 + i$.
29. true
31. true
33. $-2 + 2i$

OOPS! There is approximately a 0.098% probability that the parents would have five Rh negative children.

Tying It All Together $G = \frac{2wh + 2lh + wr}{175}$

Since about 12.1 gallons of paint would be needed, Rafael will need to purchase 13 gallons of paint, which will cost \$325.

Review Exercises
1. **a.** 5.22 **b.** 5.21
3. 4
5. -2
7. $-\frac{x}{12}$
9. 9
11. $-8z^3$
13. -4
15. 2.15×10^{-6}
17. $\frac{9x^{2/3}}{16}$
19. $-3 - \sqrt{5}$
21. $14z^2 + 3z - 2$
23. $15x^2 y^2 - 20xy^3$
25. $x^2 + 2x - 63$
27. $4x^2 - 12x + 9$
29. $x^4 + 2x^2 + 1$
31. $2xy^2(7x - 5y)$
33. $(x + 5)(2x - 1)$
35. $(4x - 5)(4x + 5)$
37. $2x(x - 3)(x + 5)$
39. $(x^2 - 2)(x + 1)$
41. $x \neq \pm 3$
43. $x + 2$ $x \neq 2$
45. $\frac{t + 3}{t + 1}$ $t \neq -1, 2$
47. $\frac{(x + 5)(x + 2)}{(x + 3)^2}$ $x \neq -3, 1, 2$
49. $\frac{2}{(x + 1)(x + 3)}$ $x \neq -1, -3$
51. $z = -5$
53. $y = 15$
55. $x = \pm 12$
57. $x = 2, 4$
59. $x = 1, \pm 2$
61. $13i$
63. $-i$
65. 12
67. $\frac{2}{5} + \frac{1}{5}i$
69. $\frac{28}{13} - \frac{3}{13}i$

Practice Test
1. 4
3. -2
5. i
7. $2y^2 - 12y + 20$
9. $(2x + 1)(x - 1)$
11. $t(2t - 3)(t + 1)$
13. $-\frac{1}{x + 1}$ $x \neq \frac{1}{2}, \pm 1$
15. $x = 9$
17. $-8 - 26i$
19. $(x^{2n} + y^{2n})(x^{2n} - y^{2n}) = (x^{2n} + y^{2n})(x^n - y^n)(x^n + y^n)$

*Answers that involve a proof, graph, or otherwise lengthy solution are not included.

Chapter 1

Section 1.1

1. $x = 4$ **3.** $m = 2$ **5.** $t = \dfrac{7}{5}$

7. $x = -10$ **9.** $n = 2$ **11.** $x = -1$

13. $p = -\dfrac{9}{2}$ **15.** $a = -8$ **17.** $x = -15$

19. $c = -\dfrac{35}{13}$ **21.** $m = \dfrac{60}{11}$ **23.** $x = 36$

25. $p = 8$ **27.** $y = -2$ **29.** $p = 2$

31. $y = \dfrac{3}{10}$ $y = 0$ must be excluded from the solution set

33. $x = \dfrac{1}{2}$ $x = 0$ must be excluded from the solution set

35. $a = \dfrac{1}{6}$ $a = 0$ must be excluded from the solution set

37. no solution $n = -1, 0$ must be excluded from the solution set

39. no solution $a = -3, 0$ must be excluded from the solution set

41. $n = \dfrac{53}{11}$ $n = 1$ must be excluded from the solution set

43. $x = -3$ $x = \dfrac{1}{2}, -\dfrac{1}{5}$ must be excluded from the solution set

45. no solution $t = 1$ must be excluded from the solution set

47. no solution $x = -1, 0$ must be excluded from the solution set

49. $\dfrac{5}{9}F - \dfrac{160}{9} = C$

51. $\lambda = 0$ must be eliminated.

53. The error is forgetting to do to both sides of any equation what is done to one side. The correct answer is $x = 5$.

55. You cannot cross multiply with the -3 in the problem. You must find a common denominator first. The correct solution is $p = \frac{6}{5}$.

57. false   **59.** true   **61.** $x = \dfrac{ab}{c}$ $x \neq \pm a$

63. $x = \dfrac{ac - cy - by}{y - a}$ $x \neq \pm c, -c - b$

65. Conditional   **67.** Identity

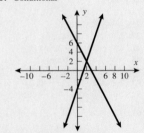

69. Contradiction (inconsistent)

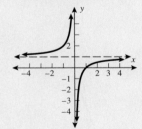

Section 1.2

1. 242.17 **3.** 13.76 **5.** $147,058.82 ($22,058.82 saved)

7. 12 miles   **9.** 9 hours of sleep   **11.** 270 units

13. 24   **15.** 8, 10   **17.** 20 inches

19. Length = 100 yards, width = 30 yards

21. $r_1 = 3$ feet, $r_2 = 6$ feet

23. 300 feet

25. The body is 63 inches or 5.25 feet.

27. $20,000 at 4% and $100,000 at 7%

29. $3,000 at 10%, $5,500 at 2%, and $5,500 at 40%

31. 6 trees, 27 shrubs

33. 70 ml of 5%, 30 ml of 15%

35. The entire bag was gummy bears.

37. 9 minutes

39. 2.3 mph

41. Jogger: 6 mph   Walker: 4 mph

43. 22.5 hours

45. 264:330:396

47. 77.5 for a B; 92.5 for an A

49. 2 field goals and 6 touchdowns

51. 3.5 feet from center

53. Fulcrum is 0.4 units from Maria and 0.6 units from Max.

55. $s_o = \dfrac{15}{2}$ or 7.5 cm (in front of the lens)

57. image distance is 3 cm, object distance is 2 cm

59. $191,983.35

61. Plan B is better for 5 or less plays/month. Plan A is better for 6 or more plays/month.

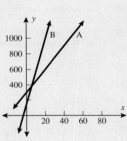

Section 1.3

1. $x = 3, x = 2$ **3.** $x = -4, x = 3$ **5.** $x = -\dfrac{1}{4}$

7. $y = 0, 2$ **9.** $p = \dfrac{2}{3}$ **11.** $x = -6, x = 2$

13. $x = 4, x = 3$ **15.** $x = -\dfrac{3}{4}, x = 2$ **17.** $p = \pm 2\sqrt{2}$

19. $x = \pm 3i$ **21.** $x = 9, x = -3$

23. $x = \dfrac{-3 + 2i}{2}$ and $x = \dfrac{-3 - 2i}{2}$

25. $x = \dfrac{2 + 3\sqrt{3}}{5}$ and $x = \dfrac{2 - 3\sqrt{3}}{5}$

27. $x = -2$ and $x = 4$ **29.** $x^2 + 6x + 9$ **31.** $x^2 - 12x + \underline{36}$

33. $x^2 - \dfrac{1}{2}x + \dfrac{1}{16}$ **35.** $x^2 + \dfrac{2}{5}x + \dfrac{1}{25}$ **37.** $x^2 - 2.4x + \underline{1.44}$

39. $x = 1$ and $x = -3$ **41.** $t = 5$ and $t = 1$ **43.** $y = 3$ and $y = 1$

45. $p = \dfrac{-4 + \sqrt{10}}{2}$ and $p = \dfrac{-4 - \sqrt{10}}{2}$

47. $x = 3$ and $x = \dfrac{1}{2}$

49. $x = \dfrac{4 + 3\sqrt{2}}{2}$ and $x = \dfrac{4 - 3\sqrt{2}}{2}$

51. $t = \dfrac{-3 + \sqrt{13}}{2}$ and $t = \dfrac{-3 - \sqrt{13}}{2}$

53. $s = \dfrac{-1 + \sqrt{3}i}{2}$ and $s = \dfrac{-1 - \sqrt{3}i}{2}$

55. $x = \dfrac{3 + \sqrt{57}}{6}$ and $x = \dfrac{3 - \sqrt{57}}{6}$

57. $x = 1 + 4i$ and $x = 1 - 4i$

59. $x = \dfrac{-7 + \sqrt{109}}{10}$ and $x = \dfrac{-7 - \sqrt{109}}{10}$

61. $x = 1.2 - \sqrt{5.96}i$ and $x = 1.2 + \sqrt{5.96}i$

63. 1 real solution   **65.** 2 real solutions
67. 2 complex solutions   **69.** $v = -2$ and $v = 10$
71. $t = 1$ and $t = -6$ **73.** $x = 1$ and $x = -7$
75. $w = \dfrac{-1 + \sqrt{167}i}{8}$ and $w = \dfrac{-1 - \sqrt{167}i}{8}$
77. $p = \dfrac{9 + \sqrt{69}}{6}$ and $p = \dfrac{9 - \sqrt{69}}{6}$
79. $t = \dfrac{10 + \sqrt{130}}{10}$ and $t = \dfrac{10 - \sqrt{130}}{10}$
81. $x = 0.4$ and $x = -0.3$
83. $t = \dfrac{\sqrt{2s}}{g}$
85. $c = \sqrt{a^2 + b^2}$
87. walker: 4 mph, jogger: 6 mph
89. 17 and 18
91. 5m, 7m
93. base is 6 and height is 20
95. $t = 8$ (August) and $t = 12$ (December) 2003.
97. 2.5 seconds after it is dropped
99. 21.2 feet   **101.** 5 ft $\times$ 5 ft
103. border is 2.3 feet wide   **105.** 10 days
107. The problem is factored incorrectly. The correction would be $t = 6$ and $t = -1$.
109. When taking the square root of both sides, the i is missing from the right side. The correction would be $a = \pm\frac{3}{4}i$.
111. false
113. true
115. $\dfrac{-b + \sqrt{b^2 - 4ac}}{2a} + \dfrac{-b - \sqrt{b^2 - 4ac}}{2a} = \dfrac{2b}{2a} = \dfrac{-b}{a}$
117. $x^2 - 6x + 4$ **119.** $ax^2 - bx + c = 0$
121. $x = -1, 2$ **123.** $x = 1.7$ (double root)

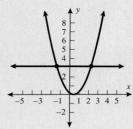

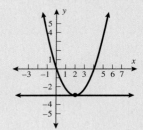

Section 1.4

1. $t = 9$ **3.** $p = 8$ **5.** no solution
7. $x = 5$ **9.** $y = -\dfrac{1}{2}$ **11.** $x = 4$
13. $y = 0, y = 25$ **15.** $s = 3, s = 6$ **17.** $x = -3, x = -1$
19. $x = 0$ **21.** $x = \dfrac{5}{2}$ **23.** no solution
25. $x = 1, x = 5$ **27.** $x = 7$ **29.** $x = 4$
31. $x = -8, x = 0$ **33.** $x = \pm\sqrt{2}, \; x = \pm 1$
35. $x = \pm\dfrac{\sqrt{6}}{2}i, \; x = \pm\sqrt{2}i$ **37.** $x = -\dfrac{5}{2}, x = -1$
39. $t = \dfrac{5}{4}, t = 3$ **41.** $x = \pm 1, \; \pm i, \; \pm\dfrac{1}{2}, \; \pm\dfrac{1}{2}i$
43. $y = -\dfrac{3}{4}, \; y = 1$ **45.** $z = 1$
47. $t = -27, t = 8$ **49.** $x = -\dfrac{4}{3}, x = 0$
51. $x = \dfrac{3}{8}, x = \dfrac{2}{3}$ **53.** $u = \pm 1, u = \pm 8$
55. $t = \sqrt{3}$ **57.** $a = 71$
59. $t = 4\sqrt{3} \approx 7$ months (Oct. 2004)   **61.** 132.66 ft
63. $t = 5$ is extraneous. No solution.
65. The error is not converting u back to x using the substitution.

67. true   **69.** false
71. $x = -1, \; x = \dfrac{1}{3}, \; x = 0, \; x = -\dfrac{2}{3}$ **73.** $x = -2$
75. $x = \dfrac{313}{64} \approx 4.891$ **77.** no solution

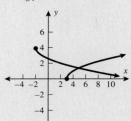

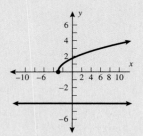

Section 1.5

1. interval notation: [3, ∞)
3. interval notation: [-2, 3)
5. interval notation: (-3, 5]
7. interval notation: [0]
9. set notation: $\{x \mid 0 \le x < 2\}$ **11.** set notation: $\{x \mid -7 < x < -2\}$
13. set notation: $\{x \mid -\infty < x \le 6\}$ **15.** set notation: $\{x \mid -\infty < x < \infty\}$
17. inequality: $-3 < x \le 7$, interval notation: $(-3, 7]$
19. inequality: $-1 \le x < \infty$, interval notation: $[-1, \infty)$
21. (-5, 3)
23. [-1, 1]
25. [-1, 2)
27. $(-\infty, -3] \cup [3, \infty)$
29. (-3, 2]
31. $(-\infty, 2) \cup [3, 5)$ **33.** $[-4, -2) \cup (3, 7]$
35. $x < 10$ $(-\infty, 10)$ **37.** $x \le 2$ $(-\infty, 2]$
39. $p \le -2$ $(-\infty, -2]$ **41.** $x \ge -2$ $[-2, \infty)$
43. $t > -3$ $(-3, \infty)$ **45.** $x < 6$ $(-\infty, 6)$
47. $x \le -8$ $(-\infty, -8]$ **49.** $y < 1$ $(-\infty, 1)$
51. $-5 < x < 2$ $(-5, 2)$ **53.** $-6 \le x < 2$ $[-6, 2)$
55. $-8 \le x < 4$ $[-8, 4)$ **57.** $-6 < y < 6$ $(-6, 6)$
59. $\dfrac{1}{2} \le y \le \dfrac{5}{4}$ $\left[\dfrac{1}{2}, \dfrac{5}{4}\right]$ **61.** $128 \le w \le 164$
63. more than 50 dresses
65. most minutes = 1013, least minutes = 878
67. $\$21,537.69 < $ invoice price $< \$24,346.96$
69. $0.9 \, r_T \le r_R \le 1.1 \, r_T$
71. $0.85 \, L \le B \le 0.95 \, L$
73. The correction should be $[-1, 4)$.
75. You must reverse the sign.
77. a and b   **79.** a and b   **81.** c
83. **a.** $x > .83582$ **b.**
c. agree.

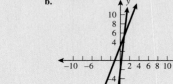

85. a. $-2 < x < 5$
c. agree.

b.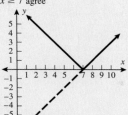

71. $x \geq 7$ agree

73. Agree

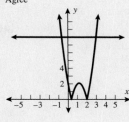

OOPS! It will take about 2.5 hours for the three people working together to pressure wash the house. Jake will profit $50.80 if two friends help him.

Tying It All Together You must score 82.6% or higher on the final exam to earn an A. If you have the two bonus points added to your course average, then you need a score of 74.6% or higher on the final to earn an A.

Section 1.6

1. $(-\infty, -2] \cup [5, \infty)$
3. $(-3, -1)$
5. $\left[-1, \dfrac{3}{2}\right]$
7. $\left(\dfrac{1}{3}, \dfrac{1}{2}\right)$
9. $\left(-\infty, -\dfrac{1}{2}\right] \cup \left[3, \infty\right)$
11. $(-\infty, -1-\sqrt{5}] \cup [-1+\sqrt{5}, \infty)$
13. $(-\infty, 0] \cup [3, \infty)$
15. $[0, 2]$
17. $(-\infty, -3) \cup (3, \infty)$
19. $\mathbb{R}$ (consistent)
21. no real solutions
23. $(0, \infty)$
25. $(-\infty, -3) \cup (0, \infty)$
27. $(-\infty, -2) \cup [-1, 2)$
29. $\left(-\dfrac{3}{2}, 1\right)$
31. $(-\infty, -5] \cup (-2, 0]$
33. $(-2, 2)$
35. no solution
37. $\mathbb{R}$ (consistent)
39. $[-3, 3) \cup (3, \infty)$
41. $30 < x < 100$ (30–100 orders will yield a profit)
43. From years 3–5, the car is worth more than you owe. In the first 3 years you owe more than the car is worth.
45. 75 seconds
47. 20 ft $\leq$ length ≤ 30 ft
49. $\$3,342 \leq$ price per acre $\leq \$3,461$
51. You cannot divide by x.
53. You must consider the $x = -2$ as a critical point.
55. false
57. $\mathbb{R}$
59. $\mathbb{R}$
61. always true, $\mathbb{R}$
63. $(-0.8960, 1.6233)$

Section 1.7

1. $x = -3$ or $x = 3$
3. no solution
5. $t = -1$ or $t = -5$
7. $y = 3$ or $y = 5$
9. $x = 3$ or $x = -3$
11. $y = 1$ or $y = -8$
13. $x = \dfrac{2}{3}$ or $x = -\dfrac{80}{21}$
15. $x = \dfrac{47}{14}$ or $x = -\dfrac{23}{14}$
17. $x = \pm\sqrt{3}$ and $x = \pm\sqrt{5}$
19. $x = \pm 2$
21. $x = a + b$ and $x = a - b$
23. $|x - 5| = 3$
25. $|x - c| = 2$
27. $-7 < x < 7$ or $(-7, 7)$
29. $y \geq 5$ or $y \leq -5$, $(-\infty, -5] \cup [5, \infty)$
31. $-10 < x < 4$, $(-10, 4)$
33. $x < 2$ or $x > 6$, $(-\infty, 2) \cup (6, \infty)$
35. $3 \leq x \leq 5$, $[3, 5]$
37. $\mathbb{R}$ (consistent)
39. $\mathbb{R}$ (consistent)
41. $\dfrac{1}{4} < x < \dfrac{3}{4}$, $\left(\dfrac{1}{4}, \dfrac{3}{4}\right)$
43. $-2.769 < x < -1.385$, $(-2.769, -1.385)$
45. $-3 \leq x \leq 3$, $[-3, 3]$
47. $|x - 2| < 7$
49. $\left|x - \dfrac{3}{2}\right| \geq \dfrac{1}{2}$
51. $|x - a| \leq 2$
53. $|T - 83| \leq 15$
55. $d = 4$ (tie) $d < 4$ (win)
57. The mistake was that $x - 3 = -7$ was not considered.
59. The 5 should have been subtracted from all three expressions.
61. $a - b < x < a + b$
63. $\mathbb{R}$
65. true
67. false
69. no solution

Review Exercises

1. $x = \dfrac{16}{7}$
3. $p = -\dfrac{8}{25}$
5. $x = 27$
7. $y = -\dfrac{17}{5}$
9. $b = \dfrac{6}{7}$
11. $x = -\dfrac{6}{17}$
13. $x = \dfrac{6 \pm \sqrt{39}}{3}$
15. $t = -\dfrac{34}{5}$
17. $x = -\dfrac{1}{2}$
19. $x = \dfrac{29}{17}$
21. $x = 8 - 7y$
23. 4.8 miles
25. 144
27. 3 inches $\times$ 7 inches
29. $\$20,000$ at 8% and $\$5,000$ at 20%
31. 60 ml of 5% and 90 ml of 10%
33. 91.5 (or 92)
35. $b = 7$ and $b = -3$
37. $x = 0$ and $x = 2$
39. $q = \pm 13$
41. $x = 2 \pm 4i$
43. $x = 6$ and $x = -2$
45. $x = \dfrac{1 \pm \sqrt{33}}{2}$
47. $t = \dfrac{7}{3}$ and $t = -1$
49. $f = \dfrac{1 \pm \sqrt{337}}{48}$
51. $q = \dfrac{3 \pm \sqrt{69}}{10}$
53. $x = \dfrac{5}{2}$ and $x = -1$
55. $x = \dfrac{2}{7}$ and $x = -3$
57. $r = \sqrt{\dfrac{S}{\pi h}}$
59. $v = \dfrac{h}{t} + 16t$
61. base $= 4$ ft, height $= 1$ ft
63. $x = 6$
65. $x = 125$
67. no solution
69. $x \approx -0.6$
71. $x \approx 5.85$
73. $x \approx -2.303$
75. $x = 2$, $x = 3$
77. $x = \dfrac{5}{4}$ and $x = \dfrac{3}{4}$
79. $x = -\dfrac{125}{8}, 1$
81. $x = \pm 2$ and $x = \pm 3i$

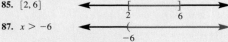

83. $(-\infty, -4]$
85. $[2, 6]$
87. $x > -6$
89. $-3 \leq x \leq 7$
91. $x \geq -4$, $[-4, \infty)$
93. $(4, \infty)$
95. $[8, 12]$
97. $x < \dfrac{5}{3}$
99. $x > -\dfrac{3}{2}$
101. $4 < x \leq 9$
103. $3 \leq x \leq \dfrac{7}{2}$

105. 74   **107.** $[-6, 6]$

109. $(-\infty, 0] \cup [4, \infty)$ **111.** $(-\infty, 0) \cup (7, \infty)$

113. $(0, 3)$ **115.** $(-\infty, -6] \cup [9, \infty)$

117. no solution   **119.** $x = 1.7$ and $x = 0.9667$

121. $(-4, 4)$ **123.** $(-\infty, -11) \cup (3, \infty)$

125. $(-\infty, -3) \cup (3, \infty)$ **127.** $\mathbb{R}$

129. $75 \leq T \leq 95$

Practice Test

1. $z = 4$ **3.** $x = \dfrac{8}{3}$ and $x = -\dfrac{1}{2}$ **5.** $y = 1$

7. $-2.24 \leq x \leq -1$ **9.** $(-\infty, \dfrac{17}{6}] \cup [\dfrac{23}{6}, \infty)$ **11.** 1000 feet

13. $627 \leq$ minutes ≤ 722

Chapter 2

Section 2.1

1. $(4, 2)$ **3.** $(-3, 0)$ **5.** $(0, -3)$

7.

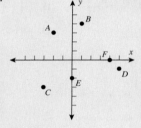

9. The line being described is $x = -3$.

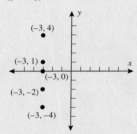

11. $d = 4$, $(3, 3)$ **13.** $d = 4\sqrt{2}$, $(1, 2)$

15. $d = 3\sqrt{10}$, $\left(-\dfrac{17}{2}, \dfrac{7}{2}\right)$ **17.** $d = 5$, $\left(\dfrac{3}{2}, \dfrac{11}{6}\right)$

19. $d = \dfrac{\sqrt{4049}}{60}$, $\left(-\dfrac{5}{24}, \dfrac{1}{15}\right)$ **21.** $d = 3.9$, $(0.3, 3.95)$

23. $d = \sqrt{1993.01} \approx 44.643$, $(1.05, -1.2)$

25. The perimeter of the triangle rounded to two decimal places is: 21.84.

27. right triangle   **29.** isosceles

31. 128.06 miles   **33.** distance ≈ 268 miles

35. midpoint $= (2003, 330)$

37. The values are misplaced. The correct distance would be $d = \sqrt{58}$.

39. The values were not used in the correct position. The correct midpoint would be: $(2, \dfrac{13}{2})$.

41. true

43. true

45. The distance is: $d = \sqrt{(b-a)^2 + (a-b)^2} = \sqrt{2(a-b)^2}$
$= \sqrt{2} \, |a - b|$. The midpoint is: $m = \left(\dfrac{a+b}{2}, \dfrac{b+a}{2}\right)$.

47. $\sqrt{\left(x_1 - \dfrac{x_1 + x_2}{2}\right)^2 + \left(y_1 - \dfrac{y_1 + y_2}{2}\right)^2}$

$\sqrt{\left(\dfrac{2x_1 - x_1 - x_2}{2}\right)^2 + \left(\dfrac{2y_1 - y_1 - y_2}{2}\right)^2}$

$\sqrt{\left(\dfrac{x_1 - x_2}{2}\right)^2 + \left(\dfrac{y_1 - y_2}{2}\right)^2}$

$\dfrac{1}{2}\sqrt{(x_1 - x_2)^2 + (y_1 - y_2)^2}$

Using (x_2, y_2) with the midpoint yields the same result.

49. The distance is $d \approx 6.357$.

51. The distance is $d \approx 3.111$.

Section 2.2

1. a. yes   **b.** no   **3. a.** yes   **b.** no

5.

x	y	(x, y)
-2	0	$(-2, 0)$
0	2	$(0, 2)$
1	3	$(1, 3)$

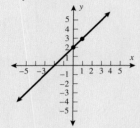

7.

x	y	(x, y)
-1	2	$(-1, 2)$
0	0	$(0, 0)$
$\dfrac{1}{2}$	$-\dfrac{1}{4}$	$\left(\dfrac{1}{2}, -\dfrac{1}{4}\right)$
1	0	$(1, 0)$
2	2	$(2, 2)$

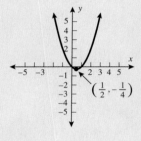

$\left(\dfrac{1}{2}, -\dfrac{1}{4}\right)$

9.

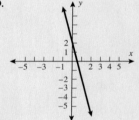

11.

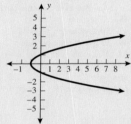

13.

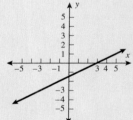

15. d

17. a

19. b

21. $(-1, -3)$

23. $(-7, 10)$

25. $(-3, -2)$, $(3, 2)$, $(-3, 2)$

27. x-axis

29. origin

31. x-axis

33. symmetric to x-axis, y-axis, and origin

35. symmetric to the y-axis

37. symmetric to y-axis

39. symmetric to origin

41.

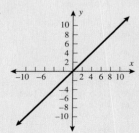

43.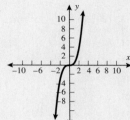

45.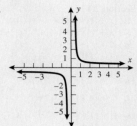

1. slope $= m = 3$

3. slope $= m = -2$

5. slope $= m = -\dfrac{19}{10}$

7. slope $= m \approx 2.379$

9. x-intercept $(0.5, 0)$, y-intercept $(0, -1)$, slope $= m = 2$ rising

11. x-intercept $(1, 0)$, y-intercept $(0, 1)$, slope $= m = -1$ falling

13. x-intercept none, y-intercept $(0, 1)$, slope $= m = 0$ horizontal

15. x-intercept $\left(\dfrac{3}{2}, 0\right)$

    y-intercept $(0, -3)$

    slope $= 2$

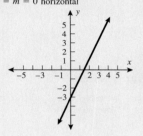

47.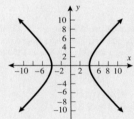

49.

17. x-intercept $(4, 0)$

    y-intercept $(0, 2)$

    slope $= -\dfrac{1}{2}$

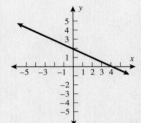

51.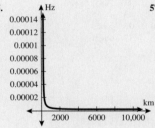

53.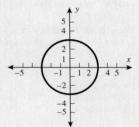

19. x-intercept $(2, 0)$

    y-intercept $\left(0, -\dfrac{4}{3}\right)$

    slope $= \dfrac{2}{3}$

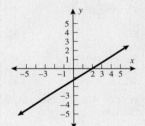

55.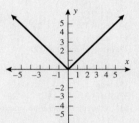

57.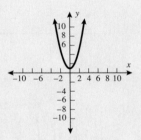

21. x-intercept $(-2, 0)$

    y-intercept $(0, -2)$

    slope $= -1$

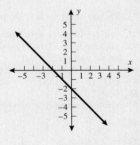

59. You are checking to see if the function is symmetric to the y-axis. Thus the substitution shown is incorrect. The correct substitution would be plugging in $-x$ for x into the function, not $-y$ for y.

61. false

63. true

65. Symmetric with respect to y-axis.

67. Symmetric with respect to x-axis, y-axis, and origin.

23. x-intercept $(-1, 0)$

    y-intercept none

    slope $=$ undefined

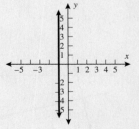

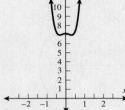

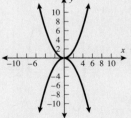

25. *x*-intercept none
y-intercept (0, 1.5)
slope = 0

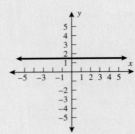

27. *x*-intercept, $\left(-\frac{7}{2}, 0\right)$

y-intercept none
slope = undefined

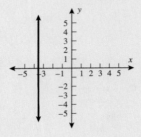

29. $y = 2x + 3$

31. $y = -\frac{1}{3}x$

33. $y = 2$

35. $x = \frac{3}{2}$

37. $y = 5x + 2$

39. $y = -3x - 4$

41. $y = \frac{3}{4}x - \frac{7}{4}$

43. $y = 4$

45. $x = -1$

47. $y = \frac{3}{5}x + \frac{1}{5}$

49. $y = -3x + 1$

51. $y = \frac{3}{2}x$

53. $x = 3$

55. $y = 7$

57. $y = \frac{6}{5}x + 6$

59. $y = 2x + 7$

61. $y = \frac{3}{2}x$

63. $y = 5$

65. $y = 2$

67. $y = \frac{3}{2}x - 4$

69. $C(h) = 1200 + 25h$ A 32-hour job will cost $2,000.

71. $375

73. $F = \frac{9}{5}C + 32$, $-40\,°C = -40\,°F$

75. The rate of change in inches per year is $\frac{1}{20}$.

77. 0.06 ounces per year. In 2040 we expect a baby to weigh 6 lb 12.4 oz.

79. The correction that needs to be made is that for the *x* intercept $y = 0$ and for the *y*-intercept, $x = 0$.

81. Values for the numerator and denominator reversed.

83. true

85. false

87. The line perpendicular is vertical and has undefined slope.

89. $y = -\frac{A}{B}x + 1$

91. perpendicular

93. perpendicular

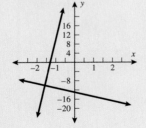

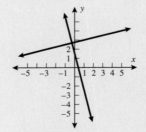

Section 2.4

1. $(x + 3)^2 + (y + 4)^2 = 100$

3. $(x - 5)^2 + (y - 7)^2 = 81$

5. $(x + 11)^2 + (y - 12)^2 = 169$

7. $x^2 + y^2 = 4$

9. $x^2 + y^2 = 2$

11. $(x - 5)^2 + (y + 3)^2 = 12$

13. $\left(x - \frac{2}{3}\right)^2 + \left(y + \frac{3}{5}\right)^2 = \frac{1}{16}$

15. $(x - 1.3)^2 + (y - 2.7)^2 = 10.24$

17. $C = (1, 3)$ $r = 5$

19. $C = (2, -5)$ $r = 7$

21. $C = (4, 9)$ $r = 2\sqrt{5}$

23. $C = \left(\frac{2}{5}, \frac{1}{7}\right)$ $r = \frac{2}{3}$

25. $C = (1.5, -2.7)$ $r = 1.3$

27. $C = (0, 0)$ $r = 5\sqrt{2}$

29. $C = (-3, -4)$ $r = 10$

31. $C = (5, 7)$ $r = 9$

33. $C = (5, -3)$ $r = 2\sqrt{3}$

35. $C = \left(\frac{1}{2}, -\frac{1}{2}\right)$ $r = \frac{1}{2}$

37. $C = (1.3, 2.7)$ $r = 3.2$

39. $(x + 1)^2 + (y + 2)^2 = 8$

41. $(x + 2)^2 + (y - 3)^2 = 41$

43. no

45. $x^2 + y^2 = 2500$

47. $x^2 + y^2 = 2,250,000$

49. Everything in this problem is correct.

51. The radius of a circle can't be imaginary.

53. true

55. true

57. the point $(-5, 3)$

59. $(x - 3)^2 + (y + 2)^2 = 20$

61. $4c = a^2 + b^2$

63. no graph (because no solution)

65. the point $(-5, 3)$

OOPS!   Alan's roof has a slope of $\frac{8}{20}$.

Tying It All Together   The coordinates of the patrol aircraft are $25\sqrt{2}$ miles east and $25\sqrt{2}$ miles north of the carrier. The equation of the outer radar coverage is $(x - 25\sqrt{2})^2 + (y - 25\sqrt{2})^2 = 200^2$. Yes, a ship 90 miles east and 150 miles south of the carrier would be visible on radar.

Review Exercises

1. quadrant II

3. quadrant III

5. $d = \sqrt{45}$

7. $d = \sqrt{205}$

9. $\left(\frac{5}{2}, 6\right)$

11. (3.85, 5.3)

13. $d \approx 52.20$

15. *y*-axis

17. origin

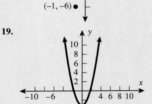

19.

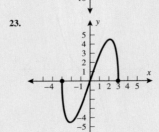

21.

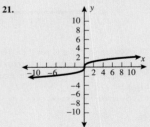

23.

25.

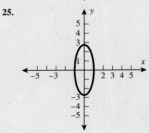

684

27. x-intercept $\left(\dfrac{5}{4}, 0\right)$

y-intercept $(0, -5)$

$m = 4$

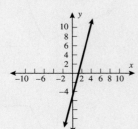

29. x-intercept $(4, 0)$

y-intercept $(0, 4)$

$m = -1$

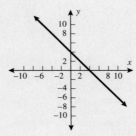

13. $y = 2x + 2$

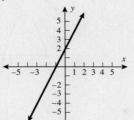

15. $(x - 6)^2 + (y + 7)^2 = 64$

17. $(x - 4)^2 + (y - 9)^2 = 20$

19. No real solutions.

31. x-intercepts: None

y-intercept $(0, 2)$

Horizontal Line

$m = 0$

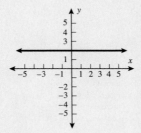

33. $y = 4x - 3$

35. $x = -3$

37. $y = -2x - 2$

39. $y = 6$

41. $y = \dfrac{5}{6}x + \dfrac{4}{3}$

43. $y = -2x - 1$

45. $y = \dfrac{2}{3}x + \dfrac{1}{3}$

47. $y = -\dfrac{3}{4}x + \dfrac{31}{16}$

49. $y = 1.2x + 100$

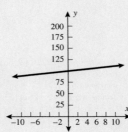

51. $(x + 2)^2 + (y - 3)^2 = 36$

53. $\left(x - \dfrac{3}{4}\right)^2 + \left(y - \dfrac{5}{2}\right)^2 = \dfrac{4}{25}$

55. $C(-2, -3)$ $r = 9$

57. $C\left(-\dfrac{3}{4}, \dfrac{1}{2}\right)$ $r = \dfrac{2}{3}$

59. Not a circle

61. $C\left(\dfrac{1}{3}, -\dfrac{2}{3}\right)$ $r = 3$

63. $(x-2)^2 + (y-7)^2 = 2$

Practice Test

1. $d = \sqrt{82}$

3. $d = \sqrt{29}$ midpoint $= \left(\dfrac{1}{2}, 5\right)$

5. $(3, 1)$ and $(3, 9)$

7. $(-1, -1)\ (1, 1)\ (-1, 1)$

9. $y = x + 5$

11. $y = -2x + 3$

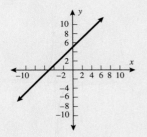

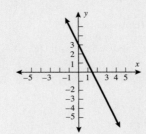

Chapter 3

Section 3.1

1. Yes, it is a function.

3. No, not a function.

5. Yes, it is a function.

7. No, not a function.

9. No, not a function.

11. Yes, it is a function.

13. No, not a function.

15. No, not a function.

17. Yes, it is a function.

19. No, not a function.

21. Yes, it is a function.

23. No, not a function.

25. $f(-2) = -7$

27. $g(1) = 6$

29. $f(-2) + g(1) = -1$

31. $2f(-1) - 2g(-3) = -33$

33. $\dfrac{f(-2)}{g(1)} = -\dfrac{7}{6}$

35. $\dfrac{f(0) - f(-2)}{g(1)} = \dfrac{2}{3}$

37. $f(x + 1) - f(x - 1) = 4$

39. $g(x + a) - f(x + a) = 8 - x - a$

41. $\dfrac{f(x + h) - f(x)}{h} = 2$

43. $\dfrac{g(t + h) - g(t)}{h} = 1$

45. $\dfrac{f(-2 + h) - f(-2)}{h} = 2$

47. $\dfrac{g(1 + h) - g(1)}{h} = 1$

49. domain: all $\mathbb{R}$, interval notation: $(-\infty, \infty)$

51. domain: all $\mathbb{R}$, interval notation: $(-\infty, \infty)$

53. domain: all $\mathbb{R}$, except 5, interval notation: $(-\infty, 5)\cup(5, \infty)$

55. domain: all $\mathbb{R}$, except -2 and 2, interval notation: $(-\infty, -2)\cup(-2, 2)\cup(2, \infty)$

57. domain: all $\mathbb{R}$, interval notation: $(-\infty, \infty)$

59. domain: $(-\infty, 7]$

61. domain: $\left[-\dfrac{5}{2}, \infty\right)$

63. domain: $(-\infty, -2]\cup[2, \infty)$

65. domain: $(3, \infty)$

67. $y = 45x$; domain: $(75, \infty)$

69. $T(12) = 90\ °F$ $T(6) = 64.8\ °F$

71. $P(x) = 10 + \sqrt{400,000 - 100x}$

when $x = 10$, $P(10) = \$641.66$

when $x = 100$, $P(100) = \$634.50$

73. $V = x(10 - 2x)^2$ domain: $0 < x < 5$

75. False, you must apply the vertical line test instead of the horizontal line test. Applying the vertical line test would show that the graph given is actually a function.

77. $f(x + 1) \neq f(x) + f(1)$

Given: $f(x) = x^2 + x$

$f(x + 1) = (x + 1)^2 - (x + 1)$

$= x^2 + 2x + 1 - x - 1$

$= x^2 + x$

79. $G(-1 + h) \neq G(-1) + G(h)$

Correct answer is $h - 2$.

81. false

83. false

85. $A = 2$

87. $C = 3$ and $D = 2$

89. Warmest at noon: 90 °F. Outside the interval [6, 18] the temperatures are too low.

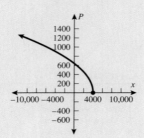

91. lowest price $10, highest $642.46 agrees.

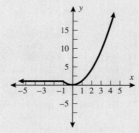

Section 3.2

1. neither

3. even

5. odd

7. neither

9. odd

11. even

13. even

15. neither

17. neither

19. neither

21. neither

23. neither

25. domain: $(-\infty, \infty)$
range: $(-\infty, 2]$
increasing: $(-\infty, 2)$
constant: $(2, \infty)$

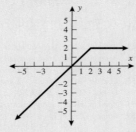

27. domain: $(-\infty, \infty)$
range: $[0, \infty)$
increasing: $(0, \infty)$
decreasing: $(-1, 0)$
constant: $(-\infty, -1)$

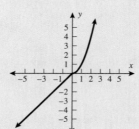

29. domain: $(-\infty, \infty)$
range: $(-\infty, \infty)$
increasing: $(-\infty, \infty)$

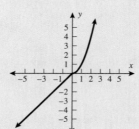

31. domain: $(-\infty, \infty)$
range: $[1, \infty)$
increasing: $(1, \infty)$
decreasing: $(-\infty, 1)$

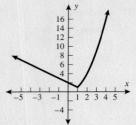

33. domain: $(-\infty, \infty)$
range: $[-1, 3]$
increasing: $(-1, 3)$
constant: $(-\infty, -1) \cup (3, \infty)$

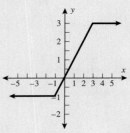

35. domain: $(-\infty, \infty)$
range: $[1, 4]$
increasing: $(1, 2)$
constant: $(-\infty, 1) \cup (2, \infty)$

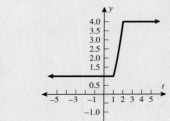

37. domain: $(-\infty, -2) \cup (-2, \infty)$
range: $(-\infty, \infty)$
increasing: $(-2, 1)$
decreasing: $(-\infty, -2) \cup (1, \infty)$

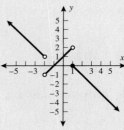

39. domain: $(-\infty, \infty)$
range: $[0, \infty)$
increasing: $(0, \infty)$
constant: $(-\infty, 0)$

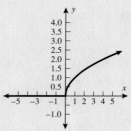

41. domain: $(-\infty, \infty)$
range: $(-\infty, \infty)$
decreasing: $(-\infty, 0) \cup (0, \infty)$

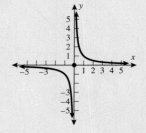

43. domain: $(-\infty, 1)\cup(1, \infty)$
range: $(-\infty, -1)\cup(-1, \infty)$
increasing: $(-1, 1)$
decreasing: $(-\infty, -1)\cup(1, \infty)$

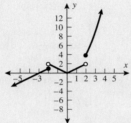

45. domain: $(-\infty, \infty)$
range: $(-\infty, 2)\cup[4, \infty)$
increasing: $(-\infty, -2)\cup(0, 2)\cup(2, \infty)$
decreasing: $(-2, 0)$

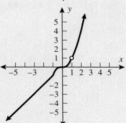

47. domain: $(-\infty, 1)\cup(1, \infty)$
range: $(-\infty, 1)\cup(1, \infty)$
increasing: $(-\infty, 1)\cup(1, \infty)$

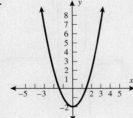

49. $C(x) = \begin{cases} 10x & x \le 50 \\ 9x & x < 50 \le 100 \\ 8x & x > 100 \end{cases}$

51. $C(x) = \begin{cases} 250x & x \le 10 \\ 175x + 750 & x > 10 \end{cases}$

53. $C(x) = \begin{cases} 1000 + 35x & x \le 100 \\ 2000 + 25x & x > 100 \end{cases}$

55. $R(x) = \begin{cases} 50,000 + 3x & x \le 100,000 \\ 4x - 50,000 & x > 100,000 \end{cases}$

57. $P(x) = 65x - 800$

59. The domain is incorrect. It should be $(-\infty, 0)\cup(0, \infty)$. The range is also incorrect and should be $(0, \infty)$.

61. $C(x) = \begin{cases} 15 & x \le 30 \\ 15 + 1(x - 30) & x > 30 \end{cases}$

63. true   **65.** false   **67.** yes, if $a = 2b$

69. The trigonometric function $\sin x$ is an odd function.

71. The trigonometric function $\tan x$ is an odd function.

Section 3.3

1. l   **3.** a   **5.** b   **7.** i   **9.** c   **11.** g

13. $y = |x| + 3$ **15.** $y = |-x|$ **17.** $y = 3|x|$

19. $y = x^3 - 4$ **21.** $y = (x + 1)^3 + 3$ **23.** $y = (-x)^3$

25. $y = -(x - 1)^2 + 2$ **27.** $y = \sqrt{-x - 1} - 2$

29. $y = \dfrac{-1}{x - 2} + 5$ **31.** $y = -x^5 - x + 2$

33.

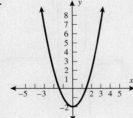

35.

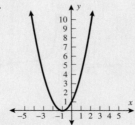

37.

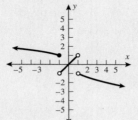

39.

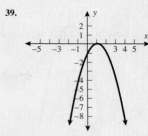

41.

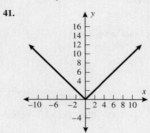

43.

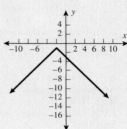

45.

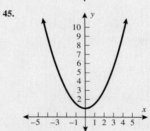

47.

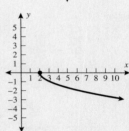

49.

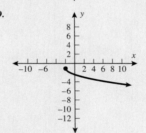

51.

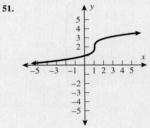

53.

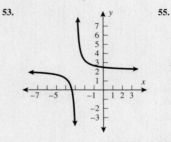

55.

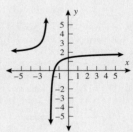

57.

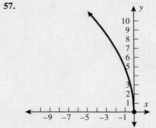

59. $f(x) = (x - 3)^2 + 2$

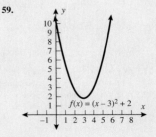

61. $f(x) = -(x+1)^2 + 1$

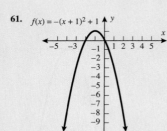

63.

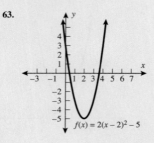

$f(x) = 2(x-2)^2 - 5$

65. $S(x) = 10x$ and $S(x) = 10x + 50$

67. $T(x) = 0.33(1.25x - 6{,}500)$

69. The following would be correct if it is shifted to the right 3.

71. $|3 - x| = |x - 3|$ Therefore, shift to the *right* 3 units.

73. true

75. true

77. $(a + 3, b + 2)$

79. Any part of the graph of $f(x)$ that is below the x-axis is reflected above it for $|f(x)|$.

81. If $0 < a < 1$, you have a vertical shrink. If $a > 1$, the graph is a vertical expansion.

Section 3.4

1. $f + g = x + 2$
$f - g = 3x$ $\Big\}$ domain: All real numbers
$f \cdot g = -2x^2 + x + 1$
$\dfrac{f}{g} = \dfrac{2x + 1}{1 - x}$ domain: $(-\infty, 1) \cup (1, \infty)$

3. $f + g = 3x^2 - x - 4$
$f - g = x^2 - x + 4$ $\Big\}$ domain: All real numbers
$f \cdot g = 2x^4 - x^3 - 8x^2 + 4x$
$\dfrac{f}{g} = \dfrac{2x^2 - x}{x^2 - 4}$ domain: $(-\infty, -2) \cup (-2, 2) \cup (2, \infty)$

5. $f + g = \dfrac{1 + x^2}{x}$
$f - g = \dfrac{1 - x^2}{x}$ $\Bigg\}$ domain: $(-\infty, 0) \cup (0, \infty)$
$f \cdot g = 1$
$\dfrac{f}{g} = \dfrac{\frac{1}{x}}{x} = \dfrac{1}{x^2}$

7. $f + g = 3\sqrt{x}$
$f - g = -\sqrt{x}$ $\Big\}$ domain: $[0, \infty)$
$f \cdot g = 2x$
$\dfrac{f}{g} = \dfrac{1}{2}$ domain: $(0, \infty)$

9. $f + g = \sqrt{4 - x} + \sqrt{x + 3}$
$f - g = \sqrt{4 - x} - \sqrt{x + 3}$ $\Big\}$ domain: $[-3, 4]$
$f \cdot g = \sqrt{4 - x}\sqrt{x + 3}$
$\dfrac{f}{g} = \dfrac{\sqrt{4 - x}}{\sqrt{x + 3}} = \dfrac{\sqrt{4 - x}\sqrt{x + 3}}{x + 3}$ domain: $(-3, 4]$

11. $f \circ g = 2x^2 - 5$ domain: $(-\infty, \infty)$
$g \circ f = 4x^2 + 4x - 2$ domain: $(-\infty, \infty)$

13. $f \circ g = \dfrac{1}{x + 1}$ domain: $(-\infty, -1) \cup (-1, \infty)$
$g \circ f = \dfrac{1}{x - 1} + 2$ domain: $(-\infty, 1) \cup (1, \infty)$

15. $f \circ g = \dfrac{1}{|x - 1|}$ domain: $(-\infty, 1) \cup (1, \infty)$
$g \circ f = \dfrac{1}{|x| - 1}$ domain: $(-\infty, -1) \cup (-1, 1) \cup (1, \infty)$

17. $f \circ g = \sqrt{x + 4}$ domain: $[-4, \infty)$
$g \circ f = \sqrt{x - 1} + 5$ domain: $[1, \infty)$

19. $f \circ g = x$ domain: $(-\infty, \infty)$
$g \circ f = x$ domain: $(-\infty, \infty)$

21. $(f + g)(2) = 15$

23. $(f \cdot g)(4) = 26\sqrt{3}$

25. $f(g(2)) = f(1) = 11$

27. not possible

29. $f(g(1)) = \dfrac{1}{3}$ $g(f(2)) = 2$

31. not possible

33. $f(g(1)) = \dfrac{1}{3}$ $g(f(2)) = 4$

41. $f(x) = 2x^2 + 5x$ $g(x) = 3x - 1$

43. $f(x) = \dfrac{2}{|x|}$ $g(x) = x - 3$

45. $f(x) = \dfrac{3}{\sqrt{x} - 2}$ $g(x) = x + 1$

47. $F = \dfrac{9}{5}(K - 273.15) + 32$

49. $A(x) = \left(\dfrac{x}{4}\right)^2$ x is the number of linear feet of fence.
$A(100) = 625$ square feet, $A(200) = 2500$ square feet

51. a. $C(p) = 62{,}000 - 20p$
b. $R(p) = 600{,}000 - 200p$
c. $P(p) = R(p) - C(p) = 538{,}000 - 180p$

53. area $= 150^2 \pi t$

55. $d(h) = \sqrt{h^2 + 4}$

57. domain: $x \neq -2$

59. The operation is composition, *not* multiplication.

61. false

63. true

65. $g \circ f = \dfrac{1}{x}$ $x \neq 0, a$

67. $g \circ f = x$ $x \geq -a$

69.

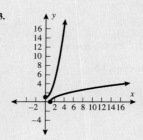

Section 3.5

1. Not one-to-one.
3. Yes, one-to-one.
5. Yes, one-to-one.
7. Not a function, so it can't be one-to-one.
9. Not one-to-one.
11. Not one-to-one.
13. Yes, one-to-one.
15. Not one-to-one.
17. Not one-to-one.
19. Yes, one-to-one.

21.

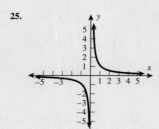

23.

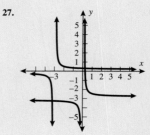

25.

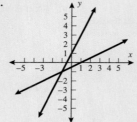

27.

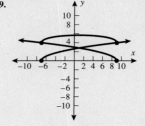

29.

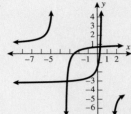

31.

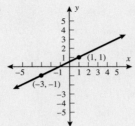

79. no

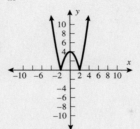

81. no

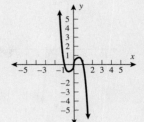

33.

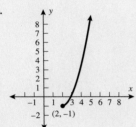

35.

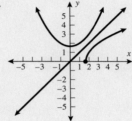

83.

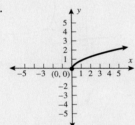

No, the functions are not inverses of each other. Had we restricted the domain of the parabola to $x > 0$, then they would be inverses.

37.

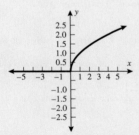

39. $f^{-1}(x) = x + 1$

41. $f^{-1}(x) = \dfrac{(x - 2)}{-3}$

43. $f^{-1}(x) = (x - 1)^{1/3}$

45. $f^{-1}(x) = x^2 + 3$

47. $f^{-1}(x) = \sqrt{x + 1}$

49. $f^{-1}(x) = \sqrt{x + 3} - 2$

51. $f^{-1}(x) = \dfrac{2}{x}$

OOPS! $y = 4000 + \sqrt{14{,}000{,}000 - 200x}$ and
$y = 4000 - \sqrt{14{,}000{,}000 - 200x}$

Tying It All Together

$$I(x) = \begin{cases} 2000 + 0.06x & 0 \le x \le 1{,}000{,}000 \\ 0.08x - 18{,}000 & x > 1{,}000{,}000 \end{cases}$$

$$P(x) = \begin{cases} 2100 + 0.06x & 0 \le x \le 1{,}200{,}000 \\ 0.09x - 33{,}900 & x > 1{,}200{,}000 \end{cases}$$

53. $f^{-1}(x) = 3 - \dfrac{2}{x}$

55. $f^{-1}(x) = \dfrac{5x - 1}{x + 7}$

57.

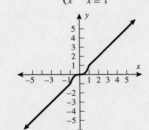

The function is NOT one-to-one.

59. $f^{-1}(x) = \begin{cases} x & x \le -1 \\ x^3 & -1 < x < 1 \\ x & x \ge 1 \end{cases}$

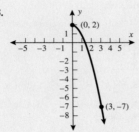

The function is one-to-one.

Review Exercises

1. yes   **3.** yes   **5.** no
7. yes   **9.** no   **11.** 5
13. -665 **15.** -2 **17.** 4
19. domain: $(-\infty, \infty)$
21. $(-\infty, -4) \cup (-4, \infty)$
23. $[4, \infty)$ **25.** $D = 18$ **27.** neither
29. odd   **31.** neither   **33.** odd
35.

35.

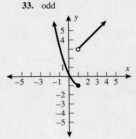

37.

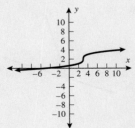

61. $f^{-1}(x) = \dfrac{5}{9}(x - 32)$

This now represents degrees Fahrenheit being turned into degrees Celsius.

63. $C^{-1}(x) = \begin{cases} \dfrac{x}{250} & x \le 2500 \\ \dfrac{x - 750}{175} & x > 2500 \end{cases}$

65. $E(x) = 5.25x$ $E^{-1}(x) = \dfrac{x}{5.25}$ tells you how many hours you will have to work to bring home x dollars.

67. No, it's not a function because it fails the vertical line test.

69. The domain of the inverse function must be given.

71. false   **73.** false   **75.** $(b, 0)$ **77.** $m \ne 0$

39. $C(x) = \begin{cases} 25 & x \le 2 \\ 25 + 10.50(x - 2) & x > 2 \end{cases}$

41.

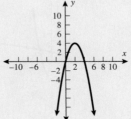

43.

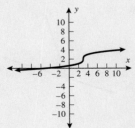

45.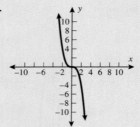

47. $y = \sqrt{x + 3}$
49. $y = \sqrt{x - 2} + 3$
51. $y = 5\sqrt{x} - 6$

53. $y = (x + 2)^2 - 12$

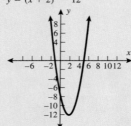

55. $g + h = -2x - 7$ domain: $(-\infty, \infty)$
$g - h = -4x - 1$ domain: $(-\infty, \infty)$
$g \cdot h = -3x^2 + 5x + 12$ domain: $(-\infty, \infty)$
$\dfrac{g}{h} = \dfrac{-3x - 4}{x - 3}$ domain: $(-\infty, 3)\cup(3, \infty)$

57. $g + h = \dfrac{1}{x^2} + \sqrt{x}$ domain: all $\mathbb{R}$ except $x = 0$, $(0, \infty)$

$g - h = \dfrac{1}{x^2} - \sqrt{x}$ domain: all $\mathbb{R}$ except $x = 0$, $(0, \infty)$

$g \cdot h = \dfrac{1}{x^{3/2}}$ domain: $(0, \infty)$

$\dfrac{g}{h} = \dfrac{1}{x^{5/2}}$ domain: $(0, \infty)$

59. $\left.\begin{array}{l} g + h = \sqrt{x - 4} + \sqrt{2x + 1} \\ g - h = \sqrt{x - 4} - \sqrt{2x + 1} \\ g \cdot h = \sqrt{x - 4}\sqrt{2x + 1} \\ \dfrac{g}{h} = \dfrac{\sqrt{x - 4}}{\sqrt{2x + 1}} \end{array}\right\}$ Domain: $[4, \infty)$ for all

61. $f \circ g = 6x - 1$ Domain: $(-\infty, \infty)$
$g \circ f = 6x - 7$ Domain: $(-\infty, \infty)$

63. $f \circ g = \dfrac{8 - 2x}{13 - 3x}$ Domain: $(-\infty, 4)\cup\left(4, \dfrac{13}{3}\right)\cup\left(\dfrac{13}{3}, \infty\right)$

$g \circ f = \dfrac{x + 3}{4x + 10}$ Domain: $(-\infty, -3)\cup\left(-3, -\dfrac{5}{2}\right)\cup\left(-\dfrac{5}{2}, \infty\right)$

65. $f(g(3)) = 857$ $g(f(-1)) = 51$

67. $f(g(3)) = \dfrac{17}{31}$ $g(f(-1)) = 1$

69. $f(x) = x - 2$ and $g(x) = 3x^2 + 4x + 7$

71. $f(x) = \dfrac{1}{\sqrt{x}}$ and $g(x) = x^2 + 7$

73. $A = 625\pi(t + 2)$ **75.** yes

77. yes   **79.** yes

81. 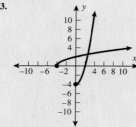   **83.**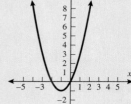

85. $f^{-1}(x) = \dfrac{x - 1}{2}$

domain f: $(-\infty, \infty)$ domain f^{-1}: $(-\infty, \infty)$
range f: $(-\infty, \infty)$ range f^{-1}: $(-\infty, \infty)$

87. $f^{-1}(x) = x^2 - 4$
domain f: $[-4, \infty)$ domain f^{-1}: $[0, \infty)$
range f: $[0, \infty)$ range: f^{-1}: $[-4, \infty)$

89. $f^{-1}(x) = \dfrac{6 - 3x}{x - 1}$

domain f: $(-\infty, -3)\cup(-3, \infty)$ domain f^{-1}: $(-\infty, 1)\cup(1, \infty)$
range f: $(-\infty, 1)\cup(1, \infty)$ range f^{-1}: $(-\infty, -3)\cup(-3, \infty)$

91. $S(x) = 22,000 + 0.08x$ $S^{-1}(x) = \dfrac{(x - 22,000)}{0.08}$.

Sales required to earn desired income.

Practice Test

1. b
3. c
5. $\left(\dfrac{f}{g}\right)(x) = \dfrac{\sqrt{x - 2}}{x^2 + 11}$ domain: $[2, \infty)$
7. $g(f(x)) = x + 9$ domain: $[2, \infty)$
9. neither
11. 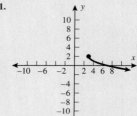   **13.**

domain: $[3, \infty)$ domain: $(-\infty, 2)\cup(2, \infty)$
range: $(-\infty, 2]$ range: $(-\infty, 3)\cup(3, \infty)$
15. $f^{-1}(x) = x^2 + 5$
The domain and range of f is domain: $[5, \infty)$, range: $[0, \infty)$.
The domain and range of f^{-1} is domain: $[0, \infty)$, range: $[5, \infty)$.
17. $f^{-1}(x) = \dfrac{-5x + 1}{-2 - x}$

The domain and range of f is domain: all $\mathbb{R}$ except $x = 5$, range: all $\mathbb{R}$ except $y = -2$.
The domain and range of f^{-1} is domain: all $\mathbb{R}$ except $x = -2$, range: all $\mathbb{R}$ except $y = 5$.
19. $x \ge 0$
21. $c(x) = (0.70)(0.60)x$, where x is the original price of the suit.
23. quadrant III, "quarter of unit circle"
25. $f(m) = \begin{cases} 15 & x \le 30 \\ -15 + x & x > 30 \end{cases}$

Chapter 4

Section 4.1

1. b   **3.** a
5. b   **7.** c
9. 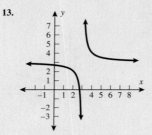   **11.**

13.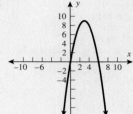

15.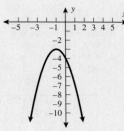

17.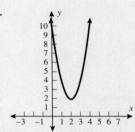

19.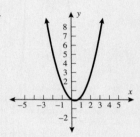

21.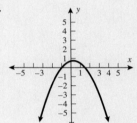

23. $(x + 3)^2 - 12$
25. $-(x + 5)^2 + 28$
27. $2(x + 2)^2 - 10$
29. $-4(x - 2)^2 + 9$
31. $\frac{1}{2}(x - 4)^2 - 5$

33.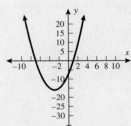

35.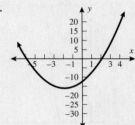

37.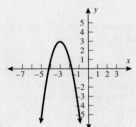

39.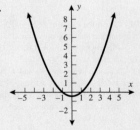

41. vertex $\left(\frac{1}{33}, \frac{494}{33}\right)$
43. vertex $\left(7, -\frac{39}{2}\right)$
45. vertex $(-75, 12.95)$
47. vertex $\left(\frac{15}{28}, \frac{829}{392}\right)$
49. $y = -2(x + 1)^2 + 4$
51. $y = -5(x - 2)^2 + 5$
53. $y = \frac{5}{9}(x + 1)^2 - 3$
55. $y = 12\left(x - \frac{1}{2}\right)^2 - \frac{3}{4}$
57. $y = \frac{5}{4}(x - 2.5)^2 - 3.5$

59. vertex $(20, 200)$
Ball will be caught on the other 40 yard line. Maximum height is 200 yards.
61. 2,083,333 square feet
63. $t = 1$ second is when rock reaches the maximum height of 116 feet. The rock hits the ground in 3.69 seconds.
65. Altitude is 26,000 feet over a horizontal distance of 18,619 feet.
67. 15 to 16 units to break even or 64 to 65 units to break even.
69. The corrections that need to be made are vertex $(-3, -1)$ x-intercepts $(-2, 0)(-4, 0)$
71. $f(x) = -(x - 1)^2 + 4$ The negative must be factored out of the x^2 and x terms.
73. true
75. false
77. $f(x) = a\left(x + \frac{b}{2a}\right)^2 + \frac{4ac - b^2}{4a}$
79. **a.** The maximum area of the rectangular fence is 62,500 square feet.
b. The maximum area of the circular fence is 79,577 square feet.
81. **a.** $(1425, 4038.25)$ **b.** $(0, -23)$
c. $(4.04, 0) (2, 845.96, 0)$ **d.** $x = 1425$

Section 4.2
1. polynomial; degree 2
3. polynomial; degree 5
5. not a polynomial
7. not a polynomial
9. not a polynomial
11. h
13. b **15.** e **17.** c
19.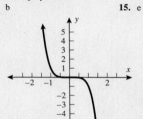
21.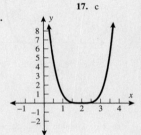

23.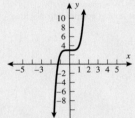
25.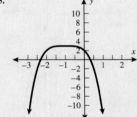

27. zero at 3, multiplicity of 1; zero at -4, multiplicity of 3
29. zero at 0, multiplicity of 2; zero at 7, multiplicity of 2; zero at -4, multiplicity of 1
31. zero at 0, multiplicity of 2; zero at 1, multiplicity of 2
33. $f(x) = 8x^3 + 6x^2 - 27x$ zero at 0; multiplicity of 1
$= x(8x^2 + 6x - 27)$ zero at $-\frac{9}{4}$; multiplicity of 1
$= x(4x + 9)(2x - 3)$ zero at $\frac{3}{2}$; multiplicity of 1
35. zero at 0; multiplicity 2 zero at -3; multiplicity 1
37. zero at 0; multiplicity 4
39. $P(x) = x(x + 3)(x - 1)(x - 2)$
41. $P(x) = x(x + 5)(x + 3)(x - 2)(x - 6)$
43. $P(x) = (2x + 1)(3x - 2)(4x - 3)$
45. $x^2 - 2x - 1$
47. $x^2(x + 2)^3$
49. $f(x) = (x + 3)^2(x - 7)^5$
51. $f(x) = x^2(x + \sqrt{3})^2(x + 1)(x - \sqrt{3})^2$

53. $f(x) = -x^2 - 6x - 9$
  a. zero at -3, multiplicity 2
  b. The graphs touches.
  c. let $x = 0$ $y = 0 - 0 - 9 = -9$
  d. Behaves like $y = -x^2$ even degree; leading coefficient is negative, graph falls without bound.
  e.

55. $f(x) = x^3 - 6x^2 + 12x - 8$
  a. zero 2; multiplicity of 3
  b. The graph crosses at $x = 2$.
  c. let $x = 0$ $y = (0 - 2)^3 = -8$
  d. Behaves like $y = x^3$ odd degree; leading coefficient is positive, graph rises to the right and falls to the left.
  e.

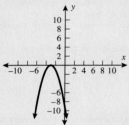

57. $f(x) = -x^4 - 3x^3$
  a. zero 0, multiplicity 3; zero -3, multiplicity 1
  b. The graph crosses at $x = 0$ and crosses at $x = -3$.
  c. $y = -x^4 - 3x^3$ let $x = 0$ then $y = 0$ y-intercept $(0, 0)$
  d. Behaves like $y = -x^4$ even degree; leading coefficient is negative.
  e.

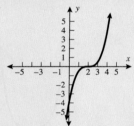

59. $f(x) = 12x^4(x - 4)(x + 1)$
  a. 0, multiplicity 4; 4 multiplicity 1; -1 multiplicity 1.
  b. The graph crosses at -1 and 4 while the graph touches at 0.
  c. let $x = 0$ $y = 0$ y-intercept $(0, 0)$
  d. Behaves like $y = x^6$ even degree with leading coefficient being positive.
  e.

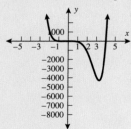

61. $f(x) = -(x + 2)^2 (x - 1)^2$
  a. zeros of -2 and 1; with both having multiplicity of 2
  b. The graph touches at each one of the zeros.
  c. let $x = 0$ $y = -4$ y-intercept $(0, -4)$
  d. Behaves like $y = -x^4$ even degree with leading coefficient being negative.
  e.

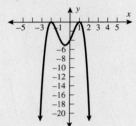

63. $f(x) = x^2(x - 2)^3(x + 3)^2$
  a. zeros of 0, 2, and -3; having multiplicities of 2, 3, and 2
  b. The graph touches at -3 and 0 but crosses at 2.
  c. let $x = 0$ $y = 0$ y-intercept $(0, 0)$
  d. Behaves like $y = x^7$ with positive leading coefficient.
  e.

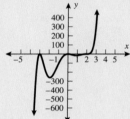

65. Sixth-degree polynomial, because there are five known turning points.

67.

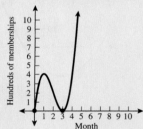

69. up

71. Given the zeros of $-2, -1, 3, 4$ the polynomial would be of degree 4, but the resulting polynomial would look like the following:
$P(x) = (x + 2)(x + 1)(x - 3)(x - 4)$

73. $f(x) = (x - 1)^2 (x + 2)^3$
  Yes, the zeros are -2 and 1. But you must remember that this is a fifth-degree polynomial. At the -2 zero the graph crosses, but it should be noted at 1 it only touches at this value. The y-intercept is also graphed incorrectly. The correct graph would be the following:

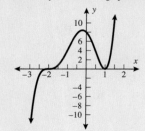

75. false   **77.** true   **79.** n

81. The zeros of the polynomial are 0, a, and $-b$.

83. no x-intercepts   **85.** $y = -x^5$

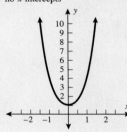

yes

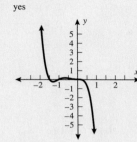

Section 4.3

1. $Q(x) = 2x - 3$ $r(x) = 0$

3. $Q(x) = 2x - 1$ $r(x) = 0$

5. $Q(x) = 3x - 28$ $r(x) = 130$

7. $Q(x) = 4x^2 + 4x + 1$ $r(x) = 0$

9. $Q(x) = 2x^2 - x - \dfrac{1}{2}$ $r(x) = \dfrac{15}{2}$

11. $Q(x) = 4x^2 - 10x - 6$ $r(x) = 0$

13. $Q(x) = -2x^2 - 3x - 9$ $r(x) = -27x^2 + 3x + 9$

15. $Q(x) = x^2 + 1$ $r(x) = 0$

17. $Q(x) = x^2 + x + \dfrac{1}{6}$ $r(x) = -\dfrac{121}{6}x + \dfrac{121}{3}$

19. $Q(x) = -3x^3 + 5.2x^2 + 3.12x - .128$ $r(x) = .9232$

21. $Q(x) = x^2 - 0.6x + 0.09$ $r(x) = 0$

23. $Q(x) = 3x + 1$ $r(x) = 0$

25. $Q(x) = 7x - 10$ $r(x) = 15$

27. $Q(x) = -x^3 + 3x - 2$ $r(x) = 0$

29. $Q(x) = x^3 - x^2 + x - 1$ $r(x) = 2$

31. $Q(x) = x^3 - 2x^2 + 4x - 8$ $r(x) = 0$

33. $Q(x) = 2x^2 - 6x + 2$ $r(x) = 0$

35. $Q(x) = 2x^3 - \dfrac{5}{3}x^2 + \dfrac{53}{9}x + \dfrac{106}{27}$ $r(x) = -\dfrac{112}{81}$

37. $Q(x) = 2x^3 + 6x^2 - 18x - 54$ $r(x) = 0$

39. $Q(x) = x^6 + x^5 + x^4 - 7x^3 - 7x^2 - 4x - 4$ $r(x) = -3$

41. $Q(x) = x^5 + \sqrt{5}x^4 - 44x^3 - 44\sqrt{5}x^2 - 69x - 69\sqrt{5}$ $r(x) = 1156$

43. $Q(x) = 2x - 7$ $r(x) = 0$

45. $Q(x) = x^2 - 9$ $r(x) = 0$

47. $Q(x) = x^4 + 2x^3 + 8x^2 + 18x + 36$ $r(x) = 71$

49. $Q(x) = x^2 + 1$ $r(x) = -24$

51. $Q(x) = x^6 + x^5 + x^4 + x^3 + x^2 + x + 1$ $r(x) = 0$

53. The width is $3x^2 + 2x + 1$ feet.

55. The trip took $x^2 + 1$ hours.

57. The correction that needs to be made is that you are adding once you have started the division. We are using long division; thus you must subtract (not add) each term.

59. The correction that needs to be made is an incorrect error with addition. Use 3, not -3.

61. true   **63.** false   **65.** yes

67. $(x - 1)(x - 3)(x + 2)$

69.

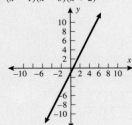

71.

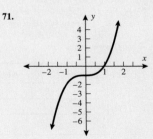

Section 4.4

1. $f(1) = 0$ **3.** $g(1) = 4$ **5.** $f(-2) = 84$

7. Yes, the given is a zero of the polynomial.

9. Yes, the given is a zero of the polynomial.

11. $P(x) = (x - 1)(x - 3)(x + 4)$

13. $P(x) = (2x - 1)(x + 3)(x - 2)$

15. $P(x) = (x - 5)(x + 3)(x - 2i)(x + 2i)$

17. $P(x) = [(x - 1) - i][(x - 1) + i]$
$[x + (1 - 2\sqrt{2})][x + (1 + 2\sqrt{2})]$

19. $P(x) = (x + 1)^2(x + 2)^2$

21.

Positive Real Zeros	Negative Real Zeros	Imaginary Zeros
1	1	2

23.

Positive Real Zeros	Negative Real Zeros	Imaginary Zeros
1	0	4

25.

Positive Real Zeros	Negative Real Zeros	Imaginary Zeros
2	1	2
0	1	4

27.

Positive Real Zeros	Negative Real Zeros	Imaginary Zeros
1	1	4

Note that $x = 0$ is also a zero.

29.

Positive Real Zeros	Negative Real Zeros	Imaginary Zeros
2	2	2
0	2	4
2	0	4
0	0	6

31.

Positive Real Zeros	Negative Real Zeros	Imaginary Zeros
4	0	0
2	0	2
0	0	4

33. $P(x) = x^4 + 3x^2 - 8x + 4$
Factors of $a_0 = 4$: ± 1, ± 2, ± 4
Factors of $a_n = 1$: ± 1
Possible rational zeros: ± 1, ± 2, ± 4

35. $P(x) = x^5 - 14x^3 + x^2 - 15x + 12$
Factors of $a_0 = 12$: ± 1, ± 2, ± 3, ± 4, ± 6, ± 12
Factors of $a_n = 1$: ± 1
Possible rational zeros: ± 1, ± 2, ± 3, ± 4, ± 6, ± 12

37. $P(x) = 2x^6 - 7x^4 + x^3 - 2x + 8$
Factors of $a_0 = 8$: ± 1, ± 2, ± 4, ± 8
Factors of $a_n = 2$: ± 1, ± 2
Possible rational zeros: ± 1, ± 2, ± 4, ± 8, $\pm \dfrac{1}{2}$

39. $P(x) = 5x^5 + 3x^4 + x^3 - x - 20$
Factors of $a_0 = -20$: ± 1, ± 2, ± 4, ± 5, ± 10, ± 20,
Factors of $a_n = 5$: ± 1, ± 5
Possible rational zeros: ± 1, ± 2, ± 4, ± 5, ± 10, ± 20, $\pm \dfrac{1}{5}$, $\pm \dfrac{2}{5}$, $\pm \dfrac{4}{5}$

41. $P(x) = x^4 + 2x^3 - 9x^2 - 2x + 8$
Factors of $a_0 = 8$: ± 1, ± 2, ± 4, ± 8
Factors of $a_n = 1$: ± 1
Possible rational zeros: ± 1, ± 2, ± 4, ± 8
Testing the zeros:
$P(1) = 1^4 + 2(1)^3 - 9(1)^2 - 2(1) + 8 = 0$
$P(-1) = 0$ $P(2) = 0$ $P(-4) = 0$

43. $P(x) = 2x^3 - 9x^2 + 10x - 3$
Factors of $a_0 = -3$: ± 1, ± 3
Factors of $a_n = 2$: ± 1, ± 2
Possible rational zeros: ± 1, ± 3, $\pm \dfrac{1}{2}$, $\pm \dfrac{3}{2}$
$P(1) = 0$ $P(3) = 0$ $P\left(\dfrac{1}{2}\right) = 0$

45. $P(x) = x^3 + 6x^2 + 11x + 6$

  a. No sign changes: 0 positive real zeros

$$P(-x) = (-x)^3 + 6(-x)^2 + 11(-x) + 6$$
$$= -x^3 + 6x^2 - 11x + 6$$

3 or 1 negative real zeros

Positive	Negative	Complex
0	3	0
0	1	2

  b. Factors of $a_0 = 6$: $\pm1, \pm2, \pm3, \pm6$

Factors of $a_n = 1$: ±1

Possible rational zeros: $\pm1, \pm2, \pm3, \pm6$

  c. $P(-1) = 0$ $P(-2) = 0$ $P(-3) = 0$

  d. $x = -1$ $x = -2$ $x = -3$

$$P(x) = (x + 1)(x + 2)(x + 3)$$

47. $P(x) = x^3 - 7x^2 - x + 7$

  a. 2 or 0 positive real zeros

$$P(-x) = (-x)^3 - 7(-x)^2 - (-x) + 7$$
$$= -x^3 - 7x^2 + x + 7$$

1 negative real zero

Positive	Negative	Complex
2	1	0
0	1	2

  b. Factors of $a_0 = 7$: $\pm1, \pm7$

Factors of $a_n = 1$: ±1

Possible rational zeros: $\pm1, \pm7$

  c. $P(1) = 0$ $P(-1) = 0$ $P(7) = 0$

  d. $P(x) = (x - 1)(x + 1)(x - 7)$

49. $P(x) = x^4 + 6x^3 + 3x^2 - 10x$

  a. 1 positive real zero

$$P(-x) = (-x)^4 + 6(-x)^3 + 3(-x)^2 - 10(-x)$$
$$= x^4 - 6x^3 + 3x^2 + 10x$$
$$= x(x^3 - 6x^2 + 3x + 10)$$

2 or 0 negative real zeros

Positive	Negative	Complex
1	2	0
1	0	2

  b. Factors of $a_0 = -10$: $\pm1, \pm2, \pm5, \pm10$

Factors of $a_n = 1$: ±1

Possible rational zeros: $\pm1, \pm2, \pm5, \pm10$

  c. $P(0) = 0$ $P(1) = 0$ $P(-2) = 0$ $P(-5) = 0$

  d. $P(x) = x(x - 1)(x + 2)(x + 5)$

51. $P(x) = x^4 - 7x^3 + 27x^2 - 47x + 26$

  a. 4, 2, or 0 positive real zeros

$$P(-x) = (-x)^4 - 7(-x)^3 + 27(-x)^2 - 47(-x) + 26$$
$$= x^4 + 7x^3 + 27x^2 + 47x + 26$$

0 negative real zero

Positive	Negative	Complex
4	0	0
2	0	2
0	0	4

  b. Factors of $a_0 = 26$: $\pm1, \pm2, \pm13, \pm26$

Factors of $a_n = 1$: ±1

Possible rational zeros: $\pm1, \pm2, \pm13, \pm26$

  c. $P(1) = 0$ $P(2) = 0$ $(x - 1)(x - 2) = x^2 - 3x + 2$

  d.
$$\begin{array}{r} x^2 - 4x + 13 \\ x^2 - 3x + 2{\overline{\smash{\big)}\,x^4 - 7x^3 + 27x^2 - 47x + 26}} \\ \underline{-(x^4 - 3x^3 + 2x^2)} \\ -4x^3 + 25x^2 - 47x \\ \underline{-(-4x^3 + 12x^2 - 8x)} \\ 13x^2 - 39x + 26 \\ \underline{-(13x^2 - 39x + 26)} \\ 0 \end{array}$$

$$P(x) = (x - 1)(x - 2)(x^2 - 4x + 13)$$
$$P(x) = (x - 1)(x - 2)(x - 2 - 3i)(x - 2 + 3i)$$

53. $P(x) = 10x^3 - 7x^2 - 4x + 1$

  a. 2 or 0 positive real zeros

$$P(-x) = 10(-x)^3 - 7(-x)^2 - 4(-x) + 1$$
$$= -10x^3 - 7x^2 + 4x + 1$$

1 negative real zero

Positive	Negative	Complex
2	1	0
0	1	2

  b. Factors of $a_0 = 1$: ±1

Factors of $a_n = 10$: $\pm1, \pm2, \pm5, \pm10$

Possible rational zeros: $\pm1, \pm\dfrac{1}{2}, \pm\dfrac{1}{5}, \pm\dfrac{1}{10}$

  c. $P(1) = 0$ $P\left(-\dfrac{1}{2}\right) = 0$ $P\left(\dfrac{1}{5}\right) = 0$

  d. $P(x) = (x - 1)\left(x + \dfrac{1}{2}\right)\left(x - \dfrac{1}{5}\right)$

$$= (x - 1)(2x + 1)(5x + 1)$$

55. $P(x) = 6x^3 + 17x^2 + x - 10$

  a. 1 positive real zero.

$$P(x) = 6(-x)^3 + 17(-x)^2 + (-x) - 10$$
$$= -6x^3 + 17x^2 - x - 10$$

2 or 0 negative real zeros.

Positive	Negative	Complex
1	2	0
1	0	2

  b. Factors of $a_0 = -10$: $\pm1, \pm2, \pm5, \pm10$

Factors of $a_n = 6$: $\pm1, \pm2, \pm3, \pm6$

Possible rational zeros: $\pm1, \pm\dfrac{1}{2}, \pm\dfrac{1}{3}, \pm\dfrac{1}{6}, \pm2, \pm\dfrac{2}{3}, \pm5,$

$\pm\dfrac{5}{2}, \pm\dfrac{5}{3}, \pm\dfrac{5}{6}, \pm10, \pm\dfrac{10}{3}$

  c. $P(-1) = 0$ $P\left(-\dfrac{5}{2}\right) = 0$ $P\left(\dfrac{2}{3}\right) = 0$

  d. $P(x) = (x + 1)\left(x - \dfrac{5}{2}\right)\left(x - \dfrac{2}{3}\right)$

$$P(x) = 6(x + 1)(2x - 5)(3x - 2)$$

57. $P(x) = x^4 - 2x^3 + 5x^2 - 8x + 4$

  a. 4, 2, or 0 positive real zeros

$$P(-x) = x^4 + 2x^3 + 5x^2 + 8x + 4$$

0 negative real zero

Positive	Negative	Complex
4	0	0
2	0	2
0	0	4

b. Factors of $a_0 = 4$: ± 1, ± 2, ± 4
   Factors of $a_n = 1$: ± 1
   Possible rational zeros: ± 1, ± 2, ± 4
c. $P(1) = 0$ $x = 1$ (multiplicity 2)
d. $P(x) = (x - 1)^2(x - 2i)(x + 2i)$

59. $P(x) = x^6 + 12x^4 + 23x^2 - 36$
a. 1 positive real zero
   $P(-x) = x^6 + 12x^4 + 23x^2 - 36$
   1 negative real zero

Positive	Negative	Complex
1	1	4

b. Factors of $a_0 = -36$: ± 1, ± 2, ± 3, ± 4, ± 6, ± 9, ± 12, ± 18, ± 36
   Factors of $a_n = 1$: ± 1
   Possible rational zeros: ± 1, ± 2, ± 3, ± 4, ± 6, ± 9, ± 12, ± 18, ± 36
c. $P(1) = 0$ $P(-1) = 0$
d. $P(x) = (x + 1)(x - 1)(x - 2i)(x + 2i)(x - 3i)(x + 3i)$

61. $P(x) = 4x^4 - 20x^3 + 37x^2 - 24x + 5$
a. 4, 2, or 0 positive real zeros
   $P(-x) = 4x^4 + 20x^3 + 37x^2 + 24x + 5$
   0 negative real zero

Positive	Negative	Complex
4	0	0
2	0	2
0	0	4

b. Factors of $a_0 = 5$: ± 1, ± 5
   Factors of $a_n = 4$: ± 1, ± 2, ± 4
   Possible rational zeros: ± 1, $\pm \dfrac{1}{2}$, $\pm \dfrac{1}{4}$, ± 5, $\pm \dfrac{5}{2}$, $\pm \dfrac{5}{4}$
c. $P\left(\dfrac{1}{2}\right) = 0$ $\dfrac{1}{2}$ has multiplicity 2.
d. $P(x) = (2x - 1)^2\big[(x - 2) - i\big]\big[(x - 2) + i\big]$

63.
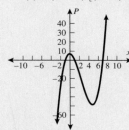

65.
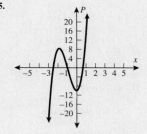

67. $R(x) = 46 - 3x^2$ $x \geq 0$
   $C(x) = 20 + 2x$
a. $P(x) = R(x) - C(x)$
         $= (46 - 3x^2) - (20 + 2x)$
         $= -3x^2 - 2x + 26$
b. Approximately 263 subscribers will break even.
69. Seventh-degree polynomial since there are six turning points.
71. The error is the fact that 1 is not a zero of the polynomial. And it is not the case that if 1 were a zero of the polynomial, then it is not necessary that -1 be a zero of the polynomial.
73. Yes, you can arrive at five negative zeros, but you must consider complex zeros also.

Positive	Negative	Complex
0	5	0
0	3	2
0	1	4

75. false
77. true
79. No, because complex zeros come in pairs.
81. No, because complex zeros come in pairs.

Section 4.5
1. domain: all $\mathbb{R}$ except $x = -3$
3. domain: all $\mathbb{R}$ except $x = -\dfrac{1}{3}$ and $x = \dfrac{1}{2}$
5. domain: all $\mathbb{R}$ except $x = -4$ and $x = 3$
7. domain: all $\mathbb{R}$
9. domain: all $\mathbb{R}$ except $x = 3$ and $x = -2$
11. HA: $y = 0$ VA: $x = -2$
13. HA: none   VA: $x = -5$
15. HA: none   VA: $x = \dfrac{1}{2}$ and $x = -\dfrac{4}{3}$
17. HA: $y = \dfrac{1}{3}$ VA: none
19. HA: $y = 1.71$ VA: $x = 0.5$ and $x = -1.5$
21. $y = x + 6$ **23.** $y = 2x + 24$ **25.** $y = 4x + \dfrac{11}{2}$
27. b   **29.** a   **31.** e

33.

35.
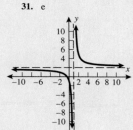

37.

39.
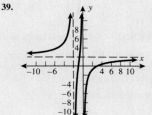

41.

43.
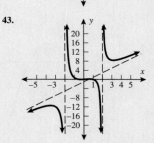

45.
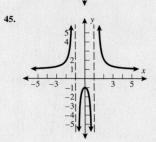

47.
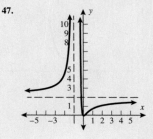

695

49.

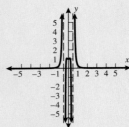

51.

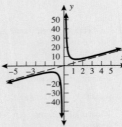

53. a. $C(1) \approx .0198$ **b.** $C(60) \approx .0324$ **c.** $C(300) \approx 0$
d. $y = 0$, after several days $C(t) \approx 0$.

55. a. $N(0) = 52$ wpm   **b.** $N(12) = 107$ wpm
c. $N(36) = 120$ wpm   **d.** 130 wpm

57. 10 ounces

59. $\dfrac{2w^2 + 1000}{w}$

61. There is a common factor in the numerator and denominator.
63. The horizontal asymptote was found incorrectly. The correct horizontal
asymptote is $y = -1$.

65. true   **67.** false
69. HA: $y = 1$ VA: $x = c$ and $x = -d$

71. $y = \dfrac{4x^2}{(x + 3)(x - 1)}$

73. yes   **75.** yes

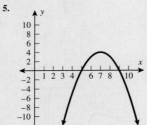

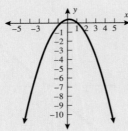

OOPS!   Car C will have the shortest braking distance.

Tying It All Together   The company should sell the game at $45 to earn the
maximum revenue of $202,500. Consumers will completely cease buying the game
when its price is $90.

Review Exercises

1. b   **3.** a

5.

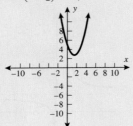

7.

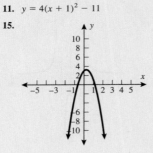

9. $y = \left(x - \dfrac{3}{2}\right)^2 - \dfrac{49}{4}$ **11.** $y = 4(x + 1)^2 - 11$

13.

15.

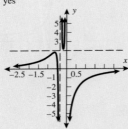

17.   vertex: $\left(\dfrac{5}{26}, \dfrac{599}{52}\right)$

19.   vertex: $\left(-\dfrac{2}{15}, \dfrac{451}{125}\right)$

21. $y = \dfrac{1}{9}(x + 2)^2 + 3$

23. $y = 5.6(x - 2.7)^2 + 3.4$

25. a. $P(x) = -2x^2 + \dfrac{35}{3}x - 14$ **c.**

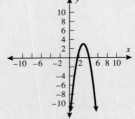

b. $x = 1.68909$ and $x = 4.1442433$
d. The range of units is $(1.689, 4.144)$
or 1,689 to 4,144.

27.   area $= -\dfrac{1}{2}x^2 + x + 4$; $x = 1$ is a maximum;

Dimensions: base $= 3$ height $= 3$

29. yes, 6   **31.** no
33. d   **35.** a

37.

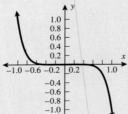

39.

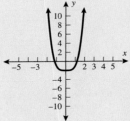

41. $x = -4$ multiplicity 2
$x = 6$ multiplicity 5
43. $x = 0$ multiplicity 1
$x = 3$ multiplicity 1
$x = -3$ multiplicity 1
$x = 2$ multiplicity 1
$x = -2$ multiplicity 1
45. $f(x) = x(x + 3)(x - 4)$
47. $f(x) = x(5x + 2)(4x - 3)$
49. $f(x) = (x + 4.2)^4(x - 3.7)^2$
51. a. 7 multiplicity 1
-2 multiplicity 1
b. The graph crosses
at $(-2, 0)$ and $(7, 0)$.
c. $(0, -14)$
d. rises right and left

e.

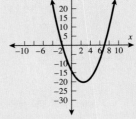

53. a. 0.8748 multiplicity 1.
b. The graph crosses
at $(0.8748, 0)$.
c. $(0, -4)$
d. falls left; rises right

e.

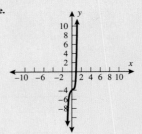

55. a.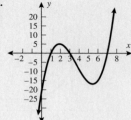

b. The real zeros occur at: $x = 1$, $x = 3$, and $x = 7$.

c. Between 1 and 3 hours or more than 7 hours is financially beneficial.

57. $Q(x) = x + 4$ $r(x) = 2$

59. $Q(x) = 2x^3 - 4x^2 - 2x - \frac{7}{2}$ $r(x) = -23$

61. $Q(x) = x^3 + 2x^2 + x - 4$ $r(x) = 0$

63. $Q(x) = x^5 - 8x^4 + 64x^3 - 512x^2 + 4096x - 32{,}768$
$r(x) = 262{,}080$

65. $Q(x) = x + 3$ $r(x) = -8 - 4x$

67. $Q(x) = x^2 - 5x + 7$ $r(x) = -15$

69. length $= 3x^3 + 2x^2 - x + 4$ $r(x) = 0$

71. $f(-2) = -207$ **73.** $g(1) = 0$

75. no   **77.** yes

79. $P(x) = x(x + 2)(x - 4)^2$ **81.** $P(x) = x^2(x + 3)(x - 2)^2$

83.

Positive	Negative	Complex
1	1	2

85.

Positive	Negative	Complex
5	2	2
5	0	4
3	2	4
3	0	6
1	2	6
1	0	8

87. The possible rational zeros are: $\pm1, \pm2, \pm3, \pm6$.

89. The possible rational zeros are: $\pm\frac{1}{2}, \pm1, \pm2, \pm4, \pm8, \pm16, \pm32, \pm64$.

91. The possible rational zeros are $\pm\frac{1}{2}, \pm1$. The rational zero is $\frac{1}{2}$.

93. The possible rational zeros are $\pm1, \pm2, \pm4, \pm8, \pm16$. The rational zeros are $1, 2, -2, 4$.

95. a.

Positive	Negative	Complex
1	0	2

b. $\pm1, \pm5$

c. -1 is a lower bound, 5 is an upper bound.

d. There are no rational zeros.

e. not possible

f.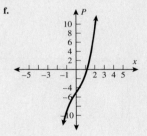

97. a.

Positive	Negative	Complex
2	0	2
0	2	2
2	2	0
0	0	4

b. $\pm1, \pm\frac{1}{2}, \pm\frac{1}{4}, \pm\frac{1}{8}$

c. $-\frac{1}{8}$ is a lower bound, 51 is an upper bound.

d. There are no rational zeros.

e. not possible

f.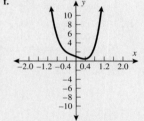

99. a.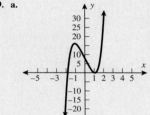

b. The turning points occur at: $(-1.07, 16.07)$ and $(1, 0)$.

c. The zeros are $(-1.824, 0)$ and $(1, 0)$.

101. VA: $x = -2$ and HA: $y = -1$

103. VA: $x = -1$, HA: none, and Oblique: $y = 4x$

105. VA: $x = -4$, HA: none, and Oblique: $y = 2x - 11$

107. 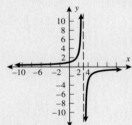   **109.**

111.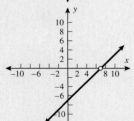

1.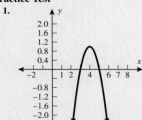

3. vertex $\left(3, \frac{1}{2}\right)$

5. $f(x) = x(x - 2)^3(x - 1)^2$

7. $Q(x) = -2x^2 - 2x - \frac{11}{2}$

$r(x) = -\frac{19}{2}x + \frac{7}{2}$

9. Yes, $x - 3$ is a factor of the polynomial.

11. Using substitution:
$P(x) = x^{21} - 2x^{18} + 5x^{12} + 7x^3 + 3x^2 + 2$
$P(-1) = (-1)^{21} - 2(-1)^{18} + 5(-1)^{12} + 7(-1)^3 + 3(-1)^2 + 2 = 0$
Since $P(-1) = 0$: Yes, -1 is a zero of the polynomial.

13. The other zeros are $x = -3i$, $x = 5$, $x = -2$.

15. $P(x) = 3x^4 - 7x^2 + 3x + 12$
Factors of $a_0 = 12$: ± 1, ± 2, ± 3, ± 4, ± 6, ± 12
Factors of $a_n = 3$: ± 1, ± 3

Possible rational zeros: ± 1, ± 2, ± 3, ± 4, ± 6, ± 12, $\pm\dfrac{1}{3}$, $\pm\dfrac{2}{3}$, $\pm\dfrac{4}{3}$

17. Given the points $(0, 300)$, $(2, 285)$, $(10, 315)$, $(52, 300)$,
You can have a polynomial of degree 3 because there are 2 turning points.

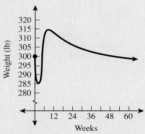

19. Given the points $(1970, 0.08)$ $(1988, 0.13)$ $(2002, 0.04)$ $(2005, 0.06)$
The lowest degree polynomial that can be represented is a third-degree polynomial.

Chapter 5

Section 5.1

1. 5.28 **3.** 0.78 **5.** 9.74
7. 24,646.66 **9.** 9.75 **11.** 1.45
13. $f(2.7) \approx 0.06$ **15.** $f\left(\dfrac{3}{4}\right) \approx 0.06$ **17.** f
19. e **21.** b
23. $f(x) = 6^x$
  y-intercept: $(0, 1)$
  horizontal asymptote: x-axis

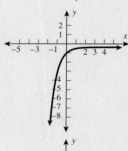

25. $f(x) = -4^{-x}$
  y-intercept: $(0, -1)$
  horizontal asymptote: x-axis

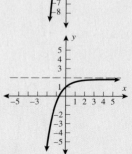

27. $f(x) = 2 - 3^{-x}$
  y-intercept: $(0, 1)$
  horizontal asymptote: $y = 2$

29. $f(x) = 1 + \left(\dfrac{1}{2}\right)^{x+2}$
  y-intercept: $(0, 1.25)$
  horizontal asymptote: $y = 1$

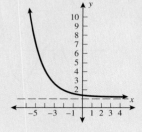

31. $f(x) = 2b^x$
  y-intercept: $(0, 2)$
  horizontal asymptote: x-axis

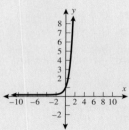

33. 10.4 million

35. $f(t) = 1500 \cdot 2^{t/5}$, where t is in years and $t = 0$ corresponds to the year the land was originally purchased. Thirty years after the initial investment the land should be worth $96,000 per acre.

37. 168 milligrams
39. \$3,031
41. \$3,448.42
43. \$13,011.03
45. $4^{-1/2} = \dfrac{1}{4^{1/2}} = \dfrac{1}{2}$
47. true
49. false
51. domain: $(-\infty, \infty)$ range: $[1, \infty)$

53.

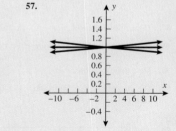

55.

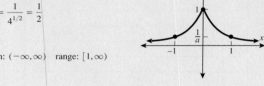

57.

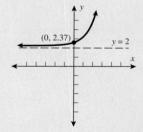

Section 5.2

1. 7.39 **3.** 3.32 **5.** 4.11
7. $(0, -1)$, $y = 0$ **9.** $(0, 2.37)$, $y = 2$

11. $(0, 2)$, $y = 0$

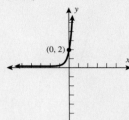

13. $(0, 1)$, $y = 0$

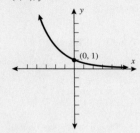

49.

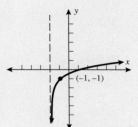

51.

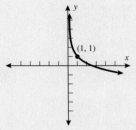

15. 805 milligrams

17. 94.4 million

19. 40,776 students

21. $4319.55

23. $13,979.42

25. 17,217 hits

27. 543 snakes

29. $A \approx 1.19$ million

31. For $t = 4$ then $A \approx 13.53$ ml. For $t = 12$ then $A \approx 0.248$ ml.

33. $r = 0$ (on axis)

35. 2.5% needs to be changed to a decimal. It should be $A = 2000e^{(0.025 \cdot 1)} \approx \$2050.63$.

37. t is still 20.
$N = 1.8$
$N_0 = .988$ million

39. false

41. true

43. $(0, a + be)$, $y = a$

45.

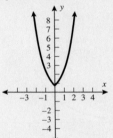

47.

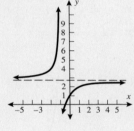

49. The graphs are nearly identical on the interval $(-1, 1.5)$.

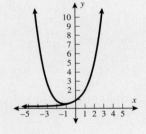

53.

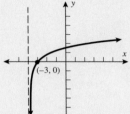

55.

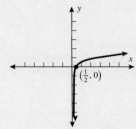

57. 60 decibels

59. 117 decibels

61. 8.5 on the Richter scale

63. 6.6 on the Richter scale

65. 3.3 pH

67. normal rainwater: 5.6 pH;   acid rain/tomato juice: 4.0 pH

69. 13,236 years old

71. 25 dB loss

73. The correction that needs to be made is:
$$2^? = 4$$
$$? = 2$$
$$\log_2 4 = 2$$

75. The correction that needs to be made is that:
$$x + 5 > 0$$
$$x > -5$$
$$\text{domain: } (-5, \infty)$$

77. false

79. true

81. domain: (a, ∞)
range: $(-\infty, \infty)$
x-intercept: $(e^b + a, 0)$

83.

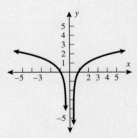

85. Symmetric with respect to $y = x$.

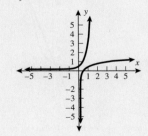

87. $(1, 0)$ and V.A. $x = 0$ are common.

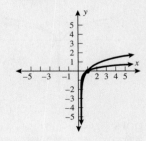

Section 5.3

1. $5^3 = 125$

3. $81^{1/4} = 3$

5. $2^{-5} = \dfrac{1}{32}$

7. $10^{-2} = 0.01$

9. $\left(\dfrac{1}{4}\right)^{-3} = 64$

11. $\log_8 512 = 3$

13. $\log 0.00001 = -5$

15. $\log_{225} 15 = \dfrac{1}{2}$

17. $\log_{2/5}\left(\dfrac{8}{125}\right) = 3$

19. $\log_{1/27} 3 = -\dfrac{1}{3}$

21. 0

23. 5

25. 7

27. -6

29. 1.46

31. 5.94

33. not possible

35. -8.11

37. domain: $(-5, \infty)$

39. domain: $(-\infty, 0) \cup (0, \infty)$

41. domain: all $\mathbb{R}$

43. b

45. c

47. d

Section 5.4

1. 0 **3.** 1 **5.** 8

7. -3 **9.** $\dfrac{3}{2}$ **11.** 5

13. $x + 5$ **15.** 8 **17.** $0.11\overline{1}$

19. $3\log_b x + 5\log_b y$ **21.** $\dfrac{1}{2}\log_b x + \dfrac{1}{3}\log_b y$

23. $\dfrac{1}{3}\log_b r - \dfrac{1}{2}\log_b s$ **25.** $\log_b x - \log_b y - \log_b z$

27. $2\log x + \dfrac{1}{2}\log(x + 5)$

29. $3\ln x + 2\ln(x - 2) - \dfrac{1}{2}\ln(x^2 + 5)$

31. $2\log(x - 1) - \log(x - 3) - \log(x + 3)$

33. $\log_b x^9 y^5$ **35.** $\log_b \dfrac{u^5}{v^2}$ **37.** $\log_b x^{1/2} y^{2/3}$

39. $\log_b \dfrac{u^2}{v^3 z^2}$ **41.** $\ln \dfrac{x^2 - 1}{(x^2 + 3)^2}$ **43.** $\ln \dfrac{(x + 3)^{1/2}}{x(x + 2)^{1/3}}$

45. 1.209 **47.** -2.322 **49.** 1.660

51. 2.011 **53.** 3.786 **55.** 110 decibels

57. 5.5 on the Richter scale (Total energy: 4.5×10^{12} joules)

59. $3 \log 5 - \log 5^2 = 3 \log 5 - 2 \log 5 = \log 5$

61. The bases are different, they must be the same in order to simplify the expression.

63. true **65.** false

67. Answers will vary. **69.** $3\log_b x - \left(\dfrac{9}{2}\log_b y - \dfrac{15}{2}\log_b z\right)$

71. yes **73.** no **75.** yes

Section 5.5

1. $x = 4$ **3.** $x = -2$ **5.** $x = \pm 2$

7. $x = -4$ **9.** $x = -\dfrac{3}{2}$ **11.** $x \approx 2.454$

13. $x \approx -0.303$ **15.** $x \approx 0.896$ **17.** $x \approx -0.904$

19. $x = 0$ **21.** $x = 50$ **23.** $x = 40$

25. $x = \pm 3$ **27.** $x = 5$ **29.** no solution

31. $x \approx \pm 12.182$ **33.** $x = 47.5$ **35.** $x \approx -1.25$

37. $x = 0.466$ **39.** $x = \dfrac{1 + \sqrt{1 + 4b^2}}{2}$ **41.** $t = 31.9$ years

43. $t \approx 19.74$ years **45.** 3.16×10^{15} joules **47.** 1 W/m^2

49. $t \approx 27.7$ years **51.** $t \approx 4.61$ hours **53.** 6.2

55. The correction that needs to be made is that the 4 should have been divided first giving the following:
$$e^x = \dfrac{9}{4}, \quad x = \ln\left(\dfrac{9}{4}\right)$$

57. The correction that needs to be made is that the solution $x = -5$ needs to be removed from the solutions. The domain of the logs cannot include a negative solution.

59. true **61.** false

63. $t = -5\ln\left(\dfrac{3000 - y}{2y}\right)$ **65.** $\ln(x + \sqrt{x^2 - 1}) \quad x \geq 1$

67. 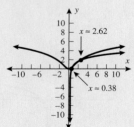 **69.**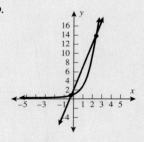

OOPS! $E \approx 1.0 \times 10^{18}$ joules is the energy released by the Indian Ocean earthquake. It was about 890 times stronger than the magnitude of the San

Francisco earthquake of 1989 and about as strong as 18,000 Hiroshima atomic bombs.

Tying It All Together $P(t) = \dfrac{500}{1 + 9e^{-0.3t}}$. The director can expect the population to reach 300 rabbits in about 9 years.

Review Exercises

1. 17,559.94 **3.** 5.52 **5.** 73.52

7. 6.25 **9.** b **11.** c

13. The y-intercept is at $(0, -1)$. The horizontal asymptote is $y = 0$.

15. The y-intercept is at $(0, 2)$. The horizontal asymptote is $y = 1$.

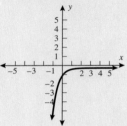

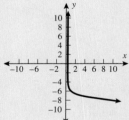

17. \$6144.68 **19.** \$18,182.60

21. 24.53 **23.** 5.89

25. The y-intercept is at: $(0, 1)$. The horizontal asymptote is $y = 0$.

27. The y-intercept is at: $(0, 3.2)$. The horizontal asymptote is $y = 0$.

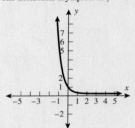

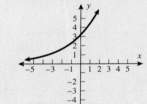

29. \$23,080.29 **31.** 343 mice

33. $4^3 = 64$ **35.** $10^{-2} = \dfrac{1}{100}$

37. $\log_6 216 = 3$ **39.** $\log_{2/13}\dfrac{4}{169} = 2$

41. 0 **43.** -4

45. 1.51 **47.** -2.08

49. domain: $(-2, \infty)$ **51.** domain: all $\mathbb{R}$

53. b **55.** d

57. 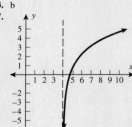 **59.**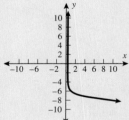

61. pH $= 6.5$ **63.** 50 dB

65. 1 **67.** 6

69. $a\log_c x + b\log_c y$ **71.** $\log_j r + \log_j s - 3\log_j t$

73. $\dfrac{1}{2}\log a - \dfrac{3}{2}\log b - \dfrac{2}{5}\log c$ **75.** 0.5283

77. 0.2939 **79.** $x = -4$

81. $x = \dfrac{4}{3}$

83. $x = -6$

85. -0.218

87. no solution

89. $x = 0$

91. $x = \dfrac{100}{3}$

93. $x = 181.02$

95. $x \approx \pm 3.004$

97. $x \approx 0.449$

99. \$28,536.88

101. $t = 16.6$ years

Practice Test

1. x^3

3. 4

5. $x \approx \pm 2.177$

7. $x \approx 7.04$

9. $x = e^2 \approx 7.389$

11. $x = e^{e^1} \approx 15.154$

13. domain: $(-1, 0) \cup (1, \infty)$

15. \$8,051.62

17. 90 dB

19. $7.9 \times 10^{11} < E < 2.5 \times 10^{13}$ joules

Chapter 6

Section 6.1

1. $x = 1, y = 0$

3. $x = 1, y = -1$

5. $u = \dfrac{32}{17}, v = \dfrac{11}{17}$

7. no solution

9. infinitely many solutions

11. $x = 3, y = 1$

13. $x = -3, y = 4$

15. $x = 9, y = \dfrac{11}{5}$

17. infinitely many solutions

19. infinitely many solutions

21. $x = \dfrac{75}{32}, y = \dfrac{7}{16}$

23. c

25. d

27. The solution is $(0, 0)$

29. The solution is $(-1, -1)$.

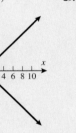

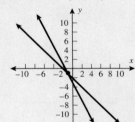

31. The solution is $(0, -6)$.

33. There are no solutions to this system of equations.

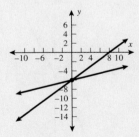

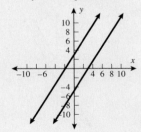

35. 16 bottles of Grey Goose and 21 bottles of Skyy

37. \$300,000 of sales

39. 169 miles highway, 180.5 miles city

41. Average plane speed is 450 mph, average wind is 50 mph.

43. \$3,500 invested at 10% and \$6,500 invested at 14%

45. Every term in Equation (1) is not multiplied correctly by -1.

47. The error in this problem is a distribution error. If the error is corrected, the correct solution should be:

$$x = \dfrac{2}{5} \text{ and } y = -\dfrac{26}{5}$$

49. false

51. false

53. $A = -4, B = 7$

55. The point of intersection is approximately $(8.9, 6.4)$.

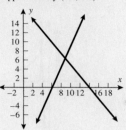

57. The lines coincide.

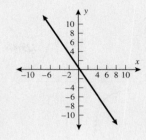

Section 6.2

1. $x = \dfrac{90}{31}, y = \dfrac{103}{31}, z = \dfrac{9}{31}$

3. $x = -\dfrac{13}{4}, y = \dfrac{1}{2}, z = -\dfrac{5}{2}$

5. $x = -2, y = -1, z = 0$

7. $x = 2, y = 5, z = -1$

9. $x = 1, y = -\dfrac{1}{2}, z = 0$

11. $x = 41 + 4t, y = 31 + 3t, z = t$

13. $x_1 = -\dfrac{1}{2}, x_2 = \dfrac{7}{4}, x_3 = -\dfrac{3}{4}$

15. no solution

17. $x_1 = 1, x_2 = -1 + t, x_3 = t$

19. $x = \dfrac{2}{3}a + \dfrac{8}{3}, y = -\dfrac{1}{3}a - \dfrac{10}{3}, z = a$

21. 8 touchdowns, 7 extra points, 1 two-point conversion, and 5 field goals

23. Three Chicken, four Tuna, six Roast Beef, and one Turkey–Bacon Wrap

25. $a = -32$ ft/sec^2, $v_0 = 52$ ft/sec, $h_0 = 0$ ft

27. $y = -0.0625x^2 + 5.25x - 50$

29. \$10,000 in the money market, \$4,000 in the mutual fund, and \$6,000 in the stock

31. $a = 4, b = -2, c = -4$

33. The correction that needs to be made is that Equation (2) and Equation (3) need to be added correctly. Also, the text tells you to begin by eliminating a variable from Equation (1).

35. true

37. $a = -\dfrac{55}{24}, b = -\dfrac{1}{4}, c = \dfrac{223}{24}, d = \dfrac{1}{4}, e = 44$

39. no solution

41. $x_1 = -2, x_2 = 1, x_3 = -4, x_4 = 5$

43. $x = 41 + 4t, y = 31 + 3t, z = t$

45. same result as Exercise 43

Section 6.3

1. d

3. a

5. b

7. $\dfrac{A}{x-5} + \dfrac{B}{x+4}$

9. $\dfrac{A}{x-4} + \dfrac{B}{x} + \dfrac{C}{x^2}$

11. $2x - 6 + \dfrac{Ax+B}{x^2+x+5}$

13. $\dfrac{Ax+B}{x^2+10} + \dfrac{Cx+D}{(x^2+10)^2}$

15. $\dfrac{1}{x} - \dfrac{1}{x+1}$

17. $\dfrac{1}{x-1}$

19. $\dfrac{2}{x-3} + \dfrac{7}{x+5}$

21. $\dfrac{4}{x+3} - \dfrac{15}{(x+3)^2}$

23. $\dfrac{3}{x+1} + \dfrac{1}{x-5} + \dfrac{2}{(x-5)^2}$

25. $\dfrac{1}{x} + \dfrac{1}{x+1} - \dfrac{1}{x^3}$

27. $\dfrac{-2}{x+4} + \dfrac{7x}{x^2+3}$

29. $\dfrac{-2}{x-7} + \dfrac{4x-3}{3x^2-7x+5}$

31. $\dfrac{x}{x^2+9} - \dfrac{9x}{(x^2+9)^2}$

33. $\dfrac{2x-3}{x^2+1} + \dfrac{5x+1}{(x^2+1)^2}$

35. $\dfrac{1}{x-1} + \dfrac{-3x-1}{2(x^2+1)} + \dfrac{1}{2(x+1)}$

37. $\dfrac{x}{x^2+1} - \dfrac{2x}{(x^2+1)^2} + \dfrac{x+2}{(x^2+1)^3}$

39. $\dfrac{1}{x-1} + \dfrac{1-x}{x^2+x+1}$

41. $\dfrac{1}{d_0} - \dfrac{1}{d_i} = \dfrac{1}{f}$

43. The correction that needs to be made is in the decomposition. Once the correct decomposition is used the correct answer will be:

$$\dfrac{1}{x} + \dfrac{3+2x}{x^2+1}$$

45. false

49. yes

47. $\dfrac{1}{x-1} - \dfrac{1}{x+2} + \dfrac{1}{x-2}$

51. no

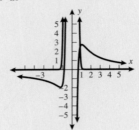

9.

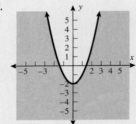

11.

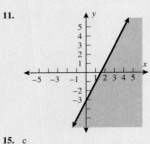

Section 6.4

1. $(2, 6), (-1, 3)$

3. $(1, 0)$

5. no solution

7. $(0, 1)$

9. $(-0.63, -1.61), (0.63, -1.61)$

11. no solution

13. $(1, 1)$

15. $(-\sqrt{2}, -2\sqrt{2}), (2\sqrt{2}, \sqrt{2}), (\sqrt{2}, 2\sqrt{2}), (-2\sqrt{2}, -\sqrt{2})$

17. $(-6, 33), (2, 1)$

19. $(3, 4)$

21. $(0, -3), \left(\dfrac{2}{5}, -\dfrac{11}{5}\right)$

23. $(-1, -1), \left(\dfrac{1}{4}, \dfrac{3}{2}\right)$

25. $(-1, -4), (4, 1)$

27. $(1, 3), (-1, -3)$

29. $(2, 4)$

31. $\left(\dfrac{1}{2}, \dfrac{1}{3}\right), \left(\dfrac{1}{2}, -\dfrac{1}{3}\right)$

33. no solution

35.

37. 3 and 7

39. 8 and 9

41. 8 centimeters by 10 centimeters

43. 400 feet by 500 feet or $333\frac{1}{3}$ feet by 600 feet

45. Professor: 2 meters per second
Jeremy: 10 meters per second

47. Can't use elimination. Should have used substitution.

49. false

51. false

53. $2n$

55. $y = x^2 + 1$ } other answers
$y = 1$ } possible

57. no solution

59. $(-1.57, -1.64)$

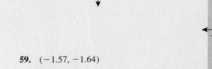

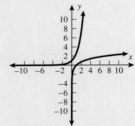

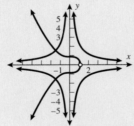

Section 6.5

1. d

3. b

5.

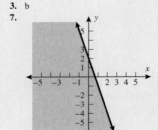

7.

13. a

15. c

17.

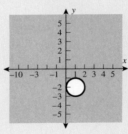

19.

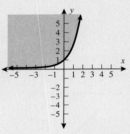

21.

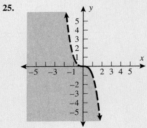

23.

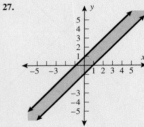

25.

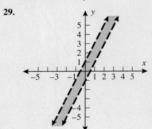

27.

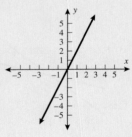

29.

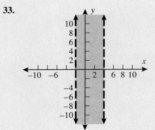

31.

33.

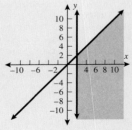

35.

37.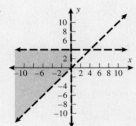

39.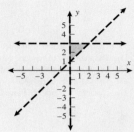

25. Minimized at $z(0, 2) = 0$.

27. Minimized at $z(6.7, 4.5) = 176.9$.

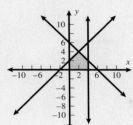

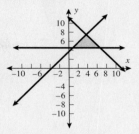

41.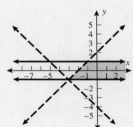

43.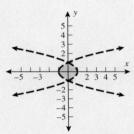

OOPS! $y = 5x^2 + 41x - 90$

Tying It All Together The company breaks even when it sells approximately 1,667 DVDs. For $x > 1,667$ DVDs, the company will show a profit.

Review Exercises

1. $r = 3, s = 0$

3. $x = \dfrac{13}{4}, y = 8$

5. $x = 2, y = 1$

7. $c = \dfrac{19}{8}, d = \dfrac{13}{8}$

9. c

11. d

13. Intersection at $(-2, 1)$.

15. Intersection at $\left(12, \dfrac{35}{6}\right)$.

45.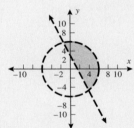

47.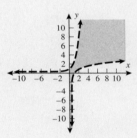

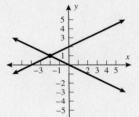

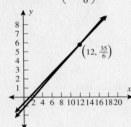

49. 4

51. $8x - 120 > 0$
$x > 15$

53. $x + 20y \le 2400$
$25x + 150y \le 6000$

55. The correction that needs to be made is that the shading should be above the line.

57. The common region is the one that is both blue and red. The graph is also shaded incorrectly.

59. true

61. false

63. The solution is a shaded rectangle. If $a > b$ and $c > d$, then there is no solution.

65. $-b \le a \le b$

Section 6.6

1. $f(x, y) = z = 2x + 3y$
$f(-1, 4) = 10$ (MAX)
$f(2, 4) = 16$
$f(-2, -1) = -7$ (MIN)
$f(1, -1) = -1$

3. $f(x, y) = z = 1.5x + 4.5y$
$f(-1, 4) = 16.5$
$f(2, 4) = 21$ (MAX)
$f(-2, -1) = -7.5$ (MIN)
$f(1, -1) = -3$

5. Minimize at $f(0, 0) = 0$

7. Maximize at $f(2, 2) = 14$

9. Minimize at $f(0, 0) = 0$

11. Maximize at $f(1, 6) = 2.65$

13. 50 Charley T-shirts, 130 Francis T-shirts (Profit = \$950)

15. 0 desktops, 25 laptops (profit = \$7,500)

17. 3 first class cars and 27 second class cars

19. You don't compare y-coordinates. You evaluate the objective function.

21. false

23. The maximum is a and occurs at $(0, a)$

17. 10.5 millimeters of 6% and 31.5 millimeters of 18%

19. $x = -1, y = -t + 2, z = t$

21. $x = -\dfrac{3}{7}t - 2, y = \dfrac{2}{7}t + 2, z = t$

23. $f(x) = -0.0050x^2 + 0.4486x - 3.8884$

25. $\dfrac{A}{x - 3} + \dfrac{B}{x + 4}$

27. $3x - 8 + \dfrac{Ax + B}{x^2 + x - 7}$

29. $\dfrac{-2}{x + 1} + \dfrac{2}{x}$

31. $\dfrac{5}{x + 2} - \dfrac{27}{(x + 2)^2}$

33. $(1, -4), (-2, -7)$

35. $(-1, 2), (1, 2)$

37. no solution

39. no solution

41. $(-3, -2), (2, 3)$

43. $\left(-\dfrac{1}{2}, -\dfrac{1}{\sqrt{7}}\right)\left(-\dfrac{1}{2}, \dfrac{1}{\sqrt{7}}\right)\left(\dfrac{1}{2}, -\dfrac{1}{\sqrt{7}}\right)\left(\dfrac{1}{2}, \dfrac{1}{\sqrt{7}}\right)$

45.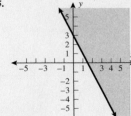

47.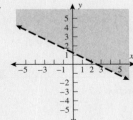

49.

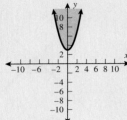

51.

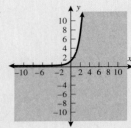

53. no solution

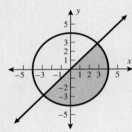

55.

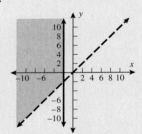

57.

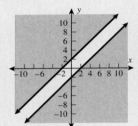

59. Minimum of z is 0, and occurs at $(0, 0)$

61. Maximum of z is 25.6, and occurs at $(0, 8)$

63. 10 watercolor, 30 geometric (Profit $390)

Practice Test

1. $x = 7, y = 3$

3. $x = 2 + t, y = t$

5. $x = 1, y = -5, z = 3$

7. $x = 8(t - 1), y = -7t + 13, z = t$

9. $\dfrac{5}{x} - \dfrac{3}{x + 1}$

11. $\dfrac{\frac{1}{3}}{x - 3} - \dfrac{\frac{2}{3}}{x + 3} + \dfrac{1}{x}$ or $\dfrac{1}{3(x - 3)} - \dfrac{2}{3(x + 3)} + \dfrac{1}{x}$

13.

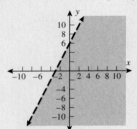

15.

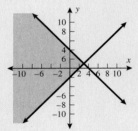

17. The minimum value occurs at $(0, 1)$. $f(0, 1) = 7$

19. $14,000 in the money market, $8,500 in the aggressive stock, and $7,500 in the conservative stock.

Chapter 7

Section 7.2

1. $\begin{bmatrix} 3 & -2 & 7 \\ -4 & 6 & -3 \end{bmatrix}$

3. $\begin{bmatrix} 2 & -3 & 4 & -3 \\ -1 & 1 & 2 & 1 \\ 5 & -2 & -3 & 7 \end{bmatrix}$

5. $\begin{bmatrix} 1 & 1 & 0 & 3 \\ 1 & 0 & -1 & 2 \\ 0 & 1 & 1 & 5 \end{bmatrix}$

7. $\begin{bmatrix} -4 & 3 & 5 & 2 \\ 2 & -3 & -2 & -3 \\ -2 & 4 & 3 & 1 \end{bmatrix}$

9. The matrix is not in reduced form. It does not satisfy Condition 3.

11. The matrix is in reduced form.

13. The matrix is not in reduced form. It does not satisfy Condition 4.

15. The matrix is in reduced form.

17. The matrix is not in reduced form. It does not satisfy Condition 2.

19. $\begin{bmatrix} 1 & -2 & -3 \\ 0 & 7 & 5 \end{bmatrix}$

21. $\begin{bmatrix} 1 & -2 & -1 & 3 \\ 0 & 5 & -1 & 0 \\ 3 & -2 & 5 & -8 \end{bmatrix}$

23. $\begin{bmatrix} 1 & -2 & 5 & -1 & 2 \\ 0 & 1 & 1 & -3 & 3 \\ 0 & -2 & 1 & -2 & 5 \\ 0 & 0 & 1 & -1 & -6 \end{bmatrix}$

25. $\begin{bmatrix} 1 & 0 & 5 & -10 & -5 \\ 0 & 1 & 2 & -3 & -2 \\ 0 & 0 & -7 & 6 & 3 \\ 0 & 0 & 8 & -10 & -9 \end{bmatrix}$

27. $\begin{bmatrix} 1 & 0 & 4 & 0 & 27 \\ 0 & 1 & 2 & 0 & -11 \\ 0 & 0 & 1 & 0 & 21 \\ 0 & 0 & 0 & 1 & -3 \end{bmatrix}$

29. $\begin{bmatrix} 1 & 0 & -8 \\ 0 & 1 & 6 \end{bmatrix}$

31. $\begin{bmatrix} 1 & 0 & 0 & -2 \\ 0 & 1 & 0 & -1 \\ 0 & 0 & 1 & 0 \end{bmatrix}$

33. $\begin{bmatrix} 1 & 0 & 0 & 2 \\ 0 & 1 & 0 & 5 \\ 0 & 0 & 1 & -1 \end{bmatrix}$

35. $\begin{bmatrix} 1 & 0 & -2 & 1 \\ 0 & 1 & -2 & 2 \end{bmatrix}$

37. $\begin{bmatrix} 1 & 0 & 1 & 1 \\ 0 & 1 & 1 & -\frac{1}{2} \\ 0 & 0 & 0 & 0 \end{bmatrix}$

39. $\begin{bmatrix} 1 & 0 & -7 \\ 0 & 1 & 5 \end{bmatrix}$ $\begin{array}{l} x = -7 \\ y = 5 \end{array}$

41. $\begin{bmatrix} 1 & -2 & -3 \\ 0 & 0 & 0 \end{bmatrix}$ $x - 2y = -3$ or $x = t, y = \dfrac{t + 3}{2}$

43. $\begin{bmatrix} 1 & \frac{1}{2} & 0 \\ 0 & 0 & 1 \end{bmatrix}$ no solution

45. $\begin{bmatrix} 1 & 0 & -4 & 41 \\ 0 & 1 & -3 & 31 \\ 0 & 0 & 0 & 0 \end{bmatrix}$ $\begin{array}{l} x = 4t + 41 \\ y = 3t + 31 \\ z = t \end{array}$

47. $\begin{bmatrix} 1 & 0 & 0 & -\frac{1}{2} \\ 0 & 1 & 0 & \frac{7}{4} \\ 0 & 0 & 1 & -\frac{3}{4} \end{bmatrix}$ $\begin{array}{l} x_1 = -\frac{1}{2} \\ x_2 = \frac{7}{4} \\ x_3 = -\frac{3}{4} \end{array}$

49. $\begin{bmatrix} 1 & 0 & -5 & 0 \\ 0 & 1 & 2 & 0 \\ 0 & 0 & 0 & 1 \end{bmatrix}$ no solution

51. $\begin{bmatrix} 1 & 0 & 0 & 1 \\ 0 & 1 & -1 & -1 \\ 0 & 0 & 0 & 0 \end{bmatrix}$ $\begin{array}{l} x_1 = 1 \\ x_2 = t - 1 \\ x_3 = t \end{array}$

53. $\begin{bmatrix} 1 & 0 & -\frac{2}{3} & \frac{8}{3} \\ 0 & 1 & \frac{1}{3} & -\frac{10}{3} \end{bmatrix}$ $\begin{array}{l} x = \frac{2}{3}t + \frac{8}{3} \\ y = -\frac{1}{3}t - \frac{10}{3} \\ z = t \end{array}$

55. $\begin{bmatrix} 1 & 0 & 0 & 0 \\ 0 & 1 & 0 & 0 \\ 0 & 0 & 1 & 0 \\ 0 & 0 & 0 & 1 \end{bmatrix}$ no solution

57. $\begin{bmatrix} 1 & 0 & 0 & 0 & -2 \\ 0 & 1 & 0 & 0 & 1 \\ 0 & 0 & 1 & 0 & -4 \\ 0 & 0 & 0 & 1 & 5 \end{bmatrix}$ $\begin{array}{l} x_1 = -2 \\ x_2 = 1 \\ x_3 = -4 \\ x_4 = 5 \end{array}$

59. 8 touchdowns, 5 extra points, 1 two-point conversion, and 2 field goals

61. 2 Chicken, 2 Tuna, 8 Roast Beef, and 2 Turkey-Bacon Wrap

63. initial height = 0 feet, initial velocity = 50 feet/second, acceleration = -32 feet/second

65. $y = -0.053x^2 + 4.58x - 34.76$

67. $5,500 in the money market, $2,500 in the mutual fund, and $2,000 in the stock

69. $a = -\dfrac{22}{17}, b = -\dfrac{44}{17}, c = -\dfrac{280}{17}$

71. The correct matrix that is needed is $\begin{bmatrix} -1 & 1 & 1 & | & 2 \\ 1 & 1 & -2 & | & -3 \\ 1 & 1 & 1 & | & 6 \end{bmatrix}$. After reducing

this matrix, the correct solution should be $\begin{bmatrix} 1 & 0 & 0 & | & 2 \\ 0 & 1 & 0 & | & 1 \\ 0 & 0 & 1 & | & 3 \end{bmatrix}$.

73. The correction that needs to be made is that row 3 is not inconsistent, $1 = 0$ is being misinterpreted. It should be $z = 0$.

75. false

77. true

79. $f(x) = -\dfrac{11}{6}x^4 + \dfrac{44}{3}x^3 - \dfrac{223}{6}x^2 + \dfrac{94}{3}x + 44$

Section 7.3

1. -2 **3.** 31

5. -28 **7.** -0.6

9. 0 **11.** $x = 5, y = -6$

13. $x = -2, y = 1$ **15.** inconsistent

17. dependent **19.** $x = \dfrac{1}{2}, y = -1$

21. $x = 1.5, y = 2.1$ **23.** $x = 0, y = 7$

25. $x = \dfrac{1}{3}, y = \dfrac{3}{4}$ **27.** 7

29. -25 **31.** -180

33. 0 **35.** abc

37. $x = 2, y = 3, z = 5$ **39.** $x = -2, y = 3, z = 5$

41. $x = 2, y = -3, z = 1$ **43.** dependent

45. inconsistent **47.** 6

49. 6 **51.** $-2x + y = 0$

53. The mistake in this problem is that a positive is used for the second term when a negative should have been used. The determinant you should obtain is -53.

55. D_x and D_y were interchanged.

57. true **59.** false **61.** -419 **65.** -180 **67.** -1019

Section 7.4

1. $\begin{bmatrix} -1 & 5 & 1 \\ 5 & 2 & 5 \end{bmatrix}$ **3.** not possible

5. $\begin{bmatrix} -2 & 6 & 0 \\ 4 & 8 & 2 \end{bmatrix}$ **7.** $\begin{bmatrix} -2 & 12 & 3 \\ 13 & 2 & 14 \end{bmatrix}$

9. $\begin{bmatrix} 2 & 5 \\ 4 & -7 \\ 11 & 2 \end{bmatrix}$ **11.** not possible

13. $\begin{bmatrix} 13 & -5 \\ 19 & 8 \end{bmatrix}$ **15.** not possible

17. $\begin{bmatrix} 0 & 1 \\ 2 & -1 \\ 3 & 1 \end{bmatrix}$ **19.** not possible

21. Yes, B is the multiplicative inverse of A.

23. Yes, B is the multiplicative inverse of A.

25. No, B is not the multiplicative inverse of A.

27. Yes, B is the multiplicative inverse of A.

29. No, B is not the multiplicative inverse of A.

31. $\begin{bmatrix} 0 & -1 \\ 1 & 2 \end{bmatrix}$ **33.** $\begin{bmatrix} -\frac{1}{13} & \frac{8}{39} \\ \frac{20}{39} & -\frac{4}{117} \end{bmatrix}$

35. $\begin{bmatrix} -0.1618 & 0.2284 \\ 0.5043 & -0.1237 \end{bmatrix}$ **37.** $\begin{bmatrix} \frac{1}{2} & \frac{1}{2} & 0 \\ \frac{1}{2} & 0 & \frac{1}{2} \\ 0 & -\frac{1}{2} & -\frac{1}{2} \end{bmatrix}$

39. A^{-1} is not possible since the determinant of A is 0.

41. $\begin{bmatrix} -\frac{1}{2} & -\frac{1}{2} & \frac{5}{2} \\ \frac{1}{2} & \frac{1}{2} & -\frac{3}{2} \\ 0 & -1 & 1 \end{bmatrix}$ **43.** $\begin{bmatrix} \frac{1}{2} & \frac{1}{2} & 0 \\ \frac{3}{4} & \frac{1}{4} & -\frac{1}{2} \\ \frac{1}{4} & \frac{3}{4} & -\frac{1}{2} \end{bmatrix}$

45. $x = 0, y = 0, z = 1$ **47.** infinitely many solutions

49. $x = -1, y = 1, z = -7$ **51.** $x = 3, y = 5, z = 4$

53. $A = \begin{bmatrix} .70 \\ .30 \end{bmatrix}$ $B = \begin{bmatrix} .89 \\ .84 \end{bmatrix}$

a. $46A = \begin{bmatrix} 32.2 \\ 13.8 \end{bmatrix}$. Out of 46 million people, 32.2 million said yes they have tried to quit smoking and 13.8 million people said they have not tried to quit smoking.

b. $46B = \begin{bmatrix} 40.94 \\ 38.64 \end{bmatrix}$. Out of 46 million people, 40.94 million people believed smoking would increase their chances of lung cancer and 38.64 million people believed smoking would shorten their lives.

55. $A = \begin{bmatrix} .589 & .628 \\ .414 & .430 \end{bmatrix}$ $B = \begin{bmatrix} 100 \text{ M} \\ 110 \text{ M} \end{bmatrix}$

$AB = \begin{bmatrix} 127.98 \text{ M} \\ 88.7 \text{ M} \end{bmatrix}$ tells us the number of registered voters and the number of voters.

57. You don't multiply corresponding elements. The correct matrix multiplication would produce the answer:
$\begin{bmatrix} 3 & 2 \\ 1 & 4 \end{bmatrix}\begin{bmatrix} -1 & 3 \\ -2 & 5 \end{bmatrix} = \begin{bmatrix} -7 & 19 \\ -9 & 23 \end{bmatrix}$

59. The mistake is that the determinant of matrix A is 0. Therefore there will be no inverse that exists. The identity matrix does not appear on the left.

61. false

63. true

67. $\begin{bmatrix} \frac{1}{a} & 0 & 0 \\ 0 & \frac{1}{b} & 0 \\ 0 & 0 & \frac{1}{c} \end{bmatrix}$

69. The determinant of the square matrix given is 0. Therefore it will not have an inverse. Additionally, one row is a constant multiple of the other row so this system is dependent.

71. $\begin{bmatrix} 33 & 35 \\ -96 & -82 \\ 31 & 19 \\ 146 & 138 \end{bmatrix}$ **73.** $\begin{bmatrix} -\frac{115}{6008} & \frac{431}{6008} & -\frac{1067}{6008} & \frac{103}{751} \\ \frac{411}{6008} & -\frac{391}{6008} & \frac{731}{6008} & -\frac{22}{751} \\ \frac{57}{751} & \frac{28}{751} & -\frac{85}{751} & \frac{3}{751} \\ -\frac{429}{6008} & \frac{145}{6008} & \frac{1035}{6008} & \frac{12}{751} \end{bmatrix}$

OOPS! The circle and parabola intersect at $(\sqrt{5}, 1), (-\sqrt{5}, 1), (\sqrt{2}, -2),$ $(-\sqrt{2}, -2)$ The matrix represents the system of equations $\begin{cases} x + y = 6 \\ x - y = 4 \end{cases}$.

Tying It All Together Sam should invest $1,190.48 into savings, $2,500 into the stock, and $1,309.52 into the mutual fund.

Review Exercises

1. $\begin{bmatrix} 5 & 7 & | & 2 \\ 3 & -4 & | & -2 \end{bmatrix}$ **3.** $\begin{bmatrix} 2 & 0 & -1 & | & 3 \\ 0 & 1 & -3 & | & -2 \\ 1 & 0 & 4 & | & -3 \end{bmatrix}$

5. No, Condition 2 is not met. **7.** No, Condition 1 is not met.

9. $\begin{bmatrix} 1 & -2 & | & 1 \\ 0 & 1 & | & -1 \end{bmatrix}$ **11.** $\begin{bmatrix} 1 & -4 & 3 & | & -1 \\ 0 & -2 & 3 & | & -2 \\ 0 & 1 & -4 & | & 8 \end{bmatrix}$

13. $\begin{bmatrix} 1 & 0 & | & \frac{3}{5} \\ 0 & 1 & | & -\frac{1}{5} \end{bmatrix}$ **15.** $\begin{bmatrix} 1 & 0 & 0 & | & -4 \\ 0 & 1 & 0 & | & 8 \\ 0 & 0 & 1 & | & -4 \end{bmatrix}$

17. $\begin{bmatrix} 1 & 0 & | & 1.25 \\ 0 & 1 & | & .875 \end{bmatrix}$ $\begin{matrix} x = 1.25 \\ y = 0.875 \end{matrix}$ **19.** $\begin{bmatrix} 1 & 0 & 0 & | & -\frac{74}{21} \\ 0 & 1 & 0 & | & -\frac{73}{21} \\ 0 & 0 & 1 & | & -\frac{3}{7} \end{bmatrix}$ $\begin{matrix} x = -\frac{74}{21} \\ y = -\frac{73}{21} \\ z = -\frac{3}{7} \end{matrix}$

21. $y = -0.005x^2 + 0.45x - 3.89$

23. -8 **25.** 5.4 **27.** $x = 3, y = 1$

29. no solution **31.** 11 **33.** $-abd$

35. $x = 1, y = 1, z = 2$ **37.** $x = -\dfrac{15}{7}, y = -\dfrac{25}{7}, z = \dfrac{19}{14}$

39. The area of the triangle = 1. **41.** not possible

43. not possible **45.** not possible

47. Yes, B is the multiplicative inverse of A.

49. Yes, B is the multiplicative inverse of A.

51. $\begin{bmatrix} 0.4 & -0.2 \\ 0.3 & 0.1 \end{bmatrix}$ **53.** $\begin{bmatrix} 0 & -\frac{1}{2} \\ 1 & 0 \end{bmatrix}$

55. $\begin{bmatrix} -\frac{1}{6} & \frac{7}{12} & -\frac{1}{12} \\ \frac{1}{2} & -\frac{1}{4} & -\frac{1}{4} \\ \frac{1}{6} & -\frac{1}{12} & -\frac{5}{12} \end{bmatrix}$ **57.** $\begin{bmatrix} 0 & -\frac{2}{5} & \frac{1}{5} \\ 1 & -\frac{2}{5} & \frac{1}{5} \\ -\frac{1}{2} & \frac{3}{10} & \frac{1}{10} \end{bmatrix}$

Practice Test

1. Augmented matrix: $\begin{bmatrix} 1 & -2 & | & 1 \\ -1 & 3 & | & 2 \end{bmatrix}$

Matrix equation: $\begin{bmatrix} 1 & 2 \\ -1 & 3 \end{bmatrix}\begin{bmatrix} x \\ y \end{bmatrix} = \begin{bmatrix} 1 \\ 2 \end{bmatrix}$

3. Augmented matrix: $\begin{bmatrix} 6 & 9 & 1 & | & 5 \\ 2 & -3 & 1 & | & 3 \\ 10 & 12 & 2 & | & 9 \end{bmatrix}$

Matrix equation: $\begin{bmatrix} 6 & 9 & 1 \\ 2 & -3 & 1 \\ 10 & 12 & 2 \end{bmatrix}\begin{bmatrix} x \\ y \\ z \end{bmatrix} = \begin{bmatrix} 5 \\ 3 \\ 9 \end{bmatrix}$

5. $\begin{bmatrix} 1 & 3 & 5 \\ 0 & 1 & -11 \\ 0 & 7 & 15 \end{bmatrix}$

7. $\begin{bmatrix} 1 & 0 & \frac{1}{3} & | & \frac{7}{6} \\ 0 & 1 & -\frac{1}{9} & | & -\frac{2}{9} \\ 0 & 0 & 0 & | & 0 \end{bmatrix}$ $\begin{array}{l} x = -\frac{1}{3}t + \frac{7}{6} \\ y = \frac{1}{9}t - \frac{2}{9} \\ z = t \end{array}$

9. Determinant $= 3$.   **11.** $x = 7, y = 3$

13. $\begin{bmatrix} -11 & 19 \\ -6 & 8 \end{bmatrix}$ **15.** $\begin{bmatrix} \frac{1}{19} & \frac{3}{19} \\ \frac{5}{19} & -\frac{4}{19} \end{bmatrix}$

17. not possible

19. The order BA is 2×2. BA tells us each person's score for each rubric.

Chapter 8

Section 8.1

1. hyperbola   **3.** circle   **5.** circle   **7.** ellipse

Section 8.2

1. c   **3.** d   **5.** c   **7.** a

9. $x^2 = 12y$ **11.** $y^2 = -20x$

13. $(x - 3)^2 = 8(y - 5)$ **15.** $(y - 4)^2 = -8(x - 2)$

17. $(x - 2)^2 = 12(y - 1)$ **19.** $(y + 1)^2 = 4(x - 2)$

21. $(y - 2)^2 = 8(x + 1)$ **23.** $(x - 2)^2 = -8(y + 1)$

25. vertex: $(0, 0)$
focus: $(0, 2)$
directrix: $y = -2$
length of latus rectum: 8

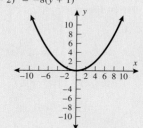

27. vertex: $(0, 0)$
focus: $\left(-\frac{1}{2}, 0\right)$
directrix: $x = \frac{1}{2}$
length of latus rectum: 2

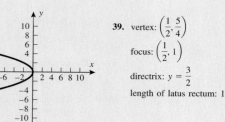

29. vertex: $(-3, 2)$
focus: $(-2, 2)$
directrix: $x = -4$
length of latus rectum: 4

31. vertex: $(3, -1)$
focus: $(3, -3)$
directrix: $y = 1$
length of latus rectum: 8

33. vertex: $(-5, 0)$
focus: $\left(-5, -\frac{1}{2}\right)$
directrix: $y = \frac{1}{2}$
length of latus rectum: 2

35. vertex: $(0, 2)$
focus: $\left(\frac{1}{2}, 2\right)$
directrix: $x = -\frac{1}{2}$
length of latus rectum: 2

37. vertex: $(-3, -1)$
focus: $(-1, -1)$
directrix: $x = -5$
length of latus rectum: 8

39. vertex: $\left(\frac{1}{2}, \frac{5}{4}\right)$
focus: $\left(\frac{1}{2}, 1\right)$
directrix: $y = \frac{3}{2}$
length of latus rectum: 1

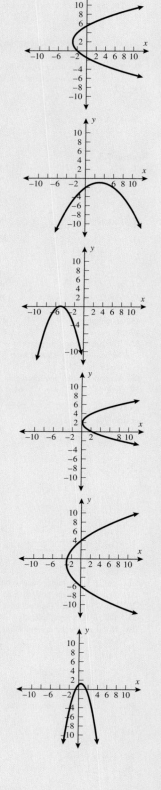

41. The focus will be at $(0, 2)$ so the receiver should be placed 2 feet from the vertex.

43. $x = \dfrac{1}{8}y^2$, $-2.5 \le y \le 2.5$

45. $x^2 = 160y$

47. Yes. The opening height is 18.75 feet, and the mast is only 17 feet.

49. $B = \dfrac{37}{800}S^2 + \dfrac{29}{80}S - \dfrac{191}{32}$

where B is the breaking distance in feet and S is the speed in miles per hour

51. The correction that needs to be made is that the formula $y^2 = 4px$ should be used.

53. true

55. false

59.

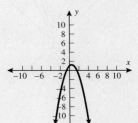

61. The vertex is located at: $(2.5, -3.5)$. The parabola opens to the right.

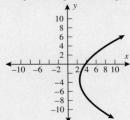

Section 8.3

1. d

3. a

5.

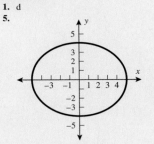

7.

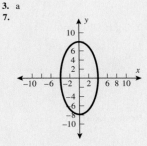

9.

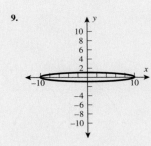

11.

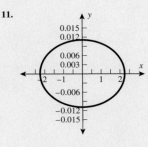

13.

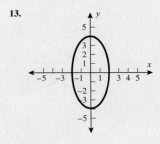

15.

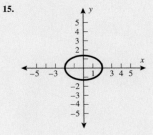

17. $\dfrac{x^2}{36} + \dfrac{y^2}{20} = 1$

19. $\dfrac{x^2}{7} + \dfrac{y^2}{16} = 1$

21. $\dfrac{x^2}{4} + \dfrac{y^2}{16} = 1$

23. $\dfrac{x^2}{9} + \dfrac{y^2}{49} = 1$

25. c

27. b

29.

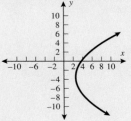

31.

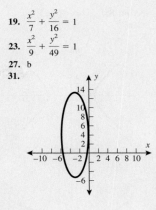

33.

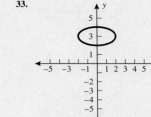

35.

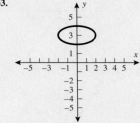

37.

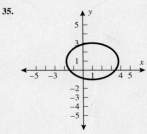

39. $\dfrac{(x - 2)^2}{25} + \dfrac{(y - 5)^2}{9} = 1$

41. $\dfrac{(x - 4)^2}{7} + \dfrac{(y + 4)^2}{16} = 1$

43. $\dfrac{(x - 3)^2}{4} + \dfrac{(y - 2)^2}{16} = 1$

45. $\dfrac{(x + 1)^2}{9} + \dfrac{(y + 4)^2}{25} = 1$

47. $\dfrac{y^2}{5625} + \dfrac{x^2}{225} = 1$

49. **a.** $\dfrac{x^2}{75^2} + \dfrac{y^2}{20^2} = 1$ **b.** Let $x = 60$ then $y = 12$. The field extends 15 feet in that direction so the track will not encompass the field.

51. $\dfrac{x^2}{5,914,000,000^2} + \dfrac{y^2}{5,729,000,000^2} = 1$

53. $\dfrac{x^2}{150,000,000^2} + \dfrac{y^2}{146,000,000^2} = 1$

55. $a = 6$ and $b = 4$ is incorrect. In the formula a and b are being squared; therefore $a = \sqrt{6}$ and $b = 2$.

57. false

59. true

61. Pluto: $e \approx 0.25$ Earth: $e \approx 0.02$

63.

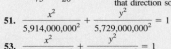

65.

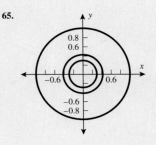

Section 8.4

1. b

3. d

5.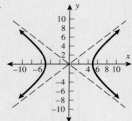

7.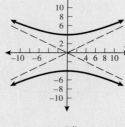

9.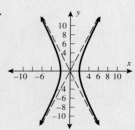

11.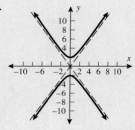

13.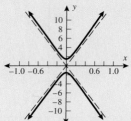

15.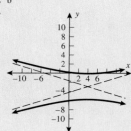

17. $\dfrac{x^2}{16} - \dfrac{y^2}{20} = 1$

19. $\dfrac{y^2}{9} - \dfrac{x^2}{7} = 1$

21. $x^2 - y^2 = a^2$

23. $\dfrac{y^2}{4} - x^2 = b^2$

25. c

27. b

29.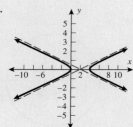

31.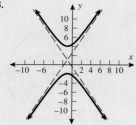

33.

35.

37.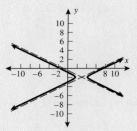

39. $\dfrac{(x-2)^2}{16} - \dfrac{(y-5)^2}{9} = 1$

41. $\dfrac{(y+4)^2}{9} - \dfrac{(x-4)^2}{7} = 1$

43. The ship will come ashore between the two stations (28.5 miles from one and 121.5 miles from the other).

45. 0.000484 seconds

47. $y^2 - \dfrac{4}{5}x^2 = 1$

49. The transverse axis is vertical. The points are $(\pm 3, 0)$. The vertices are $(0, \pm 2)$.

51. false

53. true

55. $x^2 - y^2 = a^2$ or $y^2 - x^2 = a^2$ note that $a = b$

57.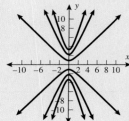

OOPS! The height of the tunnel along the sides of the truck is approximately 17.8 feet, so the truck will clear the tunnel if it straddles the centerline.

Tying It All Together

$$\frac{x^2}{100} + \frac{y^2}{64} = 1$$

The foci are located 6 feet from the center of the room at coordinates $(6, 0)$ and $(-6, 0)$.

Review Exercises

1. $y^2 = 12x$

3. $y^2 = -20x$

5. $(x-2)^2 = 8(y-3)$

7. $(x-1)^2 = -8(y-6)$

9. focus: $(0, -3)$
directrix: $y = 3$
vertex: $(0, 0)$
length of latus rectum: 12

11. focus: $\left(\dfrac{1}{4}, 0\right)$
directrix: $x = -\dfrac{1}{4}$
vertex: $(0, 0)$
length of latus rectum: 1

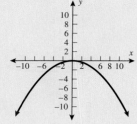

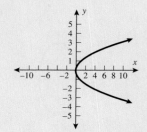

13. focus: $(3, -2)$
directrix: $x = 1$
vertex: $(2, -2)$
length of latus rectum: 4

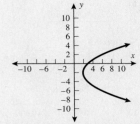

15. focus: $(-3, -1)$
directrix: $y = 3$
vertex: $(-3, 1)$
length of latus rectum: 8

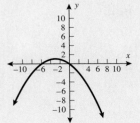

17. focus: $\left(\dfrac{5}{2}, \dfrac{71}{8}\right)$

directrix: $y = \dfrac{79}{8}$

vertex: $\left(\dfrac{5}{2}, \dfrac{75}{8}\right)$

length of latus rectum: 2

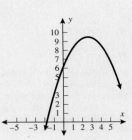

19. The focus is at $(0, 3.125)$ so the receiver should be placed 3.125 feet from the vertex.

21.

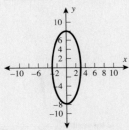

23.

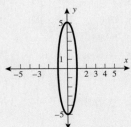

25. $\dfrac{x^2}{25} + \dfrac{y^2}{16} = 1$

27. $\dfrac{y^2}{64} + \dfrac{x^2}{9} = 1$

29.

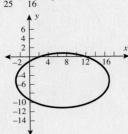

31.

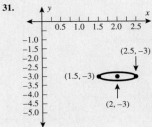
$(2.5, -3)$
$(1.5, -3)$
$(2, -3)$

33. $\dfrac{(x-3)^2}{25} + \dfrac{(y-3)^2}{9} = 1$

35. $\dfrac{x^2}{778{,}300{,}000^2} + \dfrac{y^2}{777{,}400{,}000^2} = 1$

37.

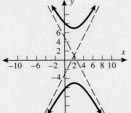

39.

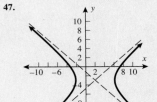

41. $\dfrac{x^2}{9} - \dfrac{y^2}{16} = 1$

43. $\dfrac{y^2}{9} - x^2 = 1$ (Other answers are possible.)

45.

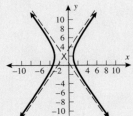

47.

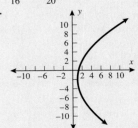

49. $\dfrac{(x-4)^2}{16} - \dfrac{(y-3)^2}{9} = 1$

51. The ship will get to shore between the two stations: 65.36 miles from one station.

Practice Test

1. c   **3.** d   **5.** f

7. $y^2 = -16x$

9. $(x+1)^2 = -12(y-5)$

11. $\dfrac{x^2}{7} + \dfrac{y^2}{16} = 1$

13. $\dfrac{(x-2)^2}{20} + \dfrac{y^2}{36} = 1$

15. $x^2 - \dfrac{y^2}{4} = 1$

17. $\dfrac{y^2}{16} - \dfrac{(x-2)^2}{20} = 1$

19.

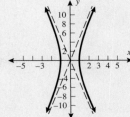

21.

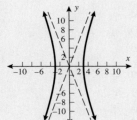

23. $x^2 = 6y$ $-2 \le x \le 2$

Chapter 9

Section 9.1

1. $a_1 = 1$ $a_2 = 2$ $a_3 = 3$ $a_4 = 4$

3. $a_1 = 1$ $a_2 = 3$ $a_3 = 5$ $a_4 = 7$

5. $a_1 = \dfrac{1}{2}$ $a_2 = \dfrac{2}{3}$ $a_3 = \dfrac{3}{4}$ $a_4 = \dfrac{4}{5}$

7. $a_1 = 2$ $a_2 = 2$ $a_3 = \dfrac{4}{3}$ $a_4 = \dfrac{2}{3}$

9. $a_1 = -x^2$ $a_2 = x^3$ $a_3 = -x^4$ $a_4 = x^5$

11. $a_1 = -\dfrac{1}{6}$ $a_2 = \dfrac{1}{12}$ $a_3 = -\dfrac{1}{20}$ $a_4 = \dfrac{1}{30}$

13. $a_9 = \dfrac{1}{512}$

15. $a_{19} = -\dfrac{1}{420}$

17. $a_{100} = 1.0201$

19. $a_{23} = 23$

21. $a_n = 2n$

23. $a_n = \dfrac{1}{n(n+1)}$

25. $a_n = \dfrac{(-1)^n 2^n}{3^n}$

27. $a_n = (-1)^{n+1}$

29. 72

31. 812

33. $\dfrac{1}{n(n+1)}$

35. $(2n+3)(2n+2)$

37. $a_1 = 7$ $a_2 = 10$ $a_3 = 13$ $a_4 = 16$

39. $a_1 = 1$ $a_2 = 2$ $a_3 = 6$ $a_4 = 24$

41. $a_1 = 100$ $a_2 = 50$ $a_3 = \dfrac{25}{3}$ $a_4 = \dfrac{25}{72}$

43. $a_1 = 1$ $a_2 = 2$ $a_3 = 2$ $a_4 = 4$

45. $a_1 = 1$ $a_2 = -1$ $a_3 = -2$ $a_4 = 5$

47. 10

49. 30

51. 36

53. 5

55. $1 - x + x^2 - x^3$

57. $\dfrac{109}{15}$

59. $1 + x + \dfrac{x^2}{2} + \dfrac{x^3}{6} + \dfrac{x^4}{24}$

61. $\dfrac{20}{9}$

63. not possible

65. $\displaystyle\sum_{n=0}^{6} \dfrac{(-1)^n}{2^n}$

67. $\displaystyle\sum_{n=1}^{\infty} (-1)^{n-1} n$

69. $\displaystyle\sum_{n=1}^{6} \dfrac{(n+1)!}{(n-1)!} = \sum_{n=1}^{5} n(n+1)$

71. $\displaystyle\sum_{n=0}^{\infty} \dfrac{(-1)^n x^n}{n!}$

73. $A_{72} \approx \$28{,}640.89$. A_{72} represents the total balance in 6 years (or 72 months).

75. $a_n = 20 + 2n$. The paralegal's salary with 20 years of experience will be $60 per hour.

77. $a_n = 1.03 a_{n-1}$ $n = 1, 2, \ldots$ and $a_0 = 30{,}000$

79. Approximately 10.6 years, $a_n = 1000 - 75n$ $n = 0, 1, 2, \ldots$

81. $A_1 = 100$ $A_2 = 200.1$ $A_3 = 300.30$ $A_4 = 400.60$ $A_{36} = 3663.72$

83. $6!$ is not equal to $(3!)(2!)$

85. $(-1)^{n+1}$ is evaluated incorrectly. The wrong sign is on each term.

87. true

89. false

91. $a_1 = C$ $a_2 = C + D$ $a_3 = C + 2D$ $a_4 = C + 3D$

93. $a_{100} = 2.705$ $a_{1000} = 2.717$ $a_{10{,}000} = 2.718$

Section 9.2

1. arithmetic; 3

3. not arithmetic

5. arithmetic; -0.03

7. arithmetic; $\dfrac{2}{3}$

9. not arithmetic

11. $a_1 = 3$ $a_2 = 1$ $a_3 = -1$ $a_4 = -3$ arithmetic; -2

13. $a_1 = 1$ $a_2 = 4$ $a_3 = 9$ $a_4 = 16$

15. $a_1 = 2$ $a_2 = 7$ $a_3 = 12$ $a_4 = 17$ arithmetic; 5

17. $a_1 = 0$ $a_2 = 10$ $a_3 = 20$ $a_4 = 30$ arithmetic; 10

19. $a_1 = -1$ $a_2 = 2$ $a_3 = -3$ $a_4 = 4$

21. $a_n = 6 + 5n$

23. $a_n = -6 + 2n$

25. $a_n = -\dfrac{2}{3} + \dfrac{2}{3} n$

27. $a_n = en - e$

29. 124

31. -684

33. $a_1 = 8$ $d = 9$ $a_n = 8 + 9(n - 1)$ or $a_n = 9n - 1$

35. $a_1 = 23$ $d = -4$ $a_n = 23 - 4(n - 1)$ or $a_n = 27 - 4n$

37. $a_1 = 1$ $d = \dfrac{2}{3}$ $a_n = 1 + \dfrac{2}{3}(n - 1)$ or $a_n = \dfrac{1}{3} + \dfrac{2}{3} n$

39. 552

41. -780

43. 51

45. 416

47. 3875

49. -66.5

51. Colin $347,500, Camden $340,000

53. 850 seats

55. 1101 glasses in the bottom row. There are 20 less glasses in every row.

57. 1600 feet

59. The correct general term is $a_n = a_1 + d(n-1)$.

61. $n = 11$ (not 10)

63. false

65. true

67. $\dfrac{(n+1)(2a + nb)}{2}$

69. 5050

71. 2500

Section 9.3

1. yes; $r = 3$

3. no

5. yes; $r = \dfrac{1}{2}$

7. yes; $r = 1.7$

9. $a_1 = 6$ $a_2 = 18$ $a_3 = 54$ $a_4 = 162$ $a_5 = 486$

11. $a_1 = 1$ $a_2 = -4$ $a_3 = 16$ $a_4 = -64$ $a_5 = 256$

13. $a_1 = 10{,}000$ $a_2 = 10{,}600$ $a_3 = 11{,}236$
$a_4 = 11{,}910.16$ $a_5 = 12{,}624.77$

15. $a_1 = \dfrac{2}{3}$ $a_2 = \dfrac{1}{3}$ $a_3 = \dfrac{1}{6}$ $a_4 = \dfrac{1}{12}$ $a_5 = \dfrac{1}{24}$

17. $a_n = 5 \cdot 2^{n-1}$

19. $a_n = 1 \cdot (-3)^{n-1}$

21. $a_n = 1000 \cdot 1.07^{n-1}$

23. $a_n = \dfrac{16}{3} \cdot \left(-\dfrac{1}{4}\right)^{n-1}$

25. $a_7 = -128$

27. $a_{13} = 1365.33$

29. $a_{15} = 6.10 \times 10^{-16}$

31. $\dfrac{8191}{3}$

33. 59,048

35. 2.222

37. 6560

39. 16,383

41. 2

43. $-\dfrac{1}{4}$

45. not possible

47. $-\dfrac{27}{2}$

49. $\dfrac{1{,}000{,}000}{95} \approx 10526$

51. $44,610.95

53. $2000(0.5)^n$; In 4 years when she graduates it will be worth $125, and after graduate school it will be worth $16.

55. On the fifth rebound the jumper will reach a height of approximately 17 feet. On the 13th rebound the jumper will essentially be at rest.

57. 58,640 students

59. 66 days; $9,618 would be paid in January

61. $3,891.48

63. 26 weeks: $13,204.46   52 weeks: $26,810.82

65. $r = -\dfrac{1}{3}$ not $\dfrac{1}{3}$

67. $a_1 = -12$ (not 4)

69. false

71. true

73. If $|b| < 1$, then the sum is $a \cdot \dfrac{1}{1 - b}$.

75. -375299968947541

77. The sum is $\dfrac{1}{1 - x}$ for $|x|$ less than or equal to $\dfrac{1}{2}$.

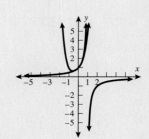

Section 9.4

25. 7

27. 31

29. false

Section 9.5

1. 35

3. 45

5. 1

7. 1

9. 17,296

11. $x^4 + 8x^3 + 24x^2 + 32x + 16$

13. $y^5 - 15y^4 + 90y^3 - 270y^2 + 405y - 243$

15. $x^5 + 5x^4 y + 10x^3 y^2 + 10x^2 y^3 + 5xy^4 + y^5$

17. $x^3 + 9x^2 y + 27xy^2 + 27y^3$

19. $125x^3 - 150x^2 + 60x - 8$

21. $\dfrac{1}{x^4} + \dfrac{20y}{x^3} + \dfrac{150y^2}{x^2} + \dfrac{500y^3}{x} + 625y^4$

23. $x^8 + 4x^6y^2 + 6x^4y^4 + 4x^2y^6 + y^8$
25. $a^5x^5 + 5a^4x^4by + 10a^3x^3b^2y^2 + 10a^2x^2b^3y^3 + 5axb^4y^4 + b^5y^5$
27. $x^3 + 12x^{5/2} + 60x^2 + 160x^{3/2} + 240x + 192x^{1/2} + 64$
29. $a^3 + 4^{9/4}b^{1/4} + 6a^{3/2}b^{1/2} + 4a^{3/4}b^{3/4} + b$
31. $x + 8x^{3/4}y^{1/2} + 24x^{1/2}y + 32x^{1/4}y^{3/2} + 16y^2$
33. $r^4 - 4r^3s + 6r^2s^2 - 4rs^3 + s^4$
35. $a^6x^6 + 6a^5x^5by + 15a^4x^4b^2y^2 + 20a^3x^3b^3y^3 + 15a^2x^2b^4y^4 + 6axb^5y^5 + b^6y^6$
37. 3360   **39.** 5670   **41.** 22,680
43. 70   **45.** 3,838,380   **47.** 2,598,960
49. $\binom{7}{5} \neq \dfrac{7!}{5!}$, $\binom{7}{5} = \dfrac{7!}{5!2!}$
51. false   **53.** true
57. The binomial expansion of
$(1 - x)^3 = 1 - 3x + 3x^2 - x^3$.

59. As each term is added, the series gets closer to $y = (1 - x)^3$. When $x > 1$ that's not true. The graphs for $-1 < x < 1$:

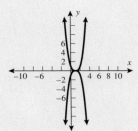

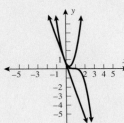

The graphs for $1 < x < 2$:

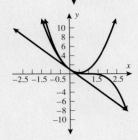

Section 9.6

1. 360   **3.** 15120   **5.** 40320
7. 1716   **9.** 252   **11.** 15,890,700
13. 1   **15.** 27,405   **17.** 215,553,195
19. 24   **21.** 12   **23.** 10,000
25. 32,760   **27.** 1.1×10^{12} **29.** 100,000; 81,000
31. 2.65×10^{32} **33.** 59,280   **35.** 997,002,000
37. 22,957,480   **39.** 2,598,960   **41.** 1326
43. $\approx 4.9 \times 10^{14}$ **45.** 256   **47.** 15,625
49. The correction that needs to be made is that the combination formula needs to be used instead of the permutation.

51. false   **53.** false   **55.** $\dfrac{{}_nC_r}{{}_nC_{r+1}} = \dfrac{r+1}{n-r}$

57. The answers are the same as exactly 1–8.

Section 9.7

1. 2, 3, 4, 5, 6, 7, 8, 9, 10, 11, 12

3. $\left\{ \begin{array}{l} BBBB, BBBG, BBGB, BBGG, BGBB, BGBG, BGGB, BGGG, \\ GBBB, GBBG, GBGB, GBGG, GGBB, GGBG, GGGB, GGGG \end{array} \right\}$

5. {RR, RB, RW, BR, BW, BB, WR, WB}

7. $\dfrac{1}{8}$ **9.** $\dfrac{7}{8}$ **11.** $\dfrac{1}{18}$ **13.** $\dfrac{1}{2}$ **15.** $\dfrac{15}{36}$

17. $\dfrac{40}{52} = \dfrac{10}{13}$ **19.** $\dfrac{4}{13}$ **21.** $\dfrac{3}{4}$ **23.** $\dfrac{3}{4}$ **25.** 0

27. a. 270,725   **b.** $\dfrac{715}{270,725} \approx 2.6\%$ **c.** $\dfrac{13}{270,725} \approx 0.005\%$

29. $\dfrac{8}{52} = \dfrac{2}{13} \approx 15.4\%$ **31.** $\dfrac{4}{663} \approx 0.6\%$ **33.** $\dfrac{1}{32} \approx 3.1\%$

35. $\dfrac{31}{32} \approx 96.9\%$ **37.** $\left(\dfrac{18}{38}\right)^4 = 5.03\%$ **39.** 20%

41. $\dfrac{1}{4} = 25\%$ **43.** $\dfrac{48}{1326} \approx 3.6\%$ **45.** $\approx 0.001526\%$

47. Subtract P (2 of spades) $= \dfrac{1}{52}$

49. true   **51.** true

53. $1 - \dfrac{(365)}{(365)} \dfrac{(364)}{(365)} = 0.0027$ **55.** $\dfrac{1}{3} \approx 0.333$

OOPS!   Solution: **a.** $\dfrac{85}{100}$ **b.** $\dfrac{37}{100}$

Tying It All Together   Solution: 582.59 feet

Review Exercises

1. $a_1 = 1$ $a_2 = 8$ $a_3 = 27$ $a_4 = 64$
3. $a_1 = 5$ $a_2 = 8$ $a_3 = 11$ $a_4 = 14$
5. $a_5 = \dfrac{32}{243} \approx 0.13$ **7.** $a_{15} = -\dfrac{1}{3600}$ **9.** $a_n = (-1)^{n+1}3n$

11. $a_n = (-1)^n$ **13.** 56   **15.** $\dfrac{1}{n+1}$

17. $a_1 = 5$ $a_2 = 3$ $a_3 = 1$ $a_4 = -1$
19. $a_1 = 1$ $a_2 = 2$ $a_3 = 4$ $a_4 = 32$

21. 15   **23.** 69   **25.** $\displaystyle\sum_{n=1}^{7} \dfrac{(-1)^n}{2^{n-1}}$ **27.** $\displaystyle\sum_{n=0}^{\infty} \dfrac{x^n}{n!}$

29. $A_{60} \approx 36,629.90$, which is the amount in the account after 5 years.

31. Yes, the sequence is arithmetic. $d = -2$

33. Yes, the sequence is arithmetic. $d = \dfrac{1}{2}$

35. Yes, the sequence is arithmetic. $d = 1$

37. $a_n = -4 + 5(n-1)$ or $a_n = 5n - 9$

39. $a_n = 1 - \dfrac{2}{3}(n-1)$ or $a_n = -\dfrac{2}{3}n + \dfrac{5}{3}$

41. $a_1 = 5$ $d = 2$ $a_n = 5 + 2(n-1)$ or $a_n = 3 + 2n$
43. $a_1 = 10$ $d = 6$ $a_n = 10 + 6(n-1)$ or $a_n = 4 + 6n$
45. 630
47. 420
49. Bob: $885,000   Tania: $990,000
51. Yes, the sequence is geometric, $r = -2$.

53. Yes, the sequence is geometric, $r = \dfrac{1}{2}$.

55. $a_1 = 3$ $a_2 = 6$ $a_3 = 12$ $a_4 = 24$ $a_5 = 48$
57. $a_1 = 100$ $a_2 = -400$ $a_3 = 1600$ $a_4 = -6400$ $a_5 = 25,600$
59. $a_n = 7 \cdot 2^{n-1}$ **61.** $a_n = (-2)^{n-1}$
63. $a_{25} = 33,554,432$ **65.** $a_{12} = -2.048 \times 10^{-6}$
67. 4,920.50   **69.** 16,400
71. 3   **73.** $59,681.97
79. 165   **81.** 1
83. $x^4 - 20x^3 + 150x^2 - 500x + 625$
85. $8x^3 - 60x^2 + 150x - 125$
87. $x^{5/2} + 5x^2 + 10x^{3/2} + 10x + 5x^{1/2} + 1$
89. $r^5 - 5r^4s + 10r^3s^2 - 10r^2s^3 + 5rs^4 - s^5$
91. 112   **93.** 37,500   **95.** 22,957,480

97. 840   **99.** 95,040   **101.** 792

103. 1   **105.** 30   **107.** 5040

109. 120 different seating orders (15 seasons)

111. 94,109,400   **113.** 20,358,520   **115.** $\dfrac{1}{16} \approx 6.25\%$

117. $\dfrac{30}{36} = \dfrac{5}{6} \approx 83.3\%$ **119.** $\dfrac{2}{3} \approx 66.7\%$ **121.** $\dfrac{7}{12} \approx 58.3\%$

123. $\dfrac{8}{52} = \dfrac{2}{13} \approx 15.4\%$ **125.** $\dfrac{31}{32} \approx 96.88\%$

Practice Test

1. $a_n = x^{n-1}$ $n = 1, 2, 3, \ldots$

3. $S_n = \dfrac{1 - x^n}{1 - x}$ **5.** $|x| < 1$ **7.** 455   **9.** 2184

11. $x^{10} + 5x^7 + 10x^4 + 10x + \dfrac{5}{x^2} + \dfrac{1}{x^5}$

13. Because combinations have no particular order and permutations do. For permutations, each combination is arranged $n!$ ways.

15. $P(\text{red}) = \dfrac{18}{38} \approx 0.47$

17. $P(\text{next roll is red}) = \dfrac{18}{38} \approx 0.47$

19. $P(\text{ace}) + P(\text{diamond}) - P(\text{ace of diamond}) = \dfrac{16}{52} \approx 0.308$

APPLICATIONS INDEX

FUNCTIONS

Constant Function	$f(x) = b$
Linear Function	$f(x) = mx + b$, where m is the slope and b is the y-intercept
Quadratic Function	$f(x) = ax^2 + bx + c$, $a \neq 0$ or $f(x) = a(x - h)^2 + k$ parabola vertex (h, k)
Polynomial Function	$f(x) = a_n x^n + a_{n-1} x^{n-1} + \cdots + a_1 x + a_0$
Rational Function	$R(x) = \dfrac{n(x)}{d(x)} = \dfrac{a_n x^n + a_{n-1} x^{n-1} + \cdots + a_1 x + a_0}{b_m x^m + a_{m-1} x^{m-1} + \cdots + b_1 x + b_0}$
Exponential Function	$f(x) = b^x$, $b > 0$, $b \neq 1$
Logarithmic Function	$f(x) = \log_b x$, $b > 0$, $b \neq 1$

GRAPHS OF COMMON FUNCTIONS

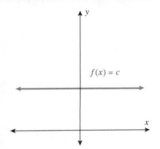

Constant Function

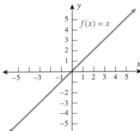

Identity Function

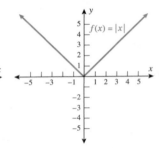

Absolute Value Function

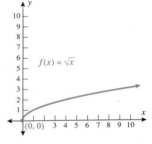

Square Root Function

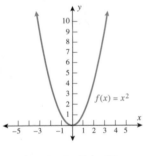

Quadratic Function

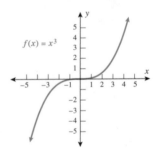

Cubic Function

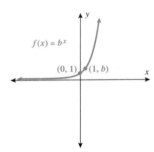

Exponential Function

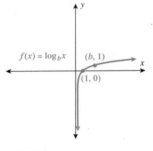

Logarithmic Function

TRANSFORMATIONS

In each case, c represents a positive real number.

Function		Draw the graph of f and:
Vertical translations	$\begin{cases} y = f(x) + c \\ y = f(x) - c \end{cases}$	Shift f upward c units. Shift f downward c units.
Horizontal translations	$\begin{cases} y = f(x - c) \\ y = f(x + c) \end{cases}$	Shift f to the right c units. Shift f to the left c units.
Reflections	$\begin{cases} y = -f(x) \\ y = f(-x) \end{cases}$	Reflect f about the x-axis. Reflect f about the y-axis.